Computer Techniques
and
Models in Power Systems

Computer Techniques and Models in Power Systems

Second Edition

K. Uma Rao
RV College of Engineering
Bangalore

I.K. International Publishing House Pvt. Ltd.
NEW DELHI

Published by

I.K. International Publishing House Pvt. Ltd.
S-25, Green Park Extension
Uphaar Cinema Market
New Delhi–110 016 (India)
E-mail: info@ikinternational.com
Website: www.ikbooks.com

ISBN 978-93-82332-31-2

Reprint 2017

Published by Krishan Makhijani for I.K. International Publishing House Pvt. Ltd., S-25, Green Park Extension, Uphaar Cinema Market, New Delhi–110 016 and Printed by Rekha Printers Pvt. Ltd., Okhla Industrial Area, Phase II, New Delhi–110 020.

Dedicated to my parents

Smt. K. Ramamani and Sri. K.S. Keshava Rao

Foreword

Modern power systems are large and complex systems extending over wide geographical areas. The system interconnections in the form of regional and national grids, while improving system reliability, add to the complexity of operation and control. The planning and operation of such large systems require system studies based on computer techniques. The use of computers for power system analysis started about 50 years ago. The growth in the computing power over the years has not yet kept pace with the growth in the size of power systems and the complexity of models.

There are several production softwares available for power flow, short-circuit and stability studies which are used routinely now by utility engineers involved with long-term and operational planning. However, it is not adequate for electrical engineers to be able to use the programs and interpret the results. It is also essential to understand the models and methodology used in the solution or simulation of the system. The training they receive during their undergraduate studies should incorporate a thorough knowledge of the principles, solution methods and implementation in the computer programs.

Dr. Uma Rao has used her extensive experience in teaching both undergraduate and graduate courses on power system analysis and computer techniques, in the preparation of this book which covers various aspects of computer methods in power system analysis starting from the fundamentals. The book covers symmetrical components, fault analysis, load-flow and stability studies and economic operation. Topics on modeling of excitation and prime-mover control systems are also included with a chapter on power system control that includes AGC (Automatic Generation Control), reactive power and voltage control. While there are many books at the undergraduate level on power system analysis, they usually do not treat computer techniques in detail. This book breaks new ground by including the advanced topics on computer techniques and explaining them at a level that the students can easily comprehend. The numerical examples are used to clarify the concepts in each topic.

It gives me great pleasure to write the Foreword for this book and I expect it will benefit immensely the student community and in particular, the undergraduate students.

K.R. Padiyar
Professor Emeritus
Indian Institute of Science
Bangalore

Preface to the Second Edition

The first edition of the book was well received by students and faculty all over India. There was a need to update the first edition. In the second edition, over 75 numerical problems have been added. A chapter on simple modeling of synchronous machines has also been included. With many universities having a laboratory course in 'Power System Simulation', there was a need to introduce a chapter on simulation. This chapter has program codes, sample data and results, and exercises to strengthen the programming skills of students. This edition is more comprehensive and covers the syllabus of a first course in power systems and also topics on computer techniques and simulation.

K. Uma Rao

Preface to the First Edition

Modern power systems have grown into large integrated, interconnected systems, spreading over expansive geographical areas. Planning, analysis, operation and control of power systems require widespread application of computers. The transition from network analyzers to digital computers, in a matter of few years, has been dramatic. Today, digital computers have become indispensable and are used extensively, not only for analysis but also for real-time control of power systems.

This book primarily deals with the application of computers for power system analysis. The major studies required for any system are fault analysis, load flows, contingency analysis, stability assessment, economic scheduling of generators and state estimation. Classic books by Stagg and El-abiad ("Computer methods in power system analysis, 1968) and G.L. Kusic ("Computer aided power system analysis,1986) have dealt with some of the above topics. The author has been teaching the subject to undergraduate and postgraduate students, for the last two decades. A strong need has been felt by the author and her students, for a book, which can cover the above topics in a systematic manner and be comprehended by the students. This book is the outcome of the teaching experience of the author.

The required mathematical background is presented at the appropriate sections in the book. A more detailed coverage of matrix algebra is given in the Appendix. It is assumed that the reader has a general understanding of elementary circuit analysis and basic knowledge of power systems.

Chapter 1 presents a brief introduction on the role of computers in power system analysis. Concepts of reactive power, directions for complex power and the per unit (pu) system are presented in detail to get the student familiar with fundamental concepts. Chapter 2 covers the concepts of network topology, vital for the understanding of the mathematical structure of the equations and the transformations involved in power systems. Chapter 3 deals with algorithms for building the bus admittance and bus impedance matrices, in a systematic manner. Here concepts of rotational and reflection symmetry are dealt with in detail. The methods presented in the chapter are developed both for single-phase representation and three-phase representations.

Chapter 4 extensively covers the concept of different reactances on the synchronous machine, symmetrical fault analysis for loaded and unloaded systems. It also covers the application of the bus-impedance matrix for fault analysis. Chapter 5 deals with the concept of symmetrical components, which transforms unbalanced phasors into set of balanced phasors. Chapter 6 uses this concept, for the analysis of unbalanced faults in practical power systems.

Chapter 7 presents the important topic of load flow analysis. Detailed algorithms have been given for the Gauss Seidel, Newton Raphson, Decoupled and Fast decoupled methods of load flows. Chapter 8 addresses the important issue of economic operation of the power system. The chapter also includes methods to compute loss coefficients and optimal operation of hydro-thermal systems. The algorithm for unit-commitment using dynamic programming, is also presented. Chapter 9 contains detailed aspects of power system stability analysis. Small signal stability and equal area criterion have been presented and illustrated with a number of examples. The algorithm for solution of the swing equation by the application of various numerical methods like Runge-Kutta, Euler, Milne's prediction correction and step-by-step method have been given and results of simulation studies presented.

Chapters 10 and 11 deal with the important topic of modeling of power system components, namely excitation systems and their accessories, turbines and loads, for digital computer applications. Detailed models of turbines, excitation systems, governors and loads have been presented. Chapter 12 covers the topic of power system control. The concepts of Automatic Load Frequency Control, Automatic Generation Control and Automatic Voltage Regulator, their models and simulation are covered in detail.

A sincere attempt has been made to include a number of solved examples in each chapter, so that the student gets a feel for the problems in practical power systems. Results from simulation are presented wherever applicable. The simulations have been carried out in MATLAB. However, the same can be done in any high-level language like FORTRAN, C, C++ etc.

The book covers more than a one semester course. It can be used for UG courses on Power System Analysis, Computer applications in power system analysis, modeling of power system components and power system operation and control. It is also useful to postgraduate students of power engineering.

K. Uma Rao

Acknowledgements

It is a pleasure to acknowledge the contributions of all those who have helped in the preparation of the book, directly and indirectly. I got a clear understanding of power systems, the complexities involved in its operation, analysis and control, as a student under Prof. K.R. Padiyar, at Indian Institute of Science, Bangalore. I am deeply indebted to him for the knowledge I have imbibed from him. I wish to acknowledge the contribution of Late Prof. P.V. Rathnam, Andhra University, Visakhapatnam, who was instrumental in me developing a keen interest in the fundamental concepts of electrical engineering. My thanks to all my students who have motivated me to write this book and have provided me with suggestions about the improvement needed in the existing books on the subject. I am extremely grateful to Ms. S. Sumathi, Assistant Professor, at RNSIT, who went through the manuscript and provided me with necessary feedback.

Special thanks to my dear friend Indira, Professor of English, Andhra University, who has always supported me. This book would have been nonexistent, but for the untiring support and encouragement of my husband Umesh, who willingly took on the reins of the family to help me complete the book. I am deeply indebted to him. Thanks to my darling children Arathi and Amrutha, who had to sacrifice their time with me, to help me finish the book on time.

K. Uma Rao

Contents

Chapter 1

Basic Principles

1.1 INTRODUCTION

A power system is an interconnected electric network which makes available, electric power generated by the generating plants to the loads, through the transmission and distribution network. The network is essentially an AC network, which supplies sinusoidal voltage at the rated voltage magnitude and frequency to the loads. The major components of the power system are generators, loads, transformers, transmission network and protection equipment called switchgear.

Modern power systems are large, highly complex and organized in the form of regional grids, which are interconnected to facilitate power transfer between areas via tie-lines. Interconnections make the system more reliable, since in the event of demand exceeding supply in an area, power can be borrowed from the neighboring area. Interconnection also helps to reduce the reserve capacity in each area and makes the system more flexible. However, stable operation of the system is more difficult in the event of faults, in large interconnected systems. Further, faults in one area, are likely to affect adjoining areas.

Due to the large size and interconnection between various regions, modern power systems are controlled from an Energy Control Center (ECC). These control centers permit a hierarchical control over the system operation, to coordinate various areas, both under normal operating conditions and under emergency conditions. Modern control centers use 'Supervisory Control and Data Acquisition' (SCADA), where the data is accessed centrally via remote acquisition systems. Computers are extensively used for data acquisition, data processing, analysis of system operating condition, control and for warning the operator of deviations from normal conditions. Since, their first application in late 40s, digital computers have become an indispensable part of the power system, both for analysis as well as for operation and control.

1.1.1 Power Systems Analysis

Some important analysis of the power system is necessary to be carried out, for continuous evaluation of the system. These studies are carried out on-line for operation and control, as well as off-line to help in future planning and expansion. Some of the important studies carried out are as following:

(i) Load flow studies: The load flow analysis provides the complex voltages at all the buses and the power flows in all lines, under specified operating conditions like generation capability, reactive power limits, taps of transformers, etc. This is a steady state solution. The study is needed for a continuous evaluation of the performance of the systems. The study also helps to identify

- Overloading of transmission lines,
- Voltage limit violations at the buses,
- Requirements of reactive power support, and
- Alternative plans for expansion to meet increased load demand.

(ii) Economic dispatch: The power system is fed from several generating plants, which have varied cost characteristics, like thermal plants, hydel plants, nuclear plants, etc. The economic dispatch problem is useful in deciding the generation schedule needed to meet the required load, so as to minimize the total cost of generation.

(iii) Fault analysis: The power system is subjected to a number of faults. The protection scheme is designed to protect the power system in the event of faults. Relays are used to sense the faults and circuit breakers isolate the faulty part of the system. Symmetrical faults affect all the three phases in a similar manner. Unsymmetrical faults, on the other hand, cause unbalance in an otherwise balanced system. Fault analysis provides information on the fault levels in various parts of the system and is vital for deciding the rating of the circuit breakers to be used. It will not be possible to protect the system against all faults. A few major contingencies are identified and the protection scheme is designed to protect the power system against these faults.

(iv) Stability analysis: Stability is the ability of the system to remain in synchronism in the event of disturbances. The power system being dynamic is constantly subject to disturbances in the form of load changes. The small signal stability study deals with the behaviour of the system under minor disturbances. Transient stability deals with the behaviour of the system under large disturbances. Stability studies are carried out on-line continuously, to assess the margin of stability. Off-line studies are also required for contingency analysis and planning of the expansion.

(v) Automatic generation control (AGC): This is an on-line control to maintain the system frequency and the net tie-line interchange between interconnected power systems. As the system load changes continuously, the generator governor is controlled so as to control the generation to restore the frequency to its rated value. During large disturbances, AGC is bypassed and other emergency controls are activated.

The mathematical formulation to perform the above studies, often leads to non-linear differential and/or algebraic equations. Non-linear equations require the application of numerical methods to solve them. Earlier, network analyzers were used but today digital computers are used widely for the solution of the equations. Some important aspects in the application of computers are next discussed.

1.1.2 Application of Computers

In application of computers for the solution of power system problems, certain steps are involved. They are:

(i) Defining the problem and the objectives of the analysis/study.

(ii) Assessment of data available and preparation of the input data in proper format. This step also involves methods to synthesize or estimate the missing data.

(iii) Development of suitable mathematical models to represent the power system and various components. This step also involves integrating the models of various components and formulating the mathematical equations to be solved.

(iv) Selection of a suitable technique to solve the mathematical equations. In most of the analysis, the equations are non-linear and a suitable numerical method has to be chosen. The choice of the method depends on the computation time involved, accuracy of the method, memory required, reliability of the method and the convergence characteristics of the method. Modern digital computers have very high computation speeds and large memory. These two aspects are no longer a major criterion in selecting the method for solution, except for real time applications where computation time is important.

(v) Development of the flow chart/algorithm for the solution of the equations. The development of the algorithm will have to take care of all the operational limitations of the power system.

(vi) The last step involves actual development of the computer programs in a high level language. FORTRAN, C++ and MATLAB are extensively used for programming. The program developed should be flexible, reliable and capable of handling diverse data.

The focus of this book is on the first five steps. Actual programming has not been discussed in detail, though a few sample programs are provided in the appendix to guide the reader. In further sections in this chapter, a few fundamental concepts necessary for the understanding of the power system and certain conventions used are also discussed.

1.2 PHASOR REPRESENTATION

The voltages at various buses are sinusoidal and have constant frequency under normal operating conditions. If we consider a linear system, the currents are also sinusoidal. Sinusoidal currents and voltages can be represented by phasors. A phasor is a complex number whose magnitude is the amplitude of the sinusoidal function and whose angle represents the phase angle of the sinusoid with respect to a reference. In electric networks, it is more common to let the magnitude of the phasor be the rms value of the voltage or current rather than the maximum value. Representation of sinusoidal voltages and currents as phasors, simplifies arithmetic operations, which would otherwise have to be carried out using cumbersome trigonometric manipulations. Further, the representation also helps to represent differential equations as algebraic equations. A sinusoidal voltage function $v(t) = V_{max} \cos(\omega t + \phi)$ can be represented as a complex number (phasor) in different ways:

- **Polar form** : $V = V\angle\phi$ where $V = V_{max}/\sqrt{2} = V_{rms}$
- **Rectangular form** : $V_1 + jV_2 = V(\cos ø + j \sin ø)$
- **Exponential form** : $Ve^{jø}$

In the equivalent circuit diagrams of the power system, polarities are marked to indicate the terminal assumed positive when the voltage is specified. An arrow indicates the assumed positive direction of current.

1.3 POWER IN SINGLE PHASE AC CIRCUITS

Here, we discuss the fundamental concept of power in AC circuits. Consider the following voltage and current functions:

$$v(t) = V_{max} \sin \omega t \quad ...(1.1)$$

$$i(t) = I_{max} \sin (\omega t - \phi) \quad ...(1.2)$$

where the radian frequency $\omega = 2\pi f$ rad/s. In (1.2) ϕ is positive when current lags the voltage and negative if current leads the voltage. The instantaneous power is given by

$$\begin{aligned} p(t) &= v(t)i(t) \\ &= V_{max} I_{max} \sin (\omega t) \sin (\omega t - \phi) \\ &= \frac{V_{max}\, I_{max}}{2} [\cos \phi - \cos (2\omega t - \phi)] \\ &= \frac{V_{max}}{\sqrt{2}} \frac{I_{max}}{\sqrt{2}} [\cos \phi - \cos (2\omega t - \phi)] \\ &= VI [\cos \phi - \cos (2\, \omega t - \phi)] \\ &= VI \cos \phi - VI \cos (2\, \omega t - \phi) \quad ...(1.3) \end{aligned}$$

From (1.3) it is evident that the power has an average value of $VI \cos \phi$ and pulsates around it at double the system radian frequency (2ω). (1.3) can be written in a different manner as given below:

$$\begin{aligned} p(t) &= VI \cos \phi - VI(\cos 2\, \omega t \cos \phi + \sin 2\, \omega t \sin \phi) \\ &= VI \cos \phi\, (1 - \cos 2\, \omega t) - VI \sin \phi \sin 2\, \omega t \quad ...(1.4) \end{aligned}$$

The first term of (1.4) has an average value of $VI \cos \phi$ which is same as the average value of the instantaneous power. The second term has an average value of zero. We now define two terms:

(i) **The real or active power:** $P \equiv VI \cos \phi$: The real or active power is defined as the average value of $p(t)$ and is therefore the useful power. The value depends on the value of the power factor angle ϕ.

(ii) **The reactive power:** $Q \equiv VI \sin \phi$: The reactive power is defined as the peak value of that component of the instantaneous power, which travels back and forth on the line at double the system frequency and results in zero average value. Hence, reactive power is not capable of any useful work.

The dimensions of P and Q are watts, but to differentiate between them, Q is given the unit of volt-ampere reactive or VAR. A more common unit in power systems is MW for P and MVAR for Q. In (1.2) ϕ is positive for lagging currents and hence Q is positive for lagging currents.

1.4 COMPLEX POWER

The complex power S is defined as

$$S \equiv VI^* = VI \cos \phi + jVI \sin \phi = P + jQ \qquad ...(1.5)$$

The magnitude of the complex power given by $\sqrt{P^2 + Q^2}$ is called as the *apparent power.* When the power flow of a system is considered, it is important to know whether the real and reactive powers are being generated or absorbed. At any bus, the power injected is considered as shown in Fig. 1.1, which shows the convention for the voltage polarity and the current direction.

The injected complex power is given by $S = P + jQ. = VI^*$. We have the following cases:

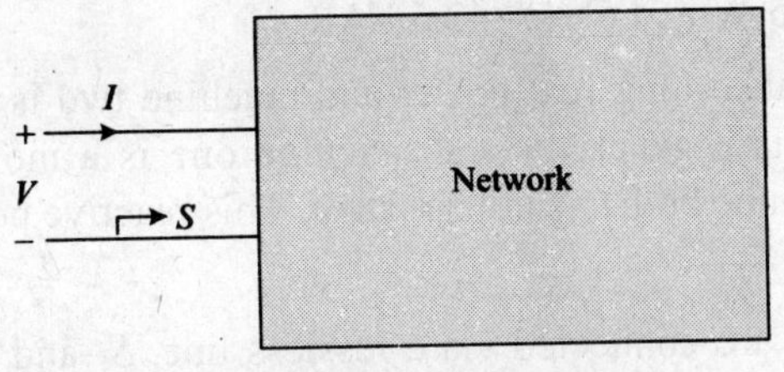

Fig. 1.1 Complex power direction.

- P is positive which means the network is absorbing real power.
- P is negative which means the network is absorbing negative real power, which means delivering real power.
- Q is positive which means network is absorbing reactive power. This means that the network is inductive in nature.
- Q is negative which means the network is generating or supplying reactive power.

The above convention can be easily illustrated with an example.

Example 1.1: Two machines are connected through a line of impedance $j2\,\Omega$ as shown in Fig. 1.2. The internal emfs are given by $E_1 = 50\angle 0°$ and $E_2 = 50\angle 15°$. Find the active and reactive powers of each machine.

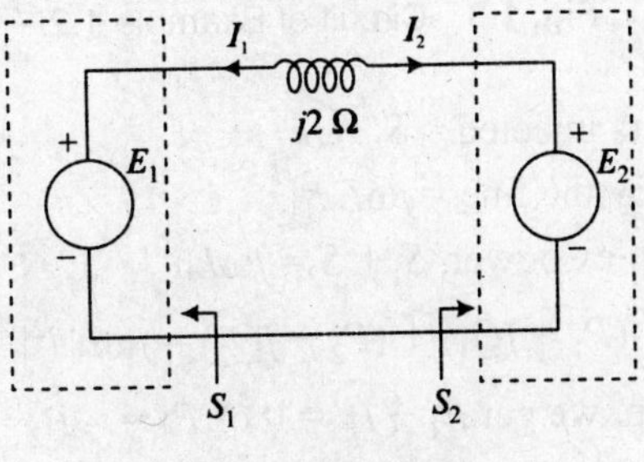

Fig. 1.2 Example 1.1.

Solution: $I_1 = \dfrac{E_2 - E_1}{j2} = \dfrac{50\angle 15° - 50\angle 0°}{j2} = 6.525\angle 7.5°$ A

$$I_2 = \frac{E_1 - E_2}{j2} = \frac{50\angle 0° - 50\angle 15°}{j2} = 6.525\angle -172.5° \text{ A}$$

$$S_1 = E_1 I_1^* = 50\angle 0° \times 6.525\angle -7.5° = 323.45 - j42.5 \text{ VA}$$

$$P_1 = 323.45 \text{ W and } Q_1 = -42.5 \text{ VAR}$$

$$S_2 = E_2 I_2^* = 50\angle 15° \times 6.525\angle 172.5° = -323.45 - j42.5$$

$$P_2 = -323.45 \text{ W and } Q_2 = -42.5 \text{ VAR}$$

Thus, machine one is absorbing real power and machine two is delivering real power. This implies that machine two is a generator and machine one is a motor. Both the machines are delivering reactive power since both Q's are negative. This reactive power is absorbed by the line and given by $|I^2|X$.

Example 1.2: Two buses are connected via a lossless line. S_1 and S_2 are the complex powers injected at the two ends. Prove that if the voltage magnitudes of both the buses are the same, $S_2 = -S_1^*$.

Solution: The circuit is shown in Fig. 1.3.

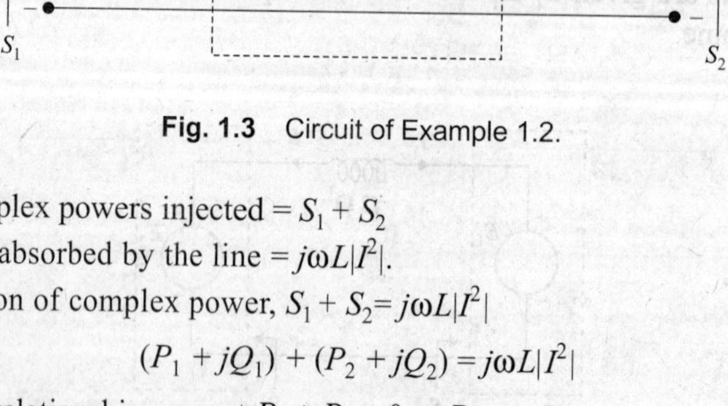

Fig. 1.3 Circuit of Example 1.2.

The sum of complex powers injected = $S_1 + S_2$

Complex power absorbed by the line = $j\omega L|I^2|$.

From conservation of complex power, $S_1 + S_2 = j\omega L|I^2|$

$$(P_1 + jQ_1) + (P_2 + jQ_2) = j\omega L|I^2|$$

From the above relationship, we get $P_1 + P_2 = 0$ or $P_1 = -P_2$

$$\text{and } Q_1 + Q_2 = \omega L|I^2|$$

Now $S_1 = V_1I_1^*$ and $S_2 = -V_2I_1^*$. Since $|V_1| = |V_2|$, $|S_1| = |S_2|$. Hence, $P_1^2 + Q_1^2 = P_2^2 + Q_2^2$. But $P_1^2 = P_2^2$. Therefore, $Q_1^2 = Q_2^2$ or $Q_1 = Q_2 = j\omega L|I^2|/2$. Thus, each end supplies half of the reactive power required by the line. We thus have $P_1 = -P_2$ and $Q_1 = Q_2$. Hence, $S_2 = -S_1^*$.

1.5 CONSERVATION OF COMPLEX POWER

The theorem of conservation of complex power states that *"For a network supplied by independent sources all at the same frequency, the sum of the complex power supplied by the independent sources equals the sum of the complex power absorbed by all other branches of the network"*. It follows that the sum of complex power injections at any bus is zero.

Example 1.3: In Fig. 1.4, find S_{12}, S_{21}, S_{23} and S_{G3}. State any assumptions made.

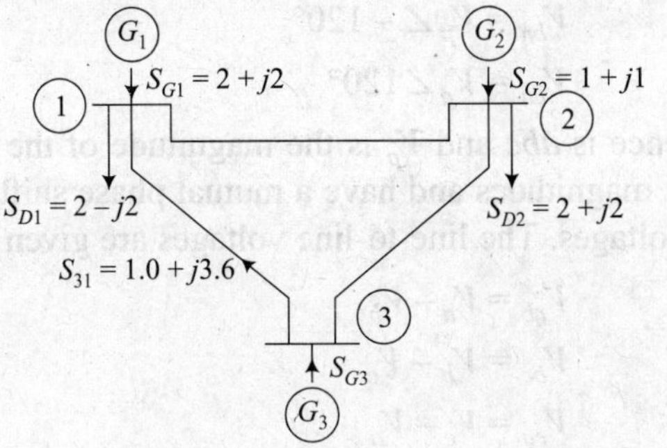

Fig. 1.4 Example 1.3.

Solution: We assume that the lines are lossless and the voltage magnitude at all the buses is same. Under these assumptions we can use the result of example 1.2. In the circuit S_{D1} and S_{D2} are the complex power demands (loads) at buses 1 and 2.

$$S_{31} = 1.0 + j3.6\ VA;\ S_{13} = -S_{31}^* = -1.0 + j3.6\ \text{VA}.$$

At bus 1, $S_{G1} = S_{D1} + S_{12} + S_{13};\ S_{12} = S_{G1} - S_{D1} - S_{13} = 1.0 + j0.4$

$$S_{21} = -S_{12}^* = -1.0 + j0.4\ \text{VA}$$

At bus 2, $S_{G2} = S_{D2} + S_{21} + S_{23};\ S_{23} = S_{G2} - S_{D2} - S_{21} = -j1.4$

$$S_{32} = -S_{23}^* = -j1.4\ \text{VA}$$

At bus 3, $S_{G3} = S_{31} + S_{32};\ S_{G3} = 1.0 + j2.2\ \text{VA}.$

1.6 THREE-PHASE SYSTEMS

From (1.4) it is obvious that the single-phase power is pulsating. A pulsating power produces a pulsating torque in motors, which is generally not acceptable. A pulsating power also produces flicker. Three-phase AC systems eliminate the power and torque pulsations. In balanced systems, there is a definite relationship between the three-phase voltages and currents.

1.6.1 Three-Phase Voltages

The three-phase generators are, in general, connected in star. We can denote the line-to-neutral voltage of phase *a* as v_{an} or simply v_a. In a balanced system the three phase line-to-neutral voltages are given by

$$v_a = \sqrt{2}\, V_p \sin \omega t$$
$$v_b = \sqrt{2}\, V_p \sin (\omega t - 120°)$$
$$v_c = \sqrt{2}\, V_p \sin (\omega t + 120°) \quad \text{...(1.6)}$$

The phasors of the line-to-neutral voltages are given by

$$V_{an} = V_p \angle 0°$$
$$V_{bn} = V_p \angle -120°$$
$$V_{cn} = V_p \angle 120° \quad \text{...(1.7)}$$

Here, the phase sequence is *abc* and V_p is the magnitude of the phase voltage (rms). Three voltages which have equal magnitudes and have a mutual phase shift of 120° form balanced and symmetrical three-phase voltages. The line-to-line voltages are given by

$$V_{ab} = V_a - V_b$$
$$V_{bc} = V_b - V_c$$
$$V_{ca} = V_c - V_a$$
$$|V_{ab}| = \sqrt{3}\, V_p \quad \text{...(1.8)}$$

Upper case letters indicate phasors. The phase and line voltage phasor relationship is shown in Fig. 1.5.

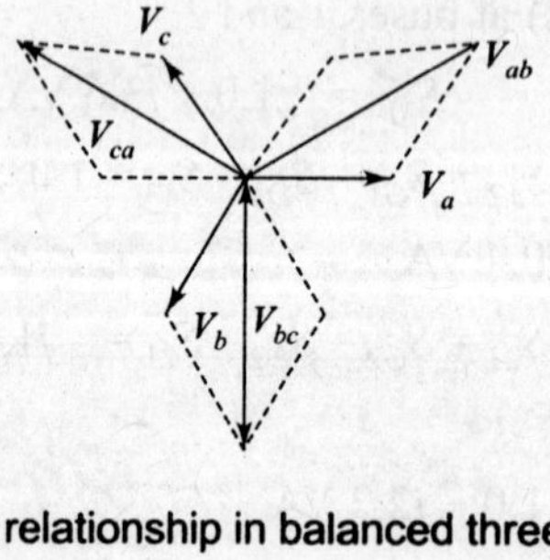

Fig. 1.5 Phasor relationship in balanced three-phase systems.

The voltage rating of a three-phase device always refers to the line-to-line voltage. If the source is connected in delta, from the phasor diagram of Fig. 1.5, we can get the phase voltages of the equivalent star connected source as

$$V_{an} = \frac{V_{ab}}{\sqrt{3}} \angle -30° \quad \text{...(1.9)}$$

The other two-phase voltages also lag the respective line voltages by 30°.

1.6.2 Current in Three-Phase Circuits

If the load is symmetrical, the three-phase currents are also balanced and symmetrical when the voltages are balanced. The currents are given by

$$i_a = \sqrt{2} I_p \sin(\omega t + \phi)$$
$$i_b = \sqrt{2} I_p \sin(\omega t + \phi - 120°)$$
$$i_c = \sqrt{2} I_p \sin(\omega t + \phi + 120°) \qquad ...(1.10)$$

I_p is the magnitude of the current (rms). The phase angle ϕ depends on the nature of the load. (1.10) in phasor form can be written as

$$I_a = I_p \angle \phi°$$
$$I_b = I_p \angle (-120° + \phi)$$
$$I_c = I_p \angle (120° + \phi) \qquad ...(1.11)$$

In (1.11) v_a is taken as the reference phasor.

1.6.3 Power Relationship in Three-Phase Circuits

The total instantaneous power is equal to the sum of the individual phase powers,

$$p(t) = v_a i_a + v_b i_b + v_c i_c \text{ W}$$

By substituting for the phase voltages and currents from (1.6) and (1.10) respectively, we get

$$P_{3\phi} = 3 V_p I_p \cos\phi \qquad ...(1.12)$$

$P_{3\phi}$ is used to indicate three-phase power. From (1.12) it can be seen that the total instantaneous three phase power is non-pulsating and is equal to three times the real power in each phase. The power rating of a three-phase device normally refers to the three-phase power. The constant power is the main advantage of the three-phase system over the single-phase system. Since, the average power of the instantaneous power is the real power itself, concept of reactive power does not make much sense in three-phase systems. However, the instantaneous power in each phase is still pulsating. Hence, reactive power exists in the phases. To maintain similarity with single-phase systems, three-phase reactive power is defined as

$$Q_{3\phi} = 3 V_p I_p \sin\phi \qquad ...(1.13)$$

The three-phase complex power is given by

$$S_{3\phi} = P_{3\phi} + jQ_{3\phi} \qquad ...(1.14)$$

1.6.4 Single-Phase Equivalent of Three-Phase Systems

The analysis of balanced three-phase circuits can be carried out on a single phase basis. Generators, motors and transformers are designed for balanced operation. Transmission lines are balanced by transposing of the conductors. Single-phase loads are distributed equally amongst all the phases. Hence, in general, the three-phase system is balanced under normal operation. Therefore, the three-phase system can be analyzed on a single-phase basis. In balanced systems, the three phases are decoupled, meaning the voltage drop in a particular phase depends only on that particular phase current. For example, if we consider phase *a*,

$$v_a = (R_a + j\omega L_a)\, I_a + j\omega M I_b + j\omega M I_c \qquad \text{...(1.15)}$$

where, R_a, L_a are the resistance and inductance of phase *a* respectively and *M* is the mutual inductance between the phases. In balanced systems, the currents are given by (1.11). From trigonometric manipulations, we can see that $I_a + I_b + I_c = 0$. Using this, (1.15) can be written as

$$v_a = (R_a + j\omega L_a)\, I_a + j\omega M\,(I_b + I_c) = (R_a + j\omega L_a)\, I_a + j\omega M\,(-I_a)$$

$$= (R_a + j\omega\,(L_a - M))I_a \qquad \text{...(1.16)}$$

From (1.16), we can see that the voltage of phase *a* depends only on the current through phase *a*. Thus, the phases are decoupled. Calculations made for one phase can be used to calculate the other phase quantities using (1.6) and (1.11). In obtaining the single-phase equivalent circuit we note the following:

- All neutrals are at the same potential.
- Phases are decoupled.
- Neutral current is zero.
- All quantities have the same phase sequence as the excitation.

All delta connected loads and sources are converted to star equivalents. A balanced delta connected load whose phase impedance is $Z_\Delta \Omega$ has an equivalent star impedance $Z_Y = \frac{Z_\Delta}{3} \Omega$ per phase. The single-phase circuit is solved for phase *a* and the other phase quantities obtained form the phase relationships.

Example 1.4: In a balanced three-phase system $V_{ab} = 400\angle 0°$. If the phase sequence is *abc*, find the other line-to-line voltages and the phase voltages.

Solution:

$$V_{bc} = 400\angle -120°;\ V_{ca} = 400\angle 120°$$

$$V_{an} = \frac{400}{\sqrt{3}}\angle -30°;\ V_{bn} = \frac{400}{\sqrt{3}}\angle -150°;\ V_{cn} = \frac{400}{\sqrt{3}}\angle 90°$$

Example 1.5: The three-phase power in a transmission system is 180 MW and 24 MVAR at the sending end. The phase impedance of the line is $1 + j10\ \Omega$. The sending end voltage is 110 kV. Find the power and voltage at the receiving end.

Solution: $S_{3\phi} = 180 + j24\text{MVA};\ S_{1\phi} = 60 + j8\,\text{MVA}$

$$Z_p = 1 + j10\Omega$$

Sending end phase voltage $V_{as} = \frac{110}{\sqrt{3}} = 63.5\angle 0°$ kV

$$S_{1\phi} = V_{as} I_a^*;\ I_a^* = \frac{60 + j8}{63.5\angle 0°} = 0.953\angle 7.59°\ \text{kA}$$

$$I_a = 0.953\angle -7.59° = 0.9446 - j0.1258 \text{ kA.}$$

Receiving end voltage phase voltage $V_{ar} = V_{as} - I_a Z_p$

$$= 63.5\angle 0° - (0.953\angle -7.59° \times 10.05\angle 84.29°)$$
$$= 61.297 - j9.3207 = 62.0\angle -8.646° \text{ kV}$$

Receiving end power $S_{1\phi r} = V_{ar} \times I_a^*$

$$= 62.0\angle -8.646° \times 0.953\angle 7.59° = 59.075 - j1.0889 \text{ MVA}$$

Three phase power at receiving end

$$= (59.075 - j1.0889) \times 3 = 177.225 - j3.2667 \text{ MVA}$$

Magnitude of line-to-line voltage at receiving end $= 62.0 \times \sqrt{3} = 107.387$ kV

Active power loss = Sending end active power – Receiving end active power

$$= 180 - 177.225 = 2.775 \text{ MW}$$

Reactive power loss = Sending end reactive power – Receiving end reactive power

$$= 24 - (-3.2667) = 27.2667 \text{ MVAR.}$$

1.7 SINGLE-LINE DIAGRAM

The single-line diagram (also called one-line diagram) is a pictorial representation of connections of various components of the power system, using standard symbols for components and lines for the transmission lines connecting them. For balanced three-phase systems, the single-line diagram shows the single phase equivalent. The single-line diagram supplies concise information of the system components. The amount and detail of information, varies depending on the purpose for which the diagram is intended. For example, the location of circuit breaker may be vital information for a stability study, whereas totally unnecessary for an expansion study. Some standard symbols published by IEEE used in single-line diagrams are shown in Fig. 1.6.

The one-line diagram (single-line diagram) of a sample power system is shown in Fig. 1.7.

The following points are to be noted while drawing the single-line diagram:

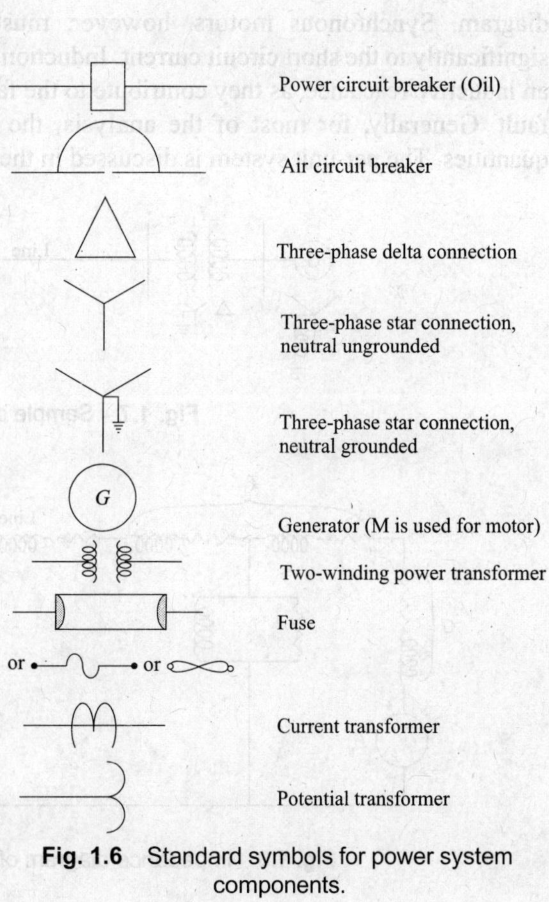

Fig. 1.6 Standard symbols for power system components.

- The location of points connected to ground has to be shown on the diagram.
- Transformer neutrals are generally solidly grounded and generator neutrals are grounded through high resistances or reactors. This information is shown on the single-line diagram and the information is vital for constructing the zero sequence network.
- Information of the equipment ratings can be given on the single-line diagram.
- In case of three phase transformers, the type of connection has to be indicated on the diagram.

The *impedance diagram* for the power system can be drawn from the single-line diagram. The impedance diagram for the sample system shown in Fig. 1.7 is shown in Fig. 1.8.

The impedance diagram shows the single phase electric circuit for three-phase systems. The impedance diagram does not include the current limiting resistors/reactors, since no current flows through the neutral under balanced conditions. Since the magnetising current of a transformer is insignificant when compared to the full load current, the shunt branch of the transformer is usually omitted in the impedance diagram. Since, in general, the reactances of the power system components like generators, motors and transmission lines are much larger than the resistances, the resistance can be omitted to obtain a *reactance diagram*. Loads which do not involve rotating machinery, do not significantly affect fault currents and hence may be omitted in the impedance diagram. Synchronous motors, however, must always be included since they contribute significantly to the short circuit current. Induction motors are represented by an emf in series with an inductive reactance, as they contribute to the fault current, immediately after occurrence of the fault. Generally, for most of the analysis, the per-unit system is used to represent various quantities. The per-unit system is discussed in the next section.

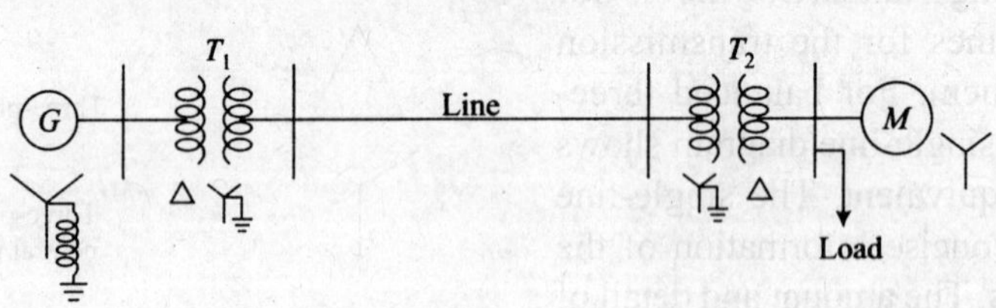

Fig. 1.7 Sample one-line diagram.

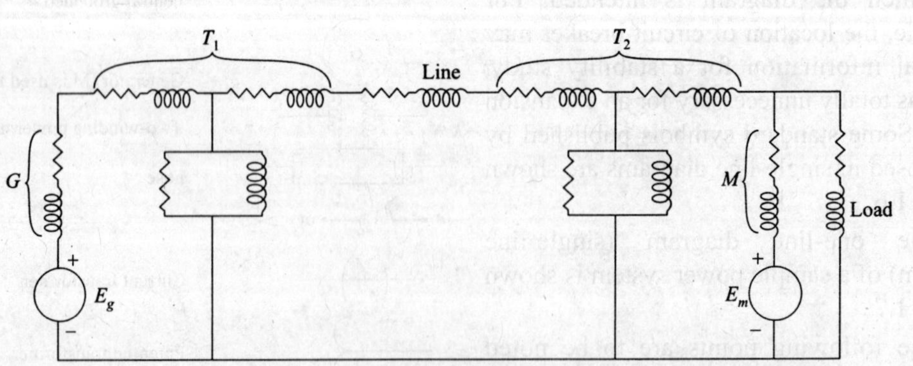

Fig. 1.8 Impedance diagram of the system shown in Fig. 1.7.

1.8 PER-UNIT SYSTEM

1.8.1 Per-unit (pu) Quantities

It is preferable in power engineering to express currents, voltages, impedances and power in per-unit rather than in actual amperes, volts, ohms or megawatts. The per-unit value of any quantity is defined as

$$\text{pu value} = \text{Actual value/Base value} \qquad ...(1.17)$$

For example, a voltage of 11 kV, has a pu value of 1 pu on a base of 11 kV and 0.1 pu on a base of 110 kV. The percentage value is the pu value multiplied by 100. Some of the advantages of the pu system are:

- Equipment impedance is normally specified by manufacturers in pu on the base of the equipment rating.
- The pu values of impedances lie in a narrow range even though the ohmic values may differ widely for machines of different ratings.
- Pu data yields information on relative magnitudes of various parameters.
- When transformers are present in the circuit, all impedances have to be referred to same side of the transformer. The pu impedance of the transformer, however is same when referred to either side, if proper bases are chosen.
- The pu impedance of transformers in three phase circuits is not affected by the connection of the transformer.

The voltage, current, power and impedance in a circuit are related to each other, so that the selection of base values of any two of them, determines the base values of the other two.

1.8.2 Selection of Base Values in Single-Phase Circuits

It is customary to select the base value of the power and voltage and obtain the base values of current and impedance from these. Suffix 1ϕ is used to denote single phase systems. The relationships between the base values are as follows:

$$\text{Base current (in amperes)} = \text{base MVA}_{1\phi} \times 10^3/\text{base kV}_{1\phi} \qquad ...(1.18)$$

$$\text{Base impedance (in Ohms)} = (\text{base kV}_{1\phi})^2/\text{base MVA}_{1\phi} \qquad ...(1.19)$$

If the above equations are used for the single phase equivalent of a three-phase system, base $\text{MVA}_{1\phi}$ stands for the base MVA per phase and $\text{kV}_{1\phi}$ stands for the line-to-neutral voltage.

Example 1.6: A 50 MVA, 11/132 kV transformer has an impedance of $0.15 + j1.0\ \Omega$ referred to the LV side. Show that by proper choice of the base values, the pu impedance is same when referred to LV side or HV side.

Solution: Let the base MVA be 50 MVA and the base kV on LV side be 11 kV and on HV side be 132 kV.

Base impedance on LV side = $11^2/50 = 2.42\ \Omega$.

Base impedance on HV side $= 132^2/50 = 348.48\ \Omega$.

pu impedance on LV side = $(0.15 + j1.0)/2.42 = 0.06198 + j0.41322$

Impedance of transformer referred to HV side = $(0.15 + j1.0)\,(132/11)^2 = 21.6 + j144\ \Omega$.

pu impedance on HV side = $(21.6 + j144)/348.48 = 0.06198 + j0.41322$

Thus, it can be seen that if the base voltages are chosen in the same ratio as the transformer voltages, the pu impedance is same on both sides.

1.8.3 Selection of Base Values in Three-Phase Circuits

Data for three-phase equipment normally specify the three-phase MVA and the line-to-line voltage. This may lead to confusion in obtaining the pu values, especially for the line-to-line voltages and the line-to-neutral voltages. If the line-to-neutral base voltage is taken to be $1/\sqrt{3}$ times that of the line-to-line base voltage, then *in balanced three-phase systems the pu value of the line-to-neutral voltage on the line-to-neutral voltage base will be equal to the pu value of the line-to-line voltage on the line-to-line voltage base.* For example, consider a 3-Φ alternator, rated at 11 kV, whose terminal voltage is 10.8 kV. If the line-to-line voltage base is taken to be 11 kV, then the line-to-neutral voltage base is $11/\sqrt{3} = 6.35$ kV. pu line-to-line voltage = 10.8/11 = 0.981. The line-to-neutral terminal voltage is $10.8/\sqrt{3} = 6.235$ kV. Hence, the pu line-to-neutral voltage is 6.235/6.35 = 0.981.

Similarly, *the pu value of the three-phase power on three phase MVA base is identical to the pu value of the single-phase power on the single-phase MVA base.* For example, consider a 3-Φ alternator, rated at 75 MVA. The per phase power is 75/3 = 25 MVA. If we choose the three-phase base MVA to be 100 MVA, then the single-phase base MVA is given by 100/3 = 33.33 MVA. The three phase pu power is 75/100 = 0.75 and single-phase pu power is 25/33.33 = 0.75.

$$\text{Base current (in amperes)} = \text{base MVA}_{3\phi} \times 10^3/\sqrt{3}\ \text{base } kV_{LL}$$

$$\text{Base impedance (in Ohms)} = (\text{base } kV_{LL})^2/\text{base MVA}_{3\phi}$$

In the above equations, kV_{LL} stands for the line-to-line voltage.

1.8.4 Change of Base

Normally, the impedances are specified on the rating of the equipment. Hence, there is a need to change the pu values from the base of the equipment rating to that of the chosen system base. Let MVA_{B1} and kV_{B1} be the base values and let Z_{pu1} be the pu value of impedance on this base. The actual impedance is given by

$$Z_{ac} = \text{pu impedance} \times \text{base impedance} = Z_{pu1} \times (kV_{B1})^2/MVA_{B1}$$

If now the new base is MVA_{B2} and kV_{B2}, the pu value on this base is given by

$$Z_{pu2} = Z_{ac}/Z_{B2} = Z_{pu1} \times (kV_{B1}{}^2/MVA_{B1}) \times (MVA_{B2}/kV_{B2}^2)$$

$$= Z_{pu1} \times (KV_{B1}/kV_{B2})^2 \times (MVA_{B2}/MVA_{B1}) \qquad ...(1.20)$$

(1.20) can be used to find the pu value of the impedance on one base given the pu value on another base.

1.8.5 Per-unit Impedance Diagram

The pu impedance diagram is obtained from the single-line diagram by the following steps:

(i) Choose a common system base MVA.
(ii) Divide the network into different sections based on the location of the transformers.
(iii) Choose a suitable voltage base in any one of the sections.
(iv) Fix the base voltages of the other sections using the voltage ratio of the transformers.
(v) Convert all impedances in pu to the base in the corresponding section.

Example 1.7: Draw the pu impedance diagram for the power system shown in Fig. 1.9.

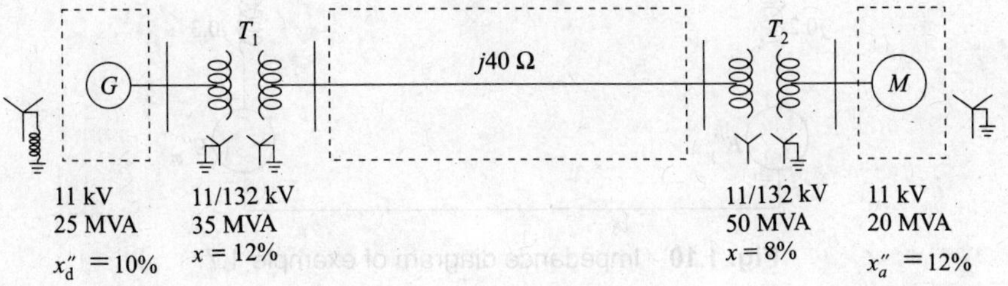

Fig. 1.9 Power system of example 1.7.

Solution: Choose a system base of 50 MVA. The system is divided into three sections as shown by the dotted line in Fig. 1.9. Let the base voltage in the generator section be 11 kV.

$$\text{Base voltage of transmission line section} = 11 \times \frac{132}{11} = 132 \text{ kV.}$$

$$\text{Base voltage on motor side} = 132 \times \frac{11}{132} = 11 \text{ kV.}$$

Per-unit impedances

Generator: $x_d'' = 10\% = 0.1$ pu (on the generator rating). The new base is 50 MVA, 11 kV.

$$\text{Therefore, pu reactance} = 0.1 \times \left(\frac{11}{11}\right)^2 \times \frac{50}{25} = 0.2 \text{ pu.}$$

Transformer T1: $x_{T1} = 12\% = 0.12$ pu (on the transformer rating). The new base is 11 kV, 50 MVA on LV side. Therefore, the pu reactance is $0.12 \times \left(\frac{11}{11}\right)^2 \times \frac{50}{35} = 0.171$ pu.

Transmission line: Base on transmission line is 132 kV, 50 MVA

Base impedance = $(132)^2/50 = 348.48\ \Omega$. Therefore, the pu reactance is $\frac{40}{348.48} = 0.115$ pu.

Transformer T2: pu reactance $= 0.08 \times \left(\frac{11}{11}\right)^2 \times \frac{50}{50} = 0.08$ pu.

Motor M: pu reactance $= 0.12 \times \left(\frac{11}{11}\right)^2 \times \frac{50}{20} = 0.3$ pu.

The impedance diagram is shown in Fig. 1.10.

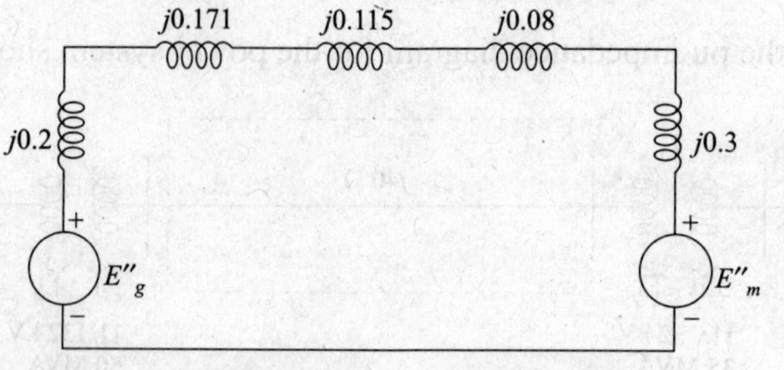

Fig. 1.10 Impedance diagram of example 1.7.

Example 1.8: Draw the pu impedance diagram of the power system shown in Fig. 1.11.

Solution: The system is divided into sections shown by the dotted lines. Let the base voltage on generator section be 22 kV. Let base MVA be 100 MVA.

$$\text{Base voltage on line } L1 = 22 \times \frac{220}{22} = 220 \text{ kV.}$$

$$\text{Base voltage on line } L2 = 22 \times \frac{110}{22} = 110 \text{ kV.}$$

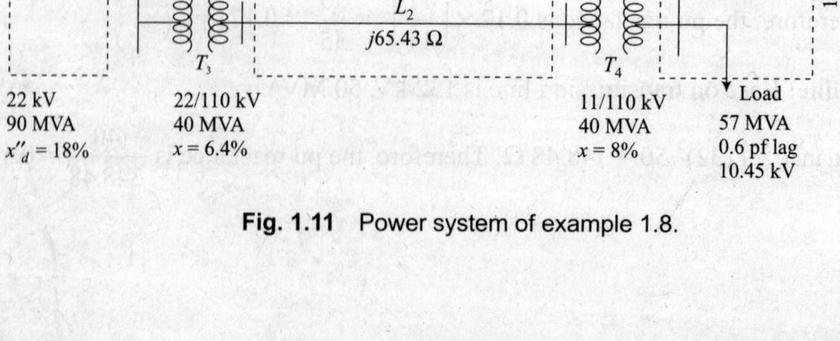

Fig. 1.11 Power system of example 1.8.

Base voltage on load section $= 110 \times \frac{11}{110} = 11$ kV.

Generator: pu reactance $= 0.18 \times \frac{100}{90} = 0.2$ pu.

Transformer T1: pu reactance $= 0.1 \times \frac{100}{50} = 0.2$ pu.

Transformer T2: pu reactance $= 0.06 \times \frac{100}{40} = 0.15$ pu.

Transformer T3: pu reactance $= 0.064 \times \frac{100}{40} = 0.16$ pu.

Transformer T4: pu reactance $= 0.08 \times \frac{100}{40} = 0.2$ pu.

Motor M: pu reactance $= 0.185 \times \left(\frac{100}{66.5}\right)\left(\frac{10.45}{11}\right)^2 = 0.25$ pu.

Base impedance on line $L1 = (220)^2/100 = 484\ \Omega$. pu reactance $= 48.4/484 = 0.1$ pu.
Base impedance on line $L2 = (110)^2/100 = 121\ \Omega$. pu reactance $= 65.43/121 = 0.54$ pu.
Complex load $S_L = 57\angle 53.13°$. Base impedance on load section $= (11)^2/100 = 1.21\ \Omega$.
Equivalent impedance $= (V_{LL})^2/S_L^* = 1.1495 + j1.533\ \Omega$

$$= (1.1495 + j1.533)/1.21 = 0.95 + j1.2667 \text{ pu}$$

The complete impedance diagram is shown in Fig. 1.12.

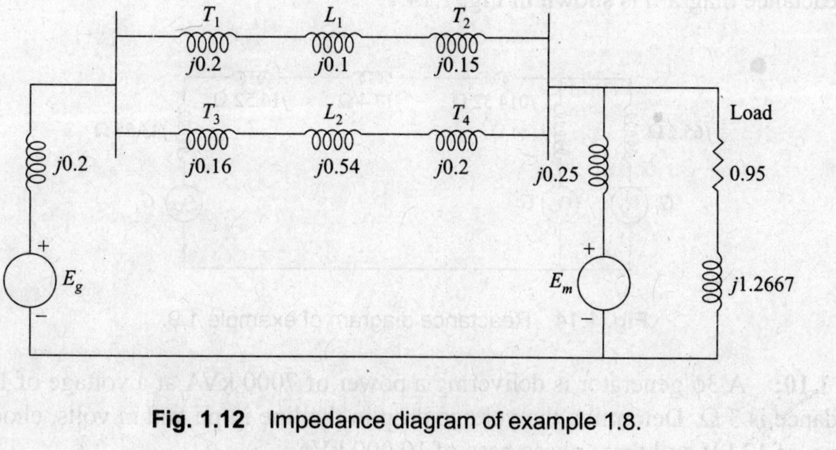

Fig. 1.12 Impedance diagram of example 1.8.

Example 1.9: Consider the system shown in Fig. 1.13.

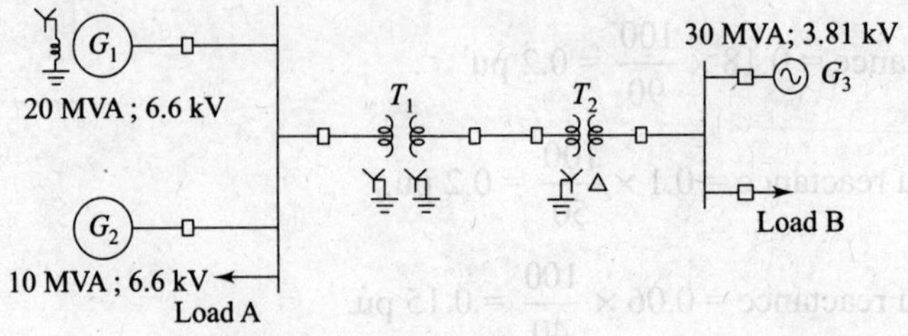

Fig. 1.13 Example 1.9.

The data is as follows: $X_{G1} = 0.652\ \Omega$; $X_{G2} = 1.31\ \Omega$; $X_{G3} = 0.1452\ \Omega$; $X_{T1} = X_{T2} = 14.52\ \Omega$; (referred to *HT*); $X_{\text{line}} = 17.4\ \Omega$; T_1 and T_2 are a bank of three single phase transformers rated 10 MVA ; 3.81/38.1 kV; draw the reactance diagram of the system.

Solution: Transformation ratio of $T_1 = KT_1 = \dfrac{V_{L2}}{V_{L1}} = \dfrac{38.1\sqrt{3}}{3.81\sqrt{3}} = 10:1$

(Note it is a λ – λ transformer)

Transformation ratio of $T_2 = KT_2 = \dfrac{38.1\sqrt{3}}{3.81} = 10\sqrt{3} = 17.32$

(Note it is a λ – Δ transformer)

X_{G1} referred to *HT* side will be $0.652 \times KT_1^{\,2} = 65.2\ \Omega$

X_{G2} referred to *HT* side $= 1.31 \times KT_1^{\,2} = 131\ \Omega$

X_{G3} referred to HT side $= 0.1452 \times (17.32)^2$

$= 43.55\ \Omega$

The reactance diagram is shown in Fig. 1.14.

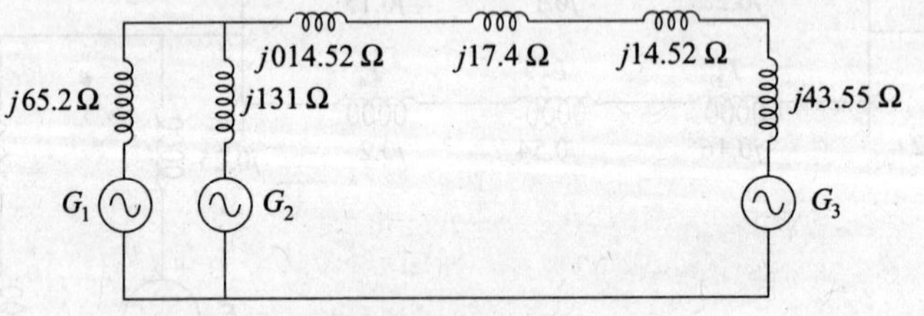

Fig. 1.14 Reactance diagram of example 1.9.

Example 1.10: A 3ϕ generator is delivering a power of 7000 kVA at a voltage of 12.9 kV. The line impedance is 3 Ω. Determine the voltage drop in the line in pu and in volts, choosing a base line voltage of 13 kV and three phase base of 10,000 kVA.

Solution: Line current $= \dfrac{3\phi\, kVA}{\sqrt{3}\, kV_{LL}} = \dfrac{7000}{\sqrt{3}\; 12.9} = 313.3$ A

Voltage drop in the line $= Z\,I$

$= 3 \times 313.3 = 939.9$ V

With a base of 10,000 kVA and 13 kV, we get generator kV in pu $= \dfrac{12\cdot 9}{13} = 0.9923$ pu

$$\text{kVA pu} = \frac{7{,}000}{10{,}000} = 0.7 \text{ pu}$$

$$\text{Base current} = \frac{\text{Base } kVA}{\sqrt{3}\ \text{Base } kV_{LL}}$$

$$= \frac{10{,}000}{\sqrt{3} \times 13} = 444.11 \text{ A}$$

$$\text{pu current} = \frac{313.3}{444.11} = 0.705 \text{ pu.}$$

$$\text{(Note that in pu, pu current} = \frac{pu\ kVA}{pu\ kV_{LL}}.)$$

$$Z_{\text{base}} = \frac{kV^2}{MVA} = \frac{13^2}{10} = 16.9\ \Omega$$

$$Z_{\text{pu}} = \frac{3}{16.9} = 0.1775 \text{ pu.}$$

Voltage drop in line in pu $= Z_{\text{pu}}\, I_{\text{pu}} = 0.1775 \times 0.705 = 0.125$ pu.

The voltage drop in the line in pu, can also be calculated as voltage drop in volts / base voltage. Here we have to take line-to-neutral voltage base, which is $13/\sqrt{3}$ kV.

Voltage drop in pu $= \dfrac{939.9}{13 \times 10^3/\sqrt{3}} = 0.125$ pu

Example 1.11: Prove that the pu impedance of the transformer is the same when referred to either side of the transformer.

Solution: Let V_1 and I_1 be the primary voltage and current; V_{1B} and I_{1B} the voltage and current base in the primary; V_2 and I_2 be the secondary voltage and current; V_{2B} and I_{2B} be the secondary base voltage and current.

Let Z_{eq1} be the impedance of the transformer referred to the primary side

Z_{eq2} be the impedance of the transformer referred to the secondary side

$$Z_{eq1}\ \text{(pu)} = \frac{Z_{eq1}}{V_{1B}/I_{1B}} = Z_{eq1}\frac{I_{1B}}{V_{1B}}$$

$$Z_{eq2}\,(\text{pu}) = \frac{Z_{eq_2}}{V_{2B}/I_{2B}} = Z_{eq2}\,\frac{I_{2B}}{V_{2B}}$$

For a transformer, $\dfrac{V_2}{V_1} = \dfrac{I_1}{I_2} = K$

$$Z_{eq2} = K^2\, Z_{eq1}$$

$$V_{2B} = KV_{1B} \text{ and } I_{1B} = K\,I_{2B}$$

$$\therefore \qquad Z_{eq2}\,(\text{pu}) = \frac{(K^2 Z_{eq_1})\, I_{1B}/K}{KV_{1B}} = Z_{eq1}\frac{I_{1B}}{V_{1B}}$$

$$= Z_{eq1}(\text{pu})$$

Thus, the pu impedance is the same, when referred to primary or secondary.

Example 1.12: Obtain the pu diagram from the following one line representation of the power system of Fig. 1.15.

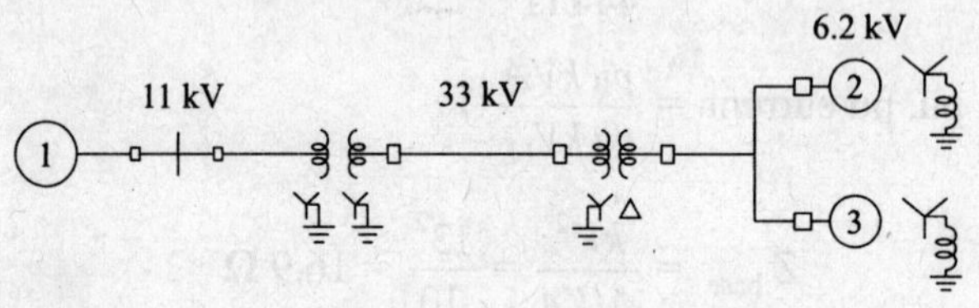

Fig. 1.15 Example 1.12.

G_1: 30 MVA; 10.5 kV; $X'' = 1.6\ \Omega$

G_2: 15 MVA; 6.6 kV; $X'' = 1.2\ \Omega$

G_3: 25 MVA; 6.6 kV; $X'' = 0.56\ \Omega$

T_1: 15 MVA; 33/11 kV; $X = 15.2\ \Omega$/ ph on HV side

Transmission line impedance is 20.5 Ω/ph.

T_2: 15 MVA; 33/6.6 kV; $X = 16\ \Omega$/ph on HV side

Choose a base of 30 MVA; 33 kV on transmission line.

Solution: $\underline{G_1}$ $Z_{pu} = \dfrac{Z \text{ in } \Omega}{Z_B} = \dfrac{Z \text{ in } \Omega}{(kV_B)^2 / MVA_B}$

Base MVA = 30; Base kV on G_1 side = 11 kV.

$$\therefore \qquad Z_{pu} = 1.6 \times \frac{30}{11^2} = 0.396 \text{ pu.}$$

$\underline{G_2}$ Base voltage is 6.2 kV.

$$Z_{pu} = 1.2 \times \frac{30}{(6.2)^2} = 0.936 \text{ pu}$$

$\underline{G_3}$ Base voltage is 6.2 kV

$$Z_{pu} = 0.56 \times \frac{30}{(6.2)^2} = 0.437$$

Transmission line
$$Z_{pu} = 20.5 \times \frac{30}{(33)^2} = 0.564$$

$$T_1 : Z_{pu} = 15.2 \times \frac{30}{(33)^2} = 0.418$$

$$T_2 : Z_{pu} = 16 \times \frac{30}{(33)^2} = 0.44$$

The pu reactance diagram is shown in Fig. 1.16.

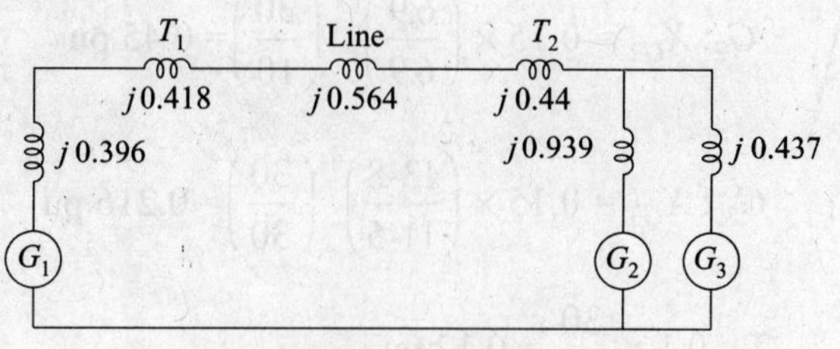

Fig. 1.16 Solution of example 1.12.

Example 1.13: Obtain the pu reactance diagram of the netwok shown in Fig. 1.17.

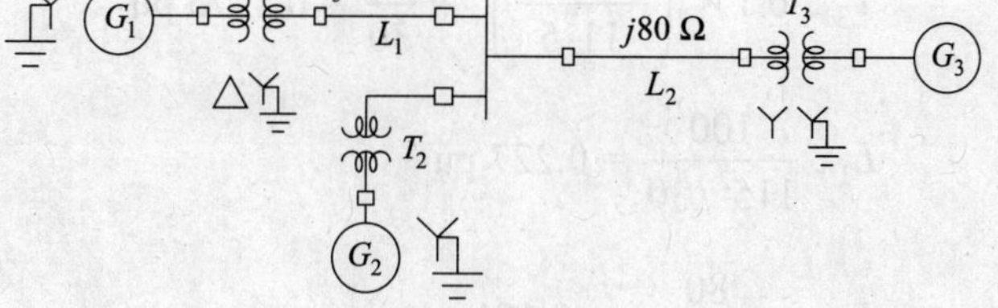

Fig. 1.17 Example 1.13.

The data is as follows:

G_1 – 20 MVA; 6.9 kV ; $X'' = 0.15$ pu
G_2 – 10 MVA; 6.9 kV $X'' = 0.15$ pu
G_3 – 30 MVA; 13.8 kV, $X'' = 0.15$ pu
T_1 – 25 MVA; 6.9 Δ – 115Y kV, $X = 10\%$

T_2 – 12 MVA; 6.9 Δ – 115Y kV, X= 10%

T_3 – 3 single phase transformers rated 10 MVA, 7.5 –75 kV; X= 10%

Draw the pu reactance diagram choosing a base of 30 MVA, 6.9 kV in the G_1 circuit.

Solution: The base voltages in various circuits are as follows:

$$G_1\text{: 6.9 kV; lines } L_1 \text{ and } L_2\text{: } 6.9 \times \frac{115}{6.9} = 115 \text{ kV}$$

$$\text{G2: } 115 \times \frac{6.9}{115} = 6.9 \text{ kV}$$

$$G_3\text{: } 115 \times \frac{7.5}{75} = 11.5 \text{ kV}$$

We calculate the pu reactances of the various components on the bases chosen, using (1.20)

$$G_1\text{: } X_{G1} = 0.15 \times \left(\frac{6.9}{6.9}\right)^2 \left(\frac{30}{20}\right) = 0.225 \text{ pu}$$

$$G_2\text{: } X_{G2} = 0.15 \times \left(\frac{6.9}{6.9}\right)^2 \left(\frac{30}{10}\right) = 0.45 \text{ pu}$$

$$G_3 : X_{G3} = 0.15 \times \left(\frac{13 \cdot 8}{11 \cdot 5}\right)^2 \left(\frac{30}{30}\right) = 0.216 \text{ pu}$$

$$\text{T}_1\text{: } 0.1 \times \frac{30}{25} = 0.12 \text{ pu}$$

$$\text{T}_2 : 0.1 \times \frac{30}{12} = 0.25 \text{ pu}$$

$$\text{T}_3 : 0.1 \times \left(\frac{7 \cdot 5\sqrt{3}}{11 \cdot 5}\right)^2 \times \frac{30}{30} = 0.1275 \text{ pu}$$

$$L_1 : \frac{100}{115^2/30} = 0.227 \text{ pu}$$

$$L_2 = \frac{80}{115^2/30} = 0.181 \text{ pu}$$

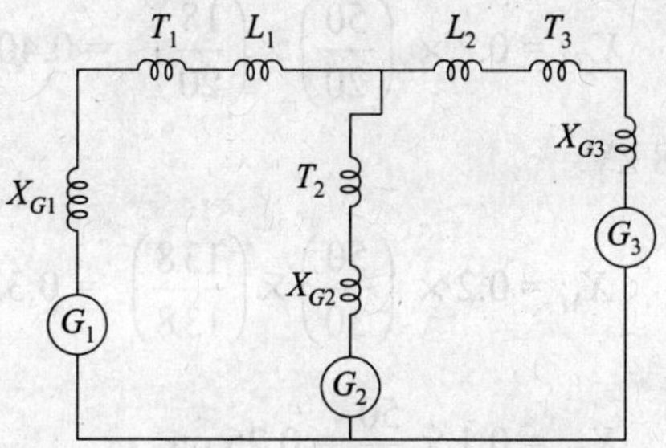

Fig. 1.18 Solution Example 1.13.

Example 1.14: Draw the pu impedance diagram of the system shown in Fig. 1.19.

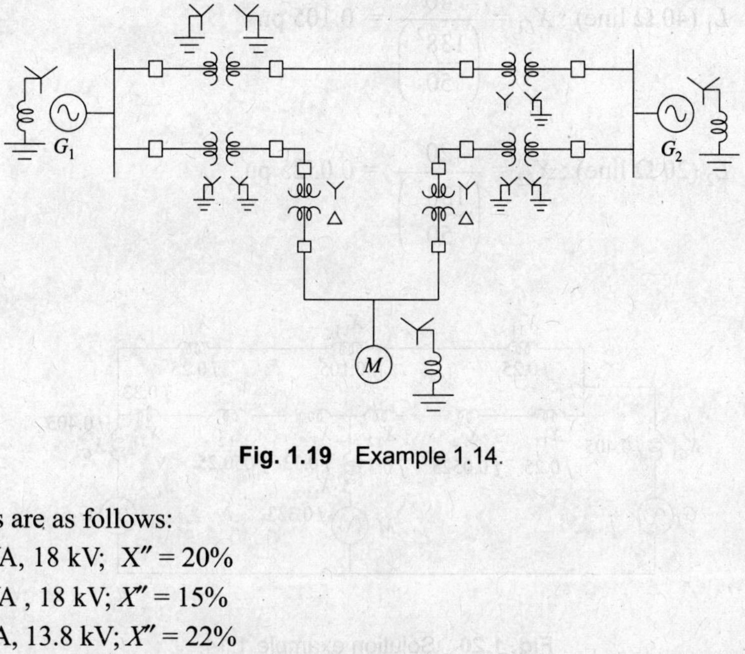

Fig. 1.19 Example 1.14.

The ratings are as follows:

G_1: 20 MVA, 18 kV; $X'' = 20\%$

G_2: 20 MVA , 18 kV; $X'' = 15\%$

M: 30 MVA, 13.8 kV; $X'' = 22\%$

Three phase Y – Y transformer: 20 MVA ,138/20 kV, $X = 10\%$

Three phase Y – Δ transformer = 15 MVA, 138/13.8 kV, $X = 10\%$

Choose a base of 50 MVA, 138 kV in the 40 Ω line.

Solution:

G_1 : Base is 50 MVA , 20 kV.

$$X_{G1} = 0.2 \times \left(\frac{50}{20}\right) \times \left(\frac{18}{20}\right)^2 = 0.405 \text{ pu}$$

G_2 : Base is 50 MVA , 20 kV.

$$X_{G2} = 0.2 \times \left(\frac{50}{20}\right) \times \left(\frac{18}{20}\right)^2 = 0.405 \text{ pu}$$

M : Base is 50 MVA , 13.8 kV.

$$X_M = 0.2 \times \left(\frac{50}{30}\right) \times \left(\frac{13.8}{13.8}\right)^2 = 0.333 \text{ pu}$$

Y – Y transformers: $X_{T1} = 0.1 \times \frac{50}{20} = 0.25$ pu

Y – Δ transformers: $X_{T2} = 0.1 \times \frac{50}{15} = 0.33$ pu

$$L_1 \text{ (40 Ω line)} : X_{L1} = \frac{40}{\left(\frac{138^2}{50}\right)} = 0.105 \text{ pu}$$

$$L_2 \text{ (20 Ω line)} : X_{L2} = \frac{20}{\left(\frac{138^2}{50}\right)} = 0.0525 \text{ pu}$$

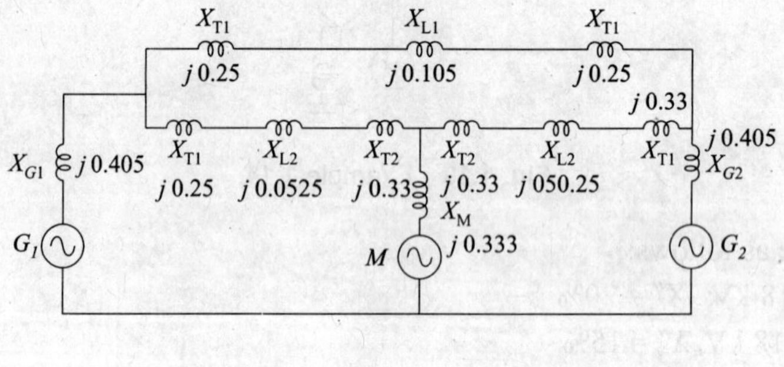

Fig. 1.20 Solution example 1.14.

Example 1.15: Figure 1.21 shows the single line diagram of a simple single phase circuit. Choosing a base of 3 kVA, 230 V on the generator section, draw the pu circuit diagram and determine the load current.

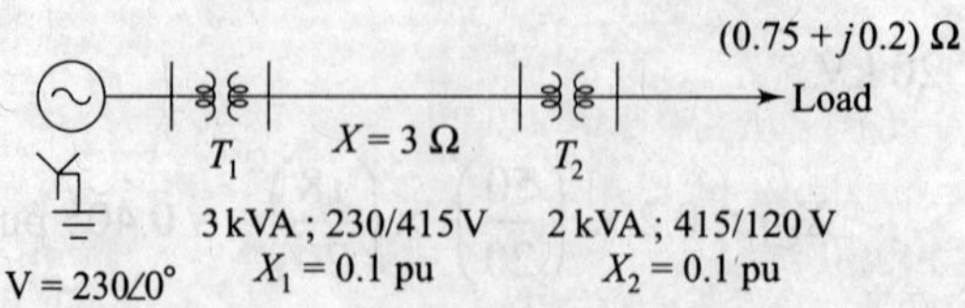

Fig. 1.21 Example 1.15.

Solution: Base on generator section is 3 kVA ; 230 V.

i.e. 3×10^{-3} MVA; 230×10^{-3} kV

Base on transmission line is 3 kVA ; 415 V.

i.e. 3×10^{-3} MVA; 415×10^{-3} kV

Base on load side is 3 kVA ; 120 V

i.e. 3×10^{-3} MVA; 120×10^{-3} kV

Z_{Bline} (Base impedance of transmission line)

$$= \frac{(0.415)^2}{0.003} = 57.4\ \Omega$$

$$Z_{line} = \frac{3}{57.4} = 0.052 \text{ pu}$$

The pu impedance of T_1 need not be changed as it is on proper base.

$$Z_{pu,T2} = 0.1 \times \frac{3}{2} = 0.15 \text{ pu}$$

$$\text{Base impedance of the load} = \frac{(0.12)^2}{0.003} = 4.8\ \Omega$$

$$Z_{load} = \frac{0.75 + j\,0.2}{4.8} = 0.15625 + j\,0.0416 \text{ pu}$$

The pu impedance diagram is shown in Fig. 1.22.

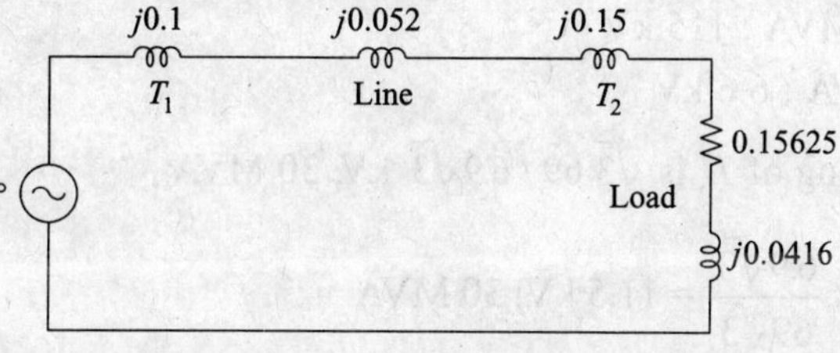

Fig. 1.22 Solution example 1.15.

$$I_L = \frac{1.0}{j\,0.1 + j\,0.052 + j\,0.15 + (0.15625 + j\,0.0416)}$$

$$= \frac{1.0}{0.15625 + j\,0.3436} = \frac{1.0}{0.377 \angle 65.54^\circ}$$

$$= 2.65 \angle -65{\cdot}54^\circ \text{ pu}$$

Base current on load side $\dfrac{3\times10^3}{120} = 25\text{A}$

$\therefore$ Load current $= 2.65\ \angle -65{\cdot}54° \times 25$

$= 66.25\ \angle -65{\cdot}54°$

Example 1.16: The data for the system of Fig 1.23, is as follows:

G_1 : 25 MVA, 6.6 kV, $X_g = 0.2$ pu

G_2 : 15 MVA, 6.6 kV, $X_g = 0.15$ pu

G_3 : 30 MVA , 13.2 kV, $X_g = 0.15$ pu

T_1 : 30 MVA , 6.6 Δ–115 Y kV, $X_T = 0.15$ pu

T_2 : 15 MVA , 6.6 Δ – 115 Y kV, $X_T = 0.2$ pu

T_3 : Bank of three single phase transformers Y – Y, 6.9 /69 kV, 10 MVA $X_{T3} = 0.1$ pu.

Draw the pu diagram taking a base of 30 MVA, 6.6 kV on the generator, G_1.

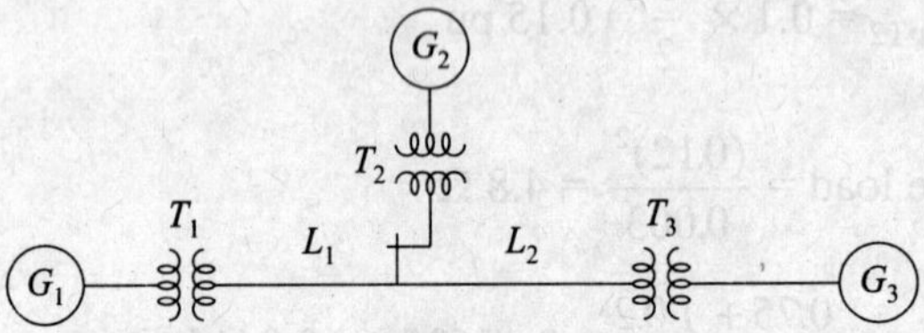

Fig.1. 23 Example 1.16.

Solution

Choose base of 30 MVA , 6.6 kV on generator, G_1.

Base on L_1, L_2: 30 MVA ; 115 kV

Base on G_2 : 30 MVA ; 6.6 kV

The three phase rating of T_3 is $\sqrt{3}\ 69 / 6.9\sqrt{3}$ kV, 30 MVA

Base on G3: $115 \times \dfrac{6.9\sqrt{3}}{69\sqrt{3}} = 11.5$ kV; 30 MVA

We now convert all values to pu.

$$X_{g1} = 0.2 \times \frac{30}{25} = 0.24 \text{ pu}$$

$$X_{g2} = 0.15 \times \frac{30}{25} = 0.30 \text{ pu}$$

$$X_{g3} = 0.15 \times \left(\frac{13{\cdot}2}{11{\cdot}5}\right)^2 = 0.2 \text{ pu}$$

$$X_{T1} = 0.1 \text{ pu}$$

$$X_{T2} = 0.1 \times \left(\frac{30}{15}\right) = 0.20 \text{ pu}$$

$$X_{T3} = 0.1 \left(\frac{69\sqrt{3}}{115}\right)^2 = 0.108 \text{ pu}$$

$$Z_{\text{Bline}} = \frac{115^2}{30} = 440\ \Omega$$

$$Z_{L1} = \frac{120}{440} = 0.27 \text{ pu}$$

$$Z_{L2} = \frac{90}{440} = 0.2045 \text{ pu}$$

The reactance diagram is shown in Fig. 1.24.

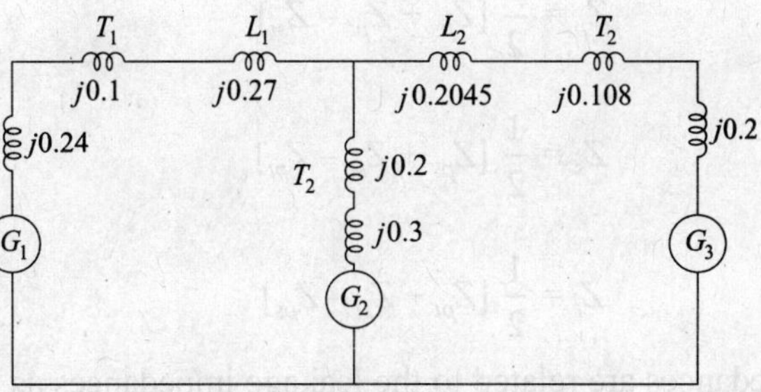

Fig. 1.24 Solution example 1.16.

1.9 PER-UNIT IMPEDANCES OF THREE-WINDING TRANSFORMER

The primary and secondary windings of a two-winding transformer have the same kVA rating. But all the three windings of a three-winding transformer may have different kVA ratings. The pu impedances of all the three windings must be expressed on the same kVA base. The schematic diagram and equivalent circuit are shown in Fig. 1.25.

The three winding are primary, secondary and tertiary. Three impedances are measured using standard short-circuit tests, defined as follows:

Z_{ps} : Leakage impedance measured in primary with secondary short-circuited and tertiary open.

Z_{pt} : Leakage impedance measured in primary with tertiary short-circuited and secondary open.

Z_{st} : Leakage impedance measured in secondary with tertiary short-circuited and primary open.

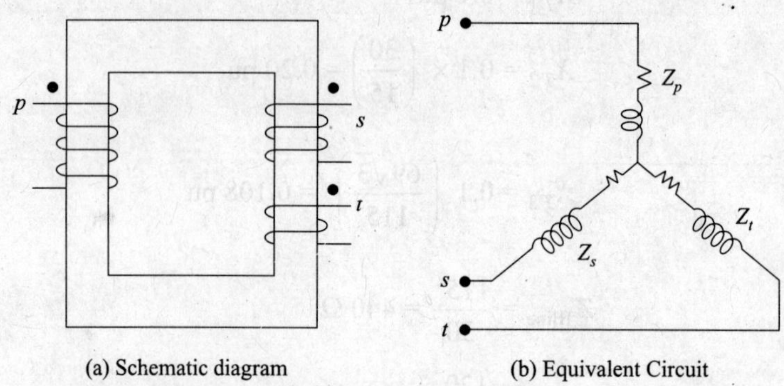

Fig. 1.25 Three-winding transformer.

If these impedances are referred to the primary circuit, and Z_p, Z_s and Z_t are impedances of the three windings referred to the primary.

$$Z_p = \frac{1}{2}\,[Z_{ps} + Z_{pt} - Z_{st}]$$

$$Z_s = \frac{1}{2}\,[Z_{ps} + Z_{st} - Z_{pt}]$$

$$Z_t = \frac{1}{2}\,[Z_{pt} + Z_{st} - Z_{ps}]$$

The individual impedances are related to the leakage impedances as follows:

$$Z_{ps} = Z_p + Z_s$$

$$Z_{pt} = Z_p + Z_t$$

$$Z_{st} = Z_s + Z_t$$

The impedances can be connected in star single-phase equivalent of the three-winding transformer, with magnetizing current neglected, as shown in Fig. 1.25 b. The neutral has no relation to the system neutral. The points *p*, *s*, *t* of the equivalent circuit, are connected to the parts in the impedance diagram connected to primary, secondary and tertiary windings. The voltage base of the three windings should be in the same ratio as the rated line-to-line voltages of the three circuits of the transformer. When three-phase, three-winding transformer is used, the primary and secondary are connected in Y and tertiary in Δ, to provide path for the third harmonic of the exciting current. Tertiary windings are provided for the following reasons:

- It reduces unbalancing in the primary due to unbalanced three-phase load.
- It redistributes the fault current.
- It can be used to supply an auxiliary load at a different voltage level.

1.10 SUMMARY

In the analysis of the power system, many studies are carried out like load flows, stability analysis, economic scheduling of generation, etc. In modern complex and integrated power systems, digital computers are extensively used for analysis, operation and control. In the analysis of power systems, we often have to solve non-linear algebraic or differential equations, which require application of numerical methods. The choice of the method depends on the computation time, complexity of programming, flexibility, reliability, memory requirements, etc.

In power system solutions, generally balanced three-phase systems are considered. The sinusoidal voltages and currents are represented as phasors using complex arithmetic. Balanced three phase systems can be solved on a single-phase basis. Hence, the three phase systems are normally pictorially represented by single-line diagrams. The Per-Unit (pu) system of representing the variables, makes the variables lie within a narrow range. The pu quantity is defined as the ratio of the actual value to the base value. It also provides an idea of relative magnitudes of different variables. The pu impedance diagram can be easily obtained from the single-line diagram of the system. The pu system also eliminates the need to transform impedances in the presence of transformers, since the pu impedance of the transformer is the same as referred to the primary or the secondary, if the base voltages on the two sides are chosen in the same ratio as the voltages.

1.11 SUPPLEMENTARY PROBLEMS

(1) If $v(t) = 325.2 \sin(\omega t + 40°)$ V and $i(t) = 23.45 \sin(\omega t - 25°)$ A, find (a) Maximum value of current and voltage (b) rms value of current and voltage (c) Phasor expression for current and voltage in rectangular and polar form.

(2) A single phase AC voltage 230 V is applied to a series circuit whose impedance is $8\angle 25°\Omega$. Find R, X, P, Q and S.

(3) A voltage source $E_{an} = -115\angle 200°$ V and current through the source is given by $I_{na} = 9\angle 60°$. Find P and Q and specify if the source is delivering or absorbing P and Q.

(4) Three impedances $8\angle -10°\ \Omega$ are Y connected to a balanced three phase supply whose line-to-line voltage is 400 V. Specify all line and phase voltages in polar form as phasors with phasor V_{bc} as the reference, for a phase sequence (i) *abc* (ii) *acb*.

(5) A three-phase generator is rated 13.2 kV, 50 MVA and has a reactance of 8%. What is the actual reactance in ohms? What is the pu value on a base of 11 kV, 100 MVA?

(6) Draw the impedance diagram of the power system whose single-line diagram is shown in Fig. P.1.6.

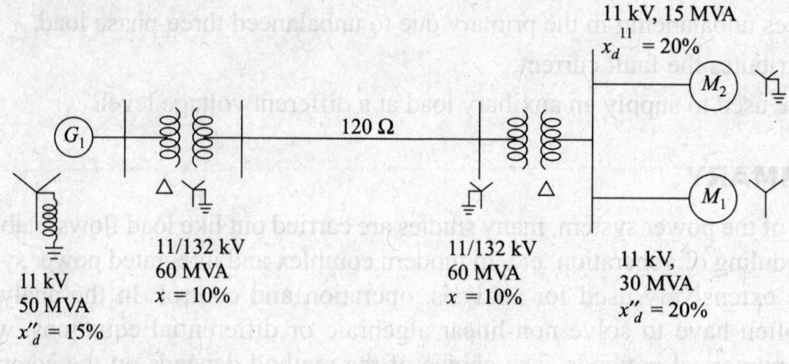

Fig. P.1.6 Single-line diagram of P6.

(7) Draw the impedance diagram of the power system whose single-line diagram is shown in Fig. P.1.7

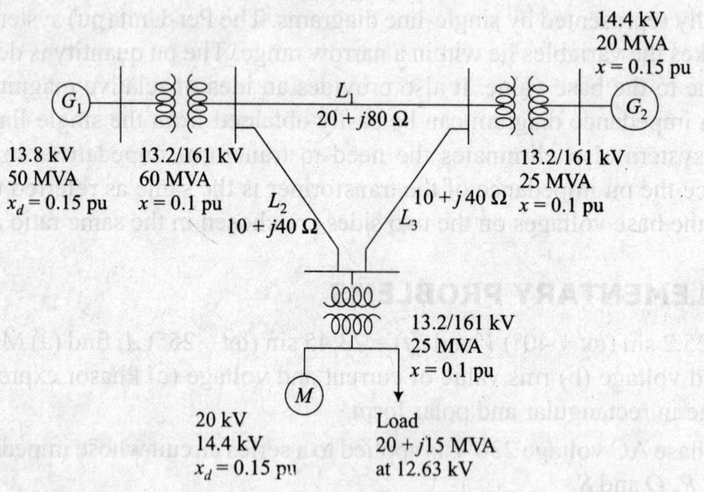

Fig. P.1.7 Single-line diagram of P7.

REVIEW QUESTIONS

1. Discuss the various studies carried out for analysis of the power system.
2. From fundamental principles, define active and reactive power.
3. ICT has become an integral part of the power system operation. Comment.
4. Explain the concept of complex power and the convention used in defining positive and negative P and Q.
5. What is the principle of conservation of complex power?
6. Prove that the total instantaneous three-phase power is non-pulsating.
7. Three-phase systems cannot always be analyzed using a single-phase equivalent.Comment.

8. Power system analysis would become virtually impossible without the pu notation. Comment.
9. Explain how bases are chosen in single-phase and three-phase circuits.
10. Why is the pu system advantageous?
11. Prove that the pu impedance of the transformer referred to the primary or secondary is the same.
12. Differentiate between reactance and impedance diagrams. Show how the pu values are changed from one base to another.

Chapter 2

Network Topology

2.1 INTRODUCTION

The solution of a given linear network problem firstly requires the formation of a set of equations describing the response of the network. The mathematical model, so derived, must describe the characteristics of the individual network components, as well as the relationship which governs the interconnection of the individual components. The formulation of the mathematical model can be based either on loop or nodal technique. In the bus frame of reference, the variables are the node voltages and node currents. In the loop frame of reference, the variables are the loop voltages and loop currents.

The independent variables in any reference frame can be either currents or voltages. Correspondingly, the coefficient matrix relating the dependent variables and the independent variables will be either an impedance matrix or an admittance matrix. The formulation of the appropriate relationships between the independent and dependent variables is an integral part of a digital computer program for the solution of power system problems. The formulation of the network equations in different reference frames requires the knowledge of graph theory. Elementary graph theory concepts are presented in the next section, followed by development of network equations in different reference frames.

2.2 ELEMENTARY GRAPH THEORY

A physical network is an assembly of interconnected branches. For example, a telephone network, a power system, a transport network, a computer network, etc. constitute physical networks. The geometrical interconnection of the various branches is called the topology of the network. The connection of the network topology, shown by replacing all physical elements by lines is called a *graph.* More precisely, a linear graph $G(V, E)$ consists of a set of objects $V = \{v_1, v_2, v_3, v_4 \ldots\}$ called *vertices* (or *nodes*) and another set $E = \{e_1, e_2, e_3, e_4 \ldots\}$ called *edges* (or *elements* or *branches*), such that each edge e_k is identified with an ordered pair of vertices (v_i, v_j). The vertices (v_i, v_j) associated with edge e_k are called the end vertices of e_k. An edge where both end vertices are same is called a *self-loop*. A graph in which a direction is assigned to each branch is called an *oriented graph* or a *directed graph*. A representative power system and its oriented graph is shown in Fig. 2.1.

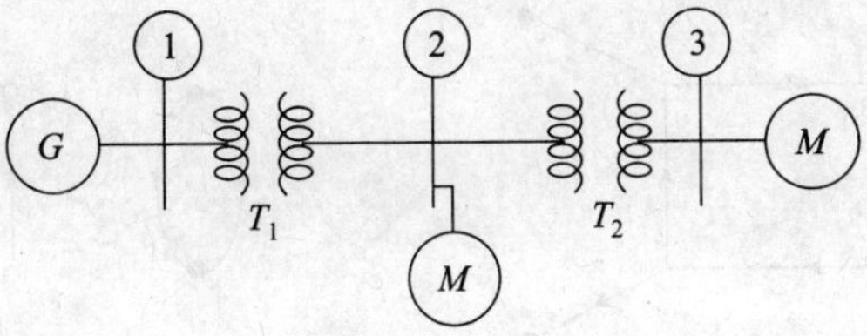

Fig. 2.1a Single-line diagram of a power system.

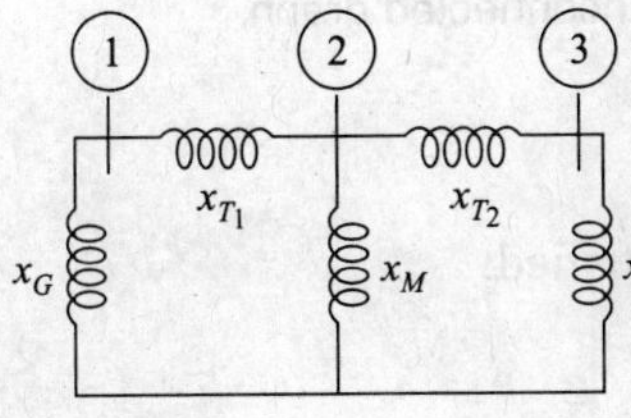

Fig. 2.1b Reactance diagram.

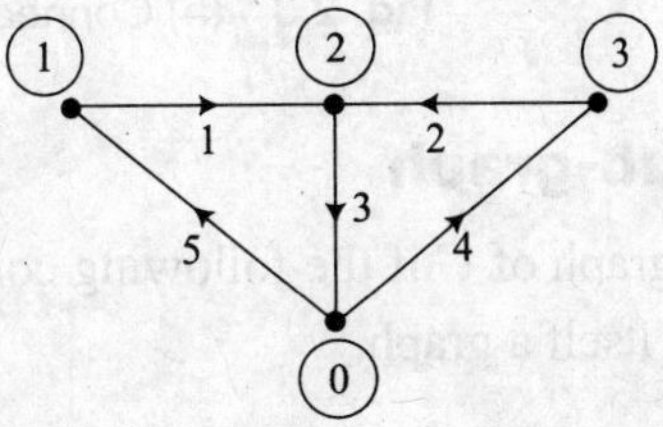

Fig. 2.1c Oriented graph.

It is to be noted that the directions of currents in various branches in an electric network are arbitrarily assigned and the network equations are derived, consistent with the assigned directions.

There are two quantities associated with each branch. The through quantity f, corresponding to the flow in the branch and an across quantity p corresponding to the potential difference across the branch. These may represent gas flow (f) and pressure difference (p) in a mechanical system or electric current (f) and voltage difference (p) in an electrical network. In linear networks, $\Delta p = r\,\Delta f$, where r is called the *branch impedance.*

In electric networks the convention is to use associated directions for the voltage drops. This means the voltage drop in a branch is taken to be in the direction of the current through the branch. Hence, we need not mark the voltage polarities in the oriented graph. This is shown in Fig. 2.2.

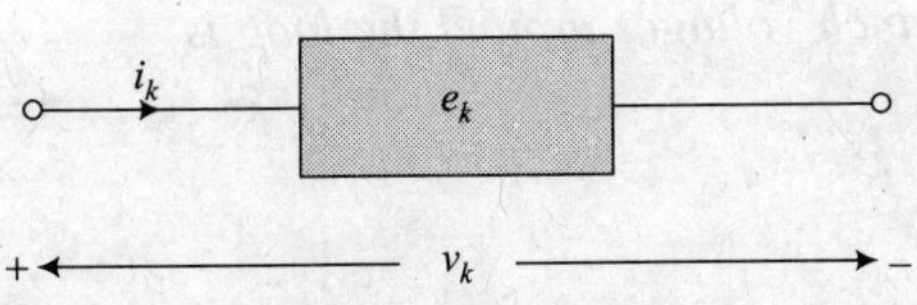

Fig. 2.2 Associated direction in a branch.

2.2.1 Connected Graph

This is a graph where at least one path (disregarding orientation) exists between any two nodes of the graph as shown in Fig. 2.3.

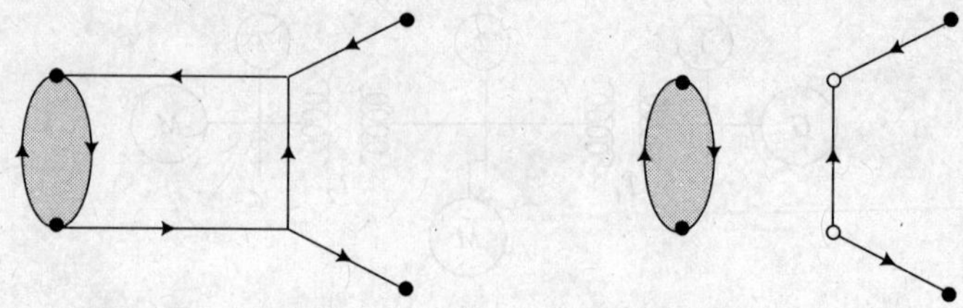

Fig. 2.3 (a) Connected graph (b) Unconnected graph.

2.2.2 Sub-graph

$G1$ is a sub-graph of G if the following conditions are satisfied:

- $G1$ is itself a graph.
- Every node of $G1$ is also a node of G.
- Every branch of $G1$ is a branch of G.

2.2.3 De-generate Sub-graph

This is a sub-graph which consists of just one node.

2.2.4 Loop

A sub-graph L of a graph G is a loop, if

- L is a connected sub-graph of G.
- Precisely two branches are incident on each node in L.

This is illustrated in Fig. 2.4.

In Fig. 2.4 the set {2, 6, 7} forms a loop, while the set {2, 6, 7, 8, 5, 4} is not a loop. The KVL (Kirchhoff's Voltage Law) for the loop is stated as follows: *"In any lumped network, for any of its loops and at any time, the algebraic sum of the branch voltages around the loop is zero."*

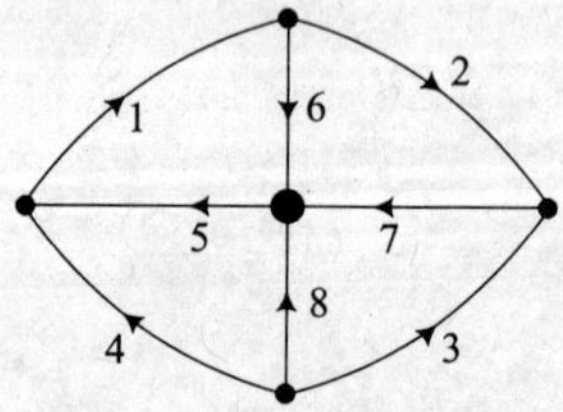

Fig. 2.4 Definition of a loop.

2.2.5 Cutset

It is a set of branches of a connected graph G which satisfies the following conditions:

- The removal of all branches of the cutset causes the remaining graph to have two separate unconnected sub-graphs.
- The removal of all but one of the branches of the set leaves the remaining graph connected.

Referring to Fig. 2.1c, the set {2, 4} constitutes a cutset since removal of them isolates node 3 from rest of the network, thus dividing the graph into two unconnected sub-graphs. The KCL

(Kirchhoff's Current Law) for the cutset is stated as follows: *"In any lumped network, for any of its cutsets and at any time, the algebraic sum of all the branch currents traversing the cutset branches is zero.*

2.2.6 Tree

It is a connected sub-graph containing all the nodes of the graph G, but without any closed paths (loops). There is one and only one path between every pair of nodes in a tree. The elements of the tree are called twigs or tree-branches*.

(*Some authors refer to the twigs simply as branches. However, the word branch is used synonymously with elements by many others. In this book branch indicates any element of the network and the branch of a tree is referred to as a twig or a tree-branch and denoted by t.). In a graph with n nodes, the number of tree-branches is given by

$$t = n - 1 \qquad ...(2.1)$$

For the graph of Fig. 2.1c, two trees are shown in Fig. 2.5.

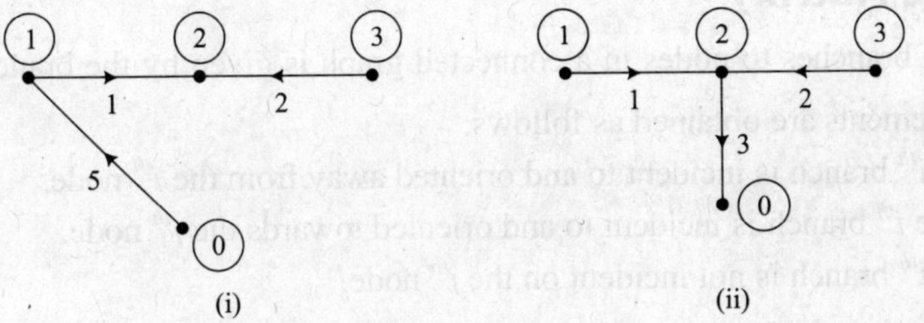

Fig. 2.5 Trees for Fig. 2.1c.

2.2.7 Co-Tree

The set of branches of the original graph G, not included in the tree is called the *co-tree*. The co-tree need not be connected. The branches of the co-tree are called *links* (or *chords* or *ties*). The co-trees corresponding to Fig. 2.5 are shown in Fig. 2.6.

The number of links l is given by

$$l = b - t = b - (n - 1) = b - n + 1 \qquad ...(2.2)$$

where, b is the total number of branches in the network.

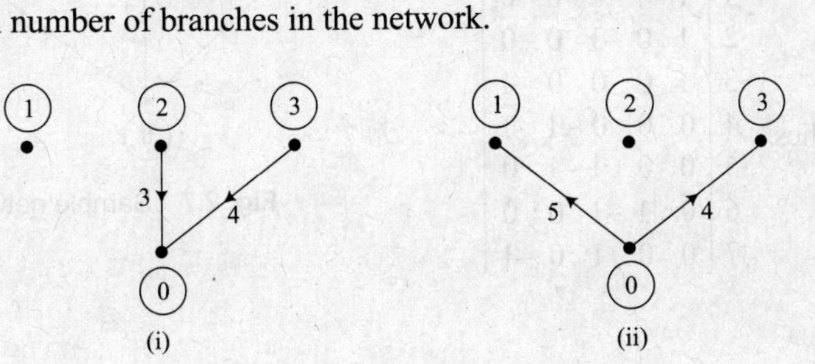

Fig. 2.6 Co-trees of Fig. 2.5.

2.2.8 Basic Loops

When a link is added to a tree it forms a loop. Addition of each subsequent link forms one or more additional loops. A loop containing only one link and remaining twigs is called a basic loop or a fundamental loop or *f*-loop. *f*-loops are defined for a particular tree. Since, each link is associated with a basic loop, the number of basic loops is equal to the number of links.

2.2.9 Basic Cutsets

Cutsets which contain only one twig and remaining links are called basic cutsets or fundamental cutsets or *f*-cutsets. They are defined for a particular tree and are equal in number to the number of twigs.

2.3 INCIDENCE MATRICES

2.3.1 Element-node (Branch-node) Incidence Matrix $\hat{A}$ (Bus-incidence Matrix)

The incidence of branches to nodes in a connected graph is given by the branch-node incidence matrix $\hat{A}$. The elements are obtained as follows:

$a_{ij} = 1$ if the ith branch is incident to and oriented away from the jth node.
$= -1$ if the ith branch is incident to and oriented towards the jth node.
$= 0$ if the ith branch is not incident on the jth node.

The dimension of $\hat{A}$ is $b \times n$, where b is the number of branches or elements and n is the number of nodes in the network. Consider the graph of Fig. 2.7.

The branch-node incidence matrix is given by

$$\hat{A} = \text{Branches} \quad \begin{array}{c|ccccc} & & & \text{nodes} & & \\ b\backslash n & 0 & 1 & 2 & 3 & 4 \\ \hline 1 & 1 & -1 & 0 & 0 & 0 \\ 2 & 1 & 0 & -1 & 0 & 0 \\ 3 & 1 & 0 & 0 & 0 & -1 \\ 4 & 0 & 0 & 0 & -1 & 1 \\ 5 & 0 & 0 & 1 & -1 & 0 \\ 6 & 0 & 1 & -1 & 0 & 0 \\ 7 & 0 & 0 & 1 & 0 & -1 \end{array}$$

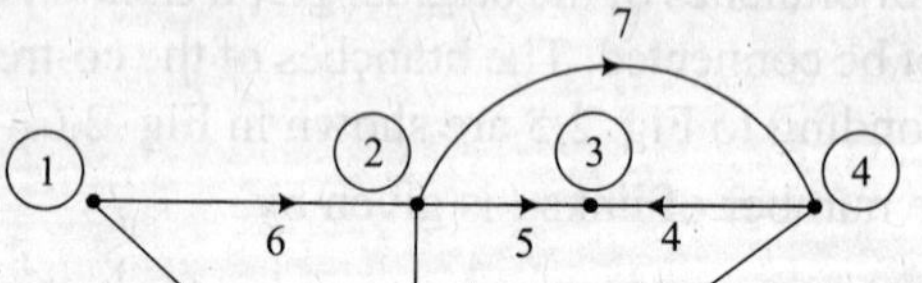

Fig. 2.7 Sample network graph.

The sum of every row is equal to zero. Hence, the rank of the matrix is less than n. Any node of the connected graph can be chosen as the reference node and the column corresponding to that node deleted from $\hat{A}$. The matrix so formed is called the bus-incidence matrix A. The dimensions of A are $b \times (n-1)$ and the rank is $n-1$. Selecting node *0* as reference node in the above example, the matrix A is obtained by deleting the column corresponding to node 0 to give

$$A = \text{Branches} \quad \begin{array}{c|cccc} b\backslash n & 1 & 2 & 3 & 4 \\ \hline 1 & -1 & 0 & 0 & 0 \\ 2 & 0 & -1 & 0 & 0 \\ 3 & 0 & 0 & 0 & -1 \\ 4 & 0 & 0 & -1 & 1 \\ 5 & 0 & 1 & -1 & 0 \\ 6 & 1 & -1 & 0 & 0 \\ 7 & 0 & 1 & 0 & 1 \end{array} \quad (\text{columns: nodes})$$

Now consider one possible tree for the graph of Fig. 2.7 as shown in Fig. 2.8.

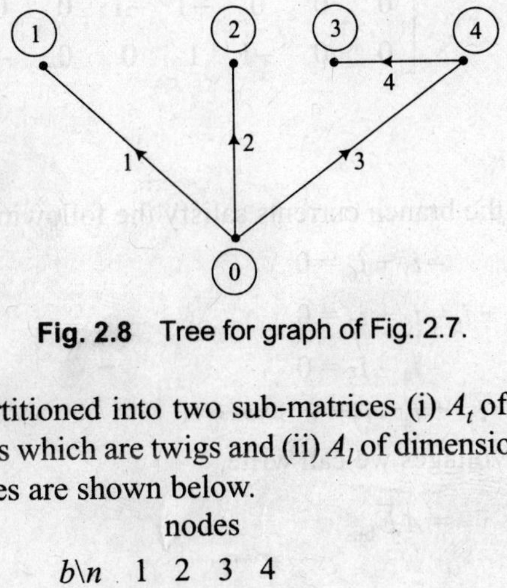

Fig. 2.8 Tree for graph of Fig. 2.7.

The matrix A can be partitioned into two sub-matrices (i) A_t of dimension $(n-1) \times (n-1)$ corresponding to the branches which are twigs and (ii) A_l of dimension $l \times (n-1)$ corresponding to links. The partitioned matrices are shown below.

$$A_t = \text{twigs} \quad \begin{array}{c|cccc} b\backslash n & 1 & 2 & 3 & 4 \\ \hline 1 & -1 & 0 & 0 & 0 \\ 2 & 0 & -1 & 0 & 0 \\ 3 & 0 & 0 & 0 & -1 \\ 4 & 0 & 0 & -1 & 1 \end{array} \quad (\text{columns: nodes})$$

$$A_l = \text{links} \begin{array}{c} \\ b\backslash n \\ 5 \\ 6 \\ 7 \end{array} \begin{array}{c} \text{nodes} \\ \begin{array}{cccc} 1 & 2 & 3 & 4 \end{array} \\ \begin{bmatrix} 0 & 1 & -1 & 0 \\ 1 & -1 & 0 & 0 \\ 0 & 1 & 0 & -1 \end{bmatrix} \end{array}$$

A is a rectangular matrix, hence it is singular. A_t is a non-singular square matrix of dimension $(n - 1)$. Since, A gives the incidence of various branches on the nodes with their direction of incidence, the KCL for the nodes can be written as

$$A^T \bar{i} = 0 \qquad \text{...(2.3)}$$

where, A^T is the transpose of matrix A and $\bar{i}$ is the vector of branch currents. For the example considered, we can write

$$A^T \bar{i} = \begin{bmatrix} -1 & 0 & 0 & 0 & 0 & 1 & 0 \\ 0 & -1 & 0 & 0 & 1 & -1 & 1 \\ 0 & 0 & 0 & -1 & -1 & 0 & 0 \\ 0 & 0 & -1 & 1 & 0 & 0 & -1 \end{bmatrix} \begin{bmatrix} i_1 \\ i_2 \\ i_3 \\ i_4 \\ i_5 \\ i_6 \\ i_7 \end{bmatrix} = 0$$

Hence, the equations for the branch currents satisfy the following:

$$\begin{aligned} -i_1 + i_6 &= 0 \\ -i_2 + i_5 - i_6 + i_7 &= 0 \\ -i_4 - i_5 &= 0 \\ -i_3 + i_4 - i_7 &= 0 \end{aligned} \qquad \text{...(2.4)}$$

Similarly, for the branch voltages we can write,

$$\bar{v} = A\,\bar{E}_{\text{bus}} \qquad \text{...(2.5)}$$

where, the vector of branch voltages $\bar{v} = \begin{bmatrix} v_1 \\ v_2 \\ .. \\ .. \\ .. \\ v_b \end{bmatrix}$, and the vector of node (bus) voltages $\bar{E}_{\text{bus}} = \begin{bmatrix} E_1 \\ E_2 \\ .. \\ .. \\ .. \\ E_n \end{bmatrix}$

The node voltages are with respect to the reference node.

For the example considered, we can write

$$\begin{bmatrix} v_1 \\ v_2 \\ v_3 \\ v_4 \\ v_5 \\ v_6 \\ v_7 \end{bmatrix} = \begin{bmatrix} -1 & 0 & 0 & 0 \\ 0 & -1 & 0 & 0 \\ 0 & 0 & 0 & -1 \\ 0 & 0 & -1 & 1 \\ 0 & 1 & -1 & 0 \\ 1 & -1 & 0 & 0 \\ 0 & 1 & 0 & -1 \end{bmatrix} \begin{bmatrix} E_1 \\ E_2 \\ E_3 \\ E_4 \end{bmatrix}$$

The equations of branch voltages in terms of node voltages are,

$$\begin{aligned} v_1 &= -E_1 \\ v_2 &= -E_2 \\ v_3 &= -E_4 \\ v_4 &= E_4 - E_3 \\ v_5 &= E_2 - E_3 \\ v_6 &= E_1 - E_2 \\ v_7 &= E_2 - E_4 \end{aligned} \qquad \text{...(2.6)}$$

2.3.2 Tree-Branch Path Incidence Matrix *K*

This matrix shows the incidence of the tree-branches (twigs) to paths in a tree, where the path is oriented from a bus to the reference bus. The elements are as follows:

$k_{ij} = 1$ if the ith tree-branch (twig) is in the path from the jth bus to reference bus and is oriented in the same direction.

= −1 if the ith tree-branch (twig) is in the path from the jth bus to reference bus and is oriented in the opposite direction.

= 0 if the ith tree-branch (twig) is not in the path from the jth bus to reference bus.

The dimension of the matrix is $t \times t$ or $(n-1) \times (n-1)$. The matrix K associated with the tree of Fig. 2.8 is

$$K = \text{twigs} \begin{array}{c} \\ t\backslash n \\ 1 \\ 2 \\ 3 \\ 4 \end{array} \begin{array}{c} \text{nodes} \\ \begin{array}{cccc} 1 & 2 & 3 & 4 \end{array} \\ \begin{bmatrix} -1 & 0 & 0 & 0 \\ 0 & -1 & 0 & 0 \\ 0 & 0 & -1 & -1 \\ 0 & 0 & -1 & 0 \end{bmatrix} \end{array}$$

The sub-matrix A_t relates the incidence of tree-branches to buses and the matrix K relates the incidence of tree-branches to paths. Since, there is a one-to-one correspondence between the paths and the buses, we can write

$$A_t K^T = I_t \quad \text{...(2.7)}$$

where, I is the identity matrix (also called unit matrix) and I_t is an identity matrix of size t (number of twigs). From (2.7)

$$K^T = [A_t]^{-1} \quad \text{...(2.8)}$$

For the example of Fig. 2.8

$$A_t K^T = \begin{bmatrix} -1 & 0 & 0 & 0 \\ 0 & -1 & 0 & 0 \\ 0 & 0 & 0 & -1 \\ 0 & 0 & -1 & 1 \end{bmatrix} \begin{bmatrix} -1 & 0 & 0 & 0 \\ 0 & -1 & 0 & 0 \\ 0 & 0 & -1 & -1 \\ 0 & 0 & -1 & 0 \end{bmatrix} = \begin{bmatrix} 1 & 0 & 0 & 0 \\ 0 & 1 & 0 & 0 \\ 0 & 0 & 1 & 0 \\ 0 & 0 & 0 & 1 \end{bmatrix} = I_t$$

2.3.3 Basic (Fundamental) Cutset Incidence Matrix B

The basic or fundamental cutset contains only one twig and remaining links. The basic cutsets for the tree of Fig. 2.8 are shown in Fig. 2.9. The orientation of the basic cutset is that of the twig in it. The elements of the basic cutset incidence matrix B are given by:

b_{ij} = 1 if the ith element is incident to and oriented in the same direction as the jth cutset.

= −1 if the ith element is incident to and oriented in an opposite direction as the jth cutset.

= 0 if the ith element is not incident to the jth cutset.

The basic cutset incidence matrix has dimensions $b \times t$. For Fig. 2.9, the matrix is given by

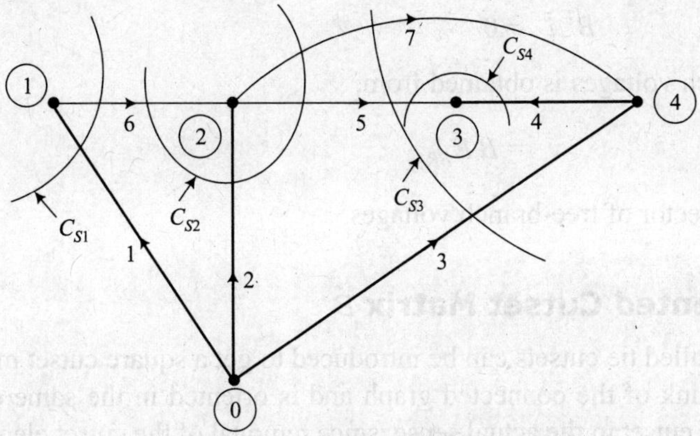

Fig. 2.9 Basic cutsets associated with Fig. 2.8.

B = branches

b\Cutset	C_{s1}	C_{s2}	C_{s3}	C_{s4}
1	1	0	0	0
2	0	1	0	0
3	0	0	1	0
4	0	0	0	1
5	0	−1	1	1
6	−1	1	0	0
7	0	−1	1	0

The matrix B can be partitioned into two sub-matrices as follows:

- B_t of dimension $t \times t$, corresponding to the twigs. Since, each twig is uniquely incident on one of the basic cutsets, this sub-matrix is an identity matrix of size t.
- B_l of dimension $l \times t$, corresponding to the elements which are links.

The matrix B is related to the bus-incidence matrix A by the following relationships:

$$B_l A_t = A_l \quad \text{...(2.9)}$$

$$B_l = A_l [A_t]^{-1} = A_l K^T \quad \text{...(2.10)}$$

For the example of Fig. 2.8 we can easily verify (2.9).

$$B_l A_t = \begin{bmatrix} 0 & -1 & 1 & 1 \\ -1 & 1 & 0 & 0 \\ 0 & -1 & 1 & 0 \end{bmatrix} \begin{bmatrix} -1 & 0 & 0 & 0 \\ 0 & -1 & 0 & 0 \\ 0 & 0 & 0 & -1 \\ 0 & 0 & -1 & 1 \end{bmatrix} = \begin{bmatrix} 0 & 1 & -1 & 0 \\ 1 & -1 & 0 & 0 \\ 0 & 1 & 0 & -1 \end{bmatrix} = A_l$$

If we apply KCL to each of the $(n - 1)$ fundamental cutsets, we obtain a system of linear homogeneous equations in b variables i_1, i_2 ... of the branch currents. Since the algebraic sum of currents traversing a cutset is zero,

$$B^T \bar{i} = 0 \qquad \text{...(2.11)}$$

The vector of branch voltages is obtained from,

$$\bar{v} = B \bar{E}_{BR} \qquad \text{...(2.12)}$$

where, $\bar{E}_{BR}$ is the vector of tree-branch voltages.

2.3.4 Augmented Cutset Matrix $\hat{B}$

Fictitious cutsets, called tie cutsets can be introduced to get a square cutset matrix. Each tie cutset contains only one link of the connected graph and is oriented in the same direction as the link. (Note that it is not a cutset in the actual sense, since removal of the cutset elements does not divide the graph into two separate parts). We get a non-singular square matrix of dimension $b \times b$. The augmented cutset for the example of Fig. 2.8 is

$$\hat{B} = \text{branches} \begin{array}{c} b\backslash b \\ 1 \\ 2 \\ 3 \\ 4 \\ 5 \\ 6 \\ 7 \end{array} \overset{\text{branches}}{\begin{array}{ccccccc} 1 & 2 & 3 & 4 & 5 & 6 & 7 \\ \end{array}} \begin{bmatrix} 1 & 0 & 0 & 0 & 0 & 0 & 0 \\ 0 & 1 & 0 & 0 & 0 & 0 & 0 \\ 0 & 0 & 1 & 0 & 0 & 0 & 0 \\ 0 & 0 & 0 & 1 & 0 & 0 & 0 \\ 0 & -1 & 1 & 1 & 1 & 0 & 0 \\ -1 & 1 & 0 & 0 & 0 & 1 & 0 \\ 0 & -1 & 1 & 0 & 0 & 0 & 1 \end{bmatrix}$$

To get elements of row 5, add the element 5 (which is a link) to the tree of Fig. 2.8. Now elements 2, 3, 4, 5 from a loop, where elements 2, 3, 4 are twigs and only element 5 is a link. Similarly, rows corresponding to links 6 and 7 are obtained.

$\hat{B}$ can be partitioned into four sub-matrices as,

$$\hat{B} = \begin{bmatrix} [I_t] & [0] \\ [B_l] & [I_l] \end{bmatrix}$$

where, I_t and I_l are identity matrices of size t and l, respectively.

2.3.5 Basic (Fundamental) Loop Incidence Matrix C

This matrix gives the incidence of the branches to the basic loops. The basic loops for the tree of Fig. 2.8 are shown in Fig. 2.10.

The dimension of the matrix is $b \times l$. The direction of the unique link in the basic loop is the orientation of the loop. The elements of the loop incidence matrix C are given by

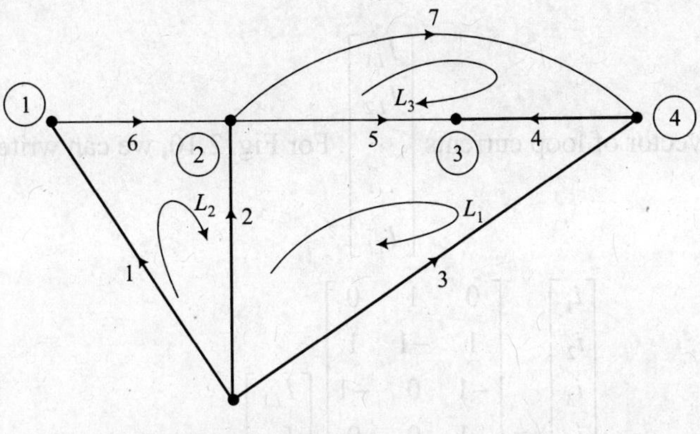

Fig. 2.10 Basic loops for Fig. 2.8.

c_{ij} = 1, if the ith branch is incident to the jth basic loop and is oriented in the same direction.
= – 1, if the ith branch is incident to the jth basic loop and is oriented in the opposite direction.
= 0, if the ith branch is not incident to the jth basic loop.

The matrix for the example of Fig. 2.10 is given by

$$C = \text{branches} \begin{array}{c} \\ b\backslash l \\ 1 \\ 2 \\ 3 \\ 4 \\ 5 \\ 6 \\ 7 \end{array} \begin{array}{c} \text{Loops} \\ \begin{array}{ccc} L_1 & L_2 & L_3 \end{array} \\ \begin{bmatrix} 0 & 1 & 0 \\ 1 & -1 & 1 \\ -1 & 0 & -1 \\ -1 & 0 & 0 \\ 1 & 0 & 0 \\ 0 & 1 & 0 \\ 0 & 0 & 1 \end{bmatrix} \end{array}$$

The matrix C can be partitioned into two sub-matrices:

- C_t of dimension $t \times l$ corresponding to the twigs.
- C_l of dimension $l \times l$ corresponding to the elements which are links. Since each link is uniquely incident on one basic loop, this sub-matrix is an identity matrix of size l.

Application of KVL to each fundamental loop, constitutes a set of l linearly independent equations, written as

$$C^T \bar{v} = 0 \qquad \text{...(2.13)}$$

Since, the loop incidence matrix contains information about the branches incident on various loops, the branch currents can be obtained in terms of the loop currents as

$$\bar{i} = C\bar{I}_{\text{Loop}} \qquad \text{...(2.14)}$$

where, $\bar{I}_{\text{Loop}}$ is the vector of loop currents $\begin{bmatrix} I_{L1} \\ I_{L2} \\ \cdots \\ \cdots \\ I_{Ll} \end{bmatrix}$. For Fig. 2.10, we can write

$$\begin{bmatrix} i_1 \\ i_2 \\ i_3 \\ i_4 \\ i_5 \\ i_6 \\ i_7 \end{bmatrix} = \begin{bmatrix} 0 & 1 & 0 \\ 1 & -1 & 1 \\ -1 & 0 & -1 \\ -1 & 0 & 0 \\ 1 & 0 & 0 \\ 0 & 1 & 0 \\ 0 & 0 & 1 \end{bmatrix} \begin{bmatrix} I_{L1} \\ I_{L2} \\ I_{L3} \end{bmatrix}$$

2.3.6 Augmented Loop Incidence Matrix $\hat{C}$

The basic loop incidence matrix can be augmented by adding open-loops corresponding to the tree-branches, to obtain a non-singular square matrix of dimension $b \times b$. An open-loop is simply a path between adjacent nodes connected by a twig. The augmented loop incidence matrix for Fig. 2.10 is given by

$$\hat{C} = \text{branches} \begin{array}{c} \\ \\ b/l \\ 1 \\ 2 \\ 3 \\ 4 \\ 5 \\ 6 \\ 7 \end{array} \begin{array}{c} \text{Open loops} \qquad\qquad \text{Basic loops} \\ \begin{array}{ccccccc} L_4 & L_5 & L_6 & L_7 & L_1 & L_2 & L_3 \end{array} \\ \begin{bmatrix} 1 & 0 & 0 & 0 & 0 & 1 & 0 \\ 0 & 1 & 0 & 0 & 1 & -1 & 1 \\ 0 & 0 & 1 & 0 & -1 & 0 & -1 \\ 0 & 0 & 0 & 1 & -1 & 0 & 0 \\ 0 & 0 & 0 & 0 & 1 & 0 & 0 \\ 0 & 1 & 0 & 0 & 0 & 1 & 0 \\ 0 & 0 & 0 & 0 & 0 & 0 & 1 \end{bmatrix} \end{array}$$

The augmented matrix can be partitioned into sub-matrices as follows:

$$\hat{C} = \begin{bmatrix} [I_t] & [C_t] \\ [0] & [I_l] \end{bmatrix}$$

2.4 CONNECTIVITY

The incidence matrices give an idea of the interconnections of the network and how removal of branches separates the network. In interconnected systems (communication, computer or power systems) the best connection is one, which ensures that the network is as invulnerable to destruction of individual nodes and branches as possible. The connectivity gives an idea about this.

Edge connectivity is defined as the number of branches in the smallest cutset i.e., one with least number of branches in it. Since, the cutset denotes the minimum number of branches, whose removal results in destroying connectivity, the cutset can be viewed as representing the vulnerability of the system. Therefore, from a reliability point of view, the edge connectivity should be high. In power systems, separation of the network into two unconnected networks, results in islanding and may be a major threat to stable system operation.

Vertex connectivity of a connected graph G is defined as the minimum number of vertices (nodes) whose removal from G, leaves the remaining graph unconnected (removal implies node and all branches incident on it).

2.5 PRIMITIVE NETWORK

In the preceding sections, the matrices of the interconnected network have been defined. These matrices contain complete information about the network connectivity, the orientation of current, the loops and cutsets. However, these matrices contain no information on the nature of the elements which form the interconnected network. The complete behaviour of the network can be obtained from the knowledge of the behaviour of the individual elements which make the network, along with the incidence matrices. An element in an electrical network is completely characterized by the relationship between the current through the element and the voltage across it.

A *primitive element* is a fundamental element which is not connected to any other element. A set of such unconnected elements is defined as a *primitive network*. To illustrate the concept, we consider two unconnected coils as shown in Fig. 2.11.

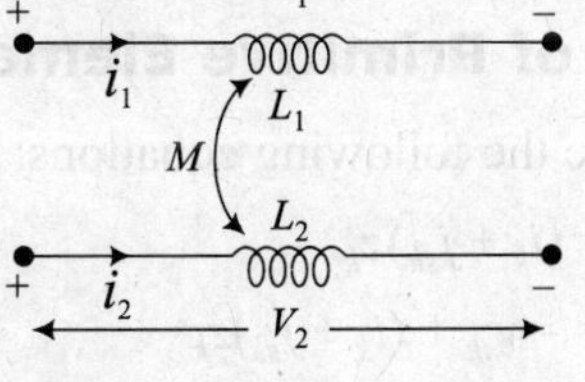

Fig. 2.11 Primitive network.

The equations describing the above elements are,

$$v_1 = L_1 \frac{di_1}{dt} + M \frac{di_2}{dt}$$

$$v_2 = M \frac{di_1}{dt} + L_2 \frac{di_2}{dt} \qquad \text{...(2.15)}$$

where, L_1 and L_2 are the self inductances of the two coils and M is the mutual inductance between them. In phasor form, (2.15) can be written as

$$V_1 = Z_{11} I_1 + Z_{12} I_2$$

$$V_2 = Z_{21} I_1 + Z_{22} I_2 \qquad ...(2.16)$$

It is important to note that (2.16) does not depend on the connectivity between the two coils and is written taking into account only the individual element currents and voltages. We can call the two coils as primitive elements and together they form a primitive network.

2.5.1 General Primitive Element

A general primitive element is shown in Fig. 2.12.

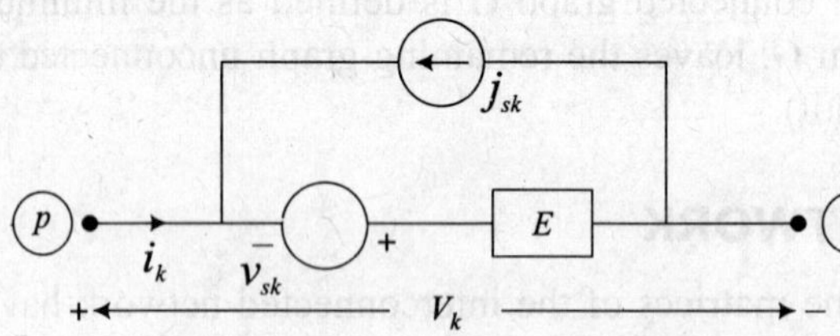

Fig. 2.12 General primitive element.

In Fig. 2.12, v_k = voltage across branch k

i_k = current through branch k

v_{sk} = independent voltage source in branch k

j_{sk} = independent current source in branch k

E = passive element specified by its impedance z_k or admittance y_k

The above element can be reduced into two simple forms.

2.5.2 Impedance Form of Primitive Element

Referring to Fig. 2.12, we can write the following equations:

$$v_k + v_{sk} = (i_k + j_{sk})z_k \qquad ...(2.17)$$

$$v_k = -v_{sk} + (i_k + j_{sk})z_k$$

$$= (j_{sk}z_k - v_{sk}) + i_k z_k \qquad ...(2.18)$$

(2.18) can be written as

$$v_k + e_k = i_k z_k \qquad ...(2.19)$$

where,

$$e_k = v_{sk} - j_{sk} i_k \qquad ...(2.20)$$

The equivalent circuit describing (2.19) is shown in Fig. 2.13.

Thus, the general primitive element is reduced to a simple voltage source, in series with an impedance. If there are b branches (elements) in a primitive network, (2.19) can be written as,

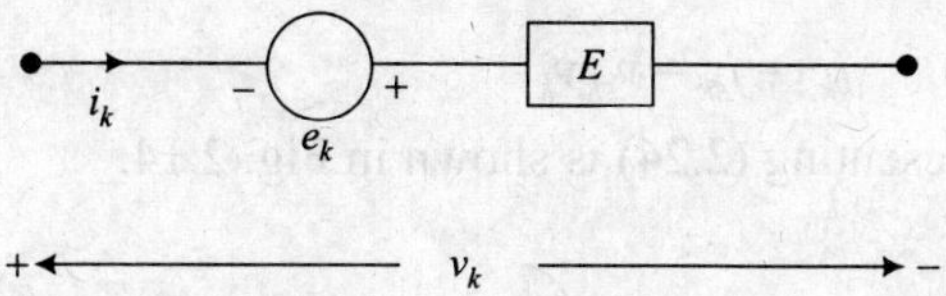

Fig. 2.13 Impedance form of the primitive element.

$$\bar{v} + \bar{e} = [z]\bar{i} \quad ...(2.21)$$

where $\bar{v}$ = vector of voltages across b branches

$\bar{i}$ = vector of currents through the b branches

$\bar{e}$ = vector of equivalent voltage sources

$[z]$ = primitive impedance matrix.

It there is no mutual coupling between the elements, $[z]$ is a diagonal matrix, whose elements are the impedances of the branches. In such a case, the primitive self admittances are equal to reciprocals of the corresponding primitive self impedances. If mutual coupling exists, then the corresponding off-diagonal elements of $[z]$ will have a non-zero entry. In such a case, the primitive admittances matrix has to be computed as inverse of the primitive impedances matrix.

Example 2.1: Consider three passive elements whose data is given in Table 2.1. Form the primitive impedance matrix.

Table 2.1

Element	*$z_k(pu)$*	*Mutual impedance*
1	$j0.6$	-
2	$j0.5$	$j0.2$ (with element 1)
3	$j0.7$	–

Solution: $[z] = \begin{bmatrix} j0.6 & j0.2 & 0.0 \\ j0.2 & j0.5 & 0.0 \\ 0.0 & 0.0 & j0.7 \end{bmatrix}$

Note:

- The size of $[z]$ is $b \times b$.
- The diagonal elements are the self impedances of the elements.
- The off-diagonal elements are the mutual impedances between the corresponding elements.

2.5.3 Admittance Form of the Primitive Element

Referring to Fig. 2.12, the describing equation can be written in a slightly different form as

$$i_k + j_{sk} = (v_k + v_{sk})y_k \quad ...(2.22)$$

$$i_k = (v_{sk}\,y_k - j_{sk}) + v_k y_k \quad ...(2.23)$$

$$i_k + j_k = v_k y_k \quad ...(2.24)$$

where, $j_k = j_{sk} - v_{sk} y_k$

The equivalent circuit representing (2.24) is shown in Fig. 2.14.

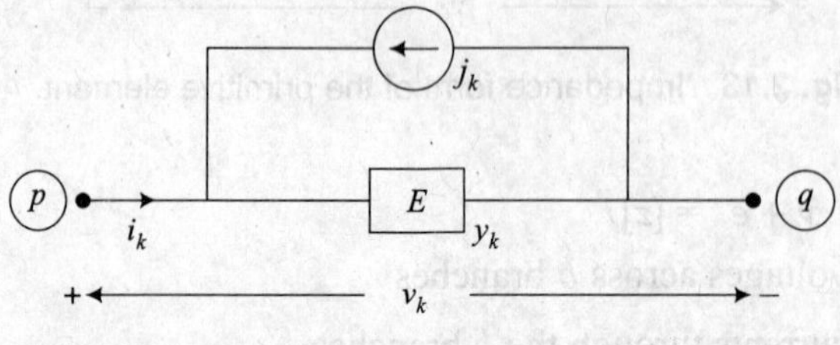

Fig. 2.14 Admittance form of the primitive element.

For the primitive network, (2.24) can be written as

$$\bar{i} + \bar{j} = [y]\bar{v} \quad ...(2.25)$$

where, $\bar{j}$ = vector of injected currents

$[y]$ = primitive admittance matrix.

In the admittance form, the primitive element is reduced to a current source in parallel with an admittance. For equivalence of (2.19) and (2.24) the following relationships must be satisfied:

$$j_k = -y_k e_k \quad ...(2.26)$$

$$y_k = \frac{1}{z_k} \quad ...(2.27)$$

$$[y] = [z]^{-1} \quad ...(2.28)$$

Example 2.2: Obtain the primitive admittance matrix of example 2.1.

Solution:

$$[y] = [z]^{-1} = \begin{bmatrix} -j1.9231 & j0.7692 & 0.0 \\ j0.7692 & -j2.3077 & 0.0 \\ 0.0 & 0.0 & -j1.4286 \end{bmatrix}$$

2.6 SINGULAR TRANSFORMATIONS

The primitive impedance matrix is the most basic matrix and depends purely on the impedance of the individual elements. However, it contains no information about the behaviour of the interconnected network variables. Hence, it is necessary to transform the primitive matrices into more meaningful matrices which can relate variables of the interconnected network.

2.6.1 Bus Admittance and Bus Impedance Matrix

In the bus frame of reference, the performance of the interconnected network is described by $(n-1)$ independent nodal equations, where n is the total number of nodes, out of which one of them is designated as the reference node. (*In power system networks, single phase equivalent of the balanced three phase network is used. In this case the reference is taken to be the ground and an n

bus system implies n nodes excluding the ground. For example, a 5-bus system will have 5 external buses and 1 ground bus). The performance equations relating the bus voltages to bus current injections are given by

$$\bar{E}_{\text{bus}} = Z_{\text{bus}} \bar{I}_{\text{bus}} \quad ...(2.29)$$

$$\bar{I}_{\text{bus}} = Y_{\text{bus}} \bar{E}_{\text{bus}} \quad ...(2.30)$$

where, $\bar{E}_{\text{bus}}$ = vector of bus voltages measured with respect to the reference bus

$\bar{I}_{\text{bus}}$ = Vector of currents injected into the bus

Z_{bus} = bus impedance matrix

Y_{bus} = bus admittance matrix

The performance equation of the primitive network in admittance form is given by

$$\bar{i} + \bar{j} = [y]\,\bar{v}$$

Premultiplying by A^T (transpose of A), we obtain

$$A^T\bar{i} + A^T\bar{j} = A^T[y]\,\bar{v} \quad ...(2.31)$$

From (2.3) we have $A^T\bar{i} = 0$. Similarly, $A^T\bar{j}$ gives the algebraic sum of all source currents incident at each bus and this is nothing but the total current injected at the bus. Hence,

$$A^T\bar{j} = \bar{I}_{\text{bus}} \quad ...(2.32)$$

Substituting (2.3) and (2.32) in (2.31) we have,

$$\bar{I}_{\text{bus}} = A^T[y]\,\bar{v} \quad ...(2.33)$$

Substituting (2.5) in (2.33) we get,

$$\bar{I}_{\text{bus}} = A^T[y]A\bar{E}_{\text{bus}} \quad ...(2.34)$$

Comparing (2.34) with (2.30), we obtain

$$Y_{bus} = A^T[y]A \quad ...(2.35)$$

The bus incidence matrix is rectangular and hence singular. Hence, (2.35) gives a singular transformation of the primitive admittance matrix $[y]$. The bus impedance matrix is given by

$$Z_{\text{bus}} = Y_{\text{bus}}^{-1} \quad ...(2.36)$$

The size of Y_{bus} and Z_{bus} are $n \times n$ where n is the number of buses in the system, excluding the ground bus (reference bus). In practical systems the reference bus is the ground. In such a case, the incidence matrix has to take into account any elements which may be connected between a bus and the ground. Typically, the line charging admittances of transmission line due to the line-to-ground capacitance are modeled as shunt capacitors between the bus and the ground at both ends.

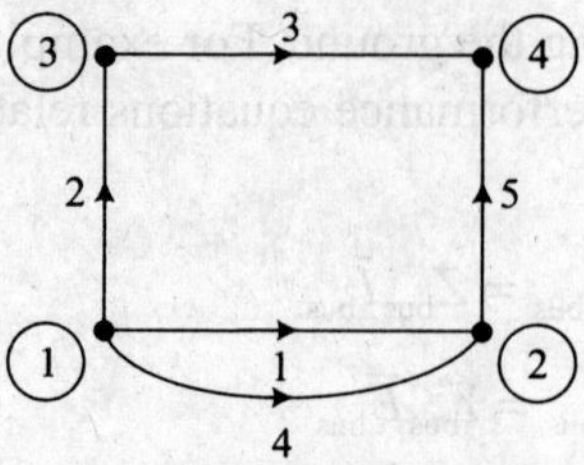

Fig. 2.15 Example 2.3.

Example 2.3: For the network of Fig. 2.15, form the primitive matrices [z] and [y] and obtain the bus admittance matrix by singular transformation. The data is given in Table 2.2.

Table 2.2 Data for example 2.3

Elements	*Self impedance*	*Mutual impedance*
1	*j* 0.6	–
2	*j* 0.5	*j* 0.1(with element 1)
3	*j* 0.5	–
4	*j* 0.4	*j* 0.2 (with element 1)
5	*j* 0.2	–

Solution: The bus incidence matrix is formed taking node 1 as the reference bus. Any other node can also be taken as reference node.

$$A = \begin{bmatrix} -1 & 0 & 0 \\ 0 & -1 & 0 \\ 0 & 1 & -1 \\ -1 & 0 & 0 \\ 1 & 0 & -1 \end{bmatrix}$$

The primitive incidence matrix is given by

$$[z] = \begin{bmatrix} j0.6 & j0.1 & 0.0 & j0.2 & 0.0 \\ j0.1 & j0.5 & 0.0 & 0.0 & 0.0 \\ 0.0 & 0.0 & j0.5 & 0.0 & 0.0 \\ j0.2 & 0.0 & 0.0 & j0.4 & 0.0 \\ 0.0 & 0.0 & 0.0 & 0.0 & j0.2 \end{bmatrix}$$

The primitive admittance matrix $[y] = [z]^{-1}$ and given by,

$$[y] = \begin{bmatrix} -j2.0833 & j0.4167 & 0.0 & j1.0417 & 0.0 \\ j0.4167 & -j2.0833 & 0.0 & -j0.2083 & 0.0 \\ 0.0 & 0.0 & -j2.0 & 0.0 & 0.0 \\ j1.0417 & -j0.2083 & 0.0 & -j3.0208 & 0.0 \\ 0.0 & 0.0 & 0.0 & 0.0 & -j5.0 \end{bmatrix}$$

The bus admittance matrix by singular transformation is obtained as

$$Y_{bus} = A^T[y]A = \begin{bmatrix} -j8.0208 & j0.2083 & j5.0 \\ j0.2083 & -j4.0833 & j2.0 \\ j5.0 & j2.0 & -j7.0 \end{bmatrix}$$

$$Z_{bus} = Y_{bus}^{-1} = \begin{bmatrix} j0.2713 & j0.1264 & j0.2299 \\ j0.1264 & j0.3437 & j0.1885 \\ j0.2299 & j0.1885 & j0.3609 \end{bmatrix}$$

Example 2.4

Consider a 4 – bus system with data as follows.

Bus code	Line Impedance (pu)	Half -line charging admittance
1 – 2	$0.2 + j\,0.6$	$j\,0.02$
2 – 3	$0.15 + j\,0.7$	$j\,0.03$
2 – 4	$0.2 + j\,0.75$	$j\,0.04$
3 – 4	$0.2 + j\,0.8$	$j\,0.02$
1 – 3	$0.1 + j\,0.4$	$j\,0.01$

Draw the network, showing the half line charging admittance also. Obtain the primitive admittance matrix Y_{bus} and Z_{bus}.

Solution The network including the half line charging admittance is shown below.

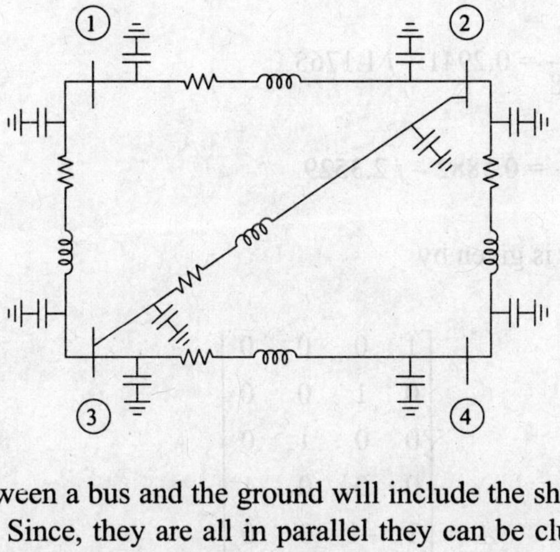

The shunt admittance between a bus and the ground will include the shunt admittances of all the lines incident on the bus. Since, they are all in parallel they can be clubbed into a single shunt admittance.

$y_{10} = j\,0.02 + j\,0.01 =$ (this is sum of half–line charging admittance of lines 1 – 2 and 1 – 3)

$y_{20} = j\,0.03 + j\,0.04 + j\,0.02 = j\,0.09$

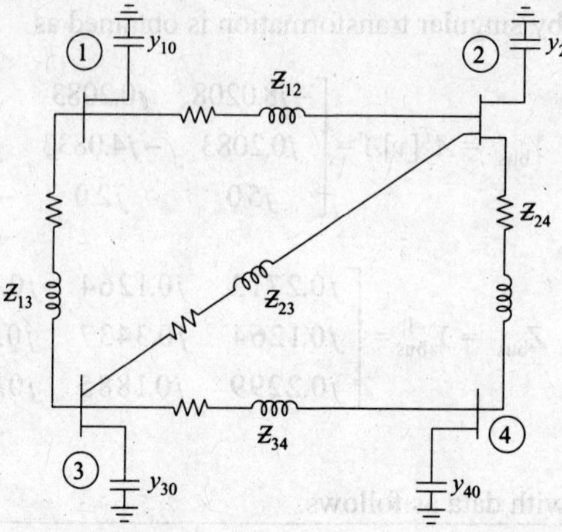

$$y_{30} = j\,0.03 + j\,0.01 + j\,0.04 = j\,0.06$$

$$y_{40} = j\,0.04 + j\,0.02 = j\,0.06$$

$$y_{12} = \frac{1}{z_{12}} = \frac{1}{0.2 + j0.6} = 0.5 - j\,1.5$$

$$y_{23} = \frac{1}{z_{23}} = \frac{1}{0.15 + j0.7} = 0.2927 - j\,1.3659$$

$$y_{24} = \frac{1}{z_{24}} = \frac{1}{0.2 + j0.75} = 0.332 - j\,1.2448$$

$$y_{34} = \frac{1}{z_{34}} = \frac{1}{0.2 + j0.8} = 0.2941 - j\,1.1765$$

$$y_{13} = \frac{1}{z_{13}} = \frac{1}{0.1 + j0.4} = 0.5882 - j\,2.3529$$

The incidence matrix is given by

$$A = \begin{bmatrix} 1 & 0 & 0 & 0 \\ 0 & 1 & 0 & 0 \\ 0 & 0 & 1 & 0 \\ 0 & 0 & 0 & 1 \\ 1 & -1 & 0 & 0 \\ 0 & 1 & -1 & 0 \\ 0 & 1 & 0 & -1 \\ 0 & 0 & 1 & -1 \\ 1 & 0 & -1 & 0 \end{bmatrix}$$

$$y \text{ (Primitive admittance matrix)} = \begin{bmatrix} y_{10} & & & & & & & & \\ & y_{20} & & & & & & & \\ & & y_{30} & & & & & & \\ & & & y_{40} & & & & & \\ & & & & y_{12} & & & & \\ & & & & & y_{23} & & & \\ & & & & & & y_{24} & & \\ & & & & & & & y_{34} & \\ & & & & & & & & y_{13} \end{bmatrix}$$

$$y_{bus} = A^T y A$$

$$= \begin{bmatrix} 1.0882 - j3.8229 & -0.5 + j1.5 & -0.5882 + j2.3529 & 0 \\ -0.5 + j1.5 & 1.1246 - j4.0207 & -0.2927 + j1.3659 & -0.332 + j1.2448 \\ -0.5882 + j2.3529 & -0.2927 + j1.3659 & 1.175 - j4.8353 & -0.2941 + j1.1765 \\ 0 & -0.332 + j1.2448 & -0.2941 + j1.1765 & 0.6261 - j2.3613 \end{bmatrix}$$

$$Z_{bus} = Y_{bus}^{-1}$$

$$= \begin{bmatrix} 0.063 - j3.9488 & -0.0094 - j4.1912 & 0.009 - j4.14 & -0.0265 - j4.2654 \\ -0.0094 - j4.1912 & 0.0253 - j4.0727 & -0.0133 - j4.2198 & -0.0199 - j4.2423 \\ 0.009 - j4.14 & -0.0133 - j4.2198 & 0.0319 - j4.0374 & -0.0165 - j4.2296 \\ -0.0265 - j4.2654 & -0.0199 - j4.2423 & -0.0165 - j4.2296 & 0.0596 - j3.9409 \end{bmatrix}$$

Note that in the absence of half line charging admittances, the sum of every row and every column of the bus admittance matrix is zero.

Example 2.5: In a 4-bus system two generators are connected to bus 1 and bus 2 respectively. The generator reactances are 0.2 and 0.4 respectively. The line data is as follows:

Bus code	Line impedance (pu)
1 – 2	$j\,0.15$
2 – 3	$j\,0.3$
2 – 4	$j\,0.6$
3 – 4	$j\,0.4$
1 – 3	$j\,0.25$

Draw the impedance diagram and compute the bus admittance matrix and bus impedance matrix.

Solution:

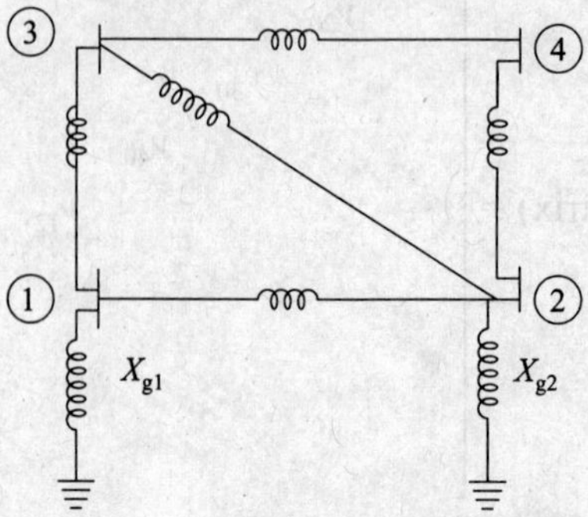

x_{g1}, x_{g2} are the reactances of the generators connected to bus 1 and bus 2.

$$y_{10} = \frac{1}{j0.2} = -j\,5.0;\; j\,y_{20} = \frac{1}{j0.4} = -j\,2.5;$$

$$y_{12} = -j\,6.667;\; y_{23} = -j\,3.33;\; y_{24} = -j\,1.667$$

$$y_{34} = -j\,2.5;\; y_{13} = -j\,4.0$$

$$A = \begin{bmatrix} 1 & 0 & 0 & 0 \\ 0 & 1 & 0 & 0 \\ 1 & -1 & 0 & 0 \\ 0 & 1 & -1 & 0 \\ 0 & 1 & 0 & -1 \\ 0 & 0 & 1 & -1 \\ 1 & 0 & -1 & 0 \end{bmatrix}$$

$$y = \begin{bmatrix} y_{10} & & & & & & \\ & y_{20} & & & & & \\ & & y_{12} & & & & \\ & & & y_{23} & & & \\ & & & & y_{24} & & \\ & & & & & y_{34} & \\ & & & & & & y_{13} \end{bmatrix}$$

$$Y_{bus} = A^T y A = \begin{bmatrix} -j15.667 & j6.667 & j4.0 & 0 \\ j6.667 & -j14.1667 & j3.33 & j1.667 \\ j4.0 & j3.33 & -j9.833 & j2.5 \\ 0 & j1.667 & j2.5 & -j4.1667 \end{bmatrix}$$

$$Z_{bus} = \begin{bmatrix} j0.144 & j0.112 & j0.1274 & j0.1212 \\ j0.112 & j0.176 & j0.1453 & j0.1576 \\ j0.1274 & j0.1453 & j0.2567 & j0.2121 \\ j0.1212 & j0.1576 & j0.2121 & j0.4303 \end{bmatrix}$$

Example 2.6: Solve example 2.5, when there is mutual coupling between line 1 – 2 and line 2 – 3, and the mutual impedance is j 0.1 pu.

Solution: Here, because of mutual coupling the primitive admittance matrix cannot be formed directly. We build the primitive impedance matrix and compute $y = z^{-1}$. The incidence matrix is same as example 2.5

$$z \text{ (primitive impedance matrix)} = j \begin{bmatrix} 0.2 & 0 & 0 & 0 & 0 & 0 & 0 \\ 0 & 0.4 & 0 & 0 & 0 & 0 & 0 \\ 0 & 0 & 0.15 & 0.1 & 0 & 0 & 0 \\ 0 & 0 & 0.1 & 0.3 & 0 & 0 & 0 \\ 0 & 0 & 0 & 0 & 0.6 & 0 & 0 \\ 0 & 0 & 0 & 0 & 0 & 0.4 & 0 \\ 0 & 0 & 0 & 0 & 0 & 0 & 0.25 \end{bmatrix}$$

Note that there are two extra entries corresponding to the mutual impedances.

We now calculate

$$Y_{bus} = A^T * [z^{-1}] * A = A^T y A$$

$$= j \begin{bmatrix} -17.5714 & 11.4286 & 1.1429 & 0 \\ 11.4286 & -22.7381 & 7.1429 & 1.6667 \\ 1.1429 & 7.1429 & -10.7857 & 2.5 \\ 0 & 1.66667 & 2.5 & -4.1667 \end{bmatrix}$$

2.6.2 Tree-branch Admittance and Tree-branch Impedance Matrices

In the tree-branch frame of reference the performance of the interconnected network is obtained by

t independent tree-branch equations, where t is the number of tree-branches. The tree-branch quantities of the interconnected network are related to the parameters of the primitive network through the basic cutset matrix B. We have the performance equation of the primitive network in admittance form as

$$\bar{i} + \bar{j} = [y]\bar{v}$$

Premultiplying by B^T, we obtain,

$$B^T\bar{i} + B^T\bar{j} = B^T[y]\bar{v} \qquad ...(2.37)$$

$B^T\bar{i}$ is the algebraic sum of the currents through the elements incident to the basic cutset and this is zero.

$$B^T\bar{i} = 0 \qquad ...(2.38)$$

Similarly, $B^T\bar{j}$ is a vector in which each element is the algebraic sum of the source currents of the elements incident to the basic cutset and can be treated as a single current source in parallel with the unique tree-branch in the basic cutset. It is denoted by $\bar{I}_{BR}$ and given by

$$B^T\bar{j} = \bar{I}_{BR} \qquad ...(2.39)$$

Substituting (2.38), (2.39) and (2.12) in (2.37), we get

$$\bar{I}_{BR} = B^T[y]B\bar{E}_{BR} \qquad ...(2.40)$$

The relationship between the tree-branch current sources and the tree-branch voltages given by (2.40) can be written as

$$\bar{I}_{BR} = Y_{BR}\bar{E}_{BR} \qquad ...(2.41)$$

where the tree-branch admittance matrix

$$Y_{BR} = B^T[y]B \qquad ...(2.42)$$

The basic cutset matrix is singular and hence (2.42) is a singular transformation of $[y]$.

The tree-branch impedance matrix Z_{BR} is given by

$$Z_{BR} = Y_{BR}^{-1}$$

Example 2.7: Obtain the tree-branch admittance and impedance matrices for the example of 2.3 using singular transformation.

Solution: The basic cutsets are shown in Fig. 2.16.

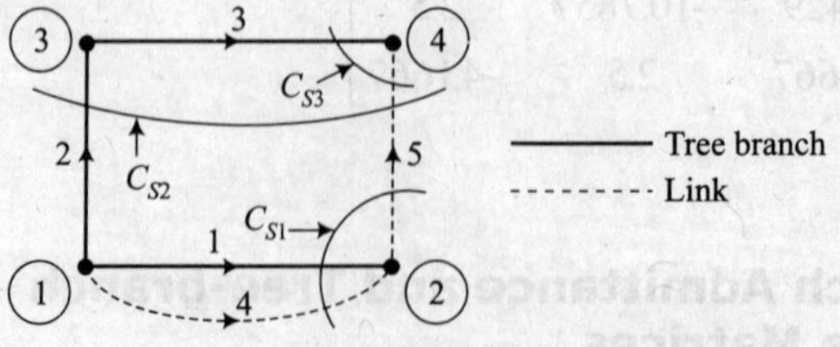

Fig. 2.16 Basic cutsets for example of 2.4.

The basic cutset incidence matrix is

$$B = \text{branches} \overset{\text{Cutsets}}{\begin{bmatrix} 1 & 0 & 0 \\ 0 & 1 & 0 \\ 0 & 0 & 1 \\ 1 & 0 & 0 \\ -1 & 1 & 1 \end{bmatrix}}$$

$$Y_{BR} = B^T[y]B = \begin{bmatrix} -j8.0208 & j5.2083 & j5.0 \\ j5.2083 & -j7.0833 & -j5.0 \\ j5.0 & -j5.0 & -j7.0 \end{bmatrix}$$

$$Z_{BR} = Y_{BR}^{-1} = \begin{bmatrix} j0.2713 & j0.1264 & j0.1034 \\ j0.1264 & j0.3437 & -j0.1552 \\ j0.1034 & -j0.1552 & j0.3276 \end{bmatrix}$$

2.6.3 Loop Impedance and Loop Admittance Matrices

In the loop frame of reference, the performance of an interconnected network is described by l independent loop equations, where l is the number of links or basic loops. The basic loop incidence matrix $[C]$ can be used to relate the parameters of the primitive network to loop quantities of the interconnected network. The performance equation of the primitive network in impedance form is given by (2.21)

$$\bar{v} + \bar{e} = [z]\bar{i}$$

Premultiplying by C^T, we obtain

$$C^T\bar{v} + C^T\bar{e} = C^T[z]\bar{i} \qquad \text{...(2.43)}$$

$C^T\bar{v}$ is the algebraic sum of all branch voltages in the loop and is equal to zero. $C^T\bar{e}$ gives the algebraic sum of source voltages around each loop. Let

$$C^T\bar{e} = \bar{E}_{\text{Loop}} \qquad \text{...(2.44)}$$

Substituting (2.13), (2.14) and (2.44) in (2.43), we get

$$\bar{E}_{\text{Loop}} = C^T[z]C\bar{I}_{\text{Loop}} \qquad \text{...(2.45)}$$

$$\bar{E}_{\text{Loop}} = Z_{\text{Loop}}\bar{I}_{\text{Loop}} \qquad \text{...(2.46)}$$

where,

$$Z_{\text{Loop}} = C^T[z]C \qquad \text{...(2.47)}$$

Since C is a singular matrix, $C^T[z]C$ is a singular transformation of $[z]$. The loop admittance matrix is given by $Y_{\text{Loop}} = Z_{\text{Loop}}^{-1}$.

Example 2.8: Obtain the loop impedance and loop admittance matrices for example 2.3, using singular transformation.

Solution: Using the same tree as example 2.4, the basic loops are shown in Fig. 2.17.

$$\text{The loop incidence matrix } [C] = \begin{bmatrix} -1 & 1 \\ 0 & -1 \\ 0 & -1 \\ 1 & 0 \\ 0 & 1 \end{bmatrix}$$

$$Z_{\text{Loop}} = C^T[z]C = \begin{bmatrix} j0.6 & -j0.3 \\ -j0.3 & j1.6 \end{bmatrix}$$

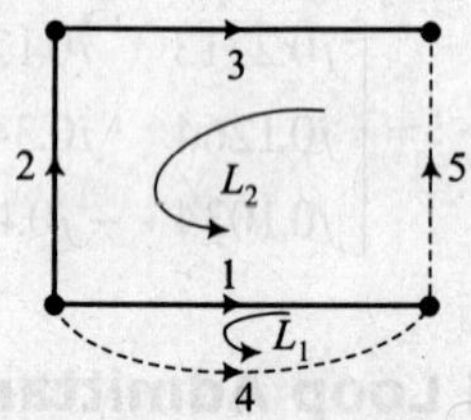

Fig. 2.17 Basic loops of example 2.5.

$$Y_{\text{Loop}} = Z_{\text{Loop}}^{-1} = \begin{bmatrix} -j1.8391 & -j0.3448 \\ -j0.3448 & -j0.6897 \end{bmatrix}$$

Example 2.9: For the network shown in Fig. 2.18, draw a tree and obtain the various incidence matrices.

Solution: Consider the tree as shown in Fig. 2.19.

Number of branches $b = 7$; Number of nodes $n = 5$; Number of tree-branches $t = 5 - 1 = 4$;

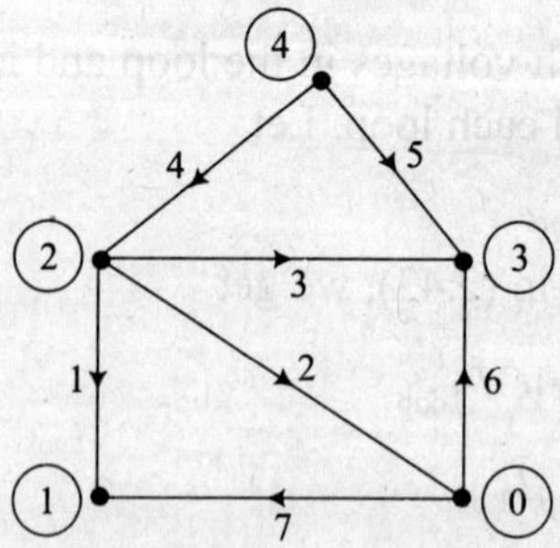

Fig. 2.18 Example 2.9.

Number of links $l = 7 - 4 = 3$; Number of basic cutsets $= t = 4$; Number of basic loops $= l = 3$. Size of $[A] = 4 \times 4$; Size of $[K] = 4 \times 4$; Size of $[B] = 7 \times 4$; Size of $[C] = 7 \times 3$.

The bus incidence matrix is formed choosing node 0 as reference node.

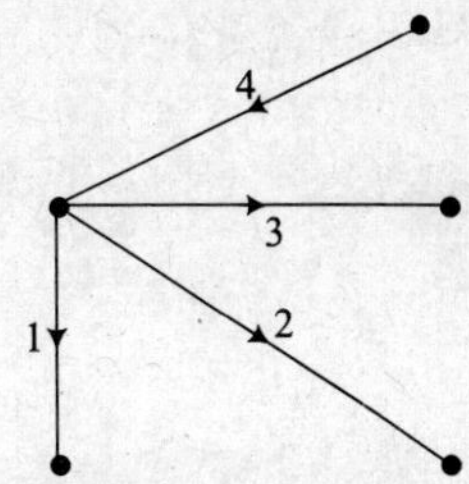

Fig. 2.19 Tree for example 2.9.

$$[A] = \begin{bmatrix} -1 & 1 & 0 & 0 \\ 0 & 1 & 0 & 0 \\ 0 & 1 & -1 & 0 \\ 0 & -1 & 0 & 1 \\ 0 & 0 & -1 & 1 \\ 0 & 0 & -1 & 0 \\ -1 & 0 & 0 & 0 \end{bmatrix}.$$

The tree-branch path incidence matrix $[K]$ is given by

$$[K] = \begin{bmatrix} -1 & 0 & 0 & 0 \\ 1 & 1 & 1 & 1 \\ 0 & 0 & -1 & 0 \\ 0 & 0 & 0 & 1 \end{bmatrix}$$

Basic cutset matrix $[B]$ is obtained from the fundamental cutsets. The basic cutsets are: $C_{S1} = \{1, 7\}$; $C_{S2} = \{2, 6, 7\}$; $C_{S3} = \{3, 5, 6\}$; $C_{S4} = \{4, 5\}$.

$$[B] = \begin{bmatrix} 1 & 0 & 0 & 0 \\ 0 & 1 & 0 & 0 \\ 0 & 0 & 1 & 0 \\ 0 & 0 & 0 & 1 \\ 0 & 0 & 1 & 1 \\ 0 & -1 & 1 & 0 \\ 1 & -1 & 0 & 0 \end{bmatrix} = \begin{bmatrix} I_t \\ B_l \end{bmatrix}$$

The basic loop incidence matrix is formed from the basic loops. The basic loops are $L_1 = \{3, 4, 5\}$; $L_2 = \{2, 3, 6\}$; $L_3 = \{1, 2, 7\}$.

$$[C] = \begin{bmatrix} 0 & 0 & -1 \\ 0 & 1 & 1 \\ -1 & -1 & 0 \\ -1 & 0 & 0 \\ 1 & 0 & 0 \\ 0 & 1 & 0 \\ 0 & 0 & 1 \end{bmatrix} = \begin{bmatrix} C_t \\ I_l \end{bmatrix}$$

2.7 NON-SINGULAR TRANSFORMATION

2.7.1 Loop Impedance and Loop Admittance Matrices

The augmented loop incidence matrix $\hat{C}$ can be used to relate the parameters of the primitive network to those of an augmented interconnected network. The augmented network is obtained by adding a fictitious link in parallel with each tree-branch of the original network as shown in Fig. 2.20.

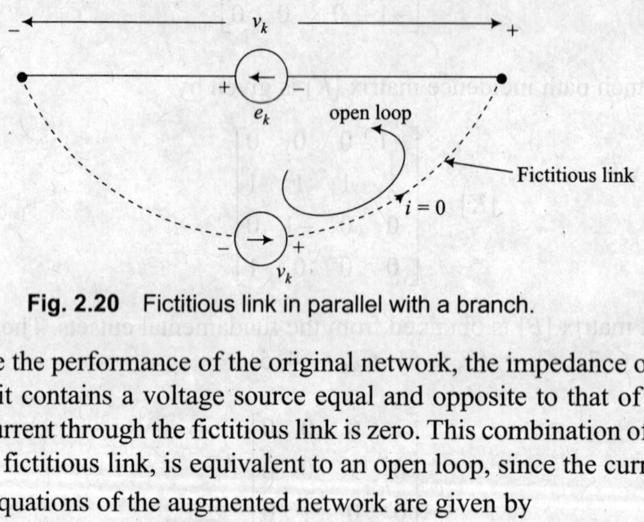

Fig. 2.20 Fictitious link in parallel with a branch.

In order to preserve the performance of the original network, the impedance of each fictitious link is set to zero and it contains a voltage source equal and opposite to that of the tree-branch voltage, such that the current through the fictitious link is zero. This combination of a loop containing a tree-branch and a fictitious link, is equivalent to an open loop, since the current is zero.

The performance equations of the augmented network are given by

$$\hat{E}_{\text{Loop}} = \hat{Z}_{\text{Loop}} \hat{I}_{\text{Loop}} \quad \text{...(2.48)}$$

Premultiplying the performance equation of the primitive network with $\hat{C}^T$, we obtain

$$\hat{C}^T \bar{v} + \hat{C}^T \bar{e} = \hat{C}^T [z] \bar{i} \quad \text{...(2.49)}$$

From the relationship derived in section 2.3.6, (2.49) can be written in partition form as

$$\begin{bmatrix} I_t & 0 \\ C_t^T & I_l \end{bmatrix} \begin{bmatrix} \bar{v}_t \\ \bar{v}_l \end{bmatrix} + \begin{bmatrix} I_t & 0 \\ C_t^T & I_l \end{bmatrix} \begin{bmatrix} \bar{e}_t \\ \bar{e}_l \end{bmatrix} = \begin{bmatrix} I_t & 0 \\ C_t^T & I_l \end{bmatrix} [z] \bar{i} \quad \text{...(2.50)}$$

where, $\bar{v}_t$, $\bar{e}_t$ are voltage vectors associated with tree-branches and $\bar{v}_l$, $\bar{e}_l$ are voltage vectors associated with links. The LHS of (2.50) is expanded to get

$$\begin{bmatrix} \bar{v}_t \\ C_t^T \bar{v}_t + \bar{v}_l \end{bmatrix} + \begin{bmatrix} \bar{e}_t \\ C_t^T \bar{e}_t + \bar{e}_l \end{bmatrix}$$

The terms can be simplified using (2.13) and (2.44) as follows:

$$\left[C_t^T \bar{v}_t + \bar{v}_l\right] = C^T \bar{v} = 0$$

$$\left[C_t^T \bar{e}_t + \bar{e}_l\right] = C^T \bar{e} = \bar{E}_{\text{Loop}}$$

Using the above relationships, the LHS of (2.50) can be simplified as

$$\begin{bmatrix} \bar{v}_t \\ 0 \end{bmatrix} + \begin{bmatrix} \bar{e}_t \\ \bar{E}_{\text{Loop}} \end{bmatrix} = \begin{bmatrix} \bar{v}_t + \bar{e}_t \\ \bar{E}_{\text{Loop}} \end{bmatrix} \equiv \hat{E}_{\text{Loop}} \qquad \text{...(2.51)}$$

From (2.49) and (2.51), we get

$$\hat{E}_{\text{Loop}} = \hat{C}^T [z] \bar{i} \qquad \text{...(2.52)}$$

Since the currents in the open loops are zero,

$$\hat{I}_{\text{Loop}} = \begin{bmatrix} 0 \\ \bar{I}_{\text{Loop}} \end{bmatrix}$$

From (2.14)

$$\bar{i} = C \bar{I}_{\text{Loop}}$$

But $C \bar{I}_{\text{Loop}} = \hat{C} \hat{I}_{\text{Loop}}$

From the above two equations $\bar{i} = \hat{C} \hat{I}_{\text{Loop}}$ ...(2.53)

Substituting (2.53) in (2.52), we get

$$\hat{E}_{\text{Loop}} = \hat{C}^T [z] \hat{C} \hat{I}_{\text{Loop}} = \hat{Z}_{\text{Loop}} \hat{I}_{\text{Loop}} \qquad \text{...(2.54)}$$

where, $\hat{Z}_{\text{Loop}} = \hat{C}^T [z] \hat{C}$ ...(2.55)

(2.55) can be partitioned as follows:

$$\begin{bmatrix} Z_1 & Z_2 \\ Z_3 & Z_4 \end{bmatrix} = \begin{bmatrix} I_t & 0 \\ C_t^T & I_l \end{bmatrix} \begin{bmatrix} z_{tt} & z_{tl} \\ z_{lt} & z_{ll} \end{bmatrix} \begin{bmatrix} I_t & C_t \\ 0 & I_l \end{bmatrix} \qquad \text{...(2.56)}$$

where,

$[z_{tt}]$ = primitive impedance sub-matrix of tree-branches.

$[z_{tl}] = [z_{lt}]^T$ = primitive impedance sub-matrix whose elements are mutual impedances between tree-branches and links.

$[z_{ll}]$ = primitive impedance sub-matrix of links.

From (2.56), $Z_4 = C_t^T[z_{tt}]C_t + [z_{lt}]C_t + C_t^T[z_{tl}] + [z_{ll}]$

From (2.47), $Z_{Loop} = C^T[z]\,C = [C_t^T\; I_l]\begin{bmatrix} z_{tt} & z_{tl} \\ z_{lt} & z_{ll} \end{bmatrix}\begin{bmatrix} C_t \\ I_l \end{bmatrix}$

$$= C_t^T[z_{tt}]C_t + [z_{lt}]C_t + C_t^T[z_{tl}] + [z_{ll}] = Z_4 \qquad ...(2.57)$$

Thus, $Z_{\text{Loop}} = Z_4$ and $Y_{\text{Loop}} = [Z_{\text{Loop}}]^{-1}$. Since $\hat{C}$ is non-singular, (2.55) can be viewed as a non-singular transformation of $[z]$.

2.7.2 Tree-branch Admittance and Tree-branch Impedance Matrices

The tree-branch admittance matrix Y_{BR} can also be obtained using the augmented cutset matrix $\hat{B}$. Here, the original network is augmented by connecting a fictitious tree-branch in series with each link of the original network as shown in Fig. 2.21.

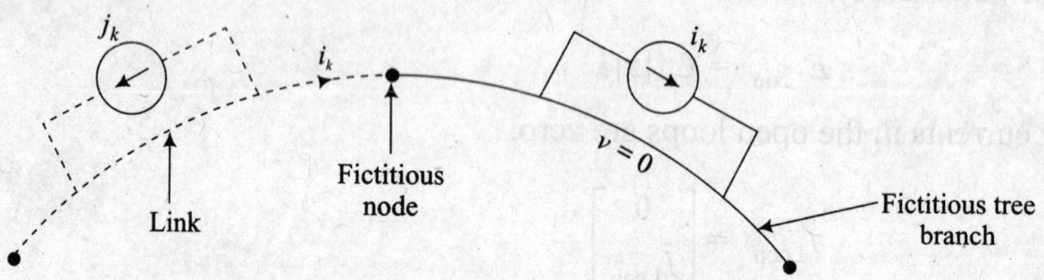

Fig. 2.21 Fictitious tree-branch in series with a link.

To preserve the performance of the original network, the admittance of the added branch is set to zero and its current source set equal to that of the associated link, so that the voltage across it is zero. The performance equation of the augmented network is given by

$$\hat{I}_{BR} = \hat{Y}_{BR}\,\hat{E}_{BR}$$

Premultiplying the performance equation of the primitive network with $\hat{B}^T$, we obtain

$$\hat{B}^T\bar{i} + \hat{B}^T\bar{j} = \hat{B}^T[y]\bar{v} \qquad ...(2.58)$$

(2.58) can be written in partitioned form as,

$$\begin{bmatrix} I_t & B_l^T \\ 0 & I_l \end{bmatrix}\begin{bmatrix} \bar{i}_t \\ \bar{i}_l \end{bmatrix} + \begin{bmatrix} I_t & B_l^T \\ 0 & I_l \end{bmatrix}\begin{bmatrix} \bar{j}_t \\ \bar{j}_l \end{bmatrix} = \begin{bmatrix} I_t & B_l^T \\ 0 & I_l \end{bmatrix}[y]\bar{v} \qquad ...(2.59)$$

where, $\bar{i}_t$ and $\bar{j}_t$ are primitive current vectors associated with tree-branches;

$\bar{i}_l$ and $\bar{j}_l$ are primitive current vectors associated with links.

The LHS of (2.59) can be written as $\begin{bmatrix} \bar{i}_t + B_l^T \bar{i}_l \\ \bar{i}_l \end{bmatrix} + \begin{bmatrix} \bar{j}_t + B_l^T \bar{j}_l \\ \bar{j}_l \end{bmatrix}$

The following relationships are used in simplifying the above arithmetic expression.

$$\bar{i}_t + B_l^T \bar{i}_l = B^T \bar{i} = 0$$

$$\bar{j}_t + B_l^T \bar{j}_l = B^T \bar{j} = \bar{I}_{BR}$$

Hence, LHS of (2.59) can be written as

$$\begin{bmatrix} 0 \\ \bar{i}_l \end{bmatrix} + \begin{bmatrix} \bar{I}_{BR} \\ \bar{j}_l \end{bmatrix} = \begin{bmatrix} \bar{I}_{BR} \\ \bar{i}_l + \bar{j}_l \end{bmatrix} = \hat{I}_{BR}$$

From the above derivation, (2.59) can be written as

$$\bar{I}_{BR} = \hat{B}^T [y] \bar{v} \qquad \text{...(2.60)}$$

Since the voltages across the fictitious tree-branches are zero, the voltage vector of the augmented network is $\hat{E}_{BR} = \begin{bmatrix} \bar{E}_{BR} \\ 0 \end{bmatrix}$. From (2.12),

$$\bar{v} = B \bar{E}_{BR} = \hat{B} \hat{E}_{BR} \qquad \text{...(2.61)}$$

Substituting (2.61) in (2.60), we get

$$\bar{I}_{BR} = \hat{B}^T [y] \hat{B} \hat{E}_{BR} = \hat{Y}_{BR} \hat{E}_{BR} \qquad \text{...(2.62)}$$

where,
$$\hat{Y}_{BR} = \hat{B}^T [y] \hat{B} \qquad \text{...(2.63)}$$

(2.63) can be partitioned as (refer section 2.3.4)

$$\begin{bmatrix} Y_1 & Y_2 \\ Y_3 & Y_4 \end{bmatrix} = \begin{bmatrix} I_t & B_l^T \\ 0 & I_l \end{bmatrix} \begin{bmatrix} y_{tt} & y_{tl} \\ y_{lt} & y_{ll} \end{bmatrix} \begin{bmatrix} I_t & 0 \\ B_l & I_l \end{bmatrix} \qquad \text{...(2.64)}$$

where, $[y_{tt}]$ = the primitive admittance sub-matrix corresponding to tree-branches.

$[y_{tl}] = [y_{lt}]^T$ = primitive admittance sub-matrix corresponding to elements which are mutual admittances between tree-branches and links.

$[y_{ll}]$ = the primitive admittance sub-matrix corresponding to links.

From (2.42), $Y_{BR} = B^T [y] B$ or in partitioned form

$$Y_{BR} = \begin{bmatrix} I_t & B_l^T \end{bmatrix} \begin{bmatrix} y_{tt} & y_{tl} \\ y_{lt} & y_{ll} \end{bmatrix} \begin{bmatrix} I_t \\ B_l \end{bmatrix}$$

$$= [y_{tt}] + B^T{}_l [y_{lt}] + [y_{tl}] B_l + B^T{}_l [y_{ll}] B_l = Y_1 \qquad \text{...(2.65)}$$

Thus,
$$Y_{BR} = Y_1$$

$$Z_{BR} = [Y_{BR}]^{-1} = [Y_1]^{-1}$$

(2.63) is a non-singular transformation of $[y]$.

2.7.3 Relationship between the Loop Admittance Matrix and Augmented Network Admittance Matrix

From (2.55) and (2.63), we get

$$\hat{Z}_{\text{Loop}}\hat{Y}_{BR} = \hat{C}^T[z]\hat{C}\hat{B}^T[y]\hat{B} \qquad \text{...(2.66)}$$

$$\hat{C}\hat{B}^T = \begin{bmatrix} I_t & C_t \\ 0 & I_l \end{bmatrix}\begin{bmatrix} I_t & B_l^T \\ 0 & I_l \end{bmatrix} = \begin{bmatrix} I_t & B_l^T + C_t \\ 0 & I_l \end{bmatrix} \qquad \text{(2.67)}$$

Premultiplying (2.14) with B^T and using the result of (2.11), we get

$$B^T\bar{i} = B^T C\bar{I}_{\text{Loop}} = 0$$

which means $(C_t + B_l^T)\,\bar{I}_{\text{Loop}} = 0$. From this, we can write

$$C_t = -B_l^T \qquad \text{...(2.68)}$$

Substituting (2.68) in (2.67), we get

$$\hat{C}\hat{B}^T = I \qquad \text{...(2.69a)}$$

Similarly,

$$\hat{C}^T\hat{B} = I \qquad \text{...(2.69b)}$$

Substituting (2.69a) into (2.66) and noting that $[z][y] = I$, we get

$$\hat{Z}_{\text{Loop}}\hat{Y}_{\text{BR}} = \hat{C}^T\hat{B} = I \qquad \text{...(2.70)}$$

(2.70) can be partitioned as

$$\begin{bmatrix} Z_1 & Z_2 \\ Z_3 & Z_4 \end{bmatrix}\begin{bmatrix} Y_1 & Y_2 \\ Y_3 & Y_4 \end{bmatrix} = \begin{bmatrix} I_t & 0 \\ 0 & I_l \end{bmatrix}$$

From the above partitioned matrices, we can write the following relationships:

$$Z_1Y_1 + Z_2Y_3 = I_t \qquad \text{...(2.71a)}$$

$$Z_1Y_2 + Z_2Y_4 = 0 \qquad \text{...(2.71b)}$$

$$Z_3Y_1 + Z_4Y_3 = 0 \qquad \text{...(2.71c)}$$

$$Z_3Y_2 + Z_4Y_4 = I_l \qquad \text{...(2.71d)}$$

From (2.71c), we get

$$Z_3 = -Z_4Y_3Y_1^{-1}$$

Substituting for Z_3 from (2.71c) in (2.71d), we get

$$Z_4(Y_4 - Y_3Y_1^{-1}Y_2) = I_l \qquad \text{...(2.72)}$$

In section 2.7.1 it has been shown that $Y_{\text{Loop}} = Z_4^{-1}$, which means that $Z_4Y_{\text{Loop}} = I_l$. Comparing this with (2.72), we can conclude that

$$Y_{\text{Loop}} = Y_4 - Y_3 Y_1^{-1} Y_2 \quad ...(2.73)$$

2.7.4 Relationship between Branch Impedance Matrix and Augmented Network Impedance Matrix

From (2.71c), we get

$$Y_3 = -Z_4^{-1} Z_3 Y_1$$

Substituting in (2.71a), we get

$$(Z_1 - Z_2 Z_4^{-1} Z_3) Y_1 = I_t \quad ...(2.74)$$

In section 2.7.2 it was shown that $Z_{\text{BR}} = Y_1^{-1}$, which means $Z_{\text{BR}} Y_1 = I_t$. Comparing this result with (2.74), we can conclude that

$$Z_{\text{BR}} = Z_1 - Z_2 Z_4^{-1} Z_3 \quad ...(2.75)$$

2.7.5 Relationship between Tree-branch Matrices and Bus Matrices

Using the tree-branch path incidence matrix K, from (2.7) and (2.10) we have $A_t K^T = I_t$ and $B_t = A_l K^T$. Post multiplying the bus incidence matrix A by K^T, we get

$$AK^T = \begin{bmatrix} A_t \\ A_l \end{bmatrix} K^T = \begin{bmatrix} A_t K^T \\ A_l K^T \end{bmatrix} = \begin{bmatrix} I_t \\ B_l \end{bmatrix} = B \quad ...(2.76a)$$

Transposing (2.76a), we get

$$KA^T = B^T \quad ...(2.76b)$$

Post multiplying (2.76b) by $[y]AK^T$, we get

$$KA^T[y]AK^T = B^T[Y]AK^T = B^T[Y]B \quad ...(2.77)$$

From singular transformations we know $A^T[y]A = Y_{\text{bus}}$ and $B^T[y]B = Y_{BR}$. Hence, (2.77) can be written as

$$KY_{\text{bus}} K^T = Y_{\text{BR}} \quad ...(2.78)$$

(2.78) relates the tree-branch admittance matrix to the bus-admittance matrix. The tree-branch impedance matrix is obtained from

$$Z_{\text{BR}} = Y_{\text{BR}}^{-1} = (K^T)^{-1} Y_{\text{bus}}^{-1} K^{-1}$$

Since $K^T = A_t^{-1}$, we can write $Z_{BR} = A_t Z_{bus} A_t^T$. Now pre-multiplying (2.78) by K^{-1} and post-multiplying by $(K^T)^{-1}$, we get

$$Y_{\text{bus}} = K^{-1} Y_{\text{BR}} (K^T)^{-1}$$

Since $$K^T = A_t^{-1},\ Y_{\text{bus}} = A^T{}_t Y_{BR} A_t \quad ...(2.79)$$

(2.79) give the bus admittance matrix in terms of the tree-branch admittance matrix.

$$Z_{\text{bus}} = Y_{\text{bus}}^{-1} = (A_t^T Y_{BR} A_t)^{-1} = K^T Z_{BR} K$$

2.7.6 Power Invariance of Singular Transformations

(i) Bus admittance matrix

From (2.33) we have

$$\bar{I}_{\text{bus}} = A^T [y]\, \bar{v}$$

The complex power into the network is $(\bar{I}^*_{\text{bus}})\, \bar{E}_{\text{bus}}$. The sum of the powers in the primitive network is $(j^*)^T \bar{v}$. For the transformation to be power invariant, the two powers must be equal.

$$(\bar{I}^*_{\text{bus}})^T\, \bar{E}_{\text{bus}} = (\bar{j}^*)^t\, \bar{v} \tag{2.80}$$

Form (2.32) $\bar{I}_{\text{bus}} = A^T \bar{j}$

$$(\bar{I}^*_{\text{bus}})^t = (\bar{j}^*)^T A \tag{2.81}$$

Substituting (2.81) in (2.80), we get

$$(\bar{j}^*)^T A\, \bar{E}_{\text{bus}} = (\bar{j}*)^T \bar{v} \tag{2.82}$$

This equation is true for all values of $\bar{j}$. Hence,

$$A\bar{E}_{\text{bus}} = \bar{v} \tag{2.83}$$

Substituting in (2.33), we get

$$\bar{I}_{\text{bus}} = A^T [y]\, A\, \bar{E}_{\text{bus}}$$

From which we get,

$$\bar{I}_{\text{bus}} = Y_{\text{bus}}\, \bar{E}_{\text{bus}}$$

where

$$Y_{\text{bus}} = A^T [y]\, A.$$

Branch Admittance Matrix

From (2.39) and (2.37), we have

$$B^T \bar{j} = \bar{I}_{BR} = B^T [y]\, \bar{v} \tag{2.84}$$

Power into the network is $(\bar{I}^*_{BR})^T\, \bar{E}_{BR}$ and this should be equal to power in primitive network.

$$(\bar{I}*_{BR})^T\, \bar{E}_{BR} = (\bar{j}*)^T\, \bar{v} \tag{2.85}$$

From (2.84)

$$(\bar{I}*_{BR})^T = (\bar{j}*)^T B* \tag{2.86}$$

$$\therefore \qquad (\bar{j}^*)^T B^* \bar{E}_{BR} = (\bar{j}^*)^T \bar{v} \tag{2.87}$$

Since this is true for all $\bar{j}$, and B is a real matrix, so that $B^* = B$, we get

$$B\bar{E}_{BR} = \bar{v} \tag{2.88}$$

Substituting in (2.84), we get

$$\bar{I}_{BR} = B^T [y] B \bar{E}_{BR} \tag{2.89}$$

or

$$Y_{BR} = B^T [y] B \tag{2.90}$$

Loop Impedance Matrix

Power into network is $(\bar{I}^*_{\text{Loop}})^T \bar{E}_{\text{Loop}}$

Power in primitive network is given by $(\bar{i}^*)^T \bar{e}$. Due to power invariance,

$$(\bar{I}^*_{\text{Loop}})^T \bar{E}_{\text{Loop}} = (\bar{i}^*)^T \bar{e} \tag{2.91}$$

Substituting for $\bar{E}_{\text{Loop}}$ form (2.44), we get

$$(\bar{I}^*_{\text{Loop}})^T C^T \bar{e} = (\bar{i}^*)^T \bar{e} \tag{2.92}$$

Since this is valid for all $\bar{e}$,

$$(\bar{i}^*)^T = (\bar{I}^*_{\text{Loop}})^T C^T \tag{2.93}$$

Hence,

$$\bar{i} = C^* \bar{I}_{\text{Loop}} \tag{2.94}$$

$$= C \bar{I}_{\text{Loop}} \text{ (as } C \text{ is a real matrix } C^* = C)$$

We now get, from (2.43)

$$\bar{E}_{\text{Loop}} = C^T\bar{e} = C^T [z] \bar{i}$$

$$= C^T [z] C \bar{I}_{\text{Loop}} \tag{2.95}$$

From (2.95) we get

$$Z_{\text{Loop}} = C^T [z] C$$

2.8 SUMMARY OF RELATIONSHIPS DERIVED

The various relationships derived in the preceding sections are presented here. In all the equations t stands for tree-branches, l stands for links and $[I]$ stands for identity matrix.

2.8.1 Incidence Matrices

(i) Bus incidence matrix $A = \begin{bmatrix} A_t \\ A_l \end{bmatrix}$

(ii) $A_t K^T = I_t$ or $K^T = [A_t]^{-1}$

(iii) Basic cutset matrix $B = \begin{bmatrix} I_t \\ B_l \end{bmatrix}$

(iv) $B_t = A_l K^T$

(v) Augmented cutset matrix $\hat{B} = \begin{bmatrix} I_t & 0 \\ B_l & I_l \end{bmatrix}$

(vi) Basic loop incidence matrix $C = \begin{bmatrix} C_t \\ I_l \end{bmatrix}$

(vii) Augmented loop incidence matrix $\hat{C} = \begin{bmatrix} I_t & C_t \\ 0 & I_l \end{bmatrix}$

2.8.2 Singular Transformations

(i) Bus admittance matrix $Y_{\text{bus}} = A^T[y]A$

(ii) Tree-branch admittance matrix $Y_{BR} = B^T[y]B$

(iii) Loop impedance matrix $Z_{\text{Loop}} = C^T[z]C$

2.8.3 Non-singular Transformations

(i) $\hat{Z}_{\text{Loop}} = \hat{C}^T[z]\hat{C} = \begin{bmatrix} Z_1 & Z_2 \\ Z_3 & Z_4 \end{bmatrix}$

(ii) $Z_{\text{Loop}} = Z_4$

(iii) $\hat{Y}_{BR} = \hat{B}^T[y]\hat{B} = \begin{bmatrix} Y_1 & Y_2 \\ Y_3 & Y_4 \end{bmatrix}$

(iv) $Y_{\text{Loop}} = Y_4 - Y_3 Y_1^{-1} Y_2$

(v) $Y_{BR} = Y_1$

(vi) $Z_{BR} = Z_1 - Z_2 Z_4^{-1} Z_3$

(vii) $Y_{BR} = K Y_{\text{bus}} K^T$

(viii) $Z_{BR} = A_t Z_{\text{bus}} A_t^T$

(ix) $Y_{\text{bus}} = A_t^T Y_{BR} A_t$

(x) $Z_{\text{bus}} = K^T Z_{BR} K$

2.8.4 Current and Voltage Relationships

(i) $\bar{i} = C\bar{I}_{\text{Loop}}$ (ii) $\bar{I}_{\text{bus}} = A^T \bar{j}$

(iii) $\bar{I}_{BR} = B^T \bar{j}$ (iv) $\hat{E}_{\text{Loop}} = C^T \bar{e}$

(v) $\bar{v} = A\bar{E}_{\text{bus}}$ (vi) $\bar{v} = B\bar{E}_{\text{BR}}$

Example 2.10: For the data given in example 2.3, obtain the following:

(i) Z_{Loop} and Z_{BR} from non-singular transformation.

(ii) Y_{Loop} and Y_{BR} from non-singular transformation.

(iii) Z_{bus} and Y_{bus} from Z_{BR} and Y_{BR}.

(iv) Y_{BR} and Z_{BR} from Y_{bus} and Z_{bus}.

Solution: (i) Z_{Loop} and Z_{BR} from non-singular transformation

$$\hat{Z}_{\text{Loop}} = \hat{C}^T [z]\hat{C}$$

$$\hat{C} = \begin{bmatrix} [I_t] & [C_t] \\ [0] & [I_l] \end{bmatrix} = \begin{bmatrix} 1 & 0 & 0 & -1 & 1 \\ 0 & 1 & 0 & 0 & -1 \\ 0 & 0 & 1 & 0 & -1 \\ 0 & 0 & 0 & 1 & 0 \\ 0 & 0 & 0 & 0 & 1 \end{bmatrix}$$

From example 2.3

$$[z] = \begin{bmatrix} j0.6 & j0.1 & 0.0 & j0.2 & 0.0 \\ j0.1 & j0.5 & 0.0 & 0.0 & 0.0 \\ 0.0 & 0.0 & j0.5 & 0.0 & 0.0 \\ j0.2 & 0.0 & 0.0 & j0.4 & 0.0 \\ 0.0 & 0.0 & 0.0 & 0.0 & j0.2 \end{bmatrix}$$

$$\hat{Z}_{\text{Loop}} = \hat{C}^T [z]\hat{C} = \begin{bmatrix} j0.6 & j0.1 & 0.0 & -j0.4 & j0.5 \\ j0.1 & j0.5 & 0.0 & -j0.1 & -j0.4 \\ 0.0 & 0.0 & j0.5 & 0.0 & -j0.5 \\ -j0.4 & -j0.1 & 0.0 & j0.6 & -j0.3 \\ j0.5 & -j0.4 & -j0.5 & -j0.3 & j1.6 \end{bmatrix} = \begin{bmatrix} Z_1 & Z_2 \\ Z_3 & Z_4 \end{bmatrix}$$

$$[Z_1] = \begin{bmatrix} j0.6 & j0.1 & 0.0 \\ j0.1 & j0.5 & 0.0 \\ 0.0 & 0.0 & j0.5 \end{bmatrix}$$

$$[Z_2] = \begin{bmatrix} -j0.4 & j0.5 \\ -j0.1 & -j0.4 \\ 0.0 & -j0.5 \end{bmatrix}$$

$$[Z_3] = \begin{bmatrix} -j0.4 & -j0.1 & 0.0 \\ j0.5 & -j0.4 & -j0.5 \end{bmatrix}$$

$$[Z_4] = \begin{bmatrix} j0.6 & -j0.3 \\ -j0.3 & j1.6 \end{bmatrix}$$

$$[Z_{\text{Loop}}] = [Z_4] = \begin{bmatrix} j0.6 & -j0.3 \\ -j0.3 & j1.6 \end{bmatrix} \text{ (compare with results of example 2.8)}$$

$$[Z_2][Z_4^{-1}][Z_3] = \begin{bmatrix} j0.3287 & -j0.0264 & -j0.1034 \\ -j0.0264 & j0.1563 & j0.1552 \\ -j0.1034 & j0.1552 & j0.1724 \end{bmatrix}$$

$$Z_{BR} = Z_1 - [Z_2][Z_4^{-1}][Z_3] = \begin{bmatrix} j0.2713 & j0.1264 & j0.1034 \\ j0.1264 & j0.3437 & -j0.1552 \\ j0.1034 & -j0.1552 & j0.3276 \end{bmatrix}$$

(compare with results of example 2.7)

(ii) Y_{Loop} and Y_{BR} from non-singular transformation

$$\hat{Y}_{BR} = \hat{B}^T[y]\hat{B}$$

$$\hat{B} = \begin{bmatrix} [I_t] & [0] \\ [B_l] & [I_l] \end{bmatrix}$$

$$\hat{B} = \begin{bmatrix} 1 & 0 & 0 & 0 & 0 \\ 0 & 1 & 0 & 0 & 0 \\ 0 & 0 & 1 & 0 & 0 \\ 1 & 0 & 0 & 1 & 0 \\ -1 & 1 & 1 & 0 & 1 \end{bmatrix}$$

From example 2.3,

$$[y] = \begin{bmatrix} -j2.0833 & j0.4167 & 0.0 & j1.0417 & 0.0 \\ j0.4167 & -j2.0833 & 0.0 & -j0.2083 & 0.0 \\ 0.0 & 0.0 & -j2.0 & 0.0 & 0.0 \\ j1.0417 & -j0.2083 & 0.0 & -j3.0208 & 0.0 \\ 0.0 & 0.0 & 0.0 & 0.0 & -j5.0 \end{bmatrix}$$

$$\hat{Y}_{BR} = \begin{bmatrix} -j8.0208 & j5.2083 & j5.0 & -j1.9792 & j5.0 \\ j5.2083 & -j7.0833 & -j5.0 & -j0.2083 & -j5.0 \\ j5.0 & -j5.0 & -j7.0 & 0.0 & -j5.0 \\ -j1.9792 & -j0.2083 & 0.0 & -j3.0208 & 0.0 \\ j5.0 & -j5.0 & -j5.0 & 0.0 & -j5.0 \end{bmatrix} = \begin{bmatrix} Y_1 & Y_2 \\ Y_3 & Y_4 \end{bmatrix}$$

$$Y_1 = \begin{bmatrix} -j8.0208 & j5.2083 & j5.0 \\ j5.2083 & -j7.0833 & -j5.0 \\ j5.0 & -j5.0 & -j7.0 \end{bmatrix}$$

$$Y_2 = \begin{bmatrix} -j1.9792 & j5.0 \\ -j0.2083 & -j5.0 \\ 0.0 & -j5.0 \end{bmatrix}$$

$$Y_3 = \begin{bmatrix} -j1.9792 & -j0.2083 & 0.0 \\ j5.0 & -j5.0 & -j5.0 \end{bmatrix}$$

$$Y_4 = \begin{bmatrix} -j3.0208 & 0.0 \\ 0.0 & -j5.0 \end{bmatrix}$$

$$Y_{BR} = Y_1 = \begin{bmatrix} -j8.0208 & j5.2083 & j5.0 \\ j5.2083 & -j7.0833 & -j5.0 \\ j5.0 & -j5.0 & -j7.0 \end{bmatrix}$$

$$Y_3 Y_1^{-1} Y_2 = \begin{bmatrix} -j1.1818 & j0.3448 \\ j0.3448 & -j4.3104 \end{bmatrix}$$

$$Y_{Loop} = Y_4 - Y_3 Y_1^{-1} Y_2 = \begin{bmatrix} -j1.8391 & -j0.3448 \\ -j0.3448 & -j0.6897 \end{bmatrix}$$

(iii) Z_{bus} and Y_{bus} from Z_{BR} and Y_{BR} (Compare with example 2.8)

For the given tree

$$[K] = \begin{bmatrix} -1 & 0 & 0 \\ 0 & -1 & -1 \\ 0 & 0 & -1 \end{bmatrix}$$

From the bus incidence matrix we obtain A_t, elements corresponding to tree-branches,

$$A_t = \begin{bmatrix} -1 & 0 & 0 \\ 0 & -1 & 0 \\ 0 & 1 & -1 \end{bmatrix}$$

$$Z_{\text{bus}} = K^T Z_{\text{BR}} K$$

$$= \begin{bmatrix} -1 & 0 & 0 \\ 0 & -1 & 0 \\ 0 & -1 & -1 \end{bmatrix} \begin{bmatrix} j0.2713 & j0.1264 & j0.1034 \\ j0.1264 & j0.3437 & -j0.1552 \\ j0.1034 & -j0.1552 & j0.3276 \end{bmatrix} \begin{bmatrix} -1 & 0 & 0 \\ 0 & -1 & -1 \\ 0 & 0 & -1 \end{bmatrix}$$

$$= \begin{bmatrix} j0.2713 & j0.1264 & j0.2299 \\ j0.1264 & j0.3437 & j0.1885 \\ j0.2299 & j0.1885 & j0.3609 \end{bmatrix}$$

$$Y_{\text{bus}} = A^T{}_t \, Y_{\text{BR}} \, A_t$$

$$= \begin{bmatrix} -1 & 0 & 0 \\ 0 & -1 & 1 \\ 0 & 0 & -1 \end{bmatrix} \begin{bmatrix} -j8.0208 & j5.2083 & j5.0 \\ j5.2083 & -j7.0833 & -j5.0 \\ j5.0 & -j5.0 & -j7.0 \end{bmatrix} \begin{bmatrix} -1 & 0 & 0 \\ 0 & -1 & 0 \\ 0 & 1 & -1 \end{bmatrix}$$

$$= \begin{bmatrix} -j8.0208 & j0.2083 & j5.0 \\ j0.2083 & -j4.0833 & j2.0 \\ j5.0 & j2.0 & -j7.0 \end{bmatrix}$$

(iv) Z_{BR} and Y_{BR} from Z_{bus} and Y_{bus}

$$Z_{\text{BR}} = A_t \, Z_{\text{bus}} \, A_t^T$$

$$= \begin{bmatrix} -1 & 0 & 0 \\ 0 & -1 & 0 \\ 0 & 1 & -1 \end{bmatrix} \begin{bmatrix} j0.2713 & j0.1264 & j0.2299 \\ j0.1264 & j0.3437 & j0.1885 \\ j0.2299 & j0.1885 & j0.3609 \end{bmatrix} \begin{bmatrix} -1 & 0 & 0 \\ 0 & -1 & 1 \\ 0 & 1 & -1 \end{bmatrix}$$

$$= \begin{bmatrix} j0.2713 & j0.1264 & j0.1034 \\ j0.1264 & j0.3437 & -j0.1552 \\ j0.1034 & -j0.1552 & j0.3276 \end{bmatrix}$$

$$Y_{\text{BR}} = K \, Y_{\text{bus}} \, K^T$$

$$= \begin{bmatrix} -1 & 0 & 0 \\ 0 & -1 & -1 \\ 0 & 0 & -1 \end{bmatrix} \begin{bmatrix} -j8.0208 & j0.2083 & j5.0 \\ j0.2083 & -j4.0833 & j2.0 \\ j5.0 & j2.0 & -j7.0 \end{bmatrix} \begin{bmatrix} -1 & 0 & 0 \\ 0 & -1 & 0 \\ 0 & -1 & -1 \end{bmatrix}$$

$$= \begin{bmatrix} -j8.0208 & j5.2083 & j5.0 \\ j5.2083 & -j7.0833 & -j5.0 \\ j5.0 & -j5.0 & -j7.0 \end{bmatrix}$$

2.9 SUMMARY

The formulation of the mathematical model is the first step in obtaining the solution of any electrical network. The independent variables can be either currents or voltages. Correspondingly, the elements of the coefficient matrix will be impedances or admittances.

Network equations can be formulated for solution of the network using graph theory, independent of the nature of elements. In the graph of a network, the tree-branches and links are distinctly identified. The complete information about the interconnection of the network, with the directions of the currents is contained in the bus incidence matrix. For a given tree, the fundamental cutset matrix and the fundamental loop-incidence matrix contain information on the basic cutsets, loops and the branches incident on them along with the orientations.

The information on the nature of the elements which form the interconnected network is contained in the primitive impedance matrix. A primitive element can be represented in impedance form or admittance form. In the bus frame of reference, the performance of the interconnected system is described by $(n-1)$ nodal equations, where n is the number of nodes. The bus admittance matrix and the bus impedance matrix relate the bus voltages and currents. These matrices can be obtained from the primitive impedance and admittance matrices.

In the tree-branch frame of reference, the performance of the interconnected network is obtained by t independent tree-branch equations, where t is the number of tree-branches. Here, the coefficient matrices are the tree-branch impedance matrix (Z_{BR}) and the tree-branch admittance matrix (Y_{BR}), which can be obtained from the primitive matrices by singular transformation. In the loop frame of reference, the performance of the interconnected network is described by l (number of links) independent loop equations. The coefficient matrices are Z_{Loop} (loop impedance matrix) and Y_{Loop} (loop admittance matrix), which can be formed from the primitive matrices by singular transformation. All the coefficient matrices can also be obtained from non-singular transformations by using augmented loop-incidence matrix and augmented cutset–incidence matrix.

2.10 SUPPLEMENTARY PROBLEMS

1. For the graph of Fig. 2.7, draw a tree different from the one given in Fig. 2.8 and obtain $\hat{A}$, A, K, B, $\hat{B}$, C, $\hat{C}$ and verify the following:

   (a) $A_t K^{\mathrm{T}} = I$

(b) $B_l = A_l K^T$

(c) $C_t = -B_t^T$

(d) $\hat{C}\hat{B}^T = I$

2. For the network of Fig. 2.7, the impedances are as follows:

Element	Impedance
1	$j\,0.4$
2	$j\,0.25$
3	$j\,0.1$
4	$-j\,0.2$
5	$j\,0.5$
6	$j\,0.3$
7	$j\,0.2$

Obtain (i) Y_{bus}, Y_{BR} and Z_{Loop} by singular transformations.

(ii) Z_{Loop}, Z_{BR} and Z_{bus} by non-singular transformations.

3. Show that if mutual coupling is not present, the elements of Y_{bus} are given by $Y_{ii} = \sum_j y_{ij}$ and $Y_{ij} = -y_{ij}$, where y_{ij} is the admittance between buses i and j.

REVIEW QUESTIONS

1. Define the following : Graph, sub-graph, connected-graph, loop, cutset, tree, co-tree.
2. How are basic loops different from other loops?
3. A cutset reflects the vulnerability in a system. Comment.
4. The reduced incidence matrix contains the information on the complete interconnectivity of the network. Comment.
5. With an example, give the incidence and the reduced incidence matrix.
6. What are f-cutsets or fundamental cutsets?
7. What is a primitive network? Give the admittance and impedance form of the primitive element.
8. What are singular transformations?
9. Derive expression for the bus admittance matrix, to be obtained from primitive admittance matrix.
10. Derive the expression for the tree-branch admittance and impedance matrices.
11. Derive expression for the loop impedance and loop admittance matrices.
12. Derive the relationship between the tree branch matrices and the bus matrices.

Chapter 3

Network Matrices

3.1 INTRODUCTION

The method of building network bus admittance and bus impedance matrices in chapter 2, involves transformations and inversion of matrices. An alternative method for bus admittance matrix is based on a simple algorithm obtained by inspection of the network interconnections. The method is applicable only if there is no mutual coupling between elements. The bus impedance matrix can be formed by using an algorithm, wherein the matrix is formed step by step by adding one element at a time to a partial network. These algorithms can be easily programmed for digital computer application and are discussed in detail in the subsequent sections.

3.2 FORMATION OF Y_{BUS} BY INSPECTION

Consider an *n* bus power system (excluding the ground). Taking ground as reference and considering the single phase equivalent of the balanced network, we can write the nodal equations at the *n* buses as

$$
\begin{aligned}
I_1 &= E_1 y_{10} + (E_1 - E_2)y_{12} \ldots + (E_1 - E_k)y_{1k} + \ldots + (E_1 - E_n)y_{1n} \\
&\vdots \\
I_k &= (E_k - E_1)y_{k1} + (E_k - E_2)y_{k2} \ldots + E_k y_{ko} + \ldots + (E_k - E_n)y_{kn} \\
&\vdots \\
I_n &= (E_n - E_1)y_{n1} + (E_n - E_2)y_{n2} \ldots + (E_n - E_k)y_{kn} + \ldots + E_n y_{no}
\end{aligned}
\qquad \ldots(3.1)
$$

In the above equations, I_k is the net current injection at bus *k*, E_k is the voltage of bus *k* with respect to ground, y_{ij} is the admittance of element connected between buses *i* and *j* and y_{ko} is the admittance of the element between bus *k* and ground. (3.1) is rewritten as follows:

$$I_1 = (y_{10} + y_{12} + \ldots + y_{1k} \ldots + y_{1n})E_1 + \ldots\ldots - y_{1k} E_k \ldots\ldots\ldots\ldots\ldots - y_{1n}E_n$$

$$\vdots$$

$$I_k = -y_{k1}E_1 \ldots + \ldots\ldots\ldots\ldots (y_{ko} + y_{k1} + \ldots y_{kn})E_k \ldots\ldots\ldots\ldots - y_{kn}E_n$$

$$\vdots$$

$$I_n = -y_{n1}E_1 \ldots\ldots\ldots\ldots\ldots\ldots\ldots\ldots -y_{nk}E_k \ldots + (y_{no} + y_{n1} \ldots)E_n$$

...(3.2)

In matrix notation it can be written as $\bar{I}_{\text{bus}} = Y_{\text{bus}}\bar{E}_{\text{bus}}$ ...(3.3)

where, $\bar{I}_{\text{bus}}$ = vector of current injections at the nodes = $\begin{bmatrix} I_1 \\ \vdots \\ I_k \\ \vdots \\ I_n \end{bmatrix}$ ...(3.4)

E_{bus} = vector of bus voltages = $\begin{bmatrix} E_1 \\ E_2 \\ \vdots \\ E_k \\ \vdots \\ E_n \end{bmatrix}$ ...(3.5)

Y_{bus} = Bus admittance matrix = $\begin{bmatrix} Y_{11} & Y_{12} & \cdots & Y_{1k} & \cdots & Y_{1n} \\ \vdots & & & & & \\ Y_{k1} & Y_{k2} & \cdots & Y_{kk} & \cdots & Y_{kn} \\ \vdots & & & & & \\ Y_{n1} & Y_{n2} & \cdots & Y_{nk} & \cdots & Y_{nn} \end{bmatrix}$ ...(3.6)

Comparing (3.6) and (3.2), we observe

(i) The diagonal element Y_{ii} of the bus admittance matrix is the sum of the admittances of all elements incident on bus i, including the element between bus i and ground.

(ii) The off diagonal element Y_{ij} is negative of the admittance of the element connected between buses i and j.

Thus, by inspecting the network, the bus admittance matrix can be formed using the above two rules. It should, however, be noted that this method cannot be used, if mutual coupling exists between the elements. If mutual coupling is present, the bus admittance matrix has to be obtained by suitable transformation of the primitive network matrices. The modeling of various components which are included in the bus admittance matrix is discussed next.

3.2.1 Modeling of Transmission Lines

Transmission lines are modeled by the nominal π model. The model is shown in Fig. 3.1. $Z_L = R_L + jX_L$ is the series impedance of the line and $\frac{y_c}{2}$ is the half line charging admittance. In the nominal π model, half of the line charging admittance is shown in shunt at the two buses, where the line is connected. Every time, a new line is added between buses i and j, the elements of the bus admittance matrix get modified as follows:

$$y_L = \frac{1}{Z_L} = \frac{1}{R_L + jX_L} \quad \text{...(3.7)}$$

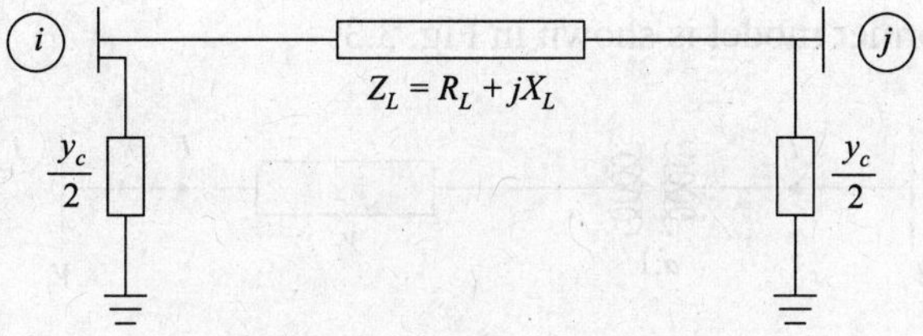

Fig. 3.1 Transmission line π model.

$$Y_{ii(\text{new})} = Y_{ii(\text{old})} + y_L + \frac{y_c}{2} \quad \text{...(3.8a)}$$

$$Y_{jj(\text{new})} = Y_{jj(\text{old})} + y_L + \frac{y_c}{2} \quad \text{...(3.8b)}$$

$$Y_{ij(\text{new})} = Y_{ij(\text{old})} - y_L \quad \text{...(3.8c)}$$

$$Y_{ji(\text{new})} = Y_{ji(\text{old})} - y_L \quad \text{...(3.8d)}$$

In (3.8a) to (3.8d), $Y_{(\text{old})}$ is the element of the bus admittance matrix before the line is included and $Y_{(\text{new})}$ is the element after line is added. The line admittance is added to the diagonal elements and the negative of the line admittance is added to the off diagonal elements. The half line charging admittance is added to the diagonal elements only, since it appears in shunt across the buses. Thus, every time a line is added, four elements of the bus admittance matrix have to be modified.

3.2.2 Modeling of Transformers

If the shunt branch of the transformer is neglected, the transformer equivalent circuit reduces to a simple series impedance as shown in Fig. 3.2.

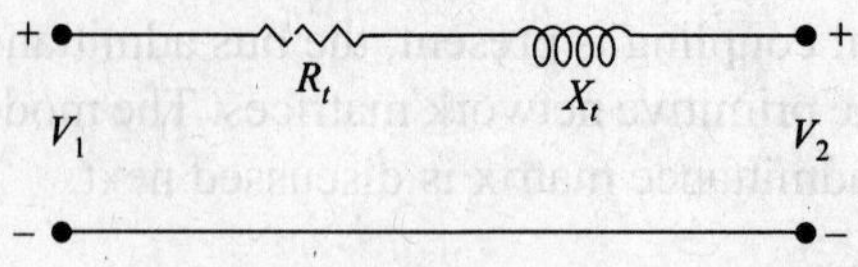

Fig. 3.2 Simplified equivalent circuit of transformer.

A transformer connected between buses *i* and *j* can be modeled as a line whose admittance is given by $\dfrac{1}{(R_t + jX_t)}$ and the bus admittance matrix modified as in section 3.2.1. However, it is frequently necessary to adjust voltages on one or both sides of the transformer to keep the voltage levels within acceptable limits. The transformers are provided with tap changing facility to adjust the voltage. If the turns ratio is equal to the ratio of rated voltages, the ratio is said to be nominal. If the turns ratio is different from the ratio of the rated voltages, the ratio is said to be off-nominal. The data of impedance of the transformer corresponds to impedance at nominal ratio. Therefore, in the presence of off-nominal ratio, the model has to be suitably modified to take into account the change in impedance. Consider a transformer connected between buses *i* and *j*. Let the off-nominal ratio be a:1. The transformer model is shown in Fig. 3.3.

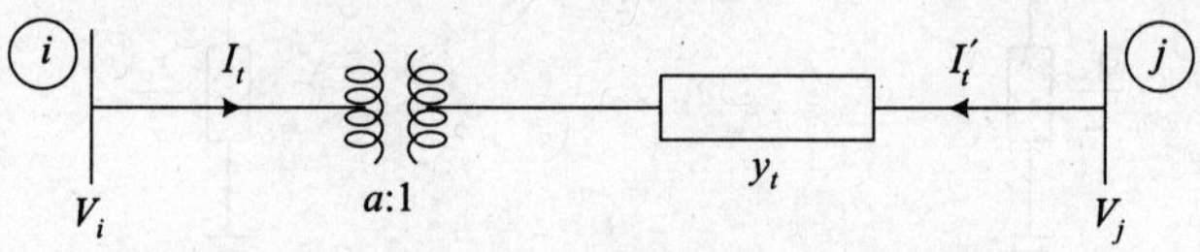

Fig. 3.3 Transformer with off-nominal turns ratio.

The non-unity side is called the tap-side. From Fig. 3.3,

$$I_t = (V_i - aV_j)\frac{y_t}{a^2} = V_i\frac{y_t}{a^2} - V_j\frac{y_t}{a} \qquad ...(3.9)$$

$$I'_t = \left(V_j - \frac{V_i}{a}\right)y_t = -\frac{V_i y_t}{a} + V_j y_t \qquad ...(3.10)$$

The transformer is modeled as an equivalent π network as shown in Fig. 3.4.

For Fig. 3.4 and Fig. 3.3 to be equivalent, the respective currents must be the same in the two networks.

From Fig. 3.4, $I_t = (V_i - V_j)y_1 + V_i y_2 = V_i(y_1 + y_2) - V_j y_1 \qquad ...(3.11)$

$$I'_t = (V_j - V_i)y_1 + V_j y_3 = -V_i y_1 + V_j(y_1 + y_3) \qquad ...(3.12)$$

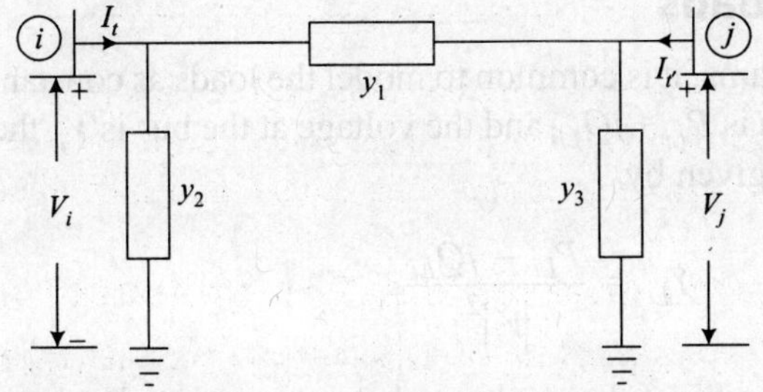

Fig. 3.4 Equivalent circuit of off-nominal ratio transformer.

Comparing (3.11) with (3.9), we get

$$y_1 = \frac{y_t}{a}; \quad \text{...(3.13)}$$

$$y_1 + y_2 = \frac{y_t}{a^2}; \text{ Hence, } y_2 = \frac{y_t}{a^2} - \frac{y_t}{a} = \frac{y_t - ay_t}{a^2} = \frac{y_t(1-a)}{a^2} \quad \text{...(3.14)}$$

Comparing (3.12) with (3.10), we get

$$y_1 + y_3 = y_t$$

$$y_3 = y_t - y_1 = \frac{y_t(a-1)}{a} \quad \text{...(3.15)}$$

Thus, the parameters of the equivalent π model are obtained from y_t and the off-nominal ratio *a*.

The elements of Y_{Bus} are now modified as follows:

$$Y_{ii(\text{new})} = Y_{ii(\text{old})} + y_1 + y_2 \quad \text{...(3.16a)}$$

$$Y_{jj(\text{new})} = Y_{jj(\text{old})} + y_1 + y_3 \quad \text{...(3.16b)}$$

$$Y_{ij(\text{new})} = Y_{ij(\text{old})} - y_1 \quad \text{...(3.16c)}$$

$$Y_{ji(\text{new})} = Y_{ji(\text{old})} - y_1 \quad \text{...(3.16d)}$$

Once again if there is a change in the tap value, four elements of the bus admittance matrix have to be modified.

3.2.3 Modeling of Shunt Elements

Shunt elements like capacitors, reactors and resistors, are connected between a bus and the ground. Let the admittance of the shunt element at bus *i* be y_{sh}. This element affects only the diagonal element of the bus admittance matrix.

$$Y_{ii(\text{new})} = Y_{ii(\text{old})} + y_{sh} \quad \text{...(3.17)}$$

3.2.4 Modeling Loads

In transient stability programs, it is common to model the loads as constant admittances at the rated voltage. If the load at bus i is $P_{Li} + jQ_{Li}$, and the voltage at the bus is V_i, the load can be replaced by an equivalent admittance given by

$$Y_{Li} = \frac{P_{Li} - jQ_{Li}}{|V_i|^2} \qquad ...(3.18)$$

This admittance can be viewed as a shunt element connected at bus i. It will therefore affect only the diagonal element. Hence,

$$Y_{ii(\text{new})} = Y_{ii(\text{old})} + Y_{Li} \qquad ...(3.19)$$

3.2.5 Modeling of Generator Internal Impedance

The generators can be modeled either as a voltage source or as a current source. The current source model is shown in Fig. 3.5.

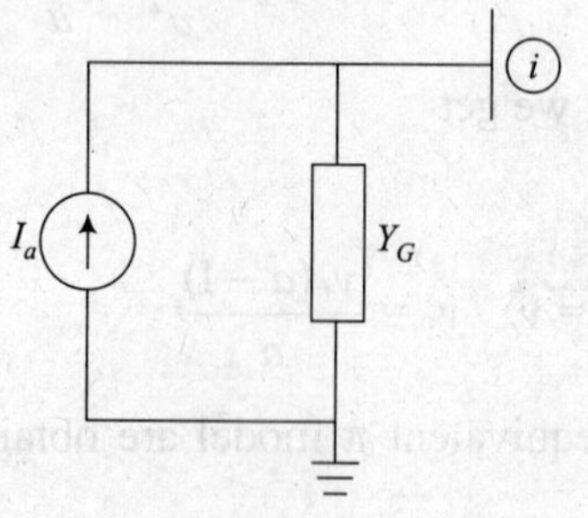

Fig. 3.5 Current source model of generator.

Here, under steady state, $y_G = \dfrac{1}{R_a + jx_d}$, where R_a is the resistance of the stator winding and x_d is the reactance.This model is used in some algorithms for transient stability evaluation. The generator admittance can be included in Y_{bus} as follows:

$$Y_{ii(\text{new})} = Y_{ii(\text{old})} + Y_G \qquad ...(3.20)$$

3.3 STEP-BY-STEP ALGORITHM FOR FORMATION OF Y_{BUS}

The Y_{bus} is formed with ground as reference. For an n bus power system the size of Y_{bus} is $n \times n$.

Step 1: Initialise all elements of Y_{bus} to zero.

Step 2: Modify the elements, by reading the data of the transmission lines one at a time. For each line added, Y_{bus} is modified using (3.8a) – (3.8d).

Step 3: Modify the elements of the Y_{bus} for transformers. For each transformer, the elements are modified using (3.16a) – (3.16d).

Step 4: Modify the elements of the Y_{bus} to include shunt elements using (3.17).

Step 5: Modify the elements to include loads and generator admittances, if necessary. This may be needed only for transient stability programs or for short circuit studies.

Example 3.1: Consider the power system shown in Fig. 3.6.

The data is given in Table 3.1.

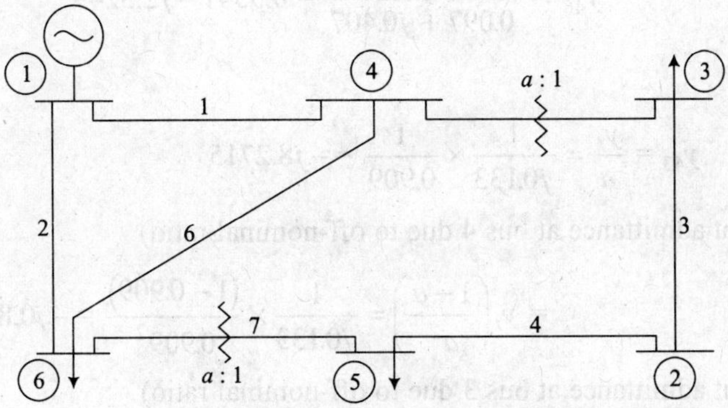

Fig 3.6 Example 3.1.

Table 3.1 Data of example 3.1 (in pu)

Line No.	Between buses	Line impedance	Half line charging admittance	Off–nominal turn ratio
1	1-4	0.08 + j0.37	0.007	—
2	1-6	0.123 + j0.518	0.010	—
3	2-3	0.723 + j1.05	0.0	—
4	2-5	0.282 + j0.64	0.0	—
5	4-3	0.0 + j0.133	0.0	0.909
6	4-6	0.097 + j0.407	0.0076	—
7	6-5	0.0 + j0.3	0.0	0.976

A static shunt capacitor is connected at bus 4 with admittance $j0.005$ pu. Formulate the Y_{bus}.

Solution: Note that the transformer data is given along with line data.

The admittances are calculated (all in pu) as follows.

(i) Transmission lines

$$y_{14} = \frac{1}{0.08 + j0.37} = 0.5583 - j2.582$$

$$y_{16} = \frac{1}{0.123 + j0.518} = 0.4339 - j1.8275$$

$$y_{23} = \frac{1}{0.723 + j1.05} = 0.4449 - j0.6461$$

$$y_{25} = \frac{1}{0.282 + j0.64} = 0.5765 - j1.3085$$

$$y_{46} = \frac{1}{0.097 + j0.407} = 0.5541 - j2.3249$$

(ii) Transformers

$$y_{43} = \frac{y_t}{a} = \frac{1}{j0.133} \times \frac{1}{0.909} = -j8.2715$$

$y_{40,3}$ **(this is the shunt admittance at bus 4 due to off-nominal ratio)**

$$= y_t\left(\frac{1-a}{a^2}\right) = \frac{1}{j0.133} \times \frac{(1-0.909)}{0.909^2} = -j0.8281$$

$y_{30,4}$ **(this is the shunt admittance at bus 3 due to off-nominal ratio)**

$$= y_t\left(\frac{a-1}{a}\right) = \frac{1}{j0.133} \times \left(\frac{0.909-1}{0.909}\right) = +j0.7527$$

Similarly,

$$y_{65} = \frac{1}{j0.3} \times \frac{1}{0.976} = -j3.4153$$

$$y_{60,5} = \frac{1}{j0.3} \times \left(\frac{1-0.976}{0.9762^2}\right) = -j0.084$$

$$y_{50,6} = \frac{1}{j0.3} \times \left(\frac{0.976-1}{0.976}\right) = j0.082$$

(iii) Half line charging admittance

$y_{10,4} = y_{40,1} = j0.007$ (This is the half line charging admittance of line 1-4 which appears between bus 1 and ground and between bus 4 and ground). Similarly, the other half line charging admittances are,

$$y_{10,6} = y_{60,1} = j0.01$$

$$y_{40,6} = y_{60,4} = j0.0076$$

(iv) Shunt element

$$y_{sh4} = j0.005 \text{ (at bus 4)}$$

Formation of Bus admittance matrix – Y_{bus}

$$Y_{11} = y_{14} + y_{16} + y_{10,4} + y_{10,6} = 0.9922 - j4.3925$$

(This is sum of all admittances incident on bus 1)

$$Y_{12} = 0.0 + j0.0$$
$$Y_{13} = 0.0 + j0.0$$
$$Y_{14} = -y_{14} = -0.5583 + j\,2.582$$

(This is negative of admittance of line between 1-4)

$$Y_{15} = 0.0 + j0.0$$
$$Y_{16} = -y_{16} = -0.4339 + j1.8275$$
$$Y_{21} = 0.0 + j0.0$$
$$Y_{22} = y_{23} + y_{25} = 1.0214 - j1.9546$$
$$Y_{23} = -y_{23} = -0.4449 + j0.6461$$
$$Y_{24} = 0.0 + j0.0$$
$$Y_{25} = -y_{25} = -0.5765 + j1.3085$$
$$Y_{26} = 0.0 + j0.0$$
$$Y_{31} = 0.0 + j0.0$$
$$Y_{32} = -y_{23} = -0.4449 + j0.6461$$
$$Y_{33} = y_{23} + y_{43} + y_{30,4} = 0.4449 - j8.1649$$
$$Y_{34} = -y_{43} = j8.2715$$
$$y_{35} = 0.0 + j0.0$$
$$Y_{36} = 0.0 + j0.0$$
$$Y_{41} = -y_{14} = -0.5583 + j2.582$$
$$Y_{42} = 0.0 + j0.0$$
$$Y_{43} = -y_{43} = j8.2715$$
$$Y_{44} = y_{14} + y_{40,1} + y_{46} + y_{40,6} + y_{43} + y_{40,3} + y_{sh4} = 1.1124 - j13.9869$$
$$Y_{45} = 0.0 + j0.0$$
$$Y_{46} = -y_{46} = -0.5541 + j2.3249$$
$$Y_{51} = 0.0 + j0.0$$
$$Y_{52} = -y_{25} = -0.5756 + j1.3085$$
$$Y_{53} = 0.0 + j0.0$$
$$Y_{54} = 0.0 + j0.0$$
$$Y_{55} = y_{25} + y_{65} + y_{50,6} = 0.5765 - j4.6418$$
$$Y_{56} = -Y_{65} = j3.4153$$
$$Y_{61} = -y_{16} = -0.4339 + j1.8275$$
$$Y_{62} = 0.0 + j0.0$$

$$Y_{63} = 0.0 + j\,0.0$$

$$Y_{64} = -y_{46} = -0.5541 + j2.3249$$

$$Y_{65} = -y_{65} = j\,3.4153$$

$$Y_{66} = y_{16} + y_{60,1} + y_{46} + y_{60,4} + y_{65} + y_{60,5} = 0.9880 - j7.6341$$

The bus admittance matrix is given below:

	1	2	3	4	5	6
1	0.9922 −*j*4.3925	0.0 + j0.0	0.0 + j0.0	−0.5583 + *j*2.5820	0.0 + j0.0	−0.4339 + *j*1.8275
2	0.0 + *j*0.0	1.0214 −j1.9546	−0.4449 + j0.6461	0.0 + *j*0.0	−0.5765 + j1.3085	0.0 + *j*0.0
3	0.0 + *j*0.0	−0.4449 + *j*0.6461	0.4449 −*j*8.1649	*j*8.2715	0.0 + *j*0.0	0.0 + *j*0.0
4	−0.5583 + *j*2.5820	0.0 + *j*0.0	*j*8.2715	1.1124 −*j*13.9869	0.0 + *j*0.0	−0.5541 +*j*2.3249
5	0.0 + j0.0	−0.5765 +*j*1.3085	0.0 + *j*0.0	0.0 + *j*0.0	0.5765 −*j*4.6418	*j*3.4153
6	−0.4339 + *j*1.8275	0.0 +j0.0	0.0 + *j*0.0	−0.5541 + j2.3249	*j*3.4153	0.9880 −*j*7.6341

Example 3.2: The one line diagram of a 4-bus system is shown in Fig. 3.7. The data is given in Table 3.2. Formulate Y_{bus} without line 1-2. If line 1-2 is added, show how Y_{bus} is modified.

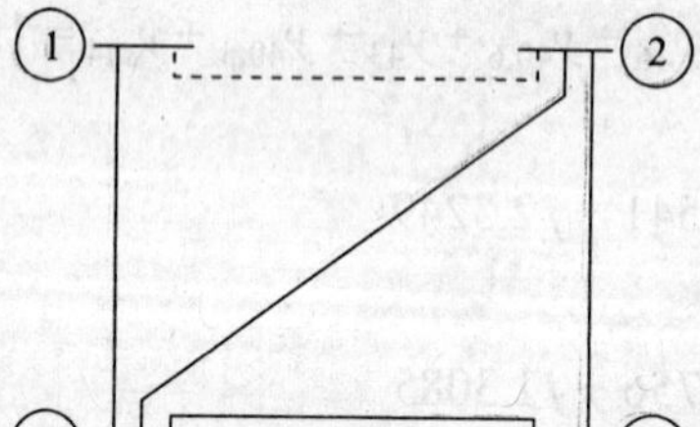

Fig. 3.7 Example 3.2.

Table 3.2 Data of example 3.2 (in pu)

Line	*R*	*X*
1 – 2	0.05	0.15
1 – 3	0.1	0.3
2 – 3	0.15	0.45
2 – 4	0.10	0.30
3 – 4	0.05	0.15

Solution: The admittances are calculated as $y_{\text{line}} = \dfrac{1}{Z_{\text{line}}}$.

The admittances are shown in Table 3.3.

Table 3.3 Admittances of example 3.2

Line	*Y = (G + jB) pu*
1 – 2	2.0 – *j*6.0
1 – 3	1.0 – *j*3.0
2 – 3	0.6667 – *j*2.0
2 – 4	1.0 – *j*3.0
3 – 4	2.0 – *j*6.0

Y_{bus} without line 1 – 2.

	1	*2*	*3*	*4*
1	1.0 – *j*3.0	0.0 + *j*0.0	–1.0 + *j*3.0	0.0 + *j*0.0
2	0.0 + *j*0.0	1.6667 – *j*5.0	–0.6667 + *j*2.0	–1.0 + *j*3.0
3	–1.0 + *j*3.0	–0.6667 + *j*2.0	3.6667 – *j*11.0	–2.0 + *j*6.0
4	0.0 + *j*0.0	–1.0 + *j*3.0	–2.0 + *j*6.0	3.0 – *j*9.0

Y_{bus} modified on addition of line 1 – 2

The four elements which get modified are:

$$Y_{11},\ Y_{12},\ Y_{21} \text{ and } Y_{22}.$$

$$Y_{11(\text{new})} = Y_{11(\text{old})} + y_{12} = (1.0 - j3.0) + (2.0 - j6.0) = 3.0 - j9.0$$

$$Y_{12(\text{new})} = Y_{12(\text{old})} - y_{12} = (0.0 + j0.0) - (2.0 - j6.0) = -2.0 + j6.0$$

$$Y_{21(\text{new})} = Y_{21(\text{old})} - y_{12} = (0.0 + j0.0) - (2.0 - j6.0) = -2.0 + j6.0$$

$$Y_{22(\text{new})} = Y_{22(\text{old})} + y_{12} = (1.667 - j5.0) + (2.0 - j6.0) = 3.6667 - j11.0$$

The other elements remain the same.

3.3.1 Network Reduction (Kron Reduction)

In a power system, the current injection is zero at buses where there are no external loads or generators connected. Such nodes can be eliminated. Let us order the buses such that the matrices can be partitioned as follows:

$$\begin{bmatrix} I_1 \\ 0 \end{bmatrix} = \begin{bmatrix} Y_1 & Y_2 \\ Y_3 & Y_4 \end{bmatrix} = \begin{bmatrix} V_1 \\ V_2 \end{bmatrix}$$

Here $[I_1]$ is vector of bus current injections.

Y_1, Y_2, Y_3, Y_4 are submatrices of the bus admittance matrix $[V_1]$ and $[V_2]$ are vector of bus voltage, suitably partitioned. We get the following equation:

$$0 = Y_3\, V_1 + Y_4 V_2$$

$$V_2 = -\,[Y_4]^{-1}\, Y_3 V_1 \qquad \text{(i)}$$

$$I_1 = Y_1\, V_1 + Y_2\, V_2$$

Substituting (i), we get

$$\begin{aligned} I_1 &= Y_1 V_1 + Y_2\,[-Y_4^{-1}\, Y_3\, V_1] \\ &= [Y_1 - Y_2\, Y_4^{-1}\, Y_3]\, V_1 \\ &= Y_{\text{Red}}\, V_1 \end{aligned}$$

Where the reduced admittance matrix is given by $Y_1 - Y_2 Y_4^{-1}\, Y_3$. The lower dimension network in which the node with zero current injection are removed is removed is called the reduced network. If in an n-bus system, node k has to be eliminated the new elements can be calculated as

$$Y_{ij}^{(\text{new})} = Y_{ij} - \frac{Y_{ik}\, Y_{kj}}{Y_{kk}};\ \begin{matrix} i, j = 1 \ldots n \\ i, j \neq k \end{matrix}$$

$Y_{ij}^{(\text{new})}$ are the elements of the new reduced $(n-1) \times (n-1)$ bus admittance matrix. This is called Kron's reduction.

3.4 FORMATION OF BUS IMPEDANCE MATRIX

The bus impedance matrix is the inverse of the bus admittance matrix. An alternative method is possible, based on an algorithm to form the bus impedance matrix directly from system parameters and the coded bus numbers. The bus impedance matrix is formed adding one element at a time to a partial network.

The equations of the network in the bus frame of reference using the currents as independent variables are given in matrix form by

$$\bar{E}_{\text{bus}} = [Z_{\text{bus}}]\bar{I}_{\text{bus}} \qquad \text{...(3.21)}$$

Expanding for an n bus system,

$$E_1 = Z_{11}I_1 + Z_{12}I_2 + ... + Z_{1k}I_k ... + Z_{1n}I_n$$

$$\vdots$$

$$\vdots$$

$$E_k = Z_{k1}I_1 + Z_{k2}I_2 + ... + Z_{kk}I_k + ... + Z_{kn}I_n$$

$$\vdots$$

$$\vdots$$

$$E_n = Z_{n1}I_1 + Z_{n2}I_2 + ... + Z_{nk}I_k + ... + Z_{nn}I_n \quad ...(3.22)$$

From (3.22), it can be seen that

$$E_k = Z_{kk}I_k \quad ...(3.23)$$

and,

$$E_i = Z_{ik}I_k \quad ...(3.24)$$

When only I_k is injected at bus k, and all other current injections are zero. If $I_k = 1.0$ pu, then

$$E_k = Z_{kk} \text{ and } E_i = Z_{ik} \quad ...(3.25)$$

The diagonal elements of $[Z_{\text{bus}}]$ are called the driving point impedance at the respective buses. For example,

$$Z_{11} = \frac{V_1}{I_1}\Big|I_2, I_3 \ldots I_n = 0$$

The driving point impedance of bus 1 is the impedance obtained by injecting a current at only bus 1 and keeping the other buses open circuited, so that their current injections are zero.

The off diagonal elements are called transfer impedances. It is important to note here that Z_{ij} is not the reciprocal of Y_{ij}.

Now assume that the bus impedance matrix Z_{bus} is known for a partial network of m buses and a reference bus 0. Let the Z_{bus} of the partial network be given by

$$Z_{\text{bus}} = \begin{bmatrix} Z_{11} & Z_{12} & \cdots & Z_{1k} & \cdots & Z_{1m} \\ \vdots & & & & & \\ Z_{k1} & Z_{k2} & \cdots & Z_{kk} & \cdots & Z_{km} \\ \vdots & & & & & \\ Z_{m1} & Z_{m2} & \cdots & Z_{mk} & \cdots & Z_{mm} \end{bmatrix} \quad ...(3.26)$$

The Z_{bus} of the partial network is of dimension $m \times m$. If now a new element is added between buses p and q, we have the following two possibilities.

(i) p is an existing bus in the partial network and q is a new bus; in this case p-q is a (tree) branch as shown in Fig. 3.8a.

(ii) both p and q are buses existing in the partial network; in this case p-q is a link as shown in Fig. 3.8b.

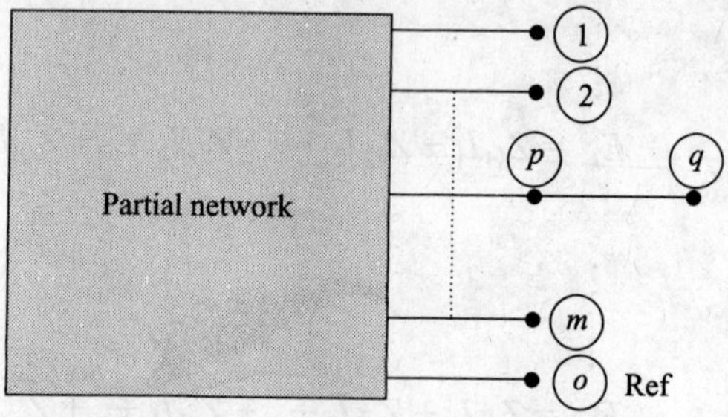

Fig. 3.8a Addition of a branch.

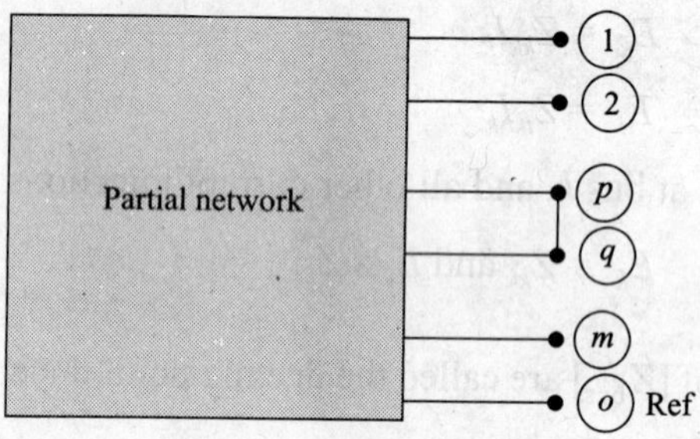

Fig. 3.8b Addition of a link.

3.4.1 Modification of Z_{bus} for Addition of a Branch

Assume that the added branch p-q is mutually coupled with some elements of the partial network. The performance equations of the network with the added branch is given by

$$\begin{bmatrix} E_1 \\ E_2 \\ \vdots \\ E_p \\ \vdots \\ E_m \\ E_q \end{bmatrix} = \begin{bmatrix} Z_{11} & Z_{12} & \cdots & Z_{1p} & \cdots & Z_{1m} & Z_{1q} \\ Z_{21} & Z_{22} & \cdots & Z_{2p} & \cdots & Z_{2m} & Z_{2q} \\ \vdots & & & & & & \\ Z_{p1} & Z_{p2} & \cdots & Z_{pp} & \cdots & Z_{pm} & Z_{pq} \\ \vdots & & & & & & \\ Z_{m1} & Z_{m2} & \cdots & Z_{mp} & \cdots & Z_{mm} & Z_{mq} \\ Z_{q1} & Z_{q2} & \cdots & Z_{qp} & \cdots & Z_{qm} & Z_{qq} \end{bmatrix} \begin{bmatrix} I_1 \\ I_2 \\ \vdots \\ I_p \\ \vdots \\ I_m \\ I_q \end{bmatrix} \quad ...(3.27)$$

If the elements of the network are bilateral passive elements, $Z_{qi} = Z_{iq}$, for all $i = 1, 2 \ldots m$ (all buses of the partial network excluding the newly added bus q). From (3.25), it is seen that Z_{qi} is nothing but the voltage at the qth bus with respect to ground if a current of 1 pu is injected at the ith bus, as shown in Fig. 3.9.

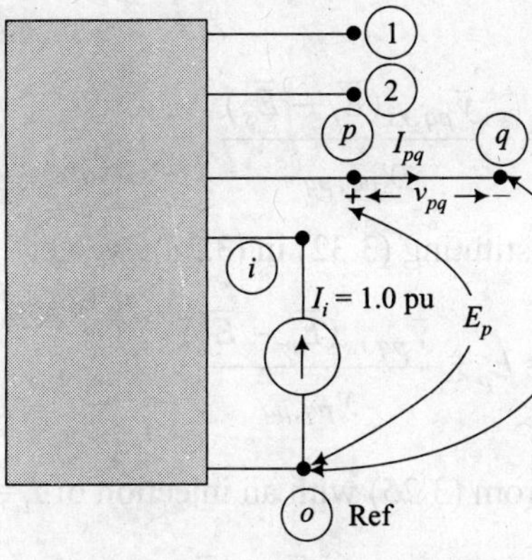

Fig. 3.9 Calculation for Z_{qi}.

The bus voltages associated with the added element and the voltage across the element are given by (refer Fig. 3.9)

$$E_p - E_q = v_{pq} \quad ...(3.28)$$

In terms of the primitive admittances and voltages across the elements, the current through the elements is given by

$$\begin{bmatrix} i_{pq} \\ \bar{i}_{rs} \end{bmatrix} = \begin{bmatrix} y_{pq,pq} & \bar{y}_{pq,rs} \\ \bar{y}_{rs,pq} & \bar{y}_{rs,rs} \end{bmatrix} \begin{bmatrix} v_{pq} \\ \bar{v}_{rs} \end{bmatrix} \quad ...(3.29)$$

where, i_{pq} is current through element *p-q*.

$\bar{i}_{rs}$ is vector of currents through elements of the partial network.

v_{pq} is voltage across element *p-q*.

$y_{pq,pq}$ is self-admittance of the added element.

$\bar{y}_{pq,rs}$ is the vector of mutual admittances between the added elements. *p-q* and elements *r-s* of the partial network.

$\bar{v}_{rs}$ is vector of voltage across elements of partial network.

$\bar{y}_{rs,pq}$ is transpose of $\bar{y}_{pq,rs}$.

$\bar{y}_{rs,rs}$ is the primitive admittance of partial network.

Since *p-q* is a branch, the current in Fig 3.9, $i_{pq} = 0$. v_{pq} need not be zero, since *p-q* is mutually coupled to some elements of the partial network. We thus have

$$i_{pq} = 0 \quad ...(3.30)$$

$$\bar{v}_{rs} = \bar{E}_r - \bar{E}_s \quad ...(3.31)$$

From (3.29), $\quad i_{pq} = y_{pq,pq}\, v_{pq} + \bar{y}_{pq,rs}\, \bar{v}_{rs} = 0$

Hence, $\quad v_{pq} = -\dfrac{\bar{y}_{pq,rs}\bar{v}_{rs}}{y_{pq,pq}}$

Substituting for $\bar{v}_{rs}$ from (3.31)

$$v_{pq} = -\frac{\bar{y}_{pq,rs}(\bar{E}_r - \bar{E}_s)}{y_{pq,pq}} \qquad ...(3.32)$$

From (3.28), $E_q = E_p - v_{pq}$. Substituting (3.32) in (3.28), we get

$$E_q = E_p + \frac{\bar{y}_{pq,rs}(\bar{E}_r - \bar{E}_s)}{y_{pq,pq}}$$

Substituting for E_q, E_p, $\bar{E}_r$, $\bar{E}_s$ from (3.25) with an injection of $I_i = 1.0$ pu at bus i, we get

$$Z_{qi} = Z_{pi} + \frac{\bar{y}_{pq,rs}(\bar{Z}_{ri} - \bar{Z}_{si})}{y_{pq,pq}} \quad i = 1, 2 \ldots m;\ i \neq q \qquad ...(3.33)$$

The element Z_{qq} can be computed by injecting a current of 1 pu at q^{th} bus i.e. $I_q = 1.0$ pu. From (3.25),

$$Z_{qq} = E_q$$

$$i_{pq} = -I_q = -1.0 \text{ (refer Fig. 3.9)}$$

From (3.29), $\quad i_{pq} = y_{pq,pq}\, v_{pq} + \bar{y}_{pq,rs}\, \bar{v}_{rs} = -1$

From which $\quad v_{pq} = -1 + \dfrac{\bar{y}_{pq,rs}\bar{v}_{rs}}{y_{pq,pq}}$

Substituting (3.31), $v_{pq} = -1 + \dfrac{\bar{y}_{pq,rs}(\bar{E}_r - \bar{E}_s)}{y_{pq,pq}}$

And from (3.28) $E_q = E_p + \dfrac{1 + \bar{y}_{pq,rs}(\bar{E}_r - \bar{E}_s)}{y_{pq,pq}}$

From (3.25) with $I_q = 1.0$ *pu*, we get

$$Z_{qq} = Z_{pq} + \frac{1 + \bar{y}_{pq,rs}(\bar{Z}_{rq} - \bar{Z}_{sq})}{y_{pq,pq}} \qquad ...(3.34)$$

Case 1. If there is no mutual coupling, then elements of $\bar{y}_{pq,rs}$ are zero and the impedance of the added element *p-q* is given by

$$Z_{pq,\,pq} = \frac{1}{y_{pq,pq}}$$

From (3.33),

$$Z_{qi} = Z_{pi},\ i = 1, 2 \ldots m;\ i \neq q \qquad \text{...(3.35a)}$$

From (3.34),

$$Z_{qq} = Z_{pq} + z_{pq,\,pq} \qquad \text{...(3.35b)}$$

Case 2. If there is no mutual coupling and p is the reference node, then $E_p = 0$ Therefore, from (3.25)

$$Z_{pi} = 0 \quad i = 1, 2 \ldots m;\ i \neq q$$

And

$$Z_{pq} = 0.$$

Hence,

$$Z_{qi} = 0 \quad i = 1, 2 \ldots m;\ i \neq q \qquad \text{...(3.36a)}$$

$$Z_{qq} = z_{pq,\,pq} \qquad \text{...(3.36b)}$$

3.4.2 Modification of Z_{bus} for Addition of a Link

The link is an element added between two existing buses. The element is first modeled as an element $p - l$ in series with a source e_l as shown in Fig. 3.10.

Note that a fictitious node l is introduced. The source e_l is between buses l and q. e_l is so chosen that the current $i_{pq} = 0$, so that the element $p - l$ can be treated as a branch. The performance equation for Fig. 3.10 is given by

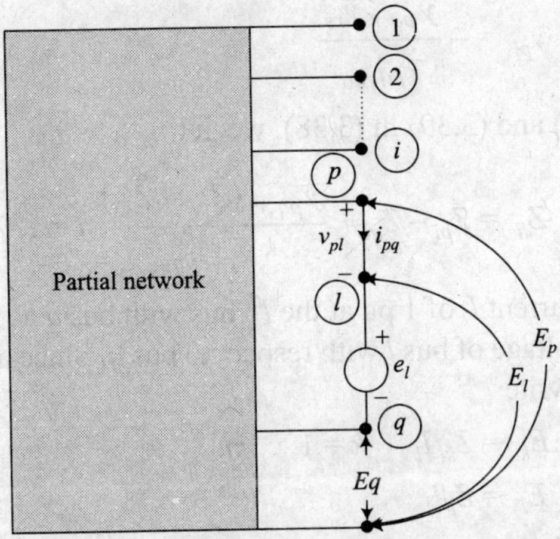

Fig. 3.10 Addition of a link.

$$\begin{bmatrix} E_1 \\ E_2 \\ \vdots \\ E_p \\ \vdots \\ E_m \\ e_l \end{bmatrix} = \begin{bmatrix} Z_{11} & Z_{12} & \cdots & Z_{1p} & \cdots & Z_{1m} & Z_{1l} \\ Z_{21} & Z_{22} & \cdots & Z_{2p} & \cdots & Z_{2m} & Z_{2l} \\ & & & & & & \\ Z_{p1} & Z_{p2} & \cdots & Z_{pp} & \cdots & Z_{pm} & Z_{pl} \\ & & & & & & \\ Z_{m1} & Z_{m2} & \cdots & Z_{mp} & \cdots & Z_{mm} & Z_{ml} \\ Z_{l1} & Z_{l2} & \cdots & Z_{lp} & \cdots & Z_{lm} & Z_{ll} \end{bmatrix} \begin{bmatrix} I_1 \\ I_2 \\ \vdots \\ I_p \\ \vdots \\ I_m \\ I_l \end{bmatrix} \qquad \text{...(3.37)}$$

In (3.37) note that e_l is the voltage of node l with reference to q and not the reference node and E_l is the voltage of node l with respect to the reference node. From Fig. 3.10,

$$e_l = E_l - E_q = E_p - E_q - v_{pl} \qquad ...(3.38)$$

Using the same steps as before, Z_{li} can be determined by injecting a current at the ith bus and calculating the voltage at the lth node with respect to bus q, with all other current injections set to zero. It follows from (3.37)

$$E_k = Z_{ki}\, I_i,\ k = 1 \ldots m \qquad ...(3.39a)$$

$$e_l = Z_{li}\, I_i \qquad ...(3.39b)$$

If $I_i = 1.0$ pu,

$$Z_{li} = e_l \qquad ...(3.39c)$$

Since $i_{pq} = 0$, $i_{pl} = 0$. Now in terms of primitive admittances,

$$i_{pl} = y_{pl,\,pl}\, v_{pl} + \bar{y}_{pl,\,rs}\, \bar{v}_{rs} = 0$$

Therefore,

$$v_{pl} = -\frac{\bar{y}_{pl,rs}\bar{v}_{rs}}{y_{pl,pl}}$$

$\bar{y}_{pl,rs} = \bar{y}_{pq,rs}$ and $y_{pl,pl} = y_{pq,pq}$ (Since l is only a fictitions node, impedance of element $p - l$ is same as the added element $p - q$)

$$v_{pl} = -\frac{\bar{y}_{pq,rs}\bar{v}_{rs}}{y_{pq,pq}} \qquad ...(3.40)$$

Substituting (3.40), (3.31) and (3.39) in (3.38), we get

$$Z_{li} = Z_{pi} - Z_{qi} + \frac{\bar{y}_{pq,rs}(\bar{Z}_{ri} - \bar{Z}_{si})}{y_{pq,pq}} \quad i = 1, 2 \ldots m;\ i \text{ ¹ } 1 \qquad ...(3.41)$$

To calculate Z_{ll}, inject a current I_l of 1 pu at the lth bus with bus q as reference (note change of reference) and calculate the voltage of bus l with respect to bus q. Since all other current injections are zero, from (3.37) we can write

$$E_k = Z_{kl}\, I_l \quad k = 1 \ldots m \qquad ...(3.42a)$$

$$E_l = Z_{ll} I_l \qquad ...(3.42b)$$

Let $I_l = 1$ pu, $i_{pl} = -I_l = -1$.

$$i_{pl} = y_{pl,\,pl}\, v_{pl} + \bar{y}_{pl,\,rs}\, \bar{v}_{rs} = -1$$

From which

$$v_{pl} = -\frac{1 + \bar{y}_{pl,rs}\bar{v}_{rs}}{y_{pl,pl}}$$

$$= -\frac{1 + \bar{y}_{pq,rs}\bar{v}_{rs}}{y_{pq,pq}} \qquad ...(3.43)$$

Substituting (3.43), (3.31) and (3.42) into (3.38), we get

$$Z_{ll} = Z_{pl} - Z_{ql} + \frac{1 + \bar{y}_{pq,rs}(\bar{Z}_{rl} - \bar{Z}_{sl})}{y_{pq,pq}} \qquad ...(3.44)$$

Case 1. If there is no mutual coupling between the added element and elements of the partial network, $\bar{y}_{pq,rs} = 0$ and $z_{pq,\,pq} = \frac{1}{y_{pq,pq}}$. In which case

$$Z_{li} = Z_{pi} - Z_{qi};\ i = 1 \ldots m \qquad ...(3.45a)$$

$$Z_{ll} = Z_{pl} - Z_{ql} + z_{pq,\,pq} \qquad ...(3.45b)$$

Case 2. Further, if there is no mutual coupling and p is the reference mode, $Z_{pi} = Z_{pl} = 0$.

$$Z_{li} = -Z_{qi};\ i = 1 \ldots m \qquad ...(3.46a)$$

$$Z_{ll} = -Z_{ql} + z_{pq,\,pq} \qquad ...(3.46b)$$

From (3.41) and (3.44), elements of the lth row and column are obtained. The elements of the bus admittance matrix, Z_{ij}, $i, j = 1 \ldots m$, should be modified to take into account the effect of the added link. Now the link has been added in series with a voltage source e_l. The node l can be eliminated by short circuiting the series voltage source. Mathematically, it is equivalent to eliminating the lth column and row from (3.37). From (3.37), the performance equations are

$$\bar{E}_{\text{bus}} = Z_{\text{bus}}\ \bar{I}_{\text{bus}} + \bar{Z}_{il} I_l \qquad ...(3.47a)$$

$$e_l = \bar{Z}_{lj}\ \bar{I}_{\text{bus}} + Z_{ll}\, I_l \qquad ...(3.47b)$$

where, Z_{bus} is the $m \times m$ bus impedance matrix.

When $e_l = 0$ (short circuited). From (3.47b), we can write

$$I_l = -\frac{\bar{Z}_{lj}\ \bar{I}_{\text{bus}}}{Z_{ll}} \qquad ...(3.48)$$

Substituting (3.48) in (3.47a), we get

$$\bar{E}_{\text{Bus}} = \left(Z_{\text{bus}} - \frac{\bar{Z}_{il}\ \bar{Z}_{lj}}{Z_{ll}} \right) \bar{I}_{\text{bus}} \qquad ...(3.49)$$

which is the performance equation of the partial network including the link p-q. Hence, the required bus impedance matrix $Z_{\text{bus (modified)}}$ can be obtained from the $Z_{\text{bus (with node } l)}$ (the impedance matrix with the fictitious node l) as

$$Z_{\text{bus (modified)}} = Z_{\text{bus (with node } l)} - \frac{\bar{Z}_{il}\ \bar{Z}_{lj}}{Z_{ll}}$$

Every element of $Z_{\text{bus (modified)}}$ is given by

$$Z_{ij\,\text{(modified)}} = Z_{ij\,\text{(with node } l)} - \frac{Z_{il}\ Z_{lj}}{Z_{ll}} \qquad ...(3.50)$$

We summarize the results below

In the results, $z_{pq,\,pq}$ is the impedance of the added branch.

1. Addition of branch *p*-*q*; *q* is new node and *p* is reference node. (This is called as **Type-1 Modification** by some authors)

$$Z_{qi} = 0 \qquad i = 1, 2 \ldots m;\ i \neq q$$
$$Z_{qq} = z_{pq,\,pq}$$

Let the original partial network have *n* buses and a branch of impedance Z_L is added between bus *q* and reference node. Let $[Z_{\text{org}}]$ be the bus impedance matrix of the original partial network Then

$$Z_{\text{new}} = \begin{bmatrix} Z_{\text{org}} & \begin{matrix} 0 \\ \vdots \\ 0 \end{matrix} \\ \begin{matrix} 0 & \cdots & 0 \end{matrix} & Z_L \end{bmatrix}$$

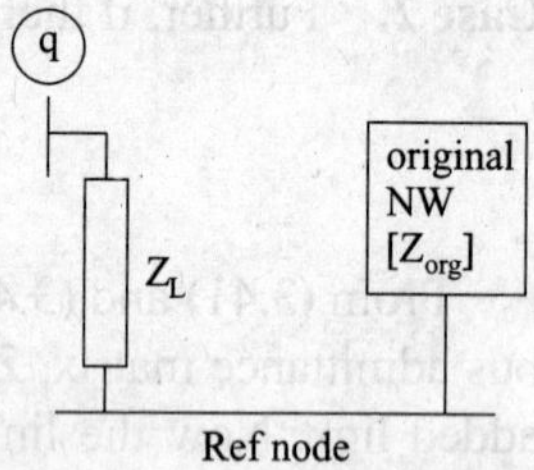

2. Addition of branch *p*-*q*; *q* is new node and *p* is a node in the existing network (This is called as **Type-2 Modification**)

$$Z_{qi} = Z_{pi} + \frac{\bar{y}_{pq,rs}(\bar{Z}_{ri} - \bar{Z}_{si})}{y_{pq,\,pq}} \qquad i = 1, 2 \ldots m;\ i\ ^{1}\ q$$

$$Z_{qq} = Z_{pq} + \frac{1 + \bar{y}_{pq,rs}(\bar{Z}_{rq} - \bar{Z}_{sq})}{y_{pq,\,pq}}$$

- If there is no mutual coupling, then

$$Z_{qi} = Z_{pi}, \qquad i = 1, 2 \ldots m;\ i \neq q$$
$$Z_{qq} = Z_{pq} + z_{pq,\,pq}$$

If there is no mutual coupling and original partial network has *n* buses and impedance matrix $[Z_{\text{org}}]$, and a branch of impedance Z_L is added between bus *q* (new bus) and bus *p* (old bus) then we can write the modified Z_{bus} as

$$Z_{\text{new}} = \left[\begin{array}{c|c} Z_{\text{org}} & \begin{matrix} Z_{p1} \\ Z_{p2} \\ \vdots \\ Z_{pn} \end{matrix} \\ \hline Z_{p1}\ Z_{p2}\ \cdots\ Z_{pn} & Z_{pq} + Z_L \end{array}\right]$$

3. Addition of link *p*-*q*; *p* is a reference node. (This is called as **Type-3 Modification**)

$$Z_{li} = -Z_{qi};\ i = 1 \ldots m$$
$$Z_{ll} = -Z_{ql} + z_{pq,\,pq}$$

We add a link between bus q and reference node.

$$Z_{new} = \begin{bmatrix} & & & & -Z_{q1} \\ & & & & -Z_{q2} \\ & Z_{org} & & & \vdots \\ & & & & -Z_{qn} \\ -Z_{q1} & -Z_{q2} & \cdots & -Z_{qn} & -Z_{ql} + Z_L \end{bmatrix}$$

We eliminate last row and column using kron's reduction.

4. Addition of link p-q; q and p are nodes in existing network. (This is called as **Type-4 Modification**)

$$Z_{li} = Z_{pi} - Z_{qi} + \frac{\bar{y}_{pq,rs}(\bar{Z}_{ri} - \bar{Z}_{si})}{y_{pq,pq}}$$

$$Z_{ll} = Z_{pl} - Z_{ql} + \frac{1 + \bar{y}_{pq,rs}(\bar{Z}_{rl} - \bar{Z}_{sl})}{y_{pq,pq}}$$

- If there is no mutual coupling, then

$$Z_{li} = Z_{pi} - Z_{qi};\ i = 1 \ldots m$$

$$Z_{ll} = Z_{pl} - Z_{ql} + z_{pq,\,pq}$$

Without mutual coupling, when we add a link between $p - q$, we get

$$Z_{new} = \left[\begin{array}{cccc|c} & & & & Z_{p1} - Z_{q1} \\ & & & & Z_{p2} - Z_{q2} \\ & Z_{org} & & & \vdots \\ & & & & Z_{pn} - Z_{qn} \\ \hline Z_{p1} - Z_{q1} & \cdots & & Z_{pn} - Z_{qn} & Z_{pl} - Z_{ql} + Z_L \end{array}\right]$$

We then eliminate the last row and column.

In using the Z_{bus} building algorithm it is important to note that, the first element to be added is always a branch between any node and the reference node. In the practical system, the ground is always the reference.

Exmple 3.3: Construct the bus impedance matrix for the network shown in Fig. 3.11. All impedances shown are in pu.

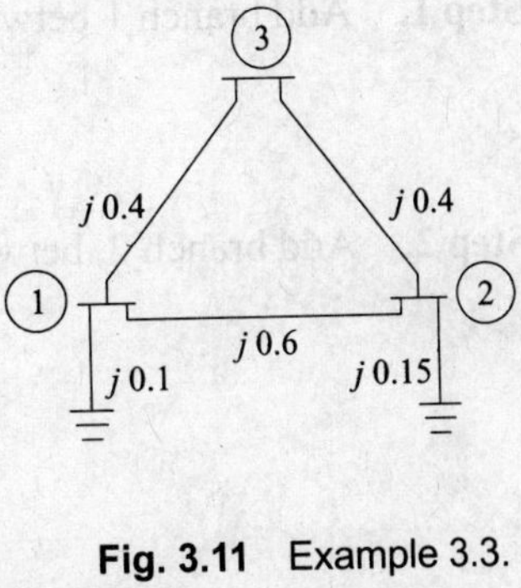

Fig. 3.11 Example 3.3.

Solution: The network is shown in Fig. 3.12.

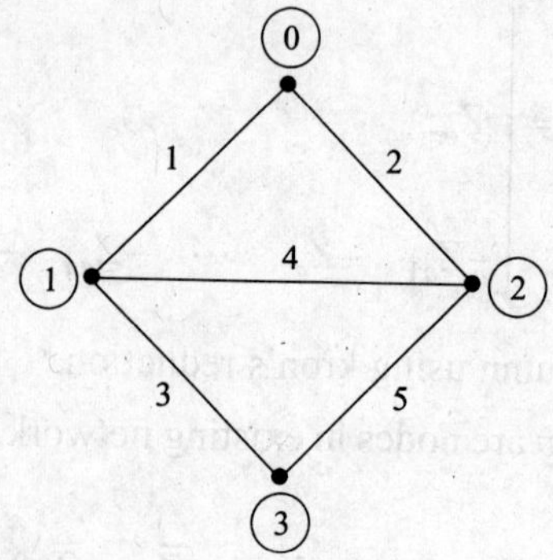

Fig. 3.12 Network of example 3.3.

Since there is no mutual coupling between elements in the network, (3.35) and (3.36) are used for addition of a branch. They are given below again for reference.

$Z_{qi} = Z_{pi}$ (Here p is node in the partial network and q is the new node added)

$$Z_{qq} = Z_{pq} + z_{pq,\,pq}$$

Similarly, (3.45) and (3.46) are used when a link is added between p-q. Both are nodes of the existing partial network

$$Z_{li} = Z_{pi} - Z_{qi}$$

$$Z_{ll} = Z_{pl} - Z_{ql} + z_{pq,\,pq}.$$

If p is the reference node,

$$Z_{li} = -Z_{qi}$$

$$Z_{ll} = -Z_{ql} + z_{pq,\,pq}.$$

Step 1. Add branch 1 between node 1 and reference node. ($q = 1, p = 0$)

$$Z_{\text{bus}} = \begin{matrix} & 1 \\ 1 & [\,j0.1\,] \end{matrix}$$

Step 2. Add branch 2, between node 2 and reference node. ($q = 2, p = 0$).

$$Z_{21} = Z_{12} = 0$$

$$Z_{22} = Z_{20,\,20} = j0.15$$

$$Z_{Bus} = \begin{matrix} & 1 & 2 \\ 1 & j0.1 & 0 \\ 2 & 0 & 0j.15 \end{matrix}$$

Step 3. Add branch 3, between node 1 and node 3 ($p = 1, q = 3$)

$$Z_{31} = Z_{11} = j0.1$$
$$Z_{32} = Z_{12} = 0$$
$$Z_{13} = Z_{31};\ Z_{23} = Z_{32}$$
$$Z_{33} = Z_{13} + z_{13,\,13} = j0.1 + j0.4 = j0.5$$

$$Z_{bus} = \begin{matrix} & 1 & 2 & 3 \\ 1 & j0.1 & 0 & j0.1 \\ 2 & 0 & j0.15 & 0 \\ 3 & j0.1 & 0 & j0.5 \end{matrix}$$

Step 4. Add element 4, which is a link between node 1 and node 2. ($p = 1, q = 2$)

$$Z_{l1} = Z_{11} - Z_{21} = j0.1 - 0.0 = j0.1$$
$$Z_{l2} = Z_{12} - Z_{22} = 0.0 - j0.15 = -j0.15$$
$$Z_{l3} = Z_{13} - Z_{23} = j0.1 - 0.0 = j0.1$$
$$Z_{1l} = Z_{l1};\ Z_{2l} = Z_{l2};\ Z_{3l} = Z_{l3};$$
$$Z_{ll} = Z_{1l} - Z_{2l} + z_{12,12} = j0.1 - (-j0.15) + j0.6 = j0.85$$

$$Z_{bus} = \begin{matrix} & 1 & 2 & 3 & 4 \\ 1 & j0.1 & 0 & j0.1 & j0.1 \\ 2 & 0 & j0.15 & 0 & -j0.15 \\ 3 & j0.1 & 0 & j0.5 & j0.1 \\ l & j0.1 & -j0.15 & j0.1 & j0.85 \end{matrix}$$

Node l has to be eliminated using (3.50)

$$Z_{11(new)} = Z_{11(old)} - \frac{Z_{1l}\, Z_{l1}}{Z_{ll}}$$

$$= j0.1 - \left(\frac{j0.1 \times j0.1}{j0.85}\right) = j0.08823$$

$$Z_{12(new)} = Z_{12(old)} - \frac{Z_{1l}\, Z_{2l}}{Z_{ll}}$$

$$= 0.0 - \left(\frac{j0.1 \times -j0.15}{j0.85} \right) = j0.01765$$

$$Z_{13(\text{new})} = Z_{13(\text{old})} - \frac{Z_{1l}\, Z_{3l}}{Z_{ll}}$$

$$= j0.1 - \left(\frac{j0.1 \times j0.1}{j0.85} \right) = j0.08823$$

$$Z_{21(\text{new})} = Z_{12(\text{new})}$$

$$Z_{22(\text{new})} = Z_{22(\text{old})} - \frac{Z_{2l}\, Z_{l2}}{Z_{ll}}$$

$$= j0.15 - \frac{(-j0.15 \times -j0.15)}{j0.85} = j0.12353$$

$$Z_{23(\text{new})} = Z_{23(\text{old})} - \frac{Z_{2l}\, Z_{l3}}{Z_{ll}}$$

$$= 0.0 - \left(\frac{-j0.15 \times j0.1}{j0.85} \right) = j0.01765$$

$$Z_{31(\text{new})} = Z_{13(\text{new})}$$

$$Z_{32(\text{new})} = Z_{23(\text{new})}$$

$$Z_{33(\text{new})} = Z_{33(\text{old})} - \frac{Z_{3l}\, Z_{l3}}{Z_{ll}}$$

$$= j0.5 - \left(\frac{j0.1 \times j0.1}{j0.85} \right) = j0.48823$$

$$Z_{\text{bus}} = \begin{array}{c} \\ \\ \\ \end{array} \overset{\begin{array}{ccc} 1 & \quad\quad 2 & \quad\quad 3 \end{array}}{\begin{bmatrix} j0.08823 & j0.01765 & j0.08823 \\ j0.01765 & j0.12353 & j0.01765 \\ j0.08823 & j0.01765 & j0.48823 \end{bmatrix}}$$

Step 5. Add link between node 2 and node 3 ($p = 2$, $q = 3$)

$$Z_{l1} = Z_{21} - Z_{31} = j0.01765 - j0.08823 = -j0.07058$$

$$Z_{l2} = Z_{22} - Z_{32} = j0.12353 - j0.01765 = j0.10588$$

$$Z_{l3} = Z_{23} - Z_{33} = j0.01765 - j0.48823 = -j0.47058$$

$$Z_{ll} = Z_{2l} - Z_{3l} + Z_{23,\,23}$$

$$= j0.10588 - (-j0.47058) + j0.4 = j0.97646$$

$$Z_{bus} = \begin{array}{c} \\ 1 \\ 2 \\ 3 \\ l \end{array} \begin{array}{c} \begin{array}{cccc} 1 & 2 & 3 & l \end{array} \\ \begin{bmatrix} j0.08823 & j0.01765 & j0.08823 & -j0.07058 \\ j0.01765 & j0.12353 & j0.01765 & j0.10588 \\ j0.08823 & j0.01765 & j0.48823 & -j0.47058 \\ -j0.07058 & j0.10588 & -j0.47058 & j0.97646 \end{bmatrix} \end{array}$$

Node l is eliminated as in previous step.

$$Z_{11} = j0.08823 - \left(\frac{-j0.07058 \times -j0.07058}{j0.97646} \right) = j0.08313$$

$$Z_{12} = j0.01765 - \left(\frac{-j0.07058 \times j0.10588}{j0.97646} \right) = j0.0253$$

$$Z_{13} = j0.08823 - \left(\frac{-j0.07058 \times -j0.47058}{j0.97646} \right) = j0.05421$$

$$Z_{22} = j0.12353 - \left(\frac{j0.10588 \times j0.10588}{j0.97646} \right) = j0.11205$$

$$Z_{23} = j0.01765 - \left(\frac{j0.10588 \times -j0.47058}{j0.97646} \right) = j0.06868$$

$$Z_{33} = j0.48823 - \left(\frac{-j0.47058 \times -j0.47058}{j0.97646} \right) = j0.26145$$

$$Z_{bus} = \begin{array}{c} \\ 1 \\ 2 \\ 3 \end{array} \begin{array}{c} \begin{array}{ccc} 1 & 2 & 3 \end{array} \\ \begin{bmatrix} j0.08313 & j0.02530 & j0.05421 \\ j0.02530 & j0.11205 & j0.06868 \\ j0.05421 & j0.06868 & j0.26145 \end{bmatrix} \end{array}$$

$$Y_{bus} = [Z_{bus}]^{-1} = \begin{array}{c} \\ 1 \\ 2 \\ 3 \end{array} \begin{array}{c} \begin{array}{ccc} 1 & 2 & 3 \end{array} \\ \begin{bmatrix} -j14.1667 & j1.6667 & j2.5 \\ j1.6667 & -j10.8334 & j2.5 \\ j2.5 & j2.5 & -j5.0 \end{bmatrix} \end{array}$$

As a check, it can be seen that the Y_{bus} can also be built by inspection to give the same answer.

Example 3.4: Form the bus impedance matrix for the network shown in Fig 3.13, using the algorithm of section 3.4. The data is given in Table 3.4.

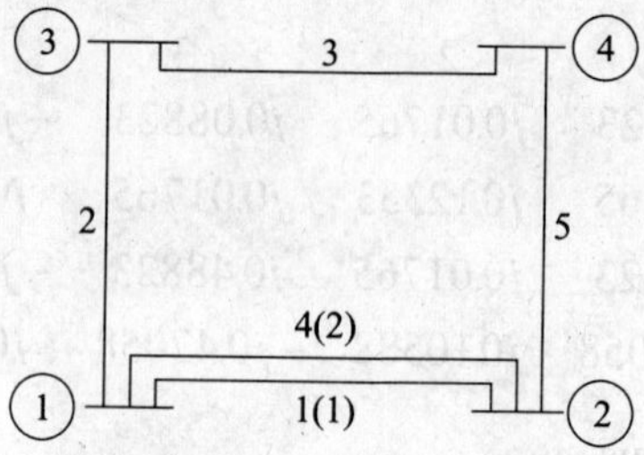

Fig. 3.13 Network for example 3.4.

Table 3.4 Data for example 3.4

Element	*Self*		*Mutual*	
	Bus p-q	*Impedance $z_{pq,\,pq}$ (pu)*	*Bus r-s*	*Impedance $z_{pq,\,rs}$ (pu)*
1	1 – 2(1)	*j*0.6		
4	1 – 2(2)	*j*0.4	1 – 2(1)	*j*0.2
2	1 – 3	*j*0.5	1 – 2(1)	*j*0.1
3	3 – 4	*j*0.5		
5	2 – 4	*j*0.2		

Solution: (3.33), (3.34), (3.41) and (3.44) are used whenever mutual coupling exists.

Step 1. Let bus 1 be reference. Add element 1 from bus 1 to 2. (1 – 2(1)). Here $p = 1$ and $q = 2$; p is the reference node

$$Z_{bus} = \begin{matrix} & 2 \\ 2 & [j0.6] \end{matrix}$$

Step 2. Add element 4, which is a link from bus 1 to 2, mutually coupled with element 1, 1 – 2(1).

$$Z_{bus} = \begin{matrix} & 2 & l \\ 2 & j0.6 & Z_{2l} \\ l & Z_{l2} & Z_{ll} \end{matrix}$$

$$Z_{2l} = Z_{l2} = -Z_{22} + \frac{y_{12(2),12(1)}(Z_{12} - Z_{22})}{y_{12(2),12(2)}}$$

$Z_{12} = Z_{1l} = 0$(as bus 1 is reference)

Primitive impedance matrix for the two elements is given by

$$[z] = \begin{matrix} & 1-2(1) & 1-2(2) \\ 1-2(1) & j0.6 & j0.2 \\ 1-2(2) & j0.2 & j0.4 \end{matrix}$$

The primitive admittance matrix is obtained by taking the inverse of $[z]$

$$[y] = \begin{array}{c} \\ 1-2(1) \\ 1-2(2) \end{array}\begin{array}{c} \begin{array}{cc} 1-2(1) & 1-2(2) \end{array} \\ \begin{bmatrix} -j2.0 & j1.0 \\ j1.0 & -j3.0 \end{bmatrix} \end{array}$$

$$y_{12(1),\,12(2)} = j1.0; \quad y_{12(2),12(2)} = -j3.0$$

$$Z_{2l} = Z_{l2} = -j0.6 + \frac{(j1.0)(-j0.6)}{-j3.0} = -j0.4$$

$$Z_{ll} = -Z_{2l} + \frac{1 + y_{12(2),12(1)}(Z_{1l} - Z_{2l})}{y_{12(2),12(2)}} = j0.4 + \frac{1 + (j1.0)(j0.4)}{-j3.0} = j0.6$$

$$Z_{\text{bus}} = \begin{array}{c} \\ 2 \\ l \end{array}\begin{array}{c} \begin{array}{cc} 2 & l \end{array} \\ \begin{bmatrix} j0.6 & -j0.4 \\ -j0.4 & j0.6 \end{bmatrix} \end{array}$$

Node l has to be eliminated.

$$Z_{22(\text{new})} = Z_{22(\text{old})} - \frac{Z_{2l}\,Z_{l2}}{Z_{ll}}$$

$$= j0.6 - \frac{(-j0.4)(-j0.4)}{j0.6} = j0.3333$$

$$Z_{\text{bus}} = \begin{array}{c} \\ 2 \end{array}\begin{array}{c} 2 \\ [j0.3333] \end{array}$$

Step 3. Add element 2 which is a branch from $p = 1$ (reference) to $q = 3$, mutually coupled with element $1(1-2(1))$.

$$Z_{32} = Z_{12} + \frac{[y_{13,12(1)} \quad y_{13,12(2)}]\begin{bmatrix} Z_{12(1)} - Z_{22} \\ Z_{12(2)} - Z_{22} \end{bmatrix}}{y_{13,13}}$$

$Z_{12} = 0$ (Since 1 is reference bus)

The primitive impedance matrix is given by

$$z_{rs,rs} = \begin{array}{c} \\ 1-2(1) \\ 1-2(2) \\ 1-3 \end{array}\begin{array}{c} \begin{array}{ccc} 1-2(1) & 1-2(2) & 1-3 \end{array} \\ \begin{bmatrix} j0.6 & j0.2 & j0.1 \\ j0.2 & j0.4 & j0.0 \\ j0.1 & j0.0 & j0.5 \end{bmatrix} \end{array}$$

$$y_{rs,rs} = z_{rs,rs}^{-1} = \begin{array}{c} \\ 1-2(1) \\ 1-2(2) \\ 1-3 \end{array}\begin{array}{c} \begin{array}{ccc} 1-2(1) & 1-2(2) & 1-3 \end{array} \\ \begin{bmatrix} -j2.0833 & j1.0417 & j0.4167 \\ j1.0417 & -j3.0208 & -j0.2083 \\ j0.4167 & -j0.2083 & -j2.0833 \end{bmatrix} \end{array}$$

$$y_{13,12(1)} = j0.4167$$

$$y_{13,12(2)} = -j0.2083$$

$$y_{13,13} = -j2.0833$$

$$Z_{32} = \frac{[j0.4167 \quad -j0.2083]\begin{bmatrix} -j0.333 \\ -j0.333 \end{bmatrix}}{-j2.0833}$$

$$= j0.0333$$

$$Z_{23} = Z_{32}$$

$$Z_{33} = Z_{13} + \frac{1 + [y_{13,12(1)} \quad y_{13,12(2)}]\begin{bmatrix} Z_{13} - Z_{23} \\ Z_{13} - Z_{23} \end{bmatrix}}{y_{13,13}}$$

$Z_{13} = 0$ (Since 1 is reference bus)

$$Z_{33} = \frac{1 + [j0.4167 \quad -j0.2083]\begin{bmatrix} -j0.0333 \\ -j0.0333 \end{bmatrix}}{-j2.0833} = j0.4833$$

$$Z_{bus} = \begin{array}{c} \\ 2 \\ 3 \end{array}\begin{array}{c} \begin{array}{cc} 2 & 3 \end{array} \\ \begin{bmatrix} j0.3333 & j0.0333 \\ j0.0333 & j0.4833 \end{bmatrix} \end{array}$$

Step 4. Add element 3, which is a branch from $p = 3$ to $q = 4$, without any coupling to elements of the partial network of step 3.

$$Z_{41} = Z_{31} = j0.0; \qquad Z_{14} = Z_{41}$$

$$Z_{42} = Z_{32} = j0.0333; \; Z_{24} = Z_{42}$$

$$Z_{43} = Z_{33} = j0.4833; \; Z_{34} = Z_{43}$$

$$Z_{44} = Z_{34} + z_{34,34} = j0.4833 + j0.5 = j0.9833.$$

$$Z_{bus} = \begin{array}{c} \\ 2 \\ 3 \\ 4 \end{array}\begin{array}{c} \begin{array}{ccc} 2 & 3 & 4 \end{array} \\ \begin{bmatrix} j0.3333 & j0.0333 & j0.0333 \\ j0.0333 & j0.4833 & j0.4833 \\ j0.0333 & j0.4833 & j0.9833 \end{bmatrix} \end{array}$$

Step 5. Add element 5 which is a link from $p = 2$ to $q = 4$. It has no coupling to elements of partial network of step 4.

$$Z_{l2} = Z_{22} - Z_{42} = j0.3333 - j0.0333 = j0.3$$

$$Z_{l3} = Z_{23} - Z_{43} = j0.0333 - j0.4833 = -j0.45$$

$$Z_{l4} = Z_{24} - Z_{44} = j0.0333 - j0.9833 = -j0.95$$

$$Z_{2l} = Z_{l2};\ Z_{3l} = Z_{l3};\ Z_{4l} = Z_{l4};$$

$$Z_{ll} = Z_{2l} - Z_{4l} + z_{24,24}$$

$$= j0.3 - (-j0.95) + j0.2 = j1.45$$

$$Z_{bus} = \begin{array}{c} \\ 2 \\ 3 \\ 4 \\ l \end{array}\begin{array}{c} \begin{array}{cccc} 2 & 3 & 4 & l \end{array} \\ \begin{bmatrix} j0.3333 & j0.0333 & j0.0333 & j0.3 \\ j0.0333 & j0.4833 & j0.4833 & -j0.45 \\ j0.0333 & j0.4833 & j0.9833 & -j0.95 \\ j0.3 & -j0.45 & -j0.95 & j1.45 \end{bmatrix} \end{array}$$

Node l is eliminated using

$$Z_{ij(\text{new})} = Z_{ij(\text{old})} - \frac{Z_{il}\, Z_{lj}}{Z_{ll}}$$

Performing the above elimination

$$Z_{bus} = \begin{array}{c} \\ 2 \\ 3 \\ 4 \end{array}\begin{array}{c} \begin{array}{ccc} 2 & 3 & 4 \end{array} \\ \begin{bmatrix} j0.2712 & j0.1263 & j0.2298 \\ j0.1263 & j0.3436 & j0.1885 \\ j0.2298 & j0.1885 & j0.3609 \end{bmatrix} \end{array}$$

Example 3.5: In example 3.4 if a new element is added between bus 2-4 (say 2-4(2)) mutually coupled with the existing line 2-4 (say 2-4(1)), modify the Z_{bus}, if self impedance of 2-4(2) is $j0.4$ and mutual impedance is $j0.15$.

Solution: Since the new element is added between existing buses, it is a link. The primitive impedance matrix is given by

$$z_{rs,\,rs} = \begin{array}{c} \\ 1-2(1) \\ 1-2(2) \\ 1-3 \\ 3-4 \\ 2-4(1) \\ 2-4(2) \end{array}\begin{array}{c} \begin{array}{cccccc} 1-2(1) & 1-2(2) & 1-3 & 3-4 & 2-4(1) & 2-4(2) \end{array} \\ \begin{bmatrix} j0.6 & j0.2 & j0.1 & 0 & 0 & 0 \\ j0.2 & j0.4 & j0.0 & 0 & 0 & 0 \\ j0.1 & j0.0 & j0.5 & 0 & 0 & 0 \\ 0 & 0 & 0 & j0.5 & 0 & 0 \\ 0 & 0 & 0 & 0 & j0.2 & j0.15 \\ 0 & 0 & 0 & 0 & j0.15 & j0.4 \end{bmatrix} \end{array}$$

Partitioning $z_{rs,rs}$, we can write $z_{rs,rs} = \begin{bmatrix} z_1 & z_2 \\ z_3 & z_4 \end{bmatrix}$, where z_2 and z_3 have all elements 0.

$$z_4 = \begin{bmatrix} j0.2 & j0.15 \\ j0.15 & j0.4 \end{bmatrix}$$

Similarly, $\quad y_{rs,rs} = z_{rs,rs}^{-1} = \begin{bmatrix} y_1 & y_2 \\ y_3 & y_4 \end{bmatrix}$

From matrix algebra $y_4 = z_4^{-1}$.

$$\therefore \quad y_4 = \begin{matrix} & 2-4(1) & 2-4(2) \\ 2-4(1) & -j6.9565 & j2.6087 \\ 2-4(2) & j2.6087 & -j3.4783 \end{matrix}$$

$$Z_{l2} = Z_{22} - Z_{42} + \frac{y_{2-4(2),2-4(1)}[Z_{22} - Z_{42}]}{y_{2-4(2),2-4(2)}}$$

$$= j0.2712 - j0.2298 + \frac{(j2.6087)\,(j0.2712 - j0.2298)}{-j3.4783} = j0.01035$$

$$Z_{l3} = Z_{23} - Z_{43} + \frac{y_{2-4(1),2-4(2)}[Z_{23} - Z_{43}]}{y_{2-4(2),2-4(2)}}$$

$$= j0.1263 - j0.1885 + \frac{(j2.6087)\,(j0.1263 - j0.1885)}{-j3.4783} = -j\,0.0155$$

$$Z_{l4} = Z_{24} - Z_{44} + \frac{y_{2-4(1),2-4(2)}[Z_{24} - Z_{44}]}{y_{2-4(2),2-4(2)}}$$

$$= j0.2298 - j0.3609 + \frac{(j2.6087)\,(j0.2298 - j0.3609)}{-j3.4783} = -j0.0328$$

$$Z_{ll} = Z_{2l} - Z_{4l} + \frac{1 + y_{2-4(1),2-4(2)}[Z_{2l} - Z_{4l}]}{y_{2-4(2),2-4(2)}}$$

$$= j0.01035 - (-j0.0328) + \frac{1 + (j2.6087)\,(j0.01035 + j0.0328)}{-j3.4783}$$

$$= j0.01035 + j0.0238 + j0.2875 + j0.0168 = j0.3475$$

$$Z_{\text{bus}} = \begin{matrix} & 2 & 3 & 4 & 1 \\ 2 & j0.2712 & j0.1263 & j0.2298 & j0.01035 \\ 3 & j0.1263 & j0.3436 & j0.1885 & -j0.0155 \\ 4 & j0.2298 & j0.1885 & j0.3609 & -j0.0328 \\ l & j0.01035 & -j0.0155 & -j0.0328 & j0.3475 \end{matrix}$$

Eliminating node *l,* we get

$$Z_{bus} = \begin{array}{c} \\ 2 \\ 3 \\ 4 \end{array}\begin{array}{c} \begin{array}{ccc} 2 & 3 & 4 \end{array} \\ \begin{bmatrix} j0.2709 & j0.1267 & j0.2307 \\ j0.1267 & j0.3429 & j0.1870 \\ j0.2307 & j0.1870 & j0.3578 \end{bmatrix} \end{array}$$

3.4.3 Modification of Z_{bus} for Removal of an Element

An element which is not coupled to any other element can be removed easily. The Z_{bus} is modified as explained in sections 3.4.1 and 3.4.2, by adding in parallel with the element (to be removed), a link whose impedance is equal to the negative of the impedance of the element to be removed.When mutually coupled elements are removed, the Z_{bus} is modified by introducing appropriate changes in the bus currents of the original network to reflect the changes introduced due to removal of the elements. Let the performance equation in terms of the new bus currents be given by

$$\bar{E}'_{bus} = Z_{bus}\,(\bar{I}_{bus} + \Delta\bar{I}_{bus}) \qquad ...(3.51)$$

In (3.51) $\Delta\bar{I}_{bus}$ is the vector of changes in bus currents such that $\bar{E}'_{bus}$, the bus voltage vector of the modified network, reflects the desired changes in the network.

From definition, Z'_{ij}, the element of the modified network, can be obtained by computing the voltage of bus *i* with a current injected at bus *j*, in the modified network. Alternatively, we can calculate the voltage at bus *i* of the original network with the same value of current injected at bus *j* and appropriate changes in currents at the buses of the elements to be removed. Let the elements *u-v* coupled to *r-s* be removed or their impedances changed. The corresponding changes in the bus currents are

$$\begin{aligned} \Delta I_u &= \Delta i_{uv} \\ \Delta I_v &= -\,\Delta i_{uv} \\ \Delta I_r &= \Delta i_{rs} \\ \Delta I_s &= -\,\Delta_{irs} \end{aligned} \qquad ...(3.52)$$

$\Delta I_k = 0$ for all other *k*.

Let $I_j = 1.0$ be injected at the *j*th bus with all other current injections equal to zero. From (3.51)

$$E'_i = \sum_{k=1}^{n} Z_{ik}\,(I_k + \Delta I_k);\; i = 1, 2 \ldots n \qquad ...(3.53)$$

Note that in (3.53) $I_k = 1.0$ for $k = j$ and zero for other values of *k*. Δi_k has non-zero values for $k = u, v, rs$ and zero for other values of *k*.

Substituting for ΔI_k and I_k, (3.53) can be written as

$$E'_i = Z_{ij} + Z_{iu}\,\overline{\Delta i}_{uv} - Z_{iv}\,\overline{\Delta i}_{uv} + Z_{ir}\,\overline{\Delta i}_{rs} - Z_{is}\,\overline{\Delta i}_{rs}$$

$$= Z_{ij} + (Z_{iu} - Z_{iv})(\overline{\Delta i}_{uv}) + (Z_{ir} - Z_{is})\overline{\Delta i}_{rs} \qquad ...(3.54)$$

Let the subscript *ab* be used for elements *u-v* and *r-s* (means *a* can take values *u,r* and *b* can take values *v* and *s*). (3.54) is now written as

$$E'_i = Z_{ij} + (Z_{ia} - Z_{ib})\overline{\Delta i}_{ab};\ i = 1, 2... n \qquad ...(3.55)$$

From the performance equation of the primitive network

$$\overline{\Delta i}_{ab} = ([y_s] - [y'_s])\overline{V}'_{cd} \qquad ...(3.56)$$

where, $[y_s]$ = square sub-matrix of the original primitive admittance matrix whose rows and columns correspond to *u-v* and *r-s*.

$[y'_s]$ = square sub-matrix of modified primitive admittance matrix whose rows and columns correspond to *u-v* and *r-s*.

Subscripts for $([y_s] - [y'_s])$ are *ab*, *cd*. $\overline{V}'_{cd}$ in (3.56) is given by

$$\overline{V}'_{cd} = \overline{E}'_c - \overline{E}'_d \qquad ...(3.57)$$

Substituting for $\overline{E}'_c$ and $\overline{E}'_d$ from (3.55), we get

$$\overline{V}'_{cd} = \overline{Z}_{cj} - \overline{Z}_{dj} + ([Z_{ca}] - [Z_{cb}] - [Z_{da}] + [Z_{db}])\overline{\Delta i}_{ab} \qquad ...(3.58)$$

Substituting (3.58) into (3.56) we can solve for $\overline{\Delta i}_{ab}$ to obtain

$$\overline{\Delta i}_{ab} = [M]^{-1}[\Delta_{ys}]\,[\overline{Z}_{cj} - \overline{Z}_{dj}] \qquad ...(3.59)$$

where,

$$[M] = \{I - [\Delta y_s]([Z_{ca}] - [Z_{cb}] - [Z_{da}] + [Z_{db}])\} \qquad ...(3.60)$$

$$\Delta y_s = [y_s] - [y'_s] \qquad ...(3.61)$$

Substituting for $\overline{\Delta i}_{ab}$ in (3.55), we get

$$E'_i = Z_{ij} + (\overline{Z}_{ia} - \overline{Z}_{ib})\left[[M]^{-1}[\Delta y_s][\overline{Z}_{cj} - \overline{Z}_{dj}]\right] \qquad ...(3.62)$$

This gives the bus voltage at bus *i* of the original network, due to a current injection of 1 pu at bus *j*, by simulating the effect of removal of the element *u-v*. Hence, this is nothing but Z'_{ij}. Therefore,

$$Z'_{ij} = Z_{ij} + (\overline{Z}_{ia} - \overline{Z}_{ib})\left[[M]^{-1}[\Delta y_s][\overline{Z}_{cj} - \overline{Z}_{dj}]\right] \qquad ...(3.63)$$

It is repeated for all $j = 1....n$.

Example 3.6: In example 3.5 remove the element connected between 2-4 (2-4(2)) mutually coupled to 2-4(1).

Solution: We use (3.63)

$$Z'_{ij} = Z_{ij} + (\overline{Z}_{ia} - \overline{Z}_{ib})\left[[M]^{-1}[\Delta y_s][\overline{Z}_{cj} - \overline{Z}_{dj}]\right]$$

$$i, j = 2, 3, 4.$$

The element removed is 2-4(2). Hence, $u = 2$; $v = 4$ (for notations used in derivation). Similarly, it is coupled to 2-4(1). Hence, $r = 2$ and $s = 4$ (element u–v coupled to r–s). Original primitive admittance sub-matrix (before removal) corresponding to u-v and r-s is given by

$$z_{24(1),24(2)} = \begin{array}{c} \\ 2-4(1) \\ 2-4(2) \end{array}\begin{array}{c} \begin{array}{cc} 2-4(1) & 2-4(2) \end{array} \\ \begin{bmatrix} j0.2 & j0.15 \\ j0.15 & j0.4 \end{bmatrix} \end{array}$$

$$[y_s] = y_{2-4(1),2-4(2)} = \begin{array}{c} \\ 2-4(1) \\ 2-4(2) \end{array}\begin{array}{c} \begin{array}{cc} 2-4(1) & 2-4(2) \end{array} \\ \begin{bmatrix} -j6.9565 & j2.6087 \\ j2.6087 & -j3.4783 \end{bmatrix} \end{array}$$

The sub-matrices after removal of the element is given by

$$z_{24(1),24(2)} = \begin{array}{c} \\ 2-4(1) \\ 2-4(2) \end{array}\begin{array}{c} \begin{array}{cc} 2-4(1) & 2-4(2) \end{array} \\ \begin{bmatrix} j0.2 & 0.0 \\ 0.0 & 0.0 \end{bmatrix} \end{array}$$

$$[y'_s] = y_{2-4(1),2-4(2)} = \begin{array}{c} \\ 2-4(1) \\ 2-4(2) \end{array}\begin{array}{c} \begin{array}{cc} 2-4(1) & 2-4(2) \end{array} \\ \begin{bmatrix} -j5.0 & 0 \\ 0 & 0 \end{bmatrix} \end{array}$$

$$\Delta y_s = [y_s] - [y'_s] = \begin{bmatrix} -j6.9565 & j2.6087 \\ j2.6087 & -j3.4783 \end{bmatrix} - \begin{bmatrix} -j5.0 & 0.0 \\ 0.0 & 0.0 \end{bmatrix}$$

$$= \begin{array}{c} \\ 2-4(1) \\ 2-4(2) \end{array}\begin{array}{c} \begin{array}{cc} 2-4(1) & 2-4(2) \end{array} \\ \begin{bmatrix} -j1.9565 & j2.6087 \\ j2.6087 & -j3.4783 \end{bmatrix} \end{array}$$

From the notations the subscripts of $[\Delta y_s]$ are ab, cd. For the sub-matrix above $a = 2,2$; $b = 4,4$; $c = 2,2$; $d = 4,4$; the values of Z_{ca}, Z_{cb}, Z_{da} and Z_{db} are taken from solution of 3.5.

$$[Z_{ca}] = \begin{array}{c} \\ 2 \\ 2 \end{array}\begin{array}{c} \begin{array}{cc} 2 & 2 \end{array} \\ \begin{bmatrix} j0.2709 & j0.2709 \\ j0.2709 & j0.2709 \end{bmatrix} \end{array}$$

$$[Z_{da}] = \begin{matrix} & 2 & 2 \\ 4 & j0.2307 & j0.2307 \\ 4 & j0.23077 & j0.2307 \end{matrix}$$

$$[Z_{cb}] = \begin{matrix} & 4 & 4 \\ 2 & j0.2307 & j0.2307 \\ 2 & j0.2307 & j0.2307 \end{matrix}$$

$$[Z_{db}] = \begin{matrix} & 4 & 4 \\ 4 & j0.3578 & j0.3578 \\ 4 & j0.3578 & j0.3578 \end{matrix}$$

$$M = \{I - [\Delta y_s]([Z_{ca}] - [Z_{cb}] - [Z_{da}] + [Z_{db}])\}$$

$$Z_{\text{temp}} = [Z_{ca}] - [Z_{cb}] - [Z_{da}] + [Z_{db}] = \begin{bmatrix} j0.1673 & j0.1673 \\ j0.1673 & j0.1673 \end{bmatrix}$$

$$[\Delta y_s] Z_{\text{temp}} = \begin{bmatrix} -j1.9565 & j2.6087 \\ j2.6087 & -j3.4783 \end{bmatrix} \begin{bmatrix} j0.1673 & j0.1673 \\ j0.1673 & j0.1673 \end{bmatrix}$$

$$= \begin{bmatrix} -0.1091 & -0.1091 \\ 0.1455 & 0.1455 \end{bmatrix}$$

$$M = [I] - [\Delta y_s] Z_{\text{temp}}$$

$$M = \begin{bmatrix} 1.0 & 0 \\ 0 & 1.0 \end{bmatrix} - \begin{bmatrix} -0.1091 & -0.1091 \\ 0.1455 & 0.1455 \end{bmatrix}$$

$$= \begin{bmatrix} 1.1091 & 0.1091 \\ -0.1455 & 0.8545 \end{bmatrix}$$

$$[M]^{-1} = \begin{bmatrix} 0.8868 & -0.1132 \\ 0.1510 & 1.1510 \end{bmatrix}$$

$$[M^{-1}]\,[\Delta y_s] = \begin{bmatrix} 0.8868 & -0.1132 \\ 0.1510 & 1.1510 \end{bmatrix} \begin{bmatrix} -j1.9565 & j2.6087 \\ j2.6087 & -j3.4783 \end{bmatrix}$$

$$= \begin{bmatrix} -j2.0303 & j2.7071 \\ j2.7071 & -j3.6096 \end{bmatrix}$$

For $i = 2$ and $j = 2$ using (3.63), ($a = 2, 2$, $b = 4, 4$, $c = 2, 2$, $d = 4,4$)

$$Z'_{22} = Z_{22} + ((Z_{22}\ Z_{22}) - (Z_{24}\ Z_{24}))\,[M]^{-1}\,[\Delta Y_s] \left[\begin{bmatrix} Z_{22} \\ Z_{22} \end{bmatrix} - \begin{bmatrix} Z_{42} \\ Z_{42} \end{bmatrix} \right]$$

$$= j0.2709 + [j0.0402 \; j0.0402]\begin{bmatrix} -j2.0303 & j2.7071 \\ j2.7071 & -j3.6096 \end{bmatrix}\begin{bmatrix} j0.0402 \\ j0.0402 \end{bmatrix}$$

$$= j0.2709 + [j0.0402 \; j0.0402]\begin{bmatrix} -0.0272 \\ 0.0363 \end{bmatrix}$$

$$= j0.2709 + j3.65 \times 10^{-4} = j0.2712$$

This value of Z'_{22} is the same as obtained in example 3.4 [the network before the addition of 2-4(2)]

Similarly, when $i = 2, j = 3$.

$$Z'_{23} = Z_{23} + [(Z_{22}\; Z_{22}) - (Z_{24}\; Z_{24})][M]^{-1}[\Delta y_s]\left[\begin{pmatrix} Z_{23} \\ Z_{23} \end{pmatrix} - \begin{pmatrix} Z_{43} \\ Z_{43} \end{pmatrix}\right]$$

$$= j0.1267 + [j0.0402 \; j0.0402]\begin{bmatrix} -j2.0303 & j2.7071 \\ j2.7071 & -j3.6096 \end{bmatrix}\begin{bmatrix} -j0.0603 \\ -j0.0603 \end{bmatrix}$$

$$= j0.1267 + [j0.0402 \; j0.0402]\begin{bmatrix} 0.0408 \\ -0.0544 \end{bmatrix}$$

$$= j0.01267 - j5.4672 \times 10^{-4} = j0.1262$$

Similarly the other elements of the Z'_{bus} are calculated.

For $i = 2$ and $j = 4$

$$Z'_{24} = Z_{24} + [(Z_{22}\; Z_{22}) - (Z_{24}\; Z_{24})]\,[M]^{-1}[\Delta y_s]\left[\begin{pmatrix} Z_{24} \\ Z_{24} \end{pmatrix} - \begin{pmatrix} Z_{44} \\ Z_{44} \end{pmatrix}\right]$$

$$= j0.2307 + [j0.0402 \; j0.0402]\begin{bmatrix} -j2.0303 & j2.7071 \\ j2.7071 & -j3.6096 \end{bmatrix}\begin{bmatrix} -j0.1271 \\ -j0.1271 \end{bmatrix}$$

$$= j0.2307 + [j0.0402 \; j0.0402]\begin{bmatrix} 0.0860 \\ -0.1147 \end{bmatrix}$$

$$= j0.2297.$$

For $i = 3$ and $j = 3$

$$Z'_{33} = Z_{33} + [(Z_{32}\; Z_{32}) - (Z_{34}\; Z_{34})]\,[M]^{-1}\,[\Delta y_s]\left[\begin{pmatrix} Z_{23} \\ Z_{23} \end{pmatrix} - \begin{pmatrix} Z_{43} \\ Z_{43} \end{pmatrix}\right]$$

$$= j0.3429 + [-j0.0603 - j0.0603]\begin{bmatrix} -j2.0303 & j2.7071 \\ j2.7071 & -j3.6096 \end{bmatrix}\begin{bmatrix} -j0.0603 \\ -j0.0603 \end{bmatrix}$$

$$= j0.3429 + [-j0.0603 - j0.0603]\begin{bmatrix} 0.0408 \\ -0.0544 \end{bmatrix}$$

$$= j0.3429 + j8.2008 \times 10^{-3} = j0.3437$$

For $i = 3$ and $j = 4$

$$Z'_{34} = Z_{34} + [(Z_{32}\ Z_{32}) - (Z_{34}\ Z_{34})][M]^{-1}\ [\Delta y_s]\left[\begin{pmatrix} Z_{24} \\ Z_{24} \end{pmatrix} - \begin{pmatrix} Z_{44} \\ Z_{44} \end{pmatrix}\right]$$

$$= j0.187 + [-j0.0603 - j0.0603]\begin{bmatrix} -j2.0303 & j2.7071 \\ j2.7071 & -j3.6096 \end{bmatrix}\begin{bmatrix} -j0.1271 \\ -j0.1271 \end{bmatrix}$$

$$= j0.187 + [-j0.0603 - j0.0603]\begin{bmatrix} 0.0860 \\ -0.1147 \end{bmatrix}$$

$$= j0.187 + j1.7306 \times 10^{-3} = j0.1886$$

For $i = 4$ and $j = 4$

$$Z'_{44} = Z_{44} + [(Z_{42}\ Z_{42}) - (Z_{44}\ Z_{44})]\,[M]^{-1}[\Delta y_s]\left[\begin{pmatrix} Z_{24} \\ Z_{24} \end{pmatrix} - \begin{pmatrix} Z_{44} \\ Z_{44} \end{pmatrix}\right]$$

$$= j0.3578 + (-j0.1271 - j0.1271)\begin{pmatrix} 0.08602 \\ -0.1147 \end{pmatrix}$$

$$= j0.3578 + j3.6477 \times 10^{-3} = j0.361$$

$$Z'_{bus} = \begin{array}{c} \\ \\ \\ \end{array}\begin{array}{ccc} 2 & 3 & 4 \\ \end{array}$$

$$Z'_{bus} = \begin{bmatrix} j0.2712 & j0.1262 & j0.2297 \\ j0.1262 & j0.3437 & j0.1886 \\ j0.2297 & j0.1886 & j0.361 \end{bmatrix}$$

(columns 2, 3, 4)

Note: This is same as Z_{bus} of example 3.4, which is before addition of line 2-4(2). Slight errors in the solution are due to numerical crunching and rounding off.

Example 3.7: Form the bus impedance matrix for the network of Fig. 3.14.

Solution:

Step 1. Add element 1, which is a branch between node 1 and reference node.

$$Z_{bus} = \begin{array}{c} \\ 1 \end{array}\begin{array}{c} 1 \\ [j1.25] \end{array}$$

Step 2. Add element 2, which is a branch between nodes 1 and 2.

$$Z_{bus} = \begin{array}{c} \\ 1 \\ 2 \end{array}\begin{array}{cc} 1 & 2 \\ \end{array}\!\!\begin{bmatrix} j1.25 & j1.25 \\ j1.25 & j1.5 \end{bmatrix}$$

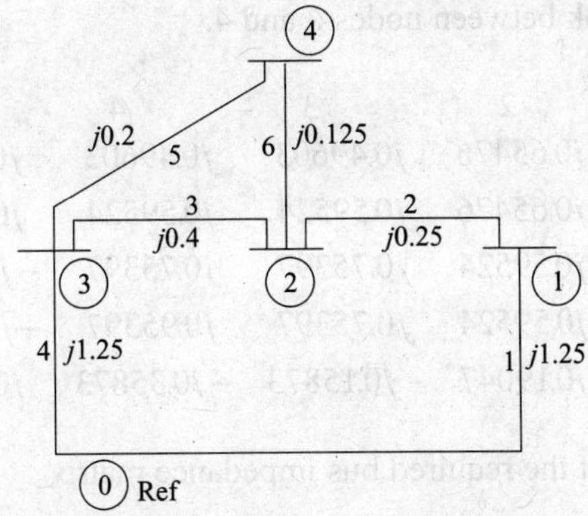

Fig. 3.14 Network of example 3.7.

Step 3. Add element 3, which is a branch between nodes 2 and 3.

$$Z_{bus} = \begin{array}{c} \\ 1 \\ 2 \\ 3 \end{array}\begin{array}{c} \begin{array}{ccc} 1 & 2 & 3 \end{array} \\ \begin{bmatrix} j1.25 & j1.25 & j1.25 \\ j1.25 & j1.5 & j1.5 \\ j1.25 & j1.5 & j1.9 \end{bmatrix} \end{array}$$

Step 4. Add element 4, which is a link from node 3 to reference node.

$$Z_{bus} = \begin{array}{c} \\ 1 \\ 2 \\ 3 \\ l \end{array}\begin{array}{c} \begin{array}{cccc} 1 & 2 & 3 & l \end{array} \\ \begin{bmatrix} j1.25 & j1.25 & j1.25 & j1.25 \\ j1.25 & j1.5 & j1.5 & j1.5 \\ j1.25 & j1.5 & j1.9 & j1.9 \\ j1.25 & j1.5 & j1.9 & j3.15 \end{bmatrix} \end{array}$$

Eliminating node l,

$$Z_{bus} = \begin{array}{c} \begin{array}{ccc} 1 & 2 & 3 \end{array} \\ \begin{bmatrix} j0.75397 & j0.65476 & j0.49603 \\ j0.65476 & j0.78571 & j0.59524 \\ j0.49603 & j0.59524 & j0.75397 \end{bmatrix} \end{array}$$

Step 5. Add element 5, a branch between nodes 3 and 4.

$$Z_{bus} = \begin{array}{c} \\ 1 \\ 2 \\ 3 \\ 4 \end{array}\begin{array}{c} \begin{array}{cccc} 1 & 2 & 3 & 4 \end{array} \\ \begin{bmatrix} j0.75397 & j0.65476 & j0.49603 & j0.49603 \\ j0.65476 & j0.65476 & j0.59524 & j0.59524 \\ j0.49603 & j0.59524 & j0.75397 & j0.75397 \\ j0.49603 & j0.59524 & j0.75397 & j0.95397 \end{bmatrix} \end{array}$$

Step 6. Add element 6, a link between nodes 2 and 4.

$$Z_{bus} = \begin{array}{c} \\ 1 \\ 2 \\ 3 \\ 4 \\ l \end{array}\begin{array}{c} \begin{array}{ccccc} 1 & 2 & 3 & 4 & l \end{array} \\ \begin{bmatrix} j0.75397 & j0.65476 & j0.49603 & j0.49603 & j0.15873 \\ j0.65476 & j0.65476 & j0.59524 & j0.59524 & j0.19047 \\ j0.49603 & j0.59524 & j0.75397 & j0.75397 & -j0.15873 \\ j0.49603 & j0.59524 & j0.75397 & j0.95397 & -j0.35873 \\ j0.15873 & j0.19047 & -j0.15873 & -j0.35873 & j0.67421 \end{bmatrix} \end{array}$$

Eliminating node l, we get the required bus impedance matrix

$$Z_{bus} = \begin{array}{c} \\ 1 \\ 2 \\ 3 \\ 4 \end{array}\begin{array}{c} \begin{array}{cccc} 1 & 2 & 3 & 4 \end{array} \\ \begin{bmatrix} j0.7166 & j0.6099 & j0.5334 & j0.5805 \\ j0.6099 & j0.7319 & j0.6401 & j0.6966 \\ j0.5334 & j0.6401 & j0.7166 & j0.6695 \\ j0.5805 & j0.6966 & j0.6695 & j0.7631 \end{bmatrix} \end{array}$$

3.5 NETWORK MATRICES FOR THREE-PHASE NETWORKS

During normal steady state operation, the three-phase system can be solved on a single phase basis, since excitation, transmission and distribution networks are balanced. However, during abnormal operations even though the three-phase network is balanced, the excitation system becomes unbalanced. With unbalanced excitation, the network can be analyzed either in terms of actual phase quantities or by suitable transformation into balanced sequence quantities.

3.5.1 Representation of Three-Phase Networks

The representation of a three-phase network component in impedance form is shown in Fig. 3.15a. and in admittance form in Fig. 3.15b.

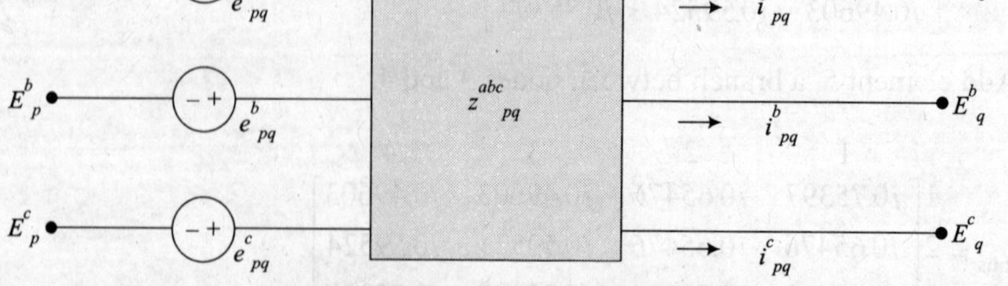

Fig. 3.15a Three phase network in impedance form.

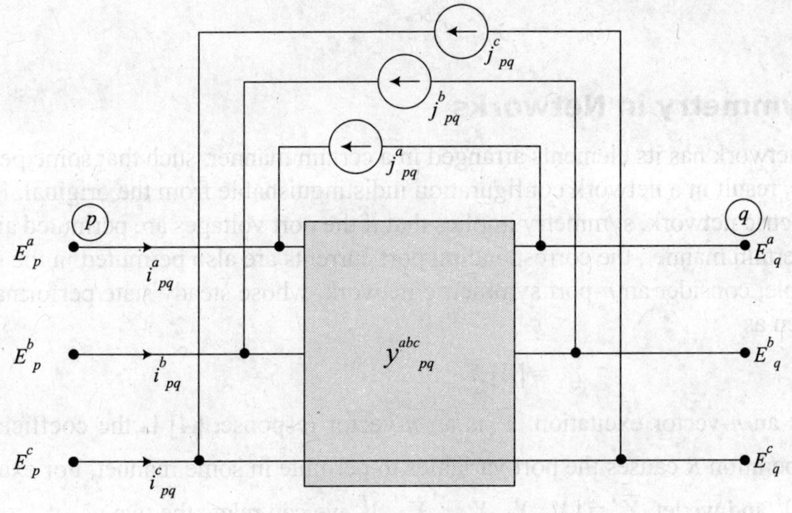

Fig. 3.15b Three-phase network in admittance form.

The variables are designated as follows:

$v_{pq}^{a}, v_{pq}^{b}, v_{pq}^{c}$, are voltages across element *p-q* for phases *a, b* and *c*, respectively.

$e_{pq}^{a}, e_{pq}^{b}, e_{pq}^{c}$ are the series voltage sources in phases *a, b* and *c*, respectively.

$i_{pq}^{a}, i_{pq}^{b}, i_{pq}^{c}$ are the currents through the elements *p-q* for phases *a, b* and *c* respectively.

$j_{pq}^{a}, j_{pq}^{b}, j_{pq}^{c}$ are the source currents in parallel with phases *a, b* and *c*, respectively.

z_{pq}^{abc} is the three phase impedance matrix for element *p-q*.

y_{pq}^{abc} is the three phase admittance matrix for element *p-q*.

The performance equation in impedance form is given by

$$\begin{bmatrix} v_{pq}^{a} \\ v_{pq}^{b} \\ v_{pq}^{c} \end{bmatrix} + \begin{bmatrix} e_{pq}^{a} \\ e_{pq}^{b} \\ e_{pq}^{c} \end{bmatrix} = \begin{bmatrix} z_{pq}^{aa} & z_{pq}^{ab} & z_{pq}^{ac} \\ z_{pq}^{ba} & z_{pq}^{bb} & z_{pq}^{bc} \\ z_{pq}^{ca} & z_{pq}^{cb} & z_{pq}^{cc} \end{bmatrix} \begin{bmatrix} i_{pq}^{a} \\ i_{pq}^{b} \\ i_{pq}^{c} \end{bmatrix} \qquad \text{...(3.64)}$$

where, $z_{pq}^{aa}, z_{pq}^{bb}, z_{pq}^{cc}$ are self impedances of phases *a, b* and *c* of the element *p-q*. z_{pq}^{ab}, z_{pq}^{ac} ...etc. are the mutual impedances between the respective phases. (3.64) can be written compactly as

$$v_{pq}^{abc} + e_{pq}^{abc} = z_{pq}^{abc}\, i_{pq}^{abc} \qquad \text{...(3.65)}$$

Similarly, the relationship in admittance form can be written as

$$i_{pq}^{abc} + j_{pq}^{abc} = y_{pq}^{abc}\, v_{pq}^{abc} \qquad \text{...(3.66)}$$

where, $$y_{pq}^{abc} = \left[z_{pq}^{abc}\right]^{-1} \qquad \text{...(3.67)}$$

$$j_{pq}^{abc} = -y_{pq}^{qbc}\, e_{pq}^{abc} \qquad ...(3.68)$$

3.5.2 Symmetry in Networks

Symmetric network has its elements arranged in a certain manner, such that some permutations of the elements, result in a network configuration indistinguishable from the original. If we consider an n-port electric network, symmetry implies that if the port voltages are permuted amongst themselves in a certain manner, the corresponding port currents are also permuted in the same manner. As an example, consider an n-port symmetric network, whose steady state performance equation can be written as

$$\overline{Y} = [A]\overline{X} \qquad ...(3.69)$$

where, $\overline{X}$ is an n-vector excitation, $\overline{Y}$ is an n-vector response, $[A]$ is the coefficient matrix. A symmetry operation R causes the port variables to permute in some manner. For example, if $\overline{X} = [X_1, X_2 \dots X_n]^T$ and we let $\overline{X}' = [X_n, X_1, X_2 \dots X_{n-1}]^T$, we can relate the two as

$$\begin{bmatrix} X_n \\ X_1 \\ X_2 \\ \vdots \\ X_{n-1} \end{bmatrix} = \begin{bmatrix} 0 & 0 & \cdots & 1 \\ 1 & 0 & \cdots & 0 \\ 0 & 1 & \cdots & 0 \\ 0 & \cdots & 1 & 0 \end{bmatrix} \begin{bmatrix} X_1 \\ X_2 \\ \vdots \\ X_n \end{bmatrix}$$

The matrix relating $\overline{X}'$ to $\overline{X}$ is called the permutation matrix, $D(R)$. Hence,

$$\overline{X}' = D(R)\overline{X} \qquad ...(3.70)$$

For any two operations $R_1, R_2, D(R_1R_2) = D(R_1)\, D(R_2)$. If the network has symmetry, it implies that if $\overline{X}$ is permutated to $\overline{X}'$ using (3.70), the output vector also permutes as

$$\overline{Y}' = D(R)\overline{Y} \qquad ...(3.71)$$

where, $$\overline{Y}' = A\overline{X}' \qquad ...(3.72)$$

Using (3.70) and (3.71), (3.72) can we written as

$$D(R)\overline{Y} = AD(R)\overline{X} \text{ or } Y = D^{-1}(R)AD(R)\overline{X} \qquad ...(3.73)$$

Comparing (3.73) with (3.69), we get

$$A = D^{-1}(R)AD(R) \qquad ...(3.74)$$

We can, therefore, conclude that the coefficient matrix of a symmetric network, satisfies (3.74).

3.5.3 Rotational Symmetry

In networks, exhibiting rotational symmetry, all operations are generated by a single operation successively applied. For example, if P is an operation, the symmetry operations of rotation are given by $R = P^0, P^1, \ldots P^k$ denoting P applied k times. This means

$$D[P^k] = [D(P)]^k$$

The symmetry operation for a three phase element p–q are rotations about an axis perpendicular to the plane of the paper through 120° (represented by R_3^1), rotation through 240° (represented by R_3^2) and rotation through 360°. Consider three coils as shown in Fig. 3.16a.

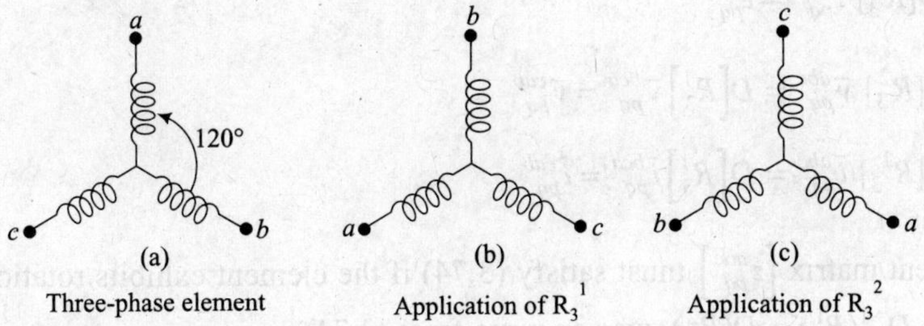

Fig. 3.16

If the operation R_3^1 is applied to the network of Fig. 3.16a, we obtain 3.16b. Now if coils a, b, c are identical, the network of Fig. 3.16b is indistinguishable from Fig. 3.16a. Application of R_3^2 to Fig. 3.16a would result in Fig. 3.16c, which is equivalent to applying R_3^1 to Fig. 3.16b. Thus, $R_3^2 = (R_3^1)^2$. If again R_3^1 is applied to Fig. 3.16c, we get back Fig. 3.16a.

These three operations constitute a permutation group of order three which can be represented as follows.

$$D(E) = \begin{bmatrix} 1 & 0 & 0 \\ 0 & 1 & 0 \\ 0 & 0 & 1 \end{bmatrix}$$

$$D(R_3^1) = \begin{bmatrix} 0 & 1 & 0 \\ 0 & 0 & 1 \\ 1 & 0 & 0 \end{bmatrix}$$

$$D(R_3^2) = \begin{bmatrix} 0 & 0 & 1 \\ 1 & 0 & 0 \\ 0 & 1 & 0 \end{bmatrix} \qquad \ldots(3.75)$$

For a three-phase element $p - q$, if

$$\bar{v}_{pq}^{abc} = \left[z_{pq}^{abc}\right]\bar{i}_{pq}^{abc}$$

Then symmetry implies

$$D[E]\,\bar{v}_{pq}^{abc} = \bar{v}_{pq}^{abc}$$

$$D[E]\,\bar{i}_{pq}^{abc} = \bar{i}_{pq}^{abc}$$

$$D[R_3^1]\,\bar{v}_{pq}^{abc} = \bar{v}_{pq}^{bca}$$

$$D[R_3^1]\,\bar{i}_{pq}^{abc} = \bar{i}_{pq}^{bca}$$

$$D[R_3^2]\,\bar{v}_{pq}^{abc} = D\left[R_3^1\right]\bar{v}_{pq}^{bca} = \bar{v}_{pq}^{cab}$$

$$D[R_3^2]\,\bar{i}_{pq}^{abc} = D\left[R_3^1\right]\bar{i}_{pq}^{bca} = \bar{i}_{pq}^{cab} \qquad ...(3.76)$$

The coefficient matrix $\left[z_{pq}^{abc}\right]$ must satisfy (3.74) if the element exhibits rotational symmetry. Recognizing that $D^{-1}(R_3^1) = D(R_3^2)$, we can write from (3.74),

$$\left[z_{pq}^{abc}\right] = D\left[R_3^1\right]\left[z_{pq}^{abc}\right]D\left[R_3^2\right] \qquad ...(3.77)$$

Substituting for z_{pq}^{abc} from (3.64) and for $D[R_3^1]$ and $D[R_3^2]$ in (3.76), we get

$$\begin{bmatrix} z_{pq}^{aa} & z_{pq}^{ab} & z_{pq}^{ac} \\ z_{pq}^{ba} & z_{pq}^{bb} & z_{pq}^{bc} \\ z_{pq}^{ca} & z_{pq}^{cb} & z_{pq}^{cc} \end{bmatrix} = \begin{bmatrix} z_{pq}^{bb} & z_{pq}^{bc} & z_{pq}^{ba} \\ z_{pq}^{cb} & z_{pq}^{cc} & z_{pq}^{ca} \\ z_{pq}^{ab} & z_{pq}^{ac} & z_{pq}^{aa} \end{bmatrix}$$

To satisfy the above equality,

$$z_{pq}^{aa} = z_{pq}^{bb} = z_{pq}^{cc} = z_{pq}^{s} \qquad ...(3.78a)$$

$$z_{pq}^{ab} = z_{pq}^{bc} = z_{pq}^{ca} = z_{pq}^{m1} \qquad ...(3.78b)$$

$$z_{pq}^{ac} = z_{pq}^{ba} = z_{pq}^{cb} = z_{pq}^{m2} \qquad ...(3.78c)$$

Thus, the coefficient matrix of a three phase element which satisfies rotational symmetry is given by

$$z_{pq}^{abc} = \begin{bmatrix} z_{pq}^{s} & z_{pq}^{m1} & z_{pq}^{m2} \\ z_{pq}^{m2} & z_{pq}^{s} & z_{pq}^{m1} \\ z_{pq}^{m1} & z_{pq}^{m2} & z_{pq}^{s} \end{bmatrix} \qquad ...(3.79)$$

Such an element is called a rotating element. Three phase balanced rotating machines exhibit rotational symmetry.

3.5.4 Reflection Symmetry

In networks possessing reflection symmetry, the network structure remains invariant for reflections of the network about axis of symmetries. In a three-phase electric network, the axes of symmetries are assumed to coincide with the magnetic axis of phases *a,b* and *c*, respectively. Consider Fig. 3.17a, showing three coils. Figure 3.17b shows the network reflected about the axis of coil *a*. If all coils are identical, 3.17b is indistinguishable from the original network of 3.17a, implying the network possesses reflection symmetry.

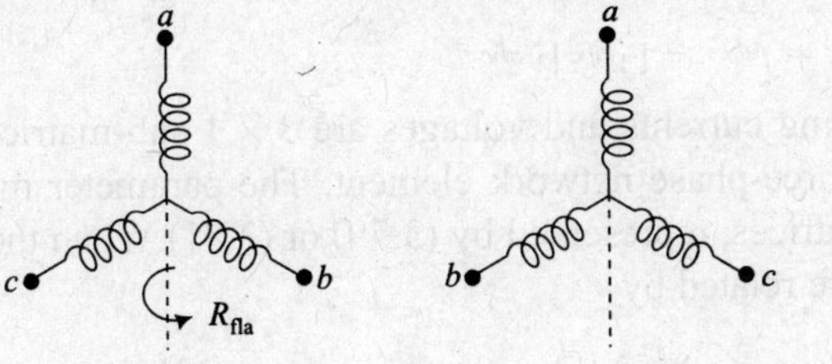

(a) Three-phase element

(b) Reflection about axis *a*

Fig. 3.17

We can thus represent three operations R_{fla}, R_{flb} anf R_{flc} to denote reflections about magnetic axis of phases *a*, *b* and *c*, respectively. These symmetry operations are represented by the following permutation matrices:

$$D(R_{fla}) = \begin{bmatrix} 1 & 0 & 0 \\ 0 & 0 & 1 \\ 0 & 1 & 0 \end{bmatrix}$$

$$D(R_{flb}) = \begin{bmatrix} 0 & 0 & 1 \\ 0 & 1 & 0 \\ 1 & 0 & 0 \end{bmatrix}$$

$$D(R_{flc}) = \begin{bmatrix} 0 & 1 & 0 \\ 1 & 0 & 0 \\ 0 & 0 & 1 \end{bmatrix} \tag{3.80}$$

The symmetry operations (3.75) and (3.80) constitute a group known as permutation group, representing three–fold rotations and three–fold reflections. The coefficient matrix z_{pq}^{abc} of a three phase element *p-q*, exhibiting both rotational and reflection symmetries will be of the form

$$z_{pq}^{abc} = \begin{bmatrix} z_{pq}^{s} & z_{pq}^{m} & z_{pq}^{m} \\ z_{pq}^{m} & z_{pq}^{s} & z_{pq}^{m} \\ z_{pq}^{m} & z_{pq}^{m} & z_{pq}^{s} \end{bmatrix} \tag{3.81}$$

Elements having this form of symmetry (both rotational and reflection) are referred to as stationary elements. The power system transmission and distribution network elements (with transposed lines) are stationary elements.

3.5.5 Balanced Excitation

The equation of the primitive three phase network is given by

$$\bar{v}^{abc} + \bar{e}^{abc} = [z^{abc}]\bar{i}^{abc} \quad ...(3.82a)$$

$$\bar{i}^{abc} + \bar{j}^{abc} = [y^{abc}]\bar{v}^{abc} \quad ...(3.82b)$$

The vectors representing currents and voltages are 3×1 sub-matrices corresponding to the variables of a particular three-phase network element. The parameter matrices z^{abc} and y^{abc} are composed of 3×3 sub-matrices, represented by (3.79) or (3.81). When the excitation is balanced, the three phase voltages are related by

$$e_{pq}^{abc} = \begin{bmatrix} e_{pq}^a \\ e_{pq}^b \\ e_{pq}^c \end{bmatrix} = \begin{bmatrix} 1 \\ a^2 \\ a \end{bmatrix} e_{pq}^a$$

$$j_{pq}^{abc} = \begin{bmatrix} j_{pq}^a \\ j_{pq}^b \\ j_{pq}^c \end{bmatrix} = \begin{bmatrix} 1 \\ a^2 \\ a \end{bmatrix} j_{pq}^a$$

where, $a = 1\angle 120°$ and the phase sequence is *abc*. The performance equation for a rotating element is given by

$$\begin{bmatrix} 1 \\ a^2 \\ a \end{bmatrix} v_{pq}^a + \begin{bmatrix} 1 \\ a^2 \\ a \end{bmatrix} e_{pq}^a = \begin{bmatrix} z_{pq}^s & z_{pq}^{m1} & z_{pq}^{m2} \\ z_{pq}^{m2} & z_{pq}^s & z_{pq}^{m1} \\ z_{pq}^{m1} & z_{pq}^{m2} & z_{pq}^s \end{bmatrix} \begin{bmatrix} 1 \\ a^2 \\ a \end{bmatrix} i_{pq}^a \quad ...(3.83)$$

Pre-multiplying by $[1 \; a \; a^2]$ on both sides, (3.83) becomes

$$v_{pq}^a + e_{pq}^a = [z_{pq}^s + a^2 z_{pq}^{m1} + a z_{pq}^{m2}] i_{pq}^a \quad ...(3.84a)$$

Similarly, $$i_{pq}^a + j_{pq}^a = [y_{pq}^s + a^2 y_{pq}^{m1} + a y_{pq}^{m2}] v_{pq}^a \quad ...(3.84b)$$

In (3.84a) $[z_{pq}^s + a^2 z_{pq}^{m1} + a z_{pq}^{m2}]$ is called the *positive sequence impedance*. It can be seen that voltage in phase *a* is dependent only on current in phase *a*. Thus, a balanced three phase rotating element with balanced excitation can be treated as a single phase element in network problems. The performance equation for a stationary element is given by

$$\begin{bmatrix} 1 \\ a^2 \\ a \end{bmatrix} v_{pq}^a + \begin{bmatrix} 1 \\ a^2 \\ a \end{bmatrix} e_{pq}^a = \begin{bmatrix} z_{pq}^s & z_{pq}^m & z_{pq}^m \\ z_{pq}^m & z_{pq}^s & z_{pq}^m \\ z_{pq}^m & z_{pq}^m & z_{pq}^s \end{bmatrix} \begin{bmatrix} 1 \\ a^2 \\ a \end{bmatrix} i_{pq}^a \quad ...(3.85)$$

Pre-multiplying by $[1\ a\ a^2]$ on both sides, we get

$$v_{pq}^a + e_{pq}^a = (z_{pq}^s - z_{pq}^m) i_{pq}^a \quad ...(3.86a)$$

Similarly,

$$i_{pq}^a + j_{pq}^a = (y_{pq}^s - y_{pq}^m) v_{pq}^a \quad ...(3.86b)$$

$z_{pq}^s - z_{pq}^m$ is called the *positive sequence impedance*. Thus, a balanced three-phase stationary element, with balanced excitation can be treated as a single-phase element.

3.5.6 Unbalanced Excitation

When excitation is unbalanced, (3.84) and (3.86) cannot be used. However, the impedance matrix can be diagonalized by suitable transformations. Using a complex transformation T, the phase variables a, b, c can be transformed into a new set of variables i, j, k as follows:

$$\begin{aligned} v_{pq}^{abc} &= T v_{pq}^{ijk} \\ i_{pq}^{abc} &= T i_{pq}^{ijk} \\ e_{pq}^{abc} &= T e_{pq}^{ijk} \\ j_{pq}^{abc} &= T j_{pq}^{ijk} \end{aligned} \quad ...(3.87)$$

The complex power in the element p-q is given by

$$S_{pq} = P_{pq} + jQ_{pq} = \left\{ \left(i_{pq}^{abc} \right)^* \right\}^T e_{pq}^{abc}$$

$$= \left\{ \left(i_{pq}^{ijk} \right)^* \right\}^T (T^*)^T T e_{pq}^{ijk} \quad ...(3.88)$$

In terms of transformed quantities,

$$S'_{pq} = \left\{ \left(i_{pq}^{ijk} \right)^* \right\}^T e_{pq}^{ijk} \quad ...(3.89)$$

For power–invariance, $S_{pq} = S'_{pq}$. Hence, from (3.88) and (3.89), we get

$$(T^*)^T T = I = T(T^*)^T$$

This implies that the transformation matrix T should be a unitary matrix. The performance equations using transferred variables are obtained by substituting (3.87) in (3.65).

$$T\left(v_{pq}^{ijk} + e_{pq}^{ijk}\right) = z_{pq}^{abc}\, T i_{pq}^{ijk}$$

Pre-multiplying by $(T^*)^T$ on both sides, we get

$$v_{pq}^{ijk} + e_{pq}^{ijk} = (T^*)^T z_{pq}^{abc} T i_{pq}^{ijk}$$

$$= z_{pq}^{ijk} \, i_{pq}^{ijk}$$

where, $$z_{pq}^{ijk} = (T^*)^T z_{pq}^{abc} T \quad ...(3.90)$$

A similar transformation can be obtained for the performance equation in admittance form.

3.5.7 Transformation Matrices

Symmetrical components

Two popular transformations are the transformation into symmetrical components and Clarke's components. Symmetrical components, originally proposed by CL Fortescue (refer chapter 5) uses a transformation

$$\begin{bmatrix} V_a \\ V_b \\ V_c \end{bmatrix} = \begin{bmatrix} 1 & 1 & 1 \\ 1 & a^2 & a \\ 1 & a & a^2 \end{bmatrix} \begin{bmatrix} V_{a0} \\ V_{a1} \\ V_{a2} \end{bmatrix} \text{ or}$$

$$[V^{abc}] = [A][V^{012}]$$

This transformation is used in chapter 5 as it is the most popularly used transformation. However, the matrix $[A]$ is not power invariant, since it is not a unitary matrix. A modified transformation matrix is used, given by

$$T_s = \frac{1}{\sqrt{3}} \begin{bmatrix} 1 & 1 & 1 \\ 1 & a^2 & a \\ 1 & a & a^2 \end{bmatrix} \quad ...(3.91)$$

which is a unitary matrix, since $(T_s^*)^T T_s = I$. For a rotating element, using (3.90)

$$z_{pq}^{012} = \frac{1}{\sqrt{3}} \begin{bmatrix} 1 & 1 & 1 \\ 1 & a^2 & a \\ 1 & a & a^2 \end{bmatrix} \begin{bmatrix} z_{pq}^{s} & z_{pq}^{m1} & z_{pq}^{m2} \\ z_{pq}^{m2} & z_{pq}^{s} & z_{pq}^{m1} \\ z_{pq}^{m1} & z_{pq}^{m2} & z_{pq}^{s} \end{bmatrix} \frac{1}{\sqrt{3}} \begin{bmatrix} 1 & 1 & 1 \\ 1 & a^2 & a \\ 1 & a & a^2 \end{bmatrix}$$

$$z_{pq}^{012} = \begin{bmatrix} z_{pq}^{s} + z_{pq}^{m1} + z_{pq}^{m2} & 0 & 0 \\ 0 & z_{pq}^{s} + a^2 z_{pq}^{m1} + a z_{pq}^{m2} & 0 \\ 0 & 0 & z_{pq}^{s} + a z_{pq}^{m1} + a^2 z_{pq}^{m2} \end{bmatrix}$$

where, $z_{pq}^{0} = z_{pq}^{s} + z_{pq}^{m1} + z_{pq}^{m2}$ = Zero sequence impedance.

$z_{pq}^{1} = z_{pq}^{s} + a^2 z_{pq}^{m1} + a^2 z_{pq}^{m2}$ = Positive sequence impedance.

$z_{pq}^{2} = z_{pq}^{s} + a z_{pq}^{m1} + a^2 z_{pq}^{m2}$ = Negative sequence impedance.

Similarly, for a stationary element

$$z_{pq}^{012} = \begin{bmatrix} z_{pq}^{s} + 2z_{pq}^{m} & 0 & 0 \\ 0 & z_{pq}^{s} - z_{pq}^{m} & 0 \\ 0 & 0 & z_{pq}^{s} - z_{pq}^{m} \end{bmatrix}$$

Both the transformed impedance matrices are diagonal. This means the voltage drops of one sequence are dependent only on the current of that sequence. The equations are decoupled.

Clarke's Components

This transforms the phase quantities into zero, alpha and beta components.

$$T_c = \frac{1}{\sqrt{3}} \begin{bmatrix} 1 & \sqrt{2} & 0 \\ 1 & -\sqrt{1/2} & \sqrt{3/2} \\ 1 & -\sqrt{1/2} & -\sqrt{3/2} \end{bmatrix} \quad ...(3.92)$$

For a rotating element

$$z_{pq}^{0\alpha\beta} = \begin{bmatrix} z_{pq}^{s} + z_{pq}^{m1} + z_{pq}^{m2} & 0 & 0 \\ 0 & z_{pq}^{s} - \frac{1}{2}(z_{pq}^{m1} + z_{pq}^{m2}) & \frac{\sqrt{3}}{2}(z_{pq}^{m1} - z_{pq}^{m2}) \\ 0 & \frac{\sqrt{3}}{2}(z_{pq}^{m2} - z_{pq}^{m1}) & z_{pq}^{s} - \frac{1}{2}(z_{pq}^{m1} + z_{pq}^{m2}) \end{bmatrix}$$

This is not a diagonal matrix. For a stationary element,

$$z_{pq}^{o\alpha\beta} = \begin{bmatrix} z_{pq}^{s} + 2z_{pq}^{m} & 0 & 0 \\ 0 & z_{pq}^{s} - z_{pq}^{m} & 0 \\ 0 & 0 & z_{pq}^{s} - z_{pq}^{m} \end{bmatrix}$$

which is a diagonal matrix. Since, rotating elements do not have a diagonal impedance matrix, we do not get independent decoupled equations, in the transformed components.

With unbalanced elements, both T_s and T_c, do not yield diagonal impedance matrix. In such a case, it may be more practical to solve the network using original three phase quantities.

3.5.8 Network Matrices for Three Phase Networks

The incidence matrices and network matrices are formed by same algorithms used for single phase networks. In all incidence matrices, the entries 1, – 1, 0 for a single phase network, are to be replaced by 3 × 3 matrices [I], [–I] and null, respectively. Further, the impedance (admittance) of each element will be a 3 × 3 matrix, represented by (3.79) (for a rotating element) and (3.81) for a stationary element. The matrices can also be composed of 3 × 3 sub-matrices whose elements refer to the sequence components. The performance equation in bus frame of reference is written as

$$\bar{E}^{abc}_{bus} = Z^{abc}_{bus}\,\bar{I}^{abc}_{bus}$$

where, $\bar{E}^{abc}_{bus}$ = Vector of three phase bus voltages measured with respect to reference bus.

$\bar{I}^{abc}_{bus}$ = Vector of three phase bus current injections.

Z^{abc}_{bus} = Three phase bus impedance matrix.

The algorithm for formation of the bus impedance matrix can be derived in a manner similar to the one presented in section 3.4. The equations are given below.

Addition of a branch

When a branch *p-q* is added where *q* is the new bus and *p* is the bus in the existing network, the elements of the bus impedance matrix are given by

$$Z^{abc}_{qi} = Z^{abc}_{pi} + (y^{abc}_{pq,pq})^{-1}\,\bar{y}^{abc}_{pq,rs}(\bar{Z}^{abc}_{ri} - Z^{adc}_{si})$$

$$Z^{abc}_{iq} = Z^{abc}_{ip} + (\bar{Z}^{abc}_{ir} - \bar{Z}^{abc}_{is})\bar{y}^{abc}_{rs,pq}(y^{abc}_{pq,pq})^{-1}$$

$$Z^{abc}_{qq} = Z^{abc}_{pq} + (y^{abc}_{pq,pq})^{-1}\{I + \bar{y}^{abc}_{pq,rs}(\bar{Z}^{abc}_{rq} - \bar{Z}^{abc}_{sq})\} \quad ...(3.93)$$

In (3.93) the element *p-q* is coupled to element *r-s* of the partial network. If there is no mutual coupling, (3.93) reduces to

$$Z^{abc}_{qi} = Z^{abc}_{pi}$$

$$Z^{abc}_{iq} = Z^{abc}_{ip}$$

$$Z^{abc}_{qq} = Z^{abc}_{pq} + z^{abc}_{pq,pq} \quad ...(3.94)$$

Further, if *p* is the reference bus,

$$Z^{abc}_{qi} = Z^{abc}_{iq} = 0$$

$$Z^{abc}_{qq} = z^{abc}_{pq,pq}$$

If network is balanced, then

$$Z^{abc}_{qi} = Z^{abc}_{iq} \quad ...(3.95)$$

Addition of a link

If the added element is a link between buses *p-q* of the partial network, we obtain

$$Z^{abc}_{li} = Z^{abc}_{pi} - Z^{abc}_{qi} + (y^{abc}_{pq,pq})^{-1}\,\bar{y}^{abc}_{pq,rs}(\bar{Z}^{abc}_{ri} - \bar{Z}^{abc}_{si})$$

$$Z^{abc}_{il} = Z^{abc}_{ip} - Z^{abc}_{iq} + (\bar{Z}^{abc}_{ir} - \bar{Z}^{abc}_{is})\bar{y}^{abc}_{rs,pq}(y^{abc}_{pq,pq})^{-1}$$

$$Z^{abc}_{il} = Z^{abc}_{pl} - Z^{abc}_{ql} + (y^{abc}_{pq,pq})^{-1}\{I + \bar{y}^{abc}_{pq,rs}(\bar{Z}^{abc}_{rl} - \bar{Z}^{abc}_{sl})\} \quad ...(3.96)$$

If there is no mutual coupling,

$$Z_{li}^{abc} = Z_{pi}^{abc} - Z_{qi}^{abc}$$

$$Z_{il}^{abc} = Z_{ip}^{abc} - Z_{iq}^{abc}$$

$$Z_{ll}^{abc} = Z_{pl}^{abc} - Z_{ql}^{abc} + z_{pq,pq}^{abc} \quad ...(3.97)$$

Further, if p is the reference node

$$Z_{li}^{abc} = -Z_{qi}^{abc}$$

$$Z_{il}^{abc} = -Z_{iq}^{abc}$$

$$Z_{ll}^{abc} = -Z_{ql}^{abc} + z_{pq,pq}^{abc} \quad ...(3.98)$$

If the network is balanced

$$Z_{li}^{abc} = Z_{il}^{abc}$$

In (3.96), (3.97) and (3.98) the node l has to be eliminated using

$$Z_{ij}^{abc} \text{ (modified)} = Z_{ij(\text{old})}^{abc} - Z_{il}^{abc} (Z_{ll}^{abc})^{-1} Z_{lj}^{abc} \quad ...(3.99)$$

Example 3.8: A simple three bus system (representing a generator supplying a load, through a connecting line) is shown in Fig. 3.18. The three phase impedance are given below.

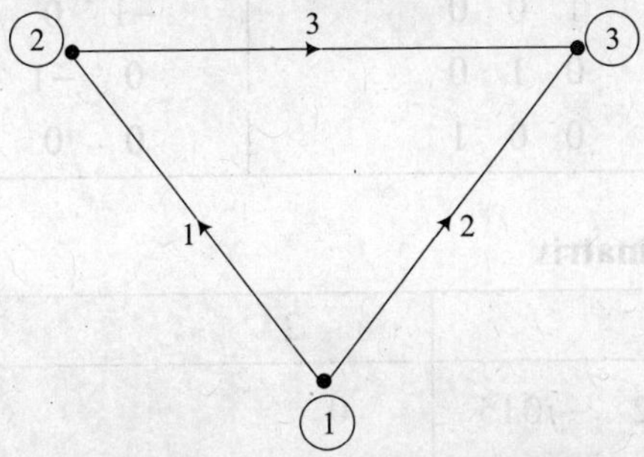

Fig. 3.18 Graph of example 3.8.

Line	*Bus code p-q*	*Impedance $z_{pq,pq}^{abc}$*	*Mutual impedance*
1	1-2	$j\begin{bmatrix} 0.06 & -0.02 & -0.15 \\ -0.15 & 0.06 & -0.02 \\ -0.02 & -0.15 & 0.06 \end{bmatrix}$	.....
2	1-3	$j\begin{bmatrix} 0.06 & -0.02 & -0.15 \\ -0.15 & 0.06 & -0.02 \\ -0.02 & -0.15 & 0.06 \end{bmatrix}$	.....
3	2-3	$j\begin{bmatrix} 1.5 & 0.5 & 0.5 \\ 0.5 & 1.5 & 0.5 \\ 0.5 & 0.5 & 1.5 \end{bmatrix}$	.....

Form

(i) The bus incidence matrix.
(ii) Primitive impedance matrix.
(iii) Primitive admittance matrix.
(iv) Build Z_{bus} using matrix building algorithm.
(v) Obtain Y_{bus} using singular transformation.

Solution: Choose bus 1 as reference.

(i) **The bus incidence matrix** is formed as in single phase networks.

	2	3
1	−1 0 0 0 −1 0 0 0 −1	0 0 0 0 0 0 0 0 0
2	0 0 0 0 0 0 0 0 0	−1 0 0 0 −1 0 0 0 −1
3	1 0 0 0 1 0 0 0 1	−1 0 0 0 −1 0 0 0 −1

(ii) **Primitive impedance matrix**

	1	2	3
1	$j0.06$ $-j0.02$ $-j0.15$ $-j0.15$ $j0.06$ $-j0.02$ $-j0.02$ $-j0.15$ $j0.06$	0	0
2	0	$j0.06$ $-j0.02$ $-j0.15$ $-j0.15$ $j0.06$ $-j0.02$ $-j0.02$ $-j0.15$ $j0.06$	0
2	0	0	$j1.5$ $j0.5$ $j0.5$ $j0.5$ $j1.5$ $j0.5$ $j0.5$ $j0.5$ $j1.5$

(iii) **Primitive admittance matrix $y = z^{-1}$**

	1	2	3
1	$\begin{matrix} j0.1619 & j6.3933 & j2.5357 \\ j2.5357 & j0.1619 & j6.3933 \\ j6.3933 & j2.5357 & j0.1619 \end{matrix}$	0	0
2	0	$\begin{matrix} j0.1619 & j6.3933 & j2.5357 \\ j2.5357 & j0.1619 & j6.3933 \\ j6.3933 & j2.5357 & j0.1619 \end{matrix}$	0
2	0	0	$\begin{matrix} -j0.8 & j0.2 & j0.2 \\ j0.2 & -j0.8 & j0.2 \\ j0.2 & j0.2 & -j0.8 \end{matrix}$

(iv) **Bus impedance matrix**

Step 1. Add element 1, which is a branch from reference bus 1 to bus 2.

$$Z_{bus}^{abc} = \begin{matrix} \\ 2 \\ \\ \end{matrix} \overset{2}{\begin{bmatrix} j0.06 & -j0.02 & -j0.15 \\ -j0.15 & j0.06 & -j0.02 \\ -j0.02 & -j0.15 & j0.06 \end{bmatrix}}$$

Step 2. Add element 2, which is a branch from reference bus 1 to bus 3.

$$Z_{bus}^{abc} = \left[\begin{array}{ccc|ccc} j0.06 & -j0.02 & -j0.15 & & & \\ -j0.15 & j0.06 & -j0.02 & & 0 & \\ -j0.02 & -j0.15 & j0.06 & & & \\ \hline & & & j0.06 & -j0.02 & -j0.15 \\ & 0 & & -j0.15 & j0.06 & -j0.02 \\ & & & -j0.02 & j0.15 & j0.06 \end{array}\right]$$

Step 3. Add element 3, which is link between bus 2 and bus 3.

$$Z_{l2}^{abc} = Z_{22}^{abc} - Z_{32}^{abc} = Z_{2l}^{abc}$$

$$Z_{l3}^{abc} = Z_{23}^{abc} - Z_{33}^{abc} = Z_{3l}^{abc}$$

$$Z_{ll}^{abc} = Z_{2l}^{abc} - Z_{3l}^{abc} + z_{23,23}^{abc}$$

The node l is finally eliminated and the elements are modified using (3.99). Using the above formulae, the final bus impedance matrix is

$$Z_{\text{bus}} = \begin{array}{c} \\ 2 \\ \\ 3 \end{array} \left[\begin{array}{c|c} \begin{bmatrix} j0.0512 & -j0.0318 & -j0.1347 \\ -j0.1347 & j0.0512 & -j0.0318 \\ -j0.0318 & -j0.1347 & j0.0512 \end{bmatrix} & \begin{bmatrix} j0.0088 & j0.0118 & -j0.0153 \\ -j0.0153 & j0.0088 & j0.0118 \\ j0.0118 & -j0.0153 & j0.0088 \end{bmatrix} \\ \hline \begin{bmatrix} j0.0088 & j0.0118 & -j0.0153 \\ -j0.0153 & j0.0088 & j0.0118 \\ j0.0118 & -j0.0153 & j0.0088 \end{bmatrix} & \begin{bmatrix} j0.0512 & -j0.0318 & -j0.1347 \\ -j0.1347 & j0.0512 & -j0.0318 \\ -j0.0318 & -j0.1347 & j0.0512 \end{bmatrix} \end{array}\right]$$

(v) Y_{bus} from singular transformation

$$Y_{\text{bus}} = A^T yA$$

$$\left[\begin{array}{c|c} \begin{bmatrix} -j0.6381 & j6.5933 & j2.7357 \\ j2.7357 & -j0.6381 & j6.5933 \\ j6.5933 & j2.7357 & -j0.6381 \end{bmatrix} & \begin{bmatrix} j0.8 & -j0.2 & -j0.2 \\ -j0.2 & j0.8 & -j0.2 \\ -j0.2 & -j0.2 & j0.8 \end{bmatrix} \\ \hline \begin{bmatrix} j0.8 & -j0.2 & -j0.2 \\ -j0.2 & j0.8 & -j0.2 \\ -j0.2 & -j0.2 & j0.8 \end{bmatrix} & \begin{bmatrix} -j0.6381 & j6.5933 & j2.7357 \\ j2.7357 & -j0.6381 & j6.5933 \\ j6.5933 & j2.7357 & -j0.6381 \end{bmatrix} \end{array}\right]$$

Note: $Z_{bus}^{abc} = [Y_{bus}^{abc}]^{-1}$

Example 3.9: Consider the system of Fig. 3.19. The data is given below. From the bus admittance matrix.

Data for example 3.9

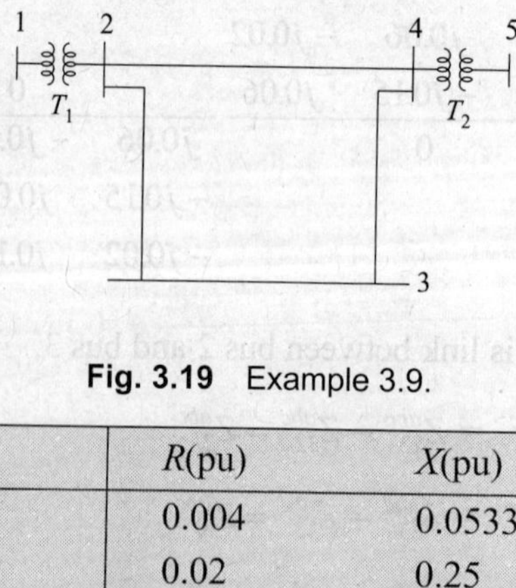

Fig. 3.19 Example 3.9.

Form Bus	To Bus	R(pu)	X(pu)	B(pu)
1	2	0.004	0.0533	0
2	3	0.02	0.25	0.22
2	4	0.01	0.15	0.11
3	4	0.02	0.25	0.22
4	5	0.006	0.08	0.0

Solution: The admittance of each line is calculated and tabulated below. The line charging admittance is divided by 2, to get the half line charging admittance.

Form Bus	To Bus	$Y_{line}^{(pu)} = \frac{1}{R + jk}$	B/2 pu
1	2	1.4 – j018.657	0
2	3	0.318 – j3.975	0.11
2	4	0.492 – j6.637	0.055
3	4	0.318 – j3.975	0.11
4	5	0.932 – j12. 43	0

The bus admittance matrix is formed by method of inspection.

Y_{Bus} =

1.4 – j18.66	–1.4 + j18.66	0	0	0
–1.4 + j18.66	2.16 – j29.10	–0.318 + j3.98	–0.442 + j6.64	0
0	–0.318 + j3.98	0.636 – j7.73	–0.318 + j3.98	0
0	–0.442 + j6.64	–0.318 + j3.98	1.692 – j22.88	–0.932 + j12.43
0	0	0	–0.932 + j12.43	0.932 – j12.43

Example 3.10: Consider the single line diagram shown in Fig. 3.20.

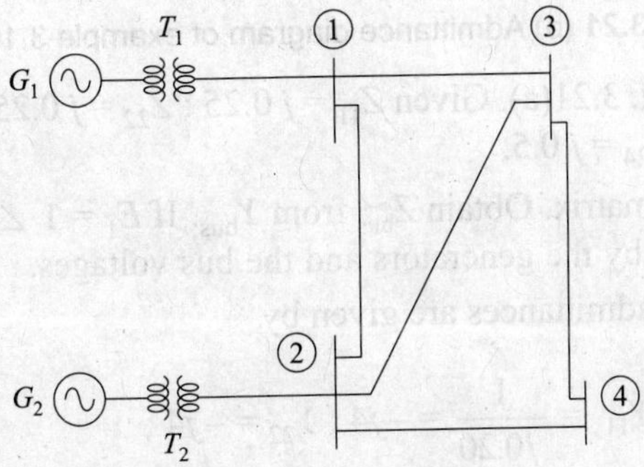

Fig. 3.20 Example 3.10.

Draw the impedance diagram and admittance diagram.

Solution The impedance diagram is shown in Fig. 3.21(a) and the corresponding admittance diagram in Fig. 3.21(b).

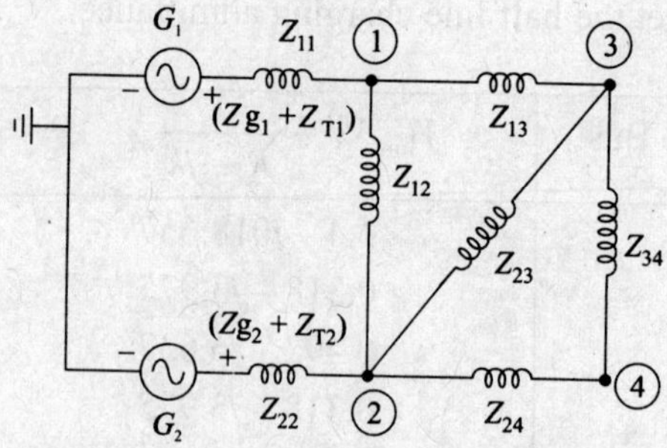

Fig. 3.21(a) Impedance Diagram of example 3.10.

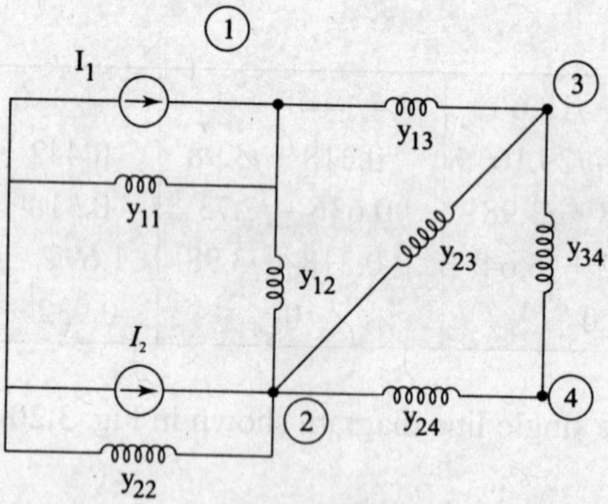

Fig. 3.21 (b) Admittance diagram of example 3.10.

Example 3.11: Consider Fig. 3.21(a). Given $Z_{11} = j\,0.25$; $Z_{22} = j\,0.25$; $Z_{12} = j\,0.2$; $Z_{13} = j\,0.25$; $Z_{23} = j\,0.4$; $Z_{34} = j\,0.4$ and $Z_{24} = j\,0.5$.

Form the bus admittance matrix. Obtain Z_{bus} from Y_{bus}. If $E_1 = 1\ \angle 30°$ pu and $E_2 = 1\angle 0°$ pu calculate the currents injected by the generators and the bus voltages.

Solution: The corresponding admittances are given by.

$$Y_{11} = \frac{1}{j0.20} = -j4\ ;\ Y_{22} = -j4\ ;$$

$$Y_{12} = -j5\ ;\ Y_{13} = -j4\ ;\ Y_{23} = Y_{34} = -j\,2.5$$

$$Y_{24} = -j2$$

We form bus admittance matrix by inspection.

$$Y_{\text{Bus}}(1, 1) = Y_{11} + Y_{12} + Y_{13} = -j13$$

$$Y_{\text{Bus}}(1, 2) = -Y_{12} = j5.$$

and so on.

$$Y_{\text{Bus}} = j\begin{bmatrix} -13 & 5 & 4 & 0 \\ 5 & -13\cdot5 & 2.5 & 2 \\ 4 & 2\cdot5 & -9 & 2\cdot5 \\ 0 & 2 & 2.5 & -4\cdot5 \end{bmatrix} \text{pu}$$

The Z_{bus} is obtained as,

$$Z_{\text{Bus}} = Y_{\text{Bus}}^{-1} = j\begin{bmatrix} 0.1531 & 0.0969 & 0.1264 & 0.1133 \\ 0.0969 & 0.1531 & 0.1236 & 0.1367 \\ 0.1264 & 0.1236 & 0.2565 & 0.1974 \\ 0.1133 & 0.1367 & 0.1974 & 0.3926 \end{bmatrix}$$

If we now consider the admittance diagram of Fig. 3.21(a), we get the currents I_1 and I_2 as,

$$I_1 = \frac{E_1}{Z_{11}} = \frac{1\angle 30°}{j0.25} = 4\angle -60° \text{ pu}$$

$$I_2 = \frac{E_2}{Z_{22}} = \frac{1\angle 0°}{j0.25} = 4\angle -90° \text{ pu}$$

We now solve for bus volatages from

$$[V_{\text{Bus}}] = [Y_{\text{Bus}}]\,[I_{\text{Bus}}] \text{ to give}$$

$$\begin{bmatrix} V_1 \\ V_2 \\ V_3 \\ V_4 \end{bmatrix} = \begin{bmatrix} 0.9677\angle 18.45° \\ 0.9677\angle 11.45° \\ 0.9659\angle 15.18° \\ 0.9662\angle 13.56° \end{bmatrix} \text{pu}$$

Example 3.12: In example 3.11 eliminate the nodes where there are no current injections.

Solutions: The current injection is zero in nodes 3 and 4. Hence, they can be eliminated. The matrices are partitioned as follows.

$$\begin{bmatrix} I_1 \\ I_2 \\ 0 \\ 0 \end{bmatrix} = \left[\begin{array}{c:c} Y_1 & Y_2 \\ \hdashline Y_3 & Y_4 \end{array}\right] = \begin{bmatrix} V_1 \\ V_2 \\ \hdashline V_3 \\ V_4 \end{bmatrix}$$

The partitioned admittance matrix is

$$j\left[\begin{array}{cc:cc} -1.3 & 5 & 4 & 0 \\ 5 & -13.5 & 2.5 & 2 \\ \hdashline 4 & 2.5 & -9 & 2.5 \\ 0 & 2 & 2.5 & -4.5 \end{array}\right]$$

$$Y_{\text{Red}} = Y_1 - Y_2\, Y_4^{-1}\, Y_3$$

$$= j\begin{bmatrix} -1.3 & 5 \\ 5 & -13.5 \end{bmatrix} - j\begin{bmatrix} 4 & 0 \\ 2.5 & 2 \end{bmatrix} j\begin{bmatrix} -9 & 2.5 \\ 2.5 & -4.5 \end{bmatrix}^{-1} j\begin{bmatrix} 4 & 2.5 \\ 0 & 2 \end{bmatrix}$$

$$= j\begin{bmatrix} -10.8978 & j6.8978 \\ j6.8978 & -j10.8978 \end{bmatrix} \text{pu.}$$

Example 3.13: Build the bus impedance matrix for the network of Fig. 3.20.

Solution: We write the data of the network as below.

S.No	From bus	To bus	Z
1	0	1	$j\,0.25$
2	0	2	$j\,0.25$
3	1	2	$j\,0.2$
4	1	3	$j\,0.25$
5	2	3	$j\,0.4$
6	2	4	$j\,0.5$
7	3	4	$j\,0.4$

Step 1. Add element 1. It is a branch to the reference node

$$Z_{\text{bus1}} = [\,j\,0.25]$$

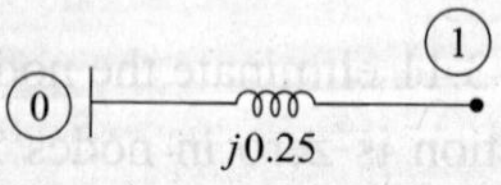

Step 2. Add element 2. This is a branch between bus 2 and reference. So Z_{bus} size increases by one.

$$Z_{\text{bus, 2}} = \begin{bmatrix} j0.25 & 0 \\ 0 & j0.25 \end{bmatrix}$$

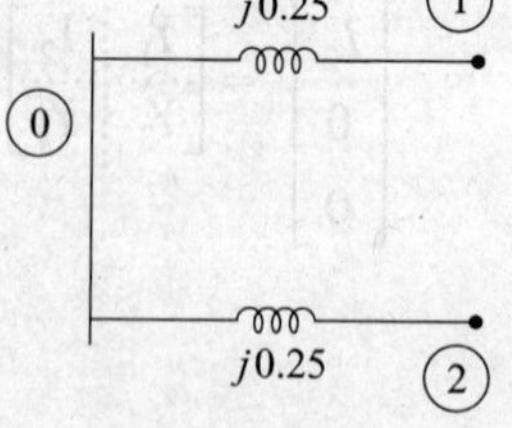

Step 3. Add element three which is a link between buses 1 and 2, we get

$$Z_{bus, 3l} = \begin{bmatrix} j0.25 & 0 & j0.25 \\ 0 & j0.25 & -j0.25 \\ j0.25 & -j0.25 & (j0.25-(-j0.25)+j0.2 = j0.7 \end{bmatrix}$$

We now eliminate the last column and last row using Kron's reduction.

$$Z_{bus, 3} = \begin{bmatrix} j0.1607 & j0.893 \\ j0.0893 & j0.1607 \end{bmatrix} =$$

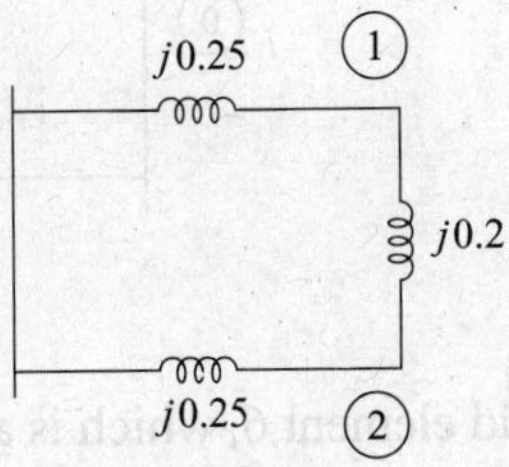

Step 4. Add element 4, which is a branch from bus 1 to bus 3 (new bus), we get

$$Z_{bus, 4} = \begin{bmatrix} j0.1607 & j0.0893 & j0.1607 \\ j0.0893 & j0.1607 & j0.0893 \\ j0.1607 & j0.0893 & (j0.1607 + j0.25 = j0.4107) \end{bmatrix}$$

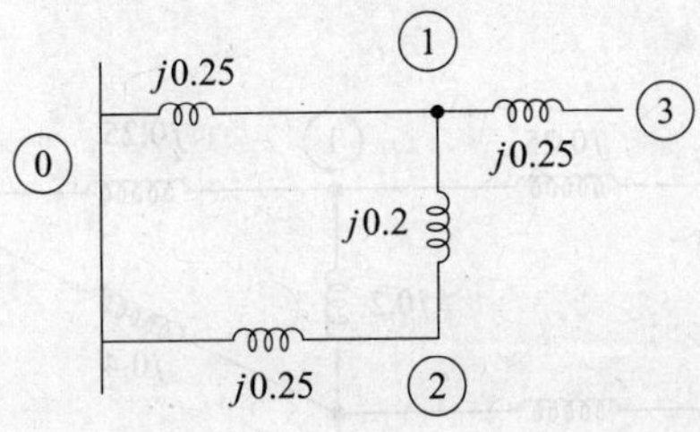

Step 5. Add element 5, which is a link between buses 2 and 3, we get

$$Z_{bus, 4l} = \begin{bmatrix} j0.1607 & j0.0893 & j0.1607 & -j0.0714 \\ j0.0893 & j0.1607 & j0.0893 & j0.0714 \\ j0.1607 & j0.0893 & j0.4107 & -j0.3214 \\ -j0.0714 & j0.0714 & -j0.3214 & (j0.0714-(-j.3214)+j0.4) = j0.7928 \end{bmatrix}$$

Reducing the last row and column we get,

$$Z_{\text{bus, 5}} = \begin{bmatrix} j0.1543 & j0.0957 & j01318 \\ j0.0957 & j0.1543 & j0.1182 \\ j0.1318 & j0.1182 & j0.2804 \end{bmatrix}$$

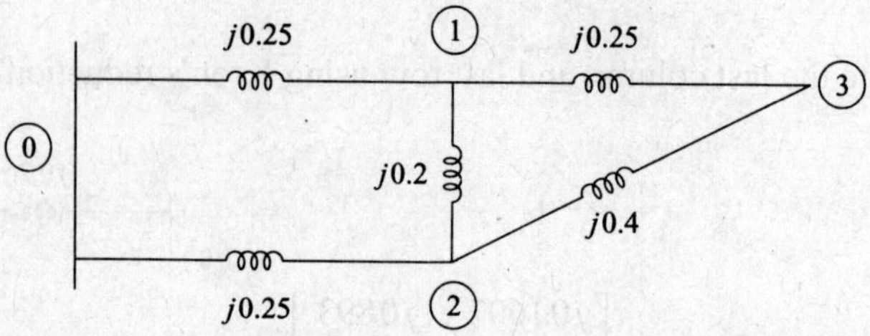

Step 6. Add element 6, which is a branch between bus 2 and bus 4 (new)

$$Z_{\text{bus, 6}} = \begin{bmatrix} j0.1543 & j0.0957 & j0.1318 & j0.0957 \\ j0.0957 & j0.1543 & j0.1182 & j0.1543 \\ j0.1318 & j0.1182 & j0.2804 & j0.1182 \\ j0.0957 & j0.1543 & j0.1182 & (j0.1543 + j0.5 = j0.6543) \end{bmatrix}$$

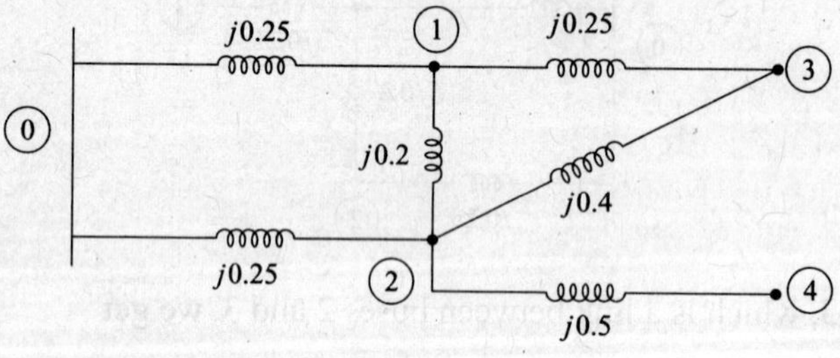

Step 7. Add link between buses 3 and 4.

$$Z_{\text{bus, 5}l} = \begin{bmatrix} j0.1543 & j0.0957 & j0.1318 & j0.0957 & j0.036 \\ j0.0957 & j0.1543 & j0.1182 & j0.1543 & -j0.0360 \\ j0.1318 & j0.1182 & j0.2804 & j0.1182 & j0.1622 \\ j0.0957 & j0.1543 & j0.1182 & j0.6543 & j0.5360 \\ j0.0360 & -j.0360 & j0.1622 & j0.5360 & j1.0982 \end{bmatrix}$$

Reducing we get

$$Z_{\text{bus, 5}} = \begin{bmatrix} j0.1531 & j0.0969 & j0.1264 & j0.1133 \\ j0.0969 & j0.1531 & j0.1236 & j0.1367 \\ j0.1264 & j0.1236 & j0.2565 & j0.1974 \\ j0.1133 & j0.1367 & j0.1974 & j0.3926 \end{bmatrix}$$

Example 3.14: Form the bus impedance matrix of the system shown in Fig. 3.22.

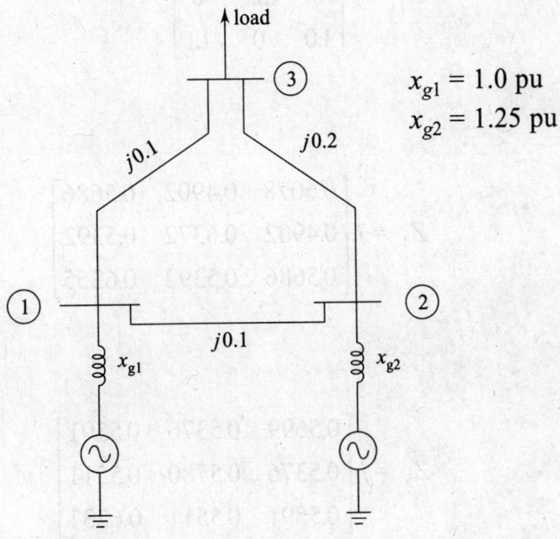

Fig. 3.22 Example 3.14.

Solution: The network equivalent is shown below.

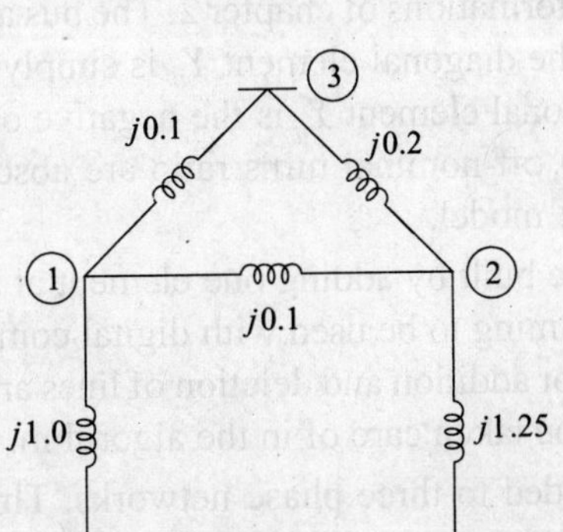

Fig. 3.23 Solution Example 3.14.

Step 1. Add element 1 – 0

$$Z_1 = [j1.0]$$

Step 2. Add element 2 – 0

$$Z_2 = j\begin{bmatrix} 1.0 & 0 \\ 0 & 1.25 \end{bmatrix}$$

Step 3. Add element 1 – 3. It is a branch.

$$Z_3 = j\begin{bmatrix} 1.0 & 0 & 1.0 \\ 0 & 1.25 & 0 \\ 1.0 & 0 & 1.1 \end{bmatrix}$$

Step 4.
Add link 2 – 3

$$Z_4 = j\begin{bmatrix} 0.6078 & 0.4902 & 0.5686 \\ 0.4902 & 0.6372 & 0.5392 \\ 0.5686 & 0.5392 & 0.6255 \end{bmatrix}$$

Step 5.
Add link 1 – 2

$$Z_5 = j\begin{bmatrix} 0.5699 & 0.5376 & 0.5591 \\ 0.5376 & 0.5780 & 0.5511 \\ 0.5591 & 0.5511 & 0.6231 \end{bmatrix}$$

3.6 SUMMARY

The bus impedance and bus admittance matrices can be built by using specific algorithms, instead of the singular and non-singular transformations of chapter 2. The bus admittance matrix is built by inspection of the interconnections. The diagonal element Y_{ii} is simply the sum of all admittances incident on the bus i and the off diagonal element Y_{ij} is the negative of the admittance of the line between buses i-j. Transformers with off-nominal turns ratio are absorbed into the bus – admittance matrix by using an equivalent π model.

The bus impedance matrix can be built by adding one element at a time to a partial network. The algorithm is suitable for programming to be used with digital computer applications. Modifications in the bus impedance matrix for addition and deletion of lines are easily carried out. Mutual coupling between elements can also be taken care of in the algorithm.

The same concepts can be extended to three phase networks. Three phase elements exhibit different types of symmetry. Rotating elements exhibit rotational symmetry and transmission network exhibits both rotational and reflection symmetries. Three phase balanced networks can be solved on a single phase basis under application of balanced three phase excitation. However, under unbalanced excitation we cannot solve the network on a single phase basis. Suitable transformations like symmetrical components and Clarke's components are possible, which transform the three phase quantities to sequence quantities to yield decoupled equations in case of unbalanced excitation.

3.7 SUPPLEMENTARY PROBLEMS

(1) In a 12 bus system, a transformer with an impedance of $0.05 + j0.4$ pu and an off-nominal ratio of 0.978 is connected between buses 8 and 12. Which elements of the bus admittance matrix are modified and how are they modified?

(2) Form the bus admittance matrix of the system shown in Fig. P.3.2.
Check your answer by obtaining the admittance matrix by using singular transformation.

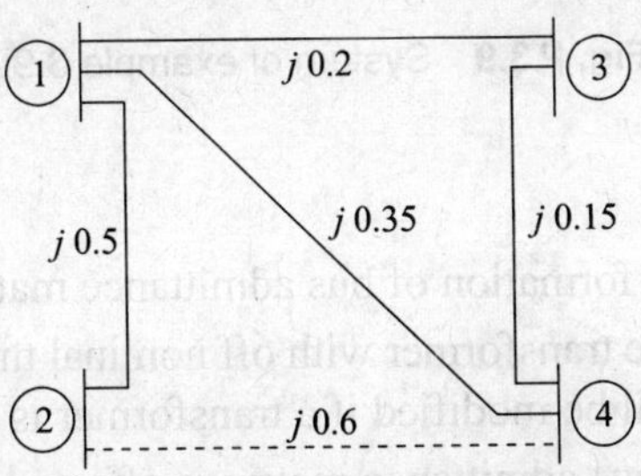

Fig. P.3.2 System for problem 3.2.

(3) In problem 2, if a new line is added between buses 2 and 4, what is the new admittance matrix?

(4) Obtain the admittance matrix of a 5 bus system with the data given in Table P3.4.

Table P3.4 Line data for problem 3.4

Starting bus	*Ending bus*	*R(pu)*	*X(pu)*	*Yc/2(pu)*
1	4	0.03	0.25	0.0065
1	5	0.026	0.108	0.0045
1	6	0.05	0.218	0.003
2	4	0.0	0.08	-
3	5	0.01	0.08	-
4	6	0.035	0.809	0.0073
5	6	0.028	0.903	0.008

(5) Modify the bus admittance matrix of problem 3.4 to include an identical parallel line between 4-6.

(6) Modify the bus admittance matrix of problem 3.4 if the line 3-5 were a transformer, with an off-nominal ratio of 1.012.

(7) In example 3.7, modify the bus impedance matrix if a new line is added between buses 1 and 3 with an impedance of $j0.5$ pu.

(8) Repeat problem 3.7 if the added element is mutually coupled to line 1-2 and the mutual impedance is $j0.12$ pu.

(9) Form the bus impedance matrix of the system shown in Fig. P.3.9. All impedances are in pu.

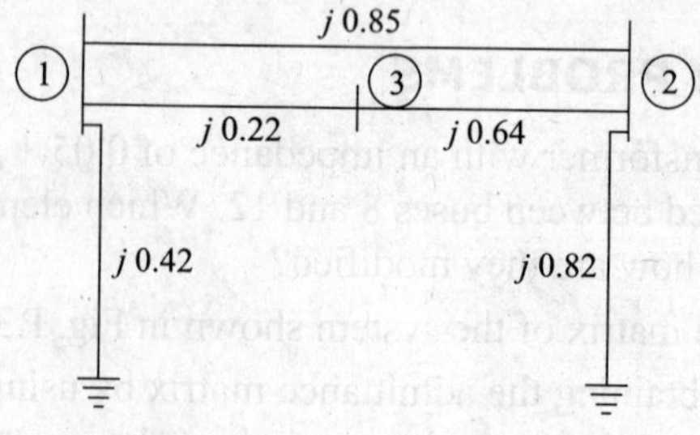

Fig. P.3.9 System of example 3.9.

REVIEW QUESTIONS

1. Explain the algorithm for formation of bus admittance matrix by method of inspection.
2. Derive the π-model for the transformer with off nominal turns ratio. Which elements of the bus admittance matrix will be modified if a transformer is connected between buses *p-q*?
3. Clearly explain how the bus admittance matrix is affected, if a new line is added between buses *i-j*?
4. If there is no branch between the ground and any of the buses in the system, then Z_{bus} building algorithm cannot be used. Comment. How would you form Z_{bus} in this case?
5. If [*z*] is the bus impedance matrix of a partial network, how will it be modified if a branch is added between buses *p-q* , if *i*) *p* is a node in the partial network and *q* is a new node ii) *p* is the reference node and *q* is the new node.
6. If [*z*] is the bus impedance matrix of a partial network, how will it be modified if a link is added between buses *p-q* , if *i*) *p* and *q* are nodes in the partial network ii) *p* is the reference node and *q* is a node in the partial network.
7. How can the Y_{bus} be modified if a line is removed, when there is no mutual coupling between the elements?
8. Explain reflection and rotation symmetry with respect to three-phase networks.
9. Prove that a balanced three-phase stationary element can be treated as a single phase element when supplied with balanced excitation.
10. If *T* is a transformation to transform the '*abc*' components, then what is the condition on *T*, such that the transformation is power invariant?
11. What is Clarke's transformation? How is it useful?
12. Explain with an example a transformation which is power variant.
13. Explain with an example a transformation which is power invariant.
14. With an example show how transformations can lead to decoupled equations.

Chapter 4

Symmetrical Faults

4.1 INTRODUCTION

Three phase power systems are designed to be symmetric and balanced. Sometimes, asymmetry can occur in currents and voltages due to unbalanced loading conditions. At higher voltage levels, the asymmetry is caused mainly due to system faults. A fault in a circuit is any event, which causes a deviation from the normal flow of currents. The two main types of faults are shunt faults, where low impedance is connected between the line and ground, and series faults, where high impedance is connected in series with the line. The most common series fault is the open circuit. This occurs when a circuit breaker is opened or a line is opened. Common shunt faults are short circuits, which are further classified as

- Balanced three phase faults
- Single line to ground faults
- Line to line faults
- Double line to ground faults.

About 75% of the faults are of single line to ground type, resulting mainly due to insulator flash over. When a lightning strikes a transmission line, the high voltage between a conductor and the grounded supporting tower causes ionization, which provides a path to ground for the charge induced by the lightning stroke. The resultant low impedance to ground allows the flow of current from one conductor to ground and through the ground to the grounded neutral of a transformer or a generator.

Protective relays are employed to trip the circuit breaker under fault conditions. The proper selection of circuit breakers depends on the current flowing immediately after the fault occurs and the current, which the breaker must interrupt. Fault analysis provides values of these currents for different type of faults at various locations in the system. The data helps to determine the settings of relays, which control the circuit breaker.

A balanced three phase fault or symmetrical fault is defined as the simultaneous short circuit across all three phases. Though infrequent, (around 5% of faults), it is the most severe type of fault. Since, we are considering a balanced network, it is solved on a single phase basis. If the fault impedance is zero, it is called a bolted fault or a solid fault. Symmetrical faults are discussed in detail in the following sections.

4.2 TRANSIENTS IN AN R-L CIRCUIT

The choice of the circuit breaker depends not only on the current, the breaker has to carry under normal conditions, but also on the maximum current it may have to carry momentarily and the current it may have to interrupt at the voltage of the line in which it is connected.

To calculate the initial current when a short circuit occurs it would be appropriate to study the transients in an *RL* circuit, to which an *AC* voltage is suddenly applied as shown in Fig. 4.1a. This situation is similar to a short circuited generator shown in Fig. 4.1b.

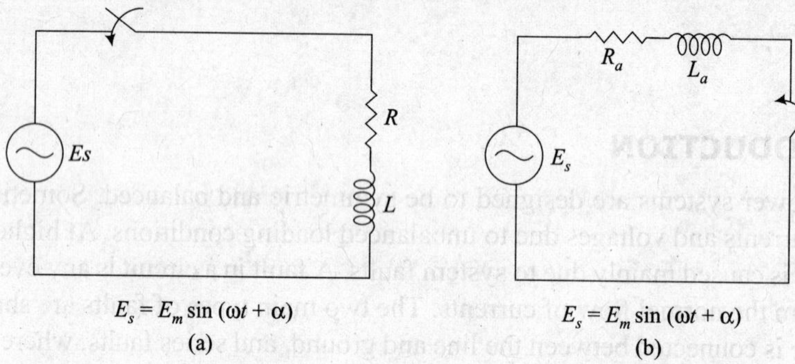

Fig. 4.1a RL circuit. **Fig. 4.1b** Short circuited generator.

Let α be the angle of the voltage wave at the instant $t = 0$, when the fault is assumed to take place, as shown in Fig. 4.2.

The instantaneous voltage equation for Fig. 4.1a can be written as

$$L\frac{di}{dt} + Ri = E_m \sin(\omega t + \alpha) \qquad ...(4.1)$$

The solution of this equation is given by

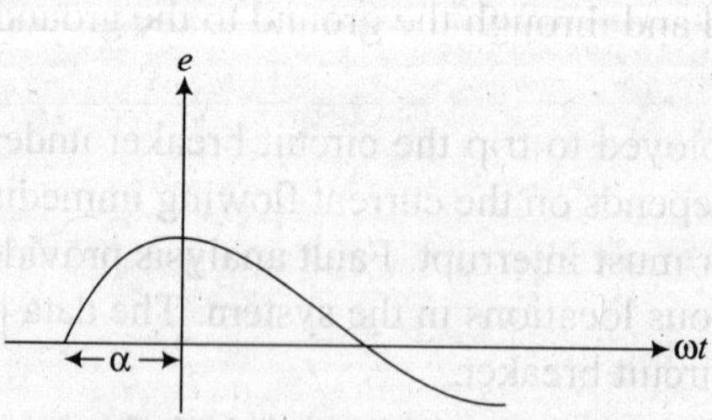

Fig 4.2 Angle α at instant of fault.

$$i(t) = \frac{E_m}{|Z|}\left[\sin(\omega t + \alpha - \theta) - \sin(\alpha - \theta)e^{\frac{-Rt}{L}}\right] \quad ...(4.2)$$

where, $$Z = |Z| \angle \theta = R + j\omega L$$

The current $i(t)$ has two components:

(i) $\frac{E_m}{|Z|}[\sin(\omega t + \alpha - \theta)]$ which is the steady state response and represents a periodic sinusoidal current with a phase difference θ with respect to the voltage wave.

(ii) $-\frac{E_m}{|Z|}\left[\sin(\alpha - \theta)e^{\frac{-Rt}{L}}\right]$ which is the transient component, non-periodic and exponentially decaying with a time constant $\frac{L}{R}$ seconds. This component is called the *DC* component of the current. It depends on the value of 'α' at the instant of closing the switch and is zero if $\alpha - \theta = 0$ or π. The current is maximum when $(\alpha - \theta) = -\frac{\pi}{2}$. These waveforms are shown in Fig. 4.3a and Fig. 4.3b. For any other value of 'α', the *DC* component may have any magnitude between 0 to $\frac{E_m}{|Z|}$.

Before the switch is closed in Fig. 4.1a, the current is zero. After switching, the current cannot change instantaneously (because of the inductor). Hence, at the instant of closing the switch, the *DC* component and the steady state component must have the same magnitude but opposite signs, so that $i(t) = 0$. The presence of the *DC* component causes asymmetry in the initial response, as seen from Fig. 4.3b. In a three phase system, the value of 'α' is different in all the three phases because of the 120° phase shift between the voltages. Hence, the transient *DC* component is differ-

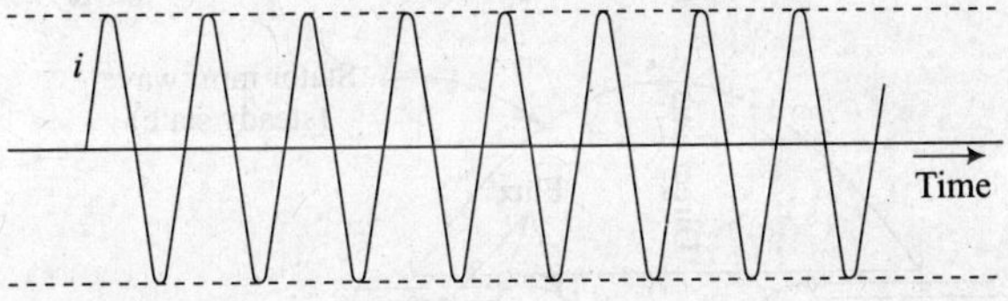

Fig. 4.3a Current in *RL* circuit when $\alpha - \theta = 0$.

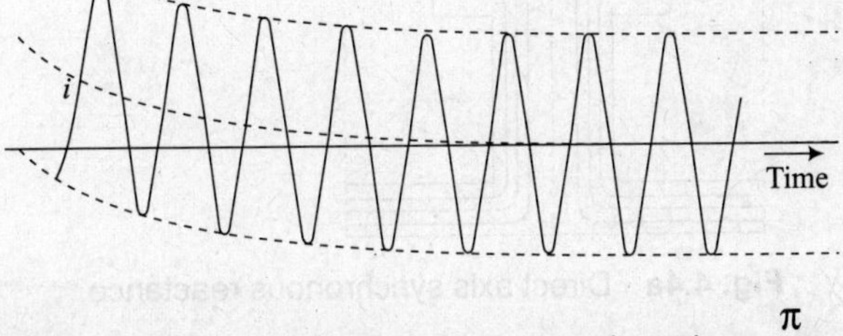

Fig. 4.3b Current in *RL* circuit when $(\alpha - \theta) = -\frac{\pi}{2}$.

ent in the three phases. A generator which is suddenly short circuited is similar to an *RL* circuit suddenly closed to an *AC* source. However, in a synchronous machine the flux across the air gap is not the same at the instant of short circuit, as it is a few cycles later. The change of flux is due to combined action of the field, the armature and the damper windings. Therefore, the synchronous machine is characterized by different reactances after the fault.

4.3 SYNCHRONOUS MACHINE REACTANCES

The reactance of the synchronous machine depends on the angular position of the rotor and whether the armature currents are applied suddenly or are steady. The impedances of the synchronous machine are classified into positive sequence, negative sequence and zero sequence impedances, according to theory of symmetrical components (Chapter 5). In the determination of these, the rotor circuits are closed but not excited, the rotor is turned forward at synchronous speed, current of the proper sequence is applied to the armature windings, and the armature voltage of the same phase sequence as the current is found. The ratio of the voltage to the current is the sequence impedance. For most purposes, the resistance can be neglected. While any machine has only one zero sequence and negative sequence reactance, it has several positive sequence reactances.

4.3.1 Direct Axis Synchronous Reactance x_d

The application of steady three-phase armature currents in a polyphase armature winding sets up a rotating magnetic field in the air gap. The space fundamental mmf and flux wave rotate forward at synchronous speed with respect to the armature and are, therefore, stationary with respect to the field. Given armature currents produce the same fundamental mmf wave regardless of the angular position of the rotor; but the fundamental flux wave varies greatly with the rotor position.

If the rotor is so rotated that the direct axis (pole axis) coincides with the crest of the mmf wave, as shown in 4.4a, a path of high permeance is offered and the fundamental flux wave has the greatest possible magnitude for the given armature current. Hence, the total flux linkage of each phase of the armature winding has the greatest possible value for the given current.

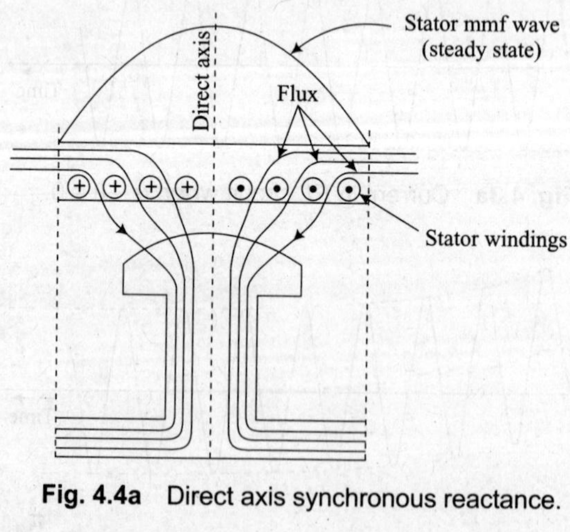

Fig. 4.4a Direct axis synchronous reactance.

Therefore, the armature inductance and reactance are greater than for any other position of the rotor. The flux linkage of an armature phase per ampere of armature current under these conditions is defined as the direct axis synchronous inductance L_d. The direct axis reactance is given by $x_d = \omega L_d$. Note that the flux is stationary with respect to the field and rotating at synchronous speed with respect to the armature. Hence, the flux linkage with each armature phase varies sinusoidally with time and induces a voltage in quadrature with the flux. The ratio of this voltage to current also gives the direct axis reactance x_d.

4.3.2 Direct Axis Transient Reactance x_d'

The conditions used in defining direct axis transient reactance are the same as that used in defining direct axis synchronous reactance. However, the armature currents are now suddenly applied and the positive sequence armature voltage is measured immediately after application of the current, rather than after the voltage has reached steady state value. Abrupt application of the armature currents causes the sudden appearance of an m.m.f opposite each field pole, tending to establish flux through the pole core. Such flux would link the field winding and is opposed by induced field current tending to maintain the flux linkage of the field winding at its initial zero value. According to constant flux linkage theorem which states that, *"the flux linkage of any closed circuit of finite resistances and sources cannot change instantaneously"*, the flux linking with the field winding is still zero (the value prior to application of sudden armature currents). Therefore, the only flux that can be established immediately is that which does not link the field winding but passes through low-permeance leakage paths, largely in air as shown in Fig. 4.4b.

The flux linkage per ampere of armature current is small and is defined as the direct axis transient inductance L_d'. The direct axis transient reactance $x_d' = \omega L_d'$ and is always less than x_d.

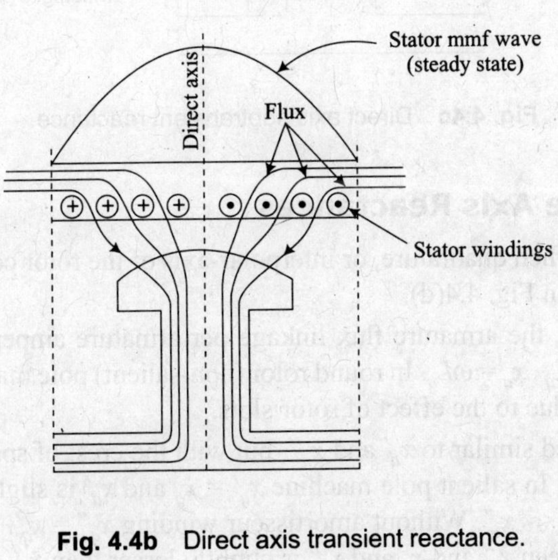

Fig. 4.4b Direct axis transient reactance.

4.3.3 Direct Axis Subtransient Reactance x''_d

In defining x'_d, the only rotor circuit considered was the main field winding. However, there may be additional rotor circuits like damper windings located on the field collar as shown in Fig. 4.4c.

The flux linkage through every rotor circuit tends to remain constant at zero value. Since, these additional rotor circuits are situated nearer the air gap than the field winding, the flux set up by the sudden application of armature currents, is initially forced into leakage paths of smaller cross-section area and lower permeance than with only field winding. The armature flux linkage per ampere under these conditions is the direct axis subtransient inductance L''_d. The direct axis subtransient reactance $x''_d = \omega L''_d$. This is always smaller than x'_d. Thus, on a short circuit the synchronous machine reactance changes from x''_d to x'_d to x_d.

The quadrature-axis steady state reactance x_q, transient reactance x'_q and subtransient reactance x''_q are defined under similar conditions, with the rotor so rotated that the inter polar axis coincides with the crest of the m.m.f wave.

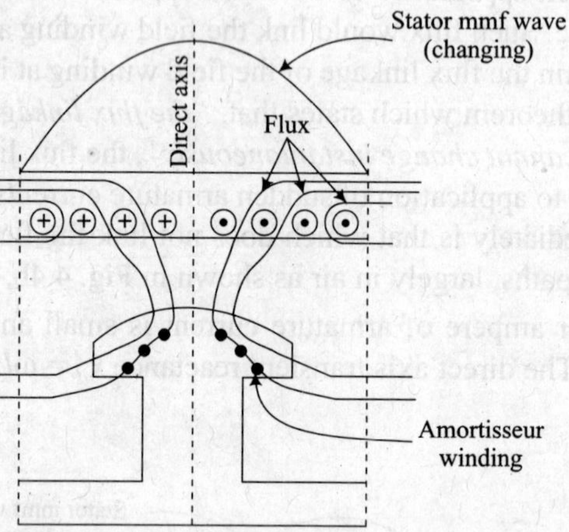

Fig. 4.4c Direct axis subtransient reactance.

4.3.4 Quadrature Axis Reactances

The flux is the smallest when quadrature, or interpolar-axis of the rotor coincides with the crest of the mmf wave as shown in Fig. 4.4(d).

Under this condition, the armature flux linkage per armature ampere is the quadrature axis synchronous inductance $L_q \cdot x_q = \omega L_q$. In round rotor (non-salient) pole machines x_q is almost equal to x_d. It is slightly lesser due to the effect of rotor slots.

x_q' and xq'' are defined similar to x_d' and x_d'', but with the crest of space-fundamental mmf in line with quadrature axis. In salient pole machine $x_q' = x_q$ and x_d' is slightly lesser than x_q'. With amortisseur windings $x_q'' << x_q'$. Without amortisseur winding $x_q'' = x_q'$. In round rotor machines the value of x_q' is in between x_d' and x_q and x_q'' is slightly lesser than x_q'.

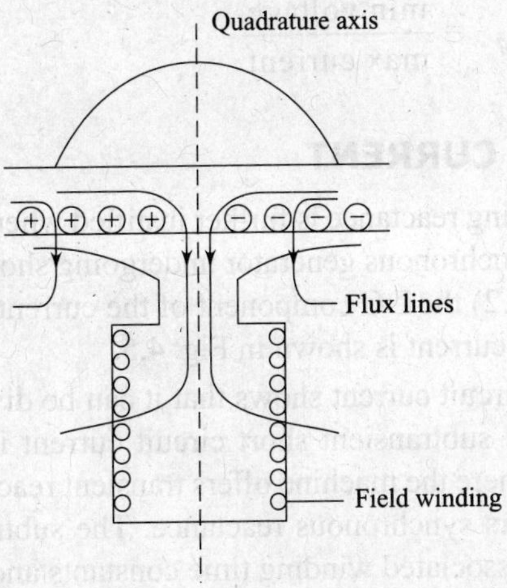

Fig. 4.4(d) Quadrature axis inductance.

4.3.5 Negative Sequence Reactance x_2

This is defined when the rotor is rotated at synchronous speed and all rotor circuits are closed, with negative-sequence currents (sequence reverse to normal) applied to the stator windings. The negative sequence currents produce an mmf rotating in reverse direction at synchronous speed with respect to the rotor. Therefore, viewed from the rotor, the mmf alternates at double the rated frequency. Hence, currents of twice the rated frequency are induced in all rotor circuits. The armature flux linkage per armature ampere under these conditions is called the negative sequence inductance L_2. The negative sequence reactance $x_2 = \omega L_2$

$$x_2 = \frac{xd'' + xq''}{2}$$

4.3.6 Zero Sequence Reactance x_o

Zero sequence currents are all in phase. So when zero sequence currents are applied to the armature, no space-fundamental mmf is produced. Very little flux is set up. Hence, x_o is the lowest reactance and even lesser than x_d''. The value depends on the pitch of the armature coils.

4.3.7 Measurement of x_d and x_q

These can be measured by running the rotor at a speed slightly less than the synchronous speed. The armature current would be modulated and vary between a maximum and minimum value. The current has a maximum amplitude when the quadrature axis is in line with the mmf wave and minimum amplitude when direct axis is in line with the mmf wave. The armature voltage is also modulated at slip frequency.

Then, $$x_d = \frac{\text{max voltage}}{\text{min crurrent}}$$

and $$x_q = \frac{\text{min voltage}}{\text{max current}}$$

4.4 SHORT CIRCUIT CURRENT

The previous concept of changing reactance is further justified when the oscillogram of the current in one phase of an unloaded synchronous generator undergoing short circuit is studied. By considering the case $\alpha - \theta = 0$ (Eq. 4.2) the DC component of the current is eliminated. The typical AC component of the short circuit current is shown in Fig. 4.5.

Observation of the short circuit current shows that it can be divided into three periods–initial subtransient period, where the subtransient short circuit current is manifest, the second period which is the transient period where the machine offers transient reactance and the final steady state period when the machine offers synchronous reactance. The subtransient and transient currents decay in accordance with the associated winding time constants and the current ultimately decays to the sustained steady state value I_s. The time constant of the damper winding is much less than that of the field winding. Hence, the subtransient period component dies down quickly. If we subtract the steady state component I_s from the AC wave, the remaining component consists of two components:

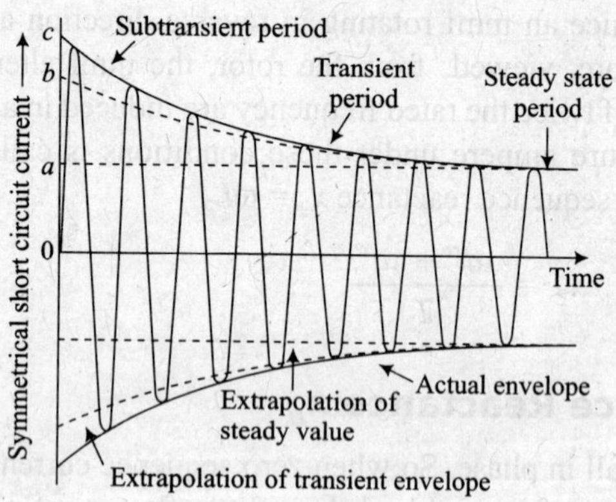

Fig. 4.5 Symmetrical short circuit current.

(i) The subtransient component decaying with a time constant T_d''.

(ii) A transient component decaying with a time constant T_d'.

The amplitude of the AC component I_{ac} is given by

$$I_{ac} = I_s + I_0' e^{-t/T_d'} + I_0'' e^{-t/T_d''} \quad \text{...(4.3)}$$

Referring to Fig. 4.5, the rms value of the steady state component is given by

$$I_s = \frac{E_g}{x_d} = \frac{oa}{\sqrt{2}} \quad \text{...(4.4a)}$$

The rms value of the transient current is

$$I' = I'_0 + I_s = \frac{E_g}{x'_d} = \frac{ob}{\sqrt{2}} \qquad \text{...(4.4b)}$$

The rms value of the initial subtransient current

$$I'' = I''_0 + I'_0 + I_s = \frac{E_g}{x''_d} = \frac{oc}{\sqrt{2}} \qquad \text{...(4.4c)}$$

From (4.4a) and (4.4b), we get

$$I'_0 = \frac{E_g}{x'_d} - \frac{E_g}{x_d} \qquad \text{...(4.5a)}$$

From (4.4b) and (4.4c), we get

$$I''_0 = \frac{E_g}{x''_d} - \frac{E_g}{x'_d} \qquad \text{...(4.5b)}$$

Substituting (4.5a) and (4.5b) in (4.3) the total AC component can be written as

$$I_{ac} = E_g\left[\frac{1}{x_d} + \left(\frac{1}{x'_d} - \frac{1}{x_d}\right)e^{-t/T'_d} + \left(\frac{1}{x''_d} - \frac{1}{x'_d}\right)e^{-t/T''_d}\right] \qquad \text{...(4.6)}$$

The DC components of the three armature currents will all be different and will decay to zero with a time constant T_a, the armature winding time constant. This typically has a value between 0.005 to 0.15 *s*. The initial values of these components, depend on the point in the voltage cycle at which the short circuit occurs. The maximum value this DC component can have is the maximum value of the initial AC subtransient current.

$$I_{dc}(\max) = \sqrt{2}\frac{E_g}{x''_d} \qquad \text{...(4.7)}$$

Most of the DC component decays during the subtransient period. The effective (rms) value of the short circuit current is given by

$$I_{sc} = \sqrt{I_{dc}^2 + I_{ac}^2} \qquad \text{...(4.8)}$$

The maximum value this can attain at the beginning of a short circuit is

$$I_{sc}(\max) = \sqrt{I''^2 + I_{dc}^2} = \sqrt{\left(\frac{E_g}{x''_d}\right)^2 + \left(\frac{\sqrt{2}E_g}{x''_d}\right)^2} = \sqrt{3}\frac{E_g}{x''_d} = \sqrt{3}I'' \qquad \text{...(4.9)}$$

This is the maximum rms value of the asymmetrical short circuit current (including AC and DC components).

Example 4.1: A 50 MVA, Y connected, 13.2 kV synchronous generator is connected to a 13.2/132 kV, 50 MVA, $\Delta - Y$ transformer. The generator reactances are: $x''_d = 0.12$ pu; $x'_d = 0.25$ pu; $x_d = 1.4$ pu; on a 50 MVA base, while the transformer reactance is 0.1 pu on the same base. The machine is unloaded when a symmetrical fault occurs at the *HT* terminals of the transformer. Find

(i) The subtransient, transient and steady state symmetrical fault currents in pu and actual values. $E_g = 1.0$ pu

(ii) What is the maximum possible value of the DC component?

(iii) What is the maximum value of the instantaneous current?

(iv) What is the maximum rms value of the asymmetrical current?

Solution: [Refer Fig. 4.6]

(i) Base current on LV side of transformer $= \dfrac{50 \times 10^6}{\sqrt{3} \times 13.2 \times 10^3} = 2{,}186.93$ A

Base current on HV side of transformer $= \dfrac{50 \times 10^6}{\sqrt{3} \times 132 \times 10^3} = 218.69$ A

Symmetrical subtransient current (rms value)

$$I'' = \frac{E_g}{x_{d''} + x_{tr}} = \frac{1.0}{j0.12 + j0.1} = 4.545 \angle -90° \text{ pu}$$

Actual subtransient current on HV side = 4.545 × 218.69 = 993.95 A

Actual subtransient current on LV side = 4.545 × 2,186.93 = 9939.5 A

Symmetrical transient current

$$= I' = \frac{E_g}{x_{d'} + x_{tr}} = \frac{1.0}{j0.25 + j0.1} = 2.857 \angle -90° \text{ pu}$$

Symmetrical steady state current

$$I_s = \frac{E_g}{x_d + x_{tr}} = \frac{1.0}{j1.4 + j0.1} = 0.666 \angle -90° \text{ pu}$$

The actual currents are found by multiplying the pu values with the respective base currents.

Fig. 4.6 Example 4.1.

(ii) Maximum DC component $= \sqrt{2}I'' = \sqrt{2} \times 4.545 = 6.427$ pu

(iii) Maximum instantaneous current = maximum value of subtransient current + DC component

$$= \sqrt{2}I'' + I_{dc} = 12.854 \text{ pu}$$

(Note that I'' is the rms value of the subtransient current)

(iv) Maximum rms value of asymmetrical AC current

$$= \sqrt{3}I'' = \sqrt{3} \times 4.545 = 7.872 \text{ pu}$$

Example 4.2: Two generators are connected in parallel to the LV side of a three phase $\Delta - Y$ transformer as shown in Fig. 4.7a. Generator 1 is rated 50 MVA, 13.8 kV and Generator 2 is rated 25 MVA, 13.8 kV. Each generator has a subtransient reactance of 25%. The transformer is rated 75 MVA, 13.8 kV (Δ) –132 kV (Y) with a reactance of 10%. Before the fault, the voltage on the HV side of the transformer is 130 kV. The transformer is unloaded when a three phase short circuit occurs on the HV side of the transformer. Find the subtransient currents in the generators.

Solution: Choose a base of 75 MVA, 132 kV on HV side of transformer. The base voltage on LV side is 13.8 kV.

Voltage on HV side = 130 kV

$$\therefore \text{ Voltage on LV side} = 130 \times \frac{13.8}{132} = 13.59 \text{ kV}$$

The generator reactances are converted to the common base, since the reactances specified are on the generator base.

G_1:

$$x''_d = 0.25 \times \frac{75}{50} = 0.375 \text{ pu}$$

$$E_{g1} = \frac{13.59}{13.8} = 0.9848 \text{ pu}$$

G_2:

$$x''_d = 0.25 \times \frac{75}{25} = 0.75 \text{ pu}$$

$$E_{g2} = \frac{13.59}{13.8} = 0.9848 \text{ pu}$$

Transformer : $x_{tr} = 0.1$ pu

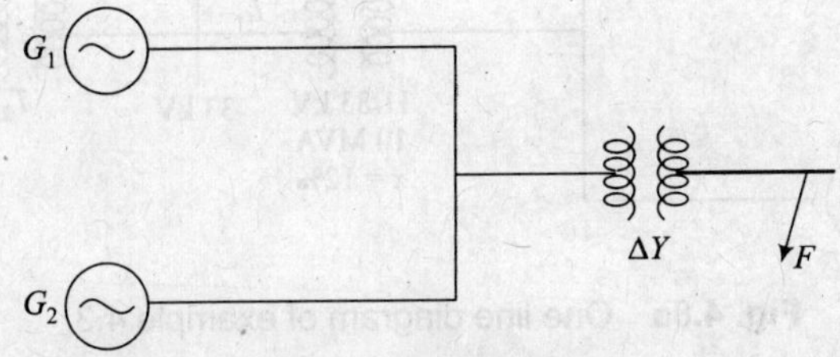

Fig. 4.7a Example 4.2.

The reactance diagram is shown in Fig. 4.7b.

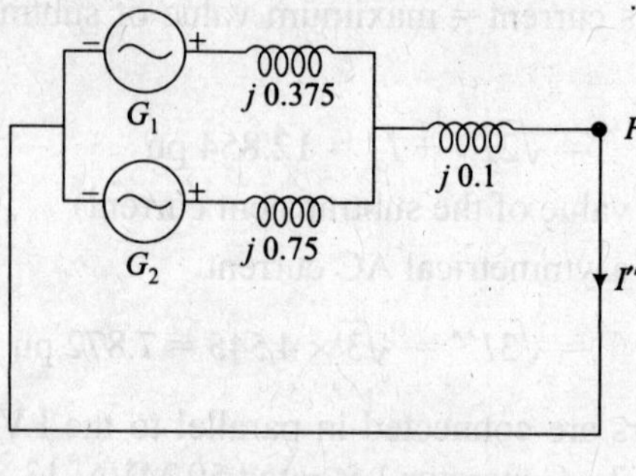

Fig. 4.7b Reactance diagram of Example 4.2.

$$\text{Voltage at } F \text{ (before the fault)} = \frac{130}{132} = 0.9848 \text{ pu.}$$

$$\text{Subtransient fault current} = I_f'' = \frac{0.9848}{(j0.375 \parallel j0.75) + j0.1} = -j2.814 \text{ pu}$$

Fault current in $G_1 = -j2.814 \text{ pu} \times \frac{0.75}{1.125} = -j1.876$(using current division in two parallel branches).

$$\text{Fault current in } G_2 = -j2.814 \times \frac{0.375}{1.125} = -j0.938 \text{ pu}$$

$$\text{Base current in } G_1 \text{ and } G_2 = \frac{75 \times 10^6}{\sqrt{3} \times 13.8 \times 10^3} = 3{,}137 \text{ A}$$

Magnitude of actual fault current in $G_1 = 1.876 \times 3{,}137 = 5{,}885$ A

Magnitude of actual fault current in $G_2 = 0.938 \times 3{,}137 = 2{,}942$ A

Example 4.3: A radial network is shown in Fig. 4.8a. The parameters of the overhead line are $Z = (0.1 + j0.4)\ \Omega$/km. L_1 has a line length of 50 km. A short circuit occurs on the line L_2 at a distance of 10 km from the 6.6 kV bus. Determine the short circuit current.

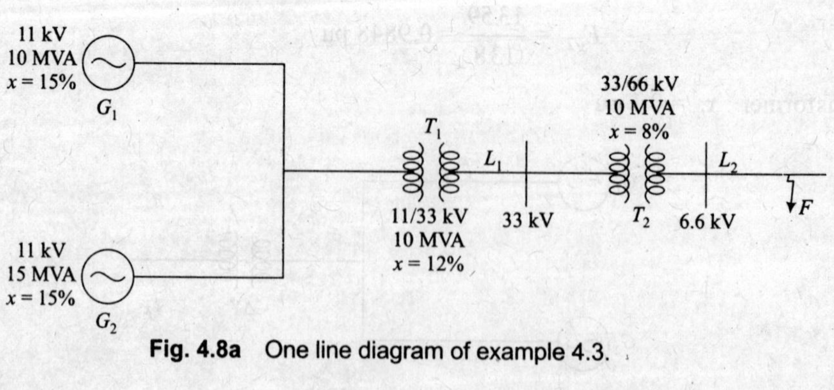

Fig. 4.8a One line diagram of example 4.3.

Solution: Choose a common base of 10 MVA. All reactances are converted to the common base

$$X_{G_1} = 0.15 \text{ pu} \quad X_{G_2} = 0.15 \times \frac{10}{15} = 0.1 \text{ pu}$$

$$X_{T_1} = 0.12 \text{ pu} \quad X_{T_2} = 0.08 \text{ pu}$$

$$Z_{\text{base}} \text{ for } L_1 = \frac{(\text{kV}_{\text{base}})^2}{(\text{MVA}_{\text{base}})} = \frac{(33)^2}{10} = 108.9\ \Omega$$

$$\text{Impedance of (length of 50 km) } L_1 = \frac{50(0.1 + j0.4)}{108.9} = 0.0459 + j0.1836 \text{ pu}$$

$$Z_{\text{base}} \text{ for } L_2 = \frac{(6.6)^2}{10} = 4.356\ \Omega$$

$$\text{Impedance of } L_2 \text{ upto point } F \text{ (10 km)} = \frac{10(0.1 + j0.4)}{4.356} = 0.229 + j0.916 \text{ pu}$$

The reactance diagram is shown in Fig. 4.8b.

Assuming no circulating current, the generators are operated in parallel with same terminal voltages. The short circuit is simulated by closing switch S in Fig. 4.8b. The total reactance upto fault point Z_F is given by

$$Z_F = (j0.15 \parallel j0.1) + j0.12 + (0.0459 + j0.1836) + j0.08 + (0.229 + j0.916)$$

$$= 0.2749 + j1.3596 = 1.387\ \angle 78.57^\circ$$

$$\text{Fault current} \quad I_F = \frac{1.0\ \angle 0^\circ}{1.387\ \angle 78.57^\circ} = 0.72\ \angle{-78.57^\circ} \text{ pu}$$

$$\text{Base current in} \quad L_2 = \frac{10 \times 10^6}{\sqrt{3} \times 6.6 \times 10^3} = 874.77 \text{ A}$$

$$\text{Fault current at} \quad F = 0.72 \times 874.77 = 629.8 \text{ A}$$

We can also determine the voltage at the 11 kV Bus. Total impedance between 11 kV bus and $F = j0.12 + (0.0459 + j0.1836) + (j0.08) + (0.229 + j0.916)$

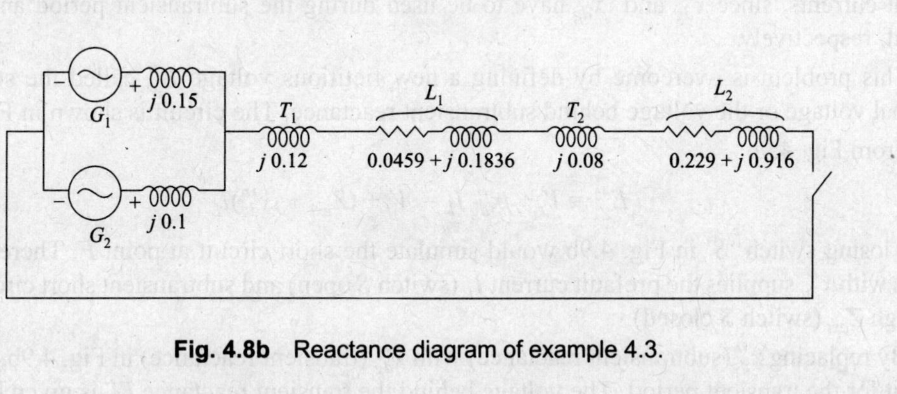

Fig. 4.8b Reactance diagram of example 4.3.

$$= 0.2749 + j1.2996 = 1.328 \angle 78.056° \text{ pu}$$

Voltage at 11 kV bus = 1.328 ∠78.056° × 0.72 ∠–78.57° = 0.956 pu

Magnitude of voltage at 11 kV bus = 0.956 × 11 = 10.516 kV

Normally, the pre-fault currents are small with respect to fault currents and hence can be neglected. However, if prefault load currents are significant, they have to be taken into account. Two approaches are possible using the internal voltages of the machine under loaded conditions and using Thevenin's theorem.

4.5 ANALYSIS OF LOADED GENERATORS

4.5.1 Internal Voltages of Machines Under Load

Consider a generator which is loaded, when a fault occurs. The steady state equivalent circuit is shown in Fig. 4.9a.

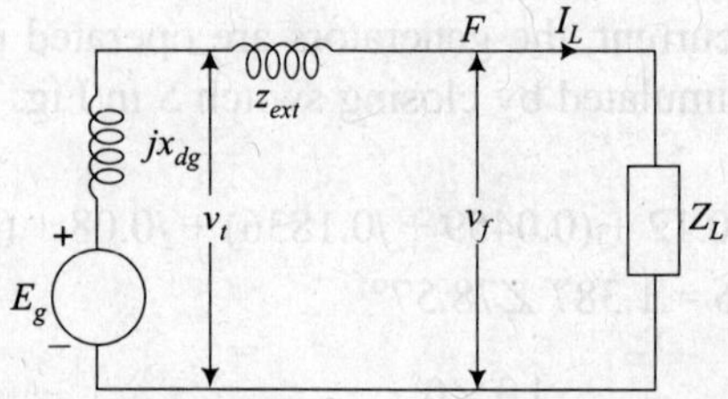

Fig. 4.9a Steady state equivalent circuit of a loaded generator.

In Fig. 4.9a,

Z_{ext} includes all external impedances between the generator and the fault point 'F'.

I_L is prefault load current.

V_t is terminal voltage of generator.

V_f is voltage at point 'F'.

Now if a short circuit occurs at F, Fig. 4.9a does not satisfy conditions for calculating the short circuit currents, since x''_{dg} and x'_{dg} have to be used during the subtransient period and transient period, respectively.

This problem is overcome by defining a new fictitious voltage E''_g, called the subtransient internal voltage or the voltage behind subtransient reactance. The circuit is shown in Fig. 4.9b.

From Fig. 4.9b

$$E''_g = V_t + jx''_{dg} I_L = V_f + (Z_{ext} + jx''_{dg}) I_L \qquad ...(4.10)$$

Closing switch 'S' in Fig. 4.9b would simulate the short circuit at point F. Therefore, E''_g in series with x''_{dg} supplies the prefault current I_L (switch S open) and subtransient short circuit current through Z_{ext} (switch S closed).

By replacing x''_{dg}(subtransient reactance) with x'_{dg} (transient reactance) in Fig. 4.9b, we get the circuit for the transient period. The voltage behind the transient reactance E'_g, is given by

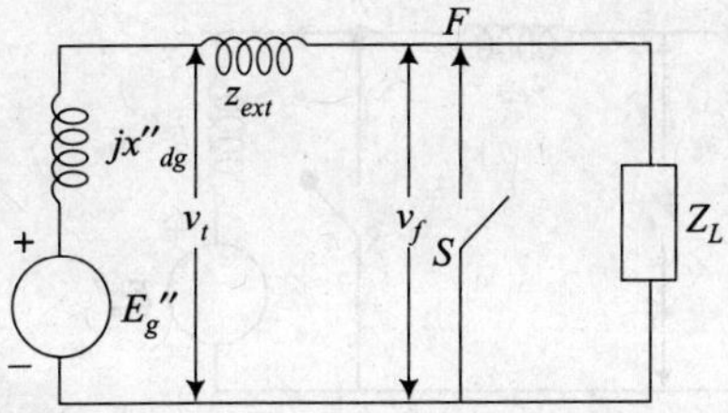

Fig. 4.9b Equivalent circuit showing voltage behind subtransient reactance

$$E_g' = V_t + jx_{dg}' I_L = V_f + (Z_{ext} + jx_{dg}')I_L \quad ...(4.11)$$

It is important to note here, that while the steady state internal voltage E_g, is determined by the excitation, E_g'' and E_g' are dependent on the prefault load current I_L. When the system is unloaded $E_g'' = E_g' = E_g = V_t$, since $I_L = 0$. The phasor diagram is shown in Fig. 4.10.

The subtransient short circuit current in the generator is obtained from Fig. 4.9b (switch S closed).

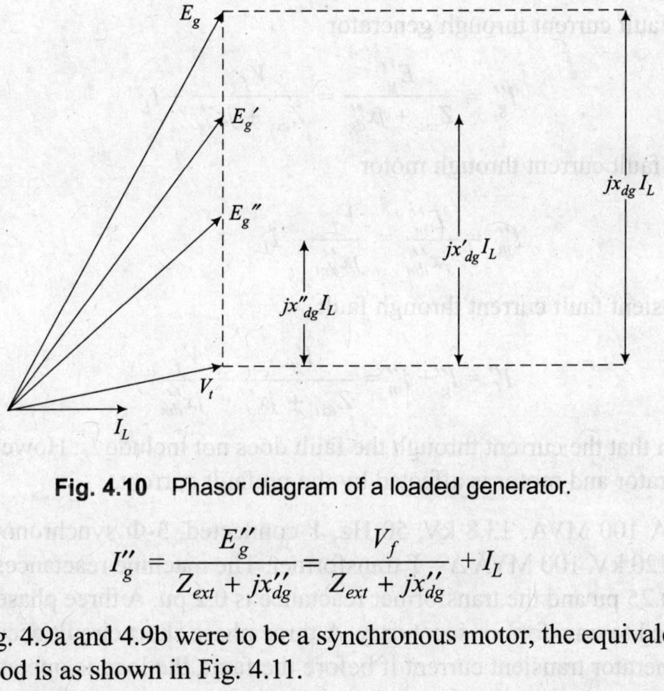

Fig. 4.10 Phasor diagram of a loaded generator.

$$I_g'' = \frac{E_g''}{Z_{ext} + jx_{dg}''} = \frac{V_f}{Z_{ext} + jx_{dg}''} + I_L \quad ...(4.12)$$

If the load in Fig. 4.9a and 4.9b were to be a synchronous motor, the equivalent circuit during the subtransient period is as shown in Fig. 4.11.

The synchronous motor has reactances similar to the synchronous generator. When its terminals are short circuited, the electric supply is cut off. However, its field is energized. Hence, it rotates for a short period of time and contributes to the fault current. The internal voltages of the synchronous motor are defined as follows:

$$E_m'' = V_{tm} - jx_{dm}'' I_L \quad ...(4.13)$$

$$E_m' = V_{tm} - jx_{dm}' I_L \quad ...(4.14)$$

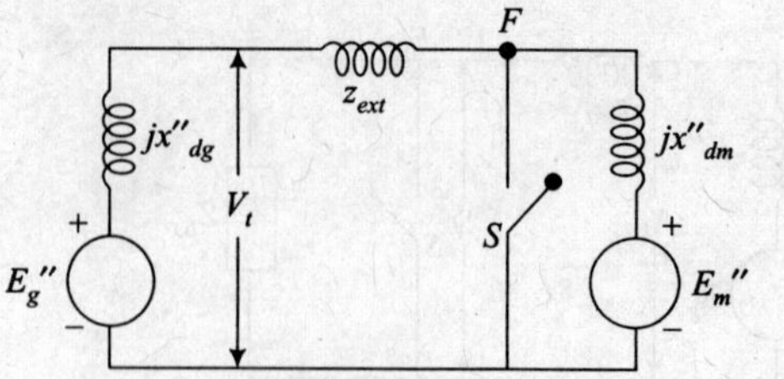

Fig. 4.11 Equivalent circuit diagram of a synchronous motor connected to a synchronous generator.

where, V_{tm} is the terminal voltage of the motor before the fault. Short circuit is simulated by closing S in Fig. 4.11. If the current before fault is I_L, and the prefault voltage at F is V_f, the subtransient short circuit current is calculated as follows: [Refer Fig 4.11]

(i) Calculate

$$E''_g = V_f + I_L\,(Z_{ext} + jx''_{dg})$$

$$E''_m = V_f - I_L\,(jx''_{dm})$$

(ii) Subtransient fault current through generator

$$I''_g = \frac{E''_g}{Z_{ext} + jx''_{dg}} = \frac{V_f}{Z_{ext} + jx''_{dg}} + I_L$$

(iii) Subtransient fault current through motor

$$I''_m = \frac{E''_m}{jx''_{dm}} = \frac{V_f}{jx''_{dm}} - I_L$$

(iv) Total subtransient fault current through fault

$$I''_f = I''_g + I''_m = \frac{V_f}{Z_{ext} + jx''_d} + \frac{V_f}{jx''_{dm}}$$

It can be seen that the current through the fault does not include I_L. However, the fault current through the generator and motor is affected by the prefault current.

Example 4.4: A 100 MVA, 13.8 kV, 50 Hz, Y connected, 3-Φ synchronous generator is connected to a 13.8/220 kV, 100 MVA $\Delta - Y$ transformer. The machine reactances on its own base are $x_d = 1.1$ pu; $x'_d = 0.25$ pu and the transformer reactance is 0.2 pu. A three phase load 100 MVA, 0.8 pf lag, is connected to transformer secondary. A three phase short circuit occurs at the load terminals. Find the generator transient current if before the fault, the load is operating at 220 kV.

Solution: The reactance diagram is shown in Fig. 4.12.

Choose a base of 220 kV, 100 MVA on HT side of transformer.

Complex load power $S_L = 100\ \angle 36.87°$ MVA

$= 1.0\ \angle 36.87°$ pu

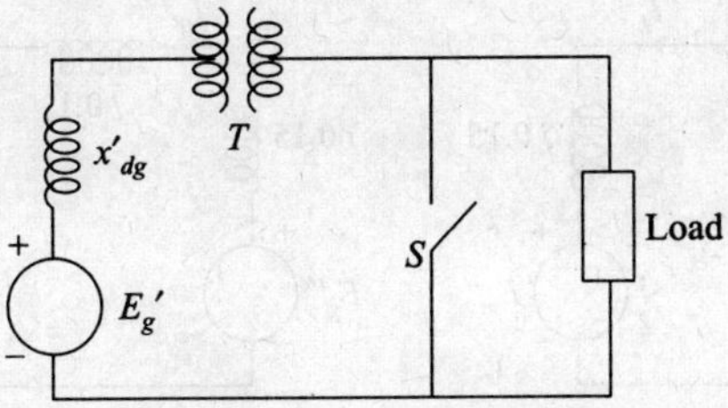

Fig. 4.12 One line diagram of example 4.4.

$$V_L = 1\ \angle 0°\ \text{pu} \quad (\text{choose } V_L \text{ as reference})$$

$$Z_L = \frac{|V_L^2|}{S_L^*} = \frac{1}{1\angle -36.87°} = 0.8 + j06 \text{ pu}$$

Before fault (S open) $I_L = \dfrac{V_L}{Z_L} = \dfrac{1.0\ \angle 0°}{1\ \angle 36.87°}$

$$= 1\ \angle{-36.87°} \text{ pu}$$

Internal voltage behind transient reactance is

$$E'_g = V_L + j(x'_{dg} + x_t)I_L$$
$$= 1\ \angle 0° + j(0.25 + 0.2) \times 1\ \angle{-36.87°}$$
$$= 1.27 + j0.36 = 1.32\ \angle 15.83° \text{ pu}$$

The short circuit is simulated by closing switch S in Fig. 4.12.

I'_g (short circuit transient current)

$$= \frac{E'_g}{j(x'_{dg} + x_t)} = \frac{1.32\ \angle 15.83°}{j(0.25 + 0.2)} = 2.93\ \angle{-74.17°} \text{ pu}$$

Example 4.5: A 25 MVA, 13.2 kV synchronous generator is connected to a synchronous motor of same rating. Both have a subtransient reactance of 15%. The line connecting them has a reactance of 10% on the machine base. The motor is drawing a power of 18 MW at 0.8 pf lead at 12.9 kV, when a short circuit occurs at its terminals. Find the subtransient currents in the motor, generator and fault.

Solution: The equivalent circuit before and after the fault is shown in Fig. 4.13a and Fig. 4.13b, respectively.

Choose a base of 25 MVA, 13.2 kV on the generator and V_f as the reference phasor.

$$V_f = \frac{12.9}{13.2} = 0.977\ \angle 0° \text{ pu}$$

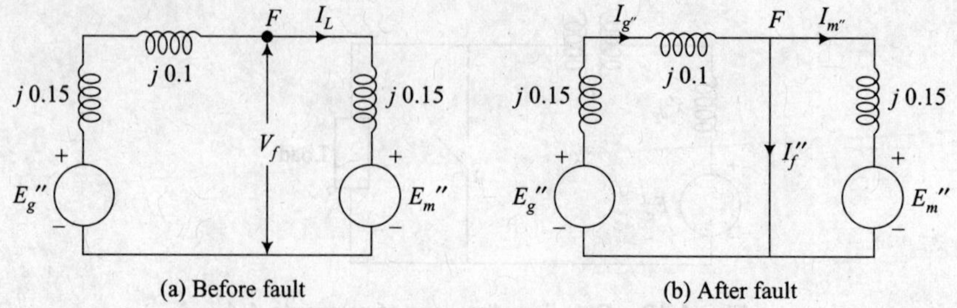

Fig. 4.13 Equivalent circuit of example 4.5.

$$\text{Base current } \frac{25\times 10^6}{\sqrt{3}\times 13.2\times 10^3} = 1093.5 \text{ A}$$

$$I_L = \frac{18\times 10^6}{\sqrt{3}\times 12.9\times 10^3\times 0.8} = 1007\ \angle 36.86° \text{ A}$$

$$= \frac{1007\ \angle 36.86°}{1093.5\ A} = 0.92\ \angle 36.86°$$

$$= 0.736 + j0.552 \text{ pu}$$

Generator

$$E_g'' = V_f + I_L\,(j0.15 + j0.1)$$

$$= 0.977\ \angle 0° + 0.92\ \angle 36.86°\ (0.25\ \angle 90°)$$

$$= 0.839 + j0.184 = 0.859\ \angle 12.37° \text{ pu}$$

$$I_g'' = \frac{E_g''}{j0.15 + j0.1} = \frac{0.859\ \angle 12.37°}{0.25\ \angle 90°} = 3.436\ \angle -77.63° \text{ pu}$$

$$= 0.736 - j3.356 \text{ pu}$$

$$= (0.736 - j3.356)\times 1093.5 = 804.8 - j3{,}670 \text{ A}$$

Motor

$$E_m'' = V_f - I_L\,(j0.15)$$

$$= 0.977\ \angle 0° - 0.92\ \angle 36.86° \times (0.15\ \angle 90°)$$

$$= 1.06 - j0.11 = 1.065\ \angle -5.92° \text{ pu}$$

$$I_m'' = \frac{E_m''}{j0.15} = \frac{1.065\ \angle -5.92°}{0.15\ \angle 90°} = 7.1\ \angle -95.92°$$

$$= -0.736 - j7.062 \text{ pu}$$

$$= (-0.736 - j7.062)\times 1093 = -804.8 - j7{,}722.3 \text{ A}$$

Fault

The fault current $I_f'' = I_g'' + I_m''$

$= (804.8 - j3,670) + (-804.8 - j7,722.3)$

$= -j11,392.3 \text{ A} = -j10.42 \text{ pu.}$

4.5.2 Application of Thevenin's Theorem

The circuits shown in Fig. 4.9a and Fig. 4.11 can be solved by the application of Thevenin's theorem to obtain the short circuit current through the fault and short circuit current in the rest of the network immediately after the fault. Considering Fig. 4.11, let V_f be the Thevenin's voltage at F before the fault. This is nothing but the voltage at point F with switch S open. Let Z_{th} be the Thevenin's impedance of the network at F. This is equal to $(jx''_{dg} + Z_{ext}) \parallel jx''_{dm}$. The equivalent circuit is now drawn as shown in Fig. 4.14.

From Fig. 4.14, the fault current at the instant of short circuit is obtained by closing the switch in Fig. 4.14 and is given by

$$I_f'' = \frac{V_f}{Z_{th}} \qquad \text{...(4.15)}$$

To calculate the current through the generator and motor at instant of short circuit, Fig. 4.11 is redrawn as shown in Fig. 4.15a and 4.15b.

Figure 4.15a shows an additional source V_f at point 'F'. This generator has no effect on current before the fault and is equivalent to Fig. 4.11 with switch 'S' open. With switch 'S' closed during the short circuit, the voltage at F is brought down to zero. This is equivalent to adding a voltage $-V_f$ in series with V_f as shown in Fig. 4.15b, so that the voltage at F is zero.

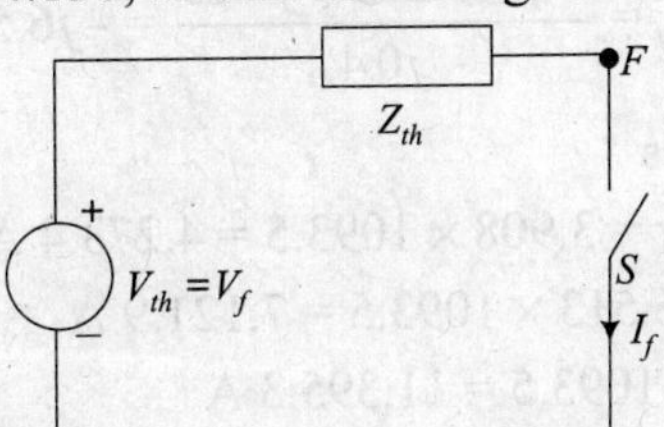

Fig. 4.14 Thevenin's equivalent circuit at fault location.

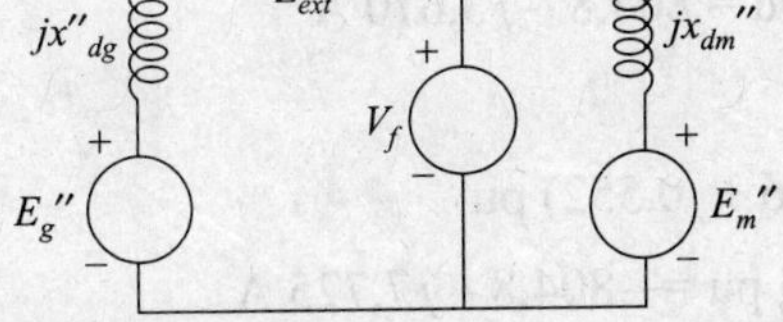

Fig. 4.15a Network before fault.

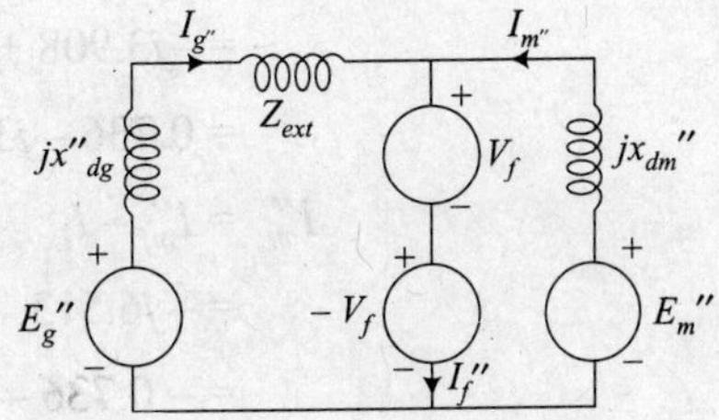

Fig. 4.15b Network during fault.

Principle of superposition is applied to Fig. 4.15b to obtain the current through the rest of the network, as follows:

(i) First short only the source $-V_f$. The network is identical to Fig. 4.15a, and prefault currents flow in the network.

(ii) Short E''_g, E''_m, and V_f to obtain the response due to $-V_f$. This is nothing but the Thevenin's voltage applied to the passive network, identical to Fig. 4.14. The fault current can now be distributed through the rest of the network.

The sum of (i) and (ii) gives the total current in each branch of the network at the instant of short circuit.

Example 4.6: Solve Example 4.5 using Thevenin's theorem.

Solution: $V_{th} = V_f = 0.977 \angle 0°$ pu

$$Z_{th} = (j0.15 + j0.1) \,||\, j0.15 = j0.09375 \text{ pu}$$

$$I''_f = \frac{0.977 \angle 0°}{j0.09375} = -j10.4213 \text{ pu} \quad \text{(Same as in Example 4.5)}$$

This fault current divides into two parallel paths (refer Fig. 4.15b – short E''_g, E''_m and V_f). One through the generator and one through the motor.

Fault current from generator: $I''_{gf} = \dfrac{-j10.4213 \times j0.15}{j0.4} = -j3.908$ pu

Fault current from motor: $I''_{mf} = \dfrac{-j10.4213 \times j0.25}{j0.4} = -j6.513$ pu

Neglecting load current gives

Fault current from Generator = 3.908 × 1093.5 = 4,273.4 A

Fault current from Motor = 6.513 × 1093.5 = 7,121.9 A

Current in fault = 10.4213 × 1093.5 = 11,395.3 A

The current in the fault is same, whether load current is considered or not. But currents in lines differ. When load current is included the currents are as follows [refer example 4.5 for I_L.]

$$\begin{aligned} I''_g &= I''_{gf} + I_L \\ &= -j3.908 + 0.736 + j0.552 \\ &= 0.736 - j3.356 \text{ pu} = 804.8 - j3{,}670 \text{ A} \\ I''_m &= I''_{mf} - I_L \\ &= -j6.513 - (0.736 + j0.552) \text{ pu} \\ &= -0.736 - j7.065 \text{ pu} = -804.8 - j7{,}725 \text{ A} \end{aligned}$$

Fault current = sum of the above

$$= -j\,10.421 \text{ pu} = 11{,}395 \text{ A}$$

The sum of magnitudes of generator and motor currents does not equal the fault current since they are not in phase when load current is included. Note that I_L is in same direction as I''_{gf}, but opposite to I''_{mf}

The fault MVA = $\sqrt{3} \times 13.2 \times 11{,}395 \times 10^{-3} = 260.5$ MVA

It can be observed that the prefault current is very small compared to the short circuit current and hence can be neglected. In the method of internal voltages, this is equivalent to assuming that the subtransient internal voltages of all machines is equal to V_f (which is the case if $I_L = 0$).

4.6 SYMMETRICAL FAULTS ON POWER SYSTEMS

In a practical power system three phase symmetrical faults can be analyzed using Thevenin's theorem discussed in section 4.5.2. However, the method is restricted to smaller systems and becomes impossible to use for large practical systems. The method is applied to a sample power system to illustrate the extension of the Thevenin's theorem method.

Example 4.7: The one line diagram of a three bus system is shown in Fig. 4.16. Each generator is represented as a voltage source behind transient reactance. Assuming the system is running on no-load and neglecting shunt capacitances, determine the fault current, bus voltages and line currents when a balanced three phase fault occurs on (a) bus 1 (b) bus 2 (c) bus 3.

The fault impedance in each case is $Z_f = j0.15$ pu.

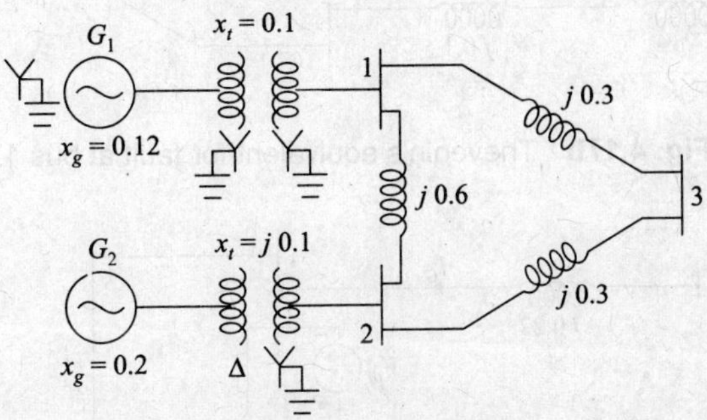

Fig. 4.16 Example 4.7.

Solution (a): Three-phase short circuit at bus 1.

The fault can be simulated by considering a switch between bus 1 and ground and closing the switch to simulate the fault as shown in Fig. 4.17a. The Thevenin's equivalent is as shown in Fig. 4.17b. The reduction stages are shown in Fig. 4.17c to 4.17e.

From Fig. 4.17d, we can obtain the Thevenin's impedance at bus 1 as

$$Z_{Th} = (j0.3 + j0.3) \,||\, j0.22 = j0.16 \text{ pu}$$

$$I_{F1} = \frac{V_{Th}}{Z_{Th} + Z_F} \quad \text{[From Fig. 4.17e]}$$

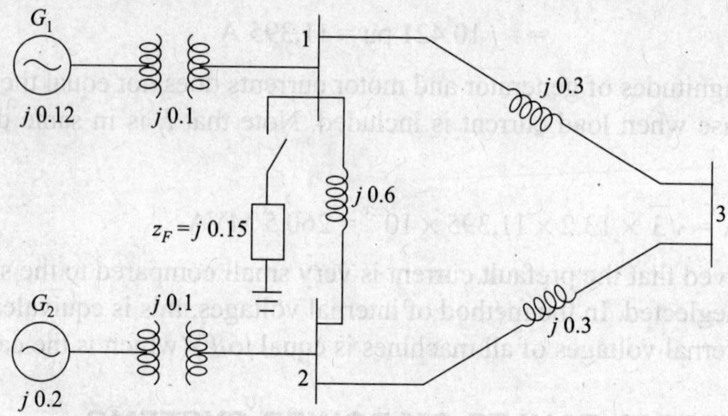

Fig. 4.17a Network to simulate fault at bus 1.

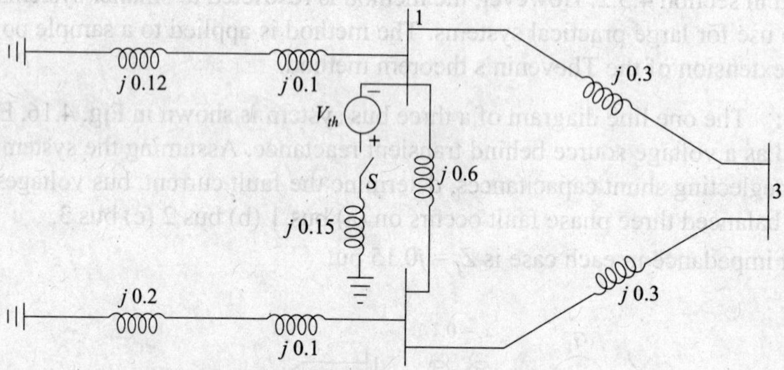

Fig. 4.17b Thevenin's equivalent for fault at bus 1.

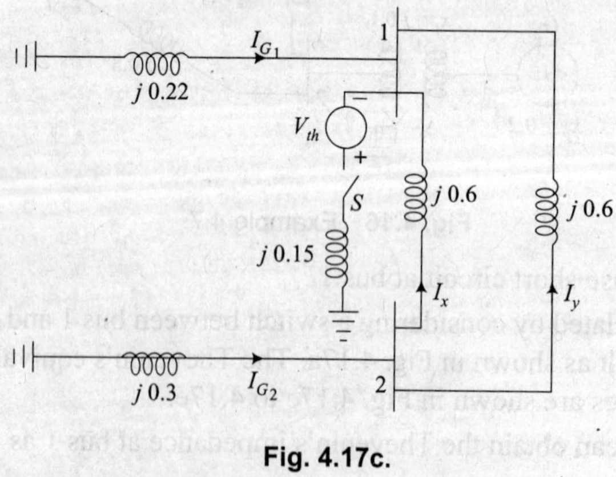

Fig. 4.17c.

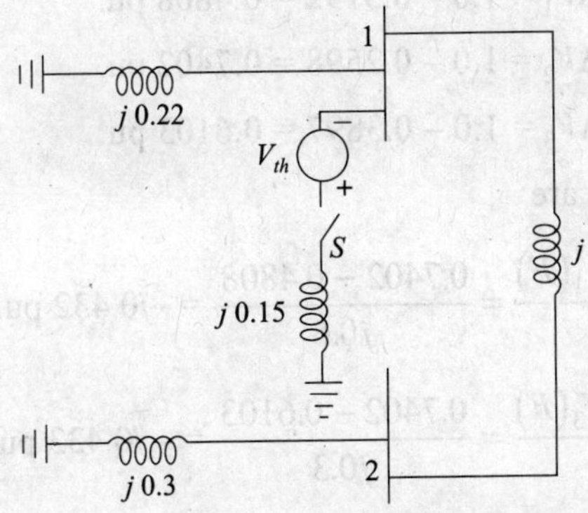

Fig. 4.17d.

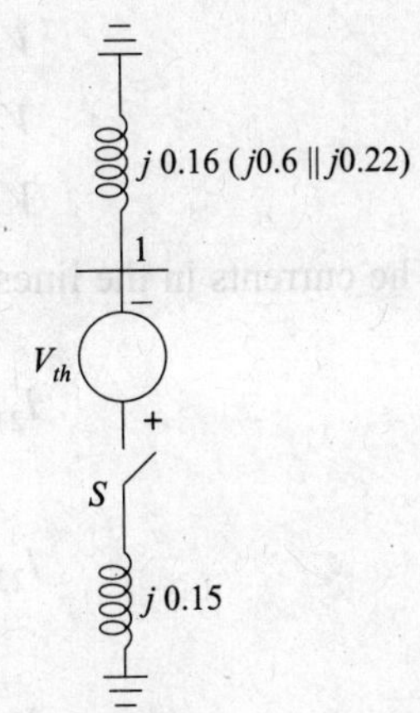

Fig. 4.17e.

Let $V_1(0)$, $V_2(0)$, and $V_3(0)$ be the prefault voltages at the buses. Since, the system is on no-load and shunt capacitances are neglected, all bus voltages before the fault can be assumed to be 1.0 pu.

The Thevenin's voltage at bus 1, V_1 (0) is therefore 1.0 pu

$$\therefore \quad I_{F1} = \frac{1.0}{j0.16 + j0.15} = -j3.226 \text{ pu}$$

Referring to Fig. 4.17d,

$$I_{G1} = \frac{I_{F1}\, j0.6}{j0.6 + j0.22} = -j2.36 \text{ pu}$$

$$I_{G2} = \frac{I_{F1}\, j0.22}{j0.6 + j0.22} = -j0.866 \text{ pu}$$

According to Thevenin's theorem, the changes in the bus voltages caused by the added branch (the fault impedance in this case) is equivalent to the voltages caused by the Thevenin's voltage with other sources shorted [refer section 4.5.2]. The change in bus voltages can be computed from Fig. 4.17b.

$$\Delta V_1 = -(I_{G1})\,(j0.22) = -(-j2.36)\,(j0.22) = -0.5192 \text{ pu.}$$

$$\Delta V_2 = -(I_{G2})\,(j0.3) = -(-j0.866)\,(j0.3) = -0.2598 \text{ pu.}$$

$$\Delta V_3 = \Delta V_2 - I_y\,(j0.3) = -0.2598 - \left[\frac{-j0.866}{2}\right][j0.3] = -0.2598 - 0.1299 = -0.3897 \text{ pu}$$

The actual bus voltages during the fault are calculated by superimposing the changes in the bus voltages to the respective prefault voltages. All prefault voltages are taken to be 1.0 pu.

$$V_1(F) = V_1(0) + \Delta V_1 = 1.0 - 0.5192 = 0.4808 \text{ pu.}$$

$$V_2(F) = V_2(0) + \Delta V_2 = 1.0 - 0.2598 = 0.7402 \text{ pu.}$$

$$V_3(F) = V_3(0) + \Delta V_3 = 1.0 - 0.3897 = 0.6103 \text{ pu.}$$

The currents in the lines during the faults are

$$I_{21}(F) = \frac{V_2(F) - V_1(F)}{z_{21}} = \frac{0.7402 - 0.4808}{j0.6} = -j0.432 \text{ pu.}$$

$$I_{23}(F) = \frac{V_2(F) - V_3(F)}{z_{23}} = \frac{0.7402 - 0.6103}{j0.3} = -j0.433 \text{ pu.}$$

$$I_{31}(F) = \frac{V_3(F) - V_1(F)}{z_{31}} = \frac{0.6103 - 0.4808}{j0.3} = -j0.432 \text{ pu.}$$

Case b. Fault at bus 2.

The equivalent network to simulate the fault, the Thevenin's equivalent and the networks used for reduction are shown from Fig. 4.18a to Fig. 4.18d.

The Thevenin's equivalent impedance at bus 2 (from Fig. 4.18c) is given by

$$Z_{22} = (j0.22 + j0.3) \parallel j0.3 \text{ pu.}$$

The fault current $$I_{2F} = \frac{V_2(0)}{Z_{22} + Z_F} = \frac{1.0}{j0.19 + j0.15} = -j2.941 \text{ pu.}$$

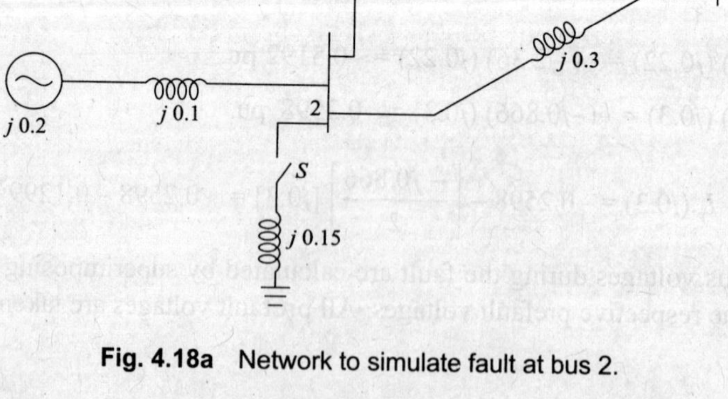

Fig. 4.18a Network to simulate fault at bus 2.

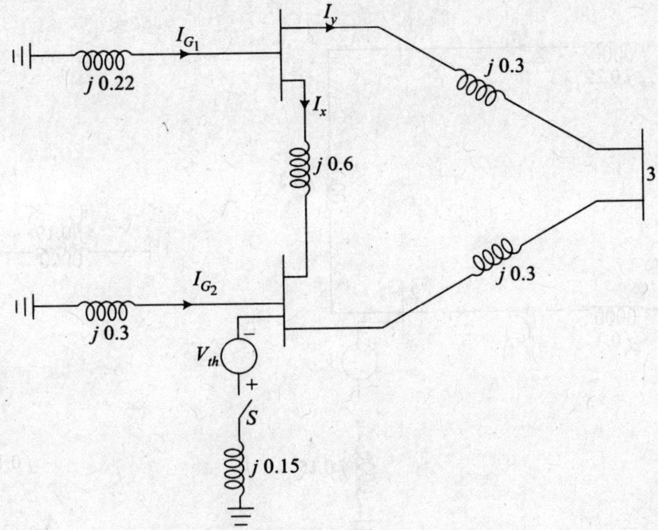

Fig. 4.18b Thevenin's equivalent for fault at bus 2.

From Fig. 4.18c, the currents are calculated as follows:

$$I_{G1} = -j2.941\frac{(j0.3)}{(j0.3+j0.52)} = -j1.076 \text{ pu.}$$

$$I_{G2} = -j2.941\frac{(j0.52)}{(j0.3+j0.52)} = -j1.865 \text{ pu.}$$

The changes in bus voltages can be computed from Fig. 4.19b.

$$\Delta V_1 = -(I_{G1})(j0.22) = -(-j1.076)(j0.22) = -0.2367 \text{ pu.}$$

$$\Delta V_2 = -(I_{G2})(j0.3) = -(-j0.3)(-j1.865) = -0.5595 \text{ pu.}$$

$$\Delta V_3 = \Delta V_1 - I_y(j0.3) = -0.2367 - (j0.3)\left(\frac{-j1.076}{2}\right) = -0.3981 \text{ pu.}$$

The bus voltages during the fault are obtained by superimposing over the prefault voltages, the changes in bus voltages caused by the fault.

$$V_1(F) = V_1(0) + \Delta V_1 = 1.0 - 0.2367 = 0.7633 \text{ pu.}$$

$$V_2(F) = V_2(0) + \Delta V_2 = 1.0 - 0.5595 = 0.4405 \text{ pu.}$$

$$V_3(F) = V_3(0) + \Delta V_3 = 1.0 - 0.3981 = 0.6019 \text{ pu.}$$

The fault currents in the lines are

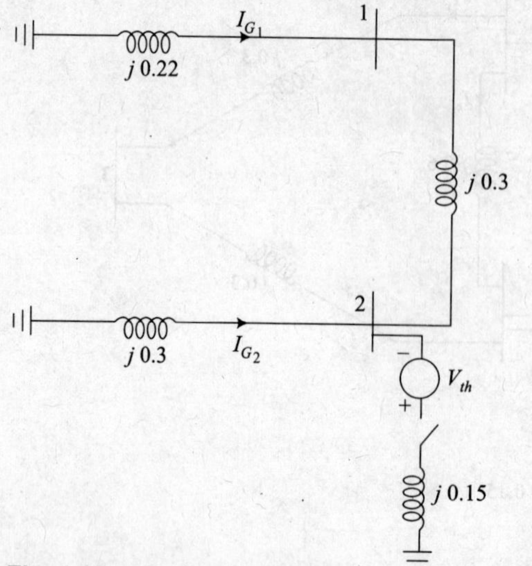

Fig. 4.18c Reduction stage 1.

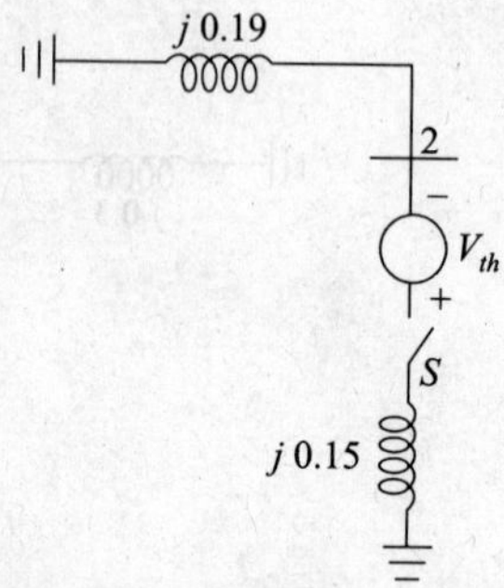

Fig. 4.18d Reduction stage 2.

$$I_{21}(F) = \frac{V_2(F) - V_1(F)}{z_{21}} = \frac{0.4405 - 0.7633}{j0.6} = j0.538 \text{ pu.}$$

$$I_{23}(F) = \frac{V_2(F) - V_3(F)}{z_{23}} = \frac{0.4405 - 0.6019}{j0.3} = j0.538 \text{ pu.}$$

$$I_{31}(F) = \frac{V_3(F) - V_1(F)}{z_{31}} = \frac{0.6019 - 0.7633}{j0.3} = j0.538 \text{ pu.}$$

Case c. Fault at bus 3.

The equivalent network, Thevenin's equivalent network and the reduced networks are shown in Fig. 4.19a to 4.19d.

From Fig. 4.19c, the Thevenin's equivalent impedance at bus 3, is given by

$$Z_{33} = (j0.37 \parallel j0.45) + j0.075 = j0.278 \text{ pu.}$$

From Fig. 4.19d, $I_{3F} = \dfrac{V_{Th}}{Z_{33} + Z_F} = \dfrac{V_3(0)}{Z_{33} + Z_F} = \dfrac{1.0}{j0.278 + j0.15} = -j2.336$ pu.

The generator currents are obtained from Fig. 4.19c.

$$I_{G1} = I_{3F} \frac{j0.45}{j0.45 + j0.37} = -j1.282 \text{ pu.}$$

$$I_{G2} = I_{3F} \frac{j0.37}{j0.45 + j0.37} = -j1.054 \text{ pu.}$$

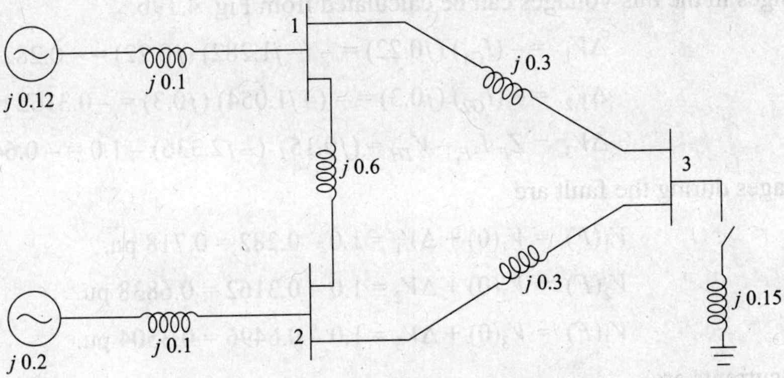

Fig. 4.19a Network to simulate fault at bus 3.

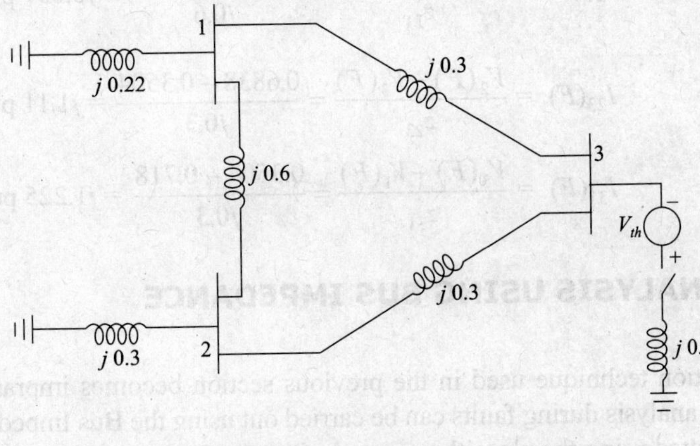

Fig. 4.19b Thevenin's Equivalent for a fault at bus 3.

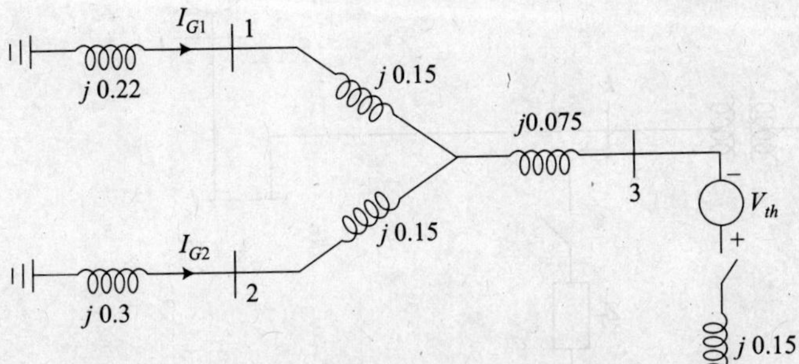

Fig. 4.19c Reduction stage 1.

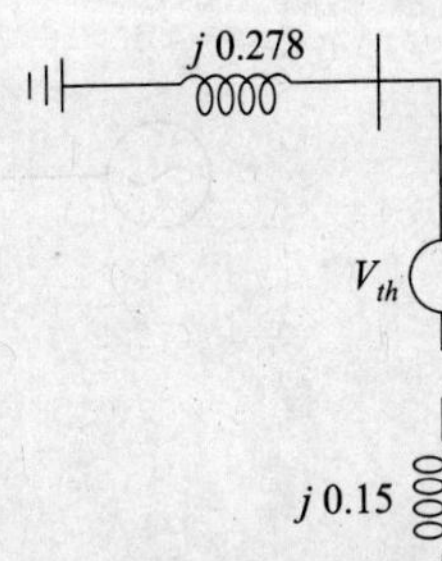

Fig. 4.19d Reduction stage 2.

The changes in the bus voltages can be calculated from Fig. 4.19b.

$$\Delta V_1 = -(I_{G1})\,(j0.22) = -\,(-j1.282)\,(j0.22) = -\,0.282 \text{ pu.}$$

$$\Delta V_2 = -(I_{G2})\,(j0.3) = -\,(-j1.054)\,(j0.3) = -\,0.3162 \text{ pu.}$$

$$\Delta V_3 = Z_F\, I_{3F} - V_{TH} = (j0.15)\,(-j2.336) - 1.0 = -\,0.6496 \text{ pu.}$$

Bus voltages during the fault are

$$V_1(F) = V_1(0) + \Delta V_1 = 1.0 - 0.282 = 0.718 \text{ pu.}$$

$$V_2(F) = V_2(0) + \Delta V_2 = 1.0 - 0.3162 = 0.6838 \text{ pu.}$$

$$V_3(F) = V_3(0) + \Delta V_3 = 1.0 - 0.6496 = 0.3504 \text{ pu.}$$

The line currents are

$$I_{21}(F) = \frac{V_2(F) - V_1(F)}{z_{21}} = \frac{0.6838 - 0.718}{j0.6} = j0.057 \text{ pu.}$$

$$I_{23}(F) = \frac{V_2(F) - V_3(F)}{z_{23}} = \frac{0.6838 - 0.3504}{j0.3} = j1.11 \text{ pu.}$$

$$I_{31}(F) = \frac{V_3(F) - V_1(F)}{z_{31}} = \frac{0.3504 - 0.718}{j0.3} = j1.225 \text{ pu.}$$

4.7 FAULT ANALYSIS USING BUS IMPEDANCE MATRIX

The network reduction technique used in the previous section becomes impractical to apply to larger systems. The analysis during faults can be carried out using the Bus Impedance Matrix Z_{bus}, which can be computed using the algorithm given in Chapter 3.

Consider an n-bus power system as shown in Fig. 4.20.

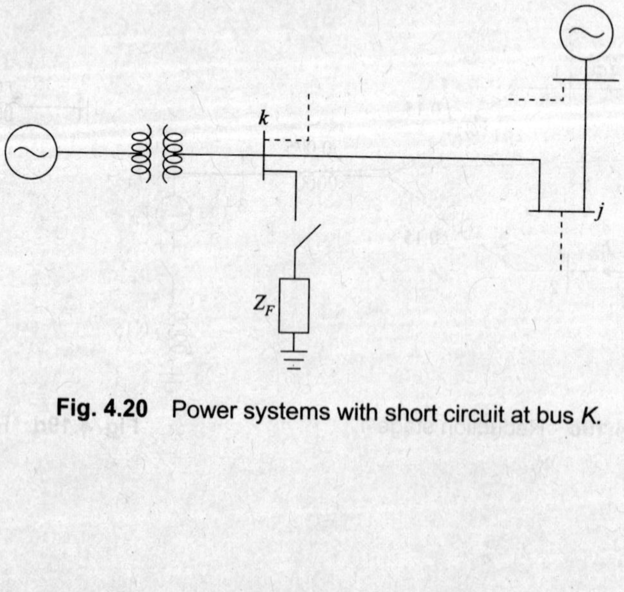

Fig. 4.20 Power systems with short circuit at bus *K*.

In the figure, the generators are represented by a constant voltage source behind a reactance (x_d, x'_d, or x''_d). Assuming the system is balanced, the prefault voltages at all the buses can be computed from the power flow (load flow) solution of the network. Let the prefault voltage at bus k be $V_k(0)$. The vector of prefault bus voltages $V_{\text{Bus}}(0)$ is given by

$$V_{\text{Bus}}(0) = \begin{bmatrix} V_1(0) \\ V_2(0) \\ \vdots \\ V_n(0) \end{bmatrix} \quad \text{...(4.16)}$$

Normally, in the formulation of the power flow solution, the loads are represented as constant power loads, based on the active and reactive power requirements of the load. In short circuit studies, the loads are represented as constant impedances evaluated at the prefault bus voltage.

The equivalent impedance Z_{iL} which draws the same complex power, is given by

$$Z_{iL} = \frac{|V_i(0)|^2}{S_{iL}^*} \quad \text{...(4.17)}$$

where, S_{iL} is complex load power at bus i.

$V_i(0)$ is the prefault voltage at bus i.

Strictly, Z_{iL} is not a constant, since the voltage during the fault does not remain at $V_i(0)$. However, this approximation causes negligible error in short circuit studies. The bus impedance matrix Z_{bus}, is formed considering these constant load impedances.

In the bus frame of reference, the network equations are given by

$$\begin{bmatrix} V_1 \\ V_2 \\ \vdots \\ V_n \end{bmatrix} = \begin{bmatrix} Z_{11} & Z_{12} & \cdots & Z_{1n} \\ Z_{21} & Z_{22} & \cdots & Z_{2n} \\ \vdots & \vdots & \vdots & \vdots \\ Z_{n1} & Z_{n2} & \cdots & Z_{nn} \end{bmatrix} \begin{bmatrix} I_1 \\ I_2 \\ \vdots \\ I_n \end{bmatrix} \quad \text{...(4.18)}$$

If now a current I_k is injected at only bus k and current injections at other buses are made zero, the power system equivalent network reduces to Fig. 4.21.

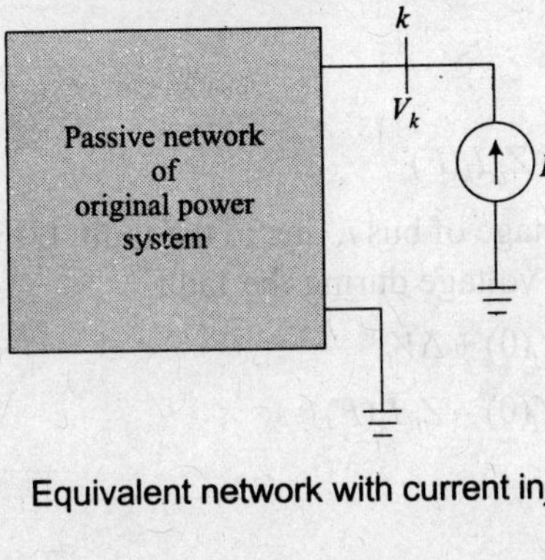

Fig. 4.21 Equivalent network with current injection only at bus k.

In Fig. 4.21, the ratio V_k/I_k is the Thevenin's impedance of the network at bus k (by definition of Thevenin's impedance). From 4.17 (with all injections, except I_k, equal to zero)

$$V_k = Z_{kk} I_k \quad \text{...(4.19)}$$

$$Z_{kk} = \frac{V_k}{I_k} \quad \text{...(4.20)}$$

Therefore, the diagonal elements of the Z_{Bus} are the Thevenin's equivalent impedance of the network at the respective buses. If now a balanced Three Phase fault occurs at bus k, we can draw the Thevenin's equivalent at bus K, as shown in Fig. 4.22.

In Fig. 4.22, $Z_{th} = Z_{kk}$ (diagonal element of Z_{Bus})

$V_{th} = V_k(0)$ (prefault voltage at bus K)

Z_F = fault impedance

The fault is simulated in Fig. 4.22, by closing the switch. The fault current is given by

$$I_k(F) = \frac{V_k(0)}{Z_{kk} + Z_F} \quad \text{...(4.20)}$$

From Fig. 4.22 it can be seen that this fault current flows out of bus k. Hence, the injection, into bus k is $-I_k(F)$. The change in the bus voltages due to this current can be obtained from (4.17) by considering a current injection $-I_k(F)$ at bus k, with other current injections remaining zero.

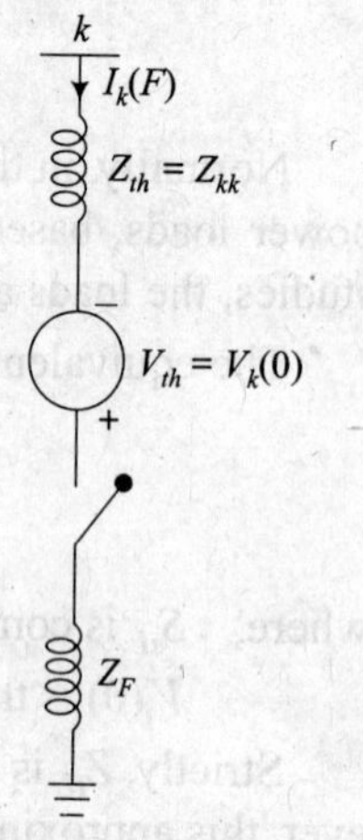

Fig. 4.22 Power system equivalent at bus K.

$$\begin{bmatrix} \Delta V_1 \\ \Delta V_2 \\ \vdots \\ \Delta V_K \\ \vdots \\ \Delta V_n \end{bmatrix} = \begin{bmatrix} Z_{11} & \cdots & Z_{1k} & Z_{1n} \\ Z_{21} & \cdots & Z_{2k} & Z_{2n} \\ \vdots & \vdots & \vdots & \vdots \\ Z_{K1} & \cdots & Z_{kk} & Z_{Kn} \\ \vdots & \vdots & \vdots & \vdots \\ Z_{n1} & \cdots & Z_{nk} & Z_{nn} \end{bmatrix} \begin{bmatrix} 0 \\ 0 \\ \vdots \\ -I_k(F) \\ \vdots \\ 0 \end{bmatrix} \quad \text{...(4.21)}$$

From (4.21), we can write

$$\Delta V_i = -Z_{ik} I_k(F) \quad \text{...(4.22)}$$

This represents the change in voltage of bus i, due to the fault. If the prefault voltage is superimposed over this, we obtain the bus voltage during the fault.

$$V_i(F) = V_i(0) + \Delta V_i$$

$$= V_i(0) - Z_{ik} I_k(F)$$

$$= V_i(0) - \frac{Z_{ik}}{Z_{kk} + Z_F} V_K(0) \quad ...(4.23)$$

The current in any line *i-j* during the fault is given by

$$I_{ij}(F) = \frac{V_i(F) - V_j(F)}{z_{ij}} \quad ...(4.24)$$

where, 'z_{ij}' is the impedance of the line *i-j*. (Not to be confused with element Z_{ij} of the bus impedance matrix). This method of fault analysis is simple and attractive for larger systems. Here too, if we neglect the load currents before fault, all the prefault bus voltages can be assumed to be 1.0 pu.

Example 4.8: Solve Example 4.7 using Z_{bus}

Solution: The bus admittance matrix for the above network can easily be formed by the method of inspection (Refer Chapter 3).

$$Y_{\text{BUS}} = \begin{matrix} 1 \\ 2 \\ 3 \end{matrix} \begin{bmatrix} -j9.5454 & j1.666 & j3.333 \\ j1.666 & -j8.333 & j3.333 \\ j3.333 & j3.333 & -j6.666 \end{bmatrix}$$

$$Z_{\text{Bus}} = Y_{\text{Bus}}^{-1} = \begin{matrix} 1 \\ 2 \\ 3 \end{matrix} \begin{bmatrix} j0.160 & j0.0805 & j0.1207 \\ j0.0805 & j0.1902 & j0.1354 \\ j0.1207 & j0.1354 & j0.2780 \end{bmatrix}$$

Case (a). Fault at bus 1

$$I_1(F) = \frac{V_1(0)}{Z_{11} + Z_F} = \frac{1.0}{j0.160 + j0.15} = -j3.226 \text{ pu.}$$

$$\begin{bmatrix} \Delta V_1 \\ \Delta V_2 \\ \Delta V_3 \end{bmatrix} = \begin{bmatrix} j0.161 & j0.0805 & j0.1207 \\ j0.0805 & j0.1902 & j0.1354 \\ j0.1207 & j0.1354 & j0.2780 \end{bmatrix} \begin{bmatrix} j3.226 \\ 0 \\ 0 \end{bmatrix}$$

$$\Delta V_1 = j0.160 \times j3.226 = -0.5161 \text{ pu}$$

$$\Delta V_2 = j0.0805 \times j3.226 = -0.2597 \text{ pu}$$

$$\Delta V_3 = j0.1207 \times j3.226 = -0.3893 \text{ pu}$$

These are same as obtained in example 4.7. The computation of bus voltages during fault and branch currents are same as before.

Case (b). Fault at bus 2

$$I_2(F) = \frac{V_2(0)}{Z_{22} + Z_F} = \frac{1.0}{j0.1902 + j0.15} = -j2.94 \text{ pu}$$

$$\begin{bmatrix} \Delta V_1 \\ \Delta V_2 \\ \Delta V_3 \end{bmatrix} = \begin{bmatrix} j0.161 & j0.0805 & j0.1207 \\ j0.0805 & j0.1902 & j0.1354 \\ j0.1207 & j0.1354 & j0.2780 \end{bmatrix} \begin{bmatrix} 0 \\ j2.94 \\ 0 \end{bmatrix}$$

$$\Delta V_1 = j0.0805 \times j2.94 = -0.237 \text{ pu}$$

$$\Delta V_2 = j0.1902 \times j2.94 = -0.5592 \text{ pu}$$

$$\Delta V_3 = j0.1354 \times j2.94 = -0.3981 \text{ pu}$$

Case (c). Fault at bus 3

$$I_3(F) = \frac{V_3(0)}{Z_{33} + Z_F} = \frac{1.0}{j0.278 + j0.15} = -j2.336 \text{ pu}$$

$$\begin{bmatrix} \Delta V_1 \\ \Delta V_2 \\ \Delta V_3 \end{bmatrix} = \begin{bmatrix} j0.161 & j0.0805 & j0.1207 \\ j0.0805 & j0.1902 & j0.1354 \\ j0.1207 & j0.1354 & j0.2780 \end{bmatrix} \begin{bmatrix} 0 \\ 0 \\ j2.336 \end{bmatrix}$$

$$\Delta V_1 = j0.1207 \times j2.336 = -0.282 \text{ pu}$$

$$\Delta V_2 = j0.1354 \times j2.336 = -0.3163 \text{ pu}$$

$$\Delta V_3 = j0.2780 \times j2.336 = -0.6494 \text{ pu}$$

(* Slight difference in the answers obtained by the two methods is due to numerical rounding-off)

4.8 SHORT CIRCUIT MVA

The short circuit MVA or short circuit capacity (SCC) at a bus k is defined as

$$\text{SCC} = \sqrt{3} V_{LL} I_k(F) \times 10^{-3} \text{ MVA} \qquad \text{...(4.25)}$$

where, V_{LL} is line to line rated voltage in kV.

$I_k(F)$ is short circuit current for a three phase fault at bus k

If the resistance is neglected, then

$$I_k(F) \text{ pu} = \frac{V_k(0)}{X_{kk}} \qquad \text{...(4.26)}$$

where, $V_k(0)$ is the prefault voltage at bus k.

X_{kk} is the pu reactance of the network at the faulted bus.

The base current is given by

$$I_B = \frac{S_B \times 10^3}{\sqrt{3} V_B} \text{ A} \qquad \text{...(4.27)}$$

where, S_B is three phase base MVA and V_B is line to line base voltage in kV. The fault current in amperes is given by

$$I_k(F) = I_k(F)pu \times I_B$$

$$= \frac{V_k(0)}{X_{kk}} \frac{S_B \times 10^3}{\sqrt{3}V_B} \qquad \text{...(4.28)}$$

Substituting (4.28) in (4.25), we get

$$\text{SCC} = \sqrt{3}V_{LL} \frac{V_k(0)}{X_{kk}} \frac{S_B \times 10^3}{\sqrt{3}V_B} \times 10^{-3} \text{ MVA}$$

$$= \frac{V_{LL}}{V_B} \frac{V_k(0)}{X_{kk}} S_B \qquad \text{...(4.29)}$$

Generally, the rated voltage is taken as the base voltage. Therefore, $V_{LL} = V_B$. Further, the prefault voltage $V_k(0)$ is close to 1.0 pu. With these assumptions (4.29) reduces to

$$\text{SCC} = \frac{S_B}{X_{kk}} \qquad \text{...(4.30)}$$

The short circuit MVA gives an indication of the required breaker capacity.

Short circuit MVA in pu = $|I_{sc}|$ in pu. Under short-circuit conditions,

$$|Z_{th}| = \frac{1.0}{|I_{sc}|} \text{ pu} = \frac{1.0}{\text{short-circuit MVA pu}}$$

4.9 RATING OF CIRCUIT BREAKER

Two factors have to be considered in selecting the circuit breakers *m*

(i) The maximum instantaneous current the breaker must withstand.

(ii) The total current which the breaker must interrupt.

The symmetrical current to be interrupted is computed using subtransient reactances for synchronous machine. However, the actual short circuit current comprises a DC component also. Inclusion of the DC component, results in a rms value of the short circuit current immediately after the fault much greater than the rms value of the symmetrical subtransient fault current (4.9). To account for this the symmetrical current is multiplied by a factor as shown in Table 4.1.

Table 4.1 Multiplication factor for circuit breakers

Breaker speed	Factor
8 Cycles	1.0
5 Cycles	1.1
3 Cycles	1.2
2 Cycles	1.4

This current is called the momentary current and the breaker must withstand the current during the first cycle after the fault occurs. The interrupting current is lower and depends on the speed of the breaker. The current which the breaker must interrupt may contain a part of the DC component. Other ratings specified normally for a breaker are the rated continuous current, rated maximum voltage, voltage range and rated maximum voltage, voltage range and the rated short circuit current at the rated maximum voltage. The rated voltage range factor K is the ratio of rated maximum voltage to the lower limit of the range of operating voltage. It is indicative of the range of voltage over which the product of the rated short circuit current and operating voltage is constant.

The interrupting rating of a circuit breaker is specified in kVA or MVA. The rated maximum voltage is the highest runs voltage for which the breaker is designed.

4.9.1 Current-Limiting Reactors

The fault levels in the power system are increasing because of expansion. Reactors are connected at suitable points in the system to reduce the fault currents levels which the circuit breaker can handle. These reactors must be designed to handle the stresses during short circuit. The reactors may be connected (i) in series with each generator (ii) in series with each feeder and (iii) between bus bar sections. Figure. 4.23(a) shows the reactors connected in series with the generator.

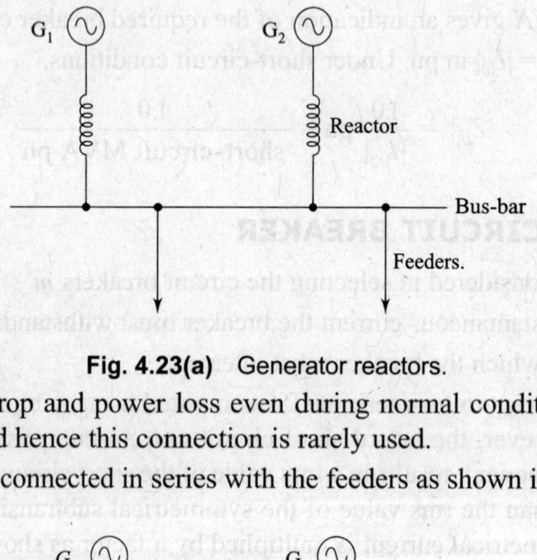

Fig. 4.23(a) Generator reactors.

These is a voltage drop and power loss even during normal conditions. Modern generators have a high reactance and hence this connection is rarely used.

The reactors may be connected in series with the feeders as shown in Fig. 4.23(b).

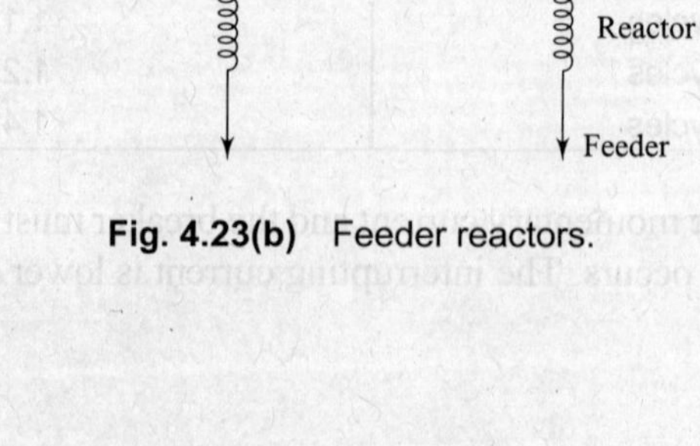

Fig. 4.23(b) Feeder reactors.

The advantage of this method is if a fault occurs on any of the feaders, the bus-bar voltage will not be affected. Here too there is a voltage drop and power loss even under normal conditions.

In bus-bar reactors, the reactors are connected in between sectionalized bus-bars. Two connections for tie-bus system and ring system are shown in Fig. 4.23(c) and 4.23(d).

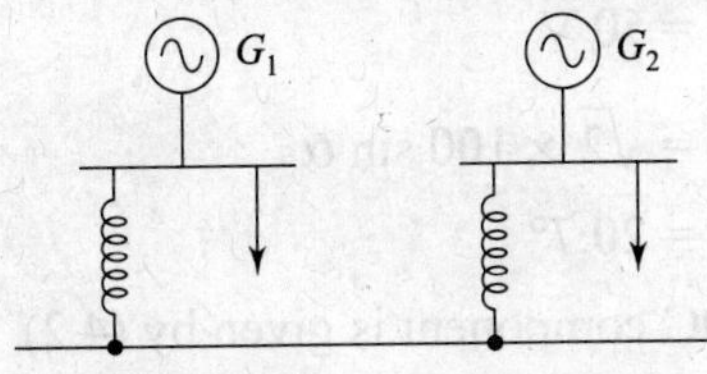

Fig. 4.23(c) Tie-bus system.

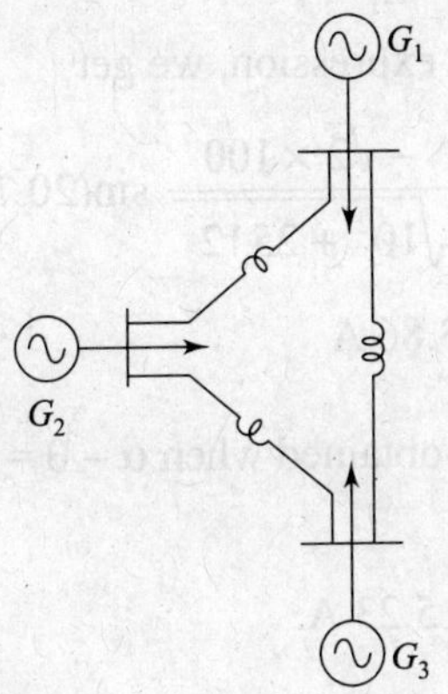

Fig. 4.23(d) Ring system.

Example 4.9: A 50 Hz AC voltage having an rms value of 100V is applied to a series RL circuit by closing a switch. The resistance is 10 Ω and the inductance is 0.08H.

(a) Find the DC component of current upon closing the switch if the instantaneous voltage is 50V at that time.

(b) What is the instantaneous value of the voltage which will produce the maximum DC component of current upon closing the switch?

(c) What is the instantaneous value of the voltage which will result in the absence of any DC component of current upon closing the switch?

(d) If the switch is closed at the instant voltage is zero, find current after 1.5 cycles.

Solution: The AC source voltage is given by $v(t) = V_m \sin(\omega t + \alpha)$ where, α is the angle after voltage zero, at the instant of closing the switch at $t = 0$. Hence, instantaneous value of voltage at $t = 0$ is given by

$$v(0) = V_m \sin \alpha$$

$$Z = R + j\omega L = 10 + j314 \times 0.08 = 10 + j25.12\ \Omega$$

$$\theta = \tan^{-1}\frac{\omega L}{R} = \tan^{-1}\frac{25.12}{10} = 68.3°$$

$$V_m = \sqrt{2} \times 100 = 141.42 \text{ V}$$

(a) If $V(0) = 50$ V

$$50 = \sqrt{2} \times 100 \sin \alpha$$

$\therefore$
$$\alpha = 20.7°$$

The instantaneous value of DC component is given by (4.2)

$$I_{dc} = -\frac{Vm}{|Z|}\sin(\alpha - \theta)e^{\frac{-R}{L}t} \quad \text{...(4.31)}$$

Substituting the values in above expression, we get

$$I_{dc}(0) = \frac{-\sqrt{2} \times 100}{\sqrt{10^2 + 25.12^2}} \sin(20.7° - 68.3°)$$

$$= 3.86 \text{ A}$$

(b) The maximum DC component is obtained when $\alpha - \theta = \pm\frac{\pi}{2}$. The magnitude of the maximum DC component is $\frac{Vm}{|Z|} = \frac{\sqrt{2} \times 100}{27.03} = 5.23$ A.

Value of α for this condition is obtained from

$$\alpha - \theta = 90°$$

$$\alpha = 90 + 68.3 = 158.3°.$$

Hence, if the switch is closed when $\alpha = 158.3°$, maximum DC component of current results at the instant of closing the switch. The voltage at this instant is

$$v(0) = \sqrt{2} \times 100 \sin(158.3°) = 52.3 \text{ V}.$$

(c) If DC component has to be eliminated $\alpha = \theta$. At the instant of closing the switch, the value of the AC voltage would be $v(0) = \sqrt{2} \times 100 \sin(68.3°) = 131.4$ V.

(d) 1.5 cycles $= \frac{1.5}{50} = 30$ ms.

Since, voltage is zero at instant of closing the switch, $\alpha = 0°$.

$$i(t) = \frac{Vm}{|Z|}\left(\sin(\omega t + \alpha - \theta) - \sin(\alpha - \theta)e^{\frac{-R}{L}t}\right)$$

$$i(30\text{ ms}) = \frac{\sqrt{2}\times 100}{27.03}\left[\sin(314\times 30\times 10^{-3} - 68.3°) - \sin(-68.3°)\times e^{\frac{-10}{0.08}\times 30\times 10^{-3}}\right]$$

$$= \frac{\sqrt{2}\times 100}{27.03}[\sin(-58.88°) - \sin(-68.3°)\,0.0235]$$

$$= \frac{\sqrt{2}\times 100}{27.03}[(-0.856) - (-0.0218)] = -4.36\text{ A.}$$

Example 4.10: A 50 Hz generator is rated 500 MVA, 20 kV with $x_d'' = 0.18$ pu. It supplies a purely resistive load of 400 MW at 20 kV, connected directly across the terminals of the generator. A three phase symmetrical short circuit occurs at the load terminals. Find the initial symmetrical rms current in pu on the generator base.

Solution: Base MVA, $S_B = 500$ MVA

Base kV, kV_B = 20 kV

Load voltage = generator terminal voltage = 20 kV = 1 pu.

$$\text{Base current} = \frac{500\times 10^3}{\sqrt{3}\times 20} = 14{,}433.7\text{ A}$$

$$\text{Load current} = \frac{400\times 10^3}{\sqrt{3}\times 20} = 11{,}547\text{ A}$$

$$= \frac{11{,}547}{14{,}433.7} = 0.8\text{ pu.}$$

$$E_g'' = V_1 + jx_{dg}''I_L$$

$$= 1.0\angle 0° + j(0.18)(0.8) = 1.0 + j0.144$$

$$= 1.0103\angle 8.19°$$

When a short circuit occurs at the generator terminals, the initial subtransient current is given by

$$|I_f''| = \frac{|E_g''|}{x_{dg}''} = \frac{1.0103}{0.18} = 5.612\text{ pu.}$$

Example 4.11: A circuit breaker having a nominal rating of 35 kV and a continuous current rating of 1400 A, has a voltage range factor K of 1.65. The rated maximum voltage is 38 kV and the rated short circuit current at that voltage is 22 kA. Find (a) The lower limit of operating voltage (b) The maximum symmetrical interrupting capability of the breaker.

Solution: (a) Voltage range factor, K = rated maximum voltage/lower limit of operating voltage

$\therefore$ Lower limit of operating voltage $= \dfrac{38}{1.65} = 23.03$ kV.

(b) Maximum symmetrical interrupting capability is given by $K \times$ rated short circuit current $= 1.65 \times 22 = 36.3$ kA.

Example 4.12: A three phase power system is shown in Fig. 4.24. The reactances indicated are pu values on a common base. For a three phase fault at bus 2, calculate (a) the fault current (b) the bus voltages during the fault (c) the line currents, and (d) the generator currents.

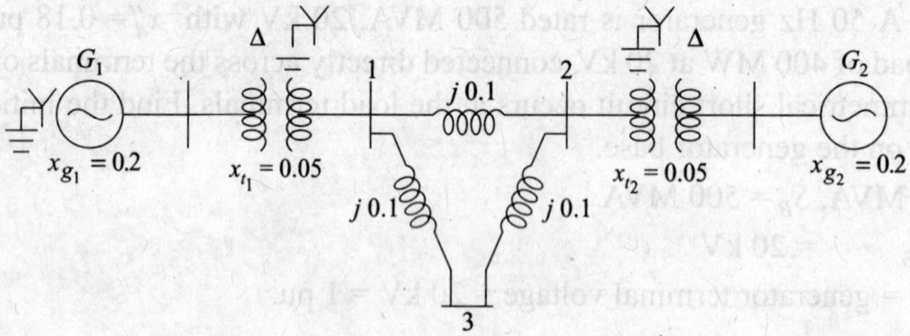

Fig. 4.24 Network of example 4.12.

Solution: The bus admittance matrix is easily obtained by inspection.

$$Y_{\text{Bus}} = \begin{matrix} 1 \\ 2 \\ 3 \end{matrix}\begin{bmatrix} -j24 & j10 & j10 \\ j10 & -j24 & j10 \\ j10 & j10 & -j20 \end{bmatrix}$$

$$Z_{\text{bus}} = Y_{\text{bus}}^{-1} = j\begin{bmatrix} 0.1397 & 0.1103 & 0.1250 \\ 0.1103 & 0.1397 & 0.1250 \\ 0.1250 & 0.1250 & 0.1750 \end{bmatrix}$$

Assume all prefault voltages are 1.0 pu.

(a) For a fault at bus 2

$$I_2(F) = \frac{V_2(0)}{Z_{22}} = \frac{1.0}{j0.1397} = -j7.158 \text{ pu}$$

$$\begin{bmatrix} \Delta V_1 \\ \Delta V_2 \\ \Delta V_3 \end{bmatrix} = j\begin{bmatrix} 0.1397 & 0.1103 & 0.1250 \\ 0.1103 & 0.1397 & 0.1250 \\ 0.1250 & 0.1250 & 0.1750 \end{bmatrix}\begin{bmatrix} 0 \\ j7.158 \\ 0 \end{bmatrix}$$

$$\Delta V_1 = (j0.1103)\,(j7.158) = -0.7895 \text{ pu.}$$
$$\Delta V_2 = (j0.1397)\,(j7.158) = -1.0 \text{ pu.}$$
$$\Delta V_3 = (j0.125)\,(j7.158) = -0.8947 \text{ pu0.}$$
$$V_1(F) = V_1(0) + \Delta V_1 = 1.0 - 0.7895 = 0.2105 \text{ pu}$$
$$V_2(F) = V_2(0) + \Delta V_2 = 1.0 - 1.0 = 0.0 \text{ pu}$$
$$V_3(F) = V_3(0) + \Delta V_3 = 1.0 - 0.8947 = 0.1053 \text{ pu}$$

(c) Currents in various lines are determined as follows:

$$I_{12}(F) = \frac{V_1(F) - V_2(F)}{z_{12}} = \frac{0.2105 - 0.0}{j0.1} = -j2.105 \text{ pu.}$$
$$I_{13}(F) = \frac{V_1(F) - V_3(F)}{z_{13}} = \frac{0.2105 - 0.1053}{j0.1} = -j1.052 \text{ pu.}$$
$$I_{23}(F) = \frac{V_2(F) - V_3(F)}{z_{23}} = \frac{0.0 - 0.1053}{j0.1} = j1.053.$$

(d) Generator currents are calculated as follows:

$$I_{G1}(F) = I_{12}(F) + I_{13}(F) = -j3.157 \text{ pu.}$$
$$I_{G2}(F) = I_{23}(F) - I_{12}(F) + I_2(F) = j1.053 + j2.105 - j7.158 = -j4 \text{ pu}$$

[note that I_2 (F), the fault current flows out of bus 2].

Example 4.13: A 25 MVA, 13.2 kV generator with $x_d'' = 20\%$ is connected to four identical motors as shown in Fig. 4.25. Each motor has $x_d'' = 25\%$; $x_d' = 35\%$ on a base of 5 MVA, 6.6 kV. The step up transformer is rated 25 MVA 13.2/66 kV with a leakage reactance of 10% and the step-down transformer is rated 25 MVA, 66/6.6 kV with a leakage reactance of 10%. The line reactance is 15% on a base of 25 MVA, 66 kV base. The bus voltage at the motors is 6.6 kV, when a three phase short circuit occurs at point F. Find

(a) The subtransient current in the fault.
(b) Subtransient current through breaker A.
(c) Momentary current in breaker A.

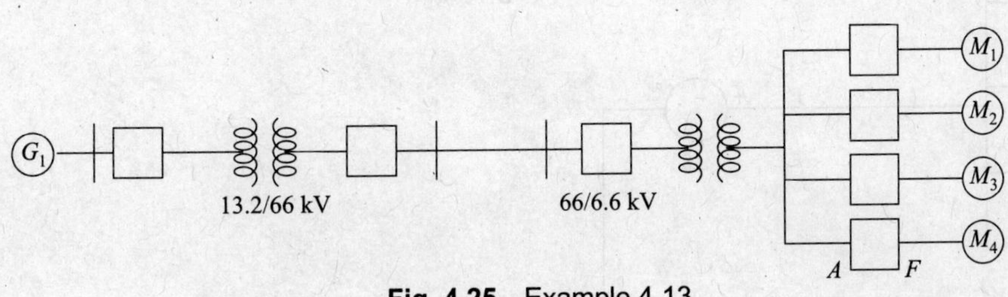

Fig. 4.25 Example 4.13.

Solution: Choose a system base of 25 MVA and 13.2 kV in generator circuit. Voltage base of line is 66 kV, voltage base of motors is 6.6 kV. For each motor

$$x''_{dm} = j0.25 \times \frac{25}{5} = j1.25 \text{ pu}$$
$$x''_{dg} = 0.2;$$
$$x_{tr} = 0.1$$

The circuit model is shown in Fig. 4.26a and the reduced models in Fig. 4.26b and Fig. 4.26c. Here we assume that the motors are on no-load.

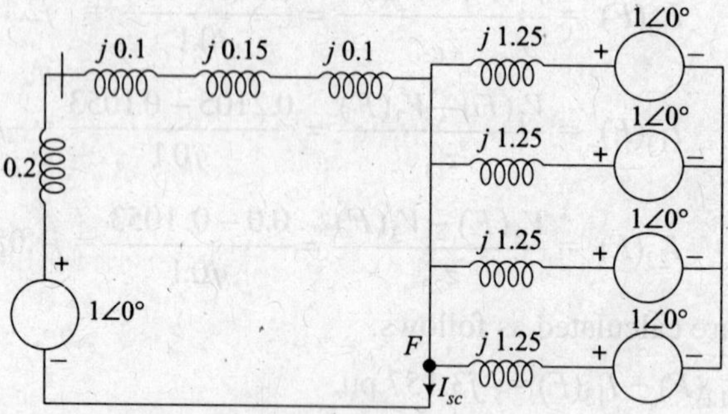

Fig. 4.26a.

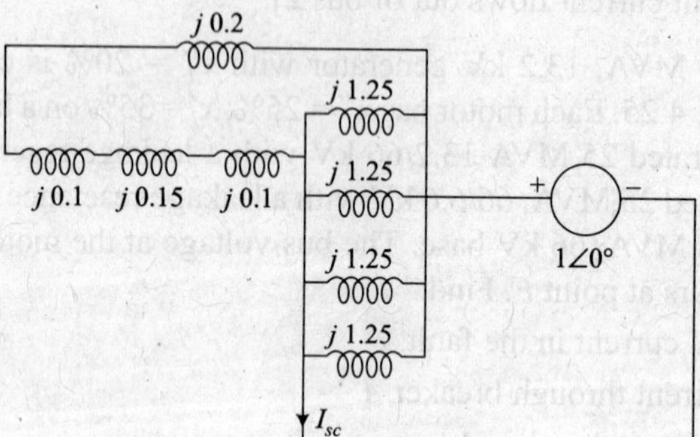

Fig. 4.26b.

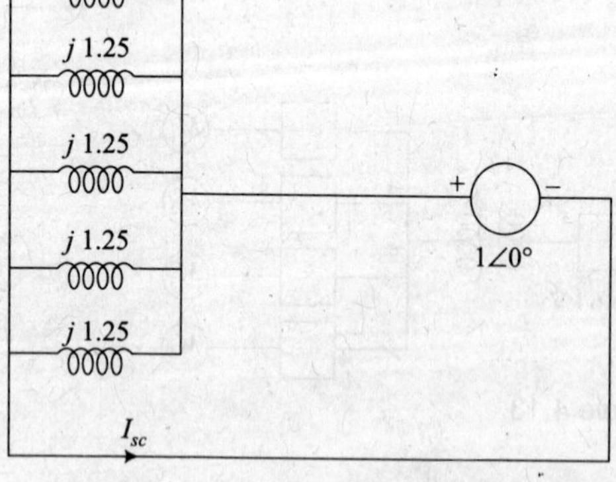

Fig. 4.26c.

(a) From Fig. 4.26c

$$I_{sc} = \frac{1\angle 0^0}{j0.1992} = -j5.018 \text{ pu.}$$

$$\text{Base current on 6.6 kV side} = \frac{25\times 10^3}{\sqrt{3}\times 6.6} = 2{,}187 \text{ A.}$$

$$I_{sc} = 5.018 \times 2{,}187 = 10{,}974 \text{ A.}$$

(b) From Fig. 4.26c, current through breaker A is $\dfrac{1}{j0.55} + 3\times\dfrac{1}{j1.25} = -j4.218$ pu $= 9{,}225$ A.

(c) To find momentary current, DC off set current must be added. On an empirical basis the momentary current is taken to be 1.6 times the symmetrical current.

Momentary current through breaker $A = 9{,}225 \times 1.6 = 14{,}760$ A.

(d) To find the current to be interrupted by breaker, the subtransient reactance is replaced by transient reactance, since the breaker normally operates in the transient period. The motor transient reactance

$$x'_{dm} = j0.35 \times \frac{25}{5} = 1.75 \text{ pu}$$

The current through breaker is now

$$3\times\frac{1}{j1.75} + \frac{1}{j0.55} = -j3.532 \text{ pu.}$$

Allowing for DC offset (Table 4.1)

Interrupting current = 3.532 × 1.1 × 2,187 = 8,497 A.

Example 4.14: Figure (4.27(a)) shows the single line diagram of a system. Determine the maximum MVA which breaker must handle for fault on (i) F_1 (ii) F_2

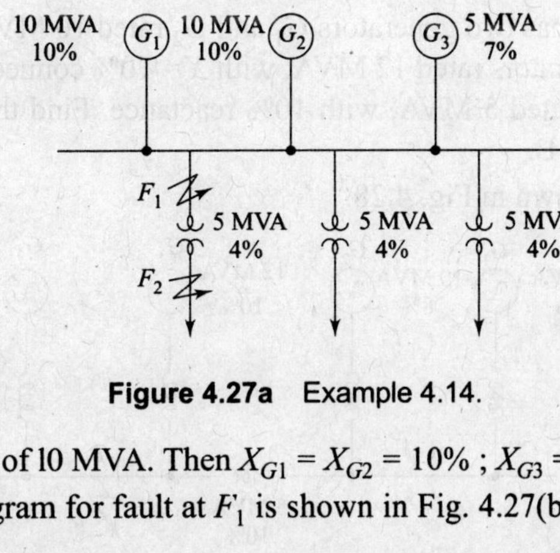

Figure 4.27a Example 4.14.

Solution: We close a base of 10 MVA. Then $X_{G1} = X_{G2} = 10\%$; $X_{G3} = 14\%$; $X_T = 8\%$

The equivalent circuit diagram for fault at F_1 is shown in Fig. 4.27(b).

Total $X = 10\% \,\|10\%\|\, 14\% = 3.68\%$

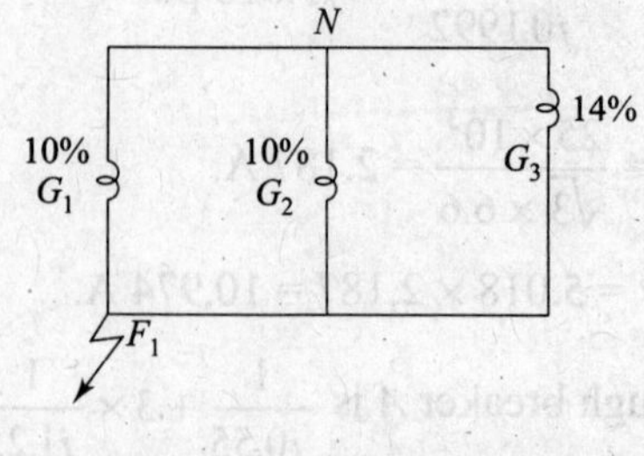

Fig. 4.27 (b) Example 4.14 fault at F_1.

$$\text{Fault MVA} = \frac{10}{0.0368} = 271.74 \text{ MVA}$$

For a fault at F_2, the equivalent circuit is shown in Fig. 4.27 (c).

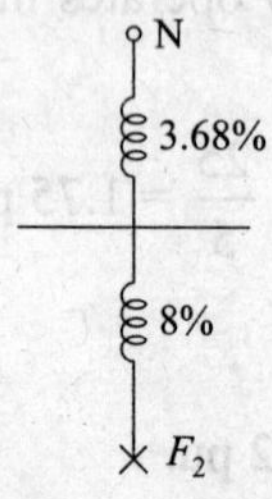

Fig. 4.27 (c) Example 4.14 fault at F_2.

$$\text{Fault MVA} = \frac{10}{0.0368 + 0.08} = 85.6 \text{ MVA}$$

It can be observed that F_2 is on the HV side of the transformer. Therefore, lower ratings of breaker is required on the HV side of the transformer.

Example 4.15: Bus bar A has two generators G_1 and G_2 rated 10 MVA, with $X = 8\%$ connected to it. Bus bar B has two generators rated 12 MVA, with $X = 10\%$ connected to it. The two bus bars are connected by a reactor rated 5 MVA, with 10% reactance. Find the short circuit MVA for a three- phase fault on bus bar B.

Solution The system is shown in Fig. 4.28.

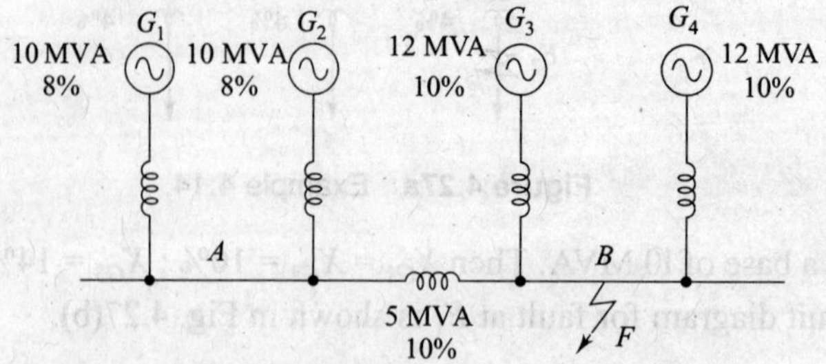

Fig. 4.28

Choose a base of 10 MVA. Reactance of G_3 and G_4 on a base of 10 MVA is $10 \times \frac{10}{12} = 8.33\%$.

Reactance of reactor on 10 MVA base is 20%. For a fault on bus bar B, the reactance from neutral to F is,

$$[(8\% \parallel 8\%) + 20\%] \parallel [8.33\% \parallel 8.33\%]$$
$$= [4\% + 20\%] \parallel [4.167\%]$$
$$= 24\% \parallel 4.167\% = 3.55\%$$

$$\text{Fault MVA} = \frac{10}{0.0355} = 281.65 \text{ MVA.}$$

Without the current limited reactor, the reactance to fault point is

$$[4\%] \parallel [4.167\%] = 2.04\%$$

$$\text{Fault MVA} = \frac{10}{0.0204} = 490 \text{ MVA}$$

4.10 SUMMARY

The current flowing through the power system elements after a fault depends on the impedances of the network branches and the synchronous machine reactances. The currents immediately after the fault, called the subtransient current, are high and generally asymmetric. This current decays into the transient current and later settles down at the steady state current value. The current through the fault itself does not depend on the prefault current. However, current in other branches of the network are determined both by the current through the fault and the prefault current. Since, the prefault current, in general, is small compared to the fault current it can be neglected.

Larger systems can be analyzed using the bus impedance matrix. The bus impedance matrix can be built using the Z_{bus} building algorithm. Changes in the bus voltages due to the fault are computed and superimposed over the prefault voltages to obtain the bus voltages during the fault.

The current through any branch ij, during the fault, can be computed as $I_{ij}(F) = \dfrac{V_i(F) - V_j(F)}{z_{ij}}$,

where the voltages are the bus voltages during the fault. Fault currents indicate the ratings of the circuit breakers to be connected in the system. The circuit breaker ratings depend both on the maximum instantaneous current they have to withstand and also the current they must interrupt.

4.11 SUPPLEMENTARY PROBLEMS

(1) A 50 MVA, 13.2 kV generator, with reactances $x_d'' = 20\%$; $x_d' = 35\%$ and $x_d = 100\%$ is connected to a 50 MVA, 13.2/132 kV transformer, with $x = 10\%$. It is operating on no-load at rated voltage. Find the sustained short circuit current, the intial symmetrical rms current and the maximum possible DC component, if the short circuit occurs on the (a) LV side of the transformer (b) HV side of the transformer.

(2) A synchronous generator and motor are rated 50 MVA, 13.2 kV with subtransient reactance of 15%. The line connecting them has a reactance of 12% on the base of machine ratings. The motor is drawing 25 MW at 0.85 *pf* lead and a terminal voltage of 13 kV, when a symmetrical three phase fault occurs at the motor terminals. Find the subtransient currents in the generator, motor and the fault by (a) application of Thevenin's theorem (b) using internal voltages of the machine.

(3) The impedance matrix of a three bus system is given by

$$Z_{bus} = j\begin{bmatrix} 0.3 & 0.2 & 0.275 \\ 0.2 & 0.4 & 0.25 \\ 0.275 & 0.250 & 0.41875 \end{bmatrix}$$

If the prefault bus voltages are all assumed to be 1.0 pu, find the short circuit current for a three phase short circuit at bus 1. What are the bus voltages during the fault?

(4) The system shown in Fig. P4.4 is initially on no-load, with generator operating at rated voltage. A balanced three phase fault occurs at the receiving end of the transmission line. Evaluate the short circuit current and short circuit MVA.

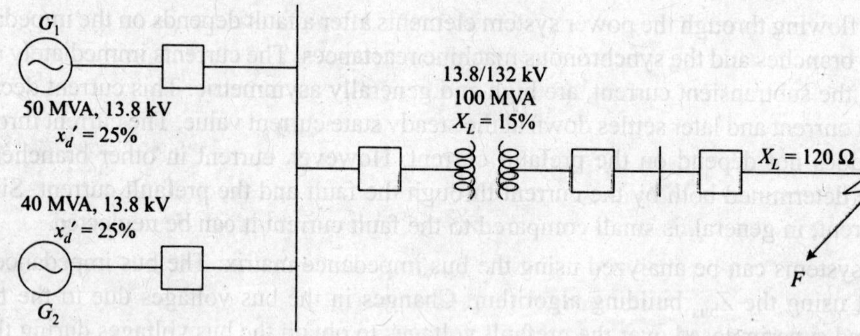

Fig. P4.4

(5) What is the short circuit current and short circuit MVA in the above problem if the fault occurs at the 13.8 kV bus.

(6) The one line diagram of a three bus system is shown in Fig. P4.6. All impedances are expressed on a common base, in pu. Assume the generators are operating on no-load at their rated voltage with their emf's in phase. A three phase fault occurs at bus 3 through an impedance $Z_f = 0.1$ pu

(a) Obtain Y_{Bus} and Z_{Bus} for the system including the generator reactances.

(b) Determine fault current.

(c) Determine the bus voltages and line currents during the fault.

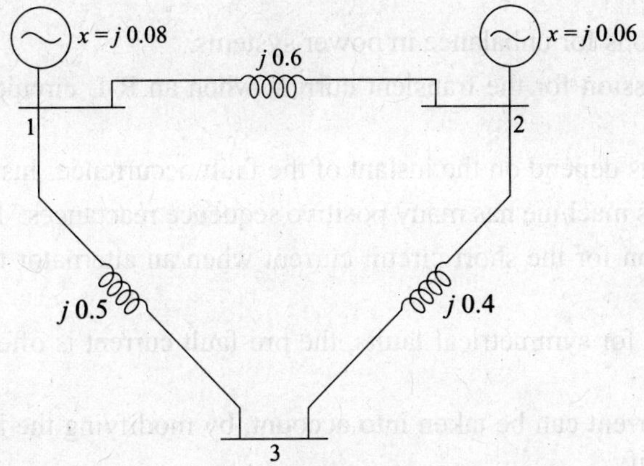

Fig. P4.6

(7) For the 4 bus system shown in Fig. P4.7, obtain the fault current, bus voltages and all branch currents during fault for a bolted fault at bus 4. All impedances are in pu on a common base. Assume all initial bus voltages are 1.0 pu. Repeat for a bolted fault at bus 3.

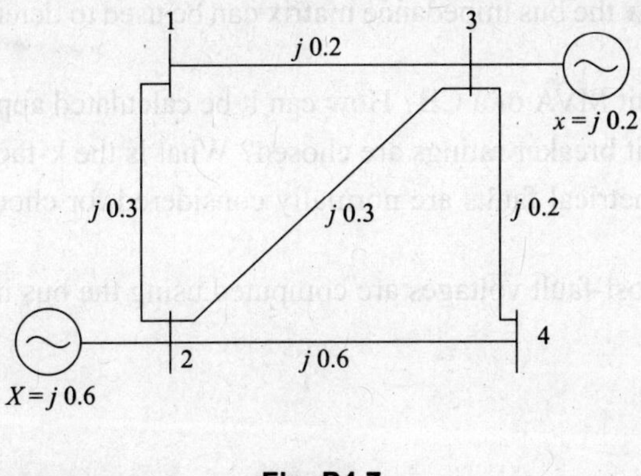

Fig. P4.7

(8) A three-phase 50 Hz, synchronous generator is initially running on no-load with the armature voltage at the rated value, when a short circuit occurs at its terminals. The machine has the following parameters:

$$x_d'' = 0.14 \text{ pu}; \; x_d' = 0.45 \text{ pu}; \; x_d = 1.2 \text{ pu}.$$

$$T_d'' = 0.03 \text{ s}; \; T_d^1 = 1.0 \text{ s}$$

Plot the fundamental frequency symmetrical AC current.

$$\left[\text{Hint: } i_{ac}(t) = E_g\left[\left(\frac{1}{x_d''} - \frac{1}{x_d'}\right)e^{\frac{-t}{T_d''}} + \left(\frac{1}{x_d'} - \frac{1}{x_d}\right)e^{\frac{-t}{T_d'}} + \frac{1}{x_d}\right]\sin \omega t\right]$$

REVIEW QUESTIONS

1. Discuss the reasons for unbalance in power systems.
2. Derive an expression for the transient current when an R-L circuit is connected to an ac source.
3. The fault currents depend on the instant of the fault occurrence. Justify.
4. The synchronous machine has many positive sequence reactances. Explain why this is so?
5. Derive expression for the short circuit current when an alternator terminals are short circuited.
6. When analyzing for symmetrical faults, the pre-fault current is often neglected. Give reasons for it.
7. The pre-fault current can be taken into account, by modifying the internal voltages to account for it. Justify.
8. When an alternator is supplying a synchronous motor, and a short circuit occurs at the terminals of the motor, prove that the fault current is independent of the pre-fault motor loading conditions.
9. Explain how Thevenin's theorem can be applied to determine the currents in different branches of the system, when a three-phase symmetrical fault occurs in any part of the network.
10. Clearly explain how the bus impedance matrix can be used to determine the fault currents in power systems. '
11. What is short circuit MVA of a CB? How can it be calculated approximately?
12. Explain how circuit breaker ratings are chosen? What is the k-factor of a breaker?
13. Three phase symmetrical faults are normally considered for choosing the breaker ratings. Why?
14. Explain how the post-fault voltages are computed using the bus impedance matrix.

Chapter 5

Symmetrical Components

5.1 INTRODUCTION

Hitherto, we have considered balanced networks, where the impedances in each phase are identical and the three-phase voltages and currents in the system are balanced. Such a system, can be solved on a single-phase basis, as has been done in the previous chapters. Unbalance in the system can be caused by

(i) Unbalance in generated voltages.
(ii) Unbalance in transmission network.
(iii) Unbalance in loads.
(iv) Unbalance created by unsymmetrical faults in an otherwise balanced system.

By design, causes (i) and (ii) are not present, since the system is designed for balanced operation. Therefore, the most common cause for unbalance in a system, is the occurrence of unsymmetrical faults. Unbalanced systems, cannot be analyzed on a single phase basis. Analysis has to be carried out on a three-phase basis.

An alternative tool to analyze unbalanced poly phase circuits, proposed by CL Fortescue, is the method of symmetrical components. Fortescue proved that an unbalanced system of n related phasors can be transformed into n systems (sets) of balanced phasors called the symmetrical components of the original phasors. If we consider a linear system, the response of the system to unbalanced voltages/currents, can be determined from the response of the system to individual symmetrical components. In a three-phase system, the three unbalanced phasors of voltages or currents are resolved into three sets of balanced phasors called the symmetrical components. If the system is originally balanced, the transformation leads to decoupled independent system networks for each component, which has greatly simplified analysis of unsymmetrical faults.

5.2 SYMMETRICAL COMPONENTS

A set of balanced three-phase phasors, with a phase sequence *abc*, are characterized by the following relationship.

$$
\begin{aligned}
V_a &= V_m \sin \omega t \\
V_b &= V_m \sin (\omega t - 120°) \\
V_c &= V_m \sin (\omega t + 120°)
\end{aligned}
\qquad ...(5.1)
$$

We define a complex number operator

$$a = e^{j120°} = 1\angle 120° \qquad ...(5.2)$$

which has the following properties:

- $a^2 = e^{j240°} = e^{-j120°} = a^*$
- $(a^2)^* = a$
- $a^3 = 1$
- $1 + a + a^2 = 0$

With this notation, (5.1) can be written as

$$V_a = V_a;\ V_b = a^2 V_a;\ V_c = aV_a \qquad ...(5.3)$$

If the phase sequence is *acb*, then

$$V_a = V_a;\ V_b = aV_a;\ V_c = a^2 V_a \qquad ...(5.4)$$

Here it is important to note that, the phase relationship is characterized by the reference phasor (V_a has been taken as *reference phasor* in (5.3) and (5.4)) and the *phase sequence.*

Three unbalanced phasors are resolved into three sets of balanced phasors called the symmetrical components. They are:

1. **Positive sequence components**, which consist of three phasors, equal in magnitude, displaced by 120° and having same phase sequence of the original phasors. If the original sequence is *abc*, the positive sequence components are

$$V_{a1};\ V_{b1} = a^2 V_{a1};\ V_{c1} = aV_{a1} \qquad ...(5.5)$$

2. **Negative sequence components**, which consist of three phasors, equal in magnitude, displaced by 120° and having a phase sequence opposite to the original phasors. Assuming original phasors have a sequence *abc*, the negative sequence components have a sequence *acb* and are given by

$$V_{a2};\ V_{b2} = aV_{a2};\ V_{c2} = a^2 V_{a2} \qquad ...(5.6)$$

3. **Zero sequence components**, which consist of three phasors equal in magnitude and with zero phase displacement with each other. The components are

$$V_{a0};\ V_{b0} = V_{a0};\ V_{c0} = V_{a0} \qquad ...(5.7)$$

Current phasors are similarly expressed in terms of their symmetrical components.

The unbalanced phasors, in terms of the symmetrical components are given by

$$
\begin{aligned}
V_a &= V_{a0} + V_{a1} + V_{a2} \\
V_b &= V_{b0} + V_{b1} + V_{b2} \\
&= V_{a0} + a^2 V_{a1} + aV_{a2}
\end{aligned}
$$

$$V_c = V_{c0} + V_{c1} + V_{c2}$$
$$= V_{a0} + aV_{a1} + a^2 V_{a2}.$$

In matrix form, we can write

$$\begin{bmatrix} V_a \\ V_b \\ V_c \end{bmatrix} = \begin{bmatrix} 1 & 1 & 1 \\ 1 & a^2 & a \\ 1 & a & a^2 \end{bmatrix} \begin{bmatrix} V_{a0} \\ V_{a1} \\ V_{a2} \end{bmatrix} \quad ...(5.8)$$

or,
$$[V_p] = [A]\,[V_s] \quad ...(5.9)$$

where, $[A]$ is the transformation matrix.

$$V_p = \begin{bmatrix} V_a \\ V_b \\ V_c \end{bmatrix} = \text{Vector of original phasors.}$$

$$V_s = \begin{bmatrix} V_{a0} \\ V_{a1} \\ V_{a2} \end{bmatrix} = \text{Vector of symmetrical components.}$$

Premultiplying both sides of (5.9) with $[A]^{-1}$, we get

$$[A]^{-1}\,[V_p] = [V_s] \quad ...(5.10)$$

Computing A^{-1}, we get

$$A^{-1} = \frac{1}{3}\begin{bmatrix} 1 & 1 & 1 \\ 1 & a & a^2 \\ 1 & a^2 & a \end{bmatrix}$$

Substituting in (5.10), we obtain

$$\begin{bmatrix} V_{a0} \\ V_{a1} \\ V_{a2} \end{bmatrix} = \frac{1}{3}\begin{bmatrix} 1 & 1 & 1 \\ 1 & a & a^2 \\ 1 & a^2 & a \end{bmatrix} \begin{bmatrix} V_a \\ V_b \\ V_c \end{bmatrix} \quad ...(5.11)$$

(5.11) can be expanded and written as

$$V_{a0} = \frac{1}{3}[V_a + V_b + V_c]$$

$$V_{a1} = \frac{1}{3}[V_a + aV_b + a^2 V_c] \quad ...(5.12)$$

$$V_{a2} = \frac{1}{3}[V_a + a^2 V_b + aV_c]$$

The above equations can be written for any set of related unbalanced phasors. The graphical representation is shown in Fig. 5.1. Fig. 5.1a shows the positive, negative and zero sequence components and Fig. 5.1b shows the phasor sum of the three components.

We can write the transformation equations for current as

$$I_p = AI_s \text{ and}$$
$$I_s = A^{-1} I_p.$$

where, $$I_P = \begin{bmatrix} I_a \\ I_b \\ I_c \end{bmatrix} \text{ and } I_s = \begin{bmatrix} I_{a0} \\ I_{a1} \\ I_{a2} \end{bmatrix}$$

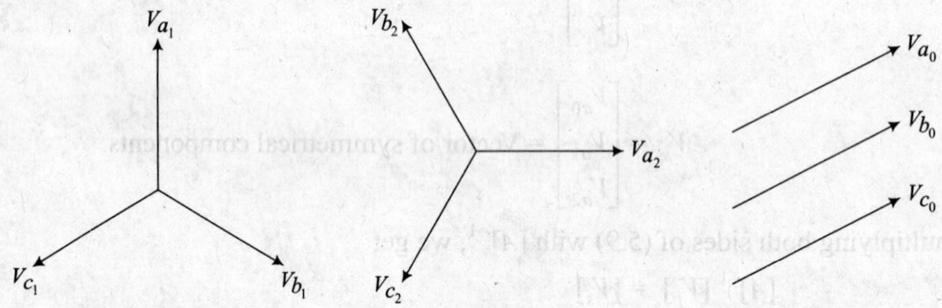

Fig. 5.1(a)

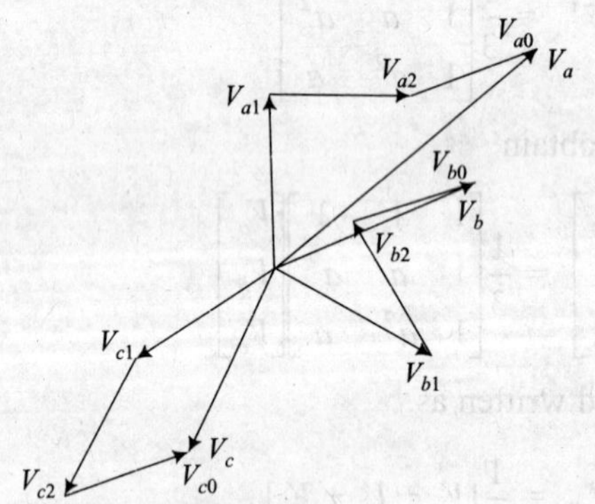

Fig. 5.1(b) Addition of symmetrical components.

Expanding, we can write

$$\begin{aligned} I_a &= I_{a0} + I_{a1} + I_{a2} \\ I_b &= I_{a0} + a^2 I_{a1} + aI_{a2} \\ I_c &= I_{a0} + aI_{a1} + a^2 I_{a2} \end{aligned} \qquad ...(5.13)$$

and,

$$I_{a0} = \frac{1}{3}[I_a + I_b + I_c]$$

$$I_{a1} = \frac{1}{3}[I_a + aI_b + a^2 I_c]$$

$$I_{a2} = \frac{1}{3}[I_a + a^2 I_b + aI_c] \quad ...(5.14)$$

5.2.1 Symmetrical Components in Δ Circuits

Relationship between symmetrical components in a Δ circuit is established with reference to Fig. 5.2.

Let branch *ab* (arbitrarily) be taken as the reference phasor. Referring to Fig. 5.2, we have

$$I_a = I_{ab} - I_{ca}$$
$$I_b = I_{bc} - I_{ab}$$
$$I_c = I_{ca} - I_{bc} \quad ...(5.15)$$

Adding the above equations, we get

$$I_a + I_b + I_c = 0.$$

Since,

$$I_{a0} = \frac{1}{3}[I_a + I_b + I_c]$$

$$I_{a0} = 0.$$

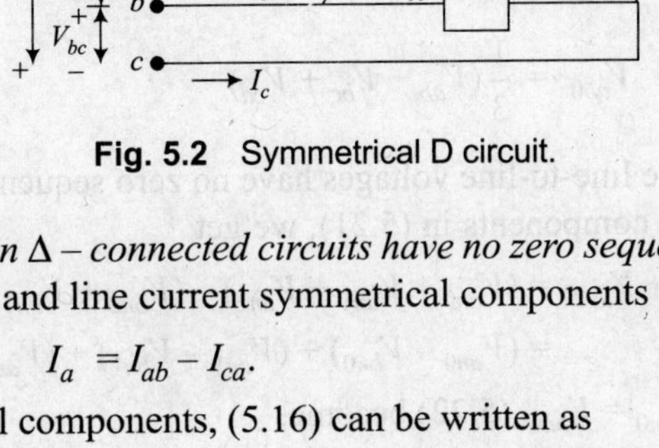

Fig. 5.2 Symmetrical D circuit.

Thus, the line currents in Δ – connected circuits have no zero sequence currents. The relationship between phase current and line current symmetrical components is derived below.

$$I_a = I_{ab} - I_{ca}. \quad ...(5.16)$$

In terms of symmetrical components, (5.16) can be written as

$$I_{a0} + I_{a1} + I_{a2} = (I_{ab0} + I_{ab1} + I_{ab2}) - (I_{ca0} + I_{ca1} + I_{ca2})$$

$$= (I_{ab0} - I_{ca0}) + (I_{ab1} - I_{ca1}) + (I_{ab2} - I_{ca2}) \quad ...(5.17)$$

Note that $I_{a0} = 0$, and $I_{ab0} = I_{ca0}$; therefore (5.17) yields

$$\begin{aligned} I_{a1} + I_{a2} &= (I_{ab1} - I_{ca1}) + (I_{ab2} - I_{ca2}) \\ &= (I_{ab1} - aI_{ab1}) + (I_{ab2} - a^2 I_{ab2}) \\ &= (1 - a)\, I_{ab1} + (1 - a^2)\, I_{ab2} \end{aligned} \quad ...(5.18)$$

Similarly, for phase b we get

$$I_{b1} + I_{b2} = (1 - a)\, I_{bc1} + (1 - a^2)\, I_{bc2}$$

or,

$$a^2 I_{a1} + aI_{a2} = (1 - a)a^2 I_{ab1} + (1 - a^2)aI_{ab2} \quad ...(5.19)$$

Solving (5.18) and (5.19), we get

$$I_{a1} = \sqrt{3}\angle -30°\, I_{ab1} \text{ and } I_{a2} = \sqrt{3}\angle 30°\, I_{ab2}$$

Similarly,

$$I_{b1} = \sqrt{3}\angle -30°\, I_{bc1} \text{ and } I_{b2} = \sqrt{3}\angle 30°\, I_{bc2}$$

$$I_{c1} = \sqrt{3}\angle -30°\, I_{ca1} \text{ and } I_{c2} = \sqrt{3}\angle 30°\, I_{ca2} \quad ...(5.20)$$

Equation (5.20) specifies the phase relationship between the delta sequence currents and the line sequence currents.

The sum of the line voltages, $V_{ab} + V_{bc} + V_{ca} = 0$. Hence, the zero sequence components of the line voltages are zero. Since, the phase voltages are same as line voltages, the phase voltages also will have no zero sequence components.

5.2.2 Symmetrical Components in *Y* Circuits

The star connected impedances, are shown in Fig. 5.3.

The line-to-line voltages in terms of the line-to-neutral voltages are given by

$$\begin{aligned} V_{ab} &= V_{an} - V_{bn} \\ V_{bc} &= V_{bn} - V_{cn} \\ V_{ca} &= V_{cn} - V_{an} \end{aligned} \quad ...(5.21)$$

Adding the three, we get $V_{ab} + V_{bc} + V_{ca} = 0$. Since,

$$V_{ab0} = \frac{1}{3}(V_{ab} + V_{bc} + V_{ca})$$

$V_{ab0} = 0$. In other words, the line-to-line voltages have no zero sequence components. Substituting the respective symmetrical components in (5.21), we get

$$\begin{aligned} V_{ab0} + V_{ab1} + V_{ab2} &= (V_{an0} + V_{an1} + V_{an2}) - (V_{bn0} + V_{bn1} + V_{bn2}) \\ &= (V_{an0} - V_{bn0}) + (V_{an1} - V_{bn1}) + (V_{an2} - V_{bn2}) \end{aligned} \quad ...(5.22)$$

Noting $V_{ab0} = 0$ and $V_{an0} = V_{bn0}$, (5.22) becomes

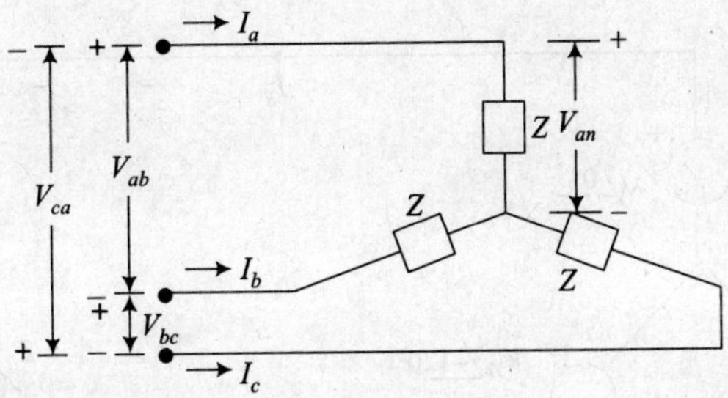

Fig. 5.3 Symmetrical star circuit.

$$V_{ab1} + V_{ab2} = (V_{an1} - V_{bn1}) + (V_{an2} - V_{bn2})$$

$$= (1 - a^2)\, V_{an1} + (1 - a)\, V_{an2} \qquad ...(5.23)$$

Similarly, $$V_{bc1} + V_{bc2} = (1 - a^2)\, V_{bn1} + (1 - a)\, V_{bn2}$$

or, $$a^2\, V_{ab1} + a V_{ab2} = (1 - a^2)\, a^2 V_{an1} + (1 - a)\, a V_{an2} \qquad ...(5.24)$$

Solving (5.23) and (5.24), we get

$$V_{ab1} = (1 - a^2)\, V_{an1} = \sqrt{3}\ \angle 30°\ V_{an1}$$

$$V_{ab2} = (1 - a)\, V_{an2} = \sqrt{3}\ \angle{-30°}\ V_{an2} \qquad ...(5.25)$$

In pu system, (5.20) and (5.25) are used as it is, if both line and phase quantities are expressed in same base.

In the star connected impedance, the sum of the three line currents is equal to the neutral current.

$$I_n = I_a + I_b + I_c$$

Also, $$I_{a0} = \frac{1}{3}[I_a + I_b + I_c] = \frac{I_n}{3} \qquad ...(5.26)$$

That is, the neutral current is three times the zero sequence line current. If there is no neutral connection, $I_{a0} = 0$. Hence, in three wire systems, where neutral does not exist, the zero sequence line current is always zero.

5.2.3 Neutral Shift in Unbalanced Star Connected Load

Consider the case of a balanced three-phase supply, supplying an unbalanced star connected load as shown in Fig. 5.3a.

The connections can be redrawn as shown in Fig. 5.3b and the phasor diagram in Fig. 5.3c.

In Fig. 5.3c, *RY*, *YB* and *BR* represent the line voltages of the supply, and applied to the load. They are equal in magnitude and displaced by 120°. *n* is the star point of the supply and is located at the centre of the equilateral triangle *RYB*. Since load is unbalanced, star point of load n' is at a different potential.

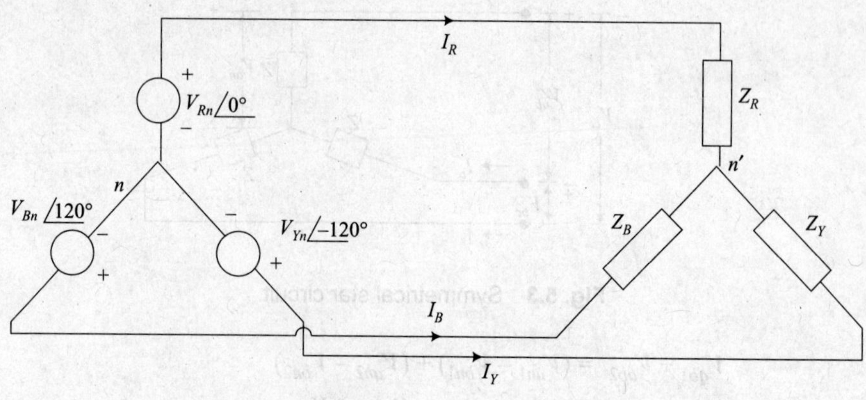

Fig. 5.3a Unbalanced star load

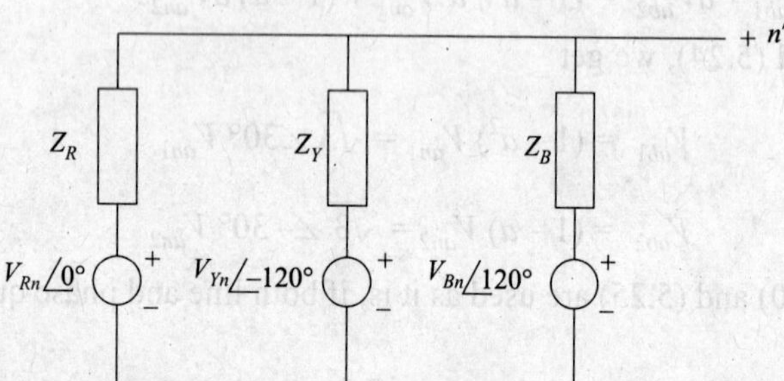

Fig. 5.3b

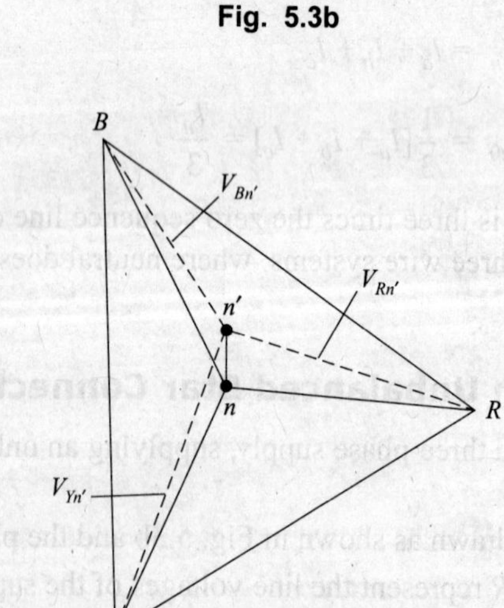

Fig. 5.3c

$V_{n'n}$ can be calculated easily using Millman's theorem. Once, $V_{n'n}$ is calculated, the load phase voltages and corresponding currents in the unbalanced star load can be determined. From Millman's theorem,

$$V_{n'n} = \frac{V_{Rn}Y_R + V_{Yn}Y_Y + V_{Bn}\ Y_B}{Y_R + Y_Y + Y_B}$$

where Y_R, Y_Y, and Y_B are the admittances of the branches of the unbalanced load. Form Fig. 5.3a, we can write

$$V_{Rn} = V_{Rn'} + V_{n'n}$$

or the load phase voltage is,

$$V_{Rn'} = V_{Rn} - V_{n'n}$$

Similarly,

$$V_{Yn'} = V_{Yn} - V_{n'n}$$

$$V_{Bn'} = V_{Bn} - V_{n'n}$$

The line currents are given by,

$$I_R = \frac{V_{Rn'}}{Z_R} = (V_{Rn} - \text{V}_{n'n})Y_R$$

$$I_Y = \frac{V_{Yn'}}{Z_y} = (V_{Yn} - \text{V}_{n'n})Y_Y$$

$$I_B = \frac{V_{Bn'}}{Z_B} = (V_{Bn} - \text{V}_{n'n})Y_B$$

Consider the following data:

$V_{RY} = 400\angle 0°$; $Z_R = 4 + j8\ \Omega$; $Z_Y = 3 + j4\ \Omega$;

$Z_B = 15 + j20\ \Omega$. Calculate the voltage of the star point of load with respect to supply neutral and the line currents.

$$V_{Rn} = \frac{400}{\sqrt{3}}\angle -30°\text{V (The phase voltage will lag line voltage by 30°)}$$

$$V_{Yn} = \frac{400}{\sqrt{3}}\angle -150°\text{ V}$$

$$Y_{Bn} = \frac{400}{\sqrt{3}}\angle -270°\text{V}$$

$$Y_R = \frac{1}{Z_R} = 0.11\ \angle -63.40°\ \mho$$

$$Y_Y = \frac{1}{Z_Y} = 0.2\angle -53.1°\ \mho$$

$$Y_B = \frac{1}{Z_B} = 0.04\ \angle -53.1°\ \mho$$

$$V_{n'n} = \frac{V_{Rn}Y_R + V_{Yn}\,Y_Y + V_{Bn}Y_B}{Y_R + Y_Y + Y_B}$$

$= 104.8\angle 238.86°$

The phase voltages of the load are

$$V_{Rn'} = V_{Rn} - V_{n'n} = 253.89\ \angle -5.67°\ \text{V}$$

$$V_{Yn'} = V_{Yn} - V_{n'n} = 149.2\ \angle\ 189.68°\ \text{V}$$

$$V_{Bn'} = V_{Bn} - V_{n'n} = 325\ \angle 80.7°\ \text{V}$$

$$I_R = V_{Rn'}\,Y_R = 253.89\angle -5.67° \times 0.11\ \angle -63.4°$$

$$= 27.93\ \angle -69.07°\ \text{A}$$

$$I_Y = V_{Yn'}\,Y_Y = 149.2\ \angle 189.68°\ V \times 0.2\ \angle -53.1°$$

$$= 29.84\ \angle 136.58°\ \text{A}$$

$$I_B = V_{Bn'}\,Y_B = 325\ \angle\ 80.7° \times 0.04\ \angle -53.1°$$

$$= 13\ \angle 27.6°$$

5.3 POWER INVARIANCE

The power expended in a 3-ϕ circuit, can be directly computed from the symmetrical components. The total complex power flowing into a three phase circuit, through lines *a,b,c* is

$$\begin{aligned} S = P + jQ &= V_a I_a^* + V_b I_b^*\ V_c I_c^* \\ &= V_p^T I_P^* \\ &= [AV_s]^T [AI_s]^* \\ &= V_s^T A^T A^* I_s^* \end{aligned} \qquad \text{...(5.27)}$$

In (5.27) V_a, V_b, V_c are the voltages to reference at the terminals and I_a, I_b, I_c are the currents flowing into the circuit in the three lines.

Now using properties of the operator *a* we get,

$$A^T A^* = \begin{bmatrix} 1 & 1 & 1 \\ 1 & a^2 & a \\ 1 & a & a^2 \end{bmatrix} \begin{bmatrix} 1 & 1 & 1 \\ 1 & a & a^2 \\ 1 & a^2 & a \end{bmatrix} = 3 \begin{bmatrix} 1 & 0 & 0 \\ 0 & 1 & 0 \\ 0 & 0 & 1 \end{bmatrix} \qquad \text{...(5.28)}$$

= 3 [*I*], where [*I*] is the identity matrix.

Substituting (5.28) in (5.27), we get

$$S = [V_{a0}\ V_{a1}\ V_{a2}]\ 3 \begin{bmatrix} 1 & 0 & 0 \\ 0 & 1 & 0 \\ 0 & 0 & 1 \end{bmatrix} \begin{bmatrix} I_{a0}\ ^* \\ I_{a1}\ ^* \\ I_{a2}\ ^* \end{bmatrix}$$

$$= 3\ V_{a0} I_{a0}^* + 3 V_{a1} I_{a1}^* + 3 V_{a2} I_{a2}^* \qquad \text{...(5.29)}$$

The power is invariant under the transformation provided the product of each sequence voltage (in volts) and complex conjugate of the corresponding sequence current (in amperes) is multiplied by 3. In pu system, care should be taken about using the correct base in (5.29). The base to be chosen is the line-to-neutral voltage.

Example 5.1: One conductor of a three phase line is open. The current flowing through line *a* is 5 A. Assuming line *c* is open, find symmetrical components of the line currents. Repeat if line *a* is open and current through line *b* is 5 A.

Solution: Fig. 5.4 shows the circuit diagram.

Taking I_a as reference, if $I_a = 5\angle 0°$, I_b must be equal to $5\angle 180°$ (since line *b* is the only return path for the current). $I_c = 0$.

$$I_{a0} = \frac{1}{3}[I_a + I_b + I_c] = \frac{1}{3}(5\angle 0° + 5\angle 180° + 0) = 0$$

$$I_{a1} = \frac{1}{3}[I_a + aI_b + a^2 I_c] = \frac{1}{3}(5\angle 0° + 5\angle(180° + 120°) + 0)$$

$$= 2.5 - j1.445 = 2.89\angle -30° \text{ A}$$

$$I_{a2} = \frac{1}{3}[I_a + a^2 I_b + aI_c] = \frac{1}{3}(5\angle 0° + 5\angle(180° - 120°) + 0)$$

$$= 2.5 + j1.445 = 2.89\angle 30° \text{ A.}$$

From (5.5) and (5.6)

$$I_{b0} = 0.$$

$$I_{b1} = a^2 I_{a1} = 2.89\angle -150° = -2.5 - j1.445 \text{ A.}$$

$$I_{b2} = aI_{a2} = 2.89\angle 150° = -2.5 + j1.445 \text{ A.}$$

$$I_{c0} = 0.$$

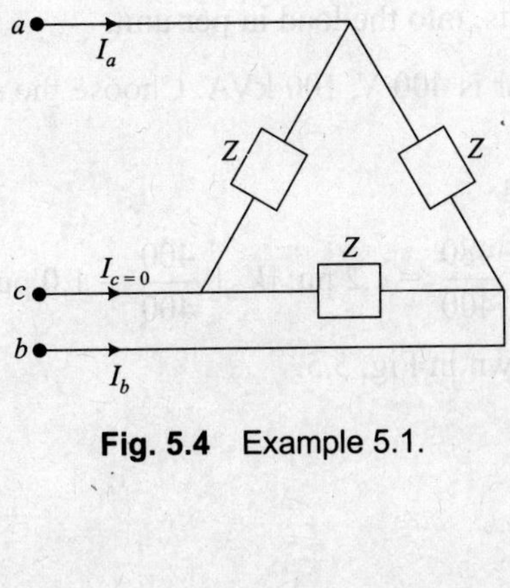

Fig. 5.4 Example 5.1.

$I_{c1} = aI_{a1} = 2.89\angle 90° = j2.89$ A.

$I_{c2} = a^2I_{a2} = 2.89\angle{-90°} = -j2.89$ A.

Check: $I_a = I_{a0} + I_{a1} + I_{a2} = 5$ A; $I_b = I_{b0} + I_{b1} + I_{b2} = -5$ A; $I_c = 0$.

When conductor a is open, then the symmetrical components change

$I_a = 0$; $I_b = 5\angle 0°$; $I_c = 5\angle 180°$

$$I_{a0} = \frac{1}{3}[I_a + I_b + I_c] = 0$$

$$I_{a1} = \frac{1}{3}[0 + 5\angle 120° + 5\angle 60°] + j2.886\,A = 2.886\angle 90°$$

$$I_{a2} = \frac{1}{3}[0 + 5\angle{-120°} + 5\angle 300°]$$

$$= -j2.886\text{A} = 2.886\angle{-90°}$$

From (5.5) and (5.6)

$$I_{b0} = 0;\ I_{b1} = a^2\, I_{a1} = 2.886\angle{-30°} = 2.5 - j1.443 \text{ A.}$$

$$I_{b2} = aI_{a2} = 2.886\angle 30° = 2.5 + j1.443 \text{ A}$$

$$I_{c0} = 0;\ I_{c1} = aI_{a1} = 2.886\angle 210° = -2.5 - j1.443 \text{ A}$$

$$I_{c2} = a^2I_{a2} = 2.886\angle{-210°} = -2.5 + j1.443 \text{ A}$$

Check: $I_a = 0$; $I_b = 5$ A; $I_c = -5$ A.

Example 5.2: Balanced load unbalanced supply.

Three identical Y-connected resistors form a load bank with a three phase rating of 400 V, 100 kVA. Unbalanced voltages are applied to the load bank as follows:

$$|V_{ab}| = 320 \text{ V};\ |V_{bc}| = 480 \text{ V};\ |V_{ca}| = 400 \text{ V.}$$

The neutral of the load is not connected to neutral of the system. On a base of 400 V, 100k VA, find the line voltages and currents, into the load in per unit.

Solution: Rating of the resistor is 400 V, 100 kVA. Choose the system base also as 400 V, 100 kVA.

∴ Value of resistors in pu = 1 pu.

$$|V_{ab}| = \frac{320}{400} = 0.8 \text{ pu};\ |V_{bc}| = \frac{480}{400} = 1.2 \text{ pu};\ |V_{ca}|\ \frac{400}{400} = 1.0 \text{ pu};$$

The voltage phasors are shown in Fig. 5.5.

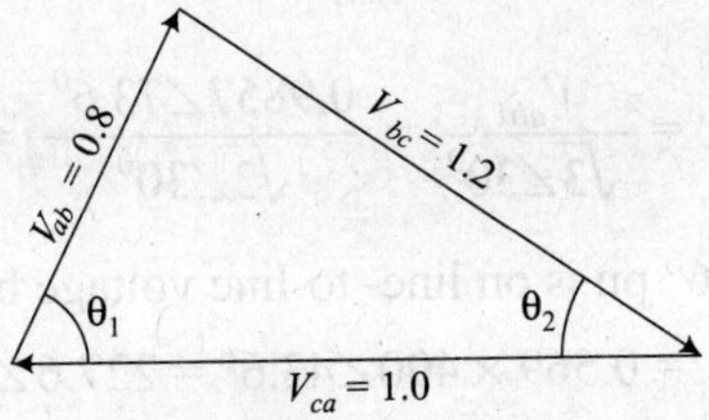

Fig. 5.5 Example 5.2.

Take V_{ca} as reference, as shown in the figure; $V_{ca} = 1.0\angle 180°$ pu.

From law of cosines,

$$V_{bc}^2 = V_{ab}^2 + V_{ca}^2 - 2V_{ab}V_{ca}\cos\theta_1.$$

$$1.2^2 = 0.8^2 + 1.0^2 - 2\times 0.8\times 1.0\times\cos\theta_1.$$

$$\cos\theta_1 = \frac{0.8^2 + 1.0^2 - 1.2^2}{2\times 0.8} = 0.125$$

$$\theta_1 = 82.8°$$

$$\therefore \quad V_{ab} = 0.8\angle 82.8° \text{ pu.}$$

Again $\quad V_{ab}^2 = V_{bc}^2 + V_{ca}^2 - 2V_{bc}V_{ca}\cos\theta_2.$

$$\cos\theta_2 = \frac{1.2^2 + 1^2 - 0.8^2}{2\times 1.2} = 0.75$$

$$\theta_2 = 41.4°$$

$$\therefore \quad V_{bc} = 1.2\angle -41.4° \text{ pu}$$

$$V_{ab1} = \frac{1}{3}[V_{ab} + aV_{bc} + a^2V_{ca}]$$

$$= \frac{1}{3}[0.8\angle 82.8° + 1.2\angle(120° - 41.4°) + 1.0\angle(-120° + 180°)]$$

$$= 0.9857\angle 73.6° \text{ pu.}$$

$$V_{ab2} = \frac{1}{3}[V_{ab} + a^2V_{bc} + aV_{ca}]$$

$$= \frac{1}{3}[0.8\angle 82.8° + 1.2\angle(-120° - 41.4°) + 1.0\angle(120° + 180°)]$$

$$= 0.2346\angle -139.7° \text{ pu.}$$

$$V_{ab0} = 0.0 \text{ pu.}$$

The phase voltages are found from (5.25). In applying (5.25) to pu values, care must be taken to specify the base voltage. From (5.25)

$$V_{ab1} = \sqrt{3}\angle 30° \, V_{an1}$$

$$\therefore \quad V_{an1} = \frac{V_{ab1}}{\sqrt{3}\angle 30^0} = \frac{0.9857\angle 73.6^0}{\sqrt{3}\angle 30^0} = 0.569\angle 42.6° \text{ pu.}$$

The value $V_{an1} = 0.569\angle 43.6°$ pu is on line- to-line voltage base (i.e., 400V).

$$\therefore \quad V_{an1} = 0.569 \times 400\angle 43.6° = 227.6\angle 43.6° \text{ V.}$$

If line-to- neutral voltage, $\frac{400}{\sqrt{3}}$, is used as base, then

$$V_{an1} = 0.9857\angle 73.6° - 30° = 0.9857\angle 43.6° \text{ pu}$$

$$= 0.9857 \times \frac{400}{\sqrt{3}}\angle 43.6° = 227.6\angle 43.6° \text{ V.}$$

Similarly, $V_{an2} = \dfrac{V_{ab2}}{\sqrt{3}\angle -30^0} = 0.135\angle -109.7°$ pu. (*on line-to-line voltage base*)

or, $V_{an2} = \dfrac{V_{ab2}}{1\angle -30^0} = 0.2345\angle -109.7°$ pu. (*on line-to-neutral voltage base*)

$$= 0.2345 \times \frac{400}{\sqrt{3}}\angle -109.7° = 54.18\angle -109.7° \text{ V.}$$

$$V_{an0} = 0.0$$

$I_{a1} = \dfrac{V_{an1}}{R} = 0.9857\angle 43.6°$ pu. (line-to-neutral voltage base)

$I_{a2} = \dfrac{V_{an2}}{R} = 0.2346\angle -109.7°$ pu (line-to-neutral voltage base)

Positive direction of current is chosen to be from the supply towards the load.

Example 5.3: A delta connected balanced resistive load is connected across an unbalanced supply. The line currents are as indicated in Fig. 5.6. Determine the symmetrical components of line currents and delta currents.

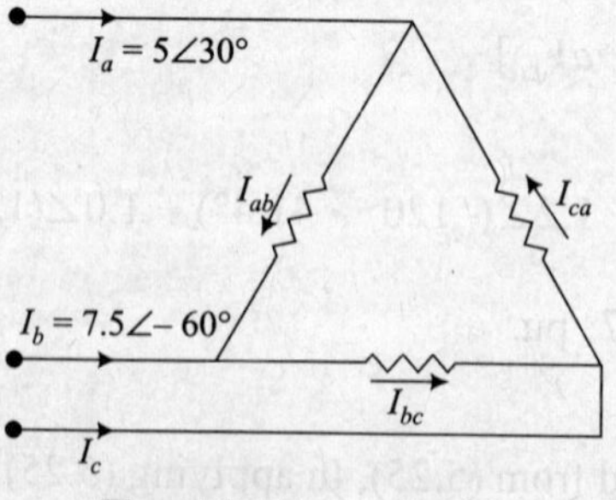

Fig. 5.6 Example 5.3.

Solution:

$$I_a + I_b + I_c = 0$$

$$I_c = -(I_a + I_b) = -(5\angle 30° + 7.5\angle -60°)$$
$$= -(4.33 + j2.5 + 3.75 - j6.5)$$
$$= -(8.08 - j4.0) = 9.0\angle 153.6° \text{ A.}$$

$$I_{a0} = \frac{1}{3}(I_a + I_b + I_c) = 0$$

$$I_{a1} = \frac{1}{3}(I_a + aI_b + a^2 I_c)$$

$$= \frac{1}{3}(5\angle 30° + 7.5\angle(-60° + 120°) + 9.0\angle(153.6° - 120°))$$

$$= 7\angle 42° \text{ A.}$$

$$I_{a2} = \frac{1}{3}(I_a + a^2 I_b + aI_c)$$

$$= \frac{1}{3}(5\angle 30° + 7.5\angle(-60° - 120°) + 9.0°\angle(153.6° + 120°))$$

$$= 2.325\angle -112° \text{ A.}$$

From (5.5) $I_{b1} = 7.0\angle 282°$; $I_{c1} = 7.0\angle 162°$ A.

From (5.6) $I_{b2} = 2.325\angle 8°$ A; $I_{c2} = 2.325\angle 128°$ A.

From (5.7) $I_{b0} = 0$; $I_{c0} = 0$;

Delta currents

From (5.20) $$I_{ab1} = \frac{7\angle 42^0}{\sqrt{3}\angle -30^0} = 4.04\angle 72° \text{ A}$$

$$I_{ab2} = \frac{2.325\angle -112^0}{\sqrt{3}\angle 30^0} = 1.34\angle -142° \text{ A}$$

$$I_{ab0} = 0$$

Since load is balanced, $I_{ab} = \dfrac{V_{ab}}{Z}$ and $I_{ab} + I_{bc} + I_{ca} = \dfrac{1}{Z}[V_{ab} + V_{bc} + V_{ca}] = 0$. Hence, $I_{ab0} = 0$. However, if load is unbalanced then we cannot conclude that I_{ab0} = o. Similarly, symmetrical components of I_{bc} and I_{ca} are found.

Example 5.4: In example 5.2 find the three phase power using symmetrical components.

Solution: The pu values on line to neutral base are taken from example 5.2.

$V_{an1} = 0.9857\angle 43.6°$ pu $I_{a1} = 0.9857\angle 43.6°$ pu

$V_{an2} = 0.2346\angle -109.7°$ pu $I_{a2} = 0.2346\angle -109.7°$ pu.

$V_{an0} = 0.0$ pu. $I_{a0} = 0.0$ pu.

$$S = 3(V_{a0} I_{a0}^* + V_{a1} I_{a1}^* + V_{a2} I_{a2}^*)$$

$$= 3(0 + 0.9857\angle 43.6° \times 0.9857\angle -43.6° + 0.2346\angle -109.7° \times 0.2346\angle 109.7°)$$

$$= 3(0.9857^2 + 0.2346^2) = 3.08 \text{ pu.}$$

$$\text{Base kVA on single phase} = \frac{100}{3}$$

$$\therefore \quad S = 3.08 \times \frac{100}{3} = 102.6 \text{ KVA}$$

Check: The Y-connected resistor ohmic value $R_Y = \dfrac{(400)^2}{100 \times 10^3} = 1.6\ \Omega$

The power can be easily computed by converting the Y to equivalent Δ.
Equivalent Δ – connected resistor $R_\Delta = 3R_Y = 4.8\ \Omega$

$$S_{3-\phi} = \frac{|V_{ab}|^2}{R_\Delta} + \frac{|V_{bc}|^2}{R_\Delta} + \frac{|V_{ca}|^2}{R_\Delta}$$

$$= \frac{320^2 + 480^2 + 400^2}{4.8} = 102.6 \text{ kVA}$$

Thus, power transformation is invariant.

5.4 SEQUENCE IMPEDANCES OF Y AND Δ CIRCUITS

Sequence impedances are the impedances offered to current of a particular sequence. Positive sequence impedance Z_1 is the impedance offered to positive sequence currents, negative sequence impedance Z_2 is the impedance offered to negative sequence current and Z_o is the impedance offered to zero sequence currents.

In (5.14) it was shown that $I_{a0} = \frac{1}{3}(I_a + I_b + I_c)$

If the neutral of a star connected impedance is grounded, the neutral current I_n is given by,

$$I_n = I_a + I_b + I_c = 3\, I_{a0} \qquad ...(5.30)$$

If an impedance Z_n is connected between neutral and ground, the neutral current I_n produces a voltage drop $V_n = 3 I_{a0} Z_n$, between the neutral and the ground. In such a case, we must differentiate between voltages to neutral and voltage to ground, under unbalanced conditions (under balanced conditions both are same, since $I_n = 0$). If V_a is the voltage of phase a with respect to ground and V_n the voltage of the neutral with respect to ground, we can write

$$V_a = V_{an} + V_n$$

Similarly,
$$V_b = V_{bn} + V_n \quad ...(5.31)$$
$$V_c = V_{cn} + V_n.$$

In matrix form

$$\begin{bmatrix} V_a \\ V_b \\ V_c \end{bmatrix} = \begin{bmatrix} V_{an} \\ V_{bn} \\ V_{cn} \end{bmatrix} + \begin{bmatrix} V_n \\ V_n \\ V_n \end{bmatrix} = Z_Y \begin{bmatrix} I_a \\ I_b \\ I_c \end{bmatrix} + 3 \begin{bmatrix} I_{a0}Z_n \\ I_{a0}Z_n \\ I_{a0}Z_n \end{bmatrix} \quad ...(5.32)$$

where, Z_Y is the per phase impedance.

In terms of symmetrical components (5.32) can be written as

$$A \begin{bmatrix} V_{a0} \\ V_{a1} \\ V_{a2} \end{bmatrix} = AZ_Y \begin{bmatrix} I_{a0} \\ I_{a1} \\ I_{a2} \end{bmatrix} + 3I_{a0}Z_n \begin{bmatrix} 1 \\ 1 \\ 1 \end{bmatrix} \quad ...(5.33)$$

Pre-multiplying (5.33) by A^{-1}, we get

$$\begin{bmatrix} V_{a0} \\ V_{a1} \\ V_{a2} \end{bmatrix} = Z_Y \begin{bmatrix} I_{a0} \\ I_{a1} \\ I_{a2} \end{bmatrix} + 3I_{a0}Z_n A^{-1} \begin{bmatrix} 1 \\ 1 \\ 1 \end{bmatrix}$$

$$= Z_Y \begin{bmatrix} I_{a0} \\ I_{a1} \\ I_{a2} \end{bmatrix} + 3I_{a0}Z_n \begin{bmatrix} 1 \\ 0 \\ 0 \end{bmatrix} \quad ...(5.34)$$

Expanding (5.34), we get

$$V_{a0} = I_{a0}\,(Z_Y + 3\,Z_n) = I_{a0}\,Z_0 \quad ...(5.35a)$$
$$V_{a1} = I_{a1}\,Z_Y = I_{a1}\,Z_1 \quad ...(5.35b)$$
$$V_{a2} = I_{a2}\,Z_Y = I_{a2}\,Z_2 \quad ...(5.35c)$$

Equations (5.35a) to (5.35c) are decoupled, meaning current of a particular sequence produces voltage drop of the same sequence, in symmetrical *balanced 3-ϕ impedances*.

The sequence networks representing (5.35) are drawn in Fig. 5.7.

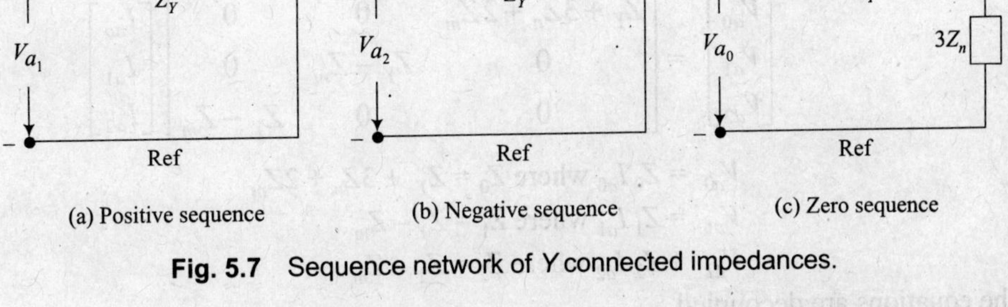

Fig. 5.7 Sequence network of *Y* connected impedances.

In Fig. 5.7, the reference is the ground. It is to be noted that in positive and negative sequence networks, the neutral and ground are at the same potential, but not in the zero sequence network. All three networks are necessary to represent the original networks.

If the neutral is not connected to ground, then $Z_n = \infty$, and no neutral current and hence zero sequence current, can flow. The zero sequence network in this case would be as shown in Fig. 5.8. It we consider a mutual impedance of Z_m between any two phases, then

$$\begin{aligned} V_a &= Z_Y I_a + Z_m I_b + Z_m I_c + Z_n I_n \\ V_b &= Z_m I_a + Z_Y I_b + Z_m I_c + Z_n I_n \\ V_c &= Z_m I_a + Z_m I_b + Z_Y I_c + Z_n I_n \end{aligned}$$

Noting $I_n = I_a + I_b + I_c$, we get

$$\begin{bmatrix} V_a \\ V_b \\ V_c \end{bmatrix} = \begin{bmatrix} Z_Y + Z_n & Z_m + Z_n & Z_m + Z_n \\ Z_m + Z_n & Z_Y + Z_n & Z_m + Z_n \\ Z_m + Z_n & Z_m + Z_n & Z_Y + Z_n \end{bmatrix} \begin{bmatrix} I_a \\ I_b \\ I_c \end{bmatrix}$$

or $V_p = Z^{\text{abc}} I_p$ where

$$Z^{\text{abc}} = \begin{bmatrix} Z_Y + Z_n & Z_m + Z_n & Z_m + Z_n \\ Z_m + Z_n & Z_Y + Z_n & Z_m + Z_n \\ Z_m + Z_n & Z_m + Z_n & Z_Y + Z_n \end{bmatrix}$$

Since $V_p = AV_s$ and $I_p = AI_s$, we get

$$AV_s = Z^{\text{abc}} AI_s$$

Premultiplying by A^{-1} we get

$$V_s = A^{-1} Z^{\text{abc}} AI_s.$$

Substituting for A and A^{-1} we get

$$Z^{012} = A^{-1} Z^{\text{abc}} A = \begin{bmatrix} Z_Y + 3Z_n + 2Z_m & 0 & 0 \\ 0 & Z_Y - Z_m & 0 \\ 0 & 0 & Z_Y - Z_m \end{bmatrix}$$

$$\therefore \quad \begin{bmatrix} V_{a0} \\ V_{a1} \\ V_{a2} \end{bmatrix} = \begin{bmatrix} Z_Y + 3Z_n + 2Z_m & 0 & 0 \\ 0 & Z_Y - Z_m & 0 \\ 0 & 0 & Z_Y - Z_m \end{bmatrix} \begin{bmatrix} I_{ao} \\ I_{a1} \\ I_{a2} \end{bmatrix}$$

or

$$\begin{aligned} V_{a0} &= Z_0 I_{a0} \text{ where } Z_0 = Z_Y + 3Z_n + 2Z_m \\ V_{a1} &= Z_1 I_{a1} \text{ where } Z_1 = Z_Y - Z_m \\ V_{a2} &= Z_2 I_{a2} \text{ where } Z_2 = Z_Y - Z_m \end{aligned}$$

Again the equations are decoupled.

In Δ-connected impedances, the sum of line voltages is always zero. Hence, we can have no zero sequence components of the line voltages. If Z_Δ is the impedance of each phase in the Δ – connected impedance,

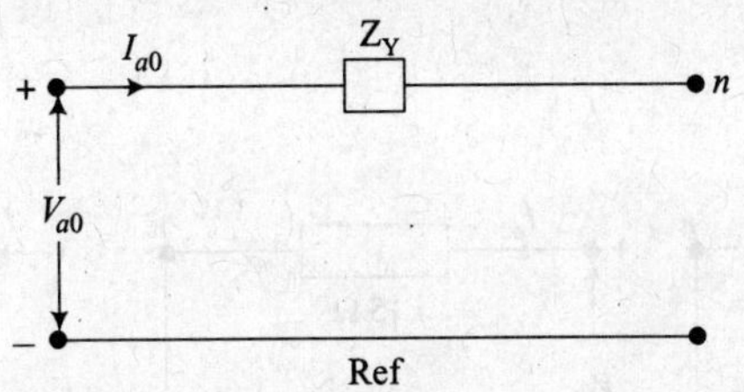

Fig. 5.8 Zero sequence network of ungrounded Y.

$$V_{ab} = Z_\Delta I_{ab};\ V_{bc} = Z_\Delta I_{bc};\ V_{ca} = Z_\Delta I_{ca}; \quad ...(5.36)$$

where, I_{ab}, I_{bc}, I_{ca} are the delta currents.

The sum of the three line voltages being zero,

$$V_{ab} + V_{bc} + V_{ca} = 3V_{ab0} = 3Z_0 I_{ab0} = 0 \quad ...(5.37)$$

From (5.37) $I_{ab0} = 0$, which means that in Δ-connected impedances without mutual coupling, there is no zero sequence component in delta currents.

The zero sequence network is shown in Fig. 5.9.

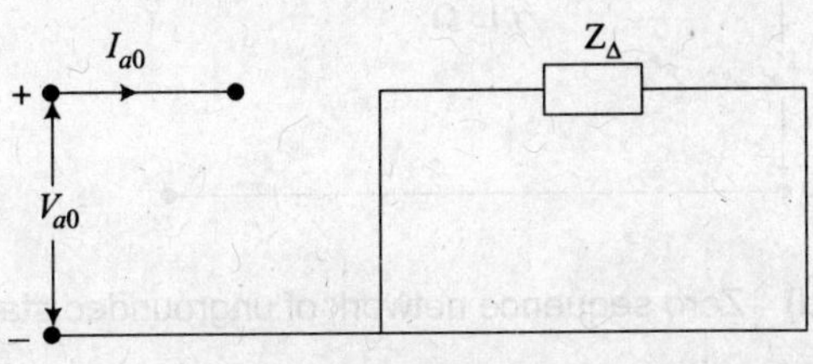

Fig. 5.9 Zero sequence network of a Δ-connected impedance.

The network represents the fact that even if zero sequence currents can flow in the delta, due to zero sequence generated voltages, these cannot exist in the line currents. The positive and negative sequence networks are identical to that shown in Fig. 5.7.

Here, Z_Y represents the equivalent star impedance given by $Z_Y = Z_\Delta/3$.

Example 5.5: Three equal impedances of $j15\ \Omega$ are connected in (a) star grounded (b) star ungrounded (c) star grounded through an impedance $j2\ \Omega$ (d) delta.

Draw the sequence networks for each case.

Solution: (a) Star grounded (Fig. 5.10 a,b,c)

(b) Star ungrounded (Fig. 5.10d)

The positive and negative sequence circuits are as shown in Fig. 5.10a and Fig. 5.10b. The zero sequence circuit is shown in Fig. 5.10d.

(c) Star grounded through an impedance $j2\ \Omega$. (Fig. 5.10e).

The positive and negative sequence circuits are unchanged. The zero sequence circuit is shown in Fig. 5.10e.

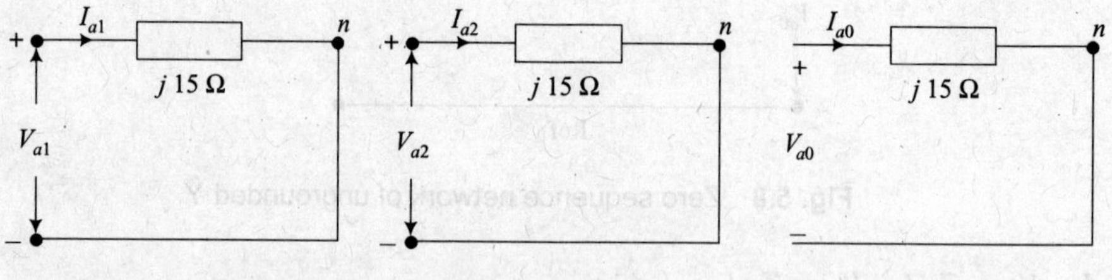

Fig. 5.10 Example 5.5 case (a).

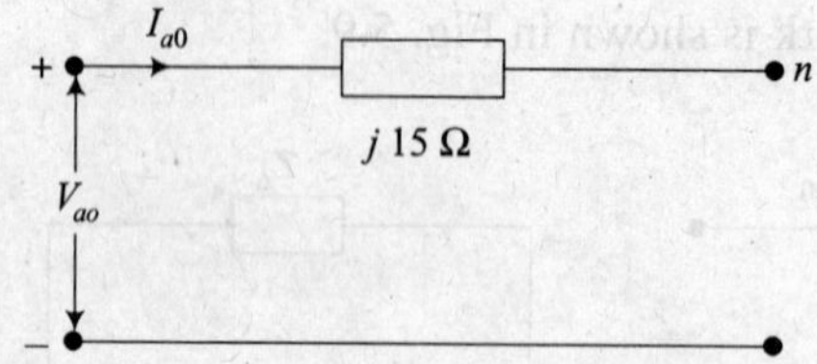

Fig. 5.10(d) Zero sequence network of ungrounded star.

Fig. 5.10e Zero sequence network of star grounded through an impedance.

(d) Delta (Fig. 5.11 a, b, c)

Equivalent star impedance is $j15/3 = j5\ \Omega$. The networks are shown in Fig. 5.11 *a,b,c*.

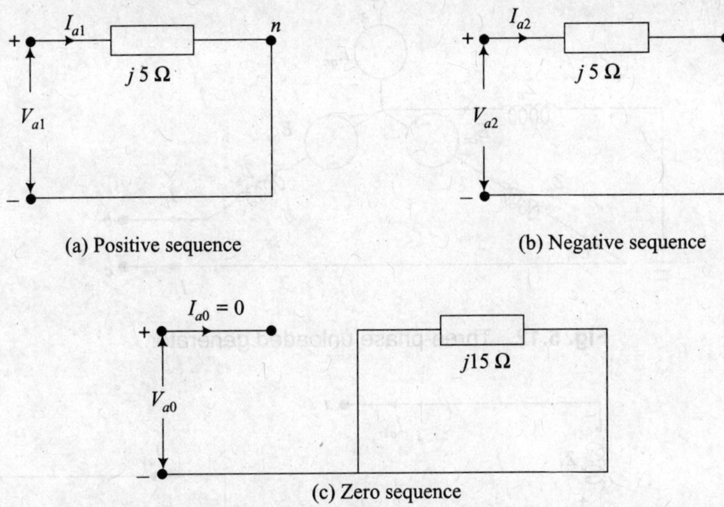

Fig. 5.11 Sequence network of Δ connected impedances.

5.5 SEQUENCE IMPEDANCES AND NETWORKS OF SYNCHRONOUS GENERATOR

The three phase synchronous generator is shown in Fig. 5.12. The neutral is grounded through an impedance Z_n. E_{an}, E_{bn}, E_{cn} are the induced emf's. When the machine is unloaded, the line currents are zero. In the event of a fault, currents flow through the lines and the neutral current $I_n = I_a + I_b + I_c$. Unbalanced line currents can be resolved into symmetrical components and sequence networks derived for the generator.

Since, the machine is designed with symmetrical windings, it induces emf's of positive sequence only. Positive sequence impedance, is the impedance offered by the windings to the flow of positive sequence currents. A machine can offer different positive sequence reactances during the sub-transient, transient and steady state periods, as discussed in detail in chapter 4. If the short circuit occurs on an unloaded generator, the terminal voltage prior to short circuit is the positive sequence voltage. If short circuit occurs on a loaded generator, the voltage behind the appropriate reactance is the positive sequence voltage.

Figure 5.13(a) shows the positive sequence network for the three phases and Fig. 5.13(b), the single phase equivalent. The reference bus is at the neutral potential. With the given directions

$$V_{a1} = E_{an} - I_{a1} Z_1 \qquad ...(5.38)$$

The negative sequence impedance is the impedance offered to flow of negative sequence currents. Since, the phase sequence of negative sequence currents is opposite to the positive sequence, the field created by negative sequence currents rotates at double the synchronous speed with respect to the rotor. While sweeping over the rotor surface, the negative sequence m.m.f. alternately crosses the direct – axis and quadrature-axis. The negative sequence impedance is defined as

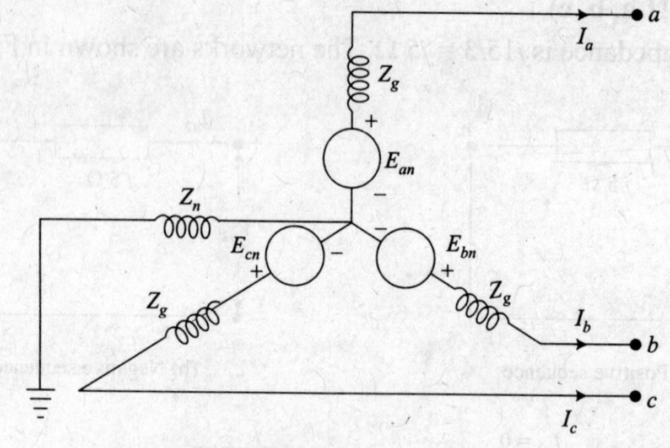

Fig. 5.12 Three-phase unloaded generator.

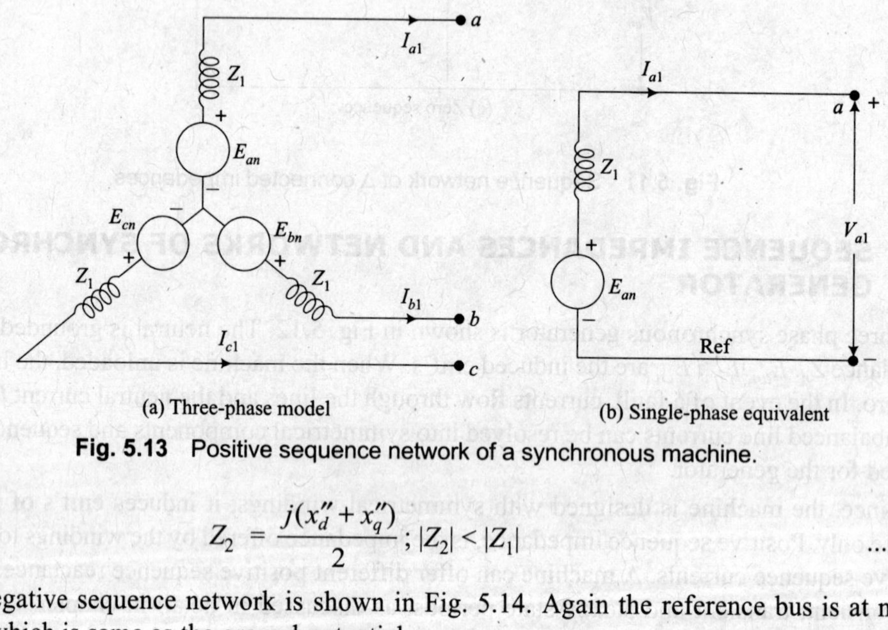

(a) Three-phase model (b) Single-phase equivalent

Fig. 5.13 Positive sequence network of a synchronous machine.

$$Z_2 = \frac{j(x_d'' + x_q'')}{2}; \; |Z_2| < |Z_1| \qquad ...(5.39)$$

The negative sequence network is shown in Fig. 5.14. Again the reference bus is at neutral potential, which is same as the ground potential.

$$V_{a2} = -Z_2 I_{a2} \qquad ...(5.40)$$

Since, the generator does not produce negative sequence voltages, the negative sequence network does not contain any source.

The impedance offered by the windings, to the flow of zero sequence currents is the zero sequence impedance Z_{0g}. The zero sequence networks for three phase and the single phase equivalent are shown in Fig. 5.15a and Fig. 5.15b respectively.

Referring to Fig. 5.15a, it can be seen that the voltage of terminal a with respect to ground is

$$V_{a0} = -3Z_n I_{a0} - Z_{0g} I_{a0} = -(3Z_n + Z_{0g}) I_{a0}. \qquad ...(5.41)$$

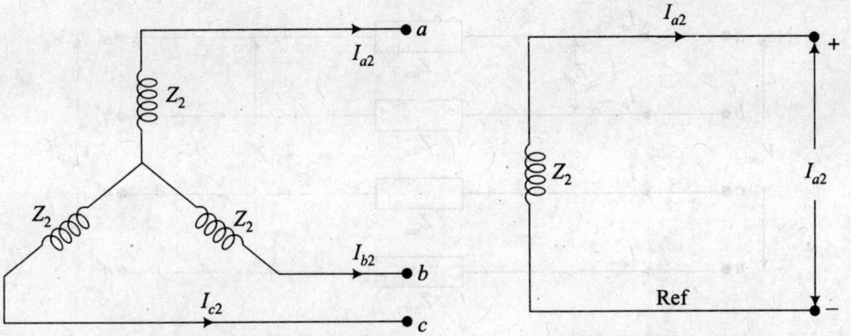

(a) Three-phase model (b) Single-phase equivalent

Fig. 5.14 Negative sequence network of a synchronous machine.

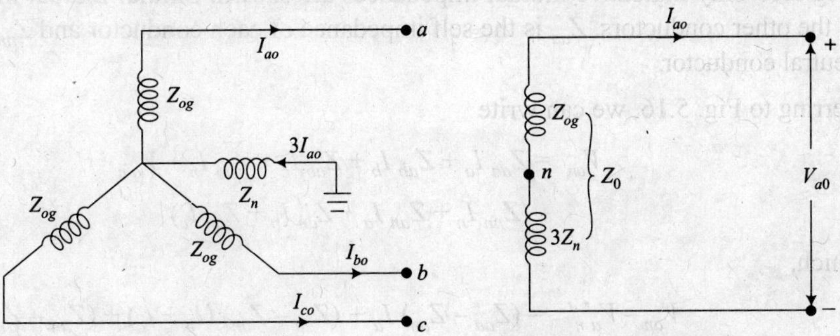

(a) Three-phase model (b) Single-phase equivalent

Fig. 5.15 Zero sequence network of a synchronous machine.

$$= -Z_0 I_{a0}$$

where, $$Z_0 = 3Z_n + Z_{0g} \qquad ...(5.42)$$

This is represented in the zero sequence network.

The reference bus is at ground potential, which is not same as the neutral potential.

When only zero sequence currents flows through the armature winding, the maximum mmf produced by one phase is displaced 120 electrical degrees in space from the point of maximum mmf of each of the other phases. If the mmf had a perfectly sinusoidal distribution, the sum of the three mmf's would be zero and no flux would be produced. In practical situation, the mmf is not perfectly sinusoidal, and a small flux is produced. Hence, the zero sequence reactance has a small value, much lower than positive and negative sequence reactances.

5.6 SEQUENCE IMPEDANCES OF TRANSMISSION LINES

Three phase transmission lines are transposed, so that the per phase impedance of the three lines have the same value. Fig. 5.16 shows the section of a transmission line, with the neutral conductor. Under balanced conditions, no current flows through the neutral conductor. When the currents I_a, I_b, I_c are unbalanced, the neutral conductor serves as a return path. Mutual coupling between the three phase conductors is assumed to be equal and designated as Z_{ab}. Similarly, the coupling be-

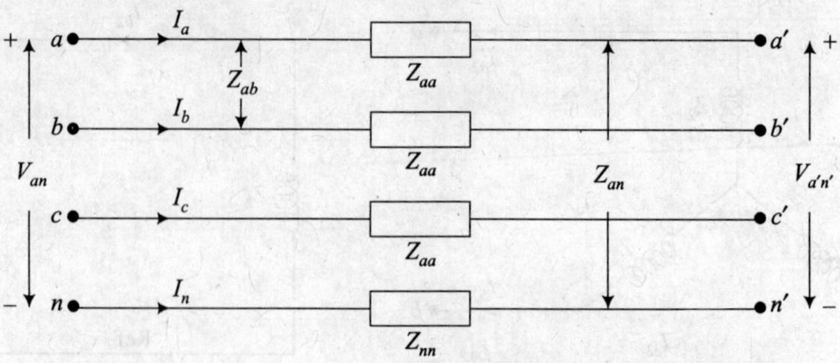

Fig. 5.16 Three-phase transmission line section.

tween the neutral conductor and each phase conductor is assumed to be identical and designated as Z_{an}. In Fig 5.16 only indicative mutual impedances are shown. Similar mutual impedances exist between the other conductors. Z_{aa} is the self impedance of each conductor and Z_{nn} the impedance of the neutral conductor.

Referring to Fig. 5.16, we can write

$$V_{an} = Z_{aa} I_a + Z_{ab} I_b + Z_{ab} I_c + Z_{an} I_n + V_{a'n'} - (Z_{nn} I_n + Z_{an} I_a + Z_{an} I_b + Z_{an} I_c) \qquad ...(5.43)$$

From which,

$$V_{an} - V_{a'n'} = (Z_{aa} - Z_{an}) I_a + (Z_{ab} - Z_{an}) (I_b + I_c) + (Z_{an} - Z_{nn}) I_n.$$

Now $I_n = -(I_a + I_b + I_c)$ with the given directions. Substituting for I_n in (5.43), we get after rearranging the terms

$$V_{an} - V_{a'n'} = (Z_{aa} + Z_{nn} - 2Z_{an})I_a + (Z_{ab} + Z_{nn} - 2Z_{an})I_b + (Z_{ab} + Z_{nn} - 2Z_{an})I_c \qquad ...(5.44)$$

Let

$$Z_s = Z_{aa} + Z_{nn} - 2Z_{an}$$

$$Z_m = Z_{ab} + Z_{nn} - 2Z_{an} \qquad ...(5.45)$$

With these definitions (5.44) can be written as

$$V_{an} - V_{a'n'} = V_{aa'} = Z_s I_a + Z_m I_b + Z_m I_c \qquad ...(5.46)$$

Similarly, we can obtain equations for $V_{bb'}$ and $V_{cc'}$.

In matrix form, we can express in the form

$$\begin{bmatrix} V_{aa'} \\ V_{bb'} \\ V_{cc'} \end{bmatrix} = \begin{bmatrix} V_{aa'} \\ V_{bb'} \\ V_{cc'} \end{bmatrix} = \begin{bmatrix} V_{an} - V_{a'n'} \\ V_{bn} - V_{b'n'} \\ V_{cn} - V_{c'n'} \end{bmatrix} = \begin{bmatrix} Z_s & Z_m & Z_m \\ Z_m & Z_s & Z_m \\ Z_m & Z_m & Z_s \end{bmatrix} \begin{bmatrix} I_a \\ I_b \\ I_c \end{bmatrix} \qquad ...(5.47)$$

$V_{aa'}$, $V_{bb'}$ and $V_{cc'}$ are the drops in the phase conductors. In terms of symmetrical component, (5.47) can be written as

$$[A]\begin{bmatrix} V_{aa'0} \\ V_{aa'1} \\ V_{aa'2} \end{bmatrix} = \begin{bmatrix} Z_s & Z_m & Z_m \\ Z_m & Z_s & Z_m \\ Z_m & Z_m & Z \end{bmatrix}[A]\begin{bmatrix} I_{a0} \\ I_{a1} \\ I_{a2} \end{bmatrix} \qquad ...(5.48)$$

Premultiplying by $[A]^{-1}$ and simplifying, we get

$$\begin{bmatrix} V_{aa'0} \\ V_{aa'1} \\ V_{aa'2} \end{bmatrix} = \begin{bmatrix} V_{aa'0} \\ V_{aa'1} \\ V_{aa'2} \end{bmatrix} = \begin{bmatrix} Z_s + 2Z_m & 0 & 0 \\ 0 & Z_s - Z_m & 0 \\ 0 & 0 & Z_s - Z_m \end{bmatrix}\begin{bmatrix} I_{a0} \\ I_{a1} \\ I_{a2} \end{bmatrix} \qquad ...(5.49)$$

The sequence impedances are defined as

$$\begin{aligned} Z_0 &= Z_s + 2Z_m \\ Z_1 &= Z_s - Z_m \\ Z_2 &= Z_s - Z_m \end{aligned} \qquad ...(5.50)$$

Thus,

$$\begin{aligned} V_{aa'0} &= Z_0 I_{a0} \\ V_{aa'1} &= Z_1 I_{a1} \\ V_{aa'2} &= Z_2 I_{a2} \end{aligned} \qquad ...(5.51)$$

From (5.51) the following observations can be made.

- The equations are decoupled, meaning the sequence currents of a particular type produce voltages of the same sequence only. Hence, we can draw separate networks, without any coupling as shown in Fig. 5.17.
- The positive and negative sequence impedances are same.It can be observed that the neutral-conductor impedances Z_{nn} and Z_{an}, affect only the zero sequence impedance Z_o. Most transmission lines have two overhead conductors called ground wires which are grounded at intervals along the length of the line. These provide return path similar to Z_{nn} and Z_{an} and hence affect the zero sequence impedance. The zero sequence impedance of transmission lines is around 2 to 3.5 times that of positive and negative sequence reactance.

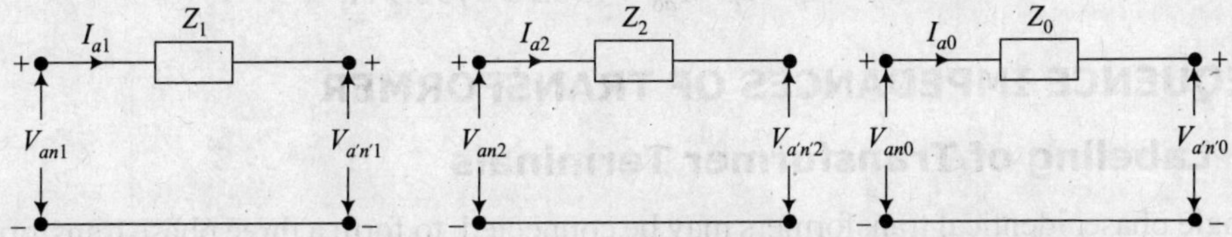

(a) Positive sequence (b) Negative sequence (c) Zero sequence

Fig. 5.17 Sequence networks of transmission lines.

Example 5.6: In Fig. 5.16, the voltages are as follows.

$V_{an} = 91 + j35$ kV $\qquad V_{a'n'} = 77 + j14$ kV

$V_{bn} = 36.12 - j16.31$ kV $\qquad V_{b'n'} = 22.12 - j37.31$ kV

$V_{cn} = -85.12 + j44.31$ kV $\qquad V_{c¢n¢} = -99.12 + j23.31$ kV

The line impedances are given by

$Z_{aa} = j50\ \Omega;\ Z_{ab} = j22\ \Omega;\ Z_{nn} = j80\ \Omega;\ Z_{an} = j30\ \Omega$

Determine line currents using symmetrical components.

Solution:

$$Z_s = Z_{aa} + Z_{nn} - 2Z_{an}$$

$$= j50 + j80 - j60 = j70\ \Omega$$

$$Z_m = Z_{ab} + Z_{nn} - 2Z_{an}$$

$$= j22 + j80 - j60 = j42\ \Omega$$

$$Z_0 = Z_s + 2Z_m = j70 + 2 \times j42 = j154\ \Omega$$

$$Z_1 = Z_2 = Z_s - Z_m = j70 - j42 = j28\ \Omega$$

$$\begin{bmatrix} V_{aa'0} \\ V_{aa'1} \\ V_{aa'2} \end{bmatrix} = A^{-1}\ \text{kV.} \begin{bmatrix} V_{an} & -V_{a'n'} \\ V_{bn} & -V_{b'n'} \\ V_{cn} & V_{c'n'} \end{bmatrix} = \frac{1}{3}\begin{bmatrix} 1 & 1 & 1 \\ 1 & a & a^2 \\ 1 & a^2 & a \end{bmatrix}\begin{bmatrix} 14 + j21.0 \\ 14 + j21.0 \\ 14 + j21.0 \end{bmatrix} = \begin{bmatrix} 14 + j21 \\ 0 \\ 0 \end{bmatrix}$$

$\therefore$

$$V_{aa'0} = 14 + j21\ \text{kV}.$$

$$V_{aa'1} = 0;$$

$$V_{aa'2} = 0;$$

$$I_{a0} = \frac{V_{aa'0}}{Z_0} = \frac{(14 + j21) \times 10^3}{j154} = 136.36 - j0.99\ \text{A}.$$

$$I_{a1} = \frac{V_{aa'1}}{Z_1} = 0;\ I_{a2} = \frac{V_{aa'2}}{Z_2} = 0;$$

$$I_a = I_b = I_c = I_{ao} = 136.36 - j90.9\ \text{A}.$$

5.7 SEQUENCE IMPEDANCES OF TRANSFORMER

5.7.1 Labeling of Transformer Terminals

Three single phase, identical transformers may be connected, to form a three phase transformer. An alternative is to build all the three phases on the same iron structure, to form a three phase transformer.

According to American standard, the high-voltage winding terminals are labeled H_1, H_2, H_3 and the low-voltage winding terminals are marked X_1, X_2, X_3. In this section A, B, C letters are used to identify phases of high-voltage side and lower case a, b, c for phases of low-voltage side. The

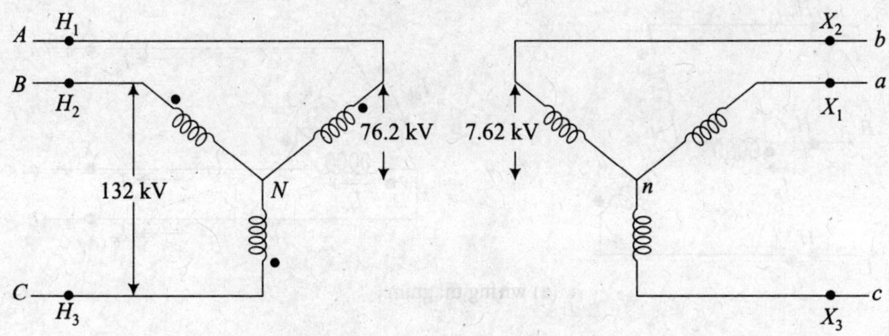

Fig. 5.18 *Y-Y* transformer.

standard markings are such that in a *Y-Y* or Δ-Δ transformer, the voltages to neutral from H_1, H_2, H_3 are in phase with the voltages to neutral from X_1, X_2, X_3, respectively. In the Δ winding, the voltage is measured with respect to the ground of the part of the system to which the Δ winding is connected. In transferring ohmic values from one voltage level to another level, the multiplying factor is the *square of the ratio of line-to-line voltages, irrespective of the connection.* In case of *Y*-Δ or Δ-*Y* transformers, the connections are such that the positive sequence voltage drop from H_1 to neutral leads the positive sequence voltage from X_1 to neutral by 30°, regardless of which (*Y* or Δ) is connected on HV side.

Consider a 132/13.2 kV *Y-Y* transformer. The wiring diagram is shown in Fig. 5.18. Windings of the primary and secondary, which are drawn parallel, are from the same single phase transformer or on the same leg of a three phase transformer. In Fig. 5.18, the flux linking the winding *AN* is same as flux linking winding *an*. Hence, V_{AN} is in phase with V_{an}. Wiring diagrams can be similarly drawn for other connections like Δ-Δ, *Y*-Δ, etc.

5.7.2 Phase Shift in Star-Delta Transformer

A phase shift occurs in star-delta transformers. Consider a *Y*-Δ transformer as shown in Fig. 5.19a. The phasor diagrams for positive and negative sequence components are shown in Fig. 5.19b and 5.19c.

From the wiring diagram, we observe V_{AN1} is in phase with V_{ab1}, according to the labeling standards. The phasor diagram is built from this step. From the phasor diagram, it can be seen that

$$V_{AB1} = \sqrt{3}\ \frac{N_1}{N_2}\ V_{ab1}\angle 30° \qquad ...(5.52)$$

where, N_1, N_2 are phase to neutral turns. If the base voltages on either side are chosen in the same ratio as the line-to-line voltages, then in per unit we get

$$V_{AB1}(\text{pu}) = V_{ab1}\angle 30°\ (\text{pu}) \qquad ...(5.53)$$

Thus, the positive sequence line voltages on star side lead the positive sequence line voltages on delta side by 30°. From Fig. 5.19c it can be seen that the negative sequence, line-to-line voltages on the star side lag the negative sequence line voltages on the delta side by 30°. If the windings are reversed on the delta side as shown in Fig. 5.19(d), the delta side positive sequence

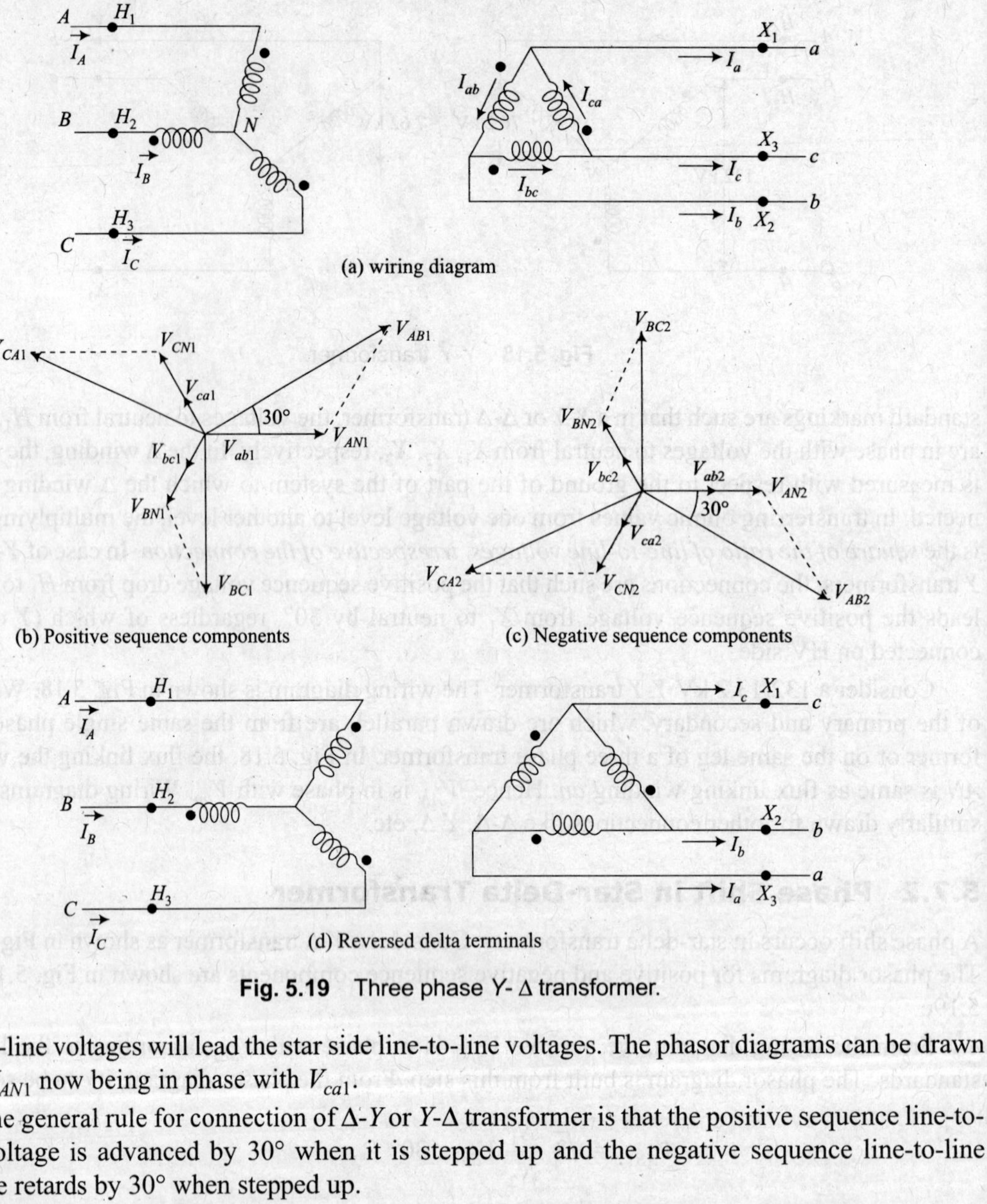

Fig. 5.19 Three phase Y- Δ transformer.

line-to-line voltages will lead the star side line-to-line voltages. The phasor diagrams can be drawn with V_{AN1} now being in phase with V_{ca1}.

The general rule for connection of Δ-Y or Y-Δ transformer is that the positive sequence line-to-line voltage is advanced by 30° when it is stepped up and the negative sequence line-to-line voltage retards by 30° when stepped up.

Equating the per unit complex power on each side we can write,

$$V_{AN1}\, I_{A1}^{\;*} = V_{an1}\angle 30° \; I_{an1}^{\;*} \angle -30° = V_{an1}\, I_{an1}^{\;*} \qquad \text{...(5.54)}$$

$$\frac{I_{an1}}{I_{A1}} = \frac{V_{AN1}^{\;*}}{V_{an1}^{\;*}} \qquad \text{...(5.55)}$$

Therefore, *the current ratio of any transformer with phase shift, is the reciprocal of the complex conjugate of the voltage ratio.*

5.7.3 Sequence Impedances of Transformers

The positive sequence impedances of a transformer is nothing but its leakage impedance. Since, the transformer is a static device, the leakage impedance does not change with the phase sequence. Therefore, $Z_1 = Z_2$. If the connections permit the flow of zero sequence currents, the zero sequence impedance may differ slightly from positive and negative sequence impedances. However, it is normal to assume all the three impedances to be equal.

To obtain the zero sequence network, we have to note the following points:

(i) If magnetizing current is neglected, the transformer primary carries current only if there is current on the secondary side.

(ii) No zero sequence currents can flow in the lines connected to a delta connected transformer, though they can flow in the delta windings.

(iii) Zero sequence currents can flow in the star connection only if the star point is grounded.

The zero sequence networks for all transformer connections are shown in Fig. 5.20. Each case is now discussed in detail.

Case 1. *Y-Y* transformer with both neutrals grounded: When both neutrals are grounded, a path exists for the zero sequence currents in both windings, via the grounded neutrals. Hence, in the zero sequence network, points *P* and *Q* in Fig. 5.20, are connected by the zero sequence impedance.

Case 2. *Y-Y* transformer bank with one neutral grounded: If only one of the neutrals is grounded, zero sequence currents cannot flow in the ungrounded star and hence cannot also flow in the grounded star. Hence, an open circuit exists in the zero sequence network,

Case 3. Δ-Δ transformer bank: Since, the zero sequence currents cannot flow on the lines connected to delta, there is an open circuit between *P* and *Q*. The zero sequence currents can however circulate in the delta windings, if zero sequence voltages are induced in either delta windings.

Case 4. *Y*-Δ transformer bank, with grounded star: If the star side is grounded, zero sequence currents can flow in the star because a path is available to the ground and the balancing zero sequence currents can flow in the delta. It cannot however flow in the lines on the delta side. Hence, an open circuit exists on the delta side and a path from reference bus, through an impedance Z_0, on the star side. If the star neutral is grounded through an impedance Z_n, $3Z_n$ is added in series with Z_0.

Case 5. *Y*-Δ transformer bank, with ungrounded star: This is similar to case (4) with $Z_n = \infty$. No zero sequence currents can flow in the transformer windings.

5.8 SEQUENCE NETWORKS OF A POWER SYSTEM

The sequence networks for various power system components have been developed. Using them the complete sequence networks of the power system can be constructed.

The positive sequence networks are constructed from the one line diagram. The generated voltages are all positive sequence voltages. The negative sequence network is similar to the positive sequence network, except for the absence of generated voltages. The negative sequence

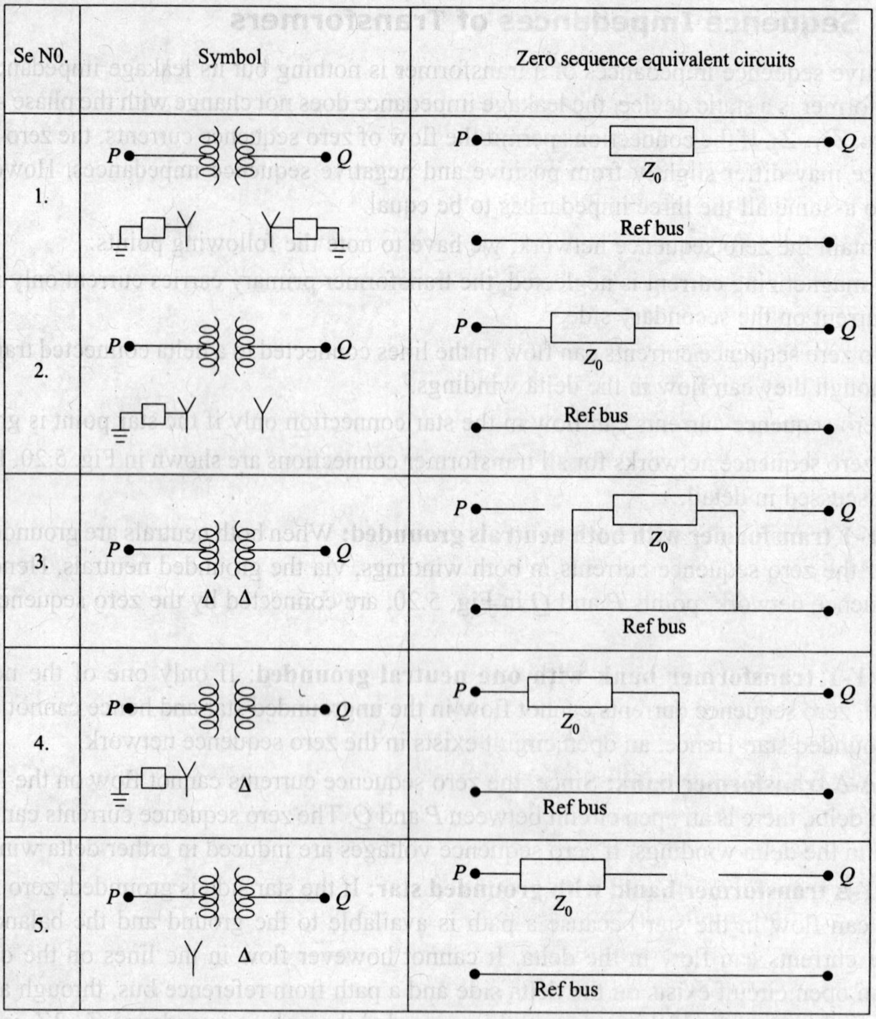

Fig. 5.20 Zero sequence networks for 3 – Φ transformers.

impedance, for static components like transmission lines and transformers, is same as the positive sequence impedance. In synchronous machines the negative sequence impedance is different. The reference bus for the positive and negative sequence networks, is the system neutral (or ground, since both are at same potential). Since, the positive sequence and negative sequence currents are balanced and displaced by 120°, they add up to zero and do not flow through the neutral. Therefore, any impedance connected between neutral and ground, is not included in these networks. Zero sequence sub-networks for various parts of a system are combined to from the complete zero sequence network. Any impedance between neutral and ground, is represented by an impedance, three times its value in the zero sequence network. The zero sequence network of the transformer must take into account the connections as discussed in section 5.7. No voltage sources are present in the zero sequence network.

Example 5.7: Draw the sequence network for the network shown in Fig. 5.21.

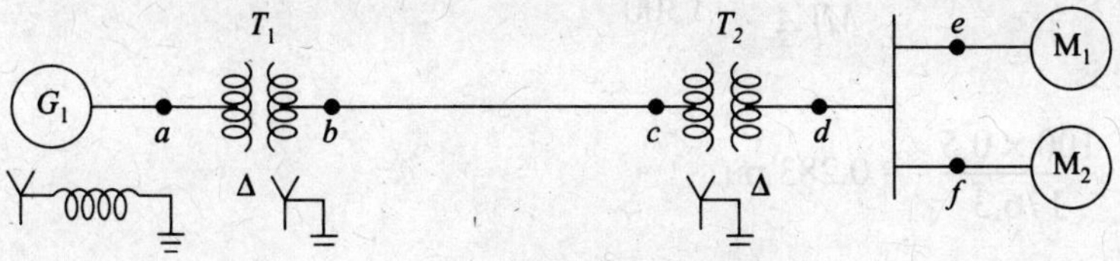

Fig. 5.21 Example 5.7

The data is as follows:

G: 300 MVA, 20 kV, $x_d'' = 15\%, x_o = 5\%; Z_n = 0.4\ \Omega$.

M_1: 13.2 kV, 200 MVA, $x_d'' = 20\%, x_o = 5\%; Z_n = 0.5\ \Omega$.

M_2: 13.2kV, 100 MVA, $x_d'' = 20\%, x_o = 5\%$.

T_1: 230/20 kV; 300 MVA; $X = 10\%$

T_2: Three single phase transformers rated 132/13.2 kV, 100 MVA, $X = 10\%$

Transmission line: 100Km; reactance = 0.5 Ω/km; $Z_o = 3Z_1$.

Solution: Positive sequence network

Three phase power rating of T_2 is

$$3 \times 100 = 300 \text{ MVA}.$$

and line-to-line voltage ratio is

$$\frac{\sqrt{3} \times 132}{13.2} = \frac{228.6}{13.2} \text{ kV}$$

The system base MVA is taken to be 300 MVA.

The generator voltage is taken as base in the generator circuit.

Voltage base in generator circuit = 20 kV

$$\text{Voltage base in transmission line} = 20 \times \frac{230}{20} = 230 \text{ kV}.$$

$$\text{Voltage base in motor circuit} = 230 \times \frac{13.2}{228.6} = 13.28 \text{ kV}.$$

The reactance's converted to the proper bases are as follows:

Generator: $x_d'' = 0.15$ pu.

T_1: $X = 0.1$ pu.

T_2: $$X = 0.1\left(\frac{13.2}{13.28}\right)^2 = 0.0988 \text{ pu}.$$

Base impedance of transmission line is

$$Z_B = \frac{kV^2}{MVA} = \frac{(230)^2}{300} = 176.3\ \Omega$$

Reactance of line is

$$\frac{100 \times 0.5}{176.3} = 0.283 \text{ pu.}$$

M_1: $$x_d'' = 0.2\left(\frac{300}{200}\right)\left(\frac{13.2}{13.28}\right)^2 \quad 0.2963 \text{ pu.}$$

M_2: $$x_d'' = 0.2\left(\frac{300}{100}\right)\left(\frac{13.2}{13.28}\right)^2 = 0.593 \text{ pu.}$$

The positive sequence network is shown in Fig 5.22a.

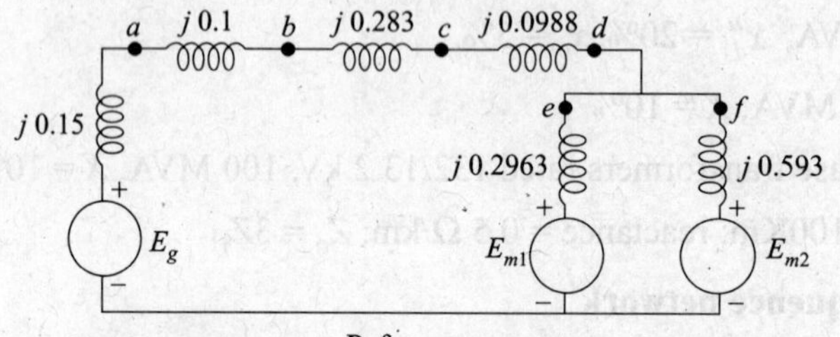

Fig. 5.22a Positive sequence network of example 5.7.

Negative sequence network: Since, all positive sequence reactances and negative sequence reactances are equal, the negative sequence network is identical to the positive sequence network, except for the voltage source omission. The network is shown in Fig. 5.22b.

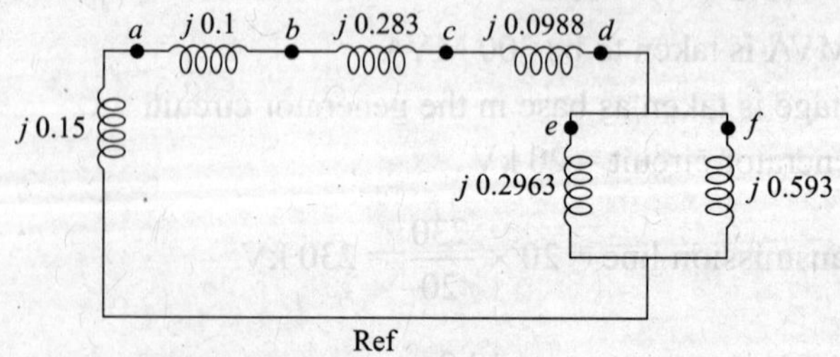

Fig. 5.22b Negative sequence network of example 5.7.

Zero sequence network

Generator: $x_o = 0.05$ pu.

$$\text{Base impedance in generator circuit} = \frac{20^2}{300} = 1.333\ \Omega.$$

$$3Z_n = 3 \times 0.4\ \Omega = \frac{1.2}{1.333} = 0.9 \text{ pu.}$$

T_1: $X_o = X_1 = 0.1$ pu.

T_2: $X_o = X_1 = 0.0988$ pu.

Line: $X_o = 3X_1 = 3 \times 0.283 = 0.849$ pu.

M_1: $X_o = 0.05$ pu.

$$\text{Base impedance} = \frac{13.28^2}{300} = 0.588\ \Omega$$

$$3Z_n = 3 \times 0.5\ \Omega = \frac{1.5}{0.588} = 2.55 \text{ pu}$$

M_2: $X_o = 0.05$ pu.

The zero sequence network is shown in Fig. 5.22c.

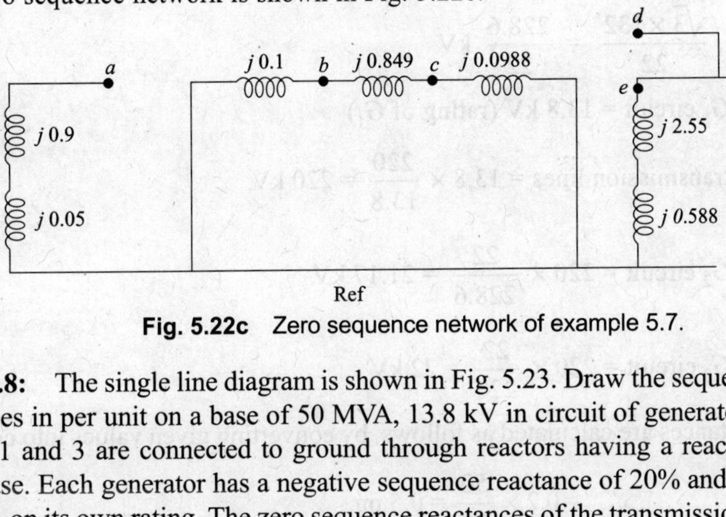

Fig. 5.22c Zero sequence network of example 5.7.

Example 5.8: The single line diagram is shown in Fig. 5.23. Draw the sequence networks. Mark all reactances in per unit on a base of 50 MVA, 13.8 kV in circuit of generator 1. The neutrals of generators 1 and 3 are connected to ground through reactors having a reactance of 6%, on the machine base. Each generator has a negative sequence reactance of 20% and zero sequence reactance of 5% on its own rating. The zero sequence reactances of the transmission lines are 2.5 times the positive sequence reactance. The other ratings are given below.

G_1: 20 MVA, 13.8 kV, $x''_d = 0.2$ pu.

G_2: 30 MVA, 20 kV, $x''_d = 0.2$ pu

G_3: 30 MVA, 20 kV, $x''_d = 0.2$ pu

T_1: 25 MVA, 220Y/13.8Δ kV, $X = 10\%$.

T_2: Single phase units, each rated 10 MVA, 132/22 kV, $X = 10\%$

T_3: 35 MVA, 220Y/22Y kV, $X = 10\%$

$X_{L1} = 80\ \Omega$

$X_{L2} = 100\ \Omega$

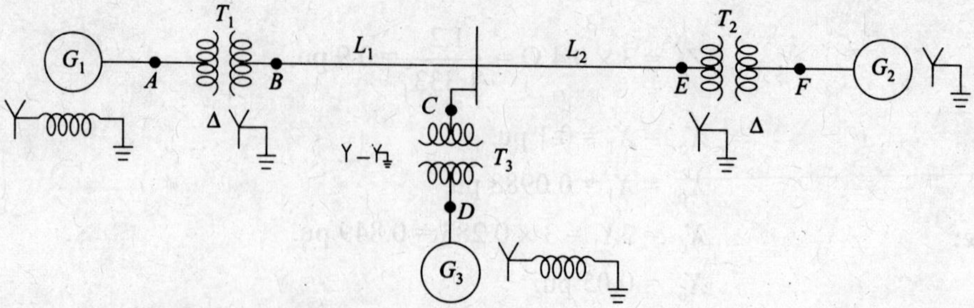

Fig. 5.23 One line diagram of example 5.8.

Solution: Choose a base of 50 MVA in the system.

Three phase power rating of T_2 is given by,

$$3 \times 10 = 30 \text{ MVA}.$$

Line-to-line voltage rating of T_2 is given by,

$$\frac{\sqrt{3} \times 132}{22} = \frac{228.6}{22} \text{ kV}$$

Base voltage in G_1 circuit = 13.8 kV (rating of G_1)

$$\text{Base voltage in transmission lines} = 13.8 \times \frac{220}{13.8} = 220 \text{ kV}.$$

$$\text{Base voltage in } G_2 \text{ circuit} = 220 \times \frac{22}{228.6} = 21.17 \text{ kV}.$$

$$\text{Base voltage in } G_3 \text{ circuit} = 220 \times \frac{22}{220} = 22 \text{ kV}$$

The various reactances are calculated as follows, by converting given values into correct base.

G_1:

$$x_d'' = 0.2 \times \frac{50}{20} = 0.5 \text{ pu.}$$

$$x_1 = x_d'' = 0.5 \text{ pu.}$$

$$x_2 = 0.2 \times \frac{50}{20} = 0.5 \text{ pu.}$$

$$x_0 = 0.05 \times \frac{50}{20} = 0.125 \text{ pu.}$$

$$3x_n = 3 \times 0.06 \times \frac{50}{20} = 0.45 \text{ pu.}$$

G$_2$:

$$x_1 = x_d'' = 0.2 \times \left(\frac{50}{30}\right) \times \left(\frac{20}{21.17}\right)^2 = 0.2975 \text{ pu.}$$

$$x_2 = 0.2 \times \left(\frac{50}{30}\right) \times \left(\frac{20}{21.17}\right)^2 = 0.2975 \text{ pu.}$$

$$x_0 = 0.05 \times \left(\frac{50}{30}\right) \times \left(\frac{20}{21.17}\right)^2 = 0.074 \text{ pu.}$$

G$_3$:

$$x_1 = x_d'' = 0.20 \times \frac{50}{30} \times \left(\frac{20}{21.17}\right)^2 = 0.2755 \text{ pu.}$$

$$x_2 = 0.2 \times \frac{50}{30} \times \left(\frac{20}{22}\right)^2 = 0.2755 \text{ pu.}$$

$$x_0 = 0.05 \times \frac{50}{30} \times \left(\frac{20}{22}\right)^2 = 0.0689 \text{ pu.}$$

$$3x_n = 3 \times 0.06 \times \frac{50}{30} \times \left(\frac{20}{22}\right)^2 = 0.2478 \text{ pu.}$$

T$_1$:

$$X = 0.1 \times \frac{50}{25} = 0.2 \text{ pu}$$

T$_2$:

$$X = 0.1 \times \frac{50}{30} \times \left(\frac{22}{21.17}\right)^2 = 0.18 \text{ pu.}$$

T$_3$:

$$X = 0.1 \times \frac{50}{35} = 0.143 \text{ pu.}$$

L$_1$: Base impedance in lines $= \dfrac{(220)^2}{50} = 968\ \Omega$

$$X_1 = X_2 = \frac{80}{968} = 0.0826 \text{ pu.}$$

$$X_o = 2.5X_1 = 2.5 \times 0.0826 = 0.2065 \text{ pu.}$$

L$_2$: Base impedance = 968 Ω.

$$X_1 = X_2 = \frac{100}{968} = 0.1033 \text{ pu.}$$

$$X_0 = 2.5X_1 = 0.2582 \text{ pu.}$$

The sequence networks are shown in Fig. 5.24.

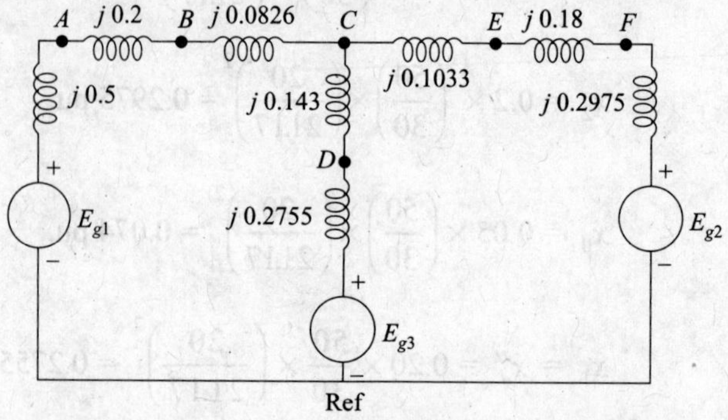

Fig. 5.24a Positive sequence network of example 5.8.

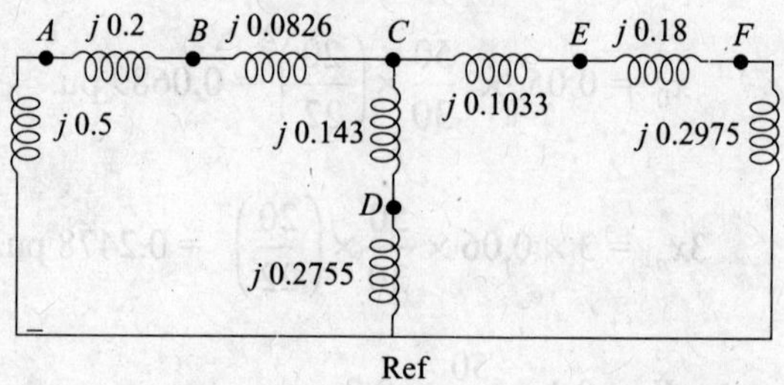

Fig. 5.24b Negative sequence network of example 5.8.

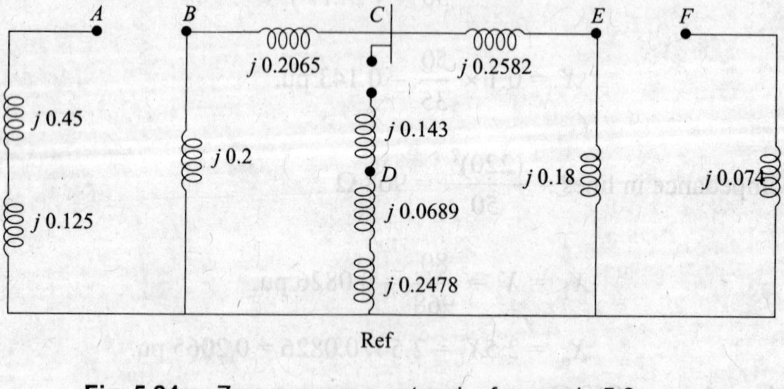

Fig. 5.24c Zero sequence network of example 5.8.

Example 5.9: Draw the zero sequence network of the one line diagram shown in Fig. 5.25.

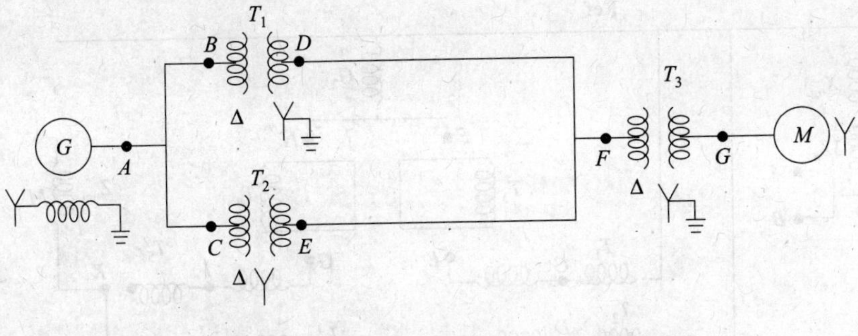

Fig. 5.25 One line diagram of example 5.9.

Solution: The zero sequence network is shown in Fig. 5.26.

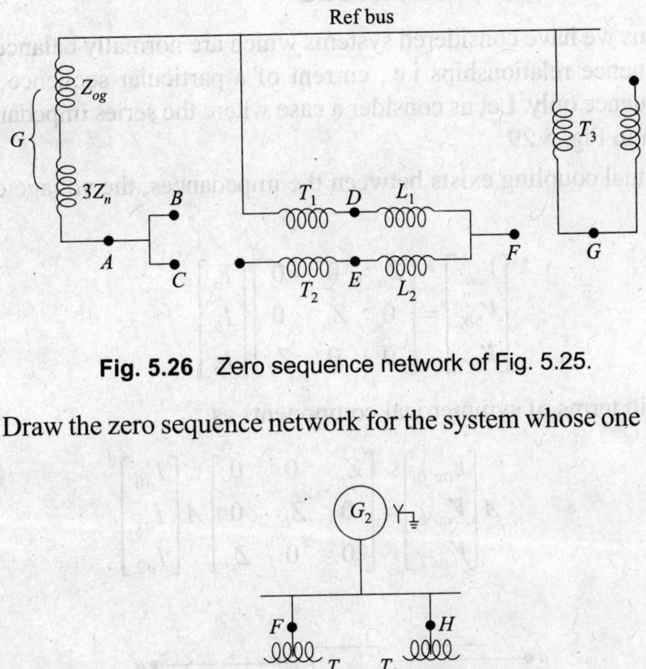

Fig. 5.26 Zero sequence network of Fig. 5.25.

Example 5.10: Draw the zero sequence network for the system whose one line diagram is shown in Fig. 5.27.

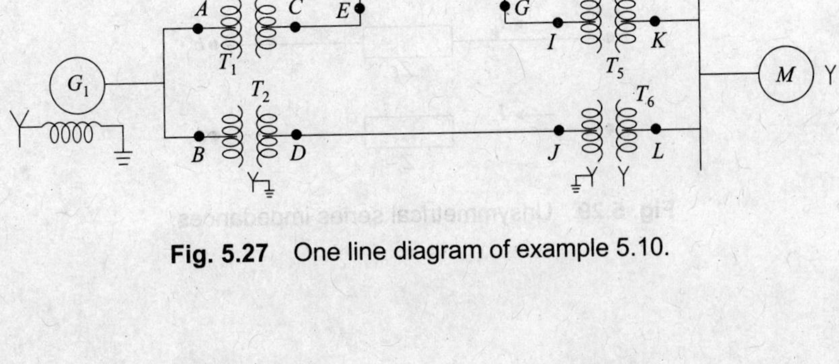

Fig. 5.27 One line diagram of example 5.10.

Solution: The zero sequence network is shown in Fig. 5.28.

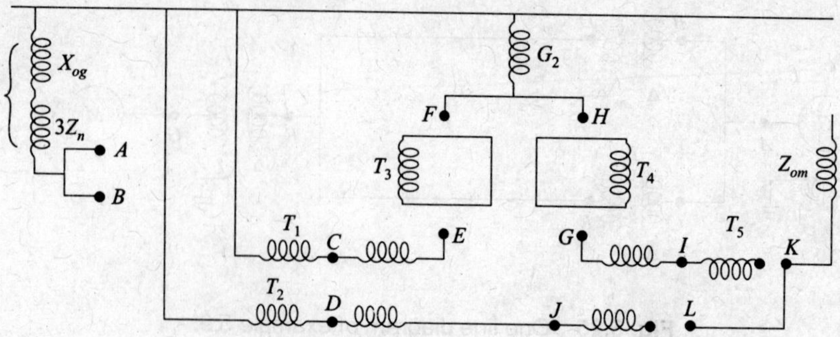

Fig. 5.28 Zero sequence network of example 5.10.

5.9 UNSYMMETRICAL IMPEDANCES

In the previous sections we have considered systems which are normally balanced. Such a system, led to decoupled sequence relationships i.e., current of a particular sequence produced voltage drops of the same sequence only. Let us consider a case where the series impedances are not equal; $Z_a \neq Z_b \neq Z_c$ as shown in Fig. 5.29.

Assuming no mutual coupling exists between the impedances, the voltage drops can be written as,

$$\begin{bmatrix} V_{aa'} \\ V_{bb'} \\ V_{cc'} \end{bmatrix} = \begin{bmatrix} Z_a & 0 & 0 \\ 0 & Z_b & 0 \\ 0 & 0 & Z_c \end{bmatrix} \begin{bmatrix} I_a \\ I_b \\ I_c \end{bmatrix} \quad \text{...(5.56)}$$

which can be written in terms of symmetrical components as

$$A \begin{bmatrix} V_{aa'0} \\ V_{aa'1} \\ V_{aa'2} \end{bmatrix} = \begin{bmatrix} Z_a & 0 & 0 \\ 0 & Z_b & 0 \\ 0 & 0 & Z_c \end{bmatrix} A \begin{bmatrix} I_{a0} \\ I_{a1} \\ I_{a2} \end{bmatrix} \quad \text{...(5.57)}$$

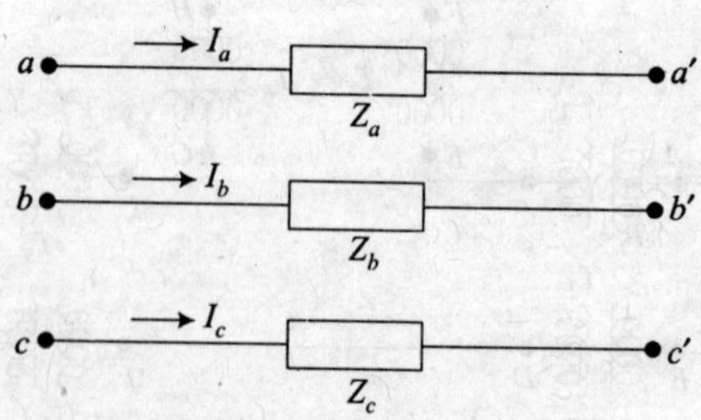

Fig. 5.29 Unsymmetrical series impedances.

Pre-multiplying by A^{-1} and expanding, we get

$$V_{aa'o} = \frac{1}{3}I_{a0}(Z_a + Z_b + Z_c) + \frac{1}{3}I_{a1}(Z_a + a^2Z_b + aZ_c) + \frac{1}{3}I_{a2}(Z_a + aZ_b + a^2Z_c)$$

$$V_{aa'1} = \frac{1}{3}I_{a0}(Z_a + aZ_b + a^2Z_c) + \frac{1}{3}I_{a1}(Z_a + Z_b + Z_c) + \frac{1}{3}I_{a2}(Z_a + a^2Z_b + aZ_c)$$

$$V_{aa'2} = \frac{1}{3}I_{a0}(Z_a + a^2Z_b + aZ_c) + \frac{1}{3}I_{a1}(Z_a + a^2Z_b + a^2Z_c) + \frac{1}{3}I_{a2}(Z_a + Z_b + Z_c)$$

...(5.58)

It can be seen that (5.58) is not decoupled, since the voltage drop of one sequence depends on currents of other sequences as well. However, under normal conditions, since transmission lines are always transposed, in spite of mutual reactance, we still get equal series impedances. Therefore, the sequence voltages depend only on that particular sequence current and we obtain decoupled equations, when symmetrical components are used.

Example 5.11: A single phase resistive load of 100 kVA is connected across lines *bc* of a balanced supply of 6 kV. Compute symmetrical components of the line currents.

Solution: The circuit is drawn in Fig. 5.30.

Let phase sequence be *abc*

$$V_{ab} = 6 \times 10^3 \angle 0° \text{ V}$$
$$V_{bc} = 6 \times 10^3 \angle -120° \text{ V}$$
$$V_{ca} = 6 \times 10^3 \angle 120° \text{ V}$$
$$I_a = 0 \text{ A}$$

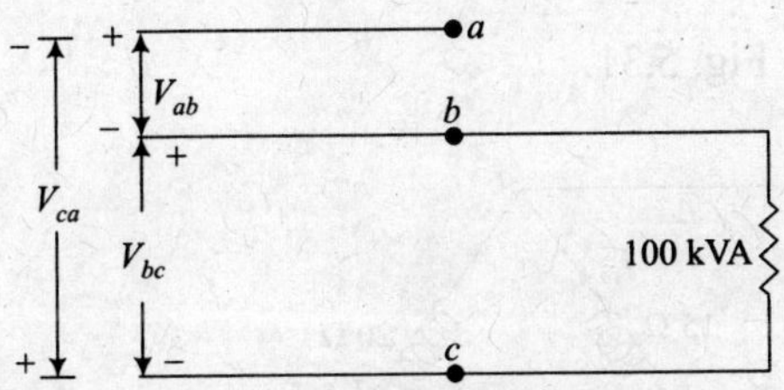

Fig. 5.30 Example 5.11.

$$|I_b| = \frac{100 \times 10^3}{6 \times 10^3} = 16.666 \text{ A}$$

I_b is in phase with V_{bc} (Since we have a resistive load).

$\therefore$ $I_b = 16.666 \angle -120°$ A.

$I_c = -I_b = 16.666\angle(-120° + 180°) = 16.666\angle 60°$ A.

$$I_{a0} = \frac{1}{3}[I_a + I_b + I_c] = 0$$

$$I_{a1} = \frac{1}{3}[I_a + aI_b + a^2I_c]$$

$$= \frac{1}{3}[0 + 16.666\angle(-120° + 120°) + 16.666\angle(60° - 120°)]$$

$$= \frac{1}{3}[16.666 + 16.666\angle{-60°}]$$

$$= \frac{1}{3}[25 - j14.43] = 8.333 - j4.81 = 9.62\angle{-30°}\ \text{A}$$

$$I_{a2} = \frac{1}{3}[I_a + a^2I_b + aI_c]$$

$$= \frac{1}{3}[0 + 16.666\angle(-120° - 120°) + 16.666\angle(60° + 120°)]$$

$$= \frac{1}{3}[-25 + j14.433] = -8.333 + j4.81 = 9.62\angle 150°\ \text{A.}$$

$$I_b = 0;\ I_{b1} = 9.62\angle{-150°}\ \text{A};\ I_{b2} = 9.62\angle{-90°}\ \text{A}$$

$$I_{c0} = 0;\ I_{c1} = 9.62\angle 90°\text{A};\ I_{c2} = 9.62\angle 30°\ \text{A.}$$

Example 5.12: A delta connected resistive load is connected across a balanced three phase supply of 415 V. $R_{ab} = 20\ \Omega$; $R_{bc} = 250\ \Omega$; $R_{ca} = 15\ \Omega$. Find symmetrical components of line currents and delta currents.

Solution: The circuit is shown in Fig. 5.31.

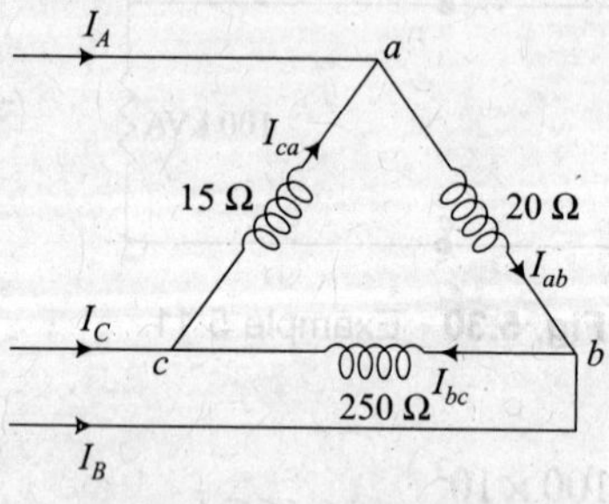

Fig. 5.31 Example 5.12.

$$V_{ab} = 415\angle 0°$$

$$V_{bc} = 415\angle{-120°}$$

$$V_{ca} = 415\angle 120°$$

The delta currents are calculated as follows:

$$I_{ab} = \frac{V_{ab}}{R_{ab}} = \frac{415\angle 0°}{20} = 20.75\angle 0° \text{ A.}$$

$$I_{bc} = \frac{V_{bc}}{R_{bc}} = \frac{415\angle -120°}{250} = 1.66\angle -120° \text{ A.}$$

$$I_{ca} = \frac{V_{ca}}{R_{ca}} = \frac{415\angle 120°}{15} = 27.66\angle 120° \text{ A.}$$

Symmetrical components of delta currents

$$I_{ab1} = \frac{1}{3}[I_{ab} + aI_{bc} + a^2 I_{ca}]$$

$$= \frac{1}{3}[20.75 + 1.66\angle(-120° + 120°) + 27.66\angle(120° - 120°)]$$

$$= 16.69 \text{ A.}$$

$$I_{ab2} = \frac{1}{3}[I_{ab} + a^2 I_{bc} + aI_{ca}]$$

$$= \frac{1}{3}[20.75 + 1.66\angle(-120° - 120°) + 27.66\angle(120° + 120°)]$$

$$= \frac{1}{3}[20.75 + 1.66\angle 120° + 27.66\angle -120°]$$

$$= 7.775\angle -75° \text{ A}$$

$$I_{ab0} = \frac{1}{3}[I_{ab} + I_{bc} + I_{ca}]$$

$$= \frac{1}{3}[20.75 + 1.66\angle -120° + 27.66\angle 120°]$$

$$= 7.775\angle 75° \text{ A.}$$

The line currents are calculated as follows:

$$I_A = I_{ab} - I_{ca} = 20.75 - 27.66\angle 120° = 34.58 - j23.954 = 42.066\angle -34.71° \text{ A}$$

$$I_B = I_{bc} - I_{ab} = 1.6\angle -120° - 20.75 = -21.55 - j1.385 = 21.594\angle -176.32° \text{ A}$$

$$I_C = I_{ca} - I_{bc} = 27.66\angle 120° - 1.66\angle -120° = -13 + j25.3916 = 28.526\angle 117.11° \text{ A}$$

The symmetrical components are:

$$I_{A1} = \frac{1}{3}[I_A + aI_B + a^2 I_C]$$

$$= \frac{1}{3}[42.066\angle{-34.71°} + 21.594\angle(-176.32° + 120°)$$

$$+ 28.526\angle(117.11° - 120°)]$$

$$= \frac{1}{3}[42.066\angle{-34.71°} + 21.594\angle{-56.32°} + 28.526\angle{-2.89°}]$$

$$= \frac{1}{3}[75.044 - j43.36] = 28.89\angle{-30°}$$

$$I_{A2} = \frac{1}{3}[I_A + a^2 I_B + aI_c]$$

$$= \frac{1}{3}[42.066\angle{-34.71°} + 21.594\angle(-176.32° - 120°)$$

$$+ 28.526\angle(117.11° + 120°)]$$

$$= \frac{1}{3}[42.066\angle{-34.71°} + 21.594\angle{-63.68°} + 28.526\angle{-122.89°}]$$

$$= \frac{1}{3}[28.664 - j28.552] = 13.486\angle{-44.88°} \text{ A}$$

$$I_{A0} = \frac{1}{3}[I_A + I_B + I_c] = 0 \text{ A}$$

Note that even though line current zero sequence component is zero, the delta current zero sequence component is not zero.

Example 5.13: The resistive Y – connected load of example 5.2 is supplied from the low – voltage Y side of a $\Delta - Y$ transformer. Find the line voltages and currents in pu on the high – voltage side of the transformer.

Solution: From example 5.2,

$$V_{ab1} = 0.9857\angle{73.6°} \text{ pu}$$

$$V_{ab2} = 0.2346\angle{-139.7°} \text{ pu}$$

$$V_{ab0} = 0.0 \text{ pu.}$$

The positive sequence line voltages are advanced by 30° on HV side and negative sequence line voltages are retarded by $30°_1$.

$$V_{AB1} = 0.9857\angle(73.6° + 30°) = 0.9857\angle{103.6°}$$

$$V_{AB2} = 0.2346\angle(-139.7° - 30°) = 0.2346\angle{-169.7°}$$

$$V_{AB0} = 0.0$$

$$V_{AB} = V_{AB1} + V_{AB2} = 0.9857\angle{103.6°} + 0.2346\angle{-169.7°}$$

$$= -0.4626 + j0.9161 = 1.026\angle{116.79°} \text{ pu}$$

$$V_{BC} = V_{BC1} + V_{BC2} = a^2 V_{AB1} + aV_{AB2}$$
$$= 0.9857\angle{-16.4°} + 0.2346\angle{-49.7°}$$
$$= 1.097 - j\,0.457 = 1.188\angle{-22.62°}\text{pu.}$$
$$V_{CA} = V_{CA1} + V_{CA2} = aV_{AB1} + a^2 V_{AB2}$$
$$= 0.9857\angle 223.6° + 0.2346\angle 70.3°$$
$$= -0.6348 - j0.4589 = 0.783\angle{-144.2°}\text{ pu.}$$

Line currents:

From example 5.2 it can be seen that I_{a1} is in phase with V_{an1} and has the same value in pu if V_{an1} is expressed in pu on a line-to-neutral voltage base. Similarly, I_{a2} is identical to V_{an2} in pu. Hence, the positive sequence line current on HV side I_{A1} will be in phase with V_{A1} and I_{A2} in phase with V_{A2}, where V_A is the line-to-neutral voltage on the delta side. The sequence components of V_A are obtained from V_{an1} and V_{an2}.

$$V_{an1} = 0.9857\angle 43.6°\text{ pu} \qquad \text{(example 5.2)}$$
$$\therefore \quad V_{A1} = 0.9857\angle(43.6° + 30°) = 0.9857\angle 73.6°\text{ pu.}$$
$$= 0.2783 + j0.9456\text{ pu.}$$
$$V_{an2} = 0.2346\angle{-109.7°}\text{ pu.}$$
$$\therefore \quad V_{A2} = 0.2346\angle(-109.7° - 30°) = 0.2346\angle{-139.7°}$$
$$= -0.1789 - j0.1517\text{ pu.}$$
$$V_A = V_{A1} + V_{A2} = 0.0994 + j0.7939 = 0.8\angle 82.8°\text{ pu.}$$
$$V_B = V_{B1} + V_{B2} = a^2 V_{A1} + aV_{A2} = 0.9857\angle{-46.4°} + 0.2346\angle{-19.7°}$$
$$= 0.9007 - j0.7929$$
$$= 1.2\angle{-41.4°}\text{ pu.}$$
$$V_C = V_{C1} + V_{C2} = aV_{A1} + a^2 V_{A2} = 0.9857\angle 193.6° + 0.2346\angle 100.3°$$
$$= -1.0 + j0 = 1.0\angle 180°\text{ pu.}$$

The line currents on HV side are now obtained as

$$I_A = 0.8\angle 82.8°\text{ pu.}$$
$$I_B = 1.20\angle{-41.4}\text{ pu.}$$
$$I_C = 1.0\angle 180°\text{ pu.}$$

Example 5.14: A three phase balanced load with a self impedance of Z_s and mutual impedance Z_m is connected in star. An impedance Z_n is connected between the neutral and the ground. Find the sequence impedance matrix.

Solution: The circuit is shown in Fig. 5.32.

$$V_a = Z_s I_a + Z_m I_b + Z_m I_c + Z_n I_n$$
$$V_b = Z_m I_a + Z_s I_b + Z_m I_c + Z_n I_n$$
$$V_c = Z_m I_a + Z_m I_b + Z_s I_c + Z_n I_n$$
$$I_n = I_a + I_b + I_c$$

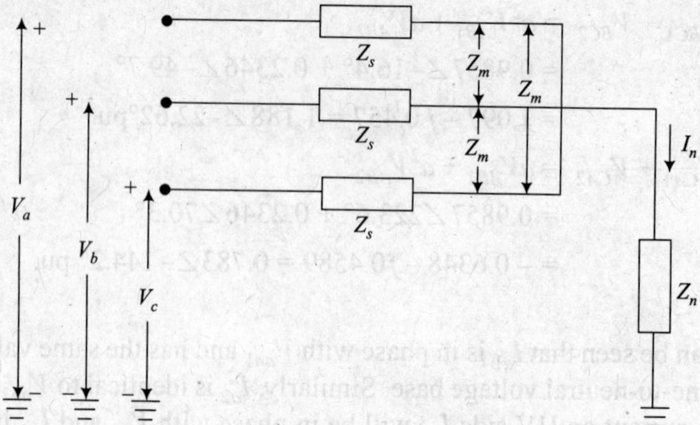

Fig. 5.32 Example 5.14.

∴ We can write

$$\begin{bmatrix} V_a \\ V_b \\ V_c \end{bmatrix} = \begin{bmatrix} Z_s + Z_n & Z_m + Z_n & Z_m + Z_n \\ Z_m + Z_n & Z_s + Z_n & Z_m + Z_n \\ Z_m + Z_n & Z_m + Z_n & Z_s + Z_n \end{bmatrix} \begin{bmatrix} I_a \\ I_b \\ I_c \end{bmatrix}.$$

i.e.
$$[V_p] = [Z_p][I_p]$$
$$A[V_s] = [Z_p]\,A[I_s]$$

Multiplying both sides by, A^{-1}.

$$[V_s] = A^{-1}[Z_p]\,A[I_s]$$

$$[Z]_s = A^{-1}[Z]_p\,[A] = \frac{1}{3}\begin{bmatrix} 1 & 1 & 1 \\ 1 & a & a^2 \\ 1 & a^2 & a \end{bmatrix} \begin{bmatrix} Z_s + Z_n & Z_m + Z_n & Z_m + Z_n \\ Z_m + Z_n & Z_s + Z_n & Z_m + Z_n \\ Z_m + Z_n & Z_m + Z_n & Z_s + Z_n \end{bmatrix} \begin{bmatrix} 1 & 1 & 1 \\ 1 & a^2 & a \\ 1 & a & a^2 \end{bmatrix}$$

$$= \begin{bmatrix} Z_s + 3Z_n + 2Z_m & 0 & 0 \\ 0 & Z_s - Z_m & 0 \\ 0 & 0 & Z_s - Z_m \end{bmatrix}$$

$$\therefore \quad \begin{bmatrix} V_{a0} \\ V_{a1} \\ V_{a2} \end{bmatrix} = \begin{bmatrix} Z_s + 3Z_n + 2Z_m & 0 & 0 \\ 0 & Z_s - Z_m & 0 \\ 0 & 0 & Z_s - Z_m \end{bmatrix} \begin{bmatrix} I_{a0} \\ I_{a1} \\ I_{a2} \end{bmatrix}$$

$[Z]_s$ is the sequence impedance matrix. Since, it has only diagonal elements, the sequence equations are decoupled.

Example 5.15: A balanced three – phase voltage of 230V, line to neutral is applied to a balanced *Y*-connected load, where each phase has a series impedance $Z_s = j12\ \Omega$ and mutual coupling between phases is $Z_m = j4\ \Omega$. Determine line currents using symmetrical components. Assume neutral is ungrounded.

Solution:

$$V_a = 230\angle 0°$$
$$V_b = 230\angle -120°$$
$$V_c = 230\angle 120°$$
$$V_{a0} = \frac{1}{3}[V_a + V_b + V_c] = 0$$

$$V_{a1} = \frac{1}{3}[230 + 230\angle(-120° + 120°) + 230\angle(120° - 120°)] = 230\text{ V.}$$

$$V_{a2} = \frac{1}{3}[230 + 230\angle(-120° - 120°) + 230\angle(120° + 120°)] = 0$$

$$[Z]_s = \begin{bmatrix} Z_s + 2Z_m + 3Z_n & 0 & 0 \\ 0 & Z_s - Z_m & 0 \\ 0 & 0 & Z_s - Z_m \end{bmatrix} = \begin{bmatrix} j20 + 3Zn & 0 & 0 \\ 0 & j8 & 0 \\ 0 & 0 & j8 \end{bmatrix} \Omega$$

[Here since neutral is not grounded, $Z_n = \infty$. It can be taken to be a very large value]

$$\begin{bmatrix} I_{a0} \\ I_{a1} \\ I_{a2} \end{bmatrix} = \begin{bmatrix} j20 + 3Z_n & 0 & 0 \\ 0 & j8 & 0 \\ 0 & 0 & j8 \end{bmatrix} \begin{bmatrix} V_{a0} \\ V_{a1} \\ V_{a2} \end{bmatrix}$$

$$= \begin{bmatrix} j20 + 3Z_n & 0 & 0 \\ 0 & j8 & 0 \\ 0 & 0 & j8 \end{bmatrix} \begin{bmatrix} 0 \\ 230 \\ 0 \end{bmatrix} = \begin{bmatrix} 0 & 0 & 0 \\ 0 & -j0.125 & 0 \\ 0 & 0 & -j0.125 \end{bmatrix} \begin{bmatrix} 0 \\ 230 \\ 0 \end{bmatrix}$$

$$I_{a0} = 0$$
$$I_{a1} = -j28.75$$
$$I_{a2} = 0$$
$$I_a = I_{a0} + I_{a1} + I_{a2} = -j28.75$$
$$I_b = I_{a0} + a^2 I_{a1} + aI_{a2} = 28.75\angle 150°$$
$$I_c = I_{a0} + a\,I_{a1} + a^2 I_{a2} = 28.75\angle 30°$$

Example 5.16: A set of three phase unbalanced voltages is applied to a load, star connected, $Z_s = 4 + j12\ \Omega$ and $Z_m = j2\ \Omega$. The load and source neutrals are solidly grounded. The phase to neutral source voltages are given by

$$[V_P] = \begin{bmatrix} 100\angle 25° \\ 50\angle -155° \\ 40\angle 100° \end{bmatrix}$$

Determine

(i) The sequence impedances.

(ii) The symmetrical components of voltage.

(iii) Symmetrical components of current.

(iv) Complex power delivered to load.

(v) Load phase currents.

(i) $$[Z]_s = \begin{bmatrix} Z_s + 2Z_m + 3Z_n & 0 & 0 \\ 0 & Z_s - Z_m & 0 \\ 0 & 0 & Z_s - Z_m \end{bmatrix} = \begin{bmatrix} 4+j16 & 0 & 0 \\ 0 & 4+j10 & 0 \\ 0 & 0 & 4+j10 \end{bmatrix}$$

(ii) $$\begin{bmatrix} V_{a0} \\ V_{a1} \\ V_{a2} \end{bmatrix} = \frac{1}{3}\begin{bmatrix} 1 & 1 & 1 \\ 1 & a & a^2 \\ 1 & a^2 & a \end{bmatrix}\begin{bmatrix} V_a \\ V_b \\ V_c \end{bmatrix}$$

$$V_{a0} = \frac{1}{3}[100\angle 25° + 50\angle -155° + 40\angle 100°]$$

$$= 12.793 + j\,20.17 = 23.88\angle 57.6°\text{ V}$$

$$V_{a1} = \frac{1}{3}[100\angle 25° + 50\angle -35° + 40\angle -20°]$$

$$= 56.39 - j0.03 = 56.39\angle -0.03°\text{ V}$$

$$V_{a2} = \frac{1}{3}[100\angle 25° + 50\angle 85° + 40\angle 220°]$$

$$= 21.449 + j22.116 = 30.8\angle 45.88°\text{ V.}$$

(iii) $$[I_s] = [Z_s]^{-1}[V_s]$$

$$= \begin{bmatrix} 4+j16 & 0 & 0 \\ 0 & 4+j10 & 0 \\ 0 & 0 & 4+j10 \end{bmatrix}^{-1} \begin{bmatrix} 23.88\angle 57.6° & 0 & 0 \\ 0 & 56.39\angle -0.03° & 0 \\ 0 & 0 & 30.8\angle 45.88° \end{bmatrix}$$

$$= \begin{bmatrix} 1.444\angle -18.34° \\ 5.236\angle -68.23° \\ 2.861\angle -22.32° \end{bmatrix}$$

$$\therefore \quad I_{a0} = 1.448\angle{-18.34°} \text{ A}$$
$$I_{a1} = 5.236\angle{-68.23°} \text{ A}$$
$$I_{a2} = 2.861\angle{-22.32°} \text{ A.}$$

(iv) Complex power delivered to loads is

$$S = 3(V_{a0}I_{a0}^* + V_{a1}I_{a1}^* + V_{a2}I_{a2}^*)$$
$$= 3(23.88\angle 57.6° \times 1.448\angle 18.34° + 56.39\angle{-0.03°} \times 5.236\angle 68.23° + 30.8\angle 45.88° \times 2.861\angle 22.32°)$$
$$= 3(34.578\angle 75.94° + 295.258\angle 68.2° + 88.11\angle 68.20)$$
$$= 3(150.769 + j389.501) = 452.307 + j1168.5 \text{ VA.}$$

(v) The load phase currents is given by

$$[I_P] = [A]\,[I_s]$$
$$\begin{bmatrix} I_a \\ I_b \\ I_c \end{bmatrix} = \begin{bmatrix} 1 & 1 & 1 \\ 1 & a^2 & a \\ 1 & a & a^2 \end{bmatrix} \begin{bmatrix} I_{a0} \\ I_{a1} \\ I_{a2} \end{bmatrix}$$
$$= \begin{bmatrix} 1 & 1 & 1 \\ 1 & 1\angle{-120°} & 1\angle 120° \\ 1 & 1\angle 120° & 1\angle{-120°} \end{bmatrix} \begin{bmatrix} 1.448\angle{-18.34°} \\ 5.236\angle{-68.23°} \\ 2.861\angle{-22.32°} \end{bmatrix}$$
$$= \begin{bmatrix} 8.75\angle{-47.04°} \\ 5.23\angle 143.24° \\ 3.03\angle 39.06° \end{bmatrix}$$

Example 5.17: A balanced three-phase voltage of 230V line-to-neutral is applied to an ungrounded star load with impedance of $j12\Omega$ per phase and mutual impedance of $j4\Omega$ between two phases. Determine the line currents.

Solution: Here $Z_n = \infty$

$$\therefore \quad Z^{012} = \begin{bmatrix} j12 + j4\times 2 + 3Z_n & 0 & 0 \\ 0 & j12 - j4 & 0 \\ 0 & 0 & j12 - j4 \end{bmatrix}$$
$$= \begin{bmatrix} j20 + 3Z_n & 0 & 0 \\ 0 & j8 & 0 \\ 0 & 0 & j8 \end{bmatrix}$$

The symmetrical components of the balanced supply voltage is given by,

$$V_{ao} = 0, V_{a1} = 230\text{V}, V_{a2} = 0$$

$$\therefore \qquad V_s = \begin{bmatrix} 0 \\ 230 \\ 0 \end{bmatrix} V$$

$$I_s = \begin{bmatrix} I_{ao} \\ I_{a1} \\ I_{a2} \end{bmatrix} = [Z^{012}]^{-1} V_s = \begin{bmatrix} 0 \\ -j28.75 \\ 0 \end{bmatrix} A$$

$$I_p = AI_s = \begin{bmatrix} 28.75\angle -90° \\ 28.75\angle 150° \\ 28.75\angle 30° \end{bmatrix}$$

Example 5.18: A three-phase unbalanced source has the following line-to-neutral voltages.

$$V_a = 100\ \angle 25° \ ; V_b = 50\ \angle -155° \ ; V_c = 40\ \angle 100°$$

It is applied to a star connected load with a series impedance of $4 + j12\Omega$ per phase and mutual impedance of $j2\Omega$. The neutral of source and load are grounded. Determine:

(i) The load sequence impedances
(ii) Symmetrical components of voltage
(iii) Symmetrical components of current
(iv) Line currents
(v) Complex power delivered to load using phase quantities
(vi) Complex power delivered to load using symmetrical components.

Solution:

(i)
$$Z_0 = Z_Y + 2Z_m = 4 + j12 + j4 = 4 + j16\Omega$$
$$Z_1 = Z_Y - Z_m = 4 + j10\Omega$$
$$Z_2 = Z_Y - Z_m = 4 + j10\Omega$$

(ii)
$$V_s = A^{-1}V_p = A^{-1}\begin{bmatrix} 100\ \angle 25° \\ 50\ \angle -155° \\ 40\ \angle 100° \end{bmatrix} = \begin{bmatrix} 23.8869\angle 57.63° \\ 56.3926\angle -0.03° \\ 30.8114\angle 45.88° \end{bmatrix}$$

(iii)
$$I_s = [Z^{012}]^{-1}\ [V_s]$$

$$= \begin{bmatrix} 4 + j16 & 0 & 0 \\ 0 & 4 + j10 & 0 \\ 0 & 0 & 4 + j10 \end{bmatrix} [V_s]$$

$$= \begin{bmatrix} 1.4484\angle -18.34° \\ 5.2359\angle -68.23° \\ 2.8608\angle -22.31° \end{bmatrix}$$

(iv)
$$I_p = AI_s = \begin{bmatrix} 8.7507\angle -47.04° \\ 5.2292\angle 143.24° \\ 3.0280\angle 39.07° \end{bmatrix}$$

(v)
$$S = V_a I_a^* + V_b I_b^* + V_c I_c^*$$
$$= 100\angle 25° \times 8.7507\angle 47.04° + 50\angle -155° \times 5.2292\angle -143.24°$$
$$+ 40\angle 100° \times 3.0280\angle 39.07°$$
$$= 452.35 + j1168.6\ VA$$

(vi)
$$S = 3[V_{ao} I_{ao}^* + V_{a1} I_{a1}^* + V_{a2} I_{a2}^*]$$
$$= 452.35 + j1168.6\ VA$$

5.10 SUMMARY

Unbalanced three phase voltages and currents can be resolved into three sets of symmetrical components. Analysis can be carried out by treating each set of components separately and superimposing the results.

In balanced networks, currents of a particular sequence produce voltage drops of the same sequence. However, in networks which are not balanced this decoupling is not present.

The three sequence impedances are not always equal. In synchronous machines, generally the positive and negative sequence reactances are equal. The zero sequence reactance is lesser. In a transformer, all the three are taken to be equal. In transmission lines, because of the presence of ground wires, the zero sequence impedance is 2.5 – 3 times the positive sequence impedance.

In obtaining the zero sequence network, care must be taken about the connections, to ensure if a path exists for the zero sequence currents.

5.11 SUPPLEMENTARY PROBLEMS

(1) Determine the symmetrical components of three voltages $V_a = 400\angle 0°$; $V_b = 400\angle 245°$; $V_c = 400\angle 105°$ V for a phase sequence *abc* and for a phase sequence *acb*.

(2) One conductor of a three phase line is open. The current flowing into the Δ-connected load through line *c* is 15 A. Using line *a* as reference, obtain the symmetrical components of the line currents and delta currents.

(3) If $V_{an1} = 100\angle 0°$; $V_{an2} = 40\angle 90°$ and $V_{an0} = 20\angle 180°$, determine V_{an}, V_{bn} and V_{cn}. Show graphically that the sum of the given symmetrical components is equal to the line-to-neutral voltages.

(4) Three identical Y– connected resistors form a load bank with a three phase rating of 3.2 kV, 750 kVA. The applied voltages are $|V_{ab}| = 3.84$ kV, $|V_{bc}| = 2.56$ kV; $|V_{ca}| = 3.2$ kV. Find the line voltages and currents in pu using symmetrical components. Also determine the power absorbed by the resistors.

(5) Draw the sequence networks for the power system shown in Fig. P5.5.

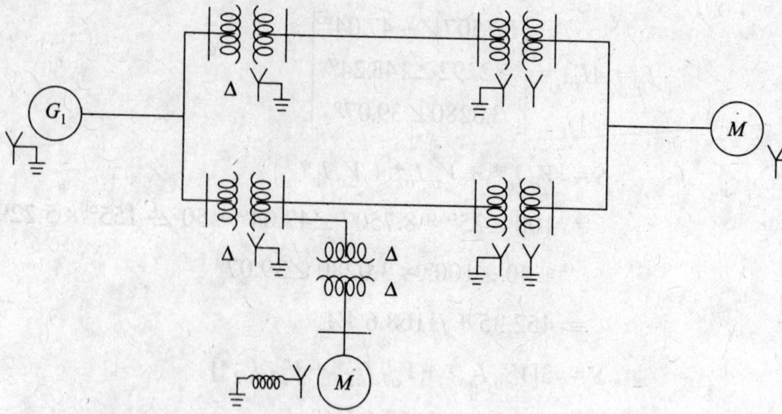

Fig. P5.5

(6) A Y connected generator has reactances $X_0 = 0.09$ pu; $X_1 = 0.2$ pu; $X_2 = 0.33$ pu. The neutral is grounded through reactance 0.08 pu. The machine is running at rated voltage no-load when an unbalanced fault occurs. The currents are $I_a = 3.75\angle 150°$; $I_b = 0$; $I_c = 3.75\angle 30°$ pu, with respect to phase a line to neutral voltage. Determine the terminal voltages with respect to ground and voltage of neutral point with respect to ground.

(7) A balanced three phase voltage 400V, line-to-line is applied to an ungrounded balanced Y connected load. The series reactance of each phase is 12 Ω and mutual coupling between phases is 5 Ω. Determine the line currents by (i) mesh analysis (ii) Using symmetrical components.

(8) An unbalanced voltage source, whose line-to-neutral voltages are given by

$$V_P = \begin{bmatrix} 150\angle -120° \\ 100\angle 90° \\ 50\angle -30° \end{bmatrix}$$

is applied to a load whose series impedance is $10 + j30$ Ω and mutual impedance between phases is $j2$ Ω. Both load and source are grounded. Find the sequence impedances, symmetrical components of voltages, currents, phase currents, three phase complex power delivered to load using symmetrical components and summing power in each phase.

(9) The line-to-line voltages of an unbalanced system are given by $V_{ab} = 480\angle 36.87°$; $V_{bc} = 640\angle 126.87°$, $V_{ca} = 800\angle -90°$ V. A Y connected load with resistance of 30 Ω is connected to the supply. Determine symmetrical components of line voltages, phase voltages and line currents.

REVIEW QUESTIONS

1. What are symmetrical components?
2. Why is there a need for symmetrical components transformation?

3. What is the symmetrical component transformation and inverse transformation matrix?
4. With phasor diagram show how the symmetrical components add up to unbalanced phasors.
5. How does the symmetrical component transformation become power invariant?
6. Prove that the zero sequence component depends on the neutral current.
7. Draw the sequence components of a (i) star (ii) delta connected impedance.
8. Obtain the sequence impedances of the transmission line. Why is the zero sequence reactance higher in case of transmission lines?
9. Explain how a phase shift occurs in a three-phase Y-Ä transformer.
10. Draw the zero sequence networks of three-phase transformers for various possible connections.
11. The zero sequence component in the line currents is zero, whereas it is present in the phase currents. Is this possible? If yes, under what conditions?
12. Prove that the equations are not decoupled if the system has unbalanced impedances in each phase.

Chapter 6

Unsymmetrical Faults

6.1 INTRODUCTION

In chapter 4, the three-phase symmetrical fault was discussed in detail. The system is balanced during a symmetrical fault and analysis can be carried out on a single-phase basis. This chapter is devoted to the analysis of unsymmetrical faults. The various types of unsymmetrical faults are:

(i) Single line–to–ground fault (*LG*)

(ii) Line–to–line fault (*LL*)

(iii) Double line–to–ground fault (*LLG*)

(iv) Open conductor fault

The first three types are *shunt* type and the last one is *series* fault.

The three phase symmetrical fault is the most severe fault and is used for choosing the rating of the circuit breaker. However, under some situations, the fault current can be greater with *LG* faults. The analysis of the unsymmetrical faults is carried out using the concept of symmetrical components presented in chapter 5. Since, symmetrical components resolve unbalanced three phase phasors into a set of balanced three phase phasors, analysis can be carried out on a single phase basis.

6.2 SYMMETRICAL COMPONENT ANALYSIS OF UNSYMMETRICAL FAULTS

A general model is developed in this section to be used in analysis of all the faults. Consider a general power system as shown in Fig. 6.1.

Assume a shunt type fault occurs at point F. Let I_a, I_b, I_c be the current flow out of the network at point F. Let V_a, V_b, V_c be the voltages of the lines with respect to the ground.

If we assume the system is operating on no load, the positive sequence voltages of all the generators will be identical. The prefault voltage at point F will also have the same value. Let us designate this voltage as E_a. The power system will present positive, negative and zero sequence impedances at the fault point. The methods of drawing the sequence networks have been discussed in detail in chapter 5. The sequence networks are shown in Fig. 6.2a to 6.2c.

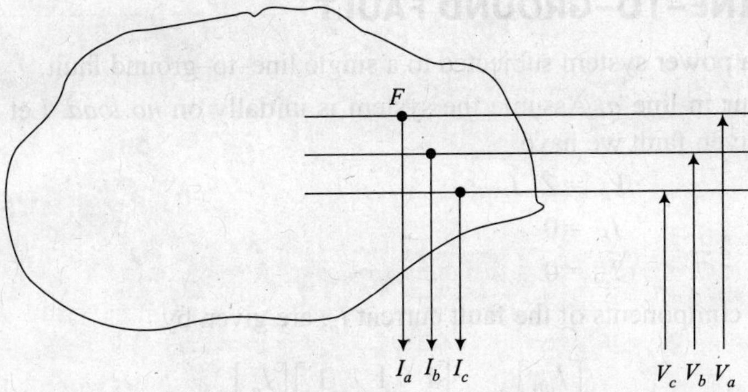

Fig. 6.1 General power system.

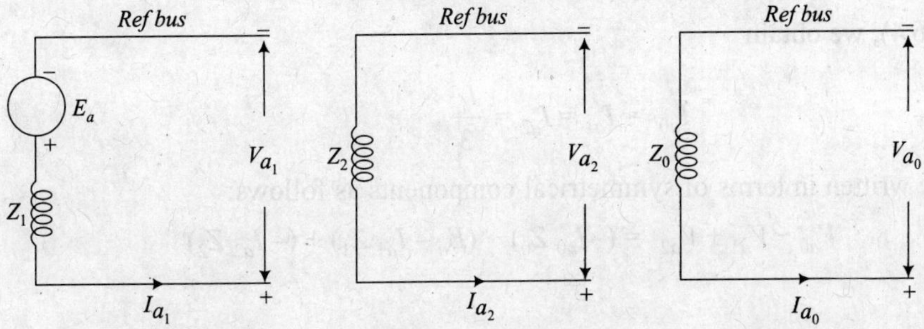

Fig. 6.2 Sequence networks.

In the figures Z_1, Z_2, Z_0 represent the Thevenin's equivalent impedances of the three sequences at the fault point. Recollect that the reference bus for the positive and negative sequence networks is the neutral (for these networks the neutral is at same potential as the ground since no current flows through the neutral) and for the zero sequence network, the reference bus is the ground. From the figures, we can write the following equations.

$$\begin{aligned} V_{a0} &= -I_{a0} Z_0 \\ V_{a1} &= E_a - I_{a1} Z_1 \\ V_{a2} &= -I_{a2} Z_2 \end{aligned} \quad \text{...(6.1)}$$

(6.1) can be written in matrix form as

$$\begin{bmatrix} V_{a0} \\ V_{a1} \\ V_{a2} \end{bmatrix} = \begin{bmatrix} 0 \\ E_a \\ 0 \end{bmatrix} - \begin{bmatrix} Z_0 & 0 & 0 \\ 0 & Z_1 & 0 \\ 0 & 0 & Z_2 \end{bmatrix} \begin{bmatrix} I_{a0} \\ I_{a1} \\ I_{a2} \end{bmatrix} \quad \text{...(6.2)}$$

Depending on the type of fault, the sequence networks are connected in a particular manner. The fault current can be computed from the connection. These concepts are discussed in detail next.

6.3 SINGLE LINE–TO–GROUND FAULT

Figure 6.3(a) shows a power system subjected to a single line–to–ground fault.

Let the fault occur in line *a*. Assume the system is initially on *no load*. Let Z_F be the fault impedance. For the given fault we have

$$V_a = Z_F I_a \quad \text{...(6.3a)}$$

$$I_b = 0 \quad \text{...(6.3b)}$$

$$I_c = 0 \quad \text{...(6.3c)}$$

The symmetrical components of the fault current I_a, are given by

$$\begin{bmatrix} I_{a0} \\ I_{a1} \\ I_{a2} \end{bmatrix} = \frac{1}{3}\begin{bmatrix} 1 & 1 & 1 \\ 1 & a & a^2 \\ 1 & a^2 & a \end{bmatrix}\begin{bmatrix} I_a \\ 0 \\ 0 \end{bmatrix} \quad \text{...(6.4)}$$

From (6.4), we obtain

$$I_{a0} = I_{a1} = I_{a2} = \frac{I_a}{3} \quad \text{...(6.5)}$$

(6.1) can be written in terms of symmetrical components as follows:

$$V_{a0} + V_{a1} + V_{a2} = (-I_{a0}\, Z_0) + (E_a - I_{a1}\, Z_1) + (-\, I_{a2}\, Z_2)$$

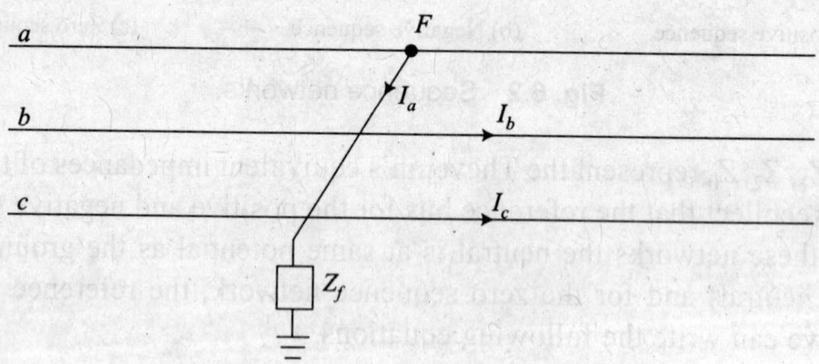

Fig. 6.3 (a) Single line–to–ground fault.

Now using (6.5)

$$V_{a0} + V_{a1} + V_{a2} = V_a = Z_F I_a = 3Z_F I_{a1}$$

Combining the two equations, we get

$$3Z_F I_{a1} = E_a - I_{a1}\,(Z_0 + Z_1 + Z_2)$$

or,

$$I_{a1} = \frac{E_a}{Z_0 + Z_1 + Z_2 + 3Z_F} \quad \text{...(6.6)}$$

Thus, in an SLG fault, all the three sequence currents of phase a (where fault has occurred) are equal and the sum of the sequence voltages is equal to $3Z_F I_{a1}$. Connecting the sequence networks in series through impedance $3Z_F$ satisfies the above equations. This is shown in Fig. 6.4.

The sequence components of the other voltages are given by

$$V_b = V_{a0} + a^2 V_{a1} + a V_{a2}$$
$$V_c = V_{a0} + a V_{a1} + a^2 V_{a2}$$

where, $a = 1\angle 120°$

The same analysis can be applied to the fault current due to an SLG fault at the terminals of an unloaded generator, as shown in Fig. 6.3(b).

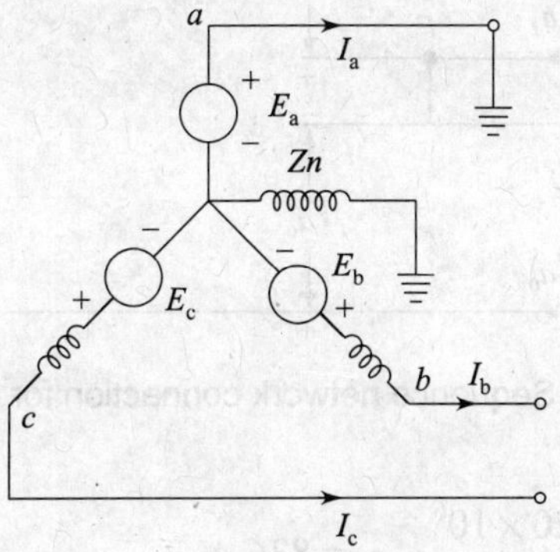

Fig. 6.3(b)

$$I_{a1} = \frac{E_a}{Z_1 + Z_2 + Z_0 + 3Z_n}$$

Example 6.1: A 20 MVA, 13.8 kV generator has a direct axis subtransient reactance of 0.25 pu. Its negative sequence reactance is 0.35 pu and zero sequence reactance is 0.1 pu. The neutral of the generator is grounded. Find the fault current, line-to-ground voltages and the line-to-line voltages for a single line-to-ground fault.

Solution: Let base kV be 13.8 kV and base MVA be 20 MVA.

$$\text{Base kV for line to neutral voltage} = \frac{13.8}{\sqrt{3}} \text{ kV}$$

The phase voltage E_a = 1.0 pu.

$$I_{a1} = \frac{E_a}{Z_1 + Z_2 + Z_0} = \frac{1.0}{j0.25 + j0.35 + j0.1} = -j1.43 \text{ pu.}$$

$$I_a = 3I_{a1} = -j4.29 \text{ pu.}$$

The fault current is I_a. The base current

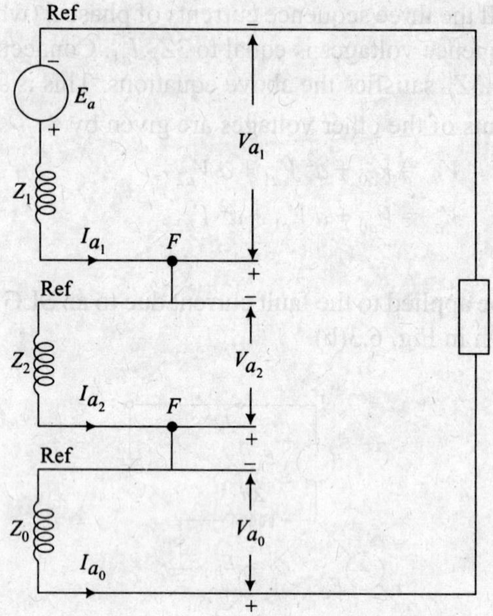

Fig. 6.4 Sequence network connection for *SLG* fault.

$$I_B = \frac{20 \times 10^6}{\sqrt{3} \times 13.8 \times 10^3} = 836 \text{ A}$$

$$\therefore \quad I_a = -j4.29 \times 836 = -j3585 \text{ A}$$

$$V_{a1} = E_a - I_{a1} Z_1 = 1.0 - (-j1.43)(j0.25) = 0.643 \text{ pu}$$

$$V_{a2} = -I_{a2} Z_2 = -(-j1.43)(j0.35) = -0.5 \text{ pu}$$

$$V_{a0} = -I_{a0} Z_0 = -(-j1.43)(j0.1) = -0.143 \text{ pu}$$

Line-to-ground voltages

$$V_a = V_{a0} + V_{a1} + V_{a2} = 0.643 - 0.5 - 0.143$$

$$= 0.0 \text{ pu.} \quad \text{(since fault is on phase } a)$$

$$V_b = V_{a0} + a^2 V_{a1} + aV_{a2}$$

$$= (-0.143) + (a^2 \times 0.643) + (a \times -0.5)$$

$$= (-0.143) + (-0.3215 - j0.5568) + (0.25 - j0.433)$$

$$= -0.215 - j0.989 \text{ pu}$$

$$V_c = V_{a0} + aV_{a1} + a^2 V_{a2}$$

$$= (-0.143) + (a \times 0.643) + (a^2 \times -0.5)$$

$$= (-0.143) + (-0.3215 + j0.5568) + (0.25 + j0.433)$$

$$= -0.215 + j0.989 \text{ pu}$$

Line-to-line voltages

$$V_{ab} = V_a - V_b = 0.0 - (-0.215 - j0.989) = 1.01\angle 77.7° \text{ pu}$$

$$= 1.01 \times \frac{13.8}{\sqrt{3}} = 8.05\angle 77.7° \text{ kV}$$

$$V_{bc} = V_b - V_c = 1.978\angle 270° \text{ pu}$$

$$= 1.978 \times \frac{13.8}{\sqrt{3}} = 15.73\angle 270° \text{ kV}$$

$$V_{ca} = V_c - V_a = 1.01\angle 102.3° \text{ pu} = 8.05\angle 102.3° \text{ kV}$$

Note that the line-to-line voltages are obtained in pu values on the line-to-neutral base. Hence, we need to multiply pu voltage by $\frac{13.8}{\sqrt{3}}$ kV.

6.3.1 Fault Under Loaded Conditions

We assume the system is balanced under loaded conditions. Hence, only positive sequence currents flow before the fault and these must be reflected only in the positive sequence network. The load current can be taken care of in two ways as presented in detail in chapter 5. In the first method, all the machines are replaced by subtransient, transient or synchronous reactance (depending on the time after the fault when currents have to be determined) and voltages behind the appropriate reactance, in the positive sequence network. This network is used to make the connection as shown in Fig. 6.4, for computing the sequence currents.

In the second method, the positive sequence network is replaced by the Thevenin's equivalent voltage, which is the prefault voltage at the fault location *F*. The Thevenin's impedance is the impedance between point F and the reference bus (with all voltage generators short circuited). The fault current computed with this model is exact. However, to compute fault current in other branches of the network, we must add the prefault steady state currents. The negative sequence and zero sequence networks do not change.

6.4 LINE–TO–LINE (*LL*) FAULT

The fault condition is shown in Fig. 6.5. The fault occurs between phases *b* and *c* through an impedance Z_F.

Assuming system is on no load initially, we can write the following relationships:

$$I_a = 0 \quad \text{...(6.7a)}$$

$$I_b = -I_c \quad \text{...(6.7b)}$$

$$V_b - V_c = I_b Z_F \quad \text{...(6.7c)}$$

The symmetrical components of the fault currents are given by

$$\begin{bmatrix} I_{a0} \\ I_{a1} \\ I_{a2} \end{bmatrix} = \frac{1}{3}\begin{bmatrix} 1 & 1 & 1 \\ 1 & a & a^2 \\ 1 & a^2 & a \end{bmatrix}\begin{bmatrix} 0 \\ I_b \\ -I_b \end{bmatrix} \quad \text{...(6.8)}$$

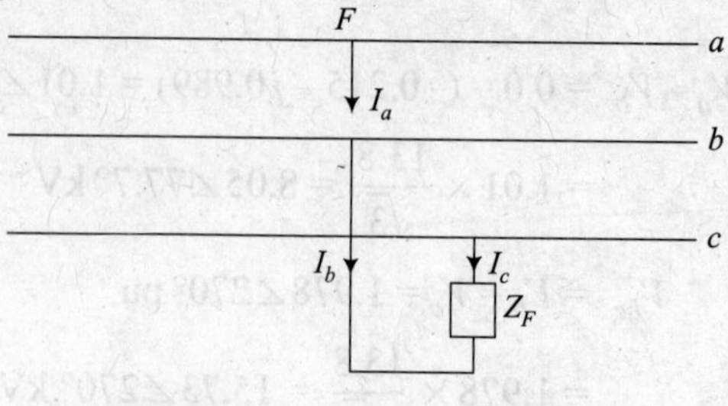

Fig. 6.5 (a) Line–to–Line fault.

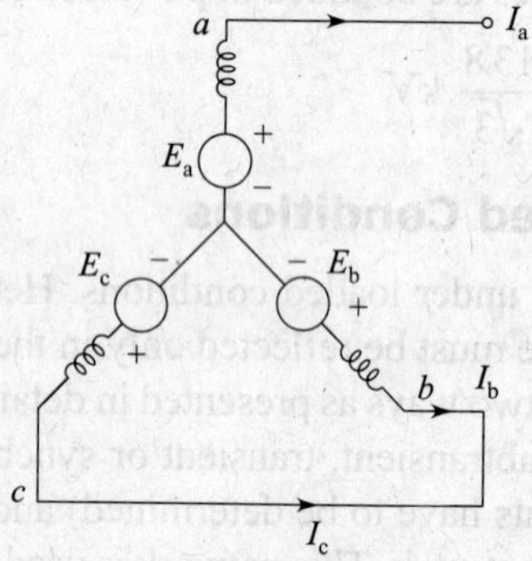

Fig. 6.5(b)

From (6.8), we get

$$I_{a1} = -I_{a2} \qquad ...(6.9)$$

$$I_{a0} = 0 \qquad ...(6.10)$$

The symmetrical components of voltages at F are given by

$$\begin{bmatrix} V_{a0} \\ V_{a1} \\ V_{a2} \end{bmatrix} = \frac{1}{3}\begin{bmatrix} 1 & 1 & 1 \\ 1 & a & a^2 \\ 1 & a^2 & a \end{bmatrix}\begin{bmatrix} V_a \\ V_b \\ V_{b-I_bZ_F} \end{bmatrix} \qquad ...(6.11)$$

From (6.11)

$$V_{a1} = \frac{1}{3}[\, V_a + aV_b + a^2V_b - a^2\, I_b\, Z_F]$$

$$3V_{a1} = V_a + (a + a^2)\, V_b - a^2 I_b Z_F \qquad ...(6.12)$$

$$V_{a2} = \frac{1}{3}[\, V_a + a^2V_b + aV_b - a\, I_b\, Z_F]$$

$$3V_{a2} = V_a + (a + a^2)\, V_b - aI_bZ_F \quad ...(6.13)$$

From (6.12) and (6.13), we can write

$$3(V_{a1} - V_{a2}) = (a - a^2)\, I_bZ_F = j\sqrt{3}\, I_bZ_F \quad ...(6.14)$$

From symmetrical components, we can write

$$I_b = I_{a0} + a^2\, I_{a1} + a\, I_{a2}$$

which using (6.9) and (6.10) becomes

$$I_b = (a^2 - a)\, I_{a1} = -j\sqrt{3}\, I_{a1} \quad ...(6.15)$$

Substituting (6.15) in (6.14), we get

$$(V_{a1} - V_{a2}) = I_{a1}\, Z_F \quad ...(6.16)$$

The network, which satisfies the equations (6.9), (6.10) and (6.16) is the parallel connection of positive and negative sequence networks through a series impedance Z_F as shown in Fig. 6.6.

From Fig. 6.6, we can see that

$$I_{a1} = \frac{E_a}{Z_1 + Z_2 + Z_F} \quad ...(6.17)$$

We can calculate sequence voltages as follows:

$$V_{a1} = E_a - I_{a1}\, Z_1$$

$$V_{a2} = -I_{a2}\, Z_2 = I_{a1}\, Z_2$$

$$V_{a0} = -I_{a0}\, Z_0 = 0$$

The *L-L* fault on an unloaded generator is shown in Fig. 6.5(b).

The fault current is given by $I_{a1} = \dfrac{E_a}{Z_1 + Z_2}$ and the sequence voltages are calculated as above.

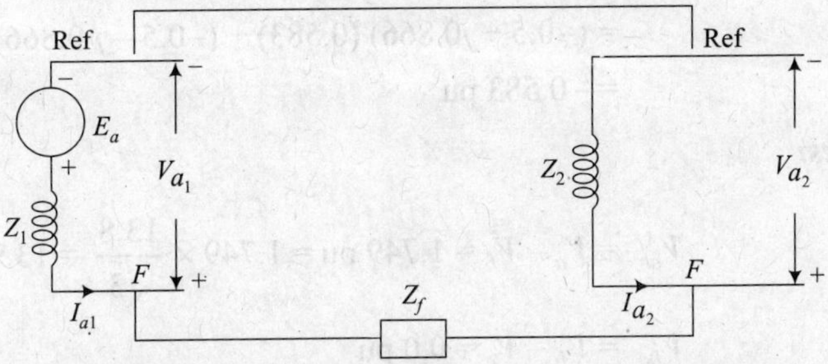

Fig. 6.6 Connection of sequence networks for *L-L* fault.

Example 6.2: Solve example 6.1, for a line–to–line fault between phases *b* and *c*.

Solution:

$$I_{a1} = \frac{E_a}{Z_1 + Z_2} = \frac{1.0}{j0.25 + j0.35} = -j1.667 \text{ pu.}$$

$$
\begin{aligned}
I_{a2} &= -I_{a1} = j1.667 \text{ pu.}\\
I_{a0} &= 0\\
I_a &= I_{a0} + I_{a1} + I_{a2} = 0\\
I_b &= I_{a0} + a^2 I_{a1} + a I_{a2}\\
&= (-0.5 - j0.866)(-j1.667) + (-0.5 + j0.866)(j1.667)\\
&= -2.887 \text{ pu}\\
I_c &= I_{a0} + a I_{a1} + a^2 I_{a2}\\
&= (-0.5 + j0.866)(-j1.667) + (-0.5 - j0.866)(j1.667)\\
&= 2.887 \text{ pu}
\end{aligned}
$$

Actual currents are obtained by multiplying with the base current

$$
\begin{aligned}
I_a &= 0\\
I_b &= -2.887 \times 836 = -2413 \text{ A}\\
I_c &= 2.887 \times 836 = 2413 \text{ A}
\end{aligned}
$$

Line-to-ground voltage:

$$
\begin{aligned}
V_{a1} &= E_a - I_{a1} Z_1 = 1.0 - (-j1.667)(j0.25) = 0.583 \text{ pu}\\
V_{a2} &= -I_{a2} Z_2 = -(j1.667)(j0.35) = 0.583 \text{ pu}\\
V_{a0} &= 0\\
V_a &= V_{a0} + V_{a1} + V_{a2} = 1.166 \text{ pu}\\
V_b &= V_{a0} + a^2 V_{a1} + a V_{a2}\\
&= (-0.5 - j0.866)(0.583) + (-0.5 + j0.866)(0.583)\\
&= -0.583 \text{ pu}\\
V_c &= V_{a0} + a V_{a1} + a^2 V_{a2}\\
&= (-0.5 + j0.866)(0.583) + (-0.5 - j0.866)(0.583)\\
&= -0.583 \text{ pu}
\end{aligned}
$$

Line-to-line voltages:

$$
\begin{aligned}
V_{ab} &= V_a - V_b = 1.749 \text{ pu} = 1.749 \times \frac{13.8}{\sqrt{3}} = 13.93 \text{ kV}\\
V_{bc} &= V_b - V_c = 0.0 \text{ pu}\\
V_{ca} &= V_c - V_a = -1.749 \text{ pu} = -13.93 \text{ kV}
\end{aligned}
$$

6.5 DOUBLE LINE–TO–GROUND (*LLG*) FAULT

The double line-to-ground fault is shown in Fig. 6.7.

The current and voltage relationships are given by

$$I_a = 0 \quad ...(6.18a)$$

$$V_b = V_c \quad ...(6.18b)$$

$$V_b = Z_F (I_b + I_c) = 3Z_F I_{a0} \quad ...(6.18c)$$

(Here we have used the fact that the ground current is $3I_{a0}$)

The symmetrical components of the voltages are

$$\begin{bmatrix} V_{a0} \\ V_{a1} \\ V_{a2} \end{bmatrix} = \frac{1}{3}\begin{bmatrix} 1 & 1 & 1 \\ 1 & a & a^2 \\ 1 & a^2 & a \end{bmatrix}\begin{bmatrix} V_a \\ V_b \\ V_b \end{bmatrix}$$

$$V_{a1} = \frac{1}{3}[\, V_a + (a + a^2)V_b] = V_{a2} \quad ...(6.19a)$$

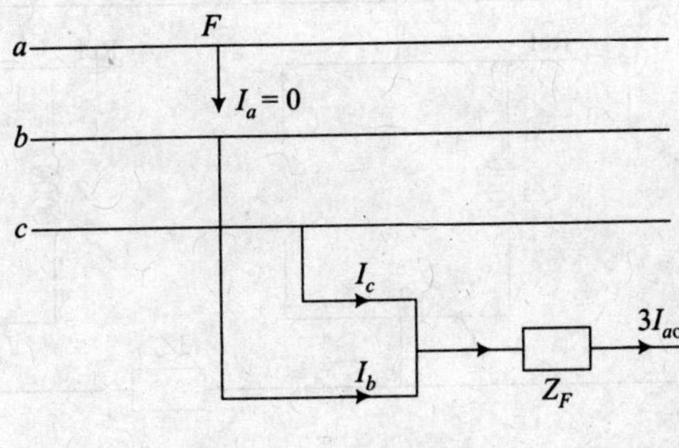

Fig. 6.7 (a) Double line-to-ground fault.

$$V_{a0} = \frac{1}{3}[\, V_a + 2V_b] \quad ...(6.19b)$$

From (6.19a) and (6.19b), we get

$$V_{a0} - V_{a1} = \frac{1}{3}[\, 2 - a - a^2]\, V_b = V_b \quad ...(6.20)$$

From (6.18c) and (6.20), we get

$$V_{a0} - V_{a1} = 3\, Z_F I_{a0}$$

or,

$$V_{a0} = 3\, Z_F I_{a0} + V_{a1} \quad ...(6.21)$$

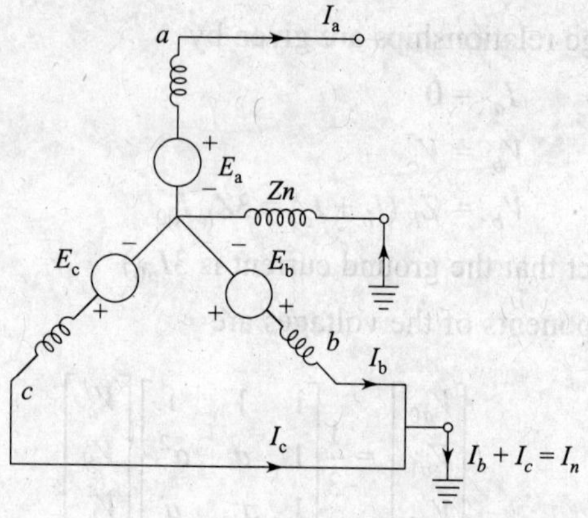

Fig. 6.7(b).

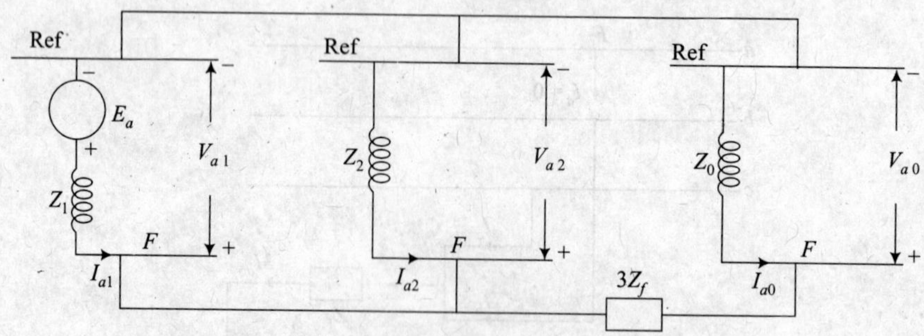

Fig. 6.8 Connection of sequence network for *LLG* fault.

The network connection satisfying (6.18a), (6.19a) and (6.21) is shown in Fig. 6.8. It can be seen that

$$I_{a1} = \frac{E_a}{Z_1 + (Z_2 \| (Z_0 + 3Z_F))} \quad ...(6.22)$$

Having obtained I_{a1} we calculate

$$V_{a1} = E_a - I_{a1} Z_1$$

$$V_{a2} = V_{a1}$$

$$I_{a2} = \frac{-V_{a2}}{Z_2}$$

$$I_{a0} = -(I_{a1} + I_{a2}) \qquad \text{(Since } I_a = 0,\ I_{a0} + I_{a1} + I_{a2} = 0\text{)}$$

$$V_{a0} = -I_{a0} Z_0$$

Thus, all sequence voltages and currents are determined.

The double line-to-ground fault on an unloaded generator is shown in Fig. 6.7(b).

Here $V_b = V_c = 0$

$$I_{a1} = \frac{E_a}{Z_1 + (Z_2 \| Z_0)}$$

$$V_{a1} = E_a - I_{a1} Z_1$$

$$V_{a2} = V_{a1} = V_{a0}$$

$$I_{a2} = -\frac{V_{a2}}{Z_2} \text{ and } I_{a0} = \frac{-V_{a0}}{Z_0}$$

Example 6.3: Solve example 6.1 for a double line-to-ground fault.

Solution: $$I_{a1} = \frac{E_a}{Z_1 + (Z_2 \| (Z_0)} = \frac{1.0}{j0.25 + \dfrac{j0.35 \times j0.1}{j0.35 + j0.1}} = -j3.05 \text{ pu}$$

Since, $Z_F = 0, V_{a1} = V_{a2} = V_{a0}$

$$V_{a1} = E_a - I_{a1} Z_1 = 1.0 - (-j0.35)(j0.25) = 0.2375 \text{ pu}$$

$$I_{a2} = \frac{-V_{a2}}{Z_2} = \frac{-0.2375}{j0.35} = j0.68 \text{ pu}$$

$$I_{a0} = \frac{-V_{a0}}{Z_0} = \frac{-0.2375}{j0.1} = j2.375 \text{ pu}$$

Phase currents:

$$I_a = I_{a0} + I_{a1} + I_{a2} = 0$$

$$I_b = I_{a0} + a^2 I_{a1} + a I_{a2}$$

$$= (j2.375) + (-0.5 - j0.866)(-j3.05) + (-0.5 + j0.866)(j0.68)$$

$$= (j2.375) + (-2.6413 + j1.525) + (-0.589 - j0.34)$$

$$= -3.23 + j3.56 = 4.807 \angle 132.2°$$

$$I_c = I_{a0} + a I_{a1} + a^2 I_{a2}$$

$$= (j2.375) + (-0.5 + j0.866)(-j3.05) + (-0.5 - j0.866)(j0.68)$$

$$= (j2.375) + (2.6413 + j1.525) + (0.589 - j0.34)$$

$$= 3.23 + j3.56 = 4.807 \angle 47.7°$$

Phase voltage:

$$V_a = V_{a0} + V_{a1} + V_{a2} = 3V_{a1} = 0.7125 \text{ pu}$$

$$V_b = V_c = 0$$

Line voltages:

$$V_{ab} = V_a - V_b = 0.7215 \text{ pu} = 0.7125 \times \frac{13.8}{\sqrt{3}} = 5.676 \text{ kV}$$

$$V_{bc} = V_b - V_c = 0.0 \text{ pu} = 0.0 \text{ kV}$$

$$V_{ca} = V_c - V_a = -0.7215 \text{ pu} = -5.676 \text{ kV}$$

Fault current:

$$3I_{a0} = 3 \times j2.375 = j7.125 \text{ pu}$$

$$= j7.125 \times 836 = 5956.5 \text{ pu}$$

Example 6.4: Consider the power system shown in Fig. 6.9. Compute the fault current for an *L–G* fault at *F*, if the system is initially on no load. The generator negative sequence reactance is 70% of positive sequence reactance.

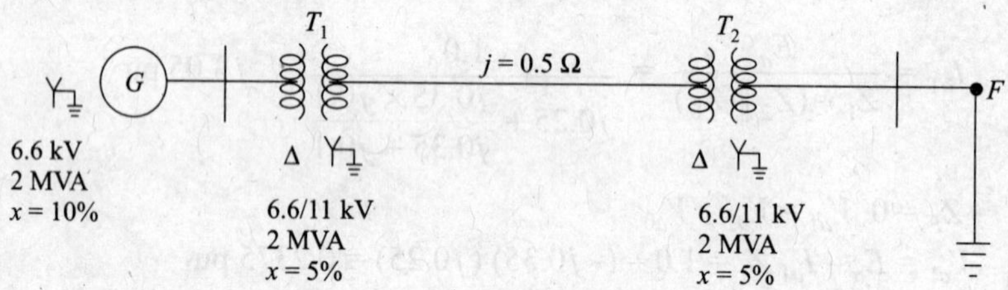

Fig. 6.9 Power system of example 6.4.

Solution: Let base MVA be 2 MVA and base on generator be 6.6 kV. All the sections are on correct bases. Hence, no change in pu values of reactance is necessary.

$$\text{Base impedance on transmission line} = \frac{(11)^2}{2} = 60.5\ \Omega$$

$$\text{pu impedance of line} = \frac{0.5}{60.5} = 0.00826 \text{ pu}$$

The positive sequence network is shown in Fig. 6.10a and the equivalent in Fig. 6.10b.

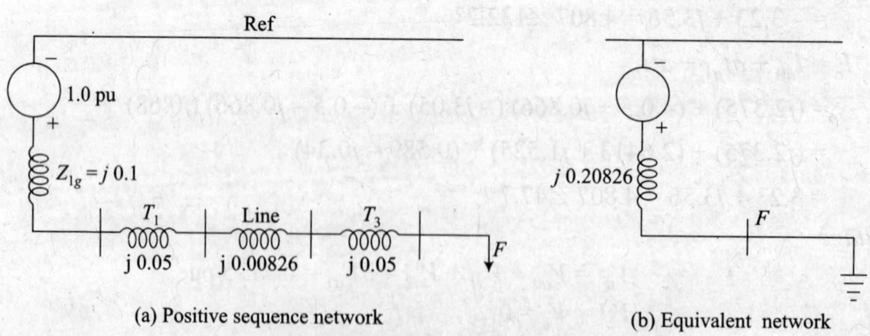

Fig. 6.10 Positive sequence network of example 6.4.

The negative sequence network is shown in Fig. 6.11a and its equivalent in 6.11b.

The zero sequence network is shown in Fig. 6.12a and its equivalent in Fig. 6.12b. (Note that the zero sequence reactance of other components are not needed)

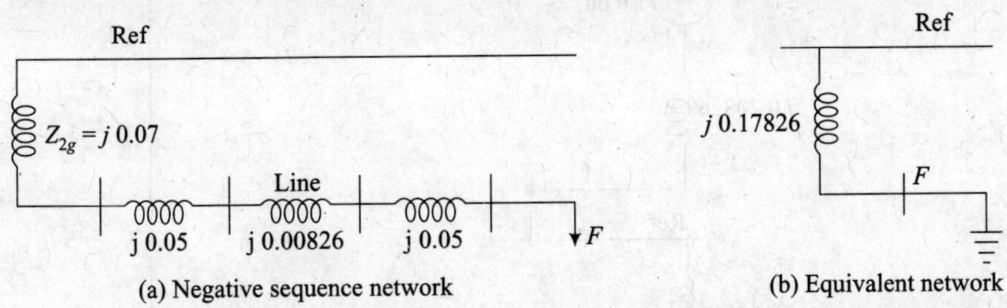

Fig. 6.11 Negative sequence network of example 6.4.

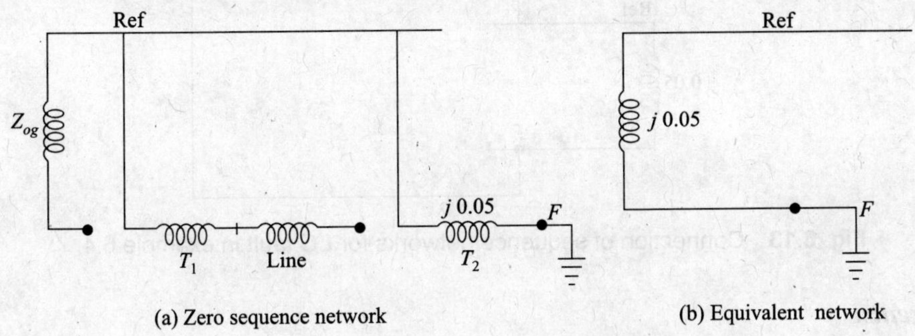

Fig. 6.12 Zero sequence network of example 6.4.

For a single line-to-ground fault the three sequence networks are connected in series as shown in Fig. 6.13.

$$I_{a1} = \frac{1.0}{j0.20826 + j0.17826 + j0.05} = -j2.29 \text{ pu.}$$

Fault current $= 3\,I_{a1} = -j\,6.8725$ pu

Base current at point $F = \dfrac{2\times10^6}{\sqrt{3}\times6.6\times10^3} = 174.95$ A

Magnitude of fault current $= 6.8725 \times 174.95 = 1202$ A

Example 6.5: Repeat example 6.4 for an LL fault.

Solution:

$$I_{a1} = \frac{E_a}{Z_1 + Z_2} = \frac{1.0}{j(0.20826 + 0.17826)} = -j\,2.587 \text{ pu}$$

$$I_{a2} = -I_{a1} = j\,2.587.$$

$$I_{a0} = 0.$$

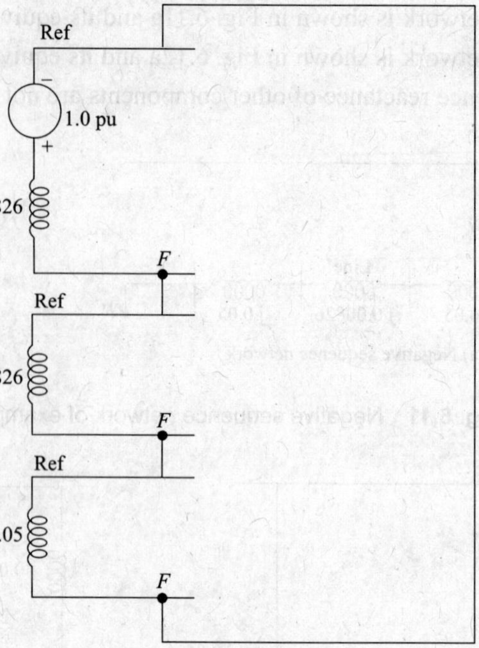

Fig. 6.13 Connection of sequence networks for *LG* fault in example 6.4.

Fault currents:

$$I_a = 0.0$$
$$I_b = I_{a0} + a^2 I_{a1} + a I_{a2}$$
$$= (a^2 - a)\, I_{a1} = (-j1.732)\,(-j2.587)$$
$$= -4.48 \text{ pu} = -784 \text{ A}$$
$$I_c = -I_b = 4.48 \text{ pu} = 784 \text{ A.}$$

Example 6.6: Derive the condition for the neutral grounding reactance, such that the *LG* fault current is less than the three phase symmetrical fault current, when fault occurs at the terminals of an unloaded generator.

Solution: The sequence network connection is shown in Fig. 6.14 for the *LG* fault. $3X_n$ is included in the zero sequence network.

For an LG fault $I_{a1} = \dfrac{E_a}{X_1 + X_2 + X_0 + 3X_n}$

Fault current $I_{FLG} = 3I_{a1}$. Assuming $X_1 = X_2$,

$$|I_{FLG}| = \frac{3|E_a|}{2X_1 + X_0 + 3X_n}$$

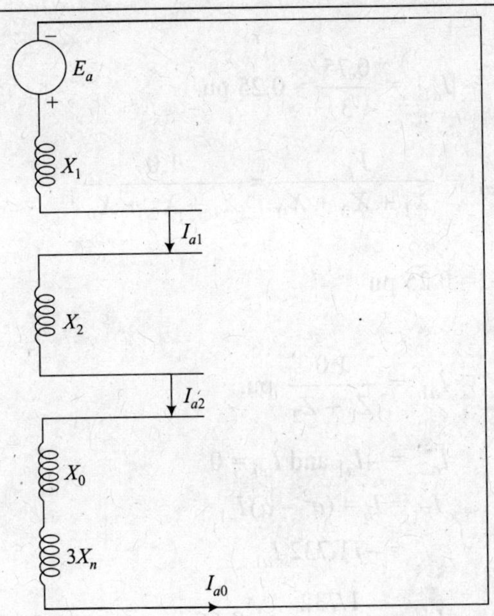

Fig. 6.14 Sequence network connection for *LG* fault.

Fault current for a three phase symmetrical fault is given by

$$|I_{F3\phi}| = \frac{E_a}{X_1} = \frac{3|E_a|}{3X_1}$$

We need $\quad |I_{FLG}| < |I_{F3\phi}|$

$$\frac{3|E_a|}{2X_1 + X_0 + 3X_n} < \frac{3|E_a|}{3X_1}$$

or, $\quad 2X_1 + X_0 + 3X_n > 3X_1$

or, $\quad X_n > \frac{1}{3}[X_1 - X_0]$

For values of grounding reactance less than $\frac{1}{3}[X_1 - X_0]$, the fault current for an *LG* fault is greater than that due to the three phase symmetrical fault.

Example 6.7: In a network, the magnitude of fault current for an *LG* fault is 0.75 pu and for an *LL* fault it is 8.66 pu. If the prefault voltage at the fault location is 1.0 pu, determine the equivalent zero sequence impedance of the network.

Solution: For an *LG* fault, $I_{FLG} = 3\,I_{a1}$.

$$\therefore \qquad |I_{a1}| = \frac{0.75}{3} = 0.25 \text{ pu.}$$

Also in an LG fault, $|I_{a1}| = \dfrac{V_F}{X_1 + X_2 + X_0} = \dfrac{1.0}{X_1 + X_2 + X_0}$

Hence, $\dfrac{1.0}{X_1 + X_2 + X_0} = 0.25$ pu

In an LL fault,
$$I_{a1} = \frac{1.0}{Z_1 + Z_2} \text{ pu.}$$

$$I_{a2} = -I_{a1} \text{ and } I_{a0} = 0$$

The fault current
$$I_F = I_b = (a^2 - a)I_{a1}$$
$$= -j1.732\, I_{a1}$$
$$|I_b| = \frac{1.732}{X_1 + X_2} = 8.66$$

We get
$$X_1 + X_2 = \frac{1.732}{8.66} = 0.2 \text{ pu}$$

$$\therefore \qquad \frac{1.0}{0.2 + X_0} = 0.25$$

$$X_0 = 3.8 \text{ pu.}$$

Example 6.8: For the network shown in Fig. 6.15, determine the fault current for

(i) Three phase symmetrical fault
(ii) LG fault
(iii) LL fault
(iv) Double line-to-ground fault

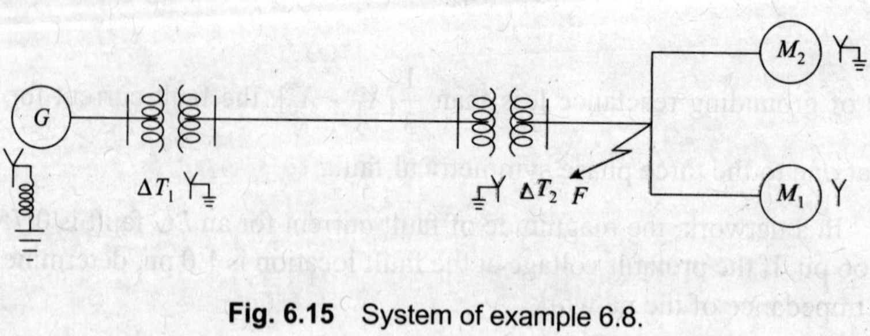

Fig. 6.15 System of example 6.8.

Data:

Generator: 11 kV, 50 MVA; x''_d = 15%; $X_2 = x''_d$; X_0 = 8%,

Grounding reactance = 2.5 Ω.

T_1 and T_2: 11/132 kV; 60 MVA; X = 10%

M_1 = 11 kV; 30 MVA; $x''_d = X_2$ = 20 %; X_0 = 8%

M_2 = 11 kV; 15 MVA; $x''_d = X_2$ = 20 %; X_0 = 8%

Grounding reactance = 2.5 Ω.

Transmission line: $X_1 = X_2$; 120 Ω reactance; $X_0 = 3X_1$

Solution: Choose a base of 11 kV, 50 MVA in generator. All reactances are now calculated in the appropriate base.

Generator:

$$X_1 = X_2 = 0.15 \text{ pu}; X_0 = 0.08 \text{ pu}.$$

$$Z_B = \frac{11^2}{50} = 2.42\ \Omega$$

$$X_n = \frac{2.5}{2.42} = 1.033 \text{ pu}$$

$$3X_n = 3.099 \text{ pu}$$

T_1 and T_2:

$$X = 0.1 \times \frac{50}{60} = 0.0833 \text{ pu}$$

Transmission line:

$$Z_B = \frac{132^2}{50} = 348.48\ \Omega$$

$$X_1 = X_2 = \frac{120}{348.48} = 0.344 \text{ pu}$$

$$X_0 = 0.344 \times 3 = 1.032 \text{ pu}$$

M_1 :

$$X_1 = X_2 = 0.2 \times \frac{50}{30} = 0.333 \text{ pu}$$

$$X_0 = 0.08 \times \frac{50}{30} = 0.133 \text{ pu}$$

M_2 :

$$X_1 = X_2 = 0.2 \times \frac{50}{15} = 0.666 \text{ pu}$$

$$X_0 = 0.08 \times \frac{50}{15} = 0.266 \text{ pu}$$

$$Z_B = \frac{11^2}{50} = 2.42\ \Omega$$

$$3X_n = \frac{2.5}{2.42} \times 3 = 3.099 \text{ pu}$$

The positive sequence network is shown in Fig. 6.16a.

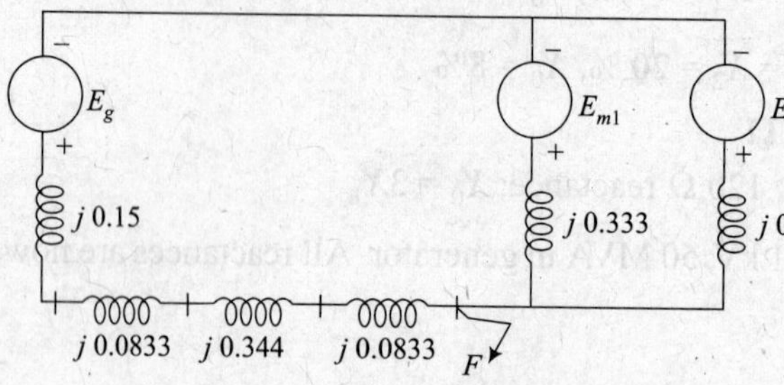

Fig. 6.16a Positive sequence network of example 6.8.

The Thevenin's equivalent impedance at fault point is given by short circuiting all the sources. X_{1F} is parallel combination of $j0.6606, j0.333$ and $j0.666$.

$$\therefore \qquad X_{1F} = j0.166 \text{ pu.}$$

The negative sequence network is shown in Fig. 6.16b.

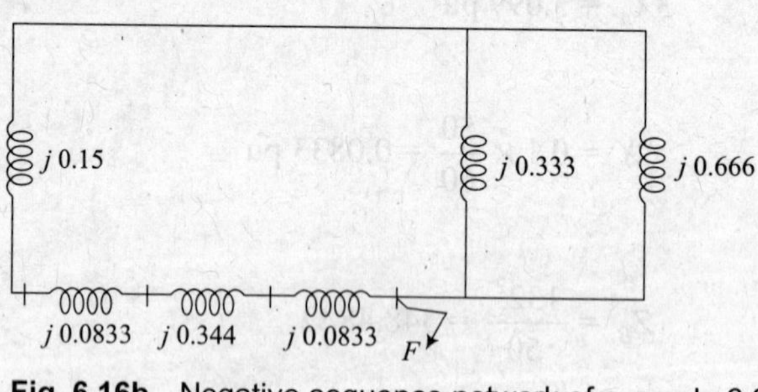

Fig. 6.16b Negative sequence network of example 6.8.

$$X_{2F} = j0.166 \text{ pu}$$

The zero sequence network is shown in Fig. 6.16c.

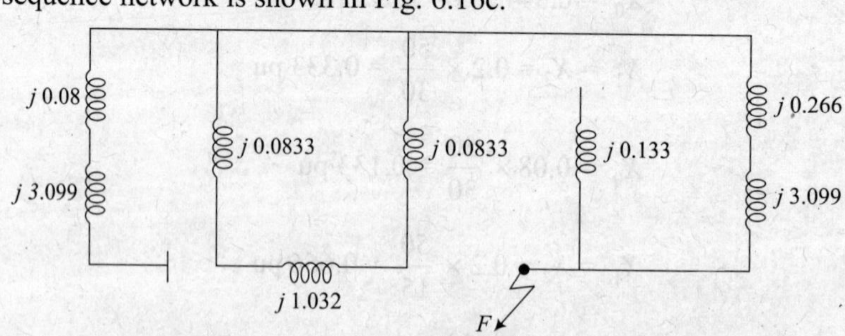

Fig. 6.16c Zero sequence network of example 6.8.

$$Z_{0F} = j3.365 \text{ pu}$$

(i) ***Three-phase symmetrical fault:***

Let us assume the voltage at fault point is 1.0 pu prior to fault.

$$I_{F3\phi} = \frac{1.0}{j0.166} = -j6.024 \text{ pu}$$

Base current at fault location

$$\frac{50 \times 10^6}{\sqrt{3}11 \times 10^3} = 2{,}624.3 \text{ A}$$

$$\therefore \qquad |I_{F3\phi}| = 15{,}808 \text{ A}$$

(ii) ***L–G Fault:***

$$I_{a1} = \frac{1.0}{j0.166 + j0.166 + j3.365}$$

$$I_{FLG} = 3\, I_{a1} = -j0.811 \text{ pu}$$

$$|I_{FLG}| = 0.811 \times 2624.3 = 2{,}128 \text{ A}$$

(iii) ***L–L Fault:***

$$I_{a1} = \frac{1.0}{j0.166 + j0.166} = 3.012$$

Fault current $\quad I_b = -j1.732 \times 3.012 = -j5.216$ pu

$$|I_{FLL}| = 5.216 \times 2624.3 = 13{,}690 \text{ A}$$

(iv) ***L–L–G Fault:***

$$I_{a1} = \frac{1.0}{j0.166 + (j0.166 \,\|\, j3.365)} = -j3.084 \text{ pu}$$

$$V_{a1} = E_a - I_{a1}\, Z_1 = 1.0 - (-j\,3.084)\,(j0.166) = 0.488 \text{ pu}$$

$$V_{a0} = V_{a1}$$

$$I_{a0} = \frac{-V_{a0}}{Z_0} = \frac{-0.488}{j3.365} = j0.145 \text{ pu}$$

Fault current $= 3I_{a0} = j0.435$ pu

$$|I_{FLLG}| = 0.435 \times 2624.3 = 1{,}142 \text{ A}$$

In this problem note that the high value of zero sequence reactance reduces the fault current levels, for faults involving ground, such as L-G and L-L-G. The L-L fault current is highest, since it does not depend on the zero sequence impedance.

Example 6.9: Consider the system shown in Fig. 6.17. Compute the fault current for different faults.

Data:

G_1 and G_2: 100 MVA; 33 kV; $X_1 = X_2 = 0.2$ pu; $X_0 = 0.05$ pu

T_1 and T_2: 100 MVA; 33/220 kV; $X = 0.05$ pu

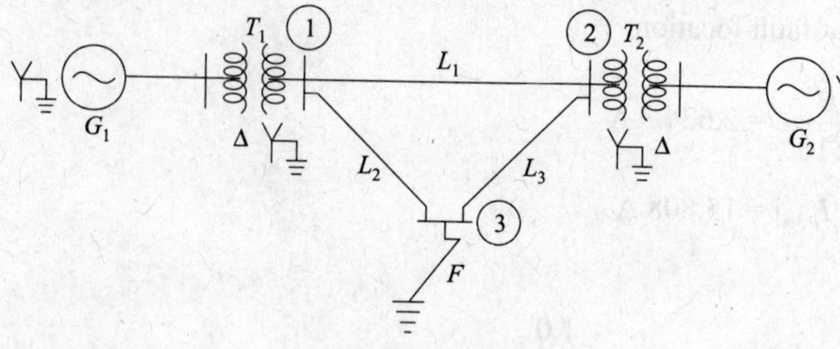

Fig. 6.17 Example 6.9.

L_1, L_2 and L_3: $X_1 = X_2 = 0.1$ pu; $X_0 = 0.3$ pu

All pu reactances are in correct base.

Solution: The positive sequence network is shown in Fig. 6.18a.

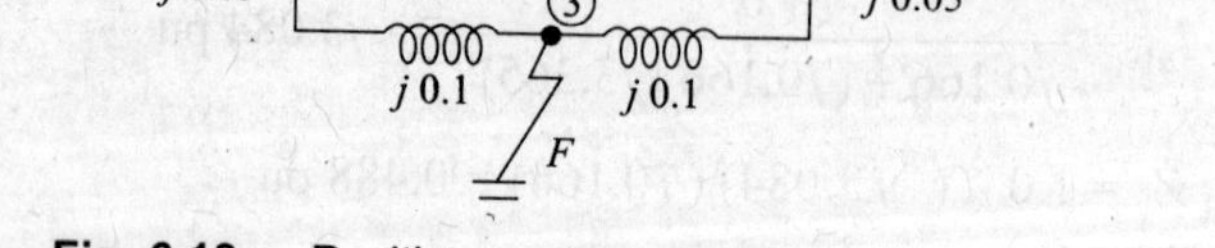

Fig. 6.18a Positive sequence network of example 6.9.

The delta is converted into equivalent star to get Fig. 6.18b.

Thevenin's equivalent voltage between F and reference bus is obtained by shorting the sources.

$$Z_1 = \{(j0.2 + j0.05 + j0.033)\} \parallel (j0.2 + j0.05 + j0.033)\} + j0.033 = j0.175 \text{ pu}$$

We can assume prefault voltage at bus 3, $V_{F0} = 1.0$ pu. The negative sequence network is identical to positive sequence network, except for the sources. Hence,

$$Z_2 = j0.175 \text{ pu}$$

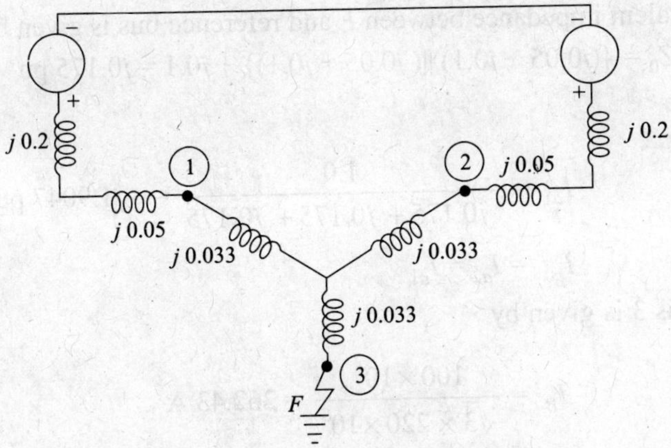

Fig. 6.18b Equivalent network.

The zero sequence network is shown in Fig. 6.18c.

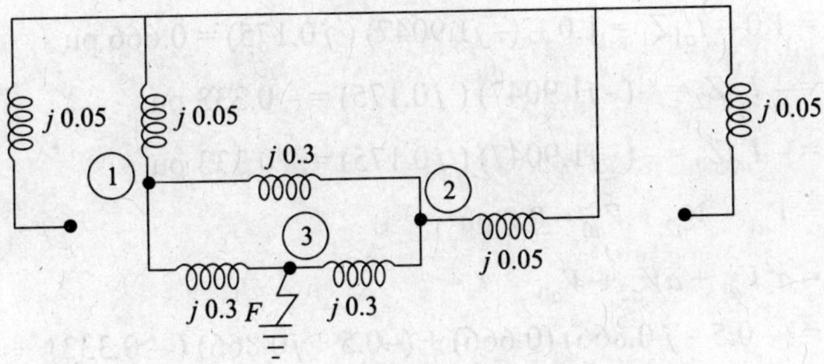

Fig. 6.18c Zero sequence of example 6.9.

The delta is converted into equivalent star and shown in Fig. 6.18d.

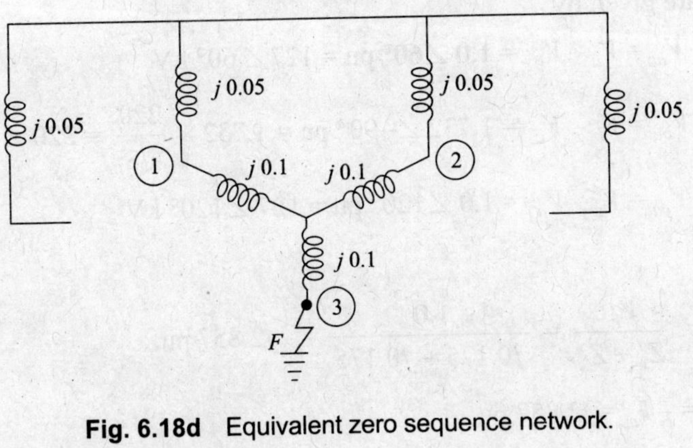

Fig. 6.18d Equivalent zero sequence network.

The Thevenin's equivalent impedance between F and reference bus is given by

$$Z_0 = \{(j0.05 + j0.1) \,||\, (j0.05 + j0.1)\} + j0.1 = j0.175 \text{ pu}$$

For an LG fault:

$$I_{a1} = \frac{1.0}{j0.175 + j0.175 + j0.175} = -j1.9047 \text{ pu.}$$

$$I_{a0} = I_{a2} = I_{a1}$$

Base current at bus 3 is given by

$$I_B = \frac{100 \times 10^6}{\sqrt{3} \times 220 \times 10^3} = 262.43 \text{ A}$$

$$I_F = 3\, I_{a1} = -j5.7141 \text{ pu}$$

$$|I_F| = 1500 \text{ A}$$

The voltages at bus 3 are computed as follows:

$$V_{a1} = 1.0 - I_{a1}Z_1 = 1.0 - (-j1.9047)\,(j0.175) = 0.666 \text{ pu}$$

$$V_{a2} = -I_{a2}Z_2 = -(-j1.9047)\,(j0.175) = -0.333 \text{ pu}$$

$$V_{a0} = -I_{a0}Z_0 = -(-j1.9047)\,(j0.175) = -0.333 \text{ pu}$$

$$V_a = V_{a1} + V_{a2} + V_{a0} = 0.0 \text{ pu}$$

$$V_b = a^2V_{a1} + aV_{a2} + V_{a0}$$

$$= (-0.5 - j\,0.866)\,(0.666) + (-0.5 + j0.866)\,(-0.333) + (-0.333)$$

$$= 1.0 \angle{-120°} \text{ pu} = 1.0 \times \frac{220}{\sqrt{3}} = 127 \angle{-120°} \text{ kV}$$

$$V_c = aV_{a1} + a^2V_{a2} + V_{a0} = 1.0 \angle 120° \text{ pu} = 127 \angle 120° \text{ kV}$$

Line voltages are given by

$$V_{ab} = V_a - V_b = 1.0 \angle 60° \text{ pu} = 127 \angle 60° \text{ kV}$$

$$V_{bc} = V_b - V_c = 1.732 \angle{-90°} \text{ pu} = 1.732 \times \frac{220}{\sqrt{3}} = 220 \angle{-90°} \text{ kV}$$

$$V_{ca} = V_c - V_a = 1.0 \angle 120° \text{ pu} = 127 \angle 120° \text{ kV}$$

For an LL fault:

$$I_{a1} = \frac{V_F}{Z_1 + Z_2} = \frac{1.0}{j0.175 + j0.175} = -j2.857 \text{ pu.}$$

$$I_{a2} = -I_{a1} = j2.857 \text{ pu}$$

$$I_{a0} = 0.0$$
$$I_a = I_{a1} + I_{a2} + I_{a0} = 0$$
$$I_b = (a^2 - a)\, I_{a1} = -4.948 = 4.948 \angle 180° \text{ pu}$$
$$I_c = (a - a^2)\, I_{a1} = 4.948 \text{ pu}$$
$$V_{a1} = V_F - I_{a1} Z_1 = 1.0 - (-j2.857)(j0.175) = 0.5 \text{ pu}$$
$$V_{a2} = -I_{a2} Z_2 = V_{a1} = 0.5 \text{ pu}$$
$$V_{a0} = 0.0$$
$$V_a = V_{a0} + V_{a1} + V_{a2} = 1.0 \text{ pu} = 127 \text{ kV}$$
$$V_b = (a^2 + a)\, V_{a1} = (-1.0)(0.5) = -0.5 \text{ pu} = 63.5 \angle 180° \text{ kV}$$
$$V_c = (a + a^2)\, Va1 = -0.5 \text{ pu} = 63.5 \angle 180° \text{ kV}$$
$$V_{ab} = V_a - V_b = 1.5 \text{ pu} = 190.5 \text{ kV}$$
$$V_{bc} = V_b - V_c = 0.0 \text{ kV}$$
$$V_{ca} = V_c - V_a = -1.5 \text{ pu} = 190.5 \angle 180° \text{ kV}$$

The voltages are the voltages at bus 3.

For an LLG Fault:

$$I_{a1} = \frac{1.0}{j0.175 + (j0.175 \,\|\, j0.175)} = -j3.809 \text{ pu}$$

$$V_{a1} = 1.0 - (-j3.809)(j0.175) = 0.333 \text{ pu}$$

$$V_{a2} = V_{a0} = V_{a1}$$

$$I_{a2} = \frac{-V_{a2}}{Z_2} = \frac{-0.333}{j0.175} = j1.9028 \text{ pu}$$

$$I_{a0} = \frac{-0.333}{j0.175} = j1.9028 \text{ pu}$$

$$\text{Fault current} = 3\, I_{a0} = j5.708 \text{ pu}$$
$$V_a = V_{a1} + V_{a2} + V_{a0} = 1.0 \text{ pu}$$
$$V_b = 0.0 \text{ pu}$$
$$V_c = 0.0 \text{ pu}$$
$$V_{ab} = V_a - V_b = 1.0 \text{ pu} = 127 \angle 0° \text{ kV}$$
$$V_{bc} = V_b - V_c = 0.0 \text{ pu}$$
$$V_{ca} = V_c - V_a = -1.0 \text{ pu} = 127 \angle 180° \text{ kV}$$

Example 6.10: Fig. 6.19 shows a system, where pu values are on a base of 1000 MVA, 20 kV, 60 Hz.

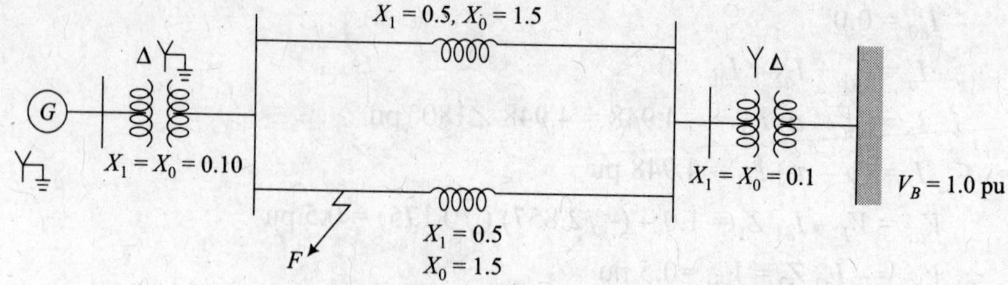

Fig. 6.19 Power system of example 6.10.

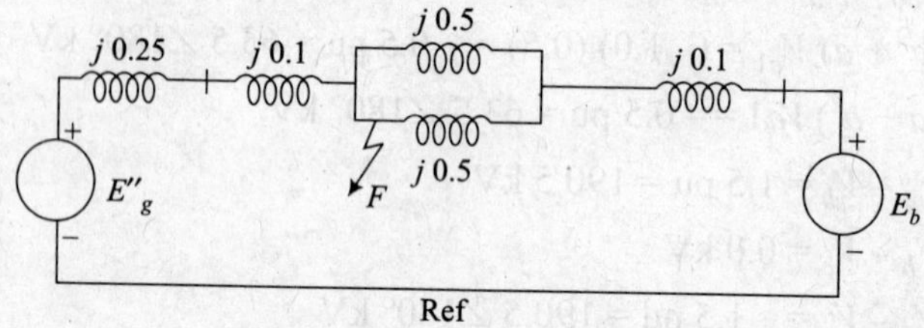

Fig. 6.20a Positive sequence network of example 6.10.

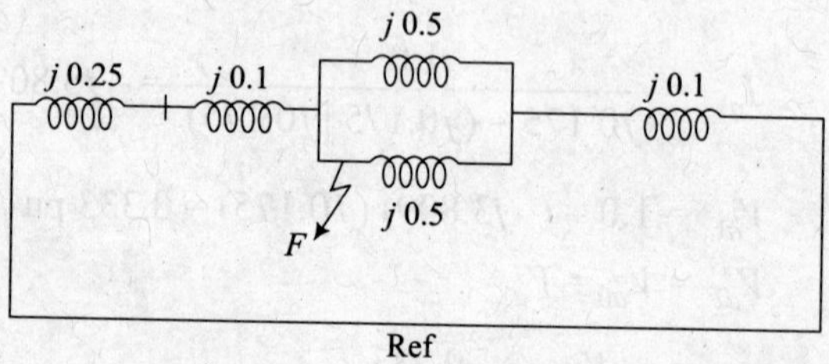

Fig. 6.20b Negative sequence network of example 6.10.

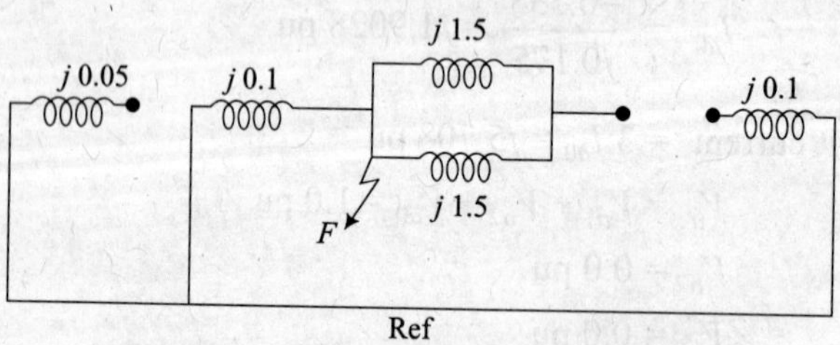

Fig. 6.20c Zero sequence network of example 6.10.

Generator $x_d'' = 0.25$; $x_0 = 0.05$. Draw the sequence networks. If a double line-to-ground fault occurs at F, find the effective fault impedance Z_{ef} which when inserted in positive sequence network, represents the fault.

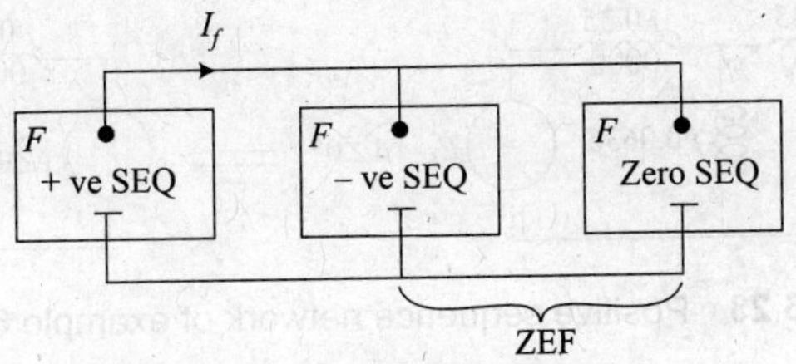

Fig. 6.21 *LLG* fault of example 6.10.

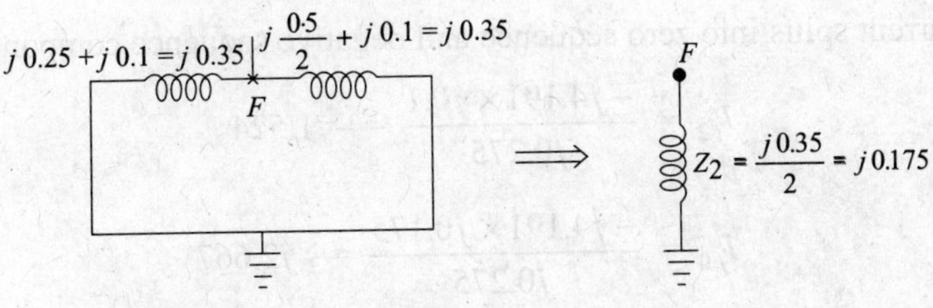

Fig. 6.22a Effective negative sequence impedance.

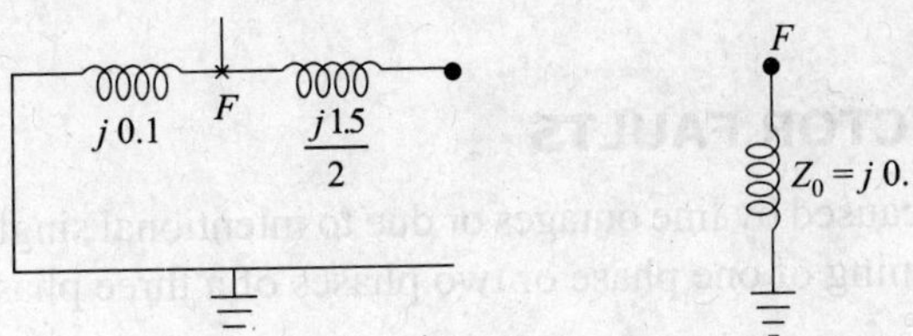

Fig. 6.22b Effective zero sequence impedance.

Solution: The positive, negative and zero sequence networks are shown in Fig. 6.20a, 6.20b and 6.20c.

The sequence network connections for *LLG* fault is shown in Fig. 6.21.

The effective negative sequence and zero sequence impedances as measured at *F* are shown in Fig. 6.22a and Fig. 6.22b, respectively.

Hence, the effective fault impedance Z_{ef} to be inserted in the positive sequence network to give same fault level as the *LLG* fault is

$$Z_{ef} = Z_2 \| Z_0 = j0.175 \,\|\, j0.1 = j0.0636 \text{ pu.}$$

Example 6.11: In example 6.10, if the system was initially on no load, find the sequence currents in the fault.

Solution: The positive sequence network with the fault impedance is shown in Fig. 6.23.

The positive sequence fault current is given by

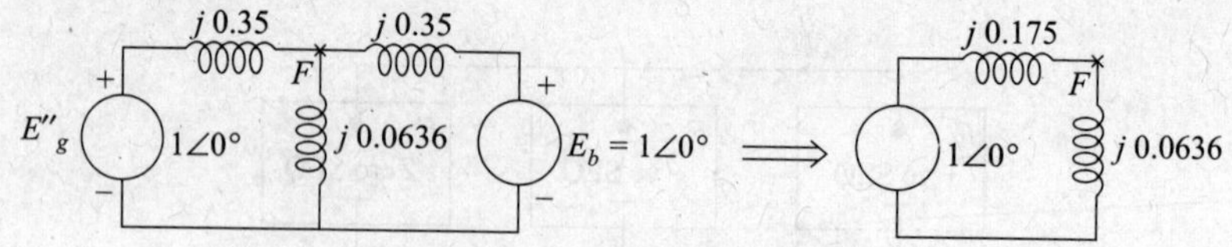

Fig. 6.23 Positive sequence network of example 6.11.

$$I_{f1} = \frac{1.0}{j0.2386} = -j4.191 \text{ pu}$$

This fault current splits into zero sequence and negative sequence components.

$$I_{f2} = \frac{-j4.191 \times j0.1}{j0.275} = -j1.524$$

$$I_{f0} = \frac{-j4.191 \times j0.175}{j0.275} = -j2.667$$

Because of network symmetry, one half of fault current of each sequence component flows from generator and other half from infinite bus. The zero sequence current is restricted to the step–up transformer.

6.6 OPEN CONDUCTOR FAULTS

Open conductor faults are caused by line outages or due to intentional single phase operation. Such faults may involve the opening of one phase or two phases of a three phase circuit.

6.6.1 One Open Conductor

Assume conductor of phase a is open, as shown in Fig. 6.24a.

$V_{aFF'}$, $V_{bFF'}$, $V_{cFF'}$ are the series voltage drops in phases between F and F' in phases a, b and c, respectively. From Fig. 6.24a, we get

$$V_{bFF'} = V_{cFF'} = 0$$

$$I_a = 0.$$

In terms of symmetrical components

$$V_{aFF_1'} = V_{aFF_2'} = V_{aFF_0'} = \frac{V_{aFF'}}{3}$$

$$I_{a1} + I_{a2} + I_{a0} = 0.$$

The connection of the sequence networks is shown in Fig. 6.24b.

6.6.2 Two Open Conductors

Let us assume phases b and c are open as shown in Fig. 6.25a.

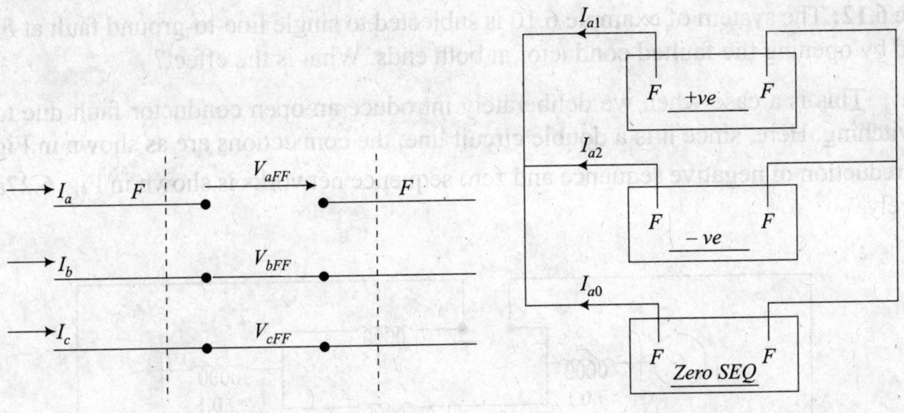

(a) Open conductor fault

(b) Sequence network connection

Fig. 6.24

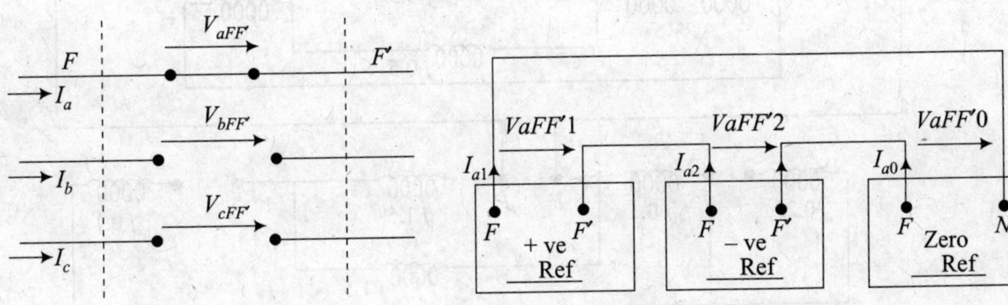

(a) Two conductors open

(b) Sequence network connection

Fig. 6.25

The equations describing Fig. 6.25a are,

$$V_{aFF'} = 0; I_b = 0; I_c = 0.$$

In terms of symmetrical components

$$V_{aFF_1'} + V_{aFF_2'} + V_{aFF_0'} = 0$$

$$I_{a1} = I_{a2} = I_{a0} = \frac{I_a}{3}$$

The sequence network connections are shown in Fig. 6.25b.

From Fig. 6.24b, we see that for a single open conductor fault, the sequence networks are connected in parallel at points F and F'. Hence, even if due to connections, the zero sequence impedance as seen from FF' is infinite, some power is transferred through the line between F and F'. When two conductors are open, the three sequence networks are connected in series. Hence, unless the zero sequence circuit provides a path of finite impedance, no power can be transferred through the line.

Example 6.12: The system of example 6.10 is subjected to single line-to-ground fault at F, which is cleared by opening the faulted conductor at both ends. What is the effect?

Solution: This is a case when we deliberately introduce an open conductor fault due to single phase switching. Here, since it is a double circuit line, the connections are as shown in Fig. 6.26.

The reduction of negative sequence and zero sequence networks is shown in Fig. 6.27a and b, respectively.

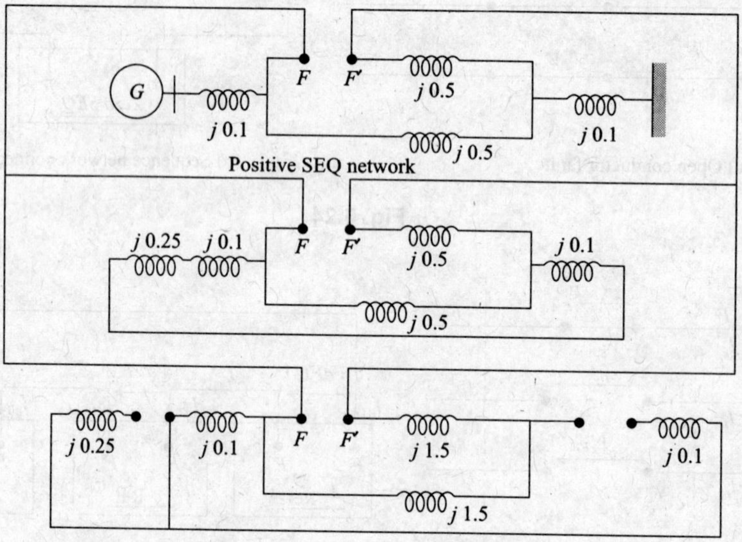

Fig. 6.26 Example 6.12.

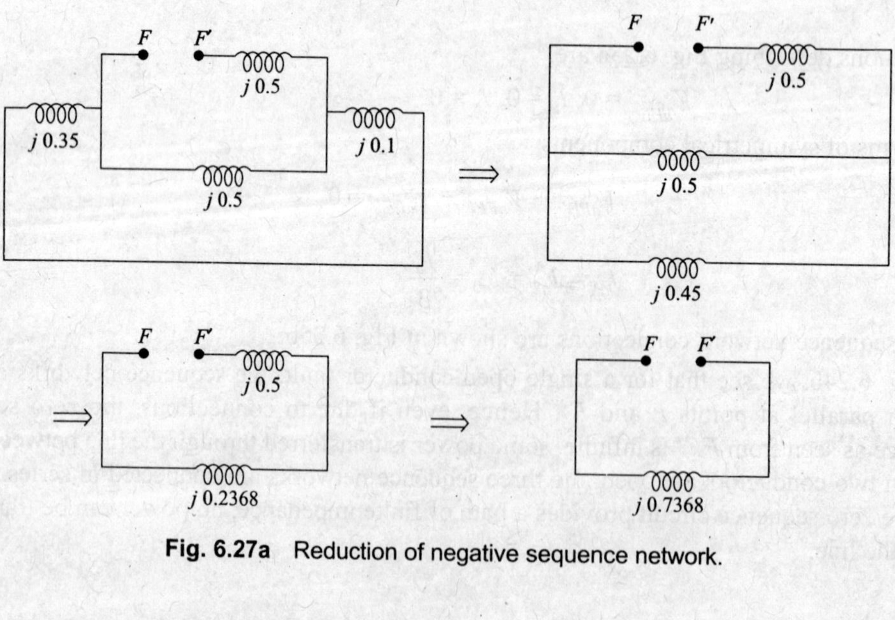

Fig. 6.27a Reduction of negative sequence network.

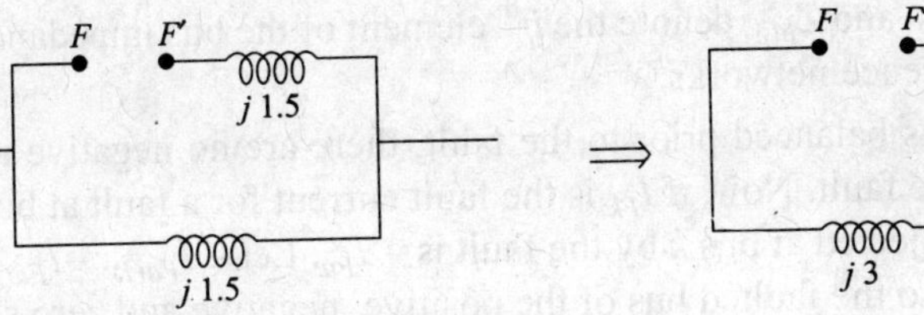

Fig. 6.27b Reduction of zero sequence network.

The net effective impedance to be inserted between F and F' of the positive sequence network is

$$Z_{ef} = j0.7368 \,||\, j3.0 = j0.5915 \text{ pu.}$$

Hence, from view point of power transfer between the generator and infinite bus, the effect of opening of one conductor in one of the double circuit lines is same as increasing the reactance of the circuit by 0.5915 pu.

6.7 ANALYSIS OF UNSYMMETRICAL FAULTS USING BUS IMPEDANCE MATRIX

In chapter 4, we used the bus impedance matrix to determine currents and voltages in all branches of the network, after the occurrence of a three phase symmetrical fault. This method can be extended to unsymmetrical faults. Here, the bus impedance matrices are built separately for the positive sequence, negative sequence and zero sequence networks. For an n bus system, we represent $Z_{bus(1)}$ as the bus impedance matrix of the positive sequence network; $Z_{bus(2)}$ as the bus impedance matrix of the negative sequence network and $Z_{bus(0)}$ as bus impedance matrix of the zero sequence network. These matrices are represented as follows:

$$Z_{bus(1)} = \begin{bmatrix} Z_{11(1)} & Z_{12(1)} & \ldots Z_{1k(1)} & \ldots Z_{1n(1)} \\ Z_{21(1)} & Z_{22(1)} & \ldots Z_{2k(1)} & \ldots Z_{2n(1)} \\ Z_{k1(1)} & Z_{k2(1)} & \ldots Z_{kk(1)} & \ldots Z_{kn(1)} \\ Z_{n1(1)} & Z_{n2(1)} & \ldots Z_{nK(1)} & \ldots Z_{nn(1)} \end{bmatrix} \quad \ldots(6.23)$$

$$Z_{bus(2)} = \begin{bmatrix} Z_{11(2)} & Z_{12(2)} & \ldots Z_{1k(2)} & \ldots Z_{1n(2)} \\ Z_{21(2)} & Z_{22(2)} & \ldots Z_{2k(2)} & \ldots Z_{2n(2)} \\ Z_{k1(2)} & Z_{k2(2)} & \ldots Z_{kk(2)} & \ldots Z_{kn(2)} \\ Z_{n1(2)} & Z_{n2(2)} & \ldots Z_{nK(2)} & \ldots Z_{nn(2)} \end{bmatrix} \quad \ldots(6.24)$$

$$Z_{bus(0)} = \begin{bmatrix} Z_{11(0)} & Z_{12(0)} & \ldots Z_{1k(0)} & \ldots Z_{1n(0)} \\ Z_{21(0)} & Z_{22(0)} & \ldots Z_{2k(0)} & \ldots Z_{2n(0)} \\ Z_{k1(0)} & Z_{k2(0)} & \ldots Z_{kk(0)} & \ldots Z_{kn(0)} \\ Z_{n1(0)} & Z_{n2(0)} & \ldots Z_{nK(0)} & \ldots Z_{nn(0)} \end{bmatrix} \quad \ldots(6.25)$$

Elements $Z_{ij(1)}$, $Z_{ij(2)}$ and $Z_{ij(0)}$ denote the ij^{th} element of the bus impedance matrices of positive, negative and zero sequence networks.

Since, the system is balanced prior to the fault, there are no negative and zero sequence currents flowing before the fault. Now, if I_{Fk} is the fault current for a fault at bus k, it flows out of bus k. Hence, the current injected at bus k by the fault is $-I_{Fa}$. Let $-I_{Fa(1)}$, $-I_{Fa(2)}$ and $-I_{Fa(0)}$ represent the injected currents into the faulted bus of the positive, negative and zero sequence networks. We can disregard the prefault currents and assume that all the system buses operate at a voltage V_0 (normally 1 pu) before the fault. Now the changes in the sequence components of the various bus voltages due to the fault current are obtained as follows.

Positive sequence voltages:

$$\begin{bmatrix} \Delta V_{1a(1)} \\ \Delta V_{2a(1)} \\ \vdots \\ \Delta V_{ka(1)} \\ \vdots \\ \Delta V_{na(1)} \end{bmatrix} = \begin{bmatrix} Z_{11(1)} & Z_{12(1)} & \ldots Z_{1k(1)} & \ldots Z_{1n(1)} \\ Z_{21(1)} & Z_{22(1)} & \ldots Z_{2k(1)} & \ldots Z_{2n(1)} \\ \vdots & & & \\ Z_{K1(1)} & Z_{K2(1)} & \ldots Z_{kk(1)} & \ldots Z_{kk(1)} \\ \vdots & & & \\ Z_{n1(1)} & Z_{n2(1)} & \ldots Z_{nk(1)} & \ldots Z_{nn(1)} \end{bmatrix} \begin{bmatrix} 0 \\ 0 \\ \vdots \\ -I_{Fa(1)} \\ \vdots \\ 0 \end{bmatrix}$$

$$= \begin{bmatrix} -Z_{1k(1)} I_{Fa(1)} \\ -Z_{2k(1)} I_{F_a(1)} \\ \vdots \\ -Z_{kk(1)} I_{Fa(1)} \\ \vdots \\ -Z_{nk(1)} I_{Fa(1)} \end{bmatrix} \quad \text{...(6.26)}$$

If this voltage is superimposed over the prefault bus voltages, we get the bus voltages during the fault.

$$\begin{bmatrix} V_{1a(1)} \\ V_{2a(1)} \\ \vdots \\ V_{Ka(1)} \\ \vdots \\ V_{na(1)} \end{bmatrix} = \begin{bmatrix} V_0 - Z_{1k(1)} I_{Fa(1)} \\ V_0 - Z_{2k(1)} I_{Fa(1)} \\ \vdots \\ V_0 - Z_{kk(1)} I_{Fa(1)} \\ \vdots \\ V_0 - Z_{nk(1)} I_{Fa(1)} \end{bmatrix} \quad \text{...(6.27)}$$

The bus voltages before the fault do not contain any negative and zero sequence components. Therefore, we get

$$\begin{bmatrix} V_{1a(2)} \\ V_{2a(2)} \\ \vdots \\ V_{ka(2)} \\ \vdots \\ V_{na(2)} \end{bmatrix} = \begin{bmatrix} -Z_{1k(2)}\, I_{Fa(2)} \\ -Z_{2k(2)}\, I_{Fa(2)} \\ \vdots \\ -Z_{kk(2)}\, I_{Fa(2)} \\ \vdots \\ -Z_{nk(2)}\, I_{Fa(2)} \end{bmatrix} \qquad ...(6.28)$$

and,

$$\begin{bmatrix} V_{1a(0)} \\ V_{2a(0)} \\ \vdots \\ V_{ka(0)} \\ \vdots \\ V_{na(0)} \end{bmatrix} = \begin{bmatrix} -Z_{1k(0)}\, I_{Fa(0)} \\ -Z_{2k(0)}\, I_{Fa(0)} \\ \vdots \\ -Z_{kk(0)}\, I_{Fa(0)} \\ \vdots \\ -Z_{nk(0)}\, I_{Fa(0)} \end{bmatrix} \qquad ...(6.29)$$

It should be noted that $I_{Fa(1)}$, $I_{Fa(2)}$ and $I_{Fa(0)}$ depend on the type of fault.

Example 6.13: Two synchronous machines are connected through three phase transformers via a line as shown in Fig. 6.28. Choosing a base of 100 MVA, 345 kV in the transmission line circuit, the line reactances are $X_1 = X_2 = 15\%$ and $X_0 = 50\%$. The other ratings are

M_1 and M_2: 100 MVA, 20 kV; $x_d'' = X_1 = X_2 = 20\%$; $X_0 = 4\%$; $X_n = 5\%$

T_1 and T_2: 100 MVA, 20/345 kV; $X = 8\%$

The system is operating on no load when an *LG* fault occurs at bus 3. Determine fault current and changes in bus voltages.

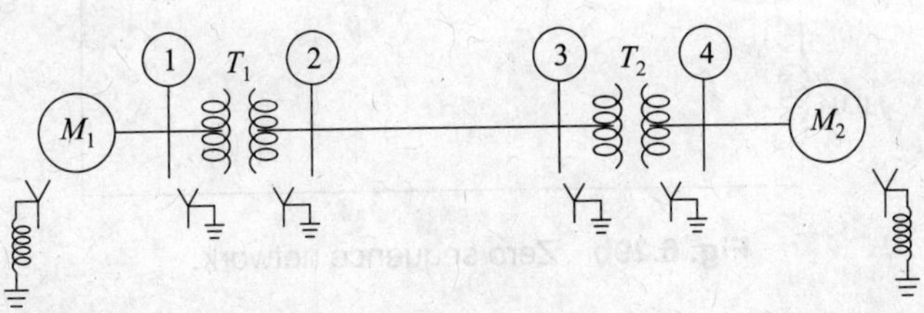

Fig. 6.28 Example 6.13.

Solution: All pu values are on proper base. Hence, they can be used directly.

Positive sequence network:

The positive sequence network is shown in Fig. 6.29a.

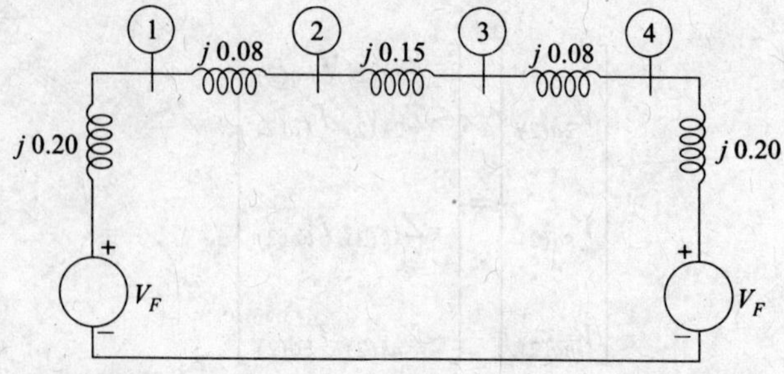

Fig. 6.29a Positive sequence network of example 6.13.

The bus impedance matrix for the positive sequence network $Z_{bus(1)}$ is built using the algorithm presented in chapter 3.

$$Z_{bus(1)} = \begin{bmatrix} j0.1437 & j0.1211 & j0.0789 & j0.0563 \\ j0.1211 & j0.1696 & j0.1104 & j0.0789 \\ j0.0789 & j0.1104 & j0.1696 & j0.1211 \\ j0.0563 & j0.0789 & j0.1211 & j0.1437 \end{bmatrix}$$

Negative sequence network:

The negative sequence network is same as positive sequence network except for the absence of the source V_F. Therefore, the negative sequence bus impedance matrix, $Z_{bus(2)} = Z_{bus(1)}$.

Zero sequence network:

The zero sequence network is shown in Fig. 6.29b.

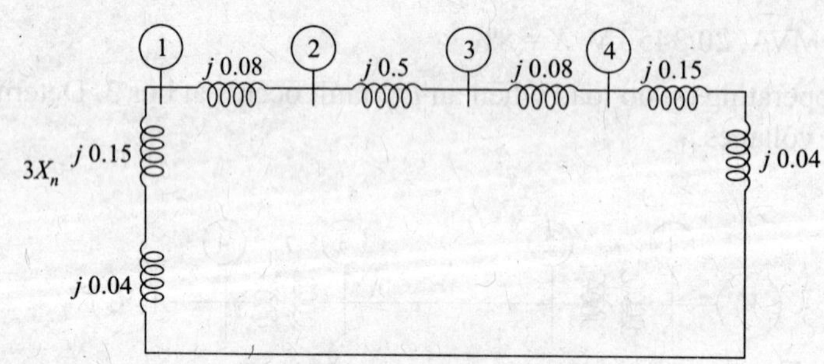

Fig. 6.29b Zero sequence network.

$$Z_{bus(0)} = \begin{bmatrix} j0.1553 & j0.1407 & j0.0493 & j0.0347 \\ j0.1407 & j0.1999 & j0.0701 & j0.0493 \\ j0.0493 & j0.0701 & j0.1999 & j0.1407 \\ j0.0347 & j0.0493 & j0.1407 & j0.1553 \end{bmatrix}$$

Since, the system is on no load we can assume all the bus voltages are 1.0 pu prior to fault. The Thevenin's equivalent impedance of the network at any bus k is Z_{kk} (refer chapter 4). Therefore, the symmetrical components of the fault current I_{Fa} are given by,

$$I_{Fa(0)} = I_{Fa(1)} = I_{Fa(2)} = \frac{V_F}{Z_{33(0)} + Z_{33(1)} + Z_{33(2)}}$$

$$= \frac{1.0}{j(0.1999 + j0.1696 + j0.1696)} = -j1.8549 \text{ pu}$$

The total fault current is $I_{Fa} = 3\ I_{Fa(1)} = -j5.5648$ pu.

Since bus 3, is on the HT side of the transmission line, the base current is

$$\frac{100 \times 10^6}{\sqrt{3} \times 345 \times 10^3} = 167.35 \text{ A}$$

Hence, fault current $I_{Fa} = -j931.3$ A

The positive sequence voltages at various buses are obtained using (6.27) with $k = 3$.

$$\begin{bmatrix} V_{1a(1)} \\ V_{2a(1)} \\ V_{3(1)} \\ V_{4a(1)} \end{bmatrix} = \begin{bmatrix} 1.0 - Z_{13(1)}\ I_{Fa(1)} \\ 1.0 - Z_{23(1)}\ I_{Fa(1)} \\ 1.0 - Z_{33(1)}\ I_{Fa(1)} \\ 1.0 - Z_{43(1)}\ I_{Fa(1)} \end{bmatrix}$$

$$= \begin{bmatrix} 1.0 - (j0.0789)(-j1.8549) \\ 1.0 - (j0.1104)(-j1.8549) \\ 1.0 - (j0.1696)(-j1.8549) \\ 1.0 - (j0.1211)(-j1.8549) \end{bmatrix} = \begin{bmatrix} 1.0 - 0.1463 \\ 1.0 - 0.2048 \\ 1.0 - 0.3146 \\ 1.0 - 0.2246 \end{bmatrix}$$

$$= \begin{bmatrix} 0.8537 \\ 0.7952 \\ 0.6854 \\ 0.7754 \end{bmatrix} \text{pu}$$

$$\begin{bmatrix} V_{1a(2)} \\ V_{2a(2)} \\ V_{3a(2)} \\ V_{4a(2)} \end{bmatrix} = \begin{bmatrix} -Z_{13(2)} I_{Fa(2)} \\ -Z_{23(2)}\ I_{Fa(2)} \\ -Z_{33(2)} I_{Fa(2)} \\ -Z_{43(2)} I_{Fa(2)} \end{bmatrix} = \begin{bmatrix} -0.1463 \\ -0.2048 \\ -0.3146 \\ -0.2246 \end{bmatrix} \text{pu}$$

$$\begin{bmatrix} V_{1a(0)} \\ V_{2a(0)} \\ V_{3a(0)} \\ V_{4a(0)} \end{bmatrix} = \begin{bmatrix} -Z_{13(0)} & I_{Fa(0)} \\ -Z_{23(0)} & I_{Fa(0)} \\ -Z_{33(0)} & I_{Fa(0)} \\ -Z_{43(0)} & I_{Fa(0)} \end{bmatrix} = \begin{bmatrix} -0.0914 \\ -0.1300 \\ -0.3708 \\ -0.2610 \end{bmatrix} \text{pu}$$

The bus voltages (phase a) are obtained adding the corresponding sequence voltages.

$$V_{1a} = V_{1a(0)} + V_{1a(1)} + V_{1a(2)} = -0.0914 - 0.1463 + 0.8537 = 0.616 \text{ pu}$$

$$V_{2a} = V_{2a(0)} + V_{2a(1)} + V_{2a(2)} = -0.1300 - 0.2048 + 0.7952 = 0.4604 \text{ pu}$$

$$V_{3a} = V_{3a(0)} + V_{3a(1)} + V_{3a(2)} = -0.3708 - 0.3146 + 0.6854 = 0.0 \text{ pu}$$

$$V_{4a} = V_{4a(0)} + V_{4a(1)} + V_{4a(2)} = -0.2610 - 0.2246 + 0.7754 = 0.2898 \text{ pu}$$

The sequence components of other phases (b and c) can be computed from the symmetrical component relationship between the three phases. To find the voltages of other buses similar procedure can be adopted. For example,

$$\begin{bmatrix} V_{4a} \\ V_{4b} \\ V_{4c} \end{bmatrix} = \begin{bmatrix} 1 & 1 & 1 \\ 1 & a^2 & a \\ 1 & a & a^2 \end{bmatrix} \begin{bmatrix} V_{4a(0)} \\ V_{4a(1)} \\ V_{4a(2)} \end{bmatrix}$$

$$= \begin{bmatrix} 1 & 1 & 1 \\ 1 & a^2 & a \\ 1 & a & a^2 \end{bmatrix} \begin{bmatrix} -0.2610 \\ 0.7754 \\ -0.2246 \end{bmatrix} = \begin{bmatrix} 0.2898 \\ -0.5364 - j0.866 \\ -0.5364 + j0.866 \end{bmatrix}$$

$$= \begin{bmatrix} 0.2898\angle 0° \\ 1.0187\angle -121.8° \\ 1.0187\angle 121.8° \end{bmatrix} \text{pu}$$

Example 6.14: A 50 MVA, 11 kV generator has the following reactances : $X_1 = X_2 = 15\%$ and $X_0 = 4\%$. The generator is connected to a 50 MVA, Δ –Y 11/132 kV transformer with reactance 10%. The generator and transformer neutrals are solidly grounded. An SLG fault occurs on the HT side of the transformer under no-load condition. Determine the currents in all the three phases of the generator.

Solution: The sequence networks are shown in Fig. 6.30.

$$I_{a1} = \frac{E_a}{Z_1 + Z_2 + Z_0} = \frac{1.0}{j0.25 + j0.25 + j0.1} = -j1.667 \text{ pu}$$

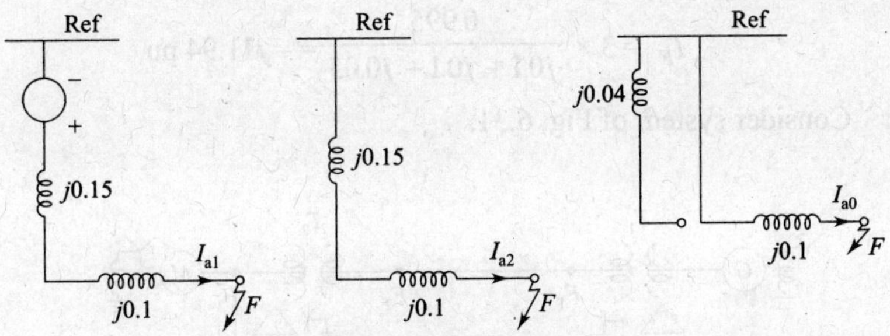

Fig. 6.30 : Example 6.14.

$$I_a = 3I_{a1} = -j5 \text{ pu}$$

$$I_b = 0\,;\, I_c = 0$$

$$\text{(Gen base current) } I_B = \frac{50 \times 10^6}{\sqrt{3} \times 11 \times 10^3} = 2624.3 \text{ A}$$

Fault current in generator = 13, 121.6A.
Fault current on HT side of transformer

$$= 13121.6 \times \frac{11}{132} = 1093.45\text{A}$$

Example 6.15: Two 100 MVA, 22 kV generators with $X_1 = X_2 = 20\%$ and $X_0 = 5\%$ are connected in parallel. The voltage developed is 21.9 kV. An SLG fault occurs on phase-a, when there is no load. Calculate the fault current if (*i*) both generator neutrals are solidly grounded, (ii) both neutrals are ungrounded and (iii) one of the neutrals is grounded.

Solution: $$E_a = \frac{21.9}{22} = 0.995 \text{ pu}$$

(i) When both neutrals are grounded, the zero sequence reactances are in parallel, as the positive sequence and negative sequence are.

$\therefore$ $$x_1 = 0.2||0.2 = 0.1\text{pu}, x_2 = 0.1 \text{ pu},$$

$$x_0 = 0.05 \;||\; 0.05 = 0.025 \text{ pu}.$$

$$I_a = \frac{0.995}{j0.1 + j0.1 + j0.025} = -j4.42 \text{ pu}$$

$$I_f = 3\text{Ia}_1 = -j\ 13.26 \text{ pu}$$

(ii) When both neutrals are ungrounded, fault current is zero, as there is no path for the fault current.

(iii) When one neutral only is grounded, $Z_o = j0.05$ pu.

$$I_F = 3 \times \frac{0.995}{j0.1 + j0.1 + j0.05} = -j11.94 \text{ pu.}$$

Example 6.16: Consider system of Fig. 6.31.

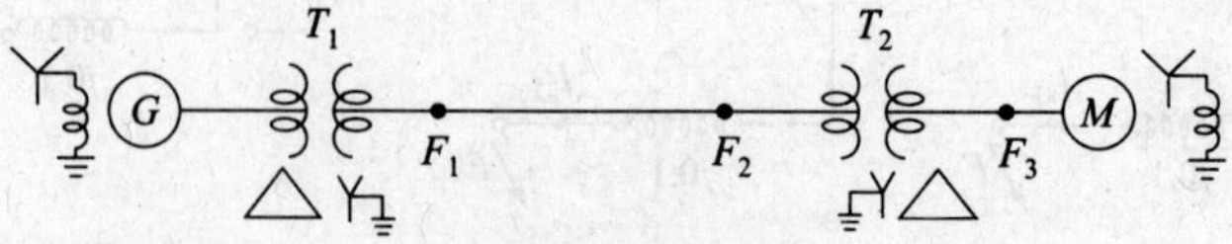

Fig. 6.31

$$G: x_1 = x_2 = 15\%, x_o = 5\%, 500 \text{ MVA}, 22 \text{ kV}$$
$$T_1, T_2: x = 10\%, 500 \text{ MVA}, 22/132 \text{ kV}$$
$$M: x_1 = x_2 = 12\%, x_o = 4\%, 500 \text{ MVA}, 22 \text{ kV}$$

Line: $x_1 = x_2 = 15\%$, $x_o = 35\%$; base 500 MVA, 32 kV

The generator and motor are grounded through reactances of 5% on their rating. Determine the fault current for a fault at F_1, F_2, jF_3 for an (i) SLG fault (ii) symmetrical three-phase fault.

Solution: The sequence networks are shown in Fig. 6.32. Since all the components are on the correct base, pu values need not be changed.

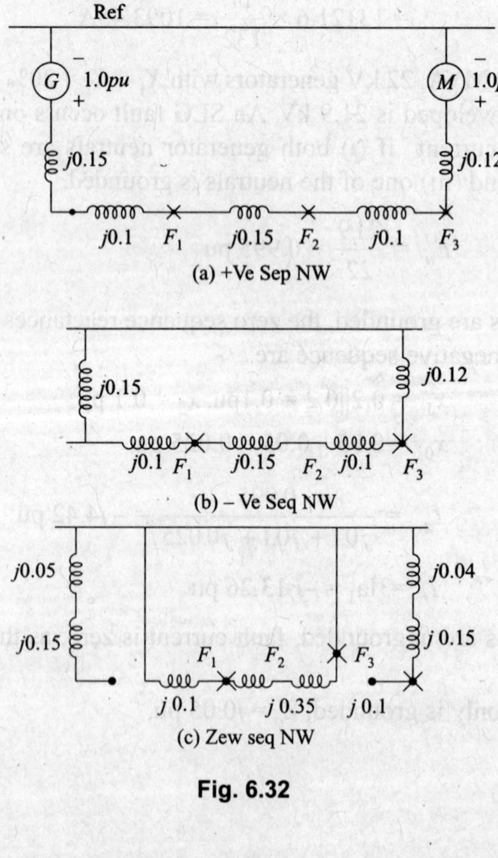

Fig. 6.32

(i) SLG Fault

Fault at F_1

$$Z_1 = j0.25 \| j0.37 = j0.149 \text{ pu}$$
$$Z_2 = Z_1 = j0.149 \text{ pu}$$
$$Z_0 = j0.1 \| j0.45 = j0.082$$
$$I_F = 3I_{a1} = 3 \times \frac{1.0}{j0.149 + j0.149 + j0.082}$$
$$= -j7.895 \text{ pu}$$

Fault at F_2

$$Z_1 = j0.4 \| j0.22 = j0.142 \text{ pu}$$
$$Z_2 = j0.142 \text{ pu}$$
$$Z_0 = j0.45 \| j0.1 = j0.082 \text{ pu}$$
$$I_F = 3 \times I_{a1} = 3 \times \frac{1.0}{j0.142 + j0.142 + j0.082}$$
$$= -j8.196 \text{ pu}$$

Fault at F_3

$$Z_1 = j0.5 \| j0.12 = j0.096 \text{ pu}$$
$$Z_2 = j0.096 \text{ pu}$$
$$Z_0 = 0$$
$$I_F = 3 \times \frac{1.0}{J0.096 + j0.096 + 0} = -j15.625 \text{ pu}$$

(ii) Symmetrical Fault

At F_1 $$I_F = \frac{1.0}{j0.149} = -j6.71 \text{pu}$$

At F_2 $$I_F = \frac{1.0}{j0.142} = -j7.042 \text{ pu}$$

At F_3 $$I_F = \frac{1.0}{j0.096} = -j10.416 \text{ pu}$$

Example 6.17: Consider the system of Fig. 6.33

G: 7500 kVA; 4.16 kV; $x_d'' = x_2 = 0.1$ pu; $x_0 = 0.05$ pu

$x_n = 0.05$ pu.

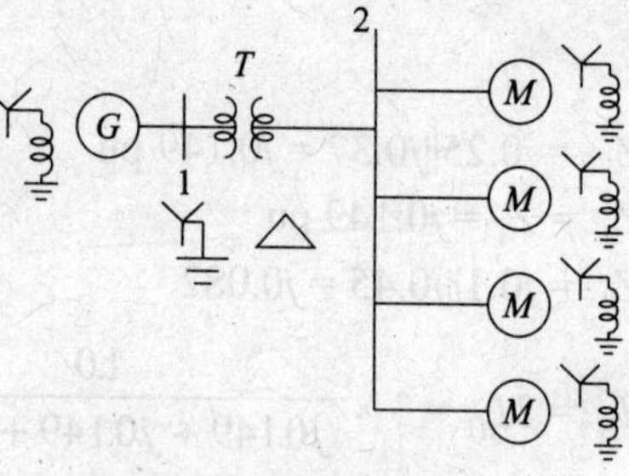

Fig. 6.33: Example 6.17.

M: 1500 HP; $\eta = 89.5\%$ at *FL* upf; $x_d'' = x_2 = 0.2$ pu,

$x_0 = 0.04$ pu; $x_n = 0.02$ pu

T: Three single-phase units, 2500 kVA, 2400/600 V

2400 V connected in λ and 600 V in $\Delta x = 0.1$ pu

The total load on motor is 5000 HP, at rated voltage, 85% pf lag and 88% efficiency.

An S-L-G fault occurs on LV side of transformer. Draw the sequence networks and determine the line current in all parts of the system (i) when pre-fault current is neglected (ii) when pre-fault current is considered.

Solution: We choose the generator rating as base in generator circuit : 7500 kVA, 4.16 kV. The transformer rating is given by

$$\sqrt{3} \times 2400 : 600 = 4157 : 600 \text{ V.}$$

Which can approximately taken to be 4160 : 600 V. Base in motor circuit is 7500 kVA; 600 V. We now replace the four motors with a single equivalent motor. The input rating of the single motor is given by

$$\frac{6000 \times 0.746}{0.895} = 5000 \text{ kVA}$$

The reactance of the equivalent motor in pu on its rating would be same as the reactance of the individual motor in pu, on the base of the individual motor rating. The reactances have to be converted to base of 7500 kVA.

$$x_d'' = x_1 = x_2 = 0.2 \times \frac{7500}{5000} = 0.3$$

$$x_0 = 0.04 \times \frac{7500}{5000} = 0.06$$

$$x_n = 0.02 \times \frac{7500}{5000} = 0.03 \text{ pu}$$

The reactance from neutral to ground would be $3x_n$ in the zero sequence network. The sequence network and connections for the S-L-G fault is shown in Fig. 6.34.

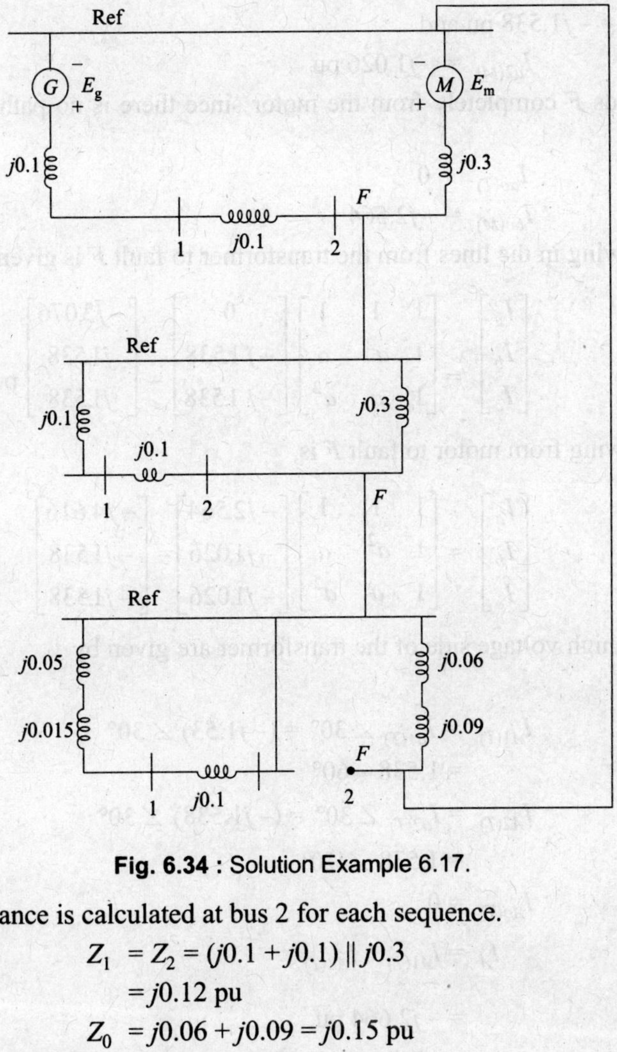

Fig. 6.34 : Solution Example 6.17.

Thevenin's impedance is calculated at bus 2 for each sequence.

$$Z_1 = Z_2 = (j0.1 + j0.1) \parallel j0.3$$

$$= j0.12 \text{ pu}$$

$$Z_0 = j0.06 + j0.09 = j0.15 \text{ pu}$$

$$I_{a1} = \frac{1.0}{j0.12 + j0.12 + j0.15} = -j2.564 \text{ pu}$$

$$I_f = 3I_{a1} = -j2.564 \times 3 = -j7.692 \text{ pu}$$

From positive sequence we find the portion of position sequence current flowing towards 'F' flow transformer by current division:

$$I_{a1(T)} = -j2.564 \times \frac{j0.3}{j0.5} = -j1.538 \text{ pu}$$

The portion of I_{a1} flowing from motor into the fault is given by

$$I_{a1(M)} = -j2.564 \times \frac{j0.2}{j0.5} = -j1.026 \text{ pu}$$

Similarly, $I_{a2(T)} = -j1.538$ pu and

$$I_{a2(M)} = -j1.026 \text{ pu}$$

I_{ao} flows towards F completely from the motor since there is no path to F, from the transformer. Therefore,

$$I_{ao(T)} = 0$$

$$I_{ao(M)} = -j2.564$$

The current flowing in the lines from the transformer to fault F is given by,

$$\begin{bmatrix} I_a \\ I_b \\ I_c \end{bmatrix} = \begin{bmatrix} 1 & 1 & 1 \\ 1 & a^2 & a \\ 1 & a & a^2 \end{bmatrix} \begin{bmatrix} 0 \\ -j1.538 \\ -j1.538 \end{bmatrix} = \begin{bmatrix} -j3.076 \\ j1.538 \\ j1.538 \end{bmatrix} \text{pu}$$

The current flowing from motor to fault F is,

$$\begin{bmatrix} I_a \\ I_b \\ I_c \end{bmatrix} = \begin{bmatrix} 1 & 1 & 1 \\ 1 & a^2 & a \\ 1 & a & a^2 \end{bmatrix} \begin{bmatrix} -j2.564 \\ -j1.026 \\ -j1.026 \end{bmatrix} = \begin{bmatrix} -j4.616 \\ -j1.538 \\ -j1.538 \end{bmatrix}$$

Now current on the high voltage side of the transformer are given by

$$I_{A1(T)} = I_{a1(T)} \angle 30° = (-j1.53) \angle 30°$$
$$= 1.538 - 60°$$
$$I_{A2(T)} = I_{a2(T)} \angle 30° = (-j1.538) \angle 30°$$
$$= 1.538 \angle 120°$$
$$I_{Ao(T)} = 0$$

$$\therefore \quad I_A = I_{A1(T)} + I_{A2(T)}$$

$$= -j2.664 \text{ pu.}$$

$$I_B = a^2 I_{A1(T)} + a^2{}_{(T)} = 0 \text{ pu}$$

$$I_C = aI_{A1(T)} + a^2{}_{IA2(T)} =$$

$$= j2.664 \text{ pu}$$

Base current on HV side of transformer

$$\frac{7{,}500{,}000}{\sqrt{3} \times 4160} = 1041 \text{A}$$

Base current on LV side of transformer and motor

$$\frac{7{,}500{,}000}{\sqrt{3}\times 600} = 7217 \text{ A}$$

Current in fault = 7.692 × 7217 = 55,500A.

Magnitude of currents in the lines between transformer and fault

$$I_a = 3.076 \times 7217 = 22{,}000\text{A} ;\ I_b = I_c = 1.538 \times 7217$$
$$= 11{,}100\text{A}$$

Magnitude of currents in lines between motor and fault

$$I_a = 4.616 \times 7271 = 33{,}300\text{A}$$
$$I_b = I_c = 1.538 \times 7217 = 11{,}100\text{A}$$

Magnitude of currents between generator bus and transformer

$$I_A = 2.664 \times 1041 = 2773\text{A}$$
$$I_B = 0$$
$$I_C = 2773\text{A}$$

When load is considered

The pre-fault load current in motor is

$$\frac{746 \times 5000}{\sqrt{3}\times 600 \times 0.88 \times 0.85} = 4798\text{A}$$

Current drawn by line a, before fault is $\frac{4798}{7217} \angle{-\cos^{-1} 0.85} = 0.665 \angle{-31.8°}$ pu

To account for this, we note that the pre-fault current from transformer to point F is in same direction as fault current flow from transformer to F. Whereas in the motor, the pre-fault current is opposite in direction to the fault current.

∴ New positive sequence current from transformer to F (taking into account pre-fault current is)

$$0.665 \angle{-31.8°} + (-j1.538) = 0.565 - j1.888 \text{ pu.}$$

New value of positive sequence current from motor to point F is

$$-j1.026 - (0.665 \angle{-31.8°}) = -0.565 - j0.676 \text{ pu}$$

The negative and zero sequence currents are not affected by pre-fault current.

The currents are shown in Fig. 6.35(a) and 6.35(b).

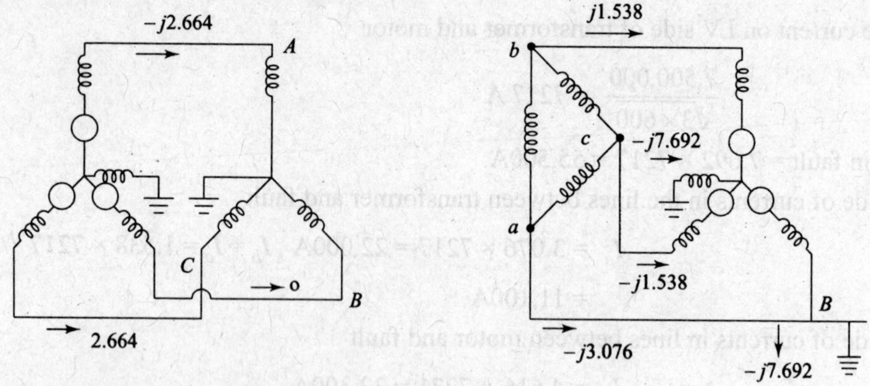

(a) Subtransient currents without pre-fault current considered.

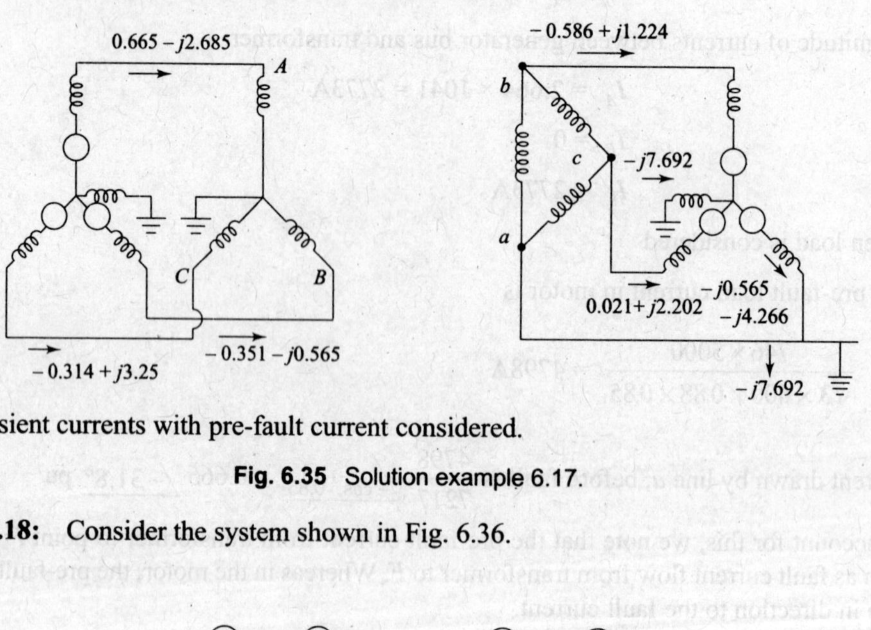

(b) Subtransient currents with pre-fault current considered.

Fig. 6.35 Solution example 6.17.

Example 6.18: Consider the system shown in Fig. 6.36.

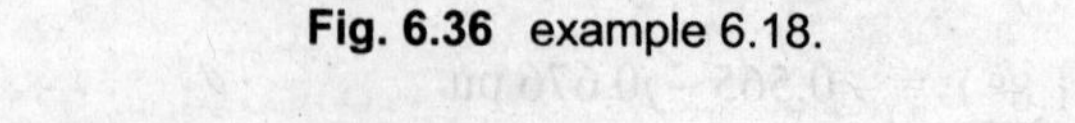

Fig. 6.36 example 6.18.

G: 100 MVA; 20 kV; $x_d'' = x_2 = 20\%$

$x_0 = 4\%$; $x_n = 5\%$

T_1 and T_2: 100 MVA ; 20/345 kV ; $x = 10\%$

Line : Base 100 MVA, 345 kV.

From bus 2 to F: $x_1 = x_2 = 20\%$; $x_0 = 50\%$

From F to bus 3: $x_1 = x_2 = 10\%$; $x_0 = 30\%$

A single line-to-ground fault occurs at F. Determine the fault current in the three lines. Draw the sequence networks. Assume that there is no load on the system.

Solution The sections are all on a base of 100 MVA, 20 kV in the generator section. The sequence networks connected for S-L-G fault is shown Fig. 6.37.

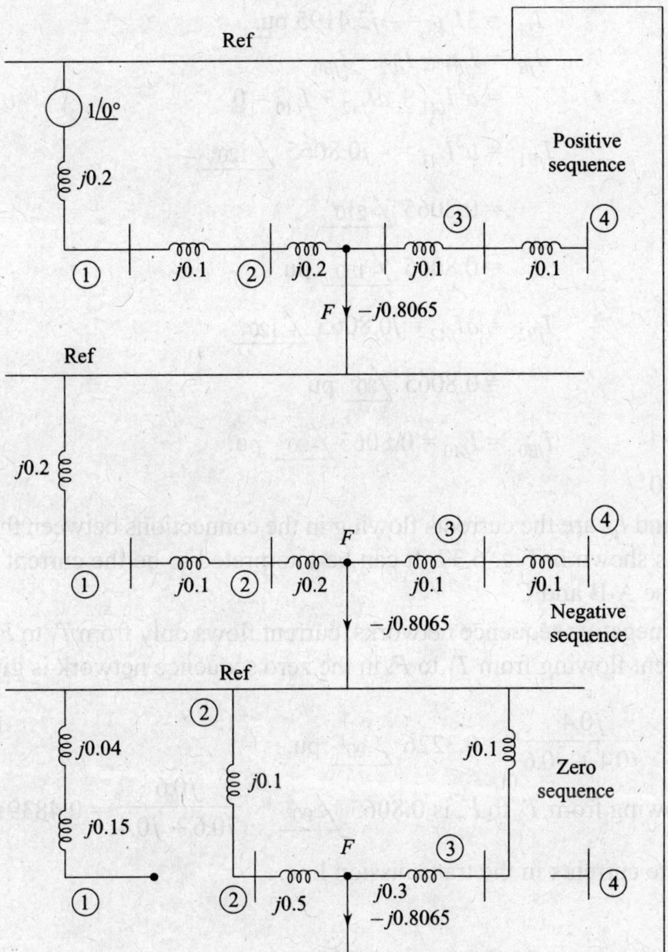

Fig. 6.37 Solution of example 6.18.

We calculate the sequence impedances at the fault point.

$$Z_{1F} = j0.2 + j0.1 + j0.2 = j0.5 \text{ pu}$$
$$Z_{2F} = Z_{1F}$$

$$Z_{OF} = \frac{(j0.6)(j0.4)}{j0.6 + j0.4} = j0.24 \text{ pu}$$

$$I_{fA1} = \frac{1.0 + j0.0}{j0.5 + j0.5 + j0.24} = -j0.8065 \text{ pu}$$

$$I_{fA2} = I_{fA0} = I_{fA1}$$

$$I_{fA} = 3I_{fA1} = -j2.4195 \text{ pu.}$$

$$I_{fB} = I_{fB1} + I_{fB2} + I_{fB0}$$

$$= a^2 I_{fA1} + aI_{fA2} + I_{fA0} = 0.$$

$$I_{fB1} = a^2 I_{fA1} = -j0.8065 \angle 120^\circ$$

$$= 0.8065 \angle 210^\circ$$

$$= 0.8065 \angle 150^\circ \text{ pu}$$

$$I_{fB2} = aI_{fA2} = j0.8065 \angle 120^\circ$$

$$= 0.8065 \angle 30^\circ \text{ pu}$$

$$I_{fB0} = I_{fA0} = 0.8065 \angle -90^\circ \text{ pu.}$$

Similarly, $I_{fc} = 0$

Hence, I_{fA}, I_{fB} and I_{fC} are the currents flowing in the connections between the fault point F and the reference bus as shown in Fig. 6.37. It can be interpreted to be the current through the hypothetical stub of phase A-B and C.

In positive and negative sequence networks, current flows only from T_1 to F as path from T_2 to F is open. The current flowing from T_1 to F, in the zero sequence network is given by,

$$0.8065 \angle -90^\circ \times \frac{j0.4}{j0.4 + j0.6} = 0.3226 \angle -90^\circ \text{ pu.}$$

The current flowing from T_2 to F, is $0.8065 \angle -90^\circ \times \dfrac{j0.6}{j0.6 + j0.4} = 0.4839 \angle -90^\circ$ pu.

We now find the currents in the transmission line.

From T_1 to F

Line A: $0.8065 \angle -90^\circ + 0.8065 \angle -90^\circ + 0.3226 \angle -90^\circ$

$$= -j1.9356 \text{ pu}$$

Line B: $0.8065 \angle 150^\circ + 0.8065 \angle 30^\circ + 0.3226 \angle -90^\circ$

$$= -j1.4839 \text{ pu}$$

Line C: $0.8065 \angle 30^\circ + 0.8065 \angle 150^\circ + 0.3226 \angle -90^\circ$

$$= j0.4839 \text{ pu.}$$

From T_2 to F

Only zero sequence currents flow

Line A: $-j0.4839$ pu

Line B: $-j0.4839$ pu

Line C: $-j0.4839$ pu

The generator is connected on the LV side.

$$\therefore \quad I_{a1} = I_{A1} \angle -30^\circ = 0.8065 \angle -90^\circ - 30^\circ$$

$$= 0.8065 \angle -120^\circ$$

$$I_{a2} = I_{A2} \angle 30^\circ = 0.8065 \angle 60^\circ$$

$$I_{a0} = 0 \text{ [No zero sequence current since path is open]}$$

The generator currents are therefore

$$I_a = I_{a1} + I_{a2} = 0.8065 \angle -120^\circ + 0.8065 \angle 60^\circ$$

$$= -j\,1.3969 \text{ pu}$$

$$I_b = a^2 I_{a1} + a I_{a2} = 0.8065 \angle -120^\circ + 240^\circ$$

$$+ 0.8065 \angle 60^\circ + 120^\circ = j1.3969 \text{ pu.}$$

$$I_c = a I_{a1} + a^2 I_{a2} = 0$$

Note the following

- In generator only positive and negative sequence currents flow.
- The Δ windings of T_1 contain positive, negative and zero sequence components.
- In the Δ windings of T_2 only zero sequence currents flow.

The currents are shown in detail in Fig. 6.38.

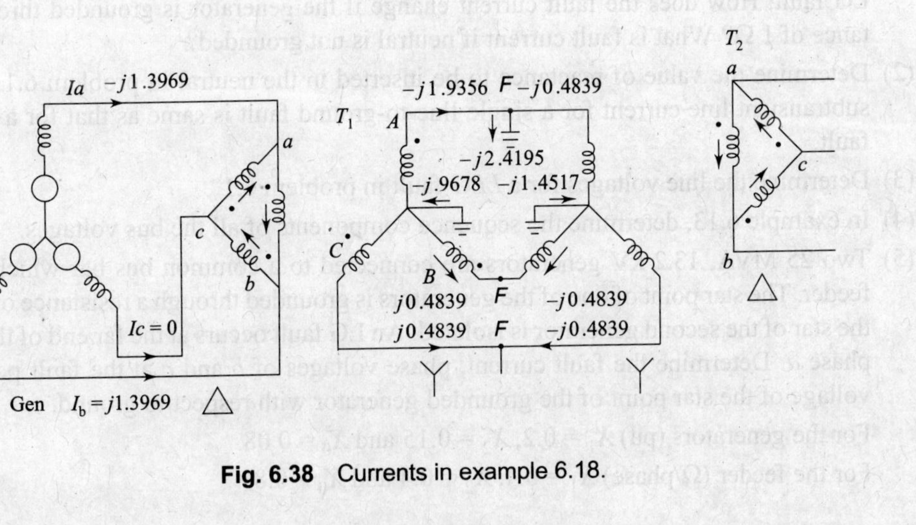

Fig. 6.38 Currents in example 6.18.

6.8 SUMMARY

This chapter deals with the analysis of unsymmetrical faults using symmetrical components. Unsymmetrical faults cause unbalance in an otherwise balanced system. The three shunt type of faults are the single line–to–ground fault (*LG*), double–line fault (*LL*) and the double line–to–ground fault (*LLG*). The conditions, which describe the fault, are derived and the positive sequence, negative sequence and zero sequence networks are connected in a manner, so as to satisfy the fault conditions.

For an *LG* fault, the three sequence networks are connected in series through an impedance $3Z_F$. The *LL* fault is unaffected by the zero sequence impedance and the positive and negative sequence networks are connected in parallel through a series impedance Z_F. For the *LLG* fault $3Z_F$ is connected in series with the zero sequence impedance and the three sequence networks are connected in parallel.

An open conductor fault is a series fault. We can have either one conductor or two conductors open. Opening of one conductor is equivalent to increasing the impedance. Hence, power transfer takes place across the line, though at a reduced level. On the other hand, with two conductors open, the power transfer across the line is reduced to zero.

The formation of the bus impedance matrices for the positive sequence, negative sequence and zero sequence networks is possible using the *Z*–bus building algorithm of chapter 3. Using these matrices, the changes in the bus voltages at all the buses of the system, during a fault, can be calculated. If parts of the circuit are separated by Δ–*Y* transformers, the phase angles of the symmetrical components have to be suitably adjusted.

6.9 SUPPLEMENTARY PROBLEMS

(1) A 50 Hz turbo generator rated 500 MVA, 22 kV is Y connected and solidly grounded. It is operating on no load. $X_d'' = X_1 = X_2 = 0.15$ and $X_0 = 0.05$ pu. Find the fault current for an LG fault. How does the fault current change if the generator is grounded through a reactance of 1 Ω? What is fault current if neutral is not grounded?

(2) Determine the value of reactance to be inserted in the neutral in problem 6.1, so that the subtransient line current for a single line-to-ground fault is same as that for a three phase fault.

(3) Determine the line voltages for a *LLG* fault in problem 6.1.

(4) In example 6.13, determine the sequence components of all the bus voltages.

(5) Two 25 MVA, 13.2 kV generators are connected to a common bus bar which supplies a feeder. The star point of one of the generators is grounded through a resistance of 1 Ω, while the star of the second generator is isolated. An LG fault occurs at the far end of the feeder on phase *a*. Determine the fault current, phase voltages of *b* and *c* at the fault point and the voltage of the star point of the grounded generator with respect to ground.
For the generators (pu) $X_1 = 0.2, X_2 = 0.15$ and $X_0 = 0.08$.
For the feeder (Ω/phase) $X_1 = 0.4, X_2 = 0.4$ and $X_0 = 0.8$.

(6) For the power system shown in Fig. P6.6, obtain the bus impedance matrices of the three sequence networks. The data is as follows:

G_1: 25 MVA, 11 kV, $X_d'' = 20\%, X_2 = X_d'', X_0 = 6\%, X_n = 2.5\ \Omega$

T_1, T_2: 30 MVA, 10.8/121 kV, $X = 10\%$

Line: $X_1 = X_2 = 100\ \Omega, X_0 = 300\ \Omega$

M_1: 15 MVA, 10 kV, $X_d'' = X_1 = X_2 = 25\%, X_0 = 6\%,$

M_1: 7.5 MVA, 10 kV, $X_d'' = X_1 = X_2 = 25\%, X_0 = 6\%, X_n = 2.5\ \Omega$

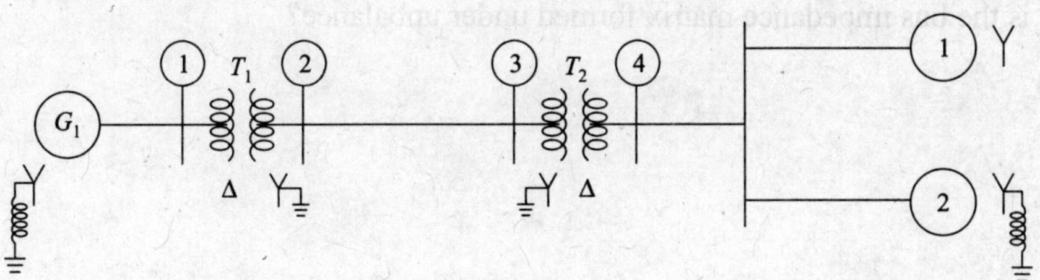

Fig. P6.6

Find the fault current for an LG fault bus 2.

(7) Repeat for an LG fault at bus 3. Find the bus voltages and the line currents during the fault.

(8) The sequence bus impedance matrices of a system are given as follows:

$$Z_{\text{bus}(1)} = Z_{\text{bus}(2)} = j\begin{bmatrix} 0.16 & 0.1 & 0.15 \\ 0.10 & 0.2 & 0.12 \\ 0.15 & 0.12 & 0.25 \end{bmatrix} \text{pu}$$

$$Z_{\text{bus}(0)} = \begin{bmatrix} 0.20 & 0.05 & 0.12 \\ 0.05 & 0.10 & 0.08 \\ 0.12 & 0.08 & 0.30 \end{bmatrix} j \text{ pu}$$

Determine the pu fault current and bus voltages for

(i) Three phase symmetrical fault at bus 2.

(ii) Three phase symmetrical fault at bus 2 through an impedance of $j0.1$ pu.

(iii) A bolted LG fault at bus 2.

(iv) A double line fault at bus 2.

(v) A LLG fault at through an impedance of $j0.1$ pu.

REVIEW QUESTIONS

1. Derive the expression for the fault current at the terminals of an unloaded generator for a (i) Single line to ground fault (ii) Line-to-line fault (iii) Double line to ground fault. Draw the connections of the sequence networks for all the cases.
2. Derive the condition under which the SLG fault current can be greater than the fault current with a three-phase symmetrical fault.
3. Derive the sequence network connections for an SLG fault in a power system.
4. What are open conductor faults. What is their effect?
5. How can unsymmetrical faults be solved with bus impedance matrix?
6. How is the bus impedance matrix formed under unbalance?

Chapter 7

Power Flow Studies

7.1 INTRODUCTION

In this chapter the problem of power flows (also called load flows) is considered. Power flow studies are important in planning, scheduling and designing future expansion. Power flow studies give steady state solutions of the complex voltages at all the buses, for a particular load condition. Different steady state solutions can be obtained for different operating conditions to help in planning, design and operation of the power system. The power flow solution also determines the active and reactive power flows in the lines.

The power flow problem can be understood by considering a sample power system, as shown in Fig. 7.1. In the figure, S_{Gi} refers to the complex power injected by the generator at bus i; S_{Di} is the complex power drawn by the load at bus i.

The generation and loads constitute the operating condition. The load flow analysis will determine the complex voltage V_i at all the buses and the line flows in all the lines, for the given operating conditions.

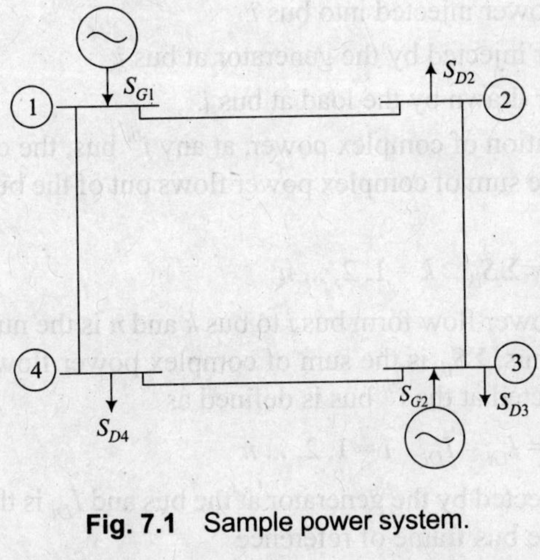

Fig. 7.1 Sample power system.

Generally, power flow studies are limited to the transmission system, which involves bulk power transmission. The load at the buses is assumed to be known *a priori*. This assumption is realistic, since the load is dictated by the customer. Load flow studies throw light on some aspects of the system operation, such as:

- Violation of voltage magnitudes at the buses.
- Overloading of lines.
- Overloading of generators.
- Stability margin reduction, indicated by power angle differences between buses linked by a line.
- Effect of contingencies like line voltages, emergency shutdown of generators, etc.

Power flow studies are required for deciding the economic operation of the power system. They are also required in transient stability studies. Hence, power flow studies play a vital role in power system studies.

7.2 POWER FLOW EQUATION

The power flow analysis can be done using the bus admittance matrix or the bus impedance matrix. In this chapter, we discuss methods using bus admittance matrix only.

The analysis is restricted to a balanced three phase power system, so that the analysis can be carried out on a single phase basis. The per unit values are used for all quantities. The first step in the analysis is the formulation of suitable equations for the power flows in the system.

The power system is a large interconnected system, where various buses are connected by transmission lines. At any bus, complex power is injected into the bus by the generators and complex power is drawn by the loads. Of course at any bus, either one of them may not be present. The power is transported from one bus to other via the transmission lines. At any bus i, the complex power S_i (injected) is defined as

$$S_i = S_{Gi} - S_{Di} \qquad ...(7.1)$$

where, S_i = net complex power injected into bus i.

S_{Gi} = complex power injected by the generator at bus i.

S_{Di} = complex power drawn by the load at bus i.

According to conservation of complex power, at any i^{th} bus, the complex power injected into the bus must be equal to the sum of complex power flows out of the bus via the transmission lines. Hence,

$$S_i = \Sigma S_{ik} \quad k = 1, 2, ... n \qquad ...(7.2)$$

where S_{ik} is the complex power flow form bus i to bus k and n is the number of buses in the system (excluding the ground). Thus, ΣS_{ik} is the sum of complex power flow over all lines connected to bus i. The bus current injected at the ith bus is defined as

$$I_i = I_{Gi} - I_{Di} \quad i = 1, 2, ... n \qquad ...(7.3)$$

where, I_{Gi} is the current injected by the generator at the bus and I_{Di} is the current drawn by the load (demand) at that bus. In the bus frame of reference

$$I = Y_{\text{bus}} V \qquad ...(7.4)$$

$$\begin{bmatrix} I_1 \\ I_2 \\ \vdots \\ I_n \end{bmatrix} = \begin{bmatrix} Y_{11} & Y_{12} & \dots & Y_{1n} \\ Y_{21} & Y_{22} & \dots & Y_{2n} \\ \vdots & & & \\ Y_{n1} & Y_{n2} & \dots & Y_{nn} \end{bmatrix} \begin{bmatrix} V_1 \\ V_2 \\ \vdots \\ V_n \end{bmatrix}$$

where, $I = \begin{bmatrix} I_1 \\ I_2 \\ . \\ . \\ I_n \end{bmatrix}$ is vector of currents injected at the buses.

Y_{bus} = Bus admittance matrix.

$V = \begin{bmatrix} V_1 \\ V_2 \\ . \\ . \\ V_n \end{bmatrix}$ vector of complex bus voltages.

From, (7.4) it can be seen that

$$I_i = \sum_{k=1}^{n} Y_{ik} V_k ; \quad i = 1, 2, \dots n \qquad \dots(7.5)$$

The complex power S_i is given by

$$S_i = V_i I_i^*$$

$$= V_i \left(\sum_{k=1}^{n} Y_{ik} V_k \right)^*$$

$$= V_i \left(\sum_{k=1}^{n} Y_{ik}^* V_k^* \right) \qquad \dots(7.6)$$

Let $V_i \underline{\underline{\Delta}} |V_i| \angle \delta_i = |V_i| (\cos \delta_i + j \sin \delta_i)$ ...(7.7a)

$$\delta_{ik} = \delta_i - \delta_k \qquad \dots(7.7b)$$

$$Y_{ik} = G_{ik} + jB_{ik} \qquad \dots(7.7c)$$

Substituting (7.7) in (7.6), we get

$$S_i = \sum_{k=1}^{n} |V_i|\,|V_k|\,(\cos\delta_{ik} + j\sin\delta_{ik})\,(G_{ik} - j\,B_{ik}) \qquad ...(7.8)$$

Separating real and imaginary parts in (7.8), we obtain

$$P_i = \sum_{k=1}^{n} |V_i|\,|V_k|\,(G_{ik}\cos\delta_{ik} + B_{ik}\sin\delta_{ik}) \qquad ...(7.9a)$$

$$Q_i = \sum_{k=1}^{n} |V_i|\,|V_k|\,(G_{ik}\sin\delta_{ik} - B_{ik}\cos\delta_{ik}) \qquad ...(7.9b)$$

An alternate form of P_i and Q_i can be obtained by representing Y_{ik} also in polar form as

$$Y_{ik} = |Y_{ik}|\,\angle\theta_{ik} \qquad ...(7.10)$$

Substituting (7.7a) and (7.10) in (7.6), we get

$$S_i = |V_i|\,\angle\delta_i \sum_{k=1}^{n} |Y_{ik}|\,\angle{-\theta_{ik}}\,|V_k|\,\angle{-\delta_k} \qquad ...(7.11)$$

The real part of (7.11) gives P_i.

$$P_i = |V_i| \sum_{k=1}^{n} |Y_{ik}|\,|V_k|\cos(-\theta_{ik} + \delta_i - \delta_k)$$

$$= |V_i| \sum_{k=1}^{n} |Y_{ik}|\,|V_k|\cos -(\theta_{ik} - \delta_i + \delta_k)$$

Noting $\cos(-\theta) = \cos\theta$,

$$P_i = \sum_{k=1}^{n} |V_i|\,|V_k|\,|Y_{ik}|\cos(\theta_{ik} - \delta_i + \delta_k) \quad i = 1, 2, \ldots n \qquad ...(7.12a)$$

Similarly, Q_i is imaginary part of (7.11) and is given by

$$Q_i = |V_i| \sum_{k=1}^{n} |Y_{ik}|\,|V_k|\sin -\,(\theta_{ik} - \delta_i + \delta_k)$$

Noting $\sin(-\theta) = -\sin\theta$

$$Q_i = -\sum_{k=1}^{n} |V_i|\,|V_k|\,|Y_{ik}|\sin(\theta_{ik} - \delta_i + \delta_k) \quad i = 1, 2, \ldots n \qquad ...(7.12b)$$

(7.9) and (7.12) are the power flow equations in two alternative forms.(*Some authors use $Y_{ik} = G_{ik} - jB_{ik}$, in which case the expressions for P_i and Q_i have to be modified.)

A simple example helps to illustrate the effect of the voltage magnitude and the phase angle on active and reactive power flows.

Example 7.1: Obtain expression for the complex power transmitted over the transmission line for the system of Fig. 7.2.

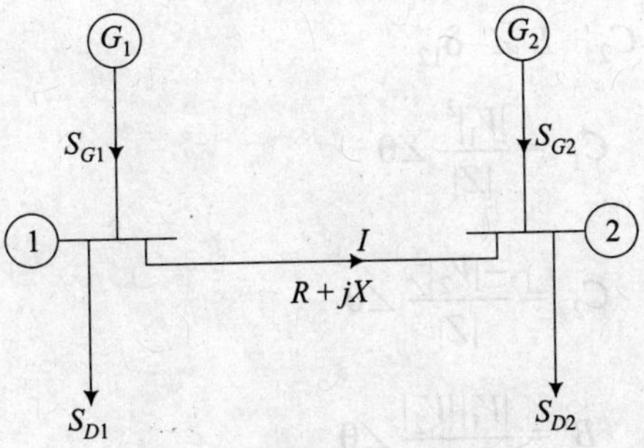

Fig. 7.2 System of example 7.1.

Solution: Let

$$V_1 = |V_1| \angle\delta_1$$

$$V_2 = |V_2| \angle\delta_2$$

$$Z = R + jX = |Z| \angle\theta$$

$$\delta_{12} = \delta_1 - \delta_2 \quad \text{(called power angle or torque angle)}$$

The complex power flow from bus 1 to bus 2 is given by

$$S_{12} = V_1 I_1^* = V_1 \left(\frac{V_1 - V_2}{Z} \right)^* = \frac{V_1 V_1{}^*}{Z^*} - \frac{V_1 V_2{}^*}{Z^*} = \frac{|V_1|^2}{Z^*} - \frac{V_1 V_2^*}{Z^*}$$

$$= \frac{|V_1|^2}{|Z|} \angle\theta - \frac{|V_1||V_2|}{|Z|} \angle(\delta_{12} + \theta) \quad \text{...(7.13a)}$$

Part of this complex power is lost in the impedance of the line.

The power flow from bus 2 to bus 1 is given by

$$S_{21} = V_2 I_2^* = \frac{|V_2|^2}{|Z|} \angle\theta - \frac{|V_2||V_1|}{|Z|} \angle(\theta - \delta_{12})$$

The power received by bus 2 via the line is $-S_{21}$,

$$-S_{21} = -\frac{|V_2|^2}{|Z|} \angle\theta + \frac{|V_2||V_1|}{|Z|} \angle\theta - \delta_{12} \quad \text{...(7.13b)}$$

If $|Z|$ is fixed (it is a constant for a transmission line), the complex power exchange depends on $|V_1|$, $|V_2|$ and δ_{12}. $|V_1|$ and $|V_2|$ are directly affected by the field currents of generators 1 and 2 and δ_{12}

by the difference in mechanical power inputs to the two generators. To increase δ_{12}, we can either increase the mechanical power input to generator 1 or decrease the mechanical power into generator 2. In (7.13a) and (7.13b) taking $|V_1|$, $|V_2|$ and Z as fixed parameters, the equations reduce to the form

$$S_{12} = C_1 - B\,\angle\delta_{12} \qquad \text{...(7.14a)}$$

$$-S_{21} = C_2 + B\,\angle{-\delta_{12}} \qquad \text{...(7.14b)}$$

where,
$$C_1 = \frac{|V_1|^2}{|Z|}\angle\theta$$

$$C_2 = \frac{-|V_2|^2}{|Z|}\angle\theta$$

$$B = \frac{|V_1||V_2|}{|Z|}\angle\theta$$

If δ_{12} is varied, (7.14a) traces out a circle in the complex plane, with center at C_1 and radius $|B|$. This circle is called the sending end circle. Similarly, (7.14b) traces out a circle with center at C_2 and radius $|B|$. This circle is called the receiving end circle. The circles are shown in Fig. 7.3. We can note the following with respect to the figure.

(i) The circles do not intersect if $|V_1| \neq |V_2|$.

(ii) As δ_{12} increases from zero, both the active power sent and received increases. The active power received reaches a maximum when $\delta_{12} = \theta$ and active power sent reaches a maximum when $\delta_{12} = 180° - \theta$.

(iii) If the line resistance is neglected, $Z = jX$ and $\theta = 90°$. Since, there is no loss in the line, the received power is equal to the sent power and the powers are given by $\frac{|V_1||V_2|}{X}\sin\delta_{12}$.

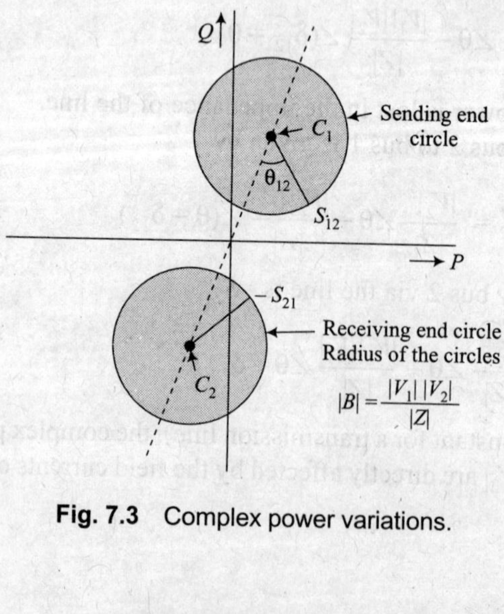

Fig. 7.3 Complex power variations.

(iv) If δ_{12} is increased beyond the transmission capability, synchronism between the generators is lost.

(v) Under normal operating conditions $|V_1| \approx |V_2| = 1.0$ pu and $\theta = 90°$. δ_{12} is small (around 10°) and good decoupling exists between P and $|V|$ and between Q and δ.

The powers are given by

$$P_{12} = -P_{21} = \frac{|V_1||V_2|}{X}\sin\delta_{12}$$

$$Q_{12} = \frac{|V_1|^2}{X} - \frac{|V_1||V_2|}{X}\cos\delta_{12}$$

$$Q_{21} = \frac{|V_2|^2}{X} - \frac{|V_2||V_1|}{X}\cos\delta_{12}$$

Example 7.2: In example 7.1, if δ_{12} is 16°, $Z = 1\angle 82°$ and $|V_1| = |V_2| = 1.0$ pu, find S_{12} and $-S_{21}$.

Solution: $S_{12} = \dfrac{|V_1|^2}{|Z|}\angle\theta - \dfrac{|V_1||V_2|}{|Z|}\angle(\delta_{12}+\theta)$

$$= 1\angle 82° - 1\angle(82° + 16°)$$

$$= (0.1392 + j\,0.9902) - (-0.1392 + j\,0.9902)$$

$$= 0.2784$$

$$-S_{21} = -\frac{|V_2|^2}{|Z|}\angle\theta + \frac{|V_2||V_1|}{|Z|}\angle(\theta - \delta_{12})$$

$$= -1\angle 82° + 1\angle 66°$$

$$= -(0.1392 + j0.9902) + (0.4067 + j\,0.9135)$$

$$= 0.2675 - j\,0.0767$$

We infer the following:

Sending end active power = 0.2784 pu

Receiving end active power = 0.2675 pu

Hence, loss in the line = 0.0109 pu

Sending end reactive power = 0.0

Receiving end reactive power = –0.0767

Hence, the reactive power absorbed by the line is completely supplied by the receiving end.

Example 7.3: Repeat example 2, if $|V_1| = 1.1$ pu and $|V_2| = 0.9$ pu.

Solution: $S_{12} = 1.21\angle 82° - 0.99\angle 98°$

$$= (0.1684 + j\,1.1982) - (-0.1378 + j\,0.9804)$$

$$= 0.3062 + j\,0.2178$$

$$-S_{21} = -0.81\ \angle 82° + 0.99\ \angle 66°$$
$$= -(0.1127 + j\,0.8021) + (0.4026 + j\,0.9044)$$
$$= 0.2899 + j\,0.1023$$

Comparing with example 2, it can be observed that the change in voltage magnitude affects the reactive power more than the active power.

Equations (7.9a and 7.9b) form $2n$ equations, where n is the number of the buses in the system (excluding the reference ground bus). Every bus is characterized by four variables: P_i, Q_i, $|V_i|$ and δ_i. Hence, there are $4n$ variables in the system. (7.9a) and (7.9b) or (7.12a) and (7.12b) can be solved for $2n$ variables, which imply that the remaining $2n$ variables must be specified. (7.9) and (7.12) are non-linear algebraic equations and can be solved only by an iterative numerical technique. Depending on which variables are specified *a priori*, the buses are classified into different categories.

7.3 CLASSIFICATION OF BUSES

1. *Voltage controlled bus (PV bus)*

At these buses the real power injected P_i and the voltage magnitude $|V_i|$ are specified. The variables to be determined are the phase angle of the voltage, δ_i, and the reactive power, Q_i. At *PV* buses, we normally have generators, whose field can be controlled to hold the magnitude of the voltage of the bus at the specified value. Otherwise, the bus must be provided with a device, such as a static var compensator (SVC) or a STATCOM, whose reactive power can be continuously controlled to hold the voltage at the specified value.

2. *Load bus*

At these buses the real power injected (P_i) and reactive power injected, Q_i, are specified. The variables to be determined are the voltage magnitude, $|V_i|$ and the phase angle of the voltage, δi. Generally, buses where only loads are connected are *PQ* buses. Buses where generators are connected but the reactive power of the generator has hit the constrained limits, are also *PQ* buses.

3. *Slack bus*

At this bus, the voltage magnitude and phase angle are specified and the active and reactive powers have to be determined. In any load flow study, the losses in the system cannot be known *a priori*, without the solution of voltages at all the buses. The slack bus, supplies the difference between the total system load plus losses and the sum of the complex powers at the remaining buses. Since, the slack bus must supply the losses, it must have a generator. In general, the bus connected to the largest generating station is normally selected as the slack bus, so that during the load flow iterations, the variations in the active and reactive powers of the slack bus is a small percentage of its capacity.

Thus, at any bus, two of the four variables are specified *a priori* and two variables have to be determined. In an n bus system, $2n$ variables have to be determined by the load flow solution. In any practical power system, all the variables lie within a certain range, dictated by the controllers and operating constraints.

7.4 OPERATING CONSTRAINTS

1. All equipment in the power system are designed to operate at fixed voltages. Further, the consumer needs necessitate, operating values of voltage magnitudes to lie in a certain range. Hence, the voltage magnitude must satisfy the inequality constraint,

$$|V_{i\,\min}| \leq |V_i| \leq |V_{i\,\max}| \quad ...(7.15a)$$

2. The mechanical power input and the rating of the generators pose a limit on the active power generation. Stable operation of prime movers and boilers dictate the lower limit of active power generation. Hence, the active power generation must satisfy the following inequality constraint,

$$P_{Gi,\,\min} \leq P_{Gi} \leq P_{Gi,\,\max} \quad ...(7.15b)$$

3. Limitations of the generator field excitation, constrains the reactive power generation so that

$$Q_{Gi,\,\min} \leq Q_{Gi} \leq Q_{Gi,\,\max} \quad ...(7.15c)$$

   This sets limit on Q_i also, so that $Q_{i,\,\min} \leq Q_i \leq Q_{i,\,\max}$

4. Considerations of stability impose constraints on the maximum permissible power angle between two buses i and j connected via a transmission line.

$$|\delta_i - \delta_j| \leq |\delta_i - \delta_j|_{\max} \quad ...(7.15d)$$

5. Since, the complex power has to be conserved at all buses,

$$\sum_i P_{Gi} = \sum_i P_{Di} + P_L \quad ...(7.15e)$$

$$\sum_i Q_{Gi} = \sum_i Q_{Di} + Q_L \quad ...(7.15f)$$

   where, P_L and Q_L are the total system real and reactive power loss, respectively.

Subject to these constraints, the power flow equations can be solved by application of standard numerical methods for solution of non-linear algebraic equations.

7. 4.1 Violation of Operating Limits

The limit on the generation P_{Gi} depends on the type of generation (thermal, hydel etc.) and on the rating of the generator. We cannot supply active power more than the sum of the maximum power generation of all the generators in the system, without a decrease in frequency. At a *PV* bus (which could be a generator bus or a load bus with variable reactive power support) if the *Q* limit is violated then it means that the voltage can no longer be held at the specified value. Similarly, at a *PQ* bus (which is a load bas) violation of voltage limits means that the specified *Q* can no longer be met within the voltage limits. Limits are violated due to limitations of the components used and due to the operating state.

7.5 DATA FOR LOAD FLOW

Irrespective of the method used for the solution, the data required is common for any load flow. These are presented below. All data is normally in pu.

7.5.1 System Data

This should give information on

- Number of buses, *n*.
- Number of *PV* buses.
- Number of loads.
- Number of transmission lines.
- Number of transformers.
- Number of shunt elements.
- The slack bus number.
- Voltage magnitude of slack bus (angle is generally taken as 0°).
- Tolerance limit.
- Base MVA.
- Maximum permissible number of iterations.

7.5.2 Generator Bus Data

For every *PV* bus *i*, the data required is

- Bus number.
- Active power generation P_{Gi}.
- The specified voltage magnitude $|V_{i, sp}|$.
- Minimum reactive power limit $Q_{i, \min}$.
- Maximum reactive power limit $Q_{i, \max}$.

7.5.3 Load Data

For all loads, the data required is

- The bus number.
- Active power demand P_{Di}.
- The reactive power demand Q_{Di}.

7.5.4 Transmission Line Data

For every transmission line connected between buses *i* and *k*, data needed is

- Starting bus number *i*.
- Ending bus number *k*.
- Resistance of the line.
- Reactance of the line.
- Half line charging admittance.

7.5.5 Transformer Data

For every transformer connected between buses i and k, the data to be given is

- Starting bus number i.
- Ending bus number k.
- Resistance of the transformer.
- Reactance of the transformer.
- Off nominal turns-ratio a.

7.5.6 Shunt Element Data

The data needed for the shunt element is

- Bus number where element is connected.
- Shunt admittance $(G_{sh} + jB_{sh})$.

The bus admittance matrix is formulated from the above data, using the algorithm discussed in chapter 3.

7.6 GAUSS–SEIDEL METHOD

The Gauss-Seidel (*GS*) method is a popular iterative algorithm for solving non-linear algebraic equations. An initial solution vector is assumed, chosen from past experiences, statistical data or from practical considerations. At every subsequent iteration, the solution is updated till convergence is reached. The *GS* method applied to power flow problem is explained in detail below.

Initially, assume all buses are *PQ* buses, except the slack bus. This means that $(n-1)$ complex bus voltages have to be determined. For ease of programming, the slack bus is generally numbered 1. *PV* buses are numbered in sequence and *PQ* buses are ordered next in sequence. This makes programming easier, compared to random ordering of buses.

Consider (7.6)

$$S_i = V_i \left(\sum_{k=1}^{n} Y_{ik} V_k \right)^*$$

This can be written as

$$S_i^* = V_i^* \left(\sum_{k=1}^{n} Y_{ik} V_k \right) \qquad \text{...(7.16)}$$

Noting $\quad S_i^* = P_i - jQ_i \quad$ (7.16) is written as

$$\frac{P_i - jQ_i}{V_i^*} = \sum_{k=1}^{n} Y_{ik} V_k$$

From which

$$\frac{P_i - jQ_i}{V_i^*} = Y_{ii} V_i + \sum_{\substack{k=1 \\ k \neq i}}^{n} Y_{ik} V_k \qquad ...(7.17)$$

The terms of (7.17) can be rearranged as

$$V_i = \frac{1}{Y_{ii}} \left[\frac{P_i - jQ_i}{V_i^*} - \sum_{\substack{k=1 \\ k \neq i}}^{n} Y_{ik} V_k \right] \quad i = 2, 3, \ldots n \qquad ...(7.18)$$

(7.18) is an implicit equation since the unknown variable appears on both sides of the equation. Hence, it needs to be solved by an iterative technique. Starting from an initial estimate of all bus voltages, in the RHS of (7.18) the most recent values of the bus voltages are substituted. One iteration of the method involves computation of all the bus voltages. In Gauss method, the values at beginning of the iteration, are used for computation of all the bus voltages in the present iteration. This leads to slow convergence. In Gauss-Seidel method, the value of the updated voltages are used in the computation of subsequent voltages in the same iteration. This speeds up convergence. Iterations are carried out till the magnitudes of all bus voltages do not change more than the tolerance value.

7.6.1 Algorithm for GS Method

The algorithm for GS method is given below.

1. Prepare data as mentioned in section 7.5.
2. Formulate the bus admittance matrix Y_{bus}. This is generally done by the method of inspection detailed in chapter 3.
3. Assume initial voltages for all buses, 2,... n. In practical power systems, the magnitude of the bus voltages is close to 1.0 pu. Hence, the complex bus voltages at all $(n - 1)$ buses (except slack bus) are taken to be $1.0\angle 0°$. This is normally called a flat start.
4. Update the voltages. In any $(r + 1)^{th}$ iteration, from (7.18) the voltages are given by

$$V_i^{(r+1)} = \frac{1}{Y_{ii}} \left[\frac{P_i - jQ_i}{(V_i^{(r)})^*} - \sum_{k=1}^{i-1} Y_{ik} V_k^{(r+1)} - \sum_{k=i+1}^{n} Y_{ik} V_k^{(r)} \right] \quad i = 2, 3, \ldots n \qquad ...(7.19)$$

Here note that when computation is carried out for i^{th} bus, updated values are already available for buses 2, 3,... $(i - 1)$ in the current $(r + 1)^{th}$ iteration. Hence, these values are used. For buses $(i + 1)$... n, values from previous, r^{th} iteration are used.

Computation time can be saved by computing constant values only once at the start of the program, instead of at every iteration. Since, P_i and Q_i are constants at PQ buses, let

$$L_i = \frac{P_i - jQ_i}{Y_{ii}} \text{ and } M_{ik} = \frac{Y_{ik}}{Y_{ii}} \text{ be computed initially.}$$

Then, (7.19) can be written as

$$V_i^{(r+1)} = \frac{L_i}{(V_i^{(r)})^*} - \sum_{k=1}^{i-1} M_{ik} V_k^{(r+1)} - \sum_{k=i+1}^{n} M_{ik} V_k^{(r)} \quad \text{...(7.20)}$$

5. Continue iterations till
$$|\Delta V_i^{(r+1)}| = |V_i^{(r+1)} - V_i^{(r)}| < \varepsilon \quad i = 2, 3,.. n$$
ε is the tolerance value. Generally, it is customary to use a value of 0.0001 pu.

6. Compute slack bus power after voltages have converged using (7.16) [assuming bus 1 is slack bus].
$$S_1^* = P_1 - jQ_1 = V_1^* \left(\sum_{k=1}^{n} Y_{1k} V_k \right) \quad \text{...(7.21)}$$

7. Compute all line flows. Consider a line (or a transformer) connected between buses i and k as shown in Fig. 7.4.

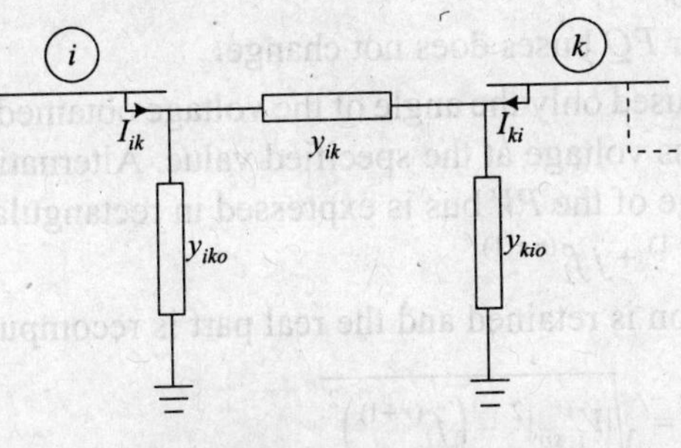

Fig. 7.4 Computation of line flow.

The figure represents the Π model of a transmission line or a transformer (refer chapter 3).

$$I_{ik} = (V_i - V_k)y_{ik} + V_i y_{ik0} \quad \text{...(7.22a)}$$
$$I_{ki} = (V_k - V_i)y_{ik} + V_k y_{ki0} \quad \text{...(7.22b)}$$
$$S_{ik} = V_i I_{ik}^* \quad \text{...(7.22c)}$$
$$S_{ki} = V_k I_{ki}^* \quad \text{...(7.22d)}$$

The complex power loss in the line is given by $S_{ik} + S_{ki}$. The total loss in the system is calculated by summing the loss over all the lines.

7.6.2 Modification of Algorithm to Include PV Buses

At the voltage controlled buses (PV buses) the magnitude of the voltage, rather than the reactive power is specified. Hence, it is necessary to first make an estimate of Q_i to be used in (7.19). From (7.16), we have

$$Q_i = -I_m \left\{ V_i^* \sum_{k=1}^{n} Y_{ik} V_k \right\}$$

where, I_m stands for imaginary part.

At any $(r+1)^{th}$ iteration, at the ith PV bus

$$Q_i^{(r+1)} = -I_m \left\{ (V_i^{(r)})^* \sum_{k=1}^{i-1} Y_{ik} V_k^{(r+1)} + (V_i^{(r)})^* \sum_{k=i}^{n} Y_{ik} V_k^{(r)} \right\} \qquad ...(7.23)$$

The steps for i^{th} PV bus are as follows:

1. Compute $Q_i^{(r+1)}$ using (7.23).
2. Calculate V_i using (7.19) with $Q_i = Q_i^{(r+1)}$.
3. Since $|V_i|$ is specified at the PV bus, the magnitude of V_i obtained in step 2 has to be modified and set to the specified value $|V_{i,sp}|$. Therefore,

$$V_i^{(r+1)} = |V_{i,sp}| \frac{V_i^{(r+1)}}{|V_i^{(r+1)}|} \qquad ...(7.24)$$

Stated another way,

$$V_i^{(r+1)} = |V_{i,sp}| \angle \delta_i^{(r+1)} \qquad ...(7.25)$$

The voltage computation for PQ buses does not change.

In the above procedure we have used only the angle of the voltage obtained in the $(r+1)^{th}$ iteration, keeping the magnitude of the bus voltage at the specified value. Alternatively, another procedure can be adopted. Here, the voltage of the PV bus is expressed in rectangular coordinates, as

$$V_i^{(r+1)} = e_i^{(r+1)} + j f_i^{(r+1)}$$

The imaginary part of the solution is retained and the real part is recomputed as,

$$e_i^{(r+1)} = \sqrt{|V_{i,sp}|^2 - \left(f_i^{(r+1)}\right)^2}$$

so that the voltage magnitude is held at the specified value.

The first algorithm is more popular.

7.6.3 Q–Limit Violations

In the previous algorithm if the Q limit at the voltage controlled bus is violated during any iteration, i.e., $Q_i^{(r+1)}$ computed using (7.23) is either less than $Q_{i,\min}$ or greater than $Q_{i,\max}$, it means that the voltage cannot be maintained at the specified value due to lack of reactive power support. This bus is then treated as a PQ bus in the $(r+1)^{th}$ iteration and the voltage is calculated with the value of Q_i set as follows:

If $\quad Q_i < Q_{i,\min}$ then

$$Q_i = Q_{i,\min}$$

If $\quad Q_i > Q_{i,\,\max}$ then

$$Q_i = Q_{i,\,\max}.$$

If in the subsequent iteration, Q_i falls within the limits, the bus can be switched back to PV status.

7.6.4 Acceleration of Convergence

It is found that in GS method of load flow, the number of iterations increase with increase in the size of the system. The number of iterations required can be reduced if the correction in voltage at each bus is accelerated, by multiplying with a constant α, called the acceleration factor. In the $(r+1)^{\text{th}}$ iteration we can let

$$V_i^{(r+1)}\,(accelerated) = V_i^{(r)} + \alpha(V_i^{(r+1)} - V_i^{(r)})$$

where, α is a real number. When $\alpha = 1$, the value of $V_i^{(r+1)}$ is the computed value. If $1 < \alpha < 2$, then the value computed is extrapolated. Generally, α is taken between 1.6 to 2.0. A wrong value may lead to divergence of the solution.

At *PQ* buses (pure load buses) if the voltage magnitude violates the limit, it simply means that the specified reactive power demand cannot be supplied, with the voltage maintained within acceptable limits.

Example 7.4: Obtain the voltage at bus 2 for the simple system shown in Fig. 7.5, using the Gauss-Seidel method, if $V_1 = 1\angle 0°$ pu.

Solution: Here the capacitor at bus 2, injects a reactive power of 1.0 pu. The complex power injection at bus 2 is

$$S_2 = j1.0 - (0.5 + j1.0) = -0.5 \text{ pu.}$$

$$V_1 = 1\angle 0°$$

$$Y_{\text{bus}} = \begin{bmatrix} -j2 & j2 \\ j2 & -j2 \end{bmatrix}$$

$$V_2^{(r+1)} = \frac{1}{Y_{22}}\left[\frac{P_2 - jQ_2}{(V_2^{(r)})^*} - Y_{21}V_1\right]$$

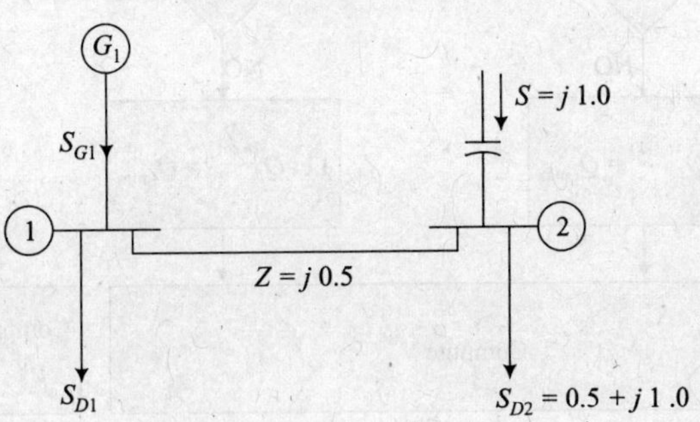

Fig. 7.5 Example 7.4.

7.6.5 Flow Chart for Gauss - Seidel Iterative Solution of Load Flow. (Assume Bus No.1 is a Block Bus)

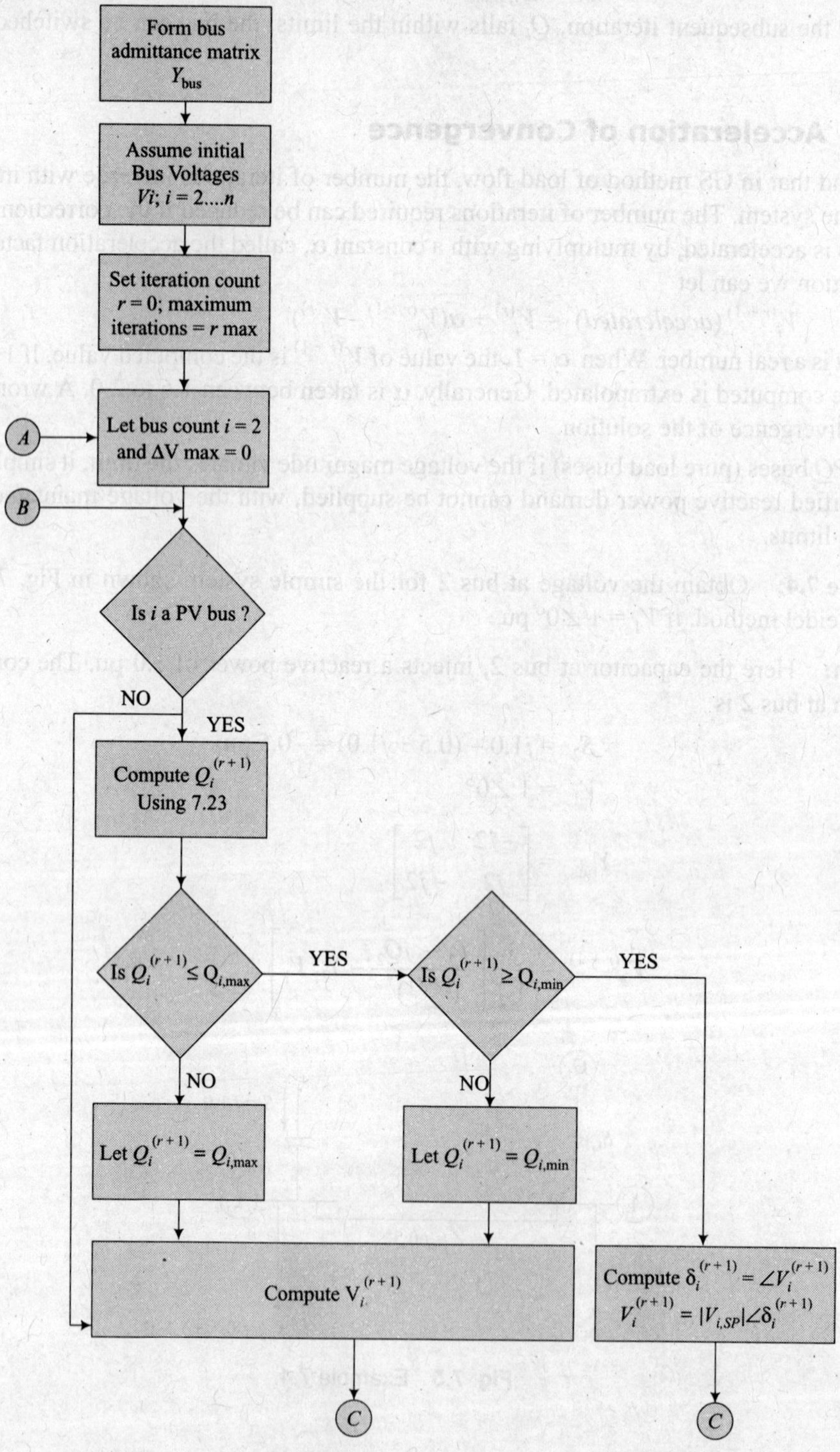

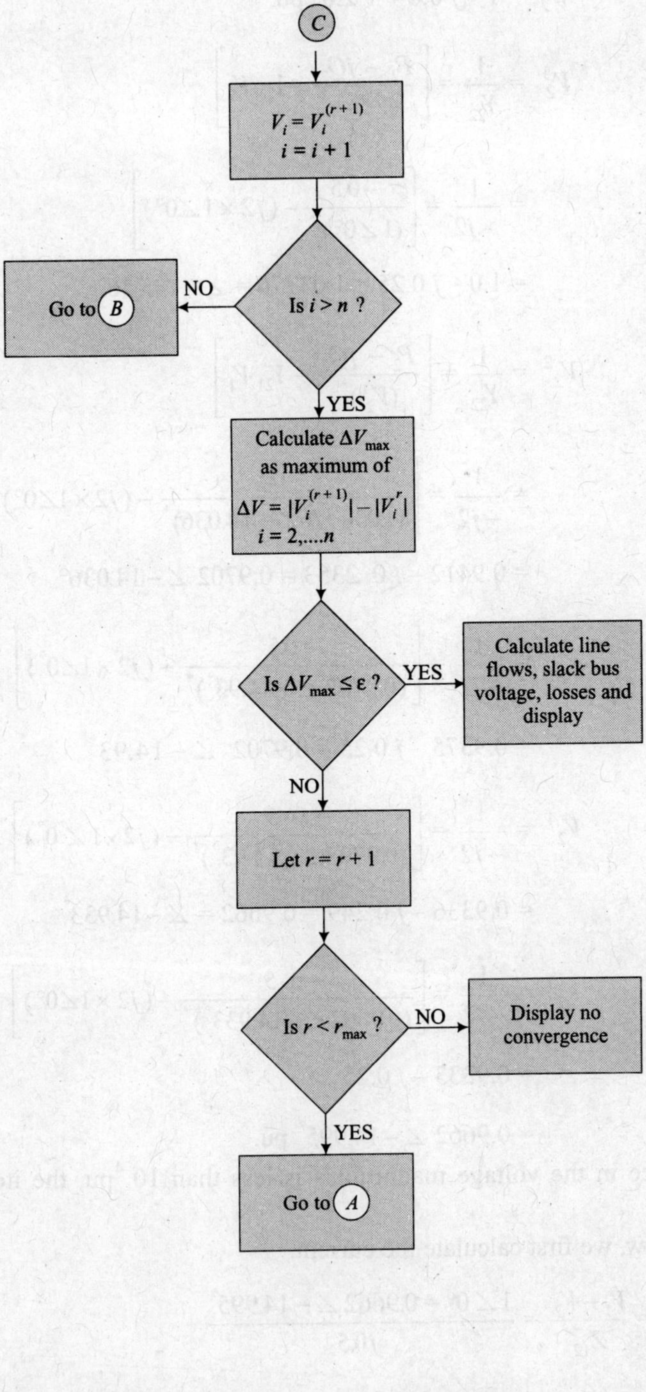
C
$V_i = V_i^{(r+1)}$
$i = i + 1$
Is $i > n$?
NO
Go to B
YES
Calculate ΔV_{max}
as maximum of
$\Delta V = |V_i^{(r+1)}| - |V_i^r|$
$i = 2,....n$
Is $\Delta V_{max} \leq \varepsilon$?
YES
Calculate line flows, slack bus voltage, losses and display
NO
Let $r = r + 1$
Is $r < r_{max}$?
NO
Display no convergence
YES
Go to A

Since, V_1 is specified, it is a constant through all the iterations. Let the initial voltage at bus 2,

$$V_2^0 = 1 + j\,0.0 = 1\ \angle 0°\ \text{pu}.$$

$$V_2^1 = \frac{1}{Y_{22}} = \left[\frac{P_2 - jQ_2}{(V_2^0)^*} - Y_{21}V_1\right]$$

$$= \frac{1}{-j2} = \left[\frac{-0.5}{(1\angle 0°)^*} - (j2 \times 1\angle 0°)\right]$$

$$= 1.0 - j\,0.25 = 1.03776 = \angle -14.036°$$

$$V_2^2 = \frac{1}{Y_{22}} = \left[\frac{P_2 - jQ_2}{(V_2^1)^*} - Y_{21}V_1\right]$$

$$= \frac{1}{-j2} = \left[\frac{-0.5}{(1.030776\angle -14.036)^*} - (j2 \times 1\angle 0°)\right]$$

$$= 0.9412 - j\,0.\,2353 = 0.9702\ \angle -14.036°$$

$$V_2^3 = \frac{1}{-j2} = \left[\frac{-0.5}{(0.9702\ \angle -14.93^{\circ})^*} - (j2 \times 1\angle 0^{\circ})\right]$$

$$= 0.9375 - j\,0.25 = 0.9702\ \ \angle -14.93^{\circ}$$

$$V_2^4 = \frac{1}{-j2} = \left[\frac{-0\cdot 5}{(0.9702\angle -14.93^{\circ})^*} - (j2 \times 1\angle 0^{\circ})\right]$$

$$= 0.9336 - j\,0.249 = 0.9662 = \angle -14.933^{\circ}$$

$$V_2^5 = \frac{1}{-j2} = \left[\frac{-0\cdot 5}{(0.9662\angle -14.933^{\circ})^*} - (j2 \times 1\angle 0°)\right]$$

$$= 0.9333 - j\,0.25$$

$$= 0.9662\ \angle -14.995^{\circ}\ \text{pu}.$$

Since the difference in the voltage magnitudes is less than 10^{-4}pu, the iterations can be stopped.

To compute line flow, we first calculate the current.

$$I_{12} = \frac{V_1 - V_2}{Z_{12}} = \frac{1\angle 0° - 0.9662\ \angle -14.995^{\circ}}{j0.5}$$

$$= 0.5 - j\,0.1333 = 0.5175\ \angle -14.928^{\circ}\ \text{pu}.$$

$$S_{12} = V_1 I_{12}^* = 1 \angle 0^\circ \times (0.5175 \angle -14.928^\circ)^*$$
$$= 1 \angle 0^\circ \times 0.5175 \angle 14.928^\circ = 0.5 + j\,0.1333 \text{ pu.}$$
$$I_{21} = -I_{12} = -0.5 + j\,0.1333 = 0.5175 \angle 165.05^\circ$$
$$S_{21} = V_2 I_{21} = 0.9662 \angle -14.995^\circ \times 0.5175 \angle -165.05^\circ$$
$$= -0.5 + j\,0.0 \text{ pu.}$$

Total loss in the line = $S_{12} + S_{21} = j\,0.1333$ pu.

This can be verified from the fact that the real power loss is zero, as the line has no resistance and the reactive power loss is $|I|^2 \times = 0.5175^2 \times 0.5 = 0.1333$ pu.

Example 7.5: Consider the power system shown in Fig. 7.6. The data for the system is given in Table 7.1 and Table 7.2. Obtain the bus voltages at the end of first iteration, by applying GS method.

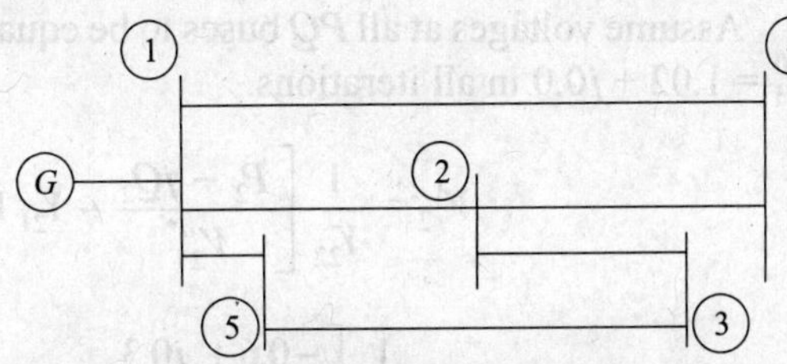

Fig. 7.6 Power system of example 7.5.

Solution: Note the following:

- Bus 1 is slack bus.
- Bus 2, 4, 5 are PQ buses.
- Bus 3 is PV bus.
- The lines do not have half line charging admittances.

$$P_2 + jQ_2 = P_{G2} + jQ_{G2} - (P_{D2} + jQ_{D2}) = -0.6 - j0.3$$
$$P_3 + jQ_3 = P_{G3} + jQ_{G3} - (P_{D3} + jQ_{D3}) = 1.0 + jQ_{G3}$$

Similarly, $P_4 + jQ_4 = -0.4 - j0.1$

$$P_5 + jQ_5 = -0.6 - j0.2$$

Table 7.1 Line data of example 7.5

SB	*EB*	*R* (pu)	*X* (pu)	$\frac{B_C}{2}$
1	2	0.10	0.40	–
1	4	0.15	0.60	–
1	5	0.05	0.20	–
2	3	0.05	0.20	–
2	4	0.10	0.40	–
3	5	0.05	0.20	–

Table 7.2 Bus data of exmple 7.5

Bus No.	P_G *(pu)*	Q_G *(pu)*	P_D *(pu)*	Q_D *(pu)*	$\|V_{SP}\|$ *(pu)*	δ
1	-	-	-	-	1.02	0°
2	-	-	0.60	0.30	-	-
3	1.0	-	-	-	1.04	-
4	-	-	0.40	0.10	-	-
5	-	-	0.60	0.20	-	-

The Y_{bus} is formed by inspection and is given by

Y_{bus} =

2.15685 − j8.62744	−0.58823 + j2.35294	0.0 + j0.0	− 0.39215 + j1.56862	− 1.17647 + j4.70588
− 0.58823 + j2.35294	2.35293 − j9.41176	− 1.17647 + j4.70588	− 0.58823 + j2.35294	0.0 + j0.0
0.0+j0.0	− 1.17647 + j4.70588	2.35294 − j9.41176	0.0+j0.0	− 1.17647 + j4.70588
−0.39215 + j1.56862	−0.58823 + j2.35294	0.0+j0.0	0.98038 − j3.92156	0.0+j0.0
− 1.17647 + j4.70588	0.0 + j0.0	− 1.17647 + j4.70588	0.0 + j 0.0	2.35294 − j9.41176

Assume voltages at all *PQ* buses to be equal to 1 + *j*0.0 pu. The slack bus voltage is taken to be $V_1^0 = 1.02 + j0.0$ in all iterations.

$$V^1{}_2 = \frac{1}{Y_{22}}\left[\frac{P_2 - jQ_2}{V_2^{o*}} - Y_{21}\,V_1^0 - Y_{23}\,V_3^0 - Y_{24}\,V_4^0 - Y_{25}\,V_5^0\right]$$

$$= \frac{1}{Y_{22}}\left[\frac{-0.6 + j0.3}{1.0 - j0.0} - \{(-0.58823 + j2.35294) \times 1.02\,\angle 0°\}\right.$$

$$\left. - \{(-1.17647 + j4.70588) \times 1.04\,\angle 0°\} - \{(-0.58823 + j2.35294) \times 1.0\,\angle 0°\}\right]$$

$$= \frac{1.81175 - j9.34705}{2.35293 - j9.41176}$$

$$= \frac{9.52101\,\angle -79.0303°}{9.70141\,\angle -75.9638°} = 0.98140\,\angle -3.0665°$$

$$= 0.97999 - j0.0525$$

Bus 3 is a PV bus. Hence, we must first calculate Q_3. This can be done using (7.9b).

$$Q_3 = |V_3|\,|V_1|\,(G_{31} \sin \delta_{31} - B_{31} \cos \delta_{31}) + |V_3|\,|V_2|\,(G_{32} \sin \delta_{32} - B_{32} \cos \delta_{32})$$
$$+ |V_3|^2\,(G_{33} \sin \delta_{33} - B_{33} \cos \delta_{33}) + |V_3|\,|V_4|\,(G_{34} \sin \delta_{34} - B_{34} \cos \delta_{34})$$
$$+ |V_3|\,|V_5|\,(G_{35} \sin \delta_{35} - B_{35} \cos \delta_{35})$$

We note that $\delta_1 = 0°$

$\delta_2 = -3.0665°$

$\delta_3 = 0°$

$\delta_4 = 0°$

$\delta_5 = 0°$

∴ $\delta_{31} = \delta_{33} = \delta_{34} = \delta_{35} = 0°$ $(\delta_{ik} = \delta_i - \delta_k)$

$\delta_{32} = 3.0665°$

$$Q_3 = 1.04\,[1.02\,(0.0 + j0.0) + 0.9814\,\{-1.17647 \times \sin(3.0665°) - 4.70588 \times \cos(3.0665°)\}$$

$$+ 1.04\begin{Bmatrix}9.41176\\ \times\cos(0°)\end{Bmatrix} + 1.0\,\{0.0 - 0.0\} + 1.0\,\{-4.70588 \times \cos(0°)\}]$$

$$= 1.04\,[-4.6735 + 9.78823 - 4.70588]$$

$$= 0.425204 \text{ pu.}$$

$$V_3^1 = \frac{1}{Y_{33}}\left[\frac{P_3 - jQ_3}{V_3^{o*}} - Y_{31}\,V_1^0 - Y_{32}\,V_2^1 - Y_{34}\,V_4^0 - Y_{35}\,V_5^0\right]$$

$Y_{31} = Y_{34} = 0.0 + j0.0$. Hence,

$$V_3^1 = \frac{1}{Y_{33}}\left[\frac{1.0 - j0.425204}{1.04 - j0.0} - \{(-1.7647 + j4.70588) \times (0.98140\,\angle -3.0665°)\}\right.$$
$$\left. - \{(-1.17647 + j4.70588) \times (1\,\angle 0°)\}\right]$$

$$= \frac{3.01385 - j9.78823}{2.35294 - j9.41176}$$

$$= \frac{10.24171\,\angle -72.886°}{9.70142\,\angle -75.963°} = 1.05569\,\angle 3.077°$$

$$= 1.0541 + j0.05666 \text{ pu.}$$

Since it is a PV bus, the voltage magnitude is adjusted to specified value and V_3^1 computed using (7.25).

$$V_3^1 = 1.04\,\angle\,3.077° \text{ pu}$$

$$V_4^1 = \frac{1}{Y_{44}}\left[\frac{P_4 - jQ_4}{V_4^{o*}} - Y_{41}\,V_1^0 - Y_{42}\,V_2^1 - Y_{43}\,V_3^1 - Y_{45}\,V_5^0\right]$$

$$Y_{43} = Y_{45} = 0.0$$

$$= \frac{1}{Y_{44}}\left[\frac{-0.4 + j0.1}{1.0 - j0.0} - \{(-0.39215 + j1.56862) \times 1.02\,\angle 0°\}\right.$$
$$\left. - \{(-0.58823 + j2.35294) \times (0.98140\,\angle 3.0665°)\}\right]$$

$$= \frac{0.45293 - j3.8366}{0.98038 - j3.92156} = \frac{3.86324\,\angle -83.267°}{4.04225\,\angle -75.964°}$$

$$= 0.955715\,\angle -7.303° \text{ pu} = 0.94796 - j0.12149$$

$$V_5^1 = \frac{1}{Y_{55}}\left[\frac{P_5 - jQ_5}{V_5^{o*}} - Y_{51}\,V_1^0 - Y_{52}\,V_2^1 - Y_{53}\,V_3^1 - Y_{54}\,V_4^1\right]$$

$$Y_{52} = Y_{54} = 0$$

$$= \frac{1}{Y_{55}}\left[\frac{-0.6 + j0.2}{1.0 - j0.0} - \{(-1.17647 + j4.70588) \times 1.02 \angle 0°\} - \{(-1.17647 + j4.70588) \times 1.04 \angle 3.077°\}\right]$$

$$= \frac{2.08445 - j9.42138}{2.35294 - j9.41176} = \frac{9.64921 \angle -77.524°}{9.70142 \angle -75.964°}$$

$$= 0.994618 \angle -1.56° = 0.994249 - j0.027$$

Thus, at end of 1st iteration

$$V_1 = 1.02 \angle 0° \text{ pu}$$
$$V_2 = 0.98140 \angle -3.066° \text{ pu}$$
$$V_3 = 1.04 \angle 3.077° \text{ pu}$$
$$V_4 = 0.955715 \angle -7.303° \text{ pu}$$
$$V_5 = 0.994618 \angle -1.56° \text{ pu}$$

Example 7.6: Obtain the load flow solution at the end of the first iteration of the power system shown in Fig. 7.7. The data is given in Table 7.4. The solution is to be obtained for following cases:

(i) All buses except bus 1 are PQ buses.

(ii) Bus 2 is a PV bus whose voltage magnitude is specified as 1.04.

(iii) Case (ii) if $0.25 \le Q_2 \le 1.0$ pu

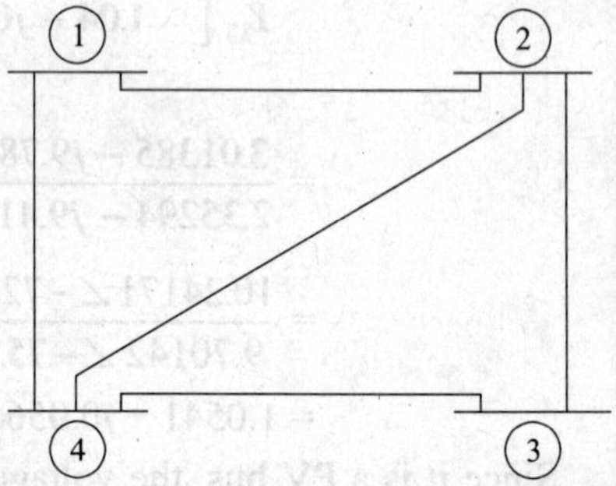

Fig. 7.7 System for example 7.6.

Table 7.3 Line data of example 7.6

SB	EB	R (pu)	X (pu)
1	2	0.05	0.15
1	3	0.10	0.30
2	3	0.15	0.45
2	4	0.10	0.30
3	4	0.05	0.15

Table 7.4 Bus data of example 7.6

Bus No.	P_i (pu)	Q_i (pu)	V_i
1	–	–	1.04 ∠0°
2	0.5	– 0.2	–
3	– 1.0	0.5	–
4	– 0.3	– 0.1	–

Solution: Note that data is directly in terms of injected powers at the buses. The bus admittance matrix is formed by inspection.

$3.0 - j9.0$	$-2.0 + j6.0$	$-1.0 + j3.0$	$0.0 + j0.0$
$-2.0 + j6.0$	$3.666 - j11.0$	$-0.666 + j2.0$	$-1.0 + j3.0$
$-1.0 + j3.0$	$-0.666 + j2.0$	$3.666 - j11.0$	$-2.0 + j6.0$
$0.0 + j0.0$	$-1.0 + j3.0$	$-2.0 + j6.0$	$3.0 - j9.0$

(Y_{bus} =)

Case(i). Assume all initial voltages to be $1.0 \angle 0°$ pu.

$$V_2^1 = \frac{1}{Y_{22}}\left[\frac{P_2 - jQ_2}{V_2^{o*}} - Y_{21} V_1^0 - Y_{23} V_3^0 - Y_{24} V_4^0\right]$$

$$= \frac{1}{Y_{22}}\left[\frac{0.5 + j0.2}{1.0 - j0.0} - \{(-2.0 + j6.0) \times (1.04 \angle 0^o)\}\right.$$

$$\left. - \{(-0.666 + j2.0) \times (1.0 \angle 0°)\} - \{(-1.0 + j3.0) \times (1.0 \angle 0°)\}\right]$$

$$= \frac{4.246 - j11.04}{3.666 - j11.0} = \frac{11.82836 \angle -68.963°}{11.5948 \angle -71.5681°}$$

$$= 1.02014 \angle 2.605° = 1.01908 + j0.04636$$

$$V_3^1 = \frac{1}{Y_{33}}\left[\frac{P_3 - jQ_3}{V_3^{o*}} - Y_{31} V_1^0 - Y_{32} V_2^1 - Y_{34} V_4^0\right]$$

$$= \frac{1}{Y_{33}}\left[\frac{-1.0 - j0.5}{1.0 - j0.0} - \{(-1.0 + j3.0) \times (1.04 \angle 0.0°)\}\right.$$

$$\left. - \{(-0.666 + j2.0) \times (1.02014 \angle 2.605°)\} - \{(-2.0 + j6.0) \times (1.0 \angle 0°)\}\right]$$

$$= \frac{2.81142 - j11.62}{3.666 - j11.0} = \frac{11.9552 \angle -76.399°}{11.5948 \angle -71.568°}$$

$$= 1.03108 \angle -4.831° = 1.027417 - j0.08683 \text{ pu.}$$

$$V_4^1 = \frac{1}{Y_{44}}\left[\frac{P_4 - jQ_4}{V_4^{o*}} - Y_{41} V_1^0 - Y_{42} V_2^1 - Y_{43} V_3^1\right]$$

$$= \frac{1}{Y_{44}}\left[\frac{0.3 + j0.1}{1.0 - j0.0} - \{(-1.0 + j3.0) \times (1.02014 \angle 2.605°)\}\right.$$

$$\left. - \{(-2.0 + j6.0) \times (1.03108 \angle -4.831°)\}\right]$$

$$= \frac{2.992 - j9.249}{3.0 - j9.0} = \frac{9.720908 \angle -72.07°}{9.48683 \angle -71.56°}$$

$$= 1.02467 \angle -0.51° = 1.02463 - j0.00912$$

Hence,

$$V_1^1 = 1.04 \angle 0° \text{ pu}$$

$$V_2^1 = 1.02014 \angle 2.605° \text{ pu}$$

$$V_3^1 = 1.03108 \angle -4.831° \text{ pu}$$

$$V_4^1 = 1.02467 \angle -0.51° \text{ pu}$$

Case (ii). If bus 2 is a PV bus and $|V_2| = 1.04$, we first compute Q_2.

$$Q_2 = |V_2|[|V_1| (G_{21} \sin \delta_{21} - B_{21} \cos \delta_{21}) + |V_2| (G_{22} \sin \delta_{22} - B_{22} \cos \delta_{22}) + |V_3| (G_{23} \sin \delta_{23} - B_{23} \cos \delta_{23}) + |V_4| (G_{24} \sin \delta_{24} - B_{24} \cos \delta_{24})]$$

$$= 1.04 [1.04 \{-6.0\} + 1.04 \{11.0\} + 1.0 \{-2.0\} + 1.0 \{-3.0\}]$$

$$= 0.208 \text{ pu.}$$

$$V_2^1 = \frac{1}{Y_{22}} \left[\frac{0.5 - j0.208}{1.04 \angle 0°} - \{(-2.0 + j6.0) \times (1.04 \angle 0°)\} - \{(-0.666 + j2.0) \times (1.0 \angle 0°)\} - \{(-1.0 + j3.0) \times (1.0 \angle 0°)\} \right]$$

$$= \frac{12.19586 \angle -69.722°}{11.5948 \angle -71.568°}$$

$$= 1.051834 \angle 1.846° = 1.051288 + j0.033883$$

The voltage magnitude is adjusted to 1.04.

Hence,

$$V_2^1 = 1.04 \angle 1.846°$$

$$V_3^1 = \frac{1}{Y_{33}} \left[\frac{-1.0 - j0.5}{1.0 \angle 0°} - \{(-1.0 + j3.0) \times (1.04 \angle 0.0°)\} - \{(-0.666 + j2.0) \times (1.04 \angle 1.846°)\} - \{(-2.0 + j6.0) \times (1.0 \angle 0°)\} \right]$$

$$= \frac{12.00743 \angle -76.519°}{11.5948 \angle -71.568°}$$

$$= 1.035587 \angle -4.951° \text{ pu.}$$

$$V_4^1 = \frac{1}{Y_{44}}\left[\frac{0.3 + j0.1}{1.0 - j0.0} - \{(-1.0 + j3.0) \times (1.04 \angle 1.846°)\}\right.$$

$$\left. - \{(-2.0 + j6.0) \times (1.035587 \angle -4.951°)\}\right]$$

$$= 0.9985 - j0.0031 = 0.9985 \angle -0.178°$$

Hence, at end of 1st iteration

$$V_1^1 = 1.04 \angle 0° \text{ pu}$$
$$V_2^1 = 1.04 \angle 1.846° \text{ pu}$$
$$V_3^1 = 1.035587 \angle -4.951° \text{ pu}$$
$$V_4^1 = 0.9985 \angle -0.178° \text{ pu}$$

Case (iii). If $0.25 \le Q_2 \le 1.0$ pu then the computed value of $Q_2 = 0.208$ is less than the lower limit. Hence, Q_2 is set equal to 0.25 pu. Iterations are carried out with this value of Q_2. The voltage magnitude at bus 2 can no longer be maintained at 1.04. Hence, there is no necessity to adjust for the voltage magnitude. Proceeding as before we obtain at the end of first iteration,

$$V_1^1 = 1.04 \angle 0° \text{ pu}$$
$$V_2^1 = 1.05645 \angle 1.849° \text{ pu}$$
$$V_3^1 = 1.038546 \angle -4.933° \text{ pu}$$
$$V_4^1 = 1.081446 \angle 4.896° \text{ pu}$$

7.7 NEWTON-RAPHSON METHOD

Newton-Raphson (NR) method is used to iteratively solve for non-linear algebraic equations of the form $f(x) = 0$. Starting from an initial guess $x°$ and a deviation $\Delta x°$ from the correct solution, Taylor's series expansion can be used to expand

$$f(x° + \Delta x°) = 0$$

The LHS when expanded gives,

$$f(x°) + \frac{df(x°)}{dx}\Delta x° + \frac{1}{2}\frac{d^2 f(x°)}{dx^2}(\Delta x)^2 + \ldots\ldots = 0$$

If we neglect second order and higher order terms, we get

$$f(x°) + \frac{df(x°)}{dx}\Delta x° = 0$$

or

$$\Delta x° = -\left[\frac{df(x°)}{dx}\right]^{-1} f(x°)$$

A new estimate can be obtained as

$$x^1 = x^\circ + \Delta x^\circ$$

$$= x^\circ - \left[\frac{df(x^\circ)}{dx}\right]^{-1} f(x^\circ)$$

Newton-Raphson (NR) method can also be used to solve a system of non-linear algebraic equations of the form $f(x) = 0$. Consider a set of n non-linear algebraic equations given by

$$f_i(x_1, x_2 \ldots x_n) = 0 \quad i = 1, 2 \ldots n \qquad \text{...(7.26)}$$

Let $x_1^0, x_2^0, \ldots x_n^0$ be the initial guess of the unknown variables and $\Delta x_1^0, \Delta x_2^0, \ldots \Delta x_n^0$ be the respective corrections. Therefore,

$$f_i(x_1^{\,0} + \Delta x_1^{\,0}, x_2^{\,0} + \Delta x_2^{\,0} \ldots x_n^{\,0} + \Delta x_n^{\,0}) = 0 \quad i = 1, 2, \ldots n \qquad \text{...(7.27)}$$

(7.27) can be expanded using Taylor's series to give

$$f_i(x_1^0, x_2^0, \ldots x_n^0) + \left[\left(\frac{\partial f_i}{\partial x_1}\right)\Delta x_1^0 + \left(\frac{\partial f_i}{\partial x_2}\right)\Delta x_2^0 + \ldots + \left(\frac{\partial f_i}{\partial x_n}\right)\Delta x_n^0\right]^0$$

$$+ \text{ Higher order terms} = 0; \quad i = 1, 2, \ldots n \qquad \text{...(7.28)}$$

$\left(\frac{\partial f_i}{\partial x_1}\right)^0, \left(\frac{\partial f_i}{\partial x_2}\right)^0 \cdots \left(\frac{\partial f_i}{\partial x_n}\right)^0$ are the partial derivatives of f_i with respect to $x_1, x_2, \ldots x_n$ respectively, evaluated at $(x_1^0, x_2^0, \ldots x_n^0)$. If the higher order terms are neglected,(7.28) can be written in matrix form as

$$\begin{bmatrix} f_1^0 \\ f_2^0 \\ \vdots \\ f_n^0 \end{bmatrix} + \begin{bmatrix} \left(\frac{\partial f_1}{\partial x_1}\right)^0 & \left(\frac{\partial f_1}{\partial x_2}\right)^0 & \cdots & \left(\frac{\partial f_1}{\partial x_n}\right)^0 \\ \left(\frac{\partial f_2}{\partial x_1}\right)^0 & \left(\frac{\partial f_2}{\partial x_2}\right)^0 & \cdots & \left(\frac{\partial f_2}{\partial x_n}\right)^0 \\ \vdots & \vdots & \cdots & \vdots \\ \left(\frac{\partial f_n}{\partial x_1}\right)^0 & \left(\frac{\partial f_n}{\partial x_2}\right)^0 & \cdots & \left(\frac{\partial f_n}{\partial x_n}\right)^0 \end{bmatrix} \begin{bmatrix} \Delta x_1^0 \\ \Delta x_2^0 \\ \vdots \\ \Delta x_n^0 \end{bmatrix} = 0 \qquad \text{...(7.29)}$$

In vector form (7.29) can be written as

$$F^0 + J^0 \, \Delta X^0 = 0$$

or,

$$F^0 = -J^0 \, \Delta X^0 \qquad \text{...(7.30)}$$

or,

$$\Delta X^0 = -[J^0]^{-1} F^0 \qquad \text{...(7.31)}$$

and,

$$X^1 = X^0 + \Delta X^0 \qquad \text{...(7.32)}$$

$[J]$ is called the Jacobian matrix. The vector of unknown variables is updated using (7.32). The process is continued till the difference between two successive iterations is less than the tolerance value.

7.7.1 NR Method for Load Flow Solution in Polar Coordinates

In application of the NR method, we have to first bring the equations to be solved, to the form $f_i(x_1, x_2, \ldots x_n) = 0$, where $x_1, x_2, \ldots x_n$ are the unknown variables to be determined. Let us assume that the power system has n_1 PV buses and n_2 *PQ* buses. In polar coordinates the unknown variables to be determined are:

(i) δ_i, the angle of the complex bus voltage at bus i, at all the PV and *PQ* buses. This gives us $n_1 + n_2$ unknown variables to be determined.

(ii) $|V_i|$, the voltage magnitude of bus i, at all the *PQ* buses. This gives us n_2 unknown variables to be determined.

Therefore, the total number of unknown variables to be computed are $n_1 + 2n_2$, for which we need $n_1 + 2n_2$ consistent equations to be solved. The equations to be solved are given by,

$$\Delta P_i = P_{i,\,sp} - P_{i,\,cal} = 0 \qquad \ldots(7.33)$$

$$\Delta Q_i = Q_{i,\,sp} - Q_{i,\,cal} = 0 \qquad \ldots(7.34)$$

where, $P_{i,\,sp}$ = specified active power at bus i.

$Q_{i,\,sp}$ = specified reactive power at bus i.

$P_{i,\,cal}$ = calculated value of active power using voltage estimates.

$Q_{i,\,cal}$ = calculated value of reactive power using voltage estimates.

ΔP = active power residue.

ΔQ = reactive power residue.

The real power is specified at all the *PV* and *PQ* buses. Hence, (7.33) is to be solved at all *PV* and *PQ* buses leading to $n_1 + n_2$ equations. Similarly, the reactive power is specified at all the *PQ* buses. Hence, (7.34) is to be solved at all *PQ* buses leading to n_2 equations.

We thus have $n_1 + 2n_2$ equations to be solved for $n_1 + 2n_2$ unknowns. (7.33) and (7.34) are of the form $F(X) = 0$. Thus, NR method can be applied to solve them. (7.33) and (7.34) can be written in the form of (7.30) as

$$\begin{bmatrix} \Delta P \\ \Delta Q \end{bmatrix} = \begin{bmatrix} J_1 & J_2 \\ J_3 & J_4 \end{bmatrix} \begin{bmatrix} \Delta\delta \\ \Delta|V| \end{bmatrix} \qquad \ldots(7.35)$$

where, J_1, J_2, J_3, J_4 are the negated partial derivatives of ΔP and ΔQ with respect to corresponding δ and $|V|$. The negated partial derivative of ΔP, is same as the partial derivative of P_{cal}, since P_{sp} is a constant (Refer (7.33) and (7.34)). The various computations involved are discussed in detail.

Computation of P_{cal} and Q_{cal}

The real and reactive powers can be computed from (7.9):

$$P_{i,\,Cal} = P_i = \sum_{k=1}^{n} |V_i|\,|V_k|\,(G_{ik}\cos\delta_{ik} + B_{ik}\sin\delta_{ik})$$

$$= G_{ii}|V_i|^2 + \sum_{\substack{k=1\\k\neq i}}^{n} |V_i|\,|V_k|\,(G_{ik}\cos\delta_{ik} + B_{ik}\sin\delta_{ik}) \qquad \text{...(7.36)}$$

$$Q_{i,\,Cal} = Q_i = \sum_{k=1}^{n} |V_i|\,|V_k|\,(G_{ik}\sin\delta_{ik} - B_{ik}\cos\delta_{ik})$$

$$= -B_{ii}|V_i|^2 + \sum_{\substack{k=1\\k\neq i}}^{n} |V_i|\,|V_k|\,(G_{ik}\sin\delta_{ik} - B_{ik}\cos\delta_{ik}) \qquad \text{...(7.37)}$$

The powers are computed at any $(r+1)^{\text{th}}$ iteration by using the voltages available from previous iteration. To compute the elements of the Jacobian, (7.36) and (7.37) are used.

Elements of J_1

Diagonal elements

The diagonals elements of J_1 are given by $\dfrac{\partial P_i}{\partial \delta_i}$. From (7.36),

$$\frac{\partial P_i}{\partial \delta_i} = \sum_{\substack{k=1\\k\neq i}}^{n} |V_i|\,|V_k|\,\{G_{ik}(-\sin\delta_{ik}) + B_{ik}\cos\delta_{ik}\}$$

$$= \sum_{\substack{k=1\\k\neq i}}^{n} -|V_i|\,|V_k|\,(G_{ik}\sin\delta_{ik} - B_{ik}\cos\delta_{ik}) \qquad \text{...(7.38a)}$$

$$= -Q_i - B_{ii}|V_i|^2 \qquad \text{...(7.38b)}$$

Off-diagonal elements

The off-diagonal elements of J_1 are $\dfrac{\partial P_i}{\partial \delta_k}$. From (7.36)

$$\frac{\partial P_i}{\partial \delta_k} = |V_i|\,|V_k|\,(G_{ik}\sin\delta_{ik} - B_{ik}\cos\delta_{ik}) \qquad \text{...(7.39)}$$

(*In deriving above recollect that $\delta_{ik} = \delta_i - \delta_k$)

Recognizing $G_{ik} = |Y_{ik}|\cos\theta_{ik}$ and $B_{ik} = |Y_{ik}|\sin\theta_{ik}$, (7.39) can be written in other forms as

$$\frac{\partial P_i}{\partial \delta_k} = |V_i|\,|V_k|\,|Y_{ik}|\,(\cos\theta_{ik}\sin\delta_{ik} - \sin\theta_{ik}\cos\delta_{ik})$$

$$= |V_i|\,|V_k|\,|Y_{ik}| \sin(\delta_{ik} - \theta_{ik}) \quad ...(7.40)$$

(7.39) can also be written in terms of real and imaginary parts of the bus voltage.

If $\quad Y_{ik} = G_{ik} + jB_{ik}$

we define, $\quad e_k + jf_k = |V_i| (\cos\delta_k + j\sin\delta_k) \quad ...(7.41)$

and,

$$(a_k + jb_k) = (G_{ik} + jB_{ik})(e_k + jf_k)$$

$$= (G_{ik} + jB_{ik})\{|V_k| (\cos\delta_k + j\sin\delta_k)\} \quad ...(7.42)$$

From (7.42), equating real and imaginary parts we obtain,

$$a_k = |V_k|\, G_{ik} \cos\delta_k - |V_k|\, B_{ik} \sin\delta_k \quad ...(7.43)$$

$$b_k = |V_k|\, G_{ik} \sin\delta_k + |V_k|\, B_{ik} \cos\delta_k \quad ...(7.44)$$

Using (7.41) and (7.43), we can write

$$a_k f_i = |V_i|\,|V_k|\, G_{ik} (\sin\delta_i \cos\delta_k) - |V_i|\,|V_k|\, B_{ik} (\sin\delta_i \sin\delta_k) \quad ..(7.45)$$

Using (7.41) and (7.44), we get

$$b_k e_i = |V_i|\,|V_k|\, G_{ik} (\sin\delta_k \cos\delta_i) - |V_i|\,|V_k|\, B_{ik} (\cos\delta_i \cos\delta_k) \quad ...(7.46)$$

From (7.45) and (7.46)

$$a_k f_i - b_k e_i = |V_i|\,|V_k| (G_{ik} \sin\delta_{ik} - B_{ik} \cos\delta_{ik}) \quad ...(7.47)$$

(7.47) is identical to (7.39). Therefore, (7.47) is an alternate method to find off diagonal elements of J_1.

Elements of J_3

Diagonal elements

The diagonal element of J_3 is $\dfrac{\partial Q_i}{\partial \delta_i}$. From (7.37)

$$\frac{\partial Q_i}{\partial \delta_i} = \sum_{\substack{k=1 \\ k \neq i}}^{n} |V_i|\,|V_k| (G_{ik} \cos\delta_{ik} + B_{ik} \sin\delta_{ik}) = P_i - G_{ii} |V_i|^2 \quad ...(7.48)$$

Off-diagonal elements

$$\frac{\partial Q_i}{\partial \delta_k} = -|V_i|\,|V_k| (G_{ik} \cos\delta_{ik} + B_{ik} \sin\delta_{ik}) \quad ...(7.49)$$

Other forms of writing (7.49) are

$$\frac{\partial Q_i}{\partial \delta_k} = -|V_i|\,|V_k|\,|Y_{ik}| \cos(\delta_{ik} - \theta_{ik}) \quad ...(7.50)$$

$$\frac{\partial Q_i}{\partial \delta_k} = -(a_k e_i + b_k f_i) \quad ...(7.51)$$

Elements of J₂

Diagonal elements

The diagonal elements are $\frac{\partial P_i}{\partial |V_i|}$. From (7.36)

$$\frac{\partial P_i}{\partial |V_i|} = 2|V_i|\, G_{ii} + \sum_{\substack{k=1\\k\neq i}}^{n} |V_k|\,(G_{ik}\cos\delta_{ik} + B_{ik}\sin\delta_{ik}) \quad ...(7.52)$$

To obtain symmetry in the elements, (7.52) is multiplied by $|V_i|$, so that the variable on RHS can be taken as $\frac{\Delta V_i}{|V_i|}$.

$$\frac{\partial P_i}{\partial |V_i|}|V_i| = 2|V_i|^2\, G_{ii} + |V_i|\sum_{\substack{k=1\\k\neq i}}^{n} |V_k|(G_{ik}\cos\delta_{ik} + B_{ik}\sin\delta_{ik}) = P_i + |V_i|^2\, G_{ii} \quad ...(7.53)$$

Off-diagonal elements

$$\frac{\partial P_i}{\partial |V_k|} = |V_i|\,(G_{ik}\cos\delta_{ik} + B_{ik}\sin\delta_{ik}) \quad ...(7.54)$$

$$\frac{\partial P_i}{\partial |V_k|}|V_k| = |V_i|\,|V_k|\,(G_{ik}\cos\delta_{ik} + B_{ik}\sin\delta_{ik}) \quad ...(7.55)$$

$$= a_k e_i + b_k f_i \quad ...(7.56)$$

Elements of J₄

Diagonal elements

The diagonal element is $\frac{\partial Q_i}{\partial |V_i|}$ and is given by

$$\frac{\partial P_i}{\partial |V_i|} = -2|V_i|B_{ii} + \sum_{\substack{k=1\\k\neq i}}^{n} |V_k|\,(G_{ik}\sin\delta_{ik} - B_{ik}\cos\delta_{ik}) \quad ...(7.57)$$

$$\frac{\partial P_i}{\partial |V_i|}|V_i| = -2|V_i|^2\, B_{ii} + \sum_{\substack{k=1\\k\neq i}}^{n} |V_i|\,|V_k|\,(G_{ik}\sin\delta_{ik} - B_{ik}\cos\delta_{ik}) = Q_i - |V_i|^2\, B_{ii} \quad ...(7.58)$$

Off-diagonal elements

$$\frac{\partial Q_i}{\partial |V_k|} = |V_i|\,(G_{ik}\sin\delta_{ik} - B_{ik}\cos\delta_{ik})$$

$$\frac{\partial Q_i}{\partial |V_k|}|V_k| = |V_i|\,|V_k|\,(G_{ik}\sin\delta_{ik} - B_{ik}\cos\delta_{ik}) = a_k f_i - b_k e_i \quad ...(7.59)$$

We can now rewrite (7.35) as follows:

$$\begin{bmatrix} \Delta P \\ \Delta Q \end{bmatrix} = \begin{bmatrix} H & N \\ M & L \end{bmatrix} \begin{bmatrix} \Delta\delta \\ \dfrac{\Delta|V|}{|V|} \end{bmatrix} \quad ...(7.60)$$

The elements are summarized below

(i) $H_{ii} = \dfrac{\partial P_i}{\partial \delta_i} = -Q_i - B_{ii}|V_i|^2$

(ii) $H_{ik} = \dfrac{\partial P_i}{\partial \delta_k} = a_k f_i - b_k e_i = |V_i|\,|V_k|\,(G_{ik}\sin\delta_{ik} - B_{ik}\cos\delta_{ik})$

(iii) $N_{ii} = \dfrac{\partial P_i}{\partial |V_i|}|V_i| = P_i + G_{ii}|V_i|^2$

(iv) $N_{ik} = \dfrac{\partial P_i}{\partial |V_k|}|V_k| = a_k e_i + b_k f_i = |V_i|\,|V_k|\,(G_{ik}\cos\delta_{ik} + B_{ik}\sin\delta_{ik})$

(v) $M_{ii} = \dfrac{\partial Q_i}{\partial \delta_i} = P_i - G_{ii}|V_i|^2$

(vi) $M_{ik} = \dfrac{\partial Q_i}{\partial \delta_k} = -(a_k e_i + b_k f_i) = -N_{ik}$

(vii) $L_{ii} = \dfrac{\partial Q_i}{\partial |V_i|}|V_i| = Q_i - B_{ii}|V_i|^2$

(viii) $L_{ik} = \dfrac{\partial Q_i}{\partial |V_k|}|V_k| = a_k f_i - b_k e_i = H_{ik}$

In the above equations,

$$Y_{ik} = G_{ik} + jB_{ik}$$

$$e_k + jf_k = |V_k|\,(\cos\delta_k + j\sin\delta_k) \text{ and}$$

$$a_k + jb_k = (G_{ik} + jB_{ik})\,(e_k + jf_k)$$

If $Y_{ik} = 0.0 + j0.0$ (if there is no line between buses i and k) then the corresponding off-diagonal elements in the Jacobian matrix will also be zero. Hence, the Jacobian is also a sparse matrix.

Size of the sub-matrices of the Jacobian

The dimensions of the various sub-matrices are as follows:

Matrix	*size*
H	$(n_1 + n_2) \times (n_1 + n_2)$
N	$(n_1 + n_2) \times (n_2)$
M	$(n_2) \times (n_1 + n_2)$
L	$(n_2) \times (n_2)$
J	$(n_1 + 2n_2) \times (n_1 + 2n_2)$
ΔP	$(n_1 + n_2) \times 1$
ΔQ	$n_2 \times 1$
$\Delta\delta$	$(n_1 + n_2) \times 1$
$\Delta\|V\|/\|V\|$	$n_2 \times 1$

7.7.2 Algorithm for NR Method in Polar Coordinates

1. Formulate the Y_{bus}.
2. Assume initial voltages as follows:

$$V_i = |V_{i,sp}| \angle 0° \text{ (at all } PV \text{ buses)}$$

$$V_i = 1 \angle 0° \text{ (at all } PQ \text{ buses)}$$

3. At $(r+1)^{th}$ iteration, calculate $P_i^{(r+1)}$ at all the PV and PQ buses and $Q_i^{(r+1)}$ at all the PQ buses, using voltages from previous iteration, $V_i^{(r)}$. The formulae to be used are:

$$P_{i,cal}^{(r+1)} = P_i = G_{ii}|V_i|^2 + \sum_{\substack{k=1\\k\neq i}}^{n} |V_i|\,|V_k|\,(G_{ik}\cos\delta_{ik} + B_{ik}\sin\delta_{ik})$$

$$Q_{i,cal}^{(r+1)} = Q_i = -B_{ii}|V_i|^2 + \sum_{\substack{k=1\\k\neq i}}^{n} |V_i|\,|V_k|\,(G_{ik}\sin\delta_{ik} - B_{ik}\cos\delta_{ik})$$

4. Calculate the power mismatches (power residues).

$$\Delta P_i^{(r)} = P_{i,sp} - P_{i,cal}^{(r+1)} \text{ (at } PV \text{ and } PQ \text{ buses)}$$

$$\Delta Q_i^{(r)} = Q_{i,sp} - Q_{i,cal}^{(r+1)} \text{ (at } PQ \text{ buses)}$$

5. Calculate the Jacobian $[J^{(r)}]$ using $V_i^{(r)}$.
6. Compute

$$\begin{bmatrix} \Delta\delta^{(r)} \\ \dfrac{\Delta|V^{(r)}|}{|V|} \end{bmatrix} = [J^{(r)}]^{-1} \begin{bmatrix} \Delta P^{(r)} \\ \Delta Q^{(r)} \end{bmatrix}$$

7. Update the variables as follows:

$$\delta_i^{(r+1)} = \delta_i^{(r)} + \Delta\delta_i^{(r)} \text{ (at all buses)}$$

$$|V_i|^{(r+1)} = |V_i|^{(r)} + \Delta|V_i|^{(r)}$$

8. Go to step 3 and iterate till the power mismatches are within acceptable tolerance.

7.7.3 NR Method in Rectangular Coordinates

The equations are formulated in a different manner to apply *NR* method to rectangular coordinates. We know,

$$S_i^* = P_i - jQ_i = V_i^* \sum_{k=1}^{n} Y_{ik} V_k \qquad ...(7.61)$$

Let

$$V_i = e_i + jf_i \qquad ...(7.62a)$$

$$V_k = e_k + jf_k \qquad ...(7.62b)$$

$$Y_{ik} = G_{ik} + jB_{ik} \qquad ...(7.62c)$$

Substituting (7.62) in (7.61), we get

$$P_i - jQ_i = (e_i - jf_i) \sum_{k=1}^{n} (G_{ik} + jB_{ik})(e_k + jf_k)$$

Equating real and imaginary parts

$$P_i = \sum_{k=1}^{n} e_i(e_k G_{ik} - f_k B_{ik}) + f_i(f_k G_{ik} + e_k B_{ik}) \qquad ...(7.63)$$

$$Q_i = \sum_{k=1}^{n} f_i(e_k G_{ik} - f_k B_{ik}) - e_i(f_k G_{ik} + e_k B_{ik}) \qquad ...(7.64)$$

Further we have $\quad |V_i|^2 = e_i^2 + f_i^2 \qquad ...(7.65)$

At the *PV* buses, even though the voltage magnitude is specified, the real and imaginary components are unknown since we do not know the voltage angle at the bus. Hence, in rectangular coordinate formulation, e_i and f_i, are unknown at all the buses and need to be solved. We thus have to solve for $2(n_1 + n_2)$ variables. The equations to be solved are:

$$\Delta P_i = P_{i,sp} - P_{i,cal} = 0 \; (n_1 + n_2 \text{ equations at all } PV \text{ and } PQ \text{ buses}) \qquad ...(7.66)$$

$$\Delta Q_i = Q_{i,sp} - Q_{i,cal} = 0 \; (n_2 \text{ equations at } PQ \text{ buses}) \qquad ...(7.67)$$

$$\Delta|V_i|^2 = (|V_{i,sp}| - |V_i|)^2 = 0 \; (n_1 \text{ equations at } PV \text{ buses}) \qquad ...(7.68)$$

The above $2(n_1 + n_2)$ are solved for the $2(n_1 + n_2)$ unknown variables which are the real and imaginary part of the bus voltages at all the buses. The equations can be written in the same form as (7.30),

$$\begin{bmatrix} \Delta P \\ \Delta Q \\ \Delta|V|^2 \end{bmatrix} = \begin{bmatrix} J_1 & J_2 \\ J_3 & J_4 \\ J_5 & J_6 \end{bmatrix} \begin{bmatrix} \Delta e \\ \Delta f \end{bmatrix} \qquad ...(7.69)$$

In rectangular coordinates the current at bus i is given by

$$I_i = c_i + jd_i = \sum_{k=1}^{n} Y_{ik}V_k = \sum_{k=1}^{n} (G_{ik} + jB_{ik})(e_k + jf_k)$$

$$= (G_{ii} + jB_{ii})(e_i + jf_i) + \sum_{\substack{k=1 \\ \neq i}}^{n} (G_{ik} + jB_{ik})(e_k + jf_k) \qquad ...(7.70)$$

Equating real and imaginary parts in (7.70), we get

$$c_i = G_{ii}e_i - B_{ii}f_i + \sum_{\substack{k=1 \\ \neq i}}^{n} (G_{ik}e_k - B_{ik}f_k) \qquad ...(7.71)$$

$$d_i = G_{ii}f_i + B_{ii}e_i + \sum_{\substack{k=1 \\ \neq i}}^{n} (G_{ik}f_k + B_{ik}e_k) \qquad ...(7.72)$$

We use (7.71) and (7.72) in simplifying the expressions for the Jacobian elements. To obtain the Jacobian elements, the expressions for the active and reactive power at a bus are written in a slightly different form in terms of rectangular coordinates of the bus voltages.

$$P_{i,\,cal} = P_i = G_{ii}e_i^2 - B_{ii}e_if_i + G_{ii}f_i^2 + B_{ii}e_if_i$$

$$+ \sum_{\substack{k=1 \\ \neq i}}^{n} \{e_i(G_{ik}e_k - B_{ik}f_k) + f_i(G_{ik}f_k + B_{ik}e_k)\} \qquad ...(7.73)$$

$$Q_{i,\,cal} = Q_i = G_{ii}f_ie_i - B_{ii}f_i^2 - G_{ii}f_ie_i - B_{ii}e_i^2$$

$$+ \sum_{\substack{k=1 \\ \neq i}}^{n} \{f_i(G_{ik}e_k - B_{ik}f_k) - e_i(G_{ik}f_k + B_{ik}e_k)\} \qquad ...(7.74)$$

Elements of J_1

Diagonal Elements

$$\frac{\partial P_i}{\partial e_i} = 2G_{ii}e_i - B_{ii}f_i + B_{ii}f_i + \sum_{\substack{k=1 \\ \neq i}}^{n} (G_{ik}e_k - B_{ik}f_k)$$

$$= G_{ii}e_i - B_{ii}f_i + \left[G_{ii}\, e_i + B_{ii}\, f_i + \sum_{\substack{k=1 \\ \neq i}}^{n} (G_{ik}\, e_k - B_{ik}\, f_k) \right]$$

$$= G_{ii}e_i - B_{ii}f_i + \sum_{\substack{k=1 \\ \neq i}}^{n} (G_{ik}\, e_k - B_{ik}\, f_k) + G_{ii}\, e_i + B_{ii} f_i$$

$$= G_{ii}e_i + B_{ii}f_i + c_i \qquad ...(7.75)$$

Off-Diagonal Elements

$$\frac{\partial P_i}{\partial e_k} = G_{ik}e_i + B_{ik}f_i \; (i \neq k) \qquad ...(7.76)$$

Elements of J_2

Diagonal Elements

$$\frac{\partial P_i}{\partial f_i} = -B_{ii}e_i + 2G_{ii}f_i + B_{ii}e_i + \sum_{\substack{k=1 \\ \neq i}}^{n} (G_{ik}f_k + B_{ik}e_k)$$

$$= -B_{ii}e_i + G_{ii}f_i + d_i \qquad ...(7.77)$$

Off-Diagonal Elements

$$\frac{\partial P_i}{\partial f_k} = G_{ik}f_i - B_{ik}e_i \; (i \neq k) \qquad ...(7.78)$$

Elements of J_3

Diagonal Elements

$$\frac{\partial Q_i}{\partial e_i} = G_{ii}f_i - G_{ii}f_i - 2B_{ii}e_i + \sum_{\substack{k=1 \\ \neq i}}^{n} - (G_{ik}f_k + B_{ik}e_k)$$

$$= G_{ii}f_i - B_{ii}e_i - \left(G_{ii}\, f_i + B_{ii}\, e_i + \sum_{\substack{k=1 \\ \neq i}}^{n} (G_{ik}\, f_k + B_{ik}\, e_k) \right)$$

$$= G_{ii}f_i - B_{ii}e_i - d_i \qquad ...(7.79)$$

Off-Diagonal Elements

$$\frac{\partial Q_i}{\partial e_k} = G_{ik}f_i - B_{ik}e_i \; (i \neq k) \qquad ...(7.80)$$

(*Note that off-diagonal elements of J_3 are same as that of J_2.)

Elements of J₄

Diagonal Elements

$$\frac{\partial Q_i}{\partial f_i} = G_{ii}e_i - 2B_{ii}f_i - G_{ii}e_i + \sum_{\substack{k=1 \\ \neq i}}^{n} (G_{ik}e_k - B_{ik}f_k)$$

$$= -G_{ii}e_i - B_{ii}f_i + c_i \qquad ...(7.81)$$

Off-Diagonal Elements

$$\frac{\partial Q_i}{\partial f_k} = -(B_{ik}f_i + G_{ik}e_i) \; (i \neq k) \qquad ...(7.82)$$

[* Note that off diagonal element of J_4 are negative of J_1]

Elements of J₅

$$|V_i|^2 = e_i^2 + f_i^2$$

$$\frac{\partial |V_i|^2}{\partial e_i} = 2e_i \qquad ...(7.83)$$

$$\frac{\partial |V_i|^2}{\partial e_k} = 0 \qquad ...(7.84)$$

Elements of J₆

$$\frac{\partial |V_i|^2}{\partial f_i} = 2f_i \qquad ...(7.85)$$

$$\frac{\partial |V_i|^2}{\partial f_k} = 0 \qquad ...(7.86)$$

7.7.4 Algorithm for NR Method in Rectangular Coordinates

1. Formulate the Y_{bus}.
2. Assume initial voltages as follows:

7.7.5 Flow chart for Gauss Newton Raphson Method of Load-flow Solution in Polal Coordinates. Assume Slack Bus is Bus No. 1.

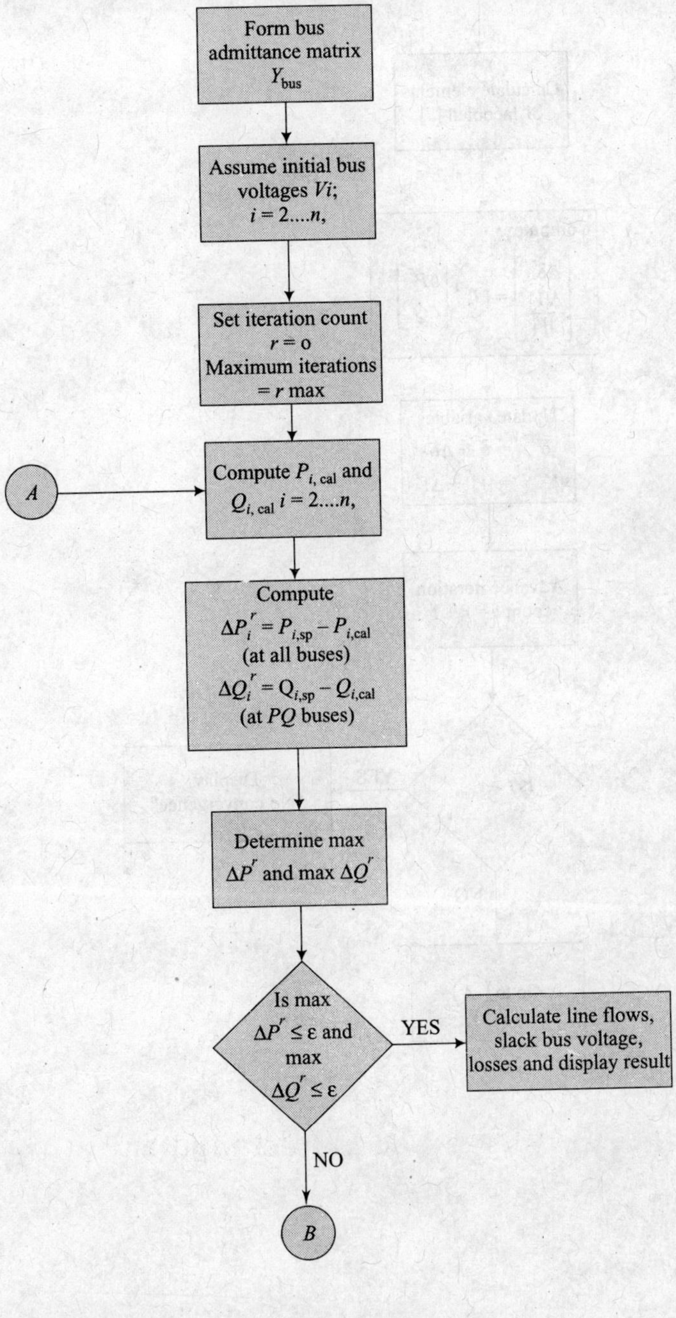

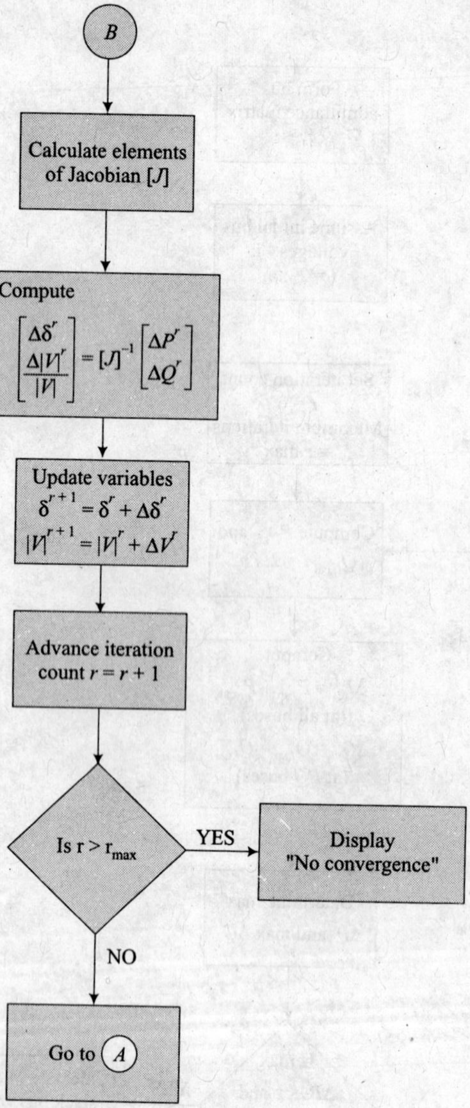
B
Calculate elements of Jacobian [J]
Compute
$\begin{bmatrix} \Delta\delta^r \\ \frac{\Delta|V|^r}{|V|} \end{bmatrix} = [J]^{-1} \begin{bmatrix} \Delta P^r \\ \Delta Q^r \end{bmatrix}$
Update variables
$\delta^{r+1} = \delta^r + \Delta\delta^r$
$|V|^{r+1} = |V|^r + \Delta V^r$
Advance iteration count $r = r + 1$
Is $r > r_{max}$
YES
Display "No convergence"
NO
Go to A

$$e_i + jf_i = |V_{i,\,sp}| + j0.0 \text{ (at all } PV \text{ buses)}$$
$$e_i + jf_i = 1 + j0.0 \text{ (at all } PQ \text{ buses)}$$

3. At $(r + 1)^{th}$ iteration, calculate $P_{i,\,cal}^{(r+1)}$ at all the PV and PQ buses, $Q_{i,\,cal}^{(r+1)}$ at all the PQ buses and $|V_i|^2$ at PV buses, using voltages from previous iteration, $V_i^{(r)}$.
4. Calculate the power mismatches
$$\Delta P_i^{(r)} = P_{i,\,sp} - P_{i,\,cal}^{(r+1)} \text{ (at } PV \text{ and } PQ \text{ buses)}$$
$$\Delta Q_i^{(r)} = Q_{i,\,sp} - Q_{i,\,cal}^{(r+1)} \text{ (at } PQ \text{ buses)}$$
$$\Delta|V_i^r|^2 = |V_{i,\,sp}|^2 - |V_i|^2 \text{ (at } PV \text{ buses)}$$
5. Calculate the Jacobian $[J^{(r)}]$ using $V_i^{(r)}$.
6. Compute
$$\begin{bmatrix} \Delta e_i^{(r)} \\ \Delta f_i^{(r)} \end{bmatrix} = [J^{(r)}]^{-1} \begin{bmatrix} \Delta P^{(r)} \\ \Delta Q^{(r)} \\ \Delta |V_i^{(r)}|^2 \end{bmatrix}$$
7. Update the variables as follows:
$$e_i^{(r+1)} = e_i^{(r)} + \Delta e_i^{(r)} \text{ (at all buses)}$$
$$f_i^{(r+1)} = f_i^{(r)} + \Delta f_i^{(r)} \text{ (at all buses)}$$
8. At PV buses the voltage magnitude is maintained at the specified value. Hence, $e_i^2 + f_i^2 = |V_{i,\,sp}|^2$ at the PV buses. Therefore, we have to adjust the voltage estimates to be used in the next iteration. The voltage angle is given by
$$\delta_i^{(r+1)} = \arctan \frac{f_i^{(r+1)}}{e_i^{(r+1)}}$$
The voltages at the PV buses are adjusted as follows:
$$e_i^{(r+1)} = |V_{i,\,sp}| \cos \delta_i^{(r+1)}$$
$$f_i^{(r+1)} = |V_{i,\,sp}| \sin \delta_i^{(r+1)}$$
9. Go to step 3 and iterate till the power mismatches are within acceptable tolerance.

7.8 DECOUPLED LOAD FLOW

In the *NR* method, the inverse of the Jacobian has to be computed at every iteration. When solving large interconnected power systems, alternative solution methods are possible, taking into account certain observations made of practical systems. These are:

- Change in voltage magnitude $|V_i|$ at a bus primarily affects the flow of reactive power Q in the lines and leaves the real power P unchanged. This observation implies that $\frac{\partial Q_i}{\partial |V_j|}$ is much larger than $\frac{\partial P_i}{\partial |V_j|}$. Hence, in the Jacobian, the elements of the sub-matrix [N], which contains terms that are partial derivatives of real power with respect to voltage magnitudes can be made zero.

- Change in voltage phase angle at a bus, primarily affects the real power flow P over the lines and the flow of Q is relatively unchanged. This observation implies that $\frac{\partial P_i}{\partial \delta_j}$ is much larger than $\frac{\partial Q_i}{\partial \delta_j}$. Hence, in the Jacobian the elements of the sub-matrix [M], which contains terms that are partial derivatives of reactive power with respect to voltage phase angles can be made zero.

These observations reduce (7.60) to

$$\begin{bmatrix} \Delta P \\ \Delta Q \end{bmatrix} = \begin{bmatrix} H & 0 \\ 0 & L \end{bmatrix} \begin{bmatrix} \Delta\delta \\ \frac{\Delta|V|}{|V|} \end{bmatrix} \qquad \text{...(7.87)}$$

From (7.87) it is obvious that the voltage angle corrections $\Delta\delta$ are obtained using real power residues ΔP and the voltage magnitude corrections $|\Delta V|$ are obtained from reactive power residues ΔQ. (7.87) can be solved in two ways.

Method I

(i) Calculate $\Delta P^{(r)}$, $\Delta Q^{(r)}$ and $J^{(r)}$.

(ii) Compute $\begin{bmatrix} \Delta\delta^{(r)} \\ \frac{\Delta|V^{(r)}|}{|V^{(r)}|} \end{bmatrix} = \left[J^{(r)}\right]^{-1} \begin{bmatrix} \Delta P^{(r)} \\ \Delta Q^{(r)} \end{bmatrix}$

(iii) Update δ and $|V|$.

(iv) Go to step (i) and iterate till convergence is reached.

Method II

(i) Compute $\Delta P^{(r)}$ and sub-matrix $H^{(r)}$. From (7.87) $\Delta\delta^{(r)} = [H^{(r)}]^{-1}\,\Delta P^{(r)}$

(ii) Up date δ using $\delta^{(r+1)} = \delta^{(r)} + \Delta\delta^{(r)}$.

(iii) Use $\delta^{(r+1)}$ to calculate $\Delta Q^{(r)}$ and $L^{(r)}$.

(iv) Compute $\frac{\Delta|V^{(r)}|}{|V^{(r)}|} = \left[L^{(r)}\right]^{-1} \Delta Q^{(r)}$

(v) Update $|V^{(r+1)}| = |V^{(r)}| + |\Delta V^{(r)}|$

(vi) Go to step (i) and iterate till convergence is reached.

In the first strategy, the variables are solved simultaneously. In the second strategy the iteration is conducted by first solving for $\Delta\delta$ and using updated values of δ to calculate $\Delta|V|$. This results in faster convergence, compared to method (i).

7.9 FAST DECOUPLED LOAD FLOW

If the coefficient matrices are constant, the need to update the Jacobian at every iteration is eliminated. This has resulted in development of Fast Decoupled Load Flow (*FDLF*) method by B. Stott in 1974. Certain assumptions are made based on observations of practical power systems. They are:

- $B_{ij} >> G_{ij}$ (Since, the X/R ratio of transmission lines is high in well designed systems)
- The voltage angle difference $(\delta_i - \delta_j)$ between two buses in the system is very small. This means $\cos(\delta_i - \delta_j) \cong 1$ and $\sin(\delta_i - \delta_j) = 0.0$
- $Q_i << B_{ii} |V_i|^2$

With these assumptions the elements of the Jacobian become

$$H_{ik} = L_{ik} = -|V_i|\,|V_k|\,B_{ik}\ (i \neq k)$$

$$H_{ii} = L_{ii} = -B_{ii}\,|V_i|^2$$

The matrix (7.87) reduces to

$$[\Delta P] = |V_i||V_k||B'_{ik}|\,[\Delta\delta] \qquad \text{...(7.88a)}$$

$$[\Delta Q] = |V_i||V_k||B''_{ik}|\left[\frac{\Delta|V|}{|V|}\right] \qquad \text{...(7.88b)}$$

where, B'_{ik} and B''_{ik} are negative of the susceptances of respective elements of the bus admittance matrix. In (7.88) if we divide LHS and RHS by $|V_i|$ and assume $|V_j| \cong 1$, we get

$$\left[\frac{\Delta P}{|V|}\right] = [B'][\Delta\delta] \qquad \text{...(7.89a)}$$

$$\left[\frac{\Delta Q}{|V|}\right] = [B''][\Delta|V|] \qquad \text{...(7.89b)}$$

Equations (7.89a) and (7.89b) constitute the Fast Decoupled load flow equations. Further simplification is possible by:

- Omitting effect of phase shifting transformers.
- Setting off-nominal turns ratio of transformers to 1.0.
- In forming B', omitting the effect of shunt reactors and capacitors which mainly affect reactive power.
- Ignoring series resistance of lines in forming the Y_{bus}.

With these assumptions we obtain a loss less network. If further, all voltage magnitudes are assumed to be 1.0 pu, we obtain a DC power flow model. This model is acceptable where only approximate solutions are required like in planning expansions and in contingency studies. In the *FDLF* method, the matrices $[B']$ and $[B'']$ are constants and need to be inverted only once at the beginning of the iterations. Separate convergence tests can be applied for real power and reactive power, as max $[\Delta P] \leq \epsilon_P$ and max $[\Delta Q] \leq \epsilon_Q$. Generally, the tolerances for power mismatch are 0.001 pu.

Equations (7.89a) and (7.89b) can be solved alternately using the most recent voltage values. $\Delta\delta$ is first evaluated using (7.89a) and the updated values of δ used in computing $\Delta|V|$ from (7.89b)

Example 7.7: Obtain the voltages at all buses for the three bus system shown in Fig. 7.8 at the end of first iteration by (a) NR method (b) Decoupled method (c) Fast decoupled method. The data is given in Table 7.5a and 7.5b.

Table 7.5a Line data for example 7.7

SB	*EB*	$R_{(pu)}$	$X_{(pu)}$	$B_C/2$
1	2	0.0	0.1	0.0
1	3	0.0	0.2	0.0
2	3	0.0	0.2	0.0

Table 7.5b Bus data for example 7.7

Bus No.	P_G	Q_G	P_L	Q_L	V_{sp}
1 (Slack)	-	-	-	-	1.0
2 (PV)	5.3217	-	-	-	1.1
3(PQ)	-	-	3.6392	0.5339	-

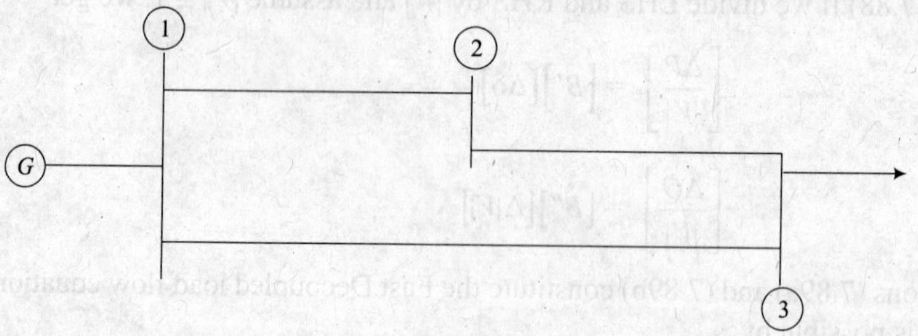

Fig. 7.8 Power system of example 7.7.

Solution: Formulate Y_{bus}:

$$Y_{bus} = \begin{bmatrix} -j15.0 & j10.0 & j5.0 \\ j10.0 & -j15.0 & j5.0 \\ j5.0 & j5.0 & -j10.0 \end{bmatrix}$$

Initial Voltages:

$$V_1 = 1.0 + j0.0 = \angle 0°$$

$$V_2 = 1.1 + j0.0 = 1.1\ \angle 0°$$

$$V_3 = 1.0 + j0.0 = 1\,0\ \angle 0°$$

(a) NR Method: $P_{2,\,sp} = 5.3217$; $P_{3,\,sp} = -3.6392$ (since it is a load P_{sp} is negative)

$$P_{2,\,cal} = P_2 = G_{22}\,|V_2|^2 + |V_2|\,|V_1|\,(G_{21}\cos\delta_{21} + B_{21}\sin\delta_{21}) + |V_2|\,|V_3|\,(G_{23}\cos\delta_{23} + B_{23}\sin\delta_{23})$$

1st Iteration

$$\delta_{21} = \delta_2 - \delta_1 = 0°;\ \delta_{23} = \delta_2 - \delta_3 = 0°;\ G_{22} = 0.0 = G_{21} = G_{23}$$

$\therefore$

$$P_{2,\,cal} = 0.0$$

$$\Delta P_2 = 5.3217 - 0.0 = 5.3217.$$

$$P_{3,\,cal} = P_3 = G_{33}\,|V_3|^2 + |V_3|\,|V_1|\,(G_{31}\cos\delta_{31} + B_{31}\sin\delta_{31}) + |V_3|\,|V_2|\,(G_{32}\cos\delta_{32} + B_{32}\sin\delta_{32})$$

$$= 0.0$$

$$\Delta P_3 = -3.6392 - 0.0 = -3.6392.$$

$$Q_{3,\,sp} = -0.5339.$$

$$Q_{3,\,cal} = Q_3 = -B_{33}\,|V_3|^2 + |V_3|\,|V_1|\,(G_{31}\sin\delta_{31} - B_{31}\cos\delta_{31}) + |V_3|\,|V_2|\,(G_{32}\sin\delta_{32} - B_{32}\cos\delta_{32})$$

$$= 10.0\,(1.0)^2 + (1.0 \times 1.0 \times -5.0) + (1.0 \times 1.1 \times -5.0)$$

$$= 10.0 - 5.0 - 5.5 = -0.5 \text{ pu.}$$

$$\Delta Q_3 = -0.5339 - (-0.5) = -0.0339 \text{ pu.}$$

The matrix for solution is given by

$$\begin{bmatrix} \Delta P_2 \\ \Delta P_3 \\ \Delta Q_3 \end{bmatrix} = \begin{bmatrix} H_{22} & H_{23} & N_{23} \\ H_{32} & H_{33} & N_{33} \\ M_{32} & M_{33} & L_{33} \end{bmatrix} \begin{bmatrix} \Delta\delta_2 \\ \Delta\delta_3 \\ \dfrac{\Delta|V_3|}{|V_3|} \end{bmatrix}$$

* *The suffixes in the Jacobian element indicate the bus numbers of the variables (for example H_{22} indicates $\dfrac{\partial P_2}{\partial \delta_2}$) and not their positions in the matrix.*

$$H_{22} = -Q_2 - B_{22}\,|V_2|^2$$

$$Q_2 = -B_{22}\,|V_2|^2 + |V_2|\,|V_1|\,(G_{21}\sin\delta_{21} - B_{21}\cos\delta_{21}) + |V_2|\,|V_3|\,(G_{23}\sin\delta_{23} - B_{23}\cos\delta_{23})$$

$$= 15\,(1.1)^2 + (1.1 \times 1.0 \times 10.0) + (1.1 \times 1.0 \times -5.0)$$

$$= 18.15 - 11.0 - 5.5 = 1.65.$$

$$H_{22} = -1.65 + 15 \times (1.1)^2 = 16.5$$

$$H_{23} = |V_2|\,|V_3|\,(G_{23} \sin \delta_{23} - B_{23} \cos \delta_{23})$$

$$= 1.1 \times 1.0 \times -5.0 = -5.5$$

$$H_{32} = |V_3|\,|V_2|\,(G_{32} \sin \delta_{32} - B_{32} \cos \delta_{32})$$

$$= 1.1 \times 1.0 \times -5.0 = -5.5$$

$$H_{33} = -Q_3 - B_{33}\,|V_3|^2 = 0.5 + 10 = 10.5$$

$$N_{23} = |V_2|\,|V_3|\,(G_{23} \cos \delta_{23} + B_{23} \sin \delta_{23}) = 0.0$$

$$N_{33} = P_3 + G_{33}\,|V_3|^2 = 0.0$$

$$M_{32} = -N_{23} = 0.0$$

$$M_{33} = P_3 - G_{33}\,|V_3|^2 = 0.0$$

$$L_{33} = Q_3 - B_{33}\,|V_3|^2 = -0.5 + 10.0 = 9.5$$

We therefore get the matrix,

$$\begin{bmatrix} 5.3217 \\ -3.6392 \\ -0.0339 \end{bmatrix} = \begin{bmatrix} 16.5 & -5.5 & 0.0 \\ -5.5 & 10.5 & 0.0 \\ 0.0 & 0.0 & 9.5 \end{bmatrix} \begin{bmatrix} \Delta\delta_2 \\ \Delta\delta_3 \\ \dfrac{\Delta|V_3|}{|V_3|} \end{bmatrix}$$

$$[J]^{-1} = \begin{bmatrix} 0.0734 & 0.0385 & 0.0 \\ 0.0385 & 0.1154 & 0.0 \\ 0.0 & 0.0 & 0.1053 \end{bmatrix}$$

$$\begin{bmatrix} \Delta\delta_2 \\ \Delta\delta_3 \\ \dfrac{\Delta|V_3|}{|V_3|} \end{bmatrix} = \begin{bmatrix} 0.0734 & 0.0385 & 0.0 \\ 0.0385 & 0.1154 & 0.0 \\ 0.0 & 0.0 & 0.1053 \end{bmatrix} \begin{bmatrix} 5.3217 \\ -3.6392 \\ -0.0339 \end{bmatrix} = \begin{bmatrix} 0.2508 \\ -0.2152 \\ -0.0036 \end{bmatrix}$$

$$\Delta\delta_2 = 0.2508 \text{ rad} = 14.37°$$

$$\delta_2 = 0.0 + 14.37° = 14.37°$$

$$\Delta\delta_3 = -0.2152 \text{ rad} = -12.33°$$

$$\delta_3 = 0.0 - 12.33° = -12.33°$$

$$\frac{\Delta|V_3|}{|V_3|} = -0.0036;\ \Delta|V_3| = -0.0036 \times 1.0 = -0.0036.$$

$$|V_3| = 1.0 - 0.0036 = 0.9964.$$

At end of Ist iteration

$$V_1 = 1.0 \angle 0° = 1.0 + j0.0$$

$$V_2 = 1.1 \angle 14.37° = 1.065584 + j0.273$$

$$V_3 = 0.9964 \angle -12.33° = 0.97342 - j0.212773$$

The results of subsequent iterations are given below.

2nd Iteration

$$P_{2,\,cal} = 5.1923;\ P_{3,\,cal} = -3.5265;\ Q_{2,\,cal} = 2.5981;\ Q_{3,\,cal} = 0.1655$$

$$J = \begin{bmatrix} 15.5519 & -4.896 & 2.4625 \\ -4896 & 9.7632 & -3.5265 \\ -2.4625 & -3.5265 & 10.0943 \end{bmatrix}$$

$$= \begin{bmatrix} \Delta\delta_2 \\ \Delta\delta_3 \\ \dfrac{\Delta|V|}{|V_3|} \end{bmatrix} = \begin{bmatrix} 0.0098 \\ -0.0353 \\ -0.0792 \end{bmatrix}$$

$$\delta_2 = 0.2508 + 0.0098 = 0.2606 = 14.93°$$

$$\delta_3 = -0.2152 - 0.0353 = -0.2505 = -14.35°$$

$$\Delta|V_3| = -0.0792 \times 0.9964 = -0.0789$$

$$|V_3| = 0.9964 - 0.0789 = 0.9175$$

The voltages at the end of 2nd iteration are

$$V_1 = 1 = \angle 0^\circ$$

$$V_2 = 1.1 \ \angle 14.93^\circ$$

$$V_3 = 0.9175 \ \angle -14.35^\circ$$

3rd Iteration

$$P_{2,\,cal} = 5.3017;\ P_{3,\,cal} = -3.6051;\ Q_{2,\,cal} = 3.1198;\ Q_{3,\,cal} = -0.4279$$

$$J = \begin{bmatrix} 15.0302 & -4.4014 & 2.4680 \\ -4.4014 & 8.8457 & -3.6051 \\ -2.4680 & -3.6051 & 7.9899 \end{bmatrix}$$

$$= \begin{bmatrix} \Delta\delta_2 \\ \Delta\delta_3 \\ \dfrac{\Delta|V_3|}{|V_3|} \end{bmatrix} = \begin{bmatrix} 0.0012 \\ -0.0105 \\ -0.0176 \end{bmatrix}$$

$$\delta_2 = 0.2606 + 0.0012 = 0.2618 = 14.99°$$

$$\delta_3 = -0.2505 - 0.0105 = -0\ 2610 = -19.95°$$

$$|\Delta V_3| = -0.0176 \times 0.9175$$

$$= -0.0162$$

$$|V_3| = 0.9175 - 0.0162 = 0.9013$$

At the end of 3rd iteration

$$V_1 = 1.0 \angle 0^{\circ}$$

$$V_2 = 1.1 \angle 14.99^{\circ}$$

$$V_3 = 0.9013 \angle -14.95^{\circ}$$

4th Iteration

$P_{2,cal} = 5.3206$; $P_{3,cal} = -3.6373$; $Q_{2,cal} = 3.2291$ $Q_{3,cal} = -0.5257$

$$J = \begin{bmatrix} 14.9209 & -4.2954 & 2.4746 \\ -4.2954 & 8.6494 & -3.6373 \\ -2.4746 & -3.6373 & 7.5979 \end{bmatrix}$$

$$\begin{bmatrix} \Delta\delta_2 \\ \Delta\delta_3 \\ \dfrac{\Delta|V_3|}{|V_3|} \end{bmatrix} = \begin{bmatrix} 0.0001 \\ -0.0008 \\ -0.0014 \end{bmatrix}$$

$$\delta_2 = 0.2618 + 0.0001 = 0.2619 = 15°$$

$$\delta_3 = -0.2610 - 0.0008 = -0.2618 = -14{,}99°$$

$$|\Delta V_3| = -0.0014 \times 0.9013 = -0.0013$$

At end of 4th iteration

$$V_1 = 1.0 \angle 0^{\circ}$$

$V_2 = 1.1 = \angle 15^\circ$

$V_3 = 0.9 = \angle -14.99^\circ$

Since, the changes have become insignificant, iterations can be stopped. More iterations can be carried out if more accuracy is needed. The converged voltages are,

$$V_1 = 1 \angle 0^\circ$$

$$V_2 = 1.1 \angle 15^\circ$$

$$V_3 = 0.9 \angle -14.99$$

The slack bus power is computed using these values.

$$P_1 = G_{11} |V_1|^2 + |V_1| |V_2| (G_{12} \cos \delta_{12} + B_{12} \sin \delta_{12}) + |V_1| |V_3| (G_{13} \cos \delta_{13} + B_{13} \sin \delta_{13})$$

$$= 1.0 \times 1.1 \times 10 \times \sin (0.0 - 15^\circ) + 1.0 \times 0.9 \times 5 \times \sin (0.0 - (- 14.99^\circ))$$

$$= - 1.6827 \text{ pu}$$

$$Q_1 = - B_{11} |V_1|^2 + |V_1| |V_2| (G_{12} \sin \delta_{12} - B_{12} \cos \delta_{12}) + |V_1| |V_3| (G_{13} \sin \delta_{13} - B_{13} \cos \delta_{13})$$

$$= 15 \times 1^2 + 1.0 \times 1.1 \times - 10 \times \cos (- 15^\circ) + 1.0 \times 0.9 \times - 5 \times \cos (14.99^\circ)$$

$$= 0.028 \text{ pu}$$

(b) Decoupled method: The equation to be solved is

$$\begin{bmatrix} \Delta P_2 \\ \Delta P_3 \\ \Delta Q_3 \end{bmatrix} = \begin{bmatrix} H_{22} & H_{23} & 0 \\ H_{32} & H_{33} & 0 \\ 0 & 0 & L_{33} \end{bmatrix} \begin{bmatrix} \Delta \delta_2 \\ \Delta \delta_3 \\ \dfrac{\Delta |V_3|}{|V_3|} \end{bmatrix}$$

(i) *In this strategy we solve for the variables simultaneously*. Therefore,

$$\begin{bmatrix} \Delta \delta_2 \\ \Delta \delta_3 \\ \dfrac{\Delta |V_3|}{|V_3|} \end{bmatrix} = [J]^{-1} \begin{bmatrix} \Delta P_2 \\ \Delta P_3 \\ \Delta Q_3 \end{bmatrix}$$

The values of the changes in the variables are as follows:

1st Iteration

$$\begin{bmatrix} \Delta \delta_2 \\ \Delta \delta_3 \\ \dfrac{\Delta |V_3|}{|V_3|} \end{bmatrix} = \begin{bmatrix} 0.2508 \\ -0.2152 \\ -0.0721 \end{bmatrix}$$

These values are used to update δ_2, δ_3 and $|V_3|$. Elements of the Jacobian and the power residues are recalculated using the updated values. The variations in subsequent iterations are given below.

2nd Iteration

$$\begin{bmatrix} \Delta\delta_2 \\ \Delta\delta_3 \\ \dfrac{\Delta|V_3|}{|V_3|} \end{bmatrix} = \begin{bmatrix} 0.0056 \\ -0.0088 \\ -0.0693 \end{bmatrix}$$

3rd Iteration

$$\begin{bmatrix} \Delta\delta_2 \\ \Delta\delta_3 \\ \dfrac{\Delta|V_3|}{|V_3|} \end{bmatrix} = \begin{bmatrix} 0.0038 \\ -0.0260 \\ -0.0110 \end{bmatrix}$$

4th Iteration

$$\begin{bmatrix} \Delta\delta_2 \\ \Delta\delta_3 \\ \dfrac{\Delta|V_3|}{|V_3|} \end{bmatrix} = \begin{bmatrix} 0.0006 \\ -0.0044 \\ -0.0126 \end{bmatrix}$$

5th Iteration

$$\begin{bmatrix} \Delta\delta_2 \\ \Delta\delta_3 \\ \dfrac{\Delta|V_3|}{|V_3|} \end{bmatrix} = \begin{bmatrix} 0.0007 \\ -0.0049 \\ -0.0024 \end{bmatrix}$$

6th Iteration

$$\begin{bmatrix} \Delta\delta_2 \\ \Delta\delta_3 \\ \dfrac{\Delta|V_3|}{|V_3|} \end{bmatrix} = \begin{bmatrix} 0.0001 \\ -0.0009 \\ -0.0025 \end{bmatrix}$$

Final voltages after 6 iterations

$V_1 = 1\angle 0°\quad V_2 = 1.1\angle 14.98°\quad V_3 = 0.9011\angle -14.85°$

(ii) *Here the equations are solved alternately in a single iteration.*

$$\begin{bmatrix}\Delta P_2\\ \Delta P_3\end{bmatrix} = \begin{bmatrix}H_{22} & H_{23}\\ H_{32} & H_{33}\end{bmatrix}\begin{bmatrix}\Delta\delta_2\\ \Delta\delta_3\end{bmatrix}$$

The values of δ_2 and δ_2 are updated. Using these values, ΔQ_3 and L_{33} are computed. Then

$$[\Delta Q_3] = [L_{33}]\left[\frac{\Delta|V_3|}{|V_3|}\right]$$

The above equation is solved for $\Delta|V_3|$. The results obtained are as follows:

1st Iteration

$$J_1 = \begin{bmatrix}H_{22} & H_{23}\\ H_{32} & H_{33}\end{bmatrix} = \begin{bmatrix}\Delta\delta_2\\ \Delta\delta_3\end{bmatrix};\ J_2 = [L_{33}]$$

$P_{2,cal} = 0\ ;\ P_{3,cal} = 0\ ;\ Q_{2,cal} = 1.65\ ;\ Q_{3,cal} = -0.5$

$$J_1 = \begin{bmatrix}16.5 & -5.5\\ -5.5 & 10.5\end{bmatrix}$$

$$= \begin{bmatrix}\Delta\delta_2\\ \Delta\delta_3\end{bmatrix} = [J_1]^{-1} = \begin{bmatrix}\Delta P_2\\ \Delta P_3\end{bmatrix} = \begin{bmatrix}0.2508\\ -0.2152\end{bmatrix}$$

We now recalculate $Q_{3,cal}$ after updating δ_2 and δ_3

$$\delta_2 = \delta_2 + \Delta\delta_2 = 0.2508$$

$$\delta_3 = \delta_3 + \Delta\delta = -0.2152$$

$$Q_{3,cal} = 0.2018$$

$$J_2 = 10.2018$$

$$= \frac{\Delta|V_3|}{|V_3|} = [J_2^{-1}]\Delta Q_3$$

$$\Delta|V_3| = -0.0721$$

$$|V_3| = V_3 + \Delta V_3 = 0.9279$$

2nd Iteration

$$J_1 = \begin{bmatrix}15.2151 & -4.5592\\ -4.5592 & 9.0915\end{bmatrix};\ J_2 = [8.2691]$$

$$= \begin{bmatrix} \Delta\delta_2 \\ \Delta\delta_3 \\ \dfrac{\Delta|V_3|}{|V_3|} \end{bmatrix} = \begin{bmatrix} 0.0093 \\ -0.0344 \\ -0.0234 \end{bmatrix}$$

$V_1 = 1.0 \angle 0°$; $V_2 = 1.1 \angle 14.9036°$; $V_3 = 0.9062 \angle -14.30°$

3rd Iteration

$$J_1 = \begin{bmatrix} 14.9803 & -4.3504 \\ -4.3504 & 8.7409 \end{bmatrix}; J_2 = [7.7195]$$

$$= \begin{bmatrix} \Delta\delta_2 \\ \Delta\delta_3 \\ \dfrac{\Delta|V_3|}{|V_3|} \end{bmatrix} = \begin{bmatrix} 0.0013 \\ -0.0094 \\ -0.0054 \end{bmatrix}$$

$$V_1 = 1.0 \angle 0°;\ V_2 = 1.1 \angle 14.98°;\ V_3 = 0.9013 \angle -14.84°$$

4th Iteration

$$J_1 = \begin{bmatrix} 14.9269 & -4.3007 \\ -4.3007 & 8.6568 \end{bmatrix}; J_2 = [7.598]$$

$$\begin{bmatrix} \Delta\delta_2 \\ \Delta\delta_3 \\ \dfrac{\Delta|V_3|}{|V_3|} \end{bmatrix} = \begin{bmatrix} 0.0003 \\ -0.0022 \\ -0.0012 \end{bmatrix}$$

$V_1 = 1.0 \angle 0°$; $V_2 = 1.1 \angle 14.99°$; $V_3 = 0.9002 \angle -14.96°$

5th Iteration

$$J_1 = \begin{bmatrix} 14.9151 & -4.2897 \\ -4.2897 & 8.6383 \end{bmatrix}; J_2 = [7.5723]$$

$$\begin{bmatrix} \Delta\delta_2 \\ \Delta\delta_3 \\ \dfrac{\Delta|V_3|}{|V_3|} \end{bmatrix} = \begin{bmatrix} 0.0605 \times 10^{-3} \\ -0.4569 \times 10^{-3} \\ -2.40 \times 10^{-4} \end{bmatrix}$$

$V_1 = 1.0 \angle 0°$; $V_2 = 1.1 \angle 15°$; $V_3 = 0.9 \angle -15°$

(c) Fast Decoupled Method: The equations to be solved are

$$\begin{bmatrix} \dfrac{\Delta P_2}{|V_2|} \\ \dfrac{\Delta P_3}{|V_3|} \end{bmatrix} = \begin{bmatrix} -B_{22} & -B_{23} \\ -B_{32} & -B_{33} \end{bmatrix} \begin{bmatrix} \Delta\delta_2 \\ \Delta\delta_3 \end{bmatrix}$$

$$\frac{\Delta Q_3}{|V_3|} = [-B_{33}] \left[\frac{\Delta|V_3|}{|V_3|} \right]$$

The Jacobian is a constant and does not require updation. The power residues are computed at every iteration. The equations are solved alternately.

1st Iteration

$$\begin{bmatrix} \Delta\delta_2 \\ \Delta\delta_3 \\ \dfrac{\Delta|V_3|}{|V_3|} \end{bmatrix} = \begin{bmatrix} 0.2415 \\ -0.2432 \\ -0.0814 \end{bmatrix}$$

2nd Iteration

$$\begin{bmatrix} \Delta\delta_2 \\ \Delta\delta_3 \\ \dfrac{\Delta|V_3|}{|V_3|} \end{bmatrix} = \begin{bmatrix} 0.0167 \\ -0.0112 \\ -0.0134 \end{bmatrix}$$

3rd Iteration

$$\begin{bmatrix} \Delta\delta_2 \\ \Delta\delta_3 \\ \dfrac{\Delta|V_3|}{|V_3|} \end{bmatrix} = \begin{bmatrix} 0.0026 \\ -0.0048 \\ -0.0045 \end{bmatrix}$$

4th Iteration

$$\begin{bmatrix} \Delta\delta_2 \\ \Delta\delta_3 \\ \dfrac{\Delta|V_3|}{|V_3|} \end{bmatrix} = \begin{bmatrix} 0.0007 \\ -0.0017 \\ -0.0015 \end{bmatrix}$$

5th Iteration

$$\begin{bmatrix} \Delta\delta_2 \\ \Delta\delta_3 \\ \dfrac{\Delta|V_3|}{|V_3|} \end{bmatrix} = \begin{bmatrix} 0.2213 \times 10^{-3} \\ -0.6024 \times 10^{-3} \\ -5.49 \times 10^{-4} \end{bmatrix}$$

The voltages after 5 iterations are

$$V_1 = 1\angle 0°$$
$$V_2 = 1.1\angle 14.99°$$
$$V_3 = 0.9007\angle -14.98°$$

Example 7.8: Solve example 7.7 using N R method in rectangular coordinates.

Solution: The network equation to be solved is given by

$$\begin{bmatrix}\Delta P_2\\ \Delta P_3\\ \Delta Q_3\\ \Delta|V_2|^2\end{bmatrix} = \begin{bmatrix}\frac{\partial P_2}{\partial e_2} & \frac{\partial P_2}{\partial e_3} & \frac{\partial P_2}{\partial f_2} & \frac{\partial P_2}{\partial f_3}\\ \frac{\partial P_3}{\partial e_2} & \frac{\partial P_3}{\partial e_3} & \frac{\partial P_3}{\partial f_2} & \frac{\partial P_3}{\partial f_3}\\ \frac{\partial Q_3}{\partial e_2} & \frac{\partial Q_3}{\partial e_3} & \frac{\partial Q_3}{\partial f_2} & \frac{\partial Q_3}{\partial f_3}\\ \frac{\partial |V_2|^2}{\partial e_2} & \frac{\partial |V_2|^2}{\partial e_3} & \frac{\partial |V_2|^2}{\partial f_2} & \frac{\partial |V_2|^2}{\partial f_3}\end{bmatrix}\begin{bmatrix}\Delta e_2\\ \Delta e_3\\ \Delta f_2\\ \Delta f_3\end{bmatrix}$$

Let initial voltages be taken as

$$V_1 = e_1 + jf_1 = 1.0 + j0.0$$
$$V_2 = e_2 + jf_2 = 1.1 + j0.0$$
$$V_3 = e_3 + jf_3 = 1.0 + j0.0$$

From example 7.7,

$$\Delta P_2 = 5.3217$$
$$\Delta P_3 = -3.6392$$
$$\Delta Q_3 = -0.0339$$
$$|V_2|^2 = e_2^{\,2} + f_2^{\,2} = 1.1$$
$$\Delta|V_2|^2 = (1.1)^2 - (1.1)^2 = 0.0$$
$$c_2 = G_{22}e_2 - B_{22}f_2 + (G_{21}e_1 - B_{21}f_1) + (G_{23}e_3 - B_{23}f_3) = 0.0$$
$$d_2 = G_{22}f_2 + B_{22}e_2 + (G_{21}f_1 + B_{21}e_1) + (G_{23}f_3 + B_{23}e_3)$$
$$= (-15\times 1.1) + (10\times 1.0) + (5\times 1.0) = -1.5$$
$$c_3 = G_{33}e_3 - B_{33}f_3 + (G_{31}e_1 - B_{31}f_1) + (G_{32}e_3 - B_{32}f_2) = 0.0$$
$$d_3 = G_{33}f_3 + B_{33}e_3 + (G_{31}f_1 + B_{31}e_1) + (G_{32}f_2 + B_{32}e_2)$$
$$= (-10\times 1.0) + (5.0\times 1.0) + (5\times 1.1) = 0.5$$

The elements are computed below:

$$\frac{\partial P_2}{\partial e_2} = G_{22}e_2 + B_{22}f_2 + c_2 = 0.0$$

$$\frac{\partial P_2}{\partial e_3} = G_{23}e_2 + B_{23}f_2 = 0.0$$

$$\frac{\partial P_2}{\partial f_2} = -B_{22}e_2 + G_{22}f_2 + d_2 = 15 \times 1.1 - 1.5 = 15.0$$

$$\frac{\partial P_2}{\partial f_3} = G_{23}f_2 - B_{23}e_2 = -5 \times 1.1 = -5.5$$

$$\frac{\partial P_3}{\partial e_2} = G_{32}e_3 + B_{32}f_3 = 0.0$$

$$\frac{\partial P_3}{\partial e_3} = G_{33}e_3 + B_{33}f_3 + c_3 = 0.0$$

$$\frac{\partial P_3}{\partial f_2} = G_{32}f_3 - B_{32}e_3 = -5 \times 1.0 = -5.0$$

$$\frac{\partial P_3}{\partial f_3} = -B_{33}e_3 + G_{33}f_3 + d_3 = 10 \times 1.0 + 0.5 = 10.5$$

$$\frac{\partial Q_3}{\partial e_2} = G_{32}f_3 - B_{32}e_3 = -5 \times 1.0 = -5.0$$

$$\frac{\partial Q_3}{\partial e_3} = G_{33}f_3 - B_{33}e_3 - d_3 = 10 \times 1.0 - 0.5 = 9.5$$

$$\frac{\partial Q_3}{\partial f_2} = -(B_{32}f_3 + G_{32}e_3) = 0.0$$

$$\frac{\partial Q_3}{\partial f_3} = -G_{33}e_3 - B_{33}f_3 + c_3 = 0.0$$

$$\frac{\partial |V_2|^2}{\partial e_2} = 2_{e2} = 2 \times 1.1 = 2.2$$

$$\frac{\partial |V_2|^2}{\partial e_3} = 0.0$$

$$\frac{\partial |V_2|^2}{\partial f_2} = 2f_2 = 0.0$$

$$\frac{\partial |V_2|^2}{\partial f_3} = 0.0$$

$$\begin{bmatrix} 5.2317 \\ -3.6392 \\ -0.0339 \\ 0.0 \end{bmatrix} = \begin{bmatrix} 0.0 & 0.0 & 15.0 & -5.5 \\ 0.0 & 0.0 & -5.0 & 10.5 \\ -5.0 & 9.5 & 0.0 & 0.0 \\ 2.2 & 0.0 & 0.0 & 0.0 \end{bmatrix} \begin{bmatrix} \Delta e_2 \\ \Delta e_3 \\ \Delta f_2 \\ \Delta f_3 \end{bmatrix}$$

$$\begin{bmatrix} \Delta e_2 \\ \Delta e_3 \\ \Delta f_2 \\ \Delta f_3 \end{bmatrix} = \begin{bmatrix} 0.0 & 0.0 & 15.0 & -5.5 \\ 0.0 & 0.0 & -5.0 & 10.5 \\ -5.0 & 9.5 & 0.0 & 0.0 \\ 2.2 & 0.0 & 0.0 & 0.0 \end{bmatrix}^{-1} \begin{bmatrix} 5.3217 \\ -3.6392 \\ -0.0339 \\ 0.0 \end{bmatrix} = \begin{bmatrix} 0.0 \\ -0.0036 \\ 0.2759 \\ -0.2152 \end{bmatrix}$$

The solution is therefore

$$\Delta e_2 = 0.0;\ e_2 = 1.1 + 0.0 = 1.1$$

$$\Delta e_3 = -0.0036;\ e_3 = 1.0 - 0.0036 = 0.9964$$

$$\Delta f_2 = 0.2759;\ f_2 = 0.0 + 0.2759 = 0.2759$$

$$\Delta f_3 = -0.2152;\ f_3 = 0.0 - 0.2152 = -0.2152$$

$$V_2 = e_2 + jf_2 = 1.1 + j0.2759$$

$$\delta_2 = \tan^{-1}\frac{0.2759}{1.1} = 14.08°$$

Since, $|V_{2,sp}| = 1.1$, and bus 2 is a *PV* bus e_2 and f_2 are estimated as follows.

$$e_2 = |V_{2,sp}| \cos \delta_2 = 1.06695$$

$$f_2 = |V_{2,sp}| \sin \delta_2 = 0.2676$$

$$V_3 = e_3 + jf_3 = 0.9964 - j0.2152 = 1.01937 \angle -12.19°$$

Hence, at end of first iteration voltages are

$$V_1 = 1 \angle 0° = 1 + j0.0$$

$$V_2 = 1.1 \angle 14.08° = 1.06695 + j02676$$

$$V_3 = 1.01937 \angle -12.19° = 0.9964 - j0.2152$$

Example 7.9: The following date pertains to a three bus system

Line Data

Line	X(Pu)
1 – 3	0.2
1 – 2	0.4
2 – 3	0.25

Bus Data

Bus No.	P_G (Po)	Q_G (Po)	P_L (Po)	Q_G (Po)	V_{SP} (Po)
1	–	–	–	–	1.04
2	–	–	1.0	0.8	–
3	1.0	–	–	–	1.005

Obtain the voltages at the end of the first iteration by Gauss-Siedel method.

Solution

$$Y_{BUS} = \begin{bmatrix} -j\,7.5 & j\,2.5 & j\,5.0 \\ j\,2.5 & -j\,6.5 & j\,4.0 \\ j\,5.0 & j\,4.0 & -j\,9.0 \end{bmatrix}$$

Here bus 1 is a slack bus, bus 2 a load bus and bus 3 *PV* bus

Initial Voltages

Let $V_1^0 = 1.04 + j\,0.0$

$$V_2^0 = 1.0 + j\,0.0$$

$$V_3^0 = 1.005 + j\,0.0$$

$$P_2 = -1.0\,;\ P_3 = +1.0\,;\ Q_2 = -0.8$$

Bus 2 is a *PQ* bus

$$V_2^1 = \frac{1}{Y_{22}}\left[\frac{P_2 - jQ_2}{(V_2^0)^*} - Y_{21}\,V_1 - Y_{23}\,V_3^0\right]$$

$$= \frac{1}{-j6.5} = \left[\frac{-1 + j0.8}{1.0 - j0.0} - (j2.5)\,(1.04 + j0.0) - (j4.0)\,(1.005 + j0.0)\right]$$

$$= 0.8954 - j\,0.\,1538 = 0.9085 \;\angle -9.74^\circ$$

Bus 3 is a *PV* bus. So we first compute Q_3 Note that all $G_{ij} = 0$

$$Q_3 = -B_{33}\,|V_3|^2 + |V_3|\,|V_1|\,(-B_{31}\cos\delta_{31}) + |V_3|\,|V_2|\,(-B_{32}\cos\delta_{32})$$

$$= 9 \times 1.005^2 + (1.005 \times 1.04 \times -5) + (1.005 \times 0.9085 \times 4 \times \cos 9\,.74^\circ)$$

$$= 0.2646 \text{ pu}$$

$$V_3^1 = \frac{1}{Y_{33}}\left[\frac{P_3 - jQ_3}{(V_3^0)^*} - Y_{31}\,V_1 - Y_{32}\,V_2^1\right]$$

$$= \frac{1}{-j9.0}\left[\frac{1.0 - j0.2646}{1.005} \begin{matrix} -(j5)\,(1.04 + j0.0) \\ -(j4.0)\,(0.8954 - j0.1538) \end{matrix}\right]$$

$$= 1.005 + \text{j}\,0.0422 = 1.0058 \;\angle 2.4^\circ$$

Thus at end of first iteration we get

$$V_1 = 1.04 + j\,0.0$$

$$V_2 = 0.8954 - j\,0.1538 = 0.9085 \;\angle -9.74^\circ$$

$$V_3 = 1.005 + j\,0.0422 = 1.0058 \;\angle 2.4^\circ$$

Since bus 3 is a PV bus, $V_3 = 1.005\angle 2.4^\circ$

Example 7.10 Solve example 7.9 using Newton. Rapson method for one iteration

Solution. Note that all $G_{ij} = 0$.we next calculate the real and reactive powers.

$$P_2 = |V_2|\,|V_1|\,B_{21} \sin \delta_{21} + |V_2|\,|V_3|\,B_{23} \sin \delta_{23}$$
$$= 0$$

(since $\delta_1 = \delta_2 = \delta_3 = 0°$ at beginning of first iteration)

$$P_3 = |V_3|\,|V_1|\,B_{31} \sin \delta_{31} + |V_3|\,|V_2|\,B_{32} \sin \delta_{32}$$
$$= 0.$$

$$Q_2 = -B_{22}\,|V_2|^2 + |V_2|\,|V_1| \times -B_{21} \cos \delta_{21} + |V_2|\,|V_3| \times -B_{23} \cos \delta_{23}$$
$$= 6.5 \times 1^2 + 1 \times 1\ .04 \times -2.5 \times \cos(0°) + 1.0 \times 1.005 \times -4.0 \times \cos(0°)$$
$$= -0.12 \text{ pu}$$

$$\Delta P_2 = P_{2,sp} - P_2 = -1.0 \text{ pu}$$
$$\Delta P_3 = P_{3,sp} - P_3 = 1.0 \text{ pu}$$
$$\Delta Q_2 = Q_{2,sp} - Q_2 = -0.8 + 0.12 = -0.68 \text{ pu.}$$

$$\begin{bmatrix} \Delta P_2 \\ \Delta P_3 \\ \Delta Q_2 \end{bmatrix} = \begin{bmatrix} \dfrac{\partial P_2}{\partial \delta_2} & \dfrac{\partial P_2}{\partial \delta_3} & \dfrac{\partial P_2}{\partial |V_2|}|V_2| \\ \dfrac{\partial P_3}{\partial \delta_2} & \dfrac{\partial P_3}{\partial \delta_3} & \dfrac{\partial P_3}{\partial |V_2|}|V_2| \\ \dfrac{\partial Q_2}{\partial \delta_2} & \dfrac{\partial Q_2}{\partial \delta_3} & \dfrac{\partial Q_2}{\partial |V_2|}|V_2| \end{bmatrix} \begin{bmatrix} \Delta \delta_2 \\ \Delta \delta_3 \\ \dfrac{\Delta |V_2|}{|V_2|} \end{bmatrix}$$

$$= \begin{bmatrix} H_{22} & H_{23} & N_{23} \\ H_{32} & H_{33} & N_{33} \\ M_{22} & M_{23} & L_{22} \end{bmatrix} \begin{bmatrix} \Delta \delta_2 \\ \Delta \delta_3 \\ \dfrac{\Delta |V_2|}{|V_2|} \end{bmatrix}$$

Jacobian elements

$$H_{22} = -Q_2 - B_{22}\,|V_2|^2$$
$$= 0.12 - (-6.5 \times 1.0^2)$$
$$= 6.62$$

$$H_{23} = |V_2|\,|V_3|\,(-B_{23} \cos \delta_{23})$$
$$= 1.0 \times 1.005\,(-4 \times \cos 0^0)$$
$$= -4.02$$

$$H_{32} = |V_3|\,|V_2|\,(-B_{32} \cos \delta_{32})$$
$$= -4\ .02$$

$$H_{33} = -Q_3 - B_{33}\,|V_3|^2$$

$$Q_3 = -B_{33}\,|V_3|^2 + |V_3|\,|V_1|\,(-B_{31}\cos\delta_{31}) + |V_3|\,|V_2|\,(-B_{32}\cos\delta_{32})$$

$$= 9 \times 1.005^2 + 1.005 \times 1.04\,(-5 \times \cos 0^0) + 1.005 \times 1.0\,(-4 \times \cos 0^0)$$

$$= -0.156$$

$$H_{33} = 0.156 + 9 \times 1.005^2 = 9.246$$

$$N_{23} = |V_2|\,|V_3|\,(B_{23}\sin\delta_{23}) = 0.0$$

$$N_{33} = P_3 + G_{33}\,|V_3|^2 = 0.0$$

$$M_{22} = P_2 - G_{22}\,|V_2|^2 = 0.0$$

$$M_{23} = -N_{23} = 0.0$$

$$L_{22} = Q_2 - B_{22}\,|V_2|^2$$

$$= -0.12 - (-6.5 \times 1^2) = 6.38$$

$$\therefore \quad \begin{bmatrix} \Delta P_2 \\ \Delta P_3 \\ \Delta Q_3 \end{bmatrix} = \begin{bmatrix} 6.62 & -4.02 & 0.0 \\ -4.02 & 9.246 & 0.0 \\ 0.0 & 0.0 & 6.38 \end{bmatrix} \begin{bmatrix} \Delta\delta_2 \\ \Delta\delta_3 \\ \dfrac{\Delta|V_2|}{|V_2|} \end{bmatrix}$$

$$\begin{bmatrix} \Delta\delta_2 \\ \Delta\delta_3 \\ \dfrac{\Delta|V_2|}{|V_2|} \end{bmatrix} = \left[J^{-1}\right] \begin{bmatrix} -1.0 \\ 1.0 \\ -0.68 \end{bmatrix} = \begin{bmatrix} -0.1160 \\ 0.0577 \\ -0.1066 \end{bmatrix}$$

Example 7.11 We next consider an example of a system with resistances as shown in Fig. 7.9.

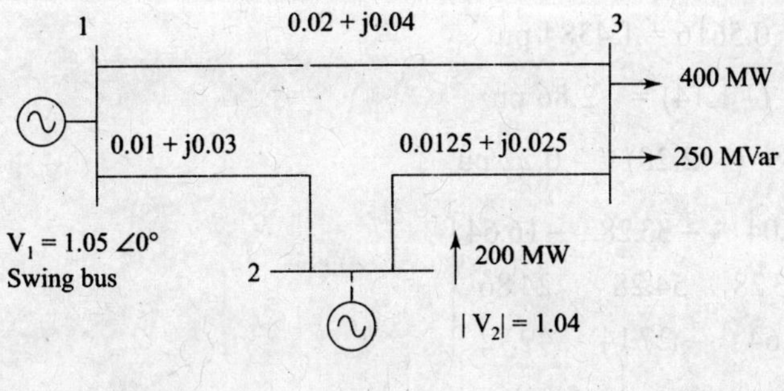

Fig. 7.9

Compute the voltages using *N-R* method.

Solution

***N–R* Method**

The bus admittance matrix is given by

$$Y_{bus} = \begin{bmatrix} 20-j50 & -10+j30 & -10+j20 \\ -10+j30 & 26-j62 & -16+j32 \\ -10+j20 & -16+j32 & 26-j52 \end{bmatrix}$$

$$P_{2,cal} = G_{22}|V_2|^2 + |V_2||V_1|(G_{21}\cos\delta_{21} + B_{21}\sin\delta_{21}) + |V_2||V_3|(G_{23}\cos\delta_{23} + B_{23}\sin\delta_{23})$$

$$P_{3,cal} = G_{33}|V_3|^2 + |V_3||V_1|(G_{31}\cos\delta_{31} + B_{31}\sin\delta_{31}) + |V_3||V_2|(G_{32}\cos\delta_{32} + B_{32}\sin\delta_{32})$$

$$Q_{2,cal} = -B_{22}|V_2|^2 + |V_2||V_1|(G_{21}\sin\delta_{21} - B_{21}\cos\delta_{21}) + |V_2||V_3|(G_{23}\sin\delta_{23} - B_{23}\cos\delta_{23})$$

$$Q_{3,cal} = -B_{33}|V_3|^2 + |V_3||V_1|(G_{31}\sin\delta_{31} - B_{31}\cos\delta_{31}) + |V_3||V_2|(G_{32}\sin\delta_{32} - B_{32}\cos\delta_{32})$$

$$\Delta P_2 = P_{2,sp} - P_{2,cal}$$

$$\Delta P_3 = P_{3,sp} - P_{3,cal}$$

$$\Delta Q_3 = Q_{3,sp} - Q_{3,cal}$$

1st Iteration

$$V_1 = 1.05\angle 0^\circ;\ V_2 = 1.04\angle 0^\circ;\ V_3 = 1.0\angle 0^\circ$$

choose base of 100 MVA

$$P_{2,sp} = \frac{200}{100} = 2\text{pu.}$$

$$P_{3,sp} = -\frac{(400+j250)}{100} = -4 - j\,2.5 \text{ pu}$$

$$P_{2,cal} = 0.5616 \text{ pu};\ P_{3,cal} = -1.14 \text{ pu};$$

$$Q_{2,cal} = 1.0192 \text{ pu};\ Q_{3,cal} = -2.28 \text{ pu}$$

$$\Delta P_2 = 2.0 - 0.5616 = 1.4384 \text{ pu}$$

$$\Delta P_3 = -4 - (-1.14) = -2.86 \text{ pu}$$

$$\Delta Q_3 = -2.5 - (-2.28) = -0.22 \text{ pu}$$

$$J = \begin{bmatrix} 66.04 & -33.28 & -16.64 \\ -33.28 & 54.28 & 24.86 \\ 16.64 & -27.14 & 49.72 \end{bmatrix}$$

$$\Delta\delta_2 = -0.0077;\ \delta_2 = \delta_2 + \Delta\delta_2 = -0.0077 \text{ rad} = -0.44220^\circ$$

$$\Delta\delta_3 = -0.0453;\ \delta_3 = \delta_3 + \Delta\delta_3 = -0.0453 \text{ rad} = -2.590^\circ$$

$$\frac{\Delta|V_3|}{|V_3|} = -0.0265 \; ; |V_3| = 1 - 0.0265 = 0.9735$$

At the end of 1st iteration

$$V_1 = 1.05 \angle 0^\circ ; V_2 = 1.04 \angle -0.4422^\circ ; V_3 = 0.9735 \angle -2{\cdot}59^\circ$$

2nd Iteration

$$J = \begin{bmatrix} 65.6564 & -32.9816 & -14.9708 \\ -31.7656 & 51.7247 & 20.7370 \\ 14.9708 & -28.5386 & 46.8265 \end{bmatrix}$$

$$V_1 = 1.05 \angle 0^\circ ; V_2 = 1.04 \angle -0.4988^\circ ; V_3 = 0.9716 \angle -2.69^\circ$$

3rd Iteration

$$J = \begin{bmatrix} 65.596 & -32.9323 & -14.9169 \\ -31.6929 & 51.5950 & 20.5466 \\ 14.9169 & -28.5462 & 46.5905 \end{bmatrix}$$

$$V_1 = 1.05 \angle 0^\circ$$

$$V_2 = 1.04 \angle -0.4988^\circ$$

$$V_3 = 0.9717 \angle -2.69^\circ$$

Since the changes are insignificant we can stop. thus the solution converges in three iterations. The other parameters are,

$$P_1 = 2.1842 \text{ pu}$$

$$Q_1 = 1.4085 \text{ pu}$$

$$Q_2 = 1.4532 \text{ pu}$$

Example 7.12

Consider a three-bus system with following data.

Bus No.	P_G (pu)	Q_G (pu)	P_L (pu)	Q_L (pu)	V_{sp} (pu)
1	–	–	–	–	$1\angle 0^\circ$
2	0.6661	–	–	–	1.05
3	–	–	2.8653	1.2244	–

There are three lines connected between 1 – 2, 2 – 3 and 1–3. The line impedances are given by $Z_{\text{line}} = j0.1$ pu and half line charging admittance is given by $y_c/2 = j\ 0.01$ pu. Determine the voltages using $N - R$ method.

Solution

We form Y_{bus}

$$Y_{\text{bus}} = j\begin{bmatrix} -19.98 & 10 & 10 \\ 10 & -19.98 & 10 \\ 10 & 10 & -19.98 \end{bmatrix}$$

The Jacobian is given by

$$J = \begin{bmatrix} \dfrac{\partial P_2}{\partial \delta_2} & \dfrac{\partial P_2}{\partial \delta_3} & \dfrac{\partial P_2}{\partial |V_3|} \\ \dfrac{\partial P_3}{\partial \delta_2} & \dfrac{\partial P_3}{\partial Q_3} & \dfrac{\partial P_3}{\partial |V_3|} \\ \dfrac{\partial Q_3}{\partial \delta_2} & \dfrac{\partial Q_3}{\partial \delta_3} & \dfrac{\partial Q_3}{\partial |V_3|} \end{bmatrix}$$

$$\begin{bmatrix} \Delta P_2 \\ \Delta P_3 \\ \Delta Q_3 \end{bmatrix} = [J] \begin{bmatrix} \Delta \delta_2 \\ \Delta \delta_3 \\ \Delta |V_3| \end{bmatrix}$$

1st Iteration

We start with $V_1 = 1 \angle 0^\circ$; $V_2 = 1.05 \angle 0^\circ$; $V_3 = 1 \angle 0^\circ$

$$J = \begin{bmatrix} 21 & -10.5 & 0 \\ -10.5 & 20.5 & 0 \\ 0 & 0 & 19.46 \end{bmatrix}; \quad \begin{bmatrix} \Delta P_2 \\ \Delta P_3 \\ \Delta Q_3 \end{bmatrix} = \begin{bmatrix} 0.6661 \\ -2.8653 \\ -0.7044 \end{bmatrix}$$

$$\begin{bmatrix} \Delta \delta_2 \\ \Delta \delta_3 \\ \Delta |V_3| \end{bmatrix} = \begin{bmatrix} -0.0513\,rad \\ -0.166\,rad \\ -0.0362 \end{bmatrix} = \begin{bmatrix} -2.939^\circ \\ -9.511^\circ \\ -0.0362 \end{bmatrix}$$

2nd Iteration

$$V_1 = 1 \angle 0^\circ;\; V_2 = 1.05;\; \angle -2.939^\circ;\; V_3 = 0.9638 \angle -9.511^\circ$$

$$J = \begin{bmatrix} 20.5396 & -10.0534 & 1.2017 \\ -10.0534 & 19.5589 & -2.8541 \\ 1.1582 & -2.7508 & 18.2199 \end{bmatrix}$$

$$\begin{bmatrix} \Delta P_2 \\ \Delta P_3 \\ \Delta q_3 \end{bmatrix} = \begin{bmatrix} 0.0463 \\ -0.1145 \\ -0.2251 \end{bmatrix}$$

$$\begin{bmatrix} \delta_2 \\ \delta_3 \\ |V_3| \end{bmatrix} = \begin{bmatrix} -3.0^\circ \\ -9.99^\circ \\ 0.9502 \end{bmatrix}$$

3rd Iteration

$$= \begin{bmatrix} \Delta P_2 \\ \Delta P_3 \\ \Delta Q_3 \end{bmatrix} = \begin{bmatrix} 0.0019 \\ -0.0023 \\ -0.0031 \end{bmatrix}$$

Since the power mismatches are very small we can stop here and take the values of the previous iteration as the converged solution.

$$V_1 = 1.0 \angle 0^\circ$$

$$V_2 = 1.05 \angle -3.0^\circ$$

$$V_3 = 0.9502 \angle -9.99^\circ$$

$$Q_{G2} = -1.6395 \text{ pu}$$

$$P_{G1} = 2.1987 \text{ (Slack bus power)}$$

$$Q_{G1} = 0.1365$$

The negative sign of P_1 indicates that there is a net real power demand at the slack bus. Since, the system is lossless it can be seen that $P_1 + P_2 + P_3 = 0$.

The reactive power generation at bus 2, the voltage controlled bus, is 3.2295 pu.

7.10 DC LOAD FLOW

The Newton Raphson method gives accurate results for the power flow equations. In the decoupled method $P - |V|$ decoupling and $Q - \delta$ decoupling is assumed. We can make further assumptions

- Neglect shunt admittances
- Neglect all line resistances
- Assume $(\delta_i - \delta_j) \cong 0 \therefore \cos(\delta_i - \delta_j) = 1$
- Drop $Q - |V|$ equations
- Assume $|V| = 1.0$ pu at all bases

With this (7.89a) reduces to

$$[\Delta P] = [B'] [\Delta\delta] \tag{7.90}$$

In (7.90) we consider the angles of all buses except the swing bus. (7.90) is called the DC load flow.

The DC load flow gives no indication of the voltage magnitudes. It is useful only to compute the active power flows in various lines. The active power flowing in each line is given by

$$P_{ik} = \frac{1}{x_{ik}} (\delta_i - \delta_k)$$

$$P_i = \sum p_k$$

Example 7.13: Consider a three-bus system with the following data.

$P_{G1} = 60$ MW; $P_{L2} = 85$ MW;

$Z_{12} = 0.05 + j0.2$ pu; $Z_{13} = 0.07 + j0.4$ pu; $Z_{23} = 0.1 + j0.25$ pu.

Bus 3 is the swing bus. Find the power flows in the lines by DC load flow.

Solution: Choose a base of 100 MVA ;

$P_{G2} = 0.6$; $P_{L2} = 0.85$ pu

We neglect the resistances $P_1 = 0.6$ pu; $P_2 = -0.85$ pu

$$B' = \begin{bmatrix} \frac{1}{0.2} + \frac{1}{0.4} & -\frac{1}{0.2} \\ -\frac{1}{0.2} & \frac{1}{0.2} + \frac{1}{0.25} \end{bmatrix}$$

Note these are the negative of the susceptances of the corresponding bus admittance matrix. B' is formed without considering the swing bus

$$\begin{bmatrix} \delta_1 \\ \delta_2 \end{bmatrix} = [B']^{-1} \begin{bmatrix} P_1 \\ P_2 \end{bmatrix}$$

$$= \begin{bmatrix} 0.2118 & 0.1176 \\ 0.1176 & 0.1765 \end{bmatrix} \begin{bmatrix} 0.6 \\ -0.85 \end{bmatrix}$$

$$= \begin{bmatrix} 0.0271 \\ -0.0794 \end{bmatrix}$$

$$P_{12} = \frac{1}{x_{12}} [\delta_1 - \delta_2] = \frac{1}{0.2} = [0.0271 - (-0.0794)]$$

$= 0.5325$ pu

$= 53.25$ MW

$$P_{13} = \frac{1}{x_{13}} [\delta_1 - \delta_3] = \frac{1}{0.4} = [0.0271 - 0]$$

$= 0.06775$ pu

$= 6.775$ MW

$$P_{23} = \frac{1}{x_{23}}[\delta_2 - \delta_3] = \frac{1}{0.25}[-0.0794 - 0]$$

$$= -0.31715 \text{ pu}$$

$$= -31.775 \text{ MW}$$

P_{23} being negative implies power flows from bus 3 to bus 2.

The slack bus power is given by

$$P_3 = P_{32} + P_{31} = 31.775 - 6.775$$

$$= 25 \text{ MW}$$

The slack bus supplies the deficit power

$$P_1 + P_3 = 85 \text{ MW (generation)}$$

$$P_2 = -85 \text{ MW (load)}$$

The *DC* load flow is not iterative and hence can give a quick solution of the power flows in various lines. It is used in contingency analysis

7.11 VOLTAGE CONTROL

In the classification of buses, certain buses are classified as *PV* buses. These are buses where voltage control is possible by location of suitable devices. Some of them are briefly discussed below.

7.11.1 Generator Excitation Control

At a bus where a generator is connected, the terminal voltage can be controlled by adjusting the generator excitation. Consider a generator modeled as a constant voltage source behind synchronous reactance, connected to the kth bus as shown in Fig. 7.10.

The power generation is given by

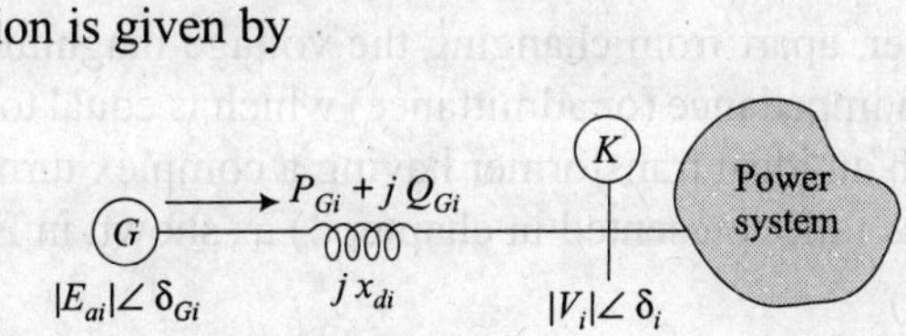

Fig. 7.10 Generator supplying a power system.

$$P_{Gi} = \frac{|E_{Gi}||V_i|}{X_{di}}\sin(\delta_{Gi} - \delta_i) \quad \text{...(7.90a)}$$

$$Q_{Gi} = \frac{|V_i|}{X_{di}}(|E_{Gi}| - |V_i|) \quad \text{...(7.90b)}$$

From given load flow solution, P_{Gi}, Q_{Gi} and $|V_1|\angle\delta_i$ are obtained. $|E_{Gi}|$ can be adjusted to give the required reactive power generation by adjusting the excitation and the input from the prime mover adjusted to obtain the required value of P_{Gi} plus losses, so that the load angle is maintained at $\delta_{Gi} - \delta_i$. If the required Q_{Gi} is beyond the generator's capacity, a reactive power generator like a capacitor or a synchronous condenser must be used. Voltages at the load buses can be obtained from the load flow solution. If the voltage at a load bus is less than the specified lower voltage limit, it means that the reactive power demand at that bus cannot be met, while holding the voltages within the limits. In such cases, it is necessary to install *VAR* generators at some load buses also. In such a case, the load bus is treated as a *PV* bus and the reactive generation at the bus obtained from the load flow solution gives the requisite reactive power injection at the bus.

In modern power system, static Var compensators (*SVC*) are used to maintain the voltage within the limits at a load bus. These devices use power electronic switches along with passive elements and can both generate and absorb reactive power. Other sophisticated Var compensators like *STACON*, (Static Condenser), *SPFC* (Series power Flow Controller), etc. are available which use advanced power electronic switches.

7.11.2 Tap Changing Under Load (TCUL)

Some transformers are provided with taps, which can be changed under load. The turns ratio is changed by a small amount Δa if

$$|V_k^{(r)}| - |V_{k,sp}| > \in$$

where, $|V_k^{(r)}|$ is the magnitude of the voltage at bus k in the rth iteration. If Δa is not very large, additional iterations need not be performed to obtain the voltage solution. The check for the voltage magnitude can be made in alternate iterations. The voltage magnitude tolerance can be taken to be 0.01 pu (for the TCUL) to give acceptable limits. Every time the tap of a transformer connected between buses i and k is changed, the elements Y_{ii}, Y_{kk}, Y_{ik} and Y_{ki} of the bus admittance matrix have to be updated, before continuing with the iterations.

7.11.3 Phase Shifting Transformers

The phase-shifting transformer, apart from changing the voltage magnitude also shifts the phase angle. It can be modeled as an impedance (or admittance) which is equal to the transformer impedance, connected in series with an ideal transformer having a complex turns ratio a (compare with modeling of off-nominal turns ratio, presented in chapter 3) as shown in Fig. 7.11. (Note a is a complex number)

Since, the transformer is ideal, complex power output is equal to complex power input.

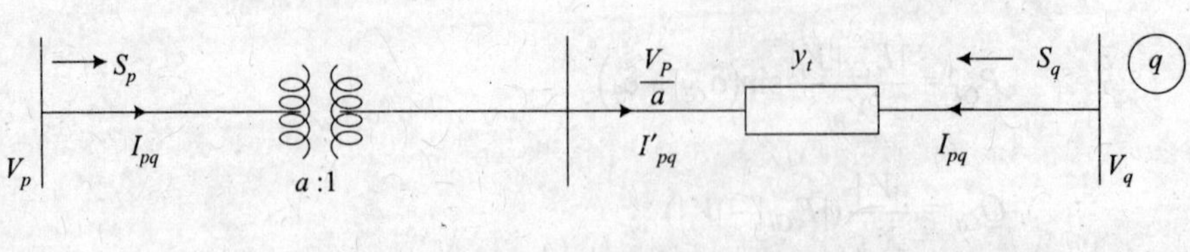

Fig. 7.11 Representation of phase shifting transformer.

Hence,

$$S_p = V_p I_{pq}^* = \frac{V_p}{a} I'^{*}_{pq}$$

From above equation we can write $I_{pq} = \dfrac{I'_{pq}}{a^*}$ ...(7.91)

For the line, $\left(\dfrac{V_p}{a} - V_q\right) y_t = I'_{pq}$

or, $$I_{pq} = \frac{I'_{pq}}{a^*} = \frac{V_p}{|a|^2} y_t - \frac{V_q y_t}{a^*} \qquad ...(7.92)$$

Also, $$I_{qp} = \left(V_q - \frac{V_p}{a}\right) y_t = \frac{-V_p y_t}{a} + V_q y_t \qquad ...(7.93)$$

Equations (7.92) and (7.93) cannot be represented by a single bilateral network. The admittance matrix representation is given as

$$y = \begin{bmatrix} \dfrac{y_t}{|a|^2} & \dfrac{-y_t}{a^*} \\ \dfrac{-yt}{a} & y_t \end{bmatrix} \qquad ...(7.94)$$

This means if there is a phase shifting transformer of complex turns ratio a and admittance Y_t between buses p and q, then the elements of bus admittance matrix are modified as follows on addition of the transformer.

$$Y_{pp\,(\text{new})} = Y_{pp\,(\text{old})} + \frac{y_t}{|a|^2}$$

$$Y_{qq\,(\text{new})} = Y_{qq\,(\text{old})} + y_t$$

$$Y_{pq\,(\text{new})} = Y_{pq\,(\text{old})} - \frac{y_t}{a^*}$$

$$Y_{qp\,(\text{new})} = Y_{qp\,(\text{old})} = -\frac{y_t}{a}$$

The complex turns ratio is calculated from,

$$a = |a| \quad (\cos\theta + j\sin\theta) \qquad ...(7.95)$$

If phase shift from bus p to q is positive, then voltage at bus p leads voltage at bus q.

7.12 COMPARISON OF LOAD FLOW METHODS

The comparison of the methods presented in the chapter should take into account

1. The computing time required for preparation of data in proper format and data processing.
2. Programming ease.
3. Storage requirements.
4. Computation time per iteration.
5. Number of iterations.
6. Ease and time required for modifying network data when operating conditions change.
   1. Since, all the methods presented are in the bus frame of reference in admittance form, the data preparation is same for all the methods. *IEEE* has proposed a standard format for data preparation for load flow studies to enable interchange of data. The bus admittance matrix can be formed using a simple algorithm, by inspection.
   2. Due to simplicity of the equations, Gauss–Seidel method is relatively easy to program. Any cases of buses switching status from a *PV* bus to a *PQ* bus or vice versa, can be easily handled. Programming of NR method is more involved. The size of the Jacobian matrix depends on the number of *PV* and *PQ* buses. Hence, with bus switching, the size of the Jacobian varies. Further, the programming becomes more complicated if the buses are randomly numbered. It is easier to program, if the *PV* buses are ordered in sequence and *PQ* buses are also ordered in sequence.
   3. The storage requirements are more for the NR method, since the Jacobian elements have to be stored. The memory is further increased for NR method using rectangular coordinates. The storage requirement can be drastically reduced by using sparse matrix techniques, since both the admittance matrix and the Jacobian are sparse matrices. Programming complication is increased if sparse matrix techniques are used. In recent times, the technology revolution in high–speed memories has effectively removed the problem of memory storage to a great extent.
   4. The time taken for a single iteration depends on the number of arithmetic and logical operations required to be performed in a full iteration. The Gauss Seidel method requires the fewest number of operations to complete an iteration. In the NR method, the computation of the Jacobian is necessary in every iteration. Further, the inverse of the Jacobian also has to be computed. Hence, the time per iteration is larger than in the GS method and is roughly about 7 times that of the GS method, in large systems. Computation time can be reduced if the Jacobian is updated once in two or three iterations. In *FDLF* method, the Jacobian is a constant and needs to be computed only once. In both NR and *FDLF* methods, the time per iteration increases directly as the number of buses.
   5. The number of iterations is determined by the convergence characteristic of the method. The GS method exhibits a linear convergence characteristic as compared to the NR method which has a quadratic convergence. Hence, the GS method requires more number of iterations to get a converged solution as compared to the NR method. In the *GS* method, the number of iterations increases directly as the size of the system increases. In contrast, the number of iterations is relatively constant in NR and *FDLF* methods. They require about 5–8 iterations for convergence in large systems. A sig-

nificant increase in rate of convergence can be obtained in the GS method if an acceleration factor is used. The number of iterations also depends on the required accuracy of the solution. Generally, a voltage tolerance of 0.0001 pu is used to obtain acceptable accuracy and the real power mismatch and reactive power mismatch can be taken as 0.001 pu.

Due to the above reasons, the NR method is faster and more reliable for large systems. The convergence of *FDLF* method is geometric and its speed is nearly 4–5 times that of NR method.

6. When operating conditions change, the data has to be modified. Network changes such as, addition or deletion of lines, changes in transformer tap position, change in impedance of a line, etc. require recalculation of some elements of the bus admittance matrix. In NR method, these changes affect the Jacobian elements also. When a series of load flow calculations have to be carried out for continuously changing operating conditions, we can use the converged values of voltages of one iteration as the initial voltage in the next iteration.

7.13 SUMMARY

In this chapter, the load flow problem, also called the power flow problem, has been discussed in detail. The load flow solution gives the complex voltages at all the buses and the complex power flows in the lines. Though, algorithms are available using the impedance form of the equations, the sparsity of the bus admittance matrix and the ease of building the bus admittance matrix have made algorithms using the admittance form of equations more popular.

The most popular methods are the Gauss-Seidel method, the Newton-Raphson method and the Fast Decoupled Load Flow method. These methods have been discussed in detail with illustrative examples. In smaller systems, the ease of programming and the memory requirements , make GS method attractive. However, the computation time increases with increase in the size of the system. Hence, in large systems NR and *FDLF* methods are more popular. There is a trade off between various requirements like speed, storage, reliability, computation time, convergence characteristics, etc. No single method has all the desirable features. However, NR method is most popular because of its versatility, reliability and accuracy.

The voltage magnitude at various buses can be controlled. The methods commonly available for voltage control are control through generator excitation adjustment, use of VAR generators, use of regulating transformers or phase shifting transformers.

7.14 LOAD FLOW TEST SYSTEMS

Coding used for buses:

Slack bus – 2

PV bus – 1

Load bus – 0

The power flow solution and line flow solutions given are obtained after running the load flow using NR method.

5–BUS SYSTEM

NUMBER OF BUSES IN THE SYSTEM	=	5
NUMBER OF PV BUSES	=	0
NUMBER OF LINES IN THE SYSTEM	=	7
NUMBER OF TRANSFORMERS PRESENT IN THE SYSTEM	=	0
NUMBER OF FIXED SHUNTS PRESENT IN THE SYSTEM	=	0
NUMBER OF THE SLACK BUS	=	1
SLACK BUS VOLTAGE	=	1.06
NUMBER OF MAXIMUM ITERATIONS ALLOWED IN LOAD FLOW	=	5
CONVERGENCE VALUE FOR ΔP and ΔQ IN LOAD FLOW	=	.00010

LINE DATA

LINE	*SB*	*EB*	*SERIES R(pu)*	*IMPEDANCE X(pu)*	*HALF LINE CHARGING ADMITTANCE(pu)*
1	1	2	.02000	.06000	.03000
2	1	3	.08000	.24000	.02500
3	2	3	.06000	.18000	.02000
4	2	4	.06000	.18000	.02000
5	2	5	.04000	.12000	.01500
6	3	4	.01000	.03000	.01000
7	4	5	.08000	.24000	.02500

BUS DATA

		GENERATION		*LOAD DEMAND*	
Bus	*Type*	*P (pu)*	*Q (pu)*	*P (pu)*	*Q (pu)*
1	2	.0000	.0000	.0000	.0000
2	0	.4000	.3000	.2000	.1000
3	0	.0000	.0000	.4500	.1500
4	0	.0000	.0000	.4000	.0500
5	0	.0000	.0000	.6000	.1000

POWER FLOW SOLUTION

		GENERATION		*LOAD DEMAND*		*BUS VOLTAGE*	
Bus	*Type*	*P (pu)*	*Q (pu)*	*P (pu)*	*Q (pu)*	*Mag*	*Angle (Deg)*
1	2	1.2959	–.0742	.0000	.0000	1.0600	.0000
2	0	.4000	.3000	.2000	.1000	1.0474	–2.8064
3	0	.0000	.0000	.4500	.1500	1.0242	–4.9970
4	0	.0000	.0000	.4000	.0500	1.0236	–5.3291
5	0	.0000	.0000	.6000	.1000	1.0179	–6.1503

SLACK BUS ACTIVE POWER	=	1.29586700 pu
SLACK BUS REACTIVE POWER	=	−0.07421447 pu
SYSTEM REAL POWER LOSS	=	0.04586673 pu
SYSTEM REACTIVE POWER LOSS	=	−0.17421100 pu
RESIDUE AFTER CONVERGENCE	=	0.00000624

ITERATIONS = 3

LINE FLOWS

LINE	*SB*	*EB*	*P–Flow*	*Q–Flow*	*P–Loss*	*Q–Loss*
1	1	2	.8886	−.0858		
1	2	1	−.8745	.0615		
					.01410419	−.02430754
2	1	3	.4072	.0116		
2	3	1	−.3953	−.0301		
					.01191953	−.01855469
3	2	3	.2469	.0355		
3	3	2	−.2434	−.0678		
					.00351517	−.03237540
4	2	4	.2794	.0296		
4	4	2	−.2749	−.0593		
					.00441328	−.02965594
5	2	5	.5482	.0734		
5	5	2	−.5370	−.0717		
					.01125234	.00175691
6	3	4	.1887	−.0520		
6	4	3	−.1884	.0321		
					.00035667	−.01989669
7	4	5	.0633	−.0228		
7	5	4	−.0630	−.0283		
					.00030718	−.05117603

14– BUS SYSTEM

NUMBER OF BUSES IN THE SYSTEM	=	14
NUMBER OF PV BUSES	=	4
NUMBER OF LINES IN THE SYSTEM	=	17
NUMBER OF TRANSFORMERS IN THE SYSTEM	=	3
NUMBER OF FIXED SHUNTS IN THE SYSTEM	=	1
NUMBER OF THE SLACK BUS	=	1
SLACK BUS VOLTAGE	=	1.06
NUMBER OF MAXIMUM ITERATIONS ALLOWED IN LOADFLOW	=	10
CONVERGENCE VALUE FOR ΔP and ΔQ IN LOADFLOW	=	.00010

LINE DATA

LINE	SB	EB	SERIES R (pu)	IMPEDANCE X (pu)	HALF LINE CHARGING ADMITTANCE (pu)
1	1	2	.01938	.05917	.02640
2	1	8	.05403	.22304	.02460
3	2	4	.04699	.19797	.02190
4	2	8	.05695	.17388	.01700
5	2	9	.05811	.17632	.01870
6	3	11	.09498	.19890	.00000
7	3	12	.12291	.25581	.00000
8	3	13	.06615	.13027	.00000
9	4	9	.06701	.17103	.01730
10	5	6	.00000	.17615	.00000
11	6	7	.00000	.11001	.00000
12	7	10	.03181	.08450	.00000
13	7	14	.12711	.27038	.00000
14	8	9	.01335	.04211	.00640
15	10	11	.08205	.19207	.00000
16	12	13	.22092	.19988	.00000
17	13	14	.17093	.34802	.00000

TRANSFORMER DATA

TRANS	SB	EB	SERIES R (pu)	IMPEDANCE X (pu)	TAP
1	8	3	.00000	.25202	.96200
2	9	6	.00000	.20912	.97800
3	9	7	.00000	.55618	.96900

SHUNT DATA

S.No	BUS	SHUNT G (pu)	ADMITTANCE B (pu)
1	7	.00000	.19000

BUS DATA

		GENERATION		LOAD DEMAND		
Bus	Type	P (pu)	Q (pu)	P (pu)	Q (pu)	Volt Mag.
1	2	.0000	.0000	.0000	.0000	1.0600
2	1	.4000	.0000	.2170	.1270	1.0450
3	1	.2000	.0000	.1120	.0750	1.0700
4	1	.0000	.0000	.9420	.1900	1.0100
5	1	.0000	.0000	.0000	.0000	1.0900
6	0	.0000	.0000	.0000	.0000	1.0000
7	0	.0000	.0000	.2950	.1660	1.0000

8	0	.0000	.0000	.0760	.0160	1.0000
9	0	.0000	.0000	.4780	–.0390	1.0000
10	0	.0000	.0000	.0900	.0580	1.0000
11	0	.0000	.0000	.0350	.0180	1.0000
12	0	.0000	.0000	.0610	.0160	1.0000
13	0	.0000	.0000	.1350	.0580	1.0000
14	0	.0000	.0000	.1490	.0500	1.0000

PV – BUS DATA

Sl No	*Bus No*	*|V| (spec)*	*Q–Min*	*Q–Max*
1	2	1.0450	–.4000	.5000
2	3	1.0700	–.0600	.2400
3	4	1.0100	.0000	.4000
4	5	1.0900	–.0600	.2400

POWER FLOW SOLUTION

		GENERATION		*LOAD DEMAND*		*BUS VOLTAGE*	
Bus	*Type*	*P (pu)*	*Q (pu)*	*P (pu)*	*Q (pu)*	*Mag*	*Angle (Deg)*
1	2	2.1061	–.1713	.0000	.0000	1.0600	.0000
2	1	.4000	.3061	.2170	.1270	1.0450	–4.5135
3	1	.2000	.1898	.1120	.0750	1.0700	–11.6653
4	1	.0000	.1985	.9420	.1900	1.0100	–11.9422
5	1	.0000	.1620	.0000	.0000	1.0900	–12.0430
6	0	.0000	.0000	.0000	.0000	1.0638	–12.0430
7	0	.0000	.0000	.2950	.1660	1.0568	–13.4160
8	0	.0000	.0000	.0760	.0160	1.0296	–7.8997
9	0	.0000	.0000	.4780	–.0390	1.0248	–9.4101
10	0	.0000	.0000	.0900	.0580	1.0514	–13.3944
11	0	.0000	.0000	.0350	.0180	1.0569	–12.6694
12	0	.0000	.0000	.0610	.0160	1.0554	–12.5909
13	0	.0000	.0000	.1350	.0580	1.0503	–12.7469
14	0	.0000	.0000	.1490	.0500	1.0359	–14.1260

SLACK BUS ACTIVE POWER	=	2.10609300 pu
SLACK BUS REACTIVE POWER	=	–0.17130510 pu
SYSTEM REAL POWER LOSS	=	0.11609210 pu
SYSTEM REACTIVE POWER LOSS	=	–0.04983985 pu
RESIDUE AFTER CONVERGENCE	=	0.00000347
ITERATIONS	=	3

LINE FLOWS

LINE	*SB*	*EB*	*P–Flow*	*Q–Flow*	*P–Loss*	*Q–Loss*
1	1	2	1.4271	–.1703		
1	2	1	–1.3917	.2201		
					.03547144	.04980880
2	1	8	.6789	–.0010		
2	8	1	–.6567	.0389		
					.02220064	.03792986
3	2	4	.7041	.0385		
3	4	2	–.6826	.0059		
					.02149987	.04432362
4	2	8	.3607	–.0331		
4	8	2	–.3539	.0172		
					.00679696	–.01583271
5	2	9	.5099	–.0464		
5	9	2	–.4960	.0484		
					.01386869	.00202145
6	3	11	.1089	.0194		
6	11	3	–.1079	–.0173		
					.00101571	.00212726
7	3	12	.0820	.0223		
7	12	3	–.0812	–.0206		
					.00077534	.00161320
8	3	13	.1955	.0642		
8	13	3	–.1930	–.0594		
					.00244613	.00481776
9	4	9	–.2594	.0027		
9	9	4	.2638	–.0271		
					.00444716	–.02446555
10	5	6	.0000	.1620		
10	6	5	.0000	–.1582		
					.00000000	.00389266
11	6	7	.2449	.0712		
11	7	6	–.2449	–.0649		
					.00000000	.00632167
12	7	10	.0176	.0599		
12	10	7	–.0175	–.0596		
					.00011031	.00029373
13	7	14	.0726	.0478		
13	14	7	–.0718	–.0459		
					.00086029	.00183032
14	8	9	.6363	–.0829		
14	9	8	–.6311	.0857		
					.00517058	.00280942
15	10	11	–.0725	.0016		
15	11	10	.0729	–.0007		

					.00039022	.00091333
16	12	13	.0202	.0046		
16	13	12	–.0202	–.0046		
					.00008560	.00007729
17	13	14	.0782	.0060		
17	14	13	–.0772	–.0041		
					.00095280	.00193954

30– BUS SYSTEM

NUMBER OF BUSES IN THE SYSTEM	=	30
NUMBER OF PV BUSES	=	5
NUMBER OF LINES IN THE SYSTEM	=	37
NUMBER OF TRANSFORMERS IN THE SYSTEM	=	4
NUMBER OF FIXED SHUNTS IN THE SYSTEM	=	2
NUMBER OF THE SLACK BUS	=	1
SLACK BUS VOLTAGE	=	1.06
NUMBER OF MAXIMUM ITERATIONS ALLOWED IN LOADFLOW	=	10
CONVERGENCE VALUE FOR ΔP and ΔQ IN LOADFLOW	=	.00010

LINE DATA

LINE	*SB*	*EB*	*SERIES R (pu)*	*IMPEDANCE X (pu)*	*HALF LINE CHARGING ADMITTANCE (pu)*
1	1	2	.01920	.05750	.02640
2	1	27	.04520	.18520	.02040
3	2	5	.04720	.19830	.02090
4	2	11	.05700	.17370	.01840
5	2	13	.05810	.17630	.01870
6	3	18	.01200	.04200	.00450
7	3	28	.06360	.20000	.02140
8	4	7	.00000	.20800	.00000
9	5	12	.04600	.11600	.01020
10	6	9	.00000	.14000	.00000
11	7	8	.00000	.11000	.00000
12	8	17	.03240	.08450	.00000
13	8	20	.09360	.20900	.00000
14	8	21	.03480	.07490	.00000
15	8	22	.07270	.14990	.00000
16	9	14	.12310	.25590	.00000
17	9	15	.06620	.13040	.00000
18	9	16	.09450	.19870	.00000
19	10	25	.10930	.20870	.00000
20	10	29	.21980	.41530	.00000
21	10	30	.32020	.60270	.00000
22	11	13	.01190	.04140	.00450

23	11	27	.01320	.03790	.00420
24	12	13	.02670	.08200	.00850
25	13	28	.01690	.05990	.00650
26	14	15	.22100	.19970	.00000
27	15	18	.10700	.21850	.00000
28	15	23	.10000	.20200	.00000
29	16	17	.08240	.19320	.00000
30	18	19	.06390	.12920	.00000
31	19	20	.03400	.06800	.00000
32	21	22	.01160	.02360	.00000
33	22	24	.11500	.17900	.00000
34	23	24	.13200	.27000	.00000
35	24	25	.18850	.32920	.00000
36	25	26	.25440	.38000	.00000
37	29	30	.23990	.45330	.00000

TRANSFORMER DATA

TRANS	*SB*	*EB*	*SERIES*	*IMPEDANCE*	*TAP*
			R (pu)	*X (pu)*	
1	13	7	.00000	.20800	.97800
2	13	8	.00000	.55600	.96900
3	11	9	.00000	.25600	.96200
4	28	10	.00000	.39600	.96800

SHUNT DATA

S.No	*BUS*	*SHUNT*	*ADMITTANCE*
		G (pu)	*B (pu)*
1	10	.00000	.19000
2	24	.00000	.04300

BUS DATA

		GENERATION		*LOAD DEMAND*		
Bus	*Type*	*P (pu)*	*Q (pu)*	*P (pu)*	*Q (pu)*	*Volt Mag.*
1	2	.0000	.0000	.0000	.0000	1.0600
2	1	.4000	.0000	.2170	.1270	1.0450
3	1	.2000	.0000	.0000	.3000	1.0100
4	1	.0000	.0000	.3000	.0000	1.0820
5	1	.0000	.0000	.9420	.1900	1.0100
6	1	.0000	.0000	.0000	.0000	1.0710
7	0	.0000	.0000	.0000	.0000	1.0000
8	0	.0000	.0000	.0580	.0200	1.0000
9	0	.0000	.0000	.1120	.0750	1.0000
10	0	.0000	.0000	.0000	.0000	1.0000
11	0	.0000	.0000	.0760	.0160	1.0000

12	0	.0000	.0000	.2280	.1090	1.0000
13	0	.0000	.0000	.0000	.0000	1.0000
14	0	.0000	.0000	.0620	.0160	1.0000
15	0	.0000	.0000	.0820	.0250	1.0000
16	0	.0000	.0000	.0350	.0180	1.0000
17	0	.0000	.0000	.0900	.0580	1.0000
18	0	.0000	.0000	.0320	.0090	1.0000
19	0	.0000	.0000	.0950	.0340	1.0000
20	0	.0000	.0000	.0220	.0070	1.0000
21	0	.0000	.0000	.1750	.1120	1.0000
22	0	.0000	.0000	.0000	.0000	1.0000
23	0	.0000	.0000	.0320	.0160	1.0000
24	0	.0000	.0000	.0870	.0670	1.0000
25	0	.0000	.0000	.0000	.0000	1.0000
26	0	.0000	.0000	.0350	.0230	1.0000
27	0	.0000	.0000	.0240	.0000	1.0000
28	0	.0000	.0000	.0000	.0000	1.0000
29	0	.0000	.0000	.0240	.0090	1.0000
30	0	.0000	.0000	.1060	.0190	1.0000

PV–BUS DATA

Sl No	*Bus No*	*|V| (spec)*	*Q–Min*	*Q–Max*
1	2	1.0450	–.4000	.5000
2	3	1.0100	–.1000	.4000
3	4	1.0820	–.0600	.2400
4	5	1.0100	–.4000	.4000
5	6	1.0710	–.0600	.2400

POWER FLOW SOLUTION

		GENERATION		*LOAD DEMAND*		*BUS VOLTAGE*	
Bus	*Type*	*P (pu)*	*Q (pu)*	*P (pu)*	*Q (pu)*	*Mag*	*Angle (Deg)*
1	2	2.3873	–.1765	.0000	.0000	1.0600	.0000
2	1	.4000	.5204	.2170	.1270	1.0450	–5.0361
3	1	.2000	.1917	.0000	.3000	1.0100	–9.8046
4	1	.0000	.2271	.3000	.0000	1.0820	–18.3703
5	1	.0000	.3574	.9420	.1900	1.0100	–13.5302
6	1	.0000	.1931	.0000	.0000	1.0710	–15.0062
7	0	.0000	.0000	.0000	.0000	1.0399	–15.1913
8	0	.0000	.0000	.0580	.0200	1.0244	–16.1105
9	0	.0000	.0000	.1120	.0750	1.0458	–15.0062
10	0	.0000	.0000	.0000	.0000	1.0792	–15.5341
11	0	.0000	.0000	.0760	.0160	1.0151	–8.7828
12	0	.0000	.0000	.2280	.1090	1.0026	–12.0551
13	0	.0000	.0000	.0000	.0000	1.0105	–10.1343
14	0	.0000	.0000	.0620	.0160	1.0311	–15.9467

15	0	.0000	.0000	.0820	.0250	1.0267	–16.0779
16	0	.0000	.0000	.0350	.0180	1.0293	–15.7442
17	0	.0000	.0000	.0900	.0580	1.0204	–16.2135
18	0	.0000	.0000	.0320	.0090	1.0135	–16.8029
19	0	.0000	.0000	.0950	.0340	1.0089	–17.0388
20	0	.0000	.0000	.0220	.0070	1.0119	–16.8652
21	0	.0000	.0000	.1750	.1120	1.0154	–16.5169
22	0	.0000	.0000	.0000	.0000	1.0171	–16.4866
23	0	.0000	.0000	.0320	.0160	1.0206	–16.5169
24	0	.0000	.0000	.0870	.0670	1.0210	–16.7487
25	0	.0000	.0000	.0000	.0000	1.0518	–16.1605
26	0	.0000	.0000	.0350	.0230	1.0348	–16.5526
27	0	.0000	.0000	.0240	.0000	1.0239	–7.2915
28	0	.0000	.0000	.0000	.0000	1.0121	–10.7532
29	0	.0000	.0000	.0240	.0090	1.0605	–16.6376
30	0	.0000	.0000	.1060	.0190	1.0497	–17.4270

SLACK BUS ACTIVE POWER = 2.38734400 pu
SLACK BUS REACTIVE POWER = –0.1765148 pu
SYSTEM REAL POWER LOSS = 0.15334330 pu
SYSTEM REACTIVE POWER LOSS = 0.06322453 pu
RESIDUE AFTER CONVERGENCE = 0.00000622
ITERATIONS = 3

LINE FLOWS

LINE	*SB*	*EB*	*P-Flow*	*Q-Flow*	*P-Loss*	*Q-Loss*
1	1	2	1.6269	–.2220		
1	2	1	–1.5810	.3009		
					.04585886	.07884644
2	1	27	.7605	.0455		
2	27	1	–.7370	.0063		
					.02345335	.05178803
3	2	5	.7987	.0299		
3	5	2	–.7710	.0423		
					.02769387	.07220702
4	2	11	.4174	.0357		
4	11	2	–.4081	–.0466		
					.00925624	–.01084672
5	2	13	.5479	.0270		
5	13	2	–.5318	–.0176		
					.01609033	.00930805
6	3	13	.1260	–.0532		
6	13	3	–.1257	.0448		
					.00021444	–.00843316
7	3	28	.0740	–.0551		
7	28	3	–.0736	.0127		
					.00041092	–.04245751

8	4	7	–.3000	.2271		
8	7	4	.3000	–.2020		
					.00000000	.02515681
9	5	12	–.1710	.1251		
9	12	5	.1731	–.1404		
					.00214662	–.01524416
10	6	9	.0000	.1931		
10	9	6	.0000	–.1885		
					.00000000	.00454868
11	7	8	.1554	.1486		
11	8	7	–.1554	–.1439		
					.00000000	.00470167
12	8	17	.0355	.0346		
12	17	8	–.0354	–.0344		
					.00007618	.00019747
13	8	20	.0772	.0266		
13	20	8	–.0766	–.0253		
					.00059551	.00132958
14	8	21	.1281	.0638		
14	21	8	–.1274	–.0623		
					.00067918	.00146281
15	8	22	.0566	.0225		
15	22	8	–.0563	–.0220		
					.00025700	.00052957
16	9	14	.0797	.0220		
16	14	9	–.0790	–.0204		
					.00076959	.00159986
17	9	15	.1848	.0607		
17	15	9	–.1825	–.0562		
					.00229068	.00451248
18	9	16	.0907	.0441		
18	16	9	–.0899	–.0422		
					.00087905	.00184849
19	10	25	.1050	.0869		
19	25	10	–.1032	–.0835		
					.00174163	.00332589
20	10	29	.0618	.0164		
20	29	10	–.0610	–.0150		
					.00077084	.00145652
21	10	30	.0708	.0163		
21	30	10	–.0693	–.0136		
					.00144941	.00272804
22	11	13	.5715	–.0497		
22	13	11	–.5677	.0537		
					.00379539	.00397033
23	11	27	–.7066	.0159		
23	27	11	.7130	–.0063		
					.00640005	.00964647

24	12	13	−.4011	.0314		
24	13	12	.4054	−.0353		
					.00431645	−.00396790
25	13	28	.1644	−.0777		
25	28	13	−.1638	.0663		
					.00053124	−.01141386
26	14	15	.0170	.0044		
26	15	14	−.0169	−.0043		
					.00006381	.00005797
27	15	18	.0732	.0264		
27	18	15	−.0726	−.0252		
					.00061449	.00125458
28	15	23	.0442	.0091		
28	23	15	−.0440	−.0087		
					.00019343	.00039031
29	16	17	.0549	.0242		
29	17	16	−.0546	−.0236		
					.00027956	.00065514
30	18	19	.0406	.0162		
30	19	18	−.0405	−.0159		
					.00011878	.00024094
31	19	20	−.0545	−.0181		
31	20	19	.0546	.0183		
					.00011089	.00022179
32	21	22	−.0476	−.0497		
32	22	21	.0476	.0498		
					.00005593	.00011365
33	22	24	.0087	−.0278		
33	24	22	−.0086	.0279		
					.00009403	.00014697
34	23	24	.0120	−.0073		
34	24	23	−.0120	.0074		
					.00002502	.00005153
35	24	25	−.0664	−.0575		
35	25	24	.0678	.0599		
					.00139464	.00243585
36	25	26	.0354	.0236		
36	26	25	−.0350	−.0230		
					.00041660	.00062228
37	29	30	.0370	.0060		
37	30	29	−.0367	−.0054		
					.00029945	.00056587

REVIEW QUESTIONS

1. Obtain the expression for active and reactive power injection at a bus, in polar and rectangular forms.
2. How are buses classified for load flow studies?
3. Generator buses are modeled as PV bus in general. Under certain conditions they are modeled as PQ buses. Discuss the above statement.
4. If the Q limits are violated at a generator bus, what are its implications?
5. What is the significance of slack bus?
6. When is the load bus modeled as a PV bus?
7. What are the elements considered in forming the bus admittance matrix, for power flow studies?
8. Briefly discuss the Gauss-Seidel method of power flow solution. How are PV buses considered in GS method.
9. What is acceleration of convergence? What is its application?
10. Draw the flow chart for GS method of power flow solution.
11. Derive the Jacobian elements for solution of power flow equations by NR method, using polar coordinates.
12. What approximations are made in NR method, to develop decoupled algorithms for power flow solutions?
13. Explain the fast decoupled method for load flow solution.
14. The Jacobian need not be updated in every iteration in decoupled methods. Justify.
15. Phase shifting transformers cause unsymmetrical bus admittance matrix. Explain.
16. Derive the mathematical model of the phase shifting transformer. Show how the bus admittance matrix elements are affected with the phase shifting transformer.
17. Compare the performance of different power flow algorithms.
18. Explain how the slack bus active and reactive power is computed.
19. How can the losses in the system be calculated?
20. The sparsity of Y_{bus} matrix is also reflected in the Jacobian. Comment.

Chapter 8

Economic Operation of Power Systems

8.1 INTRODUCTION

One of the earliest applications of on-line centralized control was to provide a central facility, to operate economically, several generating plants supplying the loads of the system. Modern integrated systems have different types of generating plants, such as coal fired thermal plants, hydel plants, nuclear plants, oil and natural gas units, etc. The capital investment, operation and maintenance costs are different for different types of plants.

The operation economics can again be subdivided into two parts.

(i) Problem of *economic dispatch*, which deals with determining the power output of each plant to meet the specified load, such that the overall fuel cost is minimized.

(ii) Problem of *optimal power flow*, which deals with minimum loss delivery, where in the power flow is optimized to minimize losses in the system. In this chapter we consider the problem of economic dispatch.

During operation of the plant, a generator may be in one of the following states:

(i) Base supply without regulation: the output is a constant.

(ii) Base supply with regulation: output power is regulated based on system load.

(iii) Automatic non-economic regulation: output level changes around a base setting as area control error changes.

(iv) Automatic economic regulation: output level is adjusted, with the area load and area control error, while tracking an economic setting.

Regardless of the *unit's* operating state, it has a contribution to the economic operation, even though its output is changed for different reasons. The factors influencing the cost of generation are the generator efficiency, fuel cost and transmission losses. The most efficient generator may not give minimum cost, since it may be located in a place where fuel cost is high. Further, if the plant is located far from the load centres, transmission losses may be high and running the plant may become uneconomical. The economic dispatch problem basically determines the generation of different plants to minimize total operating cost.

Modern generating plants like nuclear plants, geothermal plants, etc., may require capital investment of millions of rupees. The economic dispatch is however determined in terms of fuel cost per unit power generated and does not include capital investment, maintenance, depreciation, start-up and shutdown costs, etc.

8.2 PERFORMANCE CURVES

8.2.1 Input-output Curve

This is the fundamental curve for a thermal plant and is a plot of the input in British thermal units (Btu) per hour versus the power output of the plant in MW as shown in Fig. 8.1. One Btu is equal to to 1055.05585 joules.

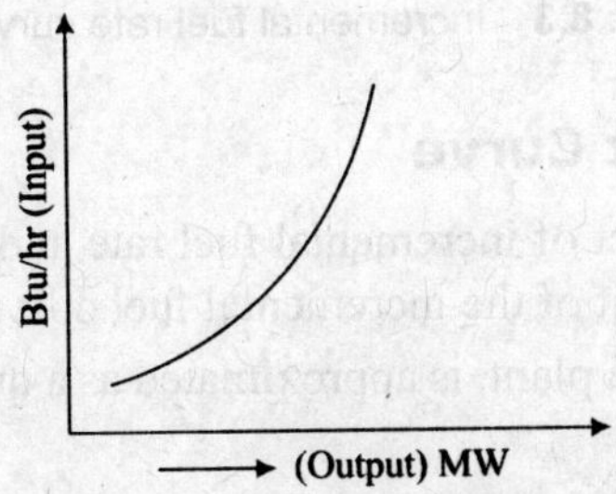

Fig. 8.1 Input-output curve.

8.2.2 Heat Rate Curve

The heat rate is the ratio of fuel input in Btu/hr to corresponding output in kW. It is the slope of the input-output curve at any point. The reciprocal of heat rate is called fuel-efficiency. The heat rate curve is a plot of heat rate versus output in MW. A typical plot is shown in Fig. 8.2.

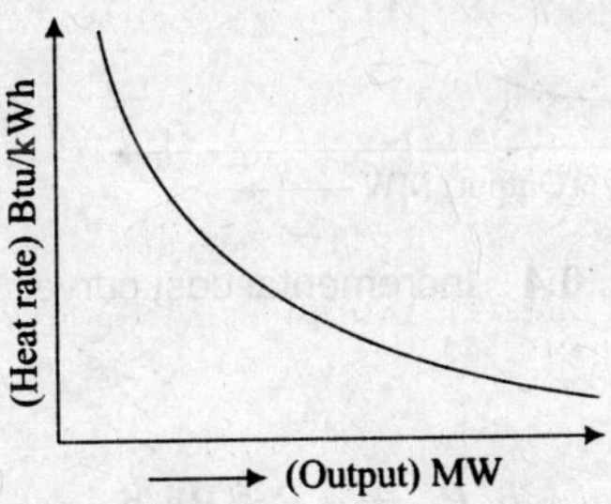

Fig. 8.2 Heat rate curve.

8.2.3 Incremental Fuel Rate Curve

The incremental fuel rate is equal to a small change in input divided by the corresponding change in output.

$$\text{Incremental fuel rate} = \frac{\Delta\,\text{Input}}{\Delta\,\text{Output}} \qquad ...(8.1)$$

The unit is again Btu/kWh. A plot of incremental fuel rate versus the output is shown in Fig. 8.3.

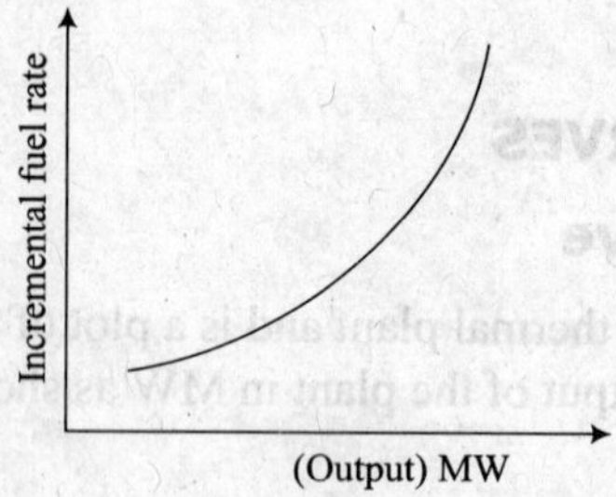

Fig. 8.3 Incremental fuel rate curve.

8.2.4 Incremental Cost Curve

The incremental cost is the product of incremental fuel rate and fuel cost (Rs/Btu or \$/Btu). The curve is shown in Fig. 8.4. The unit of the incremental fuel cost is Rs/MWh or \$/MWh.

In general, the fuel cost F_i for a plant, is approximated as a quadratic function of the generated output P_{Gi}.

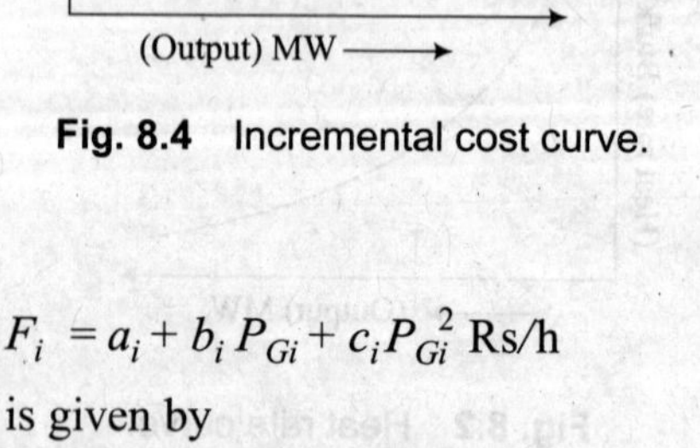

Fig. 8.4 Incremental cost curve.

$$F_i = a_i + b_i P_{Gi} + c_i P_{Gi}^2 \text{ Rs/h} \qquad ...(8.2)$$

The incremental fuel cost is given by

$$\frac{dF_i}{dP_{Gi}} = b_i + 2c_i P_{Gi} \text{ Rs/MWh} \qquad ...(8.3)$$

The incremental fuel cost is a measure of how costly it will be produce an increment of power.

The incremental production cost, is made up of incremental fuel cost plus the incremental cost of labour, water, maintenance, etc., which can be taken to be some percentage of the incremental fuel cost, instead of resorting to a rigorous mathematical model. The cost curve can be approximated by a linear curve. While there is negligible operating cost for a hydel plant, there is a limitation on the power output possible. In any plant, all units normally operate between $P_{G\,min}$, the minimum loading limit, below which it is technically infeasible to operate a unit and $P_{G\,max}$, which is the maximum output limit.

8.3 ECONOMIC GENERATION SCHEDULING NEGLECTING LOSSES AND GENERATOR LIMITS

The simplest case of economic dispatch is the case when transmission losses are neglected. The model does not consider the system configuration or line impedances. Since, losses are neglected, the total generation is equal to the total demand P_D.

Consider a system with $\mathbf{n}_g$ number of generating plants supplying the total demand P_D. If F_i is the cost of plant $\boldsymbol{i}$ in Rs/h, the mathematical formulation of the problem of economic scheduling can be stated as follows:

Minimize $$F_T = \sum_{i=1}^{n_g} F_i \qquad \text{...(8.4)}$$

such that $$\sum_{i=1}^{n_g} P_{Gi} = P_D \qquad \text{...(8.5)}$$

where, F_T = total cost.
P_{Gi} = generation of plant i.
P_D = total demand.

This is a constrained optimization problem, which can be solved by Lagrange's method.

8.3.1 Constrained Parameter Optimization

Consider the following optimization problem:

Minimize $f(x_1, x_2, \dots x_n)$ ...(8.6)

such that $g_j(x_1, x_2, \dots x_n) = 0; j = 1\dots k$ ...(8.7)

Here, $x_1, x_2, \dots x_n$ are the variables and g_j is the equality constraint and f the cost function. In the Lagrange multiplier method, the cost function is augmented by a K-vector λ of undetermined quantities. The augmented, unconstrained cost function is given by

$$£ = f(x_1, x_2, \dots x_n) + \sum_{j=1}^{k} g_j \lambda_j \qquad \text{...(8.8)}$$

The necessary conditions for local minima of £ are the following:

$$\frac{\partial £}{\partial x_i} = \frac{\partial f}{\partial x_i} + \sum_{j=1}^{k} \lambda_j \frac{\partial g_j}{\partial x_i} = 0;\ i = 1 \ldots n \qquad \text{...(8.9)}$$

$$\frac{\partial £}{\partial \lambda_j} = g_j = 0 \qquad \text{...(8.10)}$$

Note that (8.10) is simply the original constraints of the problem.

8.3.2 Lagrange Method for Solution of Economic Schedule

The above method can be applied for the solution of (8.4) subject to the constraint (8.5). The problem is restated below:

Minimize $$F_T = \sum_{i=1}^{n_g} F_i$$

Such that $$P_D - \sum_{i=1}^{n_g} P_{Gi} = 0$$

The augmented cost function is given by

$$£ = F_T + \lambda\left(P_D - \sum_{i=1}^{n_g} P_{Gi}\right) \qquad \text{...(8.11)}$$

The minimum is obtained when

$$\frac{\partial £}{\partial P_{Gi}} = 0 \text{ and } \frac{\partial £}{\partial \lambda} = 0$$

From (8.11) $$\frac{\partial £}{\partial P_{Gi}} = \frac{\partial F_T}{\partial P_{Gi}} - \lambda = 0 \qquad \text{...(8.12)}$$

$$\frac{\partial £}{\partial \lambda} = P_D - \sum_{i=1}^{n_g} P_{Gi} = 0 \qquad \text{...(8.13)}$$

(8.13) is simply the original constraint of the problem. The cost of a plant F_i depends only on its own output P_{Gi}, hence

$$\frac{\partial F_T}{\partial P_{Gi}} = \frac{\partial F_i}{\partial P_{Gi}} = \frac{dF_i}{dP_{Gi}}$$

Using the above, (8.12) can be written as

$$\frac{\partial F_i}{\partial P_{Gi}} = \frac{dF_i}{dP_{Gi}} = \lambda;\quad i = 1 \ldots n_g \qquad \text{...(8.14)}$$

Using (8.3), we can write

$$b_i + 2c_i P_{Gi} = \lambda; \quad i = 1 \dots n_g \qquad \dots(8.15)$$

(8.14) is called the coordination equation. Simply stated, *for economic generation scheduling to meet a particular load demand, when transmission losses are neglected and generation limits are not imposed, all plants must operate at equal incremental production costs, subject to the constraint that the total generation be equal to the demand.* From (8.15), we have

$$P_{Gi} = \frac{\lambda - b_i}{2c_i} \qquad \dots(8.16)$$

We know in a loss less system

$$\sum_{i=1}^{n_g} P_{Gi} = P_D$$

Substituting (8.16), we get

$$\sum_{i=1}^{n_g} \frac{\lambda - b_i}{2c_i} = P_D \qquad \dots(8.17)$$

An analytical solution of λ is obtained from (8.17) as

$$\lambda = \frac{P_D + \sum_{i=1}^{n_g} \frac{b_i}{2c_i}}{\sum_{i=1}^{n_g} \frac{1}{2c_i}} \qquad \dots(8.18)$$

It can be seen that λ is dependent on the demand and the coefficients of the cost function. The value of λ obtained from (8.18) is substituted in (8.16) to obtain the optimal generation schedule. Any deviation from these optimum loading results in increase of production cost.

Example 8.1: The fuel costs of two units are given by

$$F_1 = 1.5 + 20\,P_{G1} + 0.1\,P_{G1}^2 \text{ Rs/h}$$

$$F_2 = 1.9 + 30\,P_{G2} + 0.1\,P_{G2}^2 \text{ Rs/h}$$

P_{G1}, P_{G2} are in MW. Find the optimal schedule neglecting losses, when the demand is 200 MW.

Solution:

$$\frac{dF_1}{dP_{G1}} = 20 + 0.2\,P_{G1} \text{ Rs/MWh}$$

$$\frac{dF_2}{dP_{G2}} = 30 + 0.2\,P_{G2} \text{ Rs/MWh}$$

$$P_D = P_{G1} + P_{G2} = 200 \text{ MW}$$

For economic schedule

$$\frac{dF_1}{dP_{G1}} = \frac{dF_2}{dP_{G2}} = \lambda$$

$$20 + 0.2\, P_{G1} = 30 + 0.2\,(200 - P_{G1})$$

Solving we get, $P_{G1} = 125$ MW

$$P_{G2} = 75 \text{ MW}$$

$$\lambda = 20 + 0.2\, P_{G1}$$

$$= 20 + 0.2\,(125)$$

$$= 45 \text{ Rs/MWh.}$$

Example 8.2: The fuel cost in \$/h for two 800 MW plants is given by

$$F_1 = 400 + 6.0\, P_{G1} + 0.004\, P_{G1}^2$$

$$F_2 = 500 + b_2 P_{G2} + c_2 P_{G2}^2$$

where, P_{G1}, P_{G2} are in MW.

(a)The incremental cost of power, λ is \$8/MWh when total demand is 550 MW. Determine optimal generation schedule neglecting losses.

(b)The incremental cost of power is \$10/MWh when total demand is 1300 MW. Determine optimal schedule neglecting losses.

(c)From (a) and (b) find the coefficients b_2 and c_2.

Solution: (a) $P_{G1} = \dfrac{\lambda - b_1}{2c_1} = \dfrac{8.0 - 6.0}{2 \times 0.004} = 250$ MW

$$P_{G2} = P_D - P_{G1} = 550 - 250 = 300 \text{ MW}$$

(b) $$P_{G1} = \frac{\lambda - b_1}{2c_1} = \frac{10 - 6}{2 \times 0.004} = 500 \text{ MW}$$

$$P_{G2} = P_D - P_{G1} = 1300 - 500 = 800 \text{ MW}$$

(c) $$P_{G2} = \frac{\lambda - b_2}{2c_2}$$

From (a) $$300 = \frac{8.0 - b_2}{2c_2}$$

From (b) $$800 = \frac{10.0 - b_2}{2c_2}$$

Solving, we get $b_2 = 6.8$

$$c_2 = 0.002$$

$$F_2 = 500 + 6.8\,P_{G2} + 0.002\,P_{G2}^2 \text{ \$/h}$$

Example 8.3: The fuel cost functions in \$/h for three thermal plants are given by

$$F_1 = 350 + 7.2\,P_{G1} + 0.004\,P_{G1}^2$$
$$F_2 = 500 + 7.3\,P_{G2} + 0.0025\,P_{G2}^2$$
$$F_3 = 600 + 6.74\,P_{G3} + 0.003\,P_{G3}^2$$

P_{G1}, P_{G2}, P_{G3} are in MW. Find the optimal schedule and compare the cost of this to the case when the generators share load equally if (i) $P_D = 450$ MW (ii) $P_D = 800$ MW

(i) $P_D = 450$ MW

$$\lambda = \frac{P_D + \sum_{i=1}^{n_g} \frac{b_i}{2c_i}}{\sum_{i=1}^{n_g} \frac{1}{2c_i}}$$

$$= \frac{450 + \frac{7.2}{2\times 0.004} + \frac{7.3}{2\times 0.0025} + \frac{6.74}{2\times 0.003}}{\frac{1}{2\times 0.004} + \frac{1}{2\times 0.0025} + \frac{1}{2\times 0.003}}$$

$$= \frac{450 + 3483.33}{491.66} = 8.0 \text{ \$/MWh}$$

$$P_{G1} = \frac{\lambda - b_1}{2c_1} = \frac{8.0 - 7.2}{2\times 0.004} = 100 \text{ MW}$$

$$P_{G2} = \frac{\lambda - b_2}{2c_2} = \frac{8.0 - 7.3}{2\times 0.0025} = 140 \text{ MW}$$

$$P_{G3} = \frac{\lambda - b_3}{2c_3} = \frac{8.0 - 6.74}{2\times 0.003} = 210 \text{ MW}$$

From cost functions

$$F_1 = 350 + 7.2 \times 100 + 0.004\,(100)^2 = 1110 \text{ \$/h}$$
$$F_2 = 500 + 7.3 \times 140 + 0.0025\,(140)^2 = 1571 \text{ \$/h}$$
$$F_3 = 600 + 6.74 \times 210 + 0.003\,(210)^2 = 2147.7 \text{ \$/h}$$
$$F_T = F_1 + F_2 + F_3 = 4828.7 \text{ \$/h}$$

If load is shared equally, then

$$P_{G1} = P_{G2} = P_{G3} = \frac{450}{3} = 150 \text{ MW}$$

Calculating cost as before

$$F_1 = 1520 \text{ \$/h}; F_2 = 1651.25 \text{ \$/h}; F_3 = 1678.50 \text{ \$/h}$$

$$F_T = 4849.75 \text{ \$/h}$$

Savings with coordinated economic dispatch is

$$4849.75 - 4828.7 = 21.05 \text{ \$/h}$$

(ii) $P_D = 800$ MW

$$\lambda = \frac{800 + 3483.33}{491.66} = 8.712 \text{ \$/MWh}$$

$$P_{G1} = \frac{8.712 - 7.2}{2 \times 0.004} = 189 \text{ MW}$$

$$P_{G2} = \frac{8.712 - 7.3}{2 \times 0.0025} = 282.4 \text{ MW}$$

$$P_{G3} = \frac{8.712 - 6.74}{2 \times 0.003} = 328.6 \text{ MW}$$

Calculating costs as before,

$$F_1 = 1853.68 \text{ \$/h}; F_2 = 2760.9 \text{ \$/h}; F_3 = 3138.7 \text{ \$/h}$$

$$F_T = 7753.28 \text{ \$/h}$$

If the load is shared equally,

$$P_{G1} = P_{G2} = P_{G3} = \frac{800}{3} = 266.67 \text{ MW}$$

$F_1 = 2554.4$ \$/h; $F_2 = 2624.4$ \$/h; $F_3 = 2610.6$ \$/h; $F_T = 7789.4$ \$/h

The saving in this case = 7789.4 – 7753.28

= 36.12 \$/h with economic schedule

8.4 ECONOMIC SCHEDULE INCLUDING LIMITS ON GENERATOR (NEGLECTING LOSSES)

The power output of any generator has a maximum value dependent on the rating of the generator. It also has a minimum limit set by stable boiler operation. The economic dispatch problem now is to schedule generation to minimize cost, subject to the equality constraint.

$$\sum_{i=1}^{n_g} P_{Gi} = P_D$$

and the inequality constraint

$$P_{Gi\,(min)} \leq P_{Gi} \leq P_{Gi\,(max)};\ i = 1, \ldots n_g \qquad \ldots(8.19)$$

The procedure followed is same as before i.e., the plants are operated with equal incremental fuel costs, till their limits are not violated. As soon as a plant reaches the limit (maximum or minimum) its output is fixed at that point and is maintained a constant. The other plants are operated at equal incremental costs.

Example 8.4: Incremental fuel costs in \$/MWh for two units are given below:

$$\frac{dF_1}{dP_{G1}} = 0.01P_{G1} + 2.0 \text{ \$/MWh}$$

$$\frac{dF_2}{dP_{G2}} = 0.012P_{G2} + 1.6 \text{ \$/MWh}$$

The limits on the plants are P_{min} = 20 MW, P_{max} = 125 MW. Obtain the optimal schedule if the load varies from 50 – 250 MW.

Solution: The incremental fuel costs of the two plants are evaluated at their lower limits and upper limits of generation.

At $\quad P_{G\,(min)} = 20$ MW

$$\lambda_{1\,(min)} = \frac{dF_1}{dP_{G1}} = 0.01 \times 20 + 2.0 = 2.2 \text{ \$/MWh}$$

$$\lambda_{2\,(min)} = \frac{dF_2}{dP_{G2}} = 0.012 \times 20 + 1.6 = 1.84 \text{ \$/MWh}$$

At $\quad P_{G\,(Max)} = 125$ MW

$$\lambda_{1\,(max)} = 0.01 \times 125 + 2.0 = 3.25 \text{ \$/MWh}$$

$$\lambda_{2\,(max)} = 0.012 \times 125 + 1.6 = 3.1 \text{ \$/MWh}$$

Now at light loads unit 1 has a higher incremental cost and hence will operate at its lower limit of 20 MW. Initially, additional load is taken up by unit 2, till such time its incremental fuel cost becomes equal to 2.2 \$/MWh at P_{G2} = 50 MW. Beyond this, the two units are operated with equal incremental fuel costs. The contribution of each unit to meet the demand is obtained by assuming different values of λ; When λ = 3.1 \$/MWh, unit 2 operates at its upper limit. Further loads are taken up by unit 1. The computations are shown in Table 8.1.

For a particular value of λ, P_{G1} and P_{G2} are calculated using (8.16). Fig. 8.5 shows plot of each unit output versus the total plant output.

For any particular load, the schedule for each unit for economic dispatch can be obtained from Fig. 8.5.

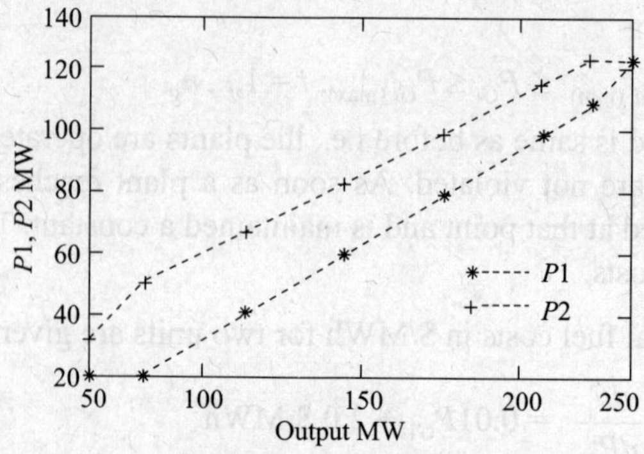

Fig. 8.5 Example 8.5

Table 8.1. Plant output and output of the two units

$\frac{dF_1}{dP_{G1}}$ $/MWh	$\frac{dF_2}{dP_{G2}}$ $/MWh	Plant λ $/MWh	P_{G1} MW	P_{G2} MW	Plant Output MW
2.2	1.96	1.96	20+	30	50
2.2	2.2	2.2	20+	50	70
2.4	2.4	2.4	40	66.7	106.7
2.6	2.6	2.6	60	83.3	143.3
2.8	2.8	2.8	80	100	180
3.0	3.0	3.0	100	116.7	216.7
3.1	3.1	3.1	110	125*	235
3.25	3.1	3.25	125*	125*	250

Example 8.5 In example 8.4, what is the saving in fuel cost for the economic schedule compared to the case where the load is shared equally. The load is 180 MW.

Solution: From Table 8.1 it is seen that for a load of 180 MW, the economic schedule is $P_{G1} = 80$ MW and $P_{G2} = 100$ MW. When load is shared equally $P_{G1} = P_{G2} = 90$ MW. Hence, the generation of unit 1 increases from 80 MW to 90 MW and that of unit 2 decreases from 100 MW to 90 MW, when the load is shared equally. There is an increase in cost of unit 1 since P_{G1} increases and decrease in cost of unit 2 since P_{G2} decreases.

$$\text{Increase in cost of unit 1} = \int_{80}^{90} \left(\frac{dF_1}{dP_{G1}}\right) dP_{G1}$$

$$= \int_{80}^{90} (0.01\, P_{G1} + 2.0) d_{PG1} = 28.5 \text{ \$/h}$$

$$\text{Decrease in cost of unit 2} = \int_{100}^{90} \left(\frac{dF_2}{dP_{G2}}\right) dP_{G2}$$

$$= \int_{100}^{90} (0.012\, P_{G2} + 1.6) dP_{G2} = -27.4 \text{ \$/h}$$

Total increase in cost if load is shared equally = 28.5 – 27.4 = 1.1 \$/h

Hence, the saving in fuel cost is 1.1 \$/h if coordinated economic schedule is used.

8.5 ITERATIVE TECHNIQUE

When losses are considered the economic schedule equations will be non-linear and have to be solved iteratively. Without losses, an analytic solution is obtained as explained in section 8.3 and 8.4. However, the iterative technique is illustrated here and later extended to the case when losses are included.

In iterative techniques, starting from an initial value of λ, better estimates are obtained in each iteration till $\Delta P = \sum_{i=1}^{n_g} P_{Gi} - P_D$ reaches the required accuracy (ideally zero). (8.17) is written as

$$f(\lambda) = P_D \qquad ...(8.20)$$

This non-linear equation can be solved by the steepest gradient technique. Expanding (8.20) in Taylor's series about an operating point $\lambda^{(r)}$ and neglecting higher order terms results in

$$f(\lambda)^{(r)} + \left(\frac{df(\lambda)}{d\lambda}\right)^{(r)} \Delta\lambda^{(r)} = P_D \qquad ...(8.21)$$

P_D is the actual demand and $f(\lambda)^{(r)}$ is the estimate in demand at the rth iteration.

$$\therefore \qquad P_D - f(\lambda)^{(r)} = \Delta P^{(r)} \qquad ...(8.22)$$

the error in estimate.

Further $\qquad f(\lambda) = P_D = \sum_{i=1}^{n_g} P_{Gi} = \sum_{i=1}^{n_g} \frac{\lambda - b_i}{2c_i}$

From which $\qquad \left(\frac{df(\lambda)}{d\lambda}\right)^r = \sum_{i=1}^{n_g} \frac{1}{2c_i} \qquad ...(8.23)$

Using (8.22) and (8.23), (8.21) can be written as

$$\Delta\lambda^{(r)} = \frac{\Delta P^{(r)}}{\sum_{i=1}^{n_g} \frac{1}{2c_i}} \qquad ...(8.24)$$

$$\lambda^{r+1} = \lambda^{(r)} + \Delta\lambda^{(r)} \qquad ...(8.25)$$

The iterative process is continued till $\Delta P^{(r)}$ reaches the required accuracy.

Example 8.6: Solve example 8.1 using an iterative technique.

Solution: Let the first estimate $\lambda^{(1)} = 30$ Rs/MWh

$$b_1 = 20;\ c_1 = 0.1;\ b_2 = 30;\ c_2 = 0.1$$

$$P_{G1}{}^{(1)} = \frac{\lambda^{(1)} - b_1}{2c_1} = \frac{30-20}{2\times 0.1} = 50 \text{ MW}$$

$$P_{G2}{}^{(1)} = \frac{\lambda^{(1)} - b_2}{2c_2} = \frac{30-30}{0.2} = 0 \text{ MW}$$

Total demand $\quad P_D = 200$ MW

$$\Delta P^{(1)} = 200 - (50 + 0) = 150 \text{ MW}$$

$$\Delta\lambda^{(1)} = \frac{\Delta P^{(1)}}{\frac{1}{2c_1} + \frac{1}{2c_2}} = \frac{150}{\frac{1}{0.2} + \frac{1}{0.2}} = 15 \text{ Rs/MWh}$$

$$\lambda^{(2)} = \lambda^{(1)} + \Delta\lambda^{(1)} = 30 + 15 = 45 \text{ Rs/MWh}$$

$$P_{G1}{}^{(2)} = \frac{\lambda^{(2)} - b_1}{2c_1} = \frac{45-20}{0.2} = 125 \text{ MW}$$

$$P_{G1}{}^{(2)} = \frac{\lambda^{(2)} - b_2}{2c_2} = \frac{45-30}{0.2} = 75 \text{ MW}$$

$$\Delta P^{(2)} = 200 - (125 + 75) = 0 \text{ MW}$$

The iterations are stopped since $\Delta P^{(2)} = 0$

Example 8.7: Solve example 8.3 using iterative technique when $P_D = 450$ MW.

Solution: Let $\lambda^{(1)} = 6$ \$/MWh

$$P_{G1}{}^{(1)} = \frac{6-7.2}{0.008} = -150 \text{ MW}$$

$$P_{G2}{}^{(1)} = \frac{6-7.3}{0.005} = -260 \text{ MW}$$

$$P_{G3}{}^{(1)} = \frac{6-6.74}{0.006} = -123.33 \text{ MW}$$

$$\Delta P^{(1)} = 450 - (-150 - 260 - 123.33) = 983.33 \text{ MW}$$

$$\Delta\lambda^{(1)} = \frac{\Delta P^{(1)}}{\sum_{i=1}^{3} \frac{1}{2c_i}} = \frac{983.33}{\frac{1}{0.008} + \frac{1}{0.005} + \frac{1}{0.006}} = 2.0 \text{ \$/MWh}$$

$$\lambda^{(2)} = \lambda^{(1)} + \Delta\lambda^{(1)} = 6 + 2 = 8.0 \text{ \$/MWh}$$

$$P_{G1}{}^{(2)} = \frac{8.0 - 7.2}{0.008} = 100 \text{ MW}$$

$$P_{G2}{}^{(2)} = \frac{8.0 - 7.3}{0.005} = 140 \text{ MW}$$

$$P_{G3}{}^{(2)} = \frac{8.0 - 6.74}{0.006} = 210 \text{ MW}$$

$$\Delta P^{(2)} = 450 - (100 + 140 + 210) = 0.0 \text{ MW}$$

Since $\Delta P = 0$, the iterations are stopped and $\lambda = 8$ \$/MWh for economic operation.

Example 8.8: In example 8.7, how does the schedule change if the limits on the generation are

$$125 \le P_G \le 300 \text{ MW}$$

Solution: In this case when $\lambda^{(2)} = 8$ \$/MWh, $P_{G1}{}^{(2)} = 100$ MW which is below the lower limit. Hence, $P_{G1}{}^{(2)}$ is pegged at 125 MW.

Then, $$\Delta P^{(2)} = 450 - (125 + 140 + 210) = -25 \text{ MW}$$

Since generator 1 is pegged, we need not consider its coefficients in further iterations.

$$\Delta\lambda^{(2)} = \frac{-25}{\left(\frac{1}{0.005} + \frac{1}{0.006}\right)} = -0.068 \text{ \$/MWh}$$

$$\lambda^{(3)} = 8 - 0.068 = 7.932 \text{ \$/MWh}$$

$$\Delta P_{G2}{}^{(3)} = \frac{7.932 - 7.3}{0.005} = 126.4 \text{ MW}$$

$$\Delta P_{G3}{}^{(3)} = \frac{7.932 - 6.74}{0.006} = 198.67 \text{ MW}$$

$$\Delta P^{(3)} = 450 - (125 + 126.4 + 198.67) = 450 - 450.07 = -0.07 \text{ MW}$$

$$\Delta\lambda^{(3)} = \frac{-0.07}{\dfrac{1}{0.005}+\dfrac{1}{0.006}} = -\ 1.9 \times 10^{-4}\ \$/\text{ MWh}$$

$$\lambda^{(4)} = 7.932 - 1.9 \times 10^{-4} = 7.9318\ \$/\text{MWh}$$

Since change in λ is very small we can stop the iterations here.

Economic schedule is

$$\lambda = 7.\ 9318\ \$/\text{MWh}$$

$$P_{G1} = 125\text{ MW}$$

$$P_{G2} = \frac{7.9318 - 7.3}{0.005} = 126.36\text{ MW}$$

$$P_{G3} = \frac{7.9318 - 6.74}{0.006} = 198.63\text{ MW}$$

8.6 ECONOMIC DISPATCH INCLUDING TRANSMISSION LOSSES

When transmission distances are large, the transmission losses are a significant part of the generation and have to be considered in the generation schedule for economic operation. The mathematical formulation is now stated as

Minimize $$F_T = \sum_{i=1}^{n_g} F_i$$

such that $$\sum_{i=1}^{n_g} P_{Gi} = P_D + P_L$$

where, P_L is the total loss.

The Lagrange function is now written as

$$£ = F_T - \lambda\left(\sum_{i=1}^{n_g} P_{Gi} - P_D - P_L\right) = 0 \quad ...(8.26)$$

The minimum point is obtained when

$$\frac{\partial £}{\partial P_{Gi}} = \frac{\partial F_T}{\partial P_{Gi}} - \lambda\left(1 - \frac{\partial P_L}{\partial P_{Gi}}\right) = 0;\ i = 1 \ldots n_g \quad ...(8.27)$$

$$\frac{\partial £}{\partial \lambda} = \sum_{i=1}^{n_g} P_{Gi} - P_D + P_L = 0 \text{ (Same as the constraint)}$$

Since, $\dfrac{\partial F_T}{\partial P_{Gi}} = \dfrac{dF_i}{dP_{Gi}}$, (8.27) can be written as

$$\frac{dF_i}{dP_{Gi}} + \lambda \frac{\partial P_L}{\partial P_{Gi}} = \lambda \qquad \text{...(8.28)}$$

$$\lambda = \frac{dF_i}{dP_{Gi}} \left(\frac{1}{1 - \dfrac{\partial P_L}{\partial P_{G_i}}} \right)$$

The term $\dfrac{1}{1 - \dfrac{\partial P_L}{\partial P_{Gi}}}$ is called the penalty factor of plant ***i***, L_i. The coordination equations including losses are given by

$$\lambda = \frac{dF_i}{dP_{Gi}} L_i;\ i = 1 \ldots n_g \qquad \text{...(8.29)}$$

The minimum operation cost is obtained when the product of the incremental fuel cost and the penalty factor of all units is the same, when losses are considered.

8.6.1 Approximate Penalty Factor

In general, $\dfrac{\partial P_L}{\partial P_{Gi}} << 1$. Hence, (8.28) can be approximated as

$$\lambda = \frac{dF_i}{dP_{Gi}} \left(1 + \frac{\partial P_L}{\partial P_{Gi}} \right) = \frac{dF_i}{dP_{Gi}} L_i' \qquad \text{...(8.30)}$$

Here, L_i' is the approximate penalty factor and is easier to use in iterative techniques. (8.30) can be written as

$$\frac{dF_i}{dP_{Gi}} + \frac{dF_i}{dP_{Gi}} \times \frac{\partial P_L}{\partial P_{Gi}} = \lambda \qquad \text{...(8.31)}$$

Thus, in the approximate penalty method, the incremental transmission loss of plant ***i*** is changed at a rate corresponding to the incremental fuel cost of plant ***i***.

8.6.2 Interpretation of Penalty Factors

The schematic representation of cost relations is shown in Fig. 8.6

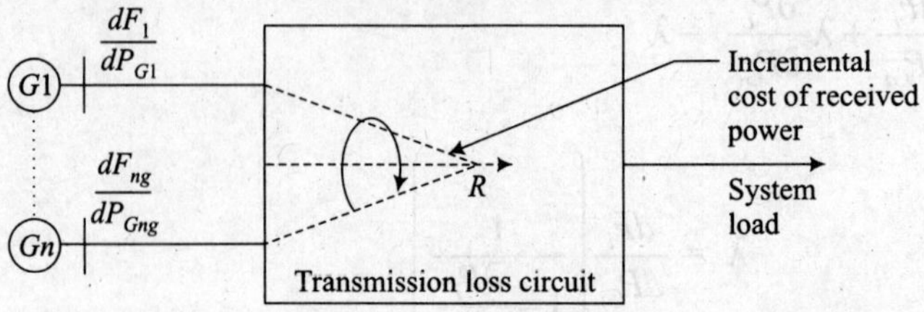

Fig. 8.6 Cost relations.

The incremental production cost of a plant ***i*** is measured at the plant bus and is given by $\frac{dF_i}{dP_{Gi}}$.

A given plant ***i***, incurs an incremental transmission loss $\frac{\partial P_L}{\partial P_{Gi}}$ in supplying the next increment of system load. If the load increases by an amount ΔP_D and this is completely met by increasing the output of plant 1 by ΔP_{G1}, the cost of this increment of power at the receiving end is given by

$$\lambda = \frac{dF_1}{dP_{G1}} \times \frac{\Delta P_{G1}}{\Delta P_D} \quad \text{...(8.32)}$$

$\Delta P_{G1} = \Delta P_D + \Delta P_L$, where ΔP_L is the increase in loss, due to the increase in demand. (8.32) can be written as

$$\lambda = \frac{dF_1}{dP_{G1}} \times \frac{\Delta P_{G1}}{\Delta P_{G1} - \Delta P_L} = \frac{dF_1}{dP_{G1}} \times \frac{1}{1 - \frac{\Delta P_L}{\Delta P_{G1}}} \quad \text{...(8.33)}$$

As $\Delta P_{G1} \to 0$

$$\lambda = \frac{dF_1}{dP_{G1}} \times \frac{1}{1 - \frac{\partial P_L}{\partial P_{G1}}} = \frac{dF_1}{dP_{G1}} L_1,$$

Thus, in general, the penalty factor of a plant ***i*** *is the ratio of the small change in power at plant* ***i*** *to a small change in receiving power, supplied by only plant* ***i***. Therefore, if ΔP_{Gi} increment leads to a larger proportion dissipated as losses, then $\frac{\Delta P_L}{\Delta P_{Gi}}$ tends to unity and the penalty factor of the plant increases. This means that a plant with a large penalty factor should operate at a low incremental cost, implying a low power output.

A rigorous general expression for the loss P_L is given by

$$P_L = \Sigma_m \Sigma_n P_{Gm} B_{mn} P_{Gn} + \Sigma_n P_{Gn} B_{no} + B_{oo} \quad \text{...(8.34)}$$

where, B_{mn}, B_{no}, B_{oo} called loss coefficients , depend on the load composition. The assumption here is that the load varies linearly between maximum and minimum values.

A simpler expression is

$$P_L = \Sigma_m \Sigma_n P_{Gm} B_{mn} P_{Gn} \quad ...(8.35)$$

The expression assumes that all load currents vary together as a constant complex fraction of the total load current. Experiences with large systems has shown that the loss of accuracy is not significant when (8.35) is used instead of (8.34).

An average set of loss coefficients may be used over the complete daily cycle in the coordination of incremental production costs and incremental transmission losses. In general, $B_{mn} = B_{nm}$ and (8.35) can be expanded for a two plant system as

$$P_L = B_{11} P_{G1}^2 + 2B_{12} P_{G1} P_{G2} + B_{22} P_{G2}^2 \quad ...(8.36)$$

Example 8.9: A generator is supplying a load. An incremental change in load of 4 MW requires generation to be increased by 6 MW. The incremental cost at the plant bus is Rs 30/MWh. What is the incremental cost at the receiving end?

Solution: $\dfrac{dF_1}{dP_{G1}} = 30$

$$\Delta P_L = \Delta P_G - \Delta P_D = 2 \text{ MW}$$

λ at receiving end is given by

$$\lambda = \frac{dF_1}{dP_{G1}} \times \frac{\Delta P_G}{\Delta P_D} = 30 \times \frac{6}{4} = 45 \text{ Rs/MWh}$$

or,

$$\lambda = \frac{dF_1}{dP_{G1}} \times \frac{1}{1 - \dfrac{\Delta P_L}{\Delta P_G}} = 30 \times \frac{1}{1 - \dfrac{2}{6}} = 45 \text{ Rs/MWh}$$

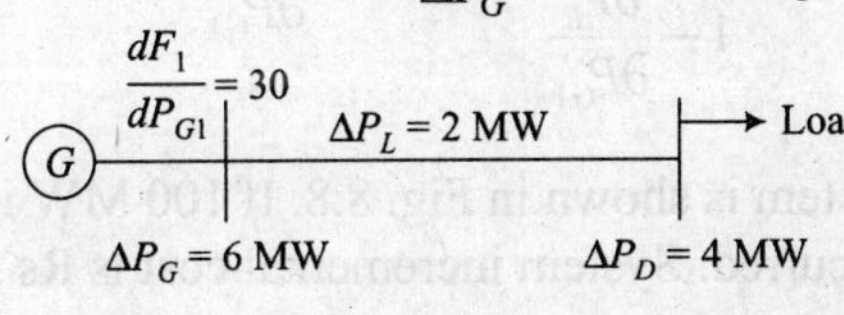

Fig. 8.7 One line diagram of example 8.9.

Note that the incremental cost at the receiving end is much higher than the incremental fuel cost of the source.

Example 8.10: In a system with two plants, the incremental fuel costs are given by

$$\frac{dF_1}{dP_{G1}} = 0.01 P_{G1} + 20 \text{ Rs/MWh}$$

$$\frac{dF_2}{dP_{G2}} = 0.015\,P_{G2} + 22.5 \text{ Rs/MWh}$$

The system is running under optimal schedule with $P_{G1} = P_{G2} = 100$ MW.

If $\dfrac{\partial P_L}{\partial P_{G2}} = 0.2$, find the plant penalty factors and $\dfrac{\partial P_L}{\partial P_{G1}}$.

Solution: For economic schedule,

$$\frac{dF_i}{dP_{Gi}} L_i = \lambda; \quad L_i = \frac{1}{1 - \dfrac{\partial P_L}{\partial P_{Gi}}};$$

For plant 2, $\quad P_{G2} = 100$ MW

$$\therefore \; (0.015 \times 100 + 22.5)\frac{1}{1-0.2} = \lambda.$$

Solving, $\quad \lambda = 30$ Rs/MWh

$$L_2 = \frac{1}{1-0.2} = 1.25$$

$$\frac{dF_1}{dP_{G1}} L_1 = \lambda \Rightarrow (0.01 \times 100 + 20)L_1 = 30$$

$$L_1 = 1.428$$

$$L_1 = \frac{1}{1 - \dfrac{\partial P_L}{\partial P_{G1}}}$$

$$1.428 = \frac{1}{1 - \dfrac{\partial P_L}{\partial P_{G1}}}; \text{ Solving } \frac{\partial P_L}{\partial P_{G1}} = 0.3$$

Example 8.11: A two bus system is shown in Fig. 8.8. If 100 MW is transmitted from plant 1 to the load, a loss of 10 MW is incurred. System incremental cost is Rs 30/MWh. Find P_{G1}, P_{G2} and power received by load if

$$\frac{dF_1}{dP_{G1}} = 0.02\,P_{G1} + 16.0 \text{ Rs/MWh}$$

$$\frac{dF_2}{dP_{G2}} = 0.04\,P_{G2} + 20.0 \text{ Rs/MWh}$$

Solution: Since, the load is connected at bus 2, no loss is incurred when plant two supplies the load.

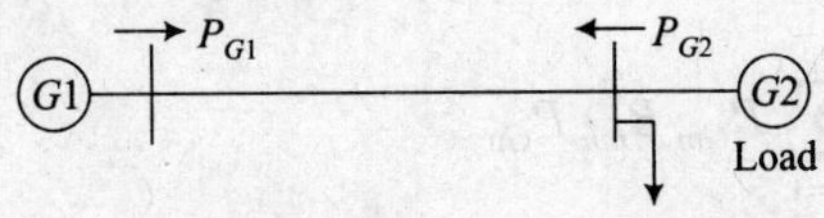

Fig. 8.8 One line diagram of example 8.11.

Therefore, in (8.36) $B_{12} = 0$ and $B_{22} = 0$

$$P_L = B_{11}\,P_{G1}^2;\quad \frac{\partial P_L}{\partial P_{G1}} = 2\,B_{11}\,P_{G1};\quad \frac{\partial P_L}{\partial P_{G2}} = 0.0$$

From data we have $P_L = 10$ MW, if $P_{G1} = 100$ MW

$$10 = B_{11}\,(100)^2$$

$$B_{11} = 0.001\ \text{MW}^{-1}$$

Coordination equation with loss is

$$\frac{dF_i}{dP_{Gi}} + \lambda\frac{\partial P_L}{\partial P_{Gi}} = \lambda$$

For plant 1 $$\frac{dF_1}{dP_{G1}} + \lambda\frac{\partial P_L}{\partial P_{G1}} = \lambda$$

$$(0.02\,P_{G1} + 16.0) + 30\,(2 \times 0.001 \times P_{G1}) = 30$$

$$0.08\,P_{G1} = 30 - 16.0. \text{ From which, } P_{G1} = 175 \text{ MW}$$

For Plant 2 $$\frac{dF_2}{dP_{G2}} + \lambda\frac{\partial P_L}{\partial P_{G2}} = \lambda$$

$$0.04\,P_{G2} + 20.0 = 30 \text{ or } P_{G2} = 250 \text{ MW}$$

$$\text{Loss} = B_{11}\,P_{G1}^2 = 0.001 \times (175)^2 = 30.625 \text{ MW}$$

$$P_D = (P_{G1} + P_{G2}) - P_L = 394.375 \text{ MW}$$

8.6.3 Iterative Technique for Solution of Economic Dispatch with Losses

A simple quadratic function for the loss as given in (8.35) is assumed

$$\frac{\partial P_L}{\partial P_{Gi}} = \frac{\partial}{\partial P_{Gi}}\left(\sum_{m=1}^{n_g}\sum_{n=1}^{n_g} P_{Gm}B_{mn}P_{Gn}\right) \qquad ...(8.37)$$

The double summation can be split into following terms.

(i) When $m \neq i$ and $n \neq i$ we get,

$$\sum_{\substack{m=1\\m\neq i}}^{n_g}\sum_{\substack{n=1\\n\neq i}}^{n_g} P_{Gm}\,B_{mn}\,P_{Gn}$$

Since, this term does not contain P_{Gi}, $\dfrac{\partial P_L}{\partial P_{Gi}} = 0$

(ii) When $m = i$, $n = 1... ng\ (\neq i)$, we get

$$\sum_{\substack{n=1\\n\neq i}}^{n_g} P_{Gi}\,B_{in}\,P_{Gn}$$

(iii) When $n = i$, $m = 1... ng\ (\neq i)$, we get

$$\sum_{\substack{m=1\\m\neq i}}^{n_g} P_{Gm}\,B_{mi}\,P_{Gi}$$

(iv) When $m = i$, $n = i$, we get

$$B_{ii}\,P_{Gi}^{\;2}$$

$$\therefore \qquad P_L = \sum_{\substack{m=1\\m\neq i}}^{n_g}\sum_{\substack{n=1\\n\neq i}}^{n_g} P_{Gm}\,B_{mn}\,P_{Gn} + \sum_{\substack{n=1\\n\neq i}}^{n_g} P_{Gi}\,B_{in}\,P_{Gn} + \sum_{\substack{m=1\\m\neq i}}^{n_g} P_{Gm}\,B_{mi}\,P_{Gi} + B_{ii}\,P_{Gi}^{\;2} \qquad ...(8.38)$$

Since, $B_{ij} = B_{ji}$, using (8.38), (8.37) now reduces to

$$\frac{\partial P_L}{\partial P_{Gi}} = \sum_{j=1}^{n_g} 2B_{ij}\,P_{Gj} \qquad ...(8.39)$$

The incremental fuel cost assuming a quadratic cost function is given by (8.3).

$$\frac{dF_i}{dP_{Gi}} = b_i + 2c_i\,P_{Gi}.$$

The coordination equation with losses for any plant is obtained by substituting (8.39) and (8.3) in (8.28).

$$\frac{dF_i}{dP_{Gi}} + \lambda\left(\frac{\partial P_L}{\partial P_{Gi}}\right) = \lambda$$

$$2c_i P_{Gi} + b_i + \lambda\left(\sum_{j=1}^{n_g} 2B_{ij} P_{Gj}\right) = \lambda \qquad ...(8.40)$$

(8.40) is rewritten as follows:

$$2c_i P_{Gi} + b_i + \lambda\,(2\,B_{ii}\,P_{Gi}) + \lambda\left(\sum_{\substack{j=1\\ j\neq i}}^{n_g} 2B_{ij}\; P_{Gj}\right) = \lambda \qquad ...(8.41)$$

$$P_{Gi} = \frac{\lambda - \lambda\left(\sum_{\substack{j=1\\ j\neq i}}^{n_g} 2B_{ij} P_{Gj}\right) - b_i}{2(c_i + \lambda B_{ii})} \qquad ...(8.42)$$

An iterative technique similar to the one presented in section 8.5 is used to iteratively solve for λ so that $\Delta P = \sum_{i=1}^{n_g} P_{Gi} - P_D - P_L$ reaches the required accuracy.

We can write an equation similar to (8.20).

$$f(\lambda) = \sum_{i=1}^{n_g} P_{Gi} = P_D + P_L \qquad ...(8.43)$$

Expanding (8.43) about an estimate $\lambda^{(r)}$ and neglecting higher order terms,

$$f(\lambda^{(r)}) + \left(\frac{df}{d\lambda}\right)^{(r)} \Delta\lambda^{(r)} = P_D + P_L^{(r)} \qquad ...(8.44)$$

From (8.43) $\left(\frac{df}{d\lambda}\right)^{(r)} = \sum_{i=1}^{n_g}\left(\frac{dP_{Gi}}{d\lambda}\right)^{(r)}$ and

From (8.42) $$\left(\frac{dP_{Gi}}{d\lambda}\right)^{(r)} = \frac{c_i + b_i B_{ii} - 2c_i \sum_{\substack{j=1\\ j\neq i}}^{n_g} B_{ij}\, P_{Gj}^{(r)}}{2(c_i + \lambda^{(r)} B_{ii})^2} \qquad ...(8.45)$$

Defining $$\Delta P^{(r)} = P_D + P_L^{(r)} - \sum_{i=1}^{n_g} P_{Gi}^{(r)} \qquad ...(8.46)$$

We obtain from (8.44)

$$\Delta\lambda^{(r)} = \frac{\Delta P^{(r)}}{\sum_{i=1}^{n_g}\left(\frac{dP_{Gi}}{d\lambda}\right)^r} \qquad ...(8.47)$$

The algorithm can now be stated as follows:

(i) Start with an estimate $\lambda^{(r)}$.

(ii) Using (8.42) solve for $P_{Gi}^{(r)}$.

(iii) Solve for $P_L^{(r)}$ using (8.35)

(iv) Compute $\left(\frac{dP_{Gi}}{d\lambda}\right)^{(r)}$ for all i, using (8.45).

(v) Compute $\Delta P^{(r)}$ from (8.46).

(vi) Compute $\Delta\lambda^{(r)}$ from (8.47).

(vii) $\lambda^{(r+1)} = \lambda^{(r)} + \Delta\lambda^{(r)}$.

(viii) Go to step (ii) till $\Delta P^{(r)} \le C$.

The above formulae can further be simplified by assuming all $B_{ij} = 0$ when $i \neq j$. Then,

$$P_L = \sum_{i=1}^{n_g} B_{ii} P_{Gi}^2 \qquad ...(8.48)$$

$$P_{Gi} = \frac{\lambda - b_i}{2(c_i + \lambda B_{ii})} \qquad ...(8.49)$$

and, $$\left(\frac{dP_{Gi}}{d\lambda}\right)^r = \frac{c_i + b_i B_{ii}}{2\left(c_i + \lambda^{(r)} B_{ii}\right)^2} \qquad ...(8.50)$$

Example 8.12: The fuel cost function of two units in \$/MWh are

$$F_1 = 320 + 6.2\,P_{G1} + 0.004\,P_{G1}^2$$
$$F_2 = 200 + 6.0\,P_{G1} + 0.003\,P_{G2}^2$$

where, P_{G1}, P_{G2} are in MW. The real power loss is given

$$P_L = 0.0125\,(P_{G1})^2 + 0.00625\,(P_{G2})^2$$

where the loss coefficients are in pu on a 100 MVA base. If the demand is 412.35 MW, find the optimal schedule.

Solution: $B_{11} = 0.0125 \text{ pu} = \dfrac{0.0125}{100} = 0.000125 \text{ MW}^{-1}$

$$B_{22} = 0.00625 \text{ pu} = \frac{0.00625}{100} = 0.0000625 \text{ MW}^{-1}$$

(B_{11} and B_{22} are multiplied by their base, $\dfrac{1}{100}$ MW^{-1})

$$\therefore \quad P_L = 0.000125\ P_{G1}^2 + 0.0000625\ P_{G2}^2 \text{ MW}$$

From data $\quad b_1 = 6.2; \quad c_1 = 0.004;$

$$b_2 = 6.0; \quad c_2 = 0.003;$$

Assume an initial value $\lambda^{(1)} = 7$ \$/MWh. From (8.49)

$$P_{G1}^{(1)} = \frac{\lambda^{(1)} - b_1}{2(c_1 + \lambda^{(1)} B_{11})} = \frac{7 - 6.2}{2(0.004 + 7 \times 0.000125)} = 82.05 \text{ MW}$$

$$P_{G2}^{(1)} = \frac{\lambda^{(1)} - b_2}{2(c_2 + \lambda^{(1)} B_{22})} = \frac{7 - 6.0}{2(0.003 + 7 \times 0.0000625)} = 145.45 \text{ MW}$$

$P_L^{(1)} = 0.000125 \times (82.05)^2 + 0.0000625\ (145.45)^2 = 2.1638$ MW.
From (8.50)

$$\sum_{i=1}^{2} \left(\frac{dP_{Gi}}{d\lambda} \right)^{(1)} = \frac{0.004 + 0.000125 \times 6.2}{2(0.004 + 7.0 \times 0.000125)^2} + \frac{0.003 + 0.0000625 \times 6.0}{2(0.003 + 7.0 \times 0.0000625)^2}$$

$$= 100.46 + 142.81 = 243.27$$

$$\Delta P^{(1)} = P_D + P_L^{(1)} - (P_{G1}^{(1)} + P_{G2}^{(1)})$$

$$= 412.35 + 2.1638 - (82.05 + 145.45)$$

$$= 187.0138 \text{ MW}$$

From (8.47) $\quad \Delta\lambda^{(1)} = \dfrac{\Delta P^{(1)}}{\displaystyle\sum_{i=1}^{n} \left(\frac{dP_{Gi}}{d\lambda} \right)^{(1)}}$

$$= \frac{187.0138}{243.27} = 0.769$$

$$\lambda^{(2)} = \lambda^{(1)} + \Delta\lambda^{(1)} = 7 + 0.769 = 7.769 \text{ \$/MWh}$$

In the second iteration, we get the values proceeding in the same manner.

$$P_{G1}^{(2)} = \frac{\lambda^{(2)} - b_1}{2(c_1 + \lambda^{(2)} B_{11})} = \frac{7.769 - 6.2}{2(0.004 + 7.769 \times 0.000125)} = 157.81 \text{ MW}$$

$$P_{G2}^{(2)} = \frac{\lambda^{(2)} - b_2}{2(c_2 + \lambda^{(2)} B_{22})} = \frac{7.769 - 6.0}{2(0.003 + 7.769 \times 0.0000625)} = 253.76 \text{ MW}$$

$$P_L^{(2)} = 0.000125 \times (157.81)^2 + 0.0000625\,(253.76)^2 = 7.1376 \text{ MW.}$$

$$\sum_{i=1}^{2}\left(\frac{dP_{Gi}}{d\lambda}\right)^{(2)} = \frac{0.004 + 0.000125 \times 6.2}{2(0.004 + 7.769 \times 0.000125)^2} + \frac{0.003 + 0.0000625 \times 6.0}{2(0.003 + 7.769 \times 0.0000625)^2}$$

$$= 96.61 + 138.89 = 235.5$$

$$\Delta P^{(2)} = P_D + P_L^{(2)} - (P_{G1}^{(2)} + P_{G2}^{(2)})$$

$$= 412.35 + 7.1376 - (157.81 + 253.76)$$

$$= 7.9176 \text{ MW}$$

$$\Delta\lambda^{(2)} = \frac{\Delta P^{(2)}}{\sum_{i=1}^{n}\left(\frac{dP_{Gi}}{d\lambda}\right)^{(2)}}$$

$$= \frac{7.9176}{235.5} = 0.034$$

$$\lambda^{(3)} = \lambda^{(2)} + \Delta\lambda^{(2)} = 7.769 + 0.034 = 7.803 \text{ \$/MWh}$$

In the third iteration, we get

$$P_{G1}^{(3)} = \frac{\lambda^{(3)} - b_1}{2(c_1 + \lambda^{(3)} B_{11})} = \frac{7.803 - 6.2}{2(0.004 + 7.803 \times 0.000125)} = 161.09 \text{ MW}$$

$$P_{G2}^{(3)} = \frac{\lambda^{(3)} - b_2}{2(c_2 + \lambda^{(3)} B_{22})} = \frac{7.803 - 6.0}{2(0.003 + 7.803 \times 0.0000625)} = 258.48 \text{ MW}$$

$$P_L^{(3)} = 0.000125 \times (161.09)^2 + 0.0000625\,(258.48)^2 = 7.42 \text{ MW.}$$

$$\sum_{i=1}^{2}\left(\frac{dP_{Gi}}{d\lambda}\right)^{(3)} = \frac{0.004 + 0.000125 \times 6.2}{2(0.004 + 7.803 \times 0.000125)^2} + \frac{0.003 + 0.0000625 \times 6.0}{2(0.003 + 7.803 \times 0.0000625)^2}$$

$$= 96.44 + 138.73 = 235.17$$

$$\Delta P^{(3)} = P_D + P_L^{(3)} - (P_{G1}^{(3)} + P_{G2}^{(3)})$$

$$= 412.35 + 7.42 - (161.09 + 258.48)$$

$$= 0.2 \text{ MW}$$

$$\Delta\lambda^{(3)} = \frac{\Delta P^{(3)}}{\sum_{i=1}^{n}\left(\frac{dP_{Gi}}{d\lambda}\right)^{(3)}}$$

$$= \frac{0.2}{235.17} = 0.00085$$

The iterations can be stopped and the following values taken for economic operation.

$$\lambda = 7.803 + 0.00085 = 7.80385 \text{ \$/MWh}$$

$$P_{G1} = 161.17 \text{ MW}$$

$$P_{G2} = 258.6 \text{ MW}$$

8.6.4 Derivation of Transmission Loss Formula

An accurate method of obtaining general loss coefficients has been presented by Kron. The method is elaborate and a simpler approach is possible by making the following assumptions:

(i) All load currents have same phase angle with respect to a common reference.
(ii) The ratio X/R is the same for all the network branches.

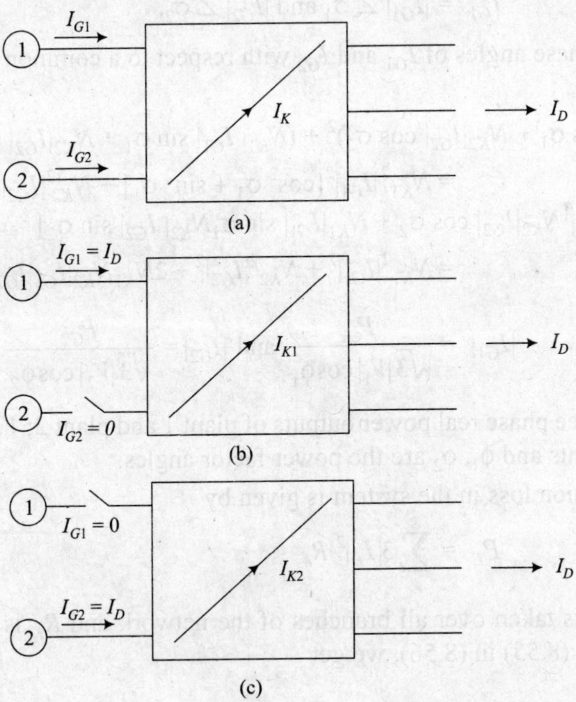

Fig. 8.9 Two plants connected to a number of loads through a transmission network.

Consider the simple case of two generating plants connected to an arbitrary number of loads through a transmission network as shown in Fig. 8.9a.

Let's assume that the total load is supplied by only generator 1 as shown in Fig. 8.9b. Let the current through a branch K in the network be I_{K1}. We define

$$N_{K1} = \frac{I_{K1}}{I_D} \qquad ...(8.51)$$

It is to be noted that $I_{G1} = I_D$ in this case. Similarly, with only plant 2 supplying the load current I_D, as shown in Fig. 8.9c, we define

$$N_{K2} = \frac{I_{K2}}{I_D} \qquad ...(8.52)$$

N_{K1} and N_{K2} are called current distribution factors and their values depend on the impedances of the lines and the network connection. They are independent of I_D. When both generators are supplying the load, then by principle of superposition

$$I_K = N_{K1} I_{G1} + N_{K2} I_{G2} \qquad ...(8.53)$$

where, I_{G1}, I_{G2} are the currents supplied by plants 1 and 2, respectively, to meet the demand I_D. Because of the assumptions made, I_{K1} and I_D have same phase angle, as do I_{K2} and I_D. Therefore, the current distribution factors are real rather than complex. Let

$$I_{G1} = |I_{G1}| \angle \sigma_1 \text{ and } |I_{G2}| \angle \sigma_2.$$

where, σ_1 and σ_2 are phase angles of I_{G1} and I_{G2} with respect to a common reference. From (8.53) we can write

$$|I_K|^2 = (N_{K1}|I_{G1}| \cos\sigma_1 + N_{K2}|I_{G2}| \cos\sigma_2)^2 + (N_{K1}|I_{G1}| \sin\sigma_1 + N_{K2}|I_{G2}| \sin\sigma_2)^2$$

$$= N_{K1}^2|I_{G1}|^2[\cos^2\sigma_1 + \sin^2\sigma_1] + N_{K2}^2|I_{G2}|^2[\cos^2\sigma_2 + \sin^2\sigma_2]$$

$$+ 2[N_{K1}|I_{G1}| \cos\sigma_1 N_{K2}|I_{G2}| \cos\sigma_2 + N_{K1}|I_{G1}| \sin\sigma_1 N_{K2}|I_{G2}| \sin\sigma_2]$$

$$= N_{K1}^2|I_{G1}|^2 + N_{K2}^2|I_{G2}|^2 + 2N_{K1}N_{K2}|I_{G1}||I_{G2}| \cos(\sigma_1 - \sigma_2) \qquad ...(8.54)$$

Now, $$|I_{G1}| = \frac{P_{G1}}{\sqrt{3}|V_1|\cos\phi_1} \text{ and } |I_{G2}| = \frac{P_{G2}}{\sqrt{3}|V_2|\cos\phi_2} \qquad ...(8.55)$$

where, P_{G1}, P_{G2} are three phase real power outputs of plant 1 and plant 2; V_1, V_2 are the line to line bus voltages of the plants and ϕ_1, ϕ_2 are the power factor angles.

The total transmission loss in the system is given by

$$P_L = \sum_k 3|I_k|^2 R_k \qquad ...(8.56)$$

where the summation is taken over all branches of the network and R_K is the branch resistance. Substituting (8.54) and (8.55) in (8.56), we get

$$P_L = \frac{P_{G1}^{\ 2}}{|V_1|^2 (\cos\phi_1)^2} \sum_K N_{K1}^{\ 2} R_K + \frac{2 P_{G1} P_{G2} \cos(\sigma_1 - \sigma_2)}{|V_1||V_2| \cos\phi_1 \cos\phi_2} \sum_K N_{K1} N_{K2} R_K$$

$$+ \frac{P_{G2}^{\ 2}}{|V_2|^2 (\cos\phi_2)^2} \sum_K N_{K2}^{\ 2} R_K \qquad ...(8.57)$$

(8.57) can be written as $P_L = P_{G1}^{\ 2} B_{11} + 2P_{G1}P_{G2} B_{12} + P_{G2}^{\ 2} B_{22}$

where,

$$B_{11} = \frac{1}{|V_1|^2 (\cos\phi_1)^2} \sum_K N_{K1}^{\ 2} R_K$$

$$B_{12} = \frac{\cos(\sigma_1 - \sigma_2)}{|V_1||V_2| \cos\phi_1 \cos\phi_2} \sum_K N_{K1} N_{K2} R_K$$

$$B_{22} = \frac{1}{|V_2|^2 (\cos\phi_2)^2} \sum_K N_{K2}^{\ 2} R_K$$

The loss coefficients are called the *B*-coefficients and have unit MW^{-1}.

For a general system with *n* plants, the transmission loss is expressed as

$$P_L = \frac{P_{G1}^{\ 2}}{|V_1|^2 (\cos\phi_1)^2} \sum_K N_{K1}^{\ 2} + ... + \frac{P_{Gn}^{\ 2}}{|V_n|^2 (\cos\phi_n)^2} \sum_K N_{Kn}^{\ 2} R_K$$

$$+ 2 \sum_{\substack{p,q=1 \\ p \neq q}}^{n} \frac{P_{GP} P_{Gq} \cos(\sigma_p - \sigma_q)}{|V_p||V_q| \cos\phi_p \cos\phi_q} \sum_k N_{KP} N_{Kq} R_K$$

In a compact form

$$P_L = \sum_{p=1}^{n} \sum_{q=1}^{n} P_{Gp} B_{Pq} P_{Gq} \qquad ...(8.58)$$

$$B_{Pq} = \frac{\cos(\sigma_p - \sigma_q)}{|V_p||V_q| \cos\phi_P \cos\phi_q} \sum_K N_{KP} N_{Kq} R_K \qquad ...(8.59)$$

B-coefficients can be treated as constants over the load cycle by computing them at average operating conditions, without significant loss of accuracy.

Example 8.13: Calculate the loss coefficients in *pu* and MW^{-1} on a base of 50 MVA for the network of Fig. 8.10. Corresponding data is given below.

$I_a = 1.2 - j\,0.4$ pu	$Z_a = 0.02 + j\,0.08$ pu
$I_b = 0.4 - j\,0.2$ pu	$Z_b = 0.08 + j\,0.32$ pu
$I_c = 0.8 - j\,0.1$ pu	$Z_c = 0.02 + j\,0.08$ pu
$I_d = 0.8 - j\,0.2$ pu	$Z_d = 0.03 + j\,0.12$ pu
$I_e = 1.2 - j\,0.3$ pu	$Z_e = 0.03 + j\,0.12$ pu

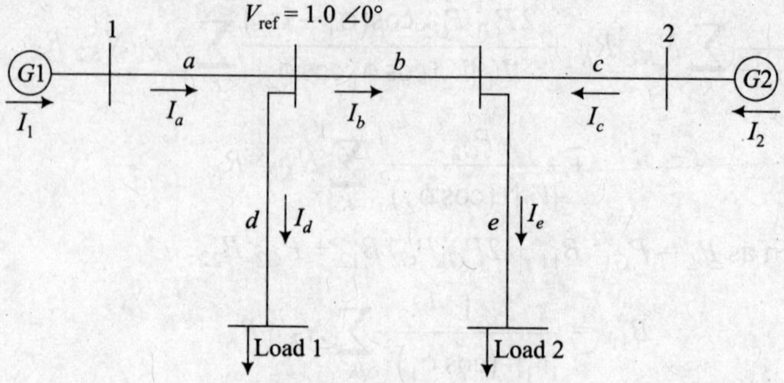

Fig. 8.10 Example 8.13.

Solution: Total load current

$$I_L = I_d + I_e = 2.0 - j0.5 = 2.061 \angle -14.03° \text{ A}$$

$$I_{L1} = I_d = 0.8 - j0.2 = 0.8246 \angle -14.03° \text{ A}$$

$$\frac{I_{L1}}{I_L} = 0.4; \ \frac{I_{L2}}{I_L} = 1.0 - 0.4 = 0.6$$

If generator 1, supplies the load then $I_1 = I_L$. The current distribution is shown in Fig. 8.11a.

$$N_{a1} = \frac{I_a}{I_L} = 1.0; N_{b1} = \frac{I_b}{I_L} = 0.6; N_{c1} = 0; N_{d1} = 0.4; N_{e1} = 0.6.$$

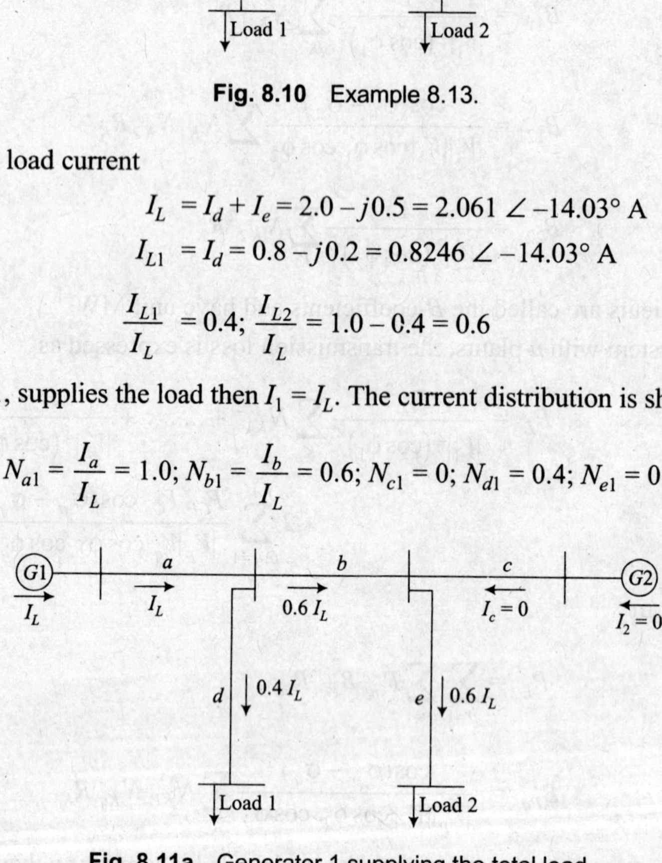

Fig. 8.11a Generator 1 supplying the total load.

Similarly, the current distribution when only generator 2 supplies the load is shown in Fig. 8.11b.

$$N_{a2} = 0; N_{b2} = -0.4; N_{c2} = 1.0; N_{d2} = 0.4; N_{e2} = 0.6$$

From Fig. 8.10,

$$V_1 = V_{ref} + Z_a I_a$$

$$= 1\angle 0° + (1.2 - j0.4)\,(0.02 + j0.08)$$

$$= 1.06 \angle 4.78° = 1.056 + j0.088 \text{ pu.}$$

$$V_2 = V_{ref} - I_b Z_b + I_c Z_c$$

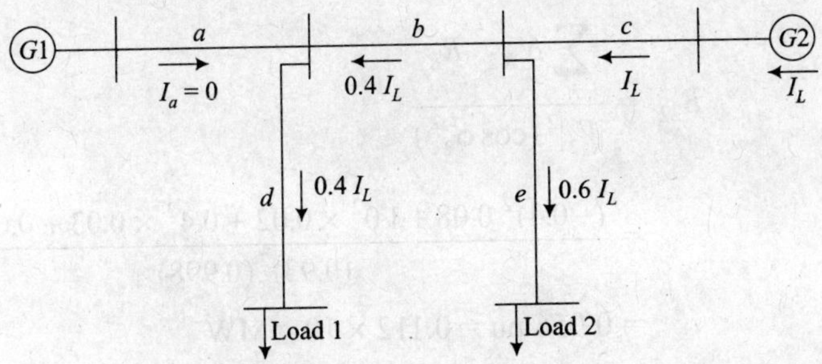

Fig. 8.11b Generator 2 supplying the total load.

$$= 1.0 \angle 0° - (0.4 - j0.2)\,(0.08 + j0.32) + (0.8 - j0.1)\,(0.02 + j0.08)$$

$$= 0.928 - j0.05 = 0.93 \angle -3.10° \text{ pu.}$$

Current phase angles

$$\sigma_1 = \text{angle of } I_1 (= I_a) = \tan^{-1}\left(\frac{-0.4}{1.2}\right) = -18.43°$$

$$\sigma_2 = \text{angle of } I_2 (= I_c) = \tan^{-1}\left(\frac{-0.1}{0.8}\right) = -7.13°$$

$$\cos(\sigma_1 - \sigma_2) = 0.98$$

Power factor angles

$$\phi_1 = 4.78° + 18.43 = 23.21°;\ \cos\phi_1 = 0.92$$

$$\phi_2 = 7.13° - 3.10° = 4.03°;\ \cos\phi_2 = 0.998$$

$$B_{11} = \frac{\sum_K N_{K1}{}^2 R_K}{|V_1|^2 (\cos\phi_1)^2} = \frac{1.0^2 \times 0.02 + 0.6^2 \times 0.08 + 0.4^2 \times 0.03 + 0.6^2 \times 0.03}{(1.06)^2 (0.920)^2}$$

$$= 0.0677 \text{ pu}$$

$$= 0.0677 \times \frac{1}{50} = 0.1354 \times 10^{-2}\ \text{MW}^{-1}$$

$$B_{12} = \frac{\cos(\sigma_1 - \sigma_2)}{|V_1||V_2|(\cos\phi_1)(\cos\phi_2)} \sum_K N_{K1} N_{K2} R_K$$

$$= \frac{0.98}{(1.06)(0.93)(0.998)(0.92)} [-0.4 \times 0.6 \times 0.08 + 0.4 \times 0.4 \times 0.03 + 0.6 \times 0.6 \times 0.03]$$

$$= -0.00389 \text{ pu}$$

$$= -0.0078 \times 10^{-2}\ \text{MW}^{-1}$$

$$B_{22} = \frac{\sum_K N_{K2}^{\ 2} R_K}{|V_2|^2 (\cos \phi_2^{\ 2})}$$

$$= \frac{(-0.4)^2\, 0.08 + 1.0^2 \times 0.02 + 0.4^2 \times 0.03 + 0.6^2 \times 0.03}{(0.93)^2 (0.998)^2}$$

$$= 0.056 \text{ pu} = 0.112 \times 10^{-2} \text{ MW}^{-1}$$

Example 8.14: Compute the loss coefficients for the network shown in Fig. 8.12 using the given data.

$$I_a = 1.0 - j\,0.15 \text{ pu} \quad Z_a = 0.02 + j\,0.15 \text{ pu}$$
$$I_b = 0.5 - j\,0.10 \text{ pu} \quad Z_b = 0.03 + j\,0.15 \text{ pu}$$
$$I_c = 0.2 - j\,0.05 \text{ pu} \quad Z_c = 0.02 + j\,0.25 \text{ pu}$$

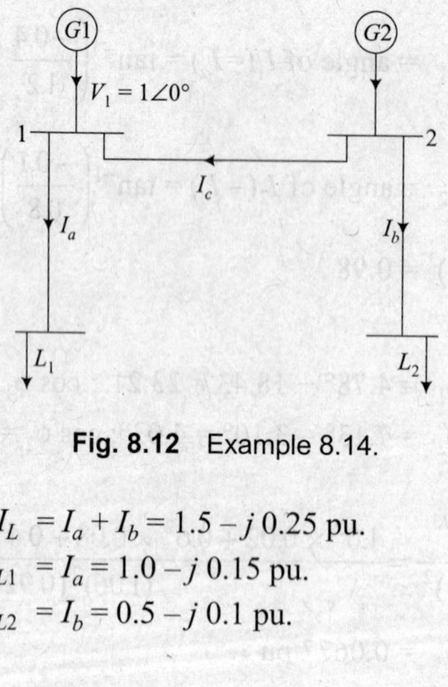

Fig. 8.12 Example 8.14.

Solution:

$$I_L = I_a + I_b = 1.5 - j\,0.25 \text{ pu.}$$
$$I_{L1} = I_a = 1.0 - j\,0.15 \text{ pu.}$$
$$I_{L2} = I_b = 0.5 - j\,0.1 \text{ pu.}$$

$$\left|\frac{I_{L1}}{I_{L1} + I_{L2}}\right| = \left|\frac{1.0 - j0.15}{1.5 - j0.25}\right| = 0.6649$$

$$\left|\frac{I_{L2}}{I_{L1} + I_{L2}}\right| = 1 - 0.6649 = 0.3351$$

$N_{a1} = 0.6649$; $N_{b1} = 0.3351$; $N_{c1} = -0.3351$

$N_{a2} = 0.6649$; $N_{b2} = 0.3351$; $N_{c3} = 0.6649$

$$V_2 = V_1 + Z_c I_c = 1 \angle 0° + (0.02 + j\,0.25)\,(0.2 - j\,0.05)$$
$$= 1.0176 \angle 2.76°$$

$$I_{G1} = I_a - I_c = 0.8 - j0.1 = 0.8062 \angle 7.12°; \sigma_1 = -7.12°$$

$$\cos \phi_1 = \cos(-7.12°) = 0.9923$$

$$I_{G2} = I_b + I_c = 0.7 - j0.15 = 0.7159 \angle -12.09°; \sigma_2 = -12.09°$$

$$\cos \phi_2 = \cos(2.76° + 12.09°) = 0.9966$$

$$B_{11} = \frac{\cos(\sigma_1 - \sigma_1)}{|V_1|^2 (\cos \phi_1)^2} [N_{a1}{}^2 R_a + N_{b1}{}^2 R_b + N_{c1}{}^2 R_c]$$

$$= \frac{1}{(1.0)^2 (0.9923)^2} [0.6649^2 \times 0.02 + 0.33512 \times 0.03 (-0.3351)^2 \, 0.02$$

$$= 0.01468 \text{ pu.}$$

$$B_{12} = \frac{\cos(\sigma_1 - \sigma_2)}{|V_1||V_2|(\cos \phi_1)(\cos \phi_2)} [N_{a1}N_{a2}R_a + N_{b1}N_{b2}R_b + N_{c1}N_{c2}R_c]$$

$$= \frac{0.996}{1 \times 1.0176 \times 0.9923 \times 0.9666} [0.6649 \times 0.6649 \times 0.02 + (0.3351)^2 \, 0.03$$

$$+ (-0.3351)(0.6649)\, 0.02] = 0.0079 \text{ pu.}$$

$$B_{22} = \frac{\cos(\sigma_2 - \sigma_2)}{|V_2|^2 (\cos \phi_2)^2} [N_{a2}{}^2 R_a + N_{b2}{}^2 R_b + N_{c2}{}^2 R_c]$$

$$= \frac{1}{(1.0176)^2 (0.9666)^2} [0.6649^2 \times 0.02 + 0.3351^2 \times 0.03 + 0.6649^2 \times 0.02]$$

$$= 0.02175 \text{ pu}$$

8.7 UNIT COMMITMENT

Generally, costs are incurred in start-up and shutdown of an unit. It is not economical to run all the units available all the time. The total load of the power system varies throughout the day and also from day-to-day. The utility has to decide in advance, the generators to start-up and shutdown, and their sequence of starting and shutting down. The computational procedure which determines the units of a plant which must operate to meet a load is called unit commitment (UC). This problem is more relevant to thermal units, since hydro-plants have negligible operating cost and start-up times.

The on-line economic dispatch problem solves to economically distribute the actual system load on-line to various plants and the UC problem decides on the best set of units to be operated to meet the plant demand. In a practical problem, the UC must be decided over the complete load cycle. As an example, consider a plant which has 4 units and has to supply 100 MW. The UC problem solution yields the units which have to be operated to meet the demand, most economically. This is determined by the production cost of each unit. If the load now changes to say 150 MW, UC decides the new set of units to be committed. This would involve not only the production cost but also the transition cost associated with changing from one combination to another, due to start-up and shutdown costs of the units. Generally, the start-up cost of a unit depends on how long the unit has been shutdown after its last operation. The shutdown cost is normally constant and does not depend on how long the unit has been running. The costs involved with changing from one combination to another are called transition costs.

Assuming a plant has K units and is never on zero load, at least one unit must be on, to supply the system load. If each unit can be considered on or off, there are $2^K - 1$ combinations of units, to be searched to determine the most economical combination. This would be computationally burdensome and hence cannot be used. If the load is assumed to increase in small but finite steps, dynamic programming can be applied to solve for computing UC.

8.7.1 Constraints in Unit Commitment

When units are committed, the constraints imposed are dependent on the system and the type of generation. Some of the constraints to be considered are discussed below.

(i) Thermal unit constraints

Thermal units have some constraints on the operating time. A thermal unit cannot be turned off suddenly and has to be ON for a minimum time after it is committed. This is called the UP time. Therefore, for any unit K

$$T_{\text{ON}, k} \geq T_{\text{UP}, k}.$$

Similarly, once it is shut down it cannot be started before a minimum time, called DOWN time, has elapsed.

(ii) Start-up costs of thermal units

A thermal unit incurs some cost during start up. When load is removed, the boiler can be cooled or banked. When a boiler is banked, it is maintained on no load at the operating temperature and pressure, by intermittent firing. The banking/cooling cost is as shown below:

If the unit is not required for up to H hours, it is cheaper to bank it rather cool it. However, if the unit is not required for more than H hours, it is cheaper to cool it.

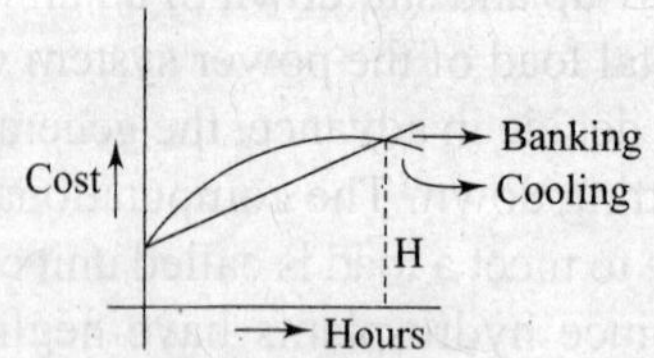

Fig. 8.13 Banking/cooling cost.

(iii) Spinning reserve constraint

Spinning reserve is the unused capacity of generators synchronized to the grid and which can be activated to supply active power. It is the sum of all generation synchronized to the grid minus the load + losses. Spinning reserve is needed to maintain system stability and also meet dynamic variations in load.

(iv) Network constraints

The network interconnection will determine which units need to be committed, due to limits imposed by transmission line capabilities. These limits, may necessitate committing more expensive units, even when cheaper units are available.

8.7.2 Dynamic Programming

The following assumptions are made in applications of dynamic programming:

- Individual production costs are known *a priori* for all the units.
- Load cycle is known *a priori.*
- The load on each unit changes in small steps.

Let's consider again a plant with 4 units, which has to supply say 4 MW, and the load varies in step sizes of 1 MW. Given the production costs of all the units the UC problem can be solved as follows:

Step 1. Consider any two units arbitrarily to be committed (ON) and determine the most economical combination for all the discrete load levels. Let's assume units 1 and 2 are committed. To supply 4 MW we have following options:

(a) Unit 1 supplies 4 MW and Unit 2, 0 MW.

(b) Unit 1 supplies 3 MW and Unit 2, 1 MW.

(c) Unit 1 supplies 2 MW and Unit 2, 2 MW.

(d) Unit 1 supplies 1 MW and Unit 2, 3 MW.

(e) Unit 1 supplies 0 MW and Unit 2, 4 MW.

For each of the combination, determine the cost of producing 4 MW and choose the combination which yields lowest cost. Likewise, determine the best option for supplying 3 MW, 2 MW and 1 MW with 2 units.

Step 2. The third unit is now committed. To meet 4 MW we now have following options:

(a) Units 1 and 2 together supply 4 MW (the most economical combination of step 1 for 4 MW) and unit 3, 0 MW.

(b) Units 1 and 2 together supply 3 MW (the most economical combination of step 1 for 3 MW) and unit 3, 1 MW and so on.

For each combination the cost of producing 4 MW (with 3 units) is computed and the one with least cost is committed.

Step 3. The fourth unit is committed and the process repeated as in steps 1 and 2, to find the optimum production.

Thus, in this approach, having obtained the optimal way of loading K units, it is easy to determine the optimal loading of $(K + 1)$ units. A systematic mathematical formulation for the above produce is presented.

We define the following quantities:

(i) The cost function $F_N(x)$ is the minimum cost in Rs/h or \$/h of generating x MW with N units. For example, $F_2(10)$ is the minimum cost of generating 10 MW with 2 units.

(*ii*) $f_N(y)$ is the cost of generating y MW with the Nth unit. For example, $f_2(10)$ is the cost of generating 10 MW by the second unit.

The application of dynamic programming results in a recursive relation.

$$F_N(x) = \min_y [f_N(y) + F_{N-1}(x-y)] \quad ...(8.60)$$

In the above formula, x MW are generated by totally N units. Out of these, y MW are generated by the Nth unit at a cost $f_N(y)$ and $(x-y)$ MW are generated by the other N-1 units. From the above recursive relation, we can easily determine the combination of units, which yields minimum operating costs for various loads in convenient steps, from the minimum permissible load of the smallest unit, to the sum of the capacities of all available units.

Example 8.15: Determine the most economical units to be committed for a load of 8 MW, if the load step is 1 MW and there are 4 units in the plant. The cost curve for any unit ***i*** at a generation level P_{Gi} is given by $F_i = a_i P_{Gi}^2 + b_i P_{Gi}$ Rs/h. The coefficients a and b are given for the various units.

Unit	*a*	*b*
1	0.35	25
2	0.8	27
3	1.0	30
4	1.2	30

Neglect transition costs.

Solution:

Step 1: Compute the cost of generation for each unit in steps of 1 MW i.e., calculate $f_i(y)$ for y varying from 1 to 8 and ***i*** varying from 1 to 4. For example, $f_1(6)$ is the cost of generating 6 MW by unit 1 and is given by

$$f_1(6) = 0.35(6)^2 + 25 \times 6 = 162.6 \text{ Rs/h.}$$

The computations are given in Table 8.2.

Table 8.2 Cost of generation in Rs/hr

Load level MW *Unit No.*	*1*	*2*	*3*	*4*	*5*	*6*	*7*	*8*
1	25.35	51.4	78.15	105.6	133.75	162.6	192.15	222.4
2	27.8	57.2	88.2	120.8	155	190.8	228.2	267.2
3	31	64	99	136	175	216	259	304
4	31.2	64.8	100.8	139.2	180	223.2	268.8	316.8

We now assume that 8 MW is generated by one unit, say the first unit. (The first unit is chosen arbitrarily)

$F_1(8) = 222.4$ Rs/h

Similarly, we compute $F_1(7), F_1(6)$...............$F_1(1)$.

Step 2. Next unit 2 is also committed and the optimum combination between units 1 and 2 is chosen.

$$F_2(8) = \text{Min}\,[(f_2(0) + F_1(8)), (f_2(1) + F_1(7)), ...(f_2(8) + F_1(0))]$$

This computation is shown in Table 8.3.

Table 8.3 Cost of generations 8 MW with 2 units

F_1	$222.4_{(8)}$	$192.15_{(7)}$	$162.6_{(6)}$	$133.75_{(5)}$	$105.6_{(4)}$	$78.15_{(3)}$	$51.4_{(2)}$	$25.35_{(1)}$
f_2	$0_{(0)}$	$27.8_{(1)}$	$57.2_{(2)}$	$88.2_{(3)}$	$120.8_{(4)}$	$155_{(5)}$	$190.8_{(6)}$	$228.2_{(7)}$
(TOTAL) F_2	222.4	219.95	**219.8**	221.95	226.4	223.15	242.2	253.55

- The number in the bracket indicates the MW generated by the unit.
- From Table 8.3 it can be seen than that the minimum F_2 (8) is 219.8 Rs/h giving the combination of 6 MW by unit 1 and 2 MW by unit 2. Similarly. $F_2(7), F_2(6)$..... are also computed.

Step 3. Unit 3 is also committed, we already know the optimal combination of units 1 and 2 for various load levels. (i.e., $F_2(8), F_2(7)$...).

$$F_3(8) = \text{Min}[(f_3(0) + F_2(8)), (f_3(1) + F_2(7)), ...\ (f_3(8) + F_2(0))]$$

From computations it is found

$$F_3(8) = f_3(0) + F_2(8) = 219.8 \text{ Rs/h.}$$

Step 4. Unit 4 is also committed.

$$F_4(8) = \text{Min}[(f_4(0) + F_3(8)), (f_4(1) + F_3(7)), ...\ (f_4(8) + F_3(0))]$$

This is found to be $F_4(8) = f_4(0) + F_3(8) = 219.8$ Rs/h.

Step 5. The final commitment is obtained by a simple backward step.

$$\begin{aligned} F_4(8) &= f_4(0) + F_3(8) \\ &= f_4(0) + f_3(0) + F_2(8) \\ &= f_4(0) + f_3(0) + f_2(2) + F_1(6) \\ &= f_4(0) + f_3(0) + f_2(2) + f_1(6) \\ &= 219.8 \text{ Rs/h.} \end{aligned}$$

Hence, we commit units 1 and 2. If we consider 1 as ON and 0 as OFF, the combination for 8 MW is 1100. If higher accuracy is desired, the step size could be reduced. The UC table is prepared for the complete load range. As the load cycle varies, it would involve starting and stopping of units according to the UC table. In general, *DP* method is used to obtain UC table for discrete load levels, while the actual load sharing among committed units is decided by the use of coordination equations. In the above problem, for a load variation from 1 to 20 MW the UC table is specified below in Table 8.4.

Table 8.4 UC table for example 8.15

Load range	*Unit Number*			
MW	*1*	*2*	*3*	*4*
01 – 05	1	0	0	0
06 – 13	1	1	0	0
14 – 18	1	1	1	0
19 – 20	1	1	1	1

* 1 (ON) 0 (OFF)

8.8 OPTIMAL SCHEDULING FOR HYDROTHERMAL PLANTS

In thermal plants, the operating cost is determined at any instant, without reference to other times. Hence, the problem reduces to one of static optimization. On the other hand, hydro-plants have negligible operating cost, but have constraints on water availability over a period of time. Hence, the problem becomes a problem of dynamic optimization. In hydrothermal systems, the problem boils down to minimizing the fuel cost of thermal plants, under constraints of water availability, over the period of optimization.

The mathematical formulation of the problem and the solution technique are illustrated with a simple hydrothermal system, with one thermal plant and one hydel plant, as shown in Fig. 8.14.

Optimization is carried out with real power generation as the control variable.

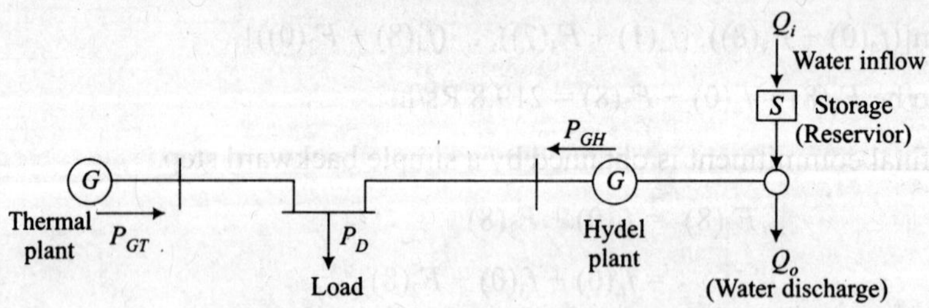

Fig. 8.14 Simple hydrothermal system.

8.8.1 Problem Formulation

The following assumptions are made over the period of operation *T*:

- Load demand as a function of time is known.
- Storage of reservoir is specified at the beginning and end of the period.

•The water inflow to the reservoir is known.

The problem is one of determining the rate of water discharge, to minimize the cost of thermal generation F_T, under certain constraints. Mathematically, we state as follows:

Minimize $$F_T = \int_0^T F'(P_{GT}(t))\,dt \qquad ...(8.61)$$

Under the following constraints

(i) **Power balance** given by

$$P_{GT}(t) + P_{GH}(t) - P_L(t) - P_D(t) = 0;\ t \in (0, T) \qquad ...(8.62)$$

where,
P_{GT} = Thermal generation
P_{GH} = Hydel generation
P_L = System loss
P_D = Demand

(ii) **Water availability**, given by

$$S(T) - S(0) - \int_0^T Q_i(t)dt + \int_0^T Q_0(t)dt = 0 \qquad ...(8.63)$$

where,
$S(T)$ = Storage at end of period.
$S(0)$ = Storage at beginning of period.
Q_i = Water inflow.
Q_0 = Water discharge.

(iii) **Hydro power generated,** is a function of water storage (head) and discharge

$$P_{GH}(t) = f(S(t), Q_0(t)) \qquad ...(8.64)$$

The solution is simplified by discretization of the optimization interval into M subintervals, of time length ΔT. It is assumed that all the variables remain fixed in value, within a subinterval.

The equations (8.61) to (8.64) are reformulated as follows:

Minimize $$\Delta T \sum_{m=1}^{M} F'(P_{GT}^m) = \sum_{m=1}^{M} F(P_{GT}^m) \qquad ...(8.65)$$

Under the following constraints

(i) Power balance is given by

$$P_{GT}^m + P_{GH}^m - P_L^m - P_D^m = 0 \qquad ...(8.66)$$

where,

P_{GT}^m = Thermal generation in the mth interval.
P_{GH}^m = Hydro generation in the mth interval.
P_L^m = Loss in the mth interval.
$= B_{TT}(P_{GT}^m)^2 + 2B_{TH}P_{GT}^m P_{GH}^m + B_{HH}(P_{GH}^m)^2$

(ii) Water availability given by

$$S^m - S^{m-1} - Q_i^m \Delta T + Q_0^m \Delta T = 0$$

Letting $S^* = \dfrac{S}{\Delta T}$, the storage in discharge units, the above equation can be written as

$$S^{*m} - S^{*m-1} - Q_i^m + Q_0^m = 0 \qquad \text{...(8.67)}$$

(iii) Hydro generation at any subinterval is given by

$$P_{GH}^{\ m} = f(S^m, Q_0^m). \text{ This is given by}$$

$$P_{GH}^{\ m} = 9.81 \times 10^{-3}\, \hbar_{av}^{\ m}(Q_0^{\ m} - \rho) \text{ MW} \qquad \text{...(8.68)}$$

where, ρ = non-effective discharge needed to run the hydel plant under no-load.

$\hbar_{av}^{\ m}$ = average head in the mth interval.

$$= \hbar'_0 + \frac{\Delta T(S^{*m} + S^{*m-1})}{2A} = \hbar'_0\,[1 + 0.5e(S^{*m} + S^{*m-1})]$$

where, A = Cross-section of reservoir at the given storage

$\hbar'_0$ = Water head corresponding to dead storage

$e = \dfrac{\Delta T}{A\hbar_0}$; the head correction factor which can be tabulated for different storage levels.

$$\therefore \quad P_{GH}^{\ m} = 9.81 \times 10^{-3}\, \hbar'_0[1 + 0.5e\,(S^{*m} + S^{*m-1})](Q_0^{\ m} - \rho)$$

$$= h^*_0[1 + 0.5e(S^{*m} + S^{*m-1})](Q_0^{\ m} - \rho) \qquad \text{...(8.69)}$$

where, $\qquad h^*_0 = 9.81 \times 10^{-3}\hbar'_0$

Dependent variables

From (8.67) by adding the equations for all values of m = 1, 2... M, we obtain (8.67)

$$S^{*m} - S^{*0} - \sum_{m=1}^{M} Q_i^m + \sum_{m=1}^{M} Q_0^{\ m} = 0 \qquad \text{...(8.70)}$$

It is obvious from (8.70) that only (M – 1) values of Q_0 can be specified independently and one of them is therefore a dependent variable. Choosing $Q_0^{\ 1}$ (arbitrarily) as the dependent variable, we get from (8.70)

$$Q_0^{\ 1} = S^{*0} - S^{*m} + \sum_{m=1}^{M} Q_i^m - \sum_{m=2}^{M} Q_0^m = 0 \qquad \text{...(8.71)}$$

The other dependent variables are the thermal generations, hydro generations and water storages at all the subintervals.

8.8.2 Solution Procedure

Non-linear programming technique is used to solve the above equations. A Lagrangian function £ is formulated by augmenting the cost function with the equality constraints using Lagrange multipliers.

$$£ = \sum_m [F(P_{GT}{}^m) - \lambda_1{}^m(P_{GT}{}^m + P_{GH}{}^m - P_L{}^m - P_D^m) + \lambda_2{}^m(S^{*m} - S^{*m-1} - Q_i^m + Q_0^m)$$
$$+ \lambda_3{}^m[P_{GH}{}^m - h*_0(1 + 0.5e(S^{*m} + S^{*m-1})(Q_0{}^m - \rho)] \qquad ...(8.72)$$

The Lagrange multipliers are obtained by equating the partial derivatives of the Lagrangian with respect to the dependent variables, to zero. We get

$$\frac{\partial £}{\partial P_{GT}{}^m} = \frac{dF(P_{GT}{}^m)}{dP_{GT}{}^m} - \lambda_1{}^m\left[1 - \frac{\partial P_L{}^m}{\partial P_{GT}{}^m}\right] = 0 \qquad ...(8.73)$$

$$\frac{\partial £}{\partial P_{GH}{}^m} = \lambda_3{}^m - \lambda_1{}^m\left(1 - \frac{\partial P_L{}^m}{\partial P_{GH}{}^m}\right) = 0 \; ... \qquad (8.74)$$

$$\left(\frac{\partial £}{\partial S^{*m}}\right)_{\substack{m\neq 0\\ m\neq M}} = \lambda_2{}^m - \lambda_2{}^{m+1} - \lambda_3{}^m[0.5\hbar *_0 e(Q_0{}^m - \rho)] - \lambda_3{}^{m+1}[0.5\hbar *_0 e(Q_0{}^{m+1} - \rho)]$$
$$= 0 \qquad ...(8.75)$$

Substituting (8.67) in (8.72) for Q_0^1 we get

$$\frac{\partial £}{\partial Q_0^1} = \lambda_2{}^1 - \lambda_3{}^1 h*_0[1 + 0.5e(2S^{*0} + Q_i{}^1 - 2Q_0{}^1 + \rho)] = 0 \qquad ...(8.76)$$

The gradient vector is the partial derivative of the Lagrangian with respect to independent variables. That is,

$$\left[\frac{\partial £}{\partial Q_0{}^m}\right]_{m\neq 1} = \lambda_2{}^m - \lambda_3{}^m \hbar *_0[1 + 0.5e(2S^{*m-1} + Q_i^m - 2Q_0^m + \rho)] \qquad ...(8.77)$$

For optimality the gradient vector should be zero.

8.8.3 Algorithm

Step 1. Assume initial set of independent variables, $Q_0^m (m \neq 1)$ for all subintervals.

Step 2. Obtain S^{*m}, $P_{GH}{}^m$, $P_{GT}{}^m$ and $Q_0{}^1$ from (8.67), (8.69), (8.66) and (8.71) respectively.

Step 3. Obtain $\lambda_1{}^m$, $\lambda_3{}^m$, $\lambda_2{}^m$ $(m \neq 1)$ and λ_2^1 from (8.73), (8.74), (8.75) and (8.76), respectively.

Step 4. Obtain the gradient vector. If zero stop, else go to Step 5.

Step 5. $$Q_0^m(\text{new}) = Q_0^m(\text{old}) - \alpha\left(\frac{\partial £}{\partial Q_0^m}\right) \quad m \neq 1 \qquad ...(8.78)$$

where, α is a positive scalar. Go to step 2.

If during the iterative process, any of the control variables cross their limits, they are made equal to the respective limits. The above method can be extended to systems with more than one thermal and hydro plants.

Example 8.16: Find the optimal water discharge, of the hydro plant, in a hydrothermal system where the load varies in 8 hour intervals, as 5 MW, 8 MW and 10 MW. There is no water inflow into the reservoir. The transmission losses can be neglected. The initial water storage is 120 m^3/s and the final water storage is 75 m^3/s. The basic head is 20 m. The head correction factor $e = 0.005$ and non-effective water discharge is 3 m^3/s. The incremental fuel cost of the thermal plant is given by

$$\frac{dF}{dP_{GT}} = 1.1P_{GT} + 30 \text{ Rs/h.}$$

Solution: From data we have

$$S^{*0} = 120 \text{ m}^3/\text{s}$$

$$S^{*M} = S^{*3} = 75 \text{ m}^3/\text{s}$$

$$\rho = 3 \text{ m}^3/\text{s}$$

$$\hbar'_0 = 20 \text{ m}$$

$$Q_i = 0$$

There are three subintervals where

$$P_D^{\,1} = 5 \text{ MW}; P_D^{\,2} = 8 \text{ MW and } P_D^{\,3} = 10 \text{ MW.}$$

Let us start the iterations with initial values of discharges in subintervals 2 and 3 as

$$Q_0^{\,2} = 15 \text{ m}^3/s$$

$$Q_0^{\,3} = 20 \text{ m}^3/s$$

From (8.71)

$$Q_0^{\,1} = S^{*0} - S^{*M} + \sum_m Q_i^{\,m} - \sum_{m=2}^{m} Q_0^{\,m}$$

$$= S^{*0} - S^{*3} - Q_0^{\,2} - Q_0^{\,3}$$

$$= 120 - 75 - 15 - 20 = 10 \text{ m}^3/\text{s}$$

From (8.67)

$$S^{*1} = S^{*0} + Q_i^{\,1} - Q_0^{\,1} = 120 - 10 = 110 \text{ m}^3/\text{s}$$

$$S^{*2} = S^{*1} + Q_i^{\,2} - Q_0^{\,2} = 110 - 15 = 95 \text{ m}^3/\text{s}$$

The hydro generations are obtained from (8.69)

$$\hbar_0 = 9.81 \times 10^{-3}\, \hbar'_0 = 9.81 \times 10^{-3} \times 20 = 0.1962.$$

$$P_{GH}^{\,1} = \hbar_0[1 + 0.5e(S^{*1} + S^{*0})](Q_0^{\,1} - \rho)$$

$$= 0.1962[1 + 0.5 \times 0.005(110 + 120)]\,(10 - 3)$$

$$= 2.163 \text{ MW}$$

$$P_{GH}^{\,2} = \hbar_0[1 + 0.5 \times e(S^{*2} + S^{*1})]\,(Q_0^{\,2} - \rho)$$

$$= 0.1962\,[1 + 0.5 \times 0.005(95 + 110)](15 - 3) = 3.561 \text{ MW}$$

$$P_{GH}^{\ 3} = \hbar_0[1 + 0.5 \times e(S^{*3} + S^{*2})]\,(Q_0^{\ 3} - \rho)$$

$$= 0.1962\,[1 + 0.5 \times 0.005(75 + 95)](20 - 3) = 4.753 \text{ MW}$$

The thermal generation is obtained from (8.66)

$$P_{GT}^{\ 1} = P_D^{\ 1} - P_{GH}^{\ 1} - P_L^{\ 1} = 5 - 2.163 - 0 = 2.837 \text{ MW}$$

$$P_{GT}^{\ 2} = P_D^{\ 2} - P_{GH}^{\ 2} - P_L^{\ 2} = 8 - 3.561 - 0 = 4.439 \text{ MW}$$

$$P_{GT}^{\ 3} = P_D^{\ 3} - P_{GH}^{\ 3} - P_L^{\ 3} = 10 - 4.753 - 0 = 5.247 \text{ MW}$$

The Lagrange multipliers are next calculated from (8.73).

$$\lambda_1^{\ m} = \frac{dF(P_{GT}^{\ m})}{dP_{GT}^{\ m}} = 1.1P_{GT}^{\ m} + 30.0$$

For the three subintervals, we have

$$\lambda_1^1 = 1.1 \times 2.837 + 30 = 33.1207$$

$$\lambda_1^2 = 1.1 \times 4.439 + 30 = 34.883$$

$$\lambda_1^3 = 1.1 \times 5.247 + 30 = 35.772$$

From (8.74) $\lambda_3^m = \lambda_1^m$ (since, losses are neglected)

$$\lambda_3^1 = 33.1207$$

$$\lambda_3^2 = 34.883$$

$$\lambda_3^3 = 35.772$$

From (8.76)

$$\lambda_2^{\ 1} = \lambda_3^{\ 1}\hbar_0[1 + 0.5e(2S^{*\circ} + Q_i^{\ 1} - 2Q_0^{\ 1} + \rho]$$

$$= 33.1207 \times 0.1962[1 + 0.5 \times 0.005(2 \times 120 - 2 \times 10 + 3)]$$

$$= 10.121$$

From (8.75)

$$\lambda_2^2 = \lambda_2^1 - \lambda_3^1\,(0.5\hbar_0 e(Q_0^{\ 1} - \rho)) - \lambda_3^2\,(0.5\hbar_0 e(Q_0^2 - \rho))$$

$$= 10.121 - 33.1207(0.5 \times 0.1962 \times 0.005\,(10/-3)$$

$$- 34.883(0.5 \times 0.1962 \times 0.005(5 - 3))$$

$$= 10.121 - 0.1137 - 0.2053 = 9.802.$$

Similarly, from (8.75)

$$\lambda_2^3 = \lambda_2^2 - \lambda_3^2(0.5\hbar_0 e(Q_0^{\ 2} - \rho)) - \lambda_3^3\,(0.5\,\hbar_0 e(Q_0^{\ 3} - \rho)$$

$$= 9.802 - 34.883(0.5 \times 0.1962 \times 0.005(15 - 3))$$

$$- 35.772\,(0.5 \times 0.1962 \times 0.005(20 - 3))$$

$$= 9.802 - 0.205 - 0.298 = 9.299$$

The gradient vector is now calculated using (8.77)

$$\frac{\partial \pounds}{\partial Q_0^2} = \lambda_2^{\ 2} - \lambda_3^{\ 2}\,\hbar_0[1 + 0.5e(2S^{*1} + Q_i^2 - 2Q_0^{\ 2} + \rho)]$$

$$= 9.802 - 34.883 \times 0.1962[1 + 0.5 \times 0.005(2 \times 110 - 2 \times 15 + 3)]$$

$$= 9.802 - 10.146 = -\,0.344.$$

$$\frac{\partial \pounds}{\partial Q_0^3} = \lambda_2^3 - \lambda_3^{\ 3}\,\hbar_0[1 + 0.5e(2S^{*2} + Q_i^3 - 2Q_0^{\ 3} + \rho)]$$

$$= 9.299 - 35.772 \times 0.1962[1 + 0.5 \times 0.055(2 \times 95 - 2 \times 20 + 3)]$$

$$= 9.299 - 9.703 = -\,0.404.$$

Since, the gradient vector has significant magnitude, convergence is not yet reached. We evaluate the new discharges, from (8.78).

$$\begin{bmatrix} Q_{onew}^{\ 2} \\ Q_{onew}^{\ 3} \end{bmatrix} = \begin{bmatrix} Q_{oold}^{\ 2} \\ Q_{oold}^{\ 3} \end{bmatrix} - \alpha \begin{bmatrix} \dfrac{\partial \pounds}{\partial Q_0^2} \\ \dfrac{\partial \pounds}{\partial Q_0^3} \end{bmatrix}$$

Take $\alpha = 0.6$

$$\begin{bmatrix} Q_{onew}^{\ 2} \\ Q_{onew}^{\ 3} \end{bmatrix} = \begin{bmatrix} 15 \\ 20 \end{bmatrix} - 0.6 \begin{bmatrix} -0.344 \\ -0.404 \end{bmatrix} = \begin{bmatrix} 15.206 \\ 20.242 \end{bmatrix}$$

$$Q_{onew}^{\ 1} = 120 - 75 - 15.206 - 20.242 = 9.552 \text{ m}^3/\text{s}$$

Hence, the initial values for the next iteration are:

$$Q_0^1 = 9.552 \text{ m}^3/\text{s}$$

$$Q_0^2 = 15.206 \text{ m}^3/s$$

$$Q_0^3 = 20.242 \text{ m}^3/s$$

Iterations are carried out till gradient vector reaches the required tolerance (generally ≤ 0.1).

Example 8.17 The fuel costs of three thermal plants in \$/*h* are given by

$$F_1 = 400 + 5.1\,P_1 + 0.004\,P_1^{\ 2}$$

$$F_2 = 450 + 5.5\,P_2 + 0.002\,P_2^{\ 2}$$

$$F_3 = 200 + 5.8\,P_3 + 0.008\,P_3^{\ 2}$$

where P_1, P_2 and P_3 are in MW. Find the optimal dispatch and total cost in \$/*h* by

(i) Analytical method (ii) graphical method

(iii) Iterative technique using gradient method for a load of 800 MW

Solution

(i) Analytic method

$$\lambda = \frac{800 + \dfrac{5.1}{2 \times 0.004} + \dfrac{5.5}{2 \times 0.002} + \dfrac{5.8}{2 \times 0.008}}{\dfrac{1}{2 \times 0.004} + \dfrac{1}{2 \times 0.002} + \dfrac{1}{2 \times 0.008}}$$

$$= \frac{3175}{437.5} = 7.257 \text{ \$/MWh}$$

$$P_1 = \frac{7.257 - 5.1}{2 \times 0.004} = 269.625 \text{ MW}$$

$$P_2 = \frac{7.257 - 5.5}{2 \times 0.002} = 439.25 \text{ MW}$$

$$P_3 = \frac{7.257 - 5.8}{2 \times 0.008} = 91.0625 \text{ MW}$$

(ii) Graphical method

We find incremental fuel costs

$$\frac{dF_1}{dP_1} = 5.1 + 0.008\, P_1 = \lambda$$

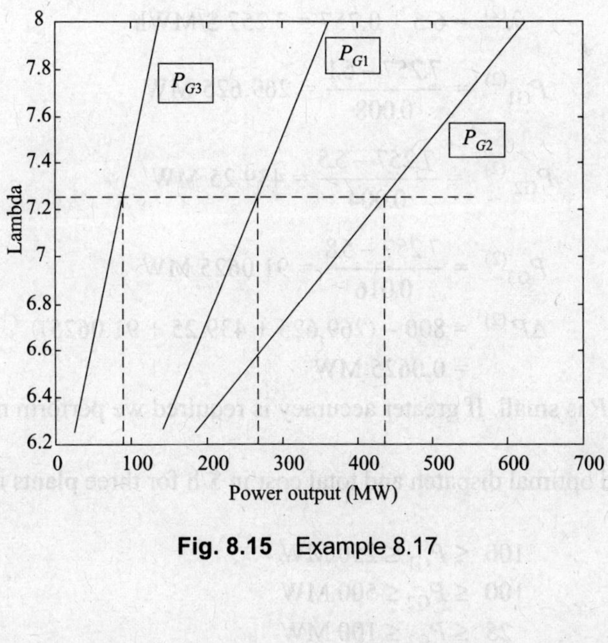

Fig. 8.15 Example 8.17

$$\frac{dF_2}{dP_2} = 5.5 + 0.004\,P_2 = \lambda$$

$$\frac{dF_3}{dP_3} = 5.8 + 0.016\,P_3 = \lambda$$

We draw three lines on the same graph as shown in Fig. 8.15.

We draw a horizontal line and drop vertical lines at the intersections to obtain the output power. By trial and error, we shift horizontal line up and down until the sum of the three outputs is equal to the demand. In this case it is approximately 7.25 \$/ MWh

(iii) Iterative method

Assume initial $\lambda^{(1)} = 6.5$ \$/MWh.

$$P_{G1}^{(1)} = \frac{6.5 - 5.1}{0.008} = 175 \text{ MW}$$

$$P_{G2}^{(1)} = \frac{6.5 - 5.5}{0.004} = 250 \text{ MW}$$

$$P_{G3}^{(1)} = \frac{6.5 - 5.8}{0.016} = 43.75 \text{ MW}$$

$$\Delta P^{(1)} = 800 - (175 + 250 + 43.75) = 331.25 \text{ MW}$$

$$\Delta\lambda = \frac{331.25}{\dfrac{1}{0.008} + \dfrac{1}{0.004} + \dfrac{1}{0.016}} = 0.757$$

$$\lambda^{(2)} = 6.5 + 0.757 = 7.257 \text{ \$/MWh}$$

$$P_{G1}^{(2)} = \frac{7.257 - 5.1}{0.008} = 269.625 \text{ MW}$$

$$P_{G2}^{(2)} = \frac{7.257 - 5.5}{0.004} = 439.25 \text{ MW}$$

$$P_{G3}^{(2)} = \frac{7.257 - 5.8}{0.016} = 91.0625 \text{ MW}$$

$$\Delta P^{(2)} = 800 - (269.625 + 439.25 + 91.0625) = 0.0625 \text{ MW.}$$

We can stop as ΔP is small. If greater accuracy is required we perform more number of iterations.

Example 8.18. Find optimal dispatch and total cost in \$/h for three plants in example 8.17 if the limits are.

$$100 \le P_{G1} \le 250 \text{ MW}$$
$$100 \le P_{G2} \le 500 \text{ MW}$$
$$25 \le P_{G3} \le 100 \text{ MW}$$

Solution We start in the same way as example 8.17. At the end of 2nd iteration through $\Delta P \,\Omega\, 0$, generator one has crossed its upper limit. So we peg P_{G1} at 250 MW.

Then we get

$$\Delta P^{(2)} = 800 - (250 + 439.25 + 91.0625)$$

$$= 19.6875.$$

$$\Delta\lambda^{(2)} = \frac{19.6875}{\frac{1}{0.004} + \frac{1}{0.016}} = 0.063$$

$$\lambda^{(3)} = 7.257 + 0.063$$

$$= 7.32 \text{ \$/MWh}.$$

$$P_{G2}^{(3)} = \frac{7.32 - 5.5}{0.004} = 455 \text{ MW}$$

$$P_{G3}^{(2)} = \frac{7.32 - 5.8}{0.016} = 95 \text{ MW}$$

$$\Delta P = 800 - (250 + 455 + 95) = 0$$

Hence, we can stop the iterations.

Example 8.19 Consider the two-bus-system shown in Fig. 8.16. Given

$$\frac{dF_1}{dP_{G1}} = 0.015\, P_{G1} + 20 \text{ Rs/ MWh}$$

$$\frac{dF_2}{dP_2} = 0.04\, P_{G2} + 18 \text{ Rs/ MWh}$$

When a load of 100 MW is transmitted from plant 1, a loss of 8 MW is incurred. The incremental cost is Rs 30/ MWh. Find P_{G1} and P_{G2} and the load.

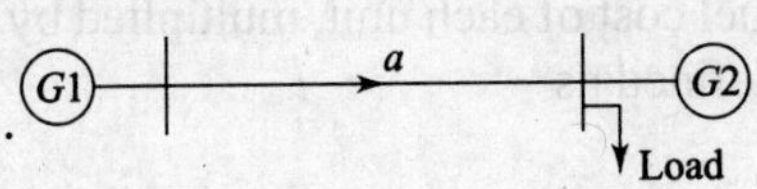

Fig. 8.16 Example 8.19.

Solution.

When only G_1 supplies load, the load current flows through line a. Therefore, $N_{a1} = 1$.

When only G_2 supplies load, no current flows through line a ∴ $N_{a2} = 0$

Since $N_{a2} = 0$, $B_{22} = 0$ and $B_{12} = 0$

$$\therefore \qquad P_L = B_{11}\, P_{G1}^{\;2} \text{ and } \frac{\partial P_L}{\partial P_{G1}} = 2\, B_{11}\, P_{G1}$$

$$\frac{\partial P_L}{\partial P_{G2}} = 0$$

When $P_L = 8$ MW, $P_{G1} = 100$ MW

$$8 = B_{11}\,(100)^2 \Rightarrow B_{11} = 8 \times 10^{-4}\ \text{MW}^{-1}$$

The coordination equation with loss is given by

$$\frac{dF_i}{dP_{Gi}} + \lambda\, \frac{\partial P_L}{\partial P_{Gi}} = \lambda$$

For plant 1

$$(0.015\, P_{G1} + 20) + 30\,(2 \times 8 \times 10^{-4} \times P_{G1}) = 30$$

$$0.063\, P_{G1} = 10$$

$$P_{G1} = 158.73 \text{ MW}$$

For plant 2,

$$0.04\, P_{G2} + 18 = 30$$

$$P_{G2} = 300 \text{ MW}$$

$$P_L = B_{11}\, P_{G1}^{\;2} = 20.156 \text{ MW}$$

$$P_D = P_{G1} + P_{G2} - P_L = 158.73 + 300 - 20.156$$

$$= 438.574 \text{ MW.}$$

8.9 SUMMARY

The classical solution of the economic dispatch problem is provided by the method of Lagrange multipliers. If losses are neglected, the most economical schedule is obtained when all the plants are operating with equal incremental costs. If losses are considered, the minimum fuel cost is obtained when the incremental fuel cost of each unit, multiplied by its penalty factor is same for all the units. The penalty factor is defined as

$$L_i = \frac{1}{1 - \dfrac{\partial P_L}{\partial P_{Gi}}}$$

where, P_L is the real power transmission loss. Iterative techniques can be used to solve the economic dispatch problem in the presence of transmission loss.

Transmission loss can be expressed in terms of B-coefficients as

$$P = \sum_{i=1}^{n_g} \sum_{j=1}^{n_g} P_m B_{mn} P_n$$

The *B*-coefficients can be calculated from knowledge of the network parameters, using current distribution factors.

The unit commitment problem addresses the issue of the generators to be committed, to meet the load demand. Dynamic programming is applied to obtain the unit commitment as presented in Sec. 8.7. The solution of hydrothermal systems has to take into account the constraints on water levels along with minimizing the fuel cost of thermal plants. A simplified approach is possible by discretizing the quantities over the period of operation as presented in Sec. 8.8.

8.10 SUPPLEMENTARY PROBLEMS

(1) The incremental fuel costs of 4 units in \$/MWh are as follows:

$$\frac{dF_1}{dP_{G1}} = 0.012P_{G1} + 10.0$$

$$\frac{dF_2}{dP_{G2}} = 0.009\,P_{G2} + 8.2$$

$$\frac{dF_3}{dP_{G3}} = 0.0075\,P_{G3} + 6.0$$

$$\frac{dF_4}{dP_{G4}} = 0.007\,P_{G4} + 9.2$$

Assuming all the four units operate to meet a demand of 100 MW, find the output of each unit and the plant λ for economic schedule.

(2) The incremental fuel costs of two units of a plant are

$$\frac{dF_1}{dP_{G1}} = 0.022\,P_{G1} + 18.0 \text{ Rs/MWh}$$

$$\frac{dF_2}{dP_{G2}} = 0.04\,P_{G2} + 20.0 \text{ Rs/MWh.}$$

If the maximum load on each unit is 500 MW and minimum load is 100 MW, plot λ versus plant output as load varies from 200 MW to 1000 MW.

(3) In problem (2) find the savings in Rs/h for economic dispatch between the units compared to sharing output equally for a load of (i) 600 MW (ii) 800 MW.

(4) A power system is supplied by three plants operating on economic dispatch. At plant 1, the incremental cost is 2 \$/MWh, at plant 2, 9 \$/MWh and at plant 3,10 \$/MW. Which plant has highest penalty factor?

Find penalty factor of plant 1, if the cost per hour to increase the total delivered load by 1 MW is 13 \$/MWh.

(5) Using data given in example 8.15, find the economical unit commitment for the daily load cycle given in Fig. P8.5.

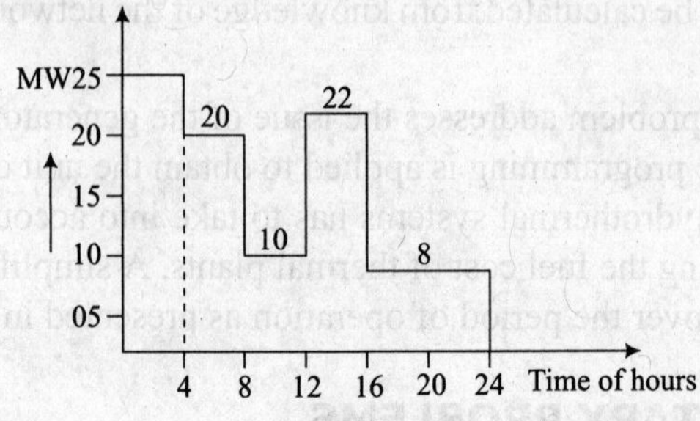

Fig. P8.5

(6) For the system shown in Fig. P8.6 obtain the *B*-Coefficients, with the following data:

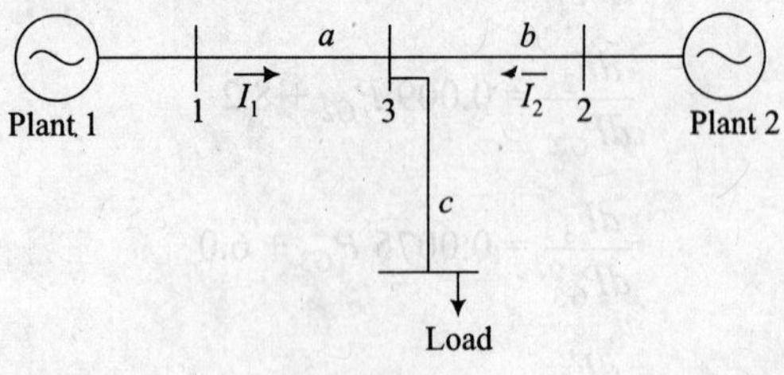

Fig. P8.6

$I_a = 1.0$ Ð $0°$ pu $\qquad Z_a = 0.04 + j0.16$ pu

$I_b = 0.8$ Ð $0°$ pu $\qquad Z_b = 0.03 + j0.12$ pu

$I_c = 1.0$ Ð $0°$ pu $\qquad Z_c = 0.02 + j0.08$ pu

(7) The fuel cost of three thermal plants is given by

$$F_1 = 400 + 8.2\, P_{G1} + 0.003\, P_{G1}^2 \text{ \$/h}$$

$$F_2 = 480 + 8.1\, P_{G2} + 0.0025\, P_{G2}^2 \text{ \$/h}$$

$$F_3 = 600 + 6.8\, P_{G3} + 0.003\, P_{G3}^2 \text{ \$/h.}$$

where P_{G1}, P_{G2}, P_{G3}, are in MW. Find the economic schedule for a load of 1000 MW, neglecting losses using

(i) analytical approach

(ii) iterative method

(8) Repeat (7) with limits on generator MW as follows:

$$120 \le P_{G1} \le 40$$

$$240 \le P_{G2} \le 600$$

$$50 \le P_{G3} \le 450.$$

Flow chart solution of incremental fuel cost by iterative technique

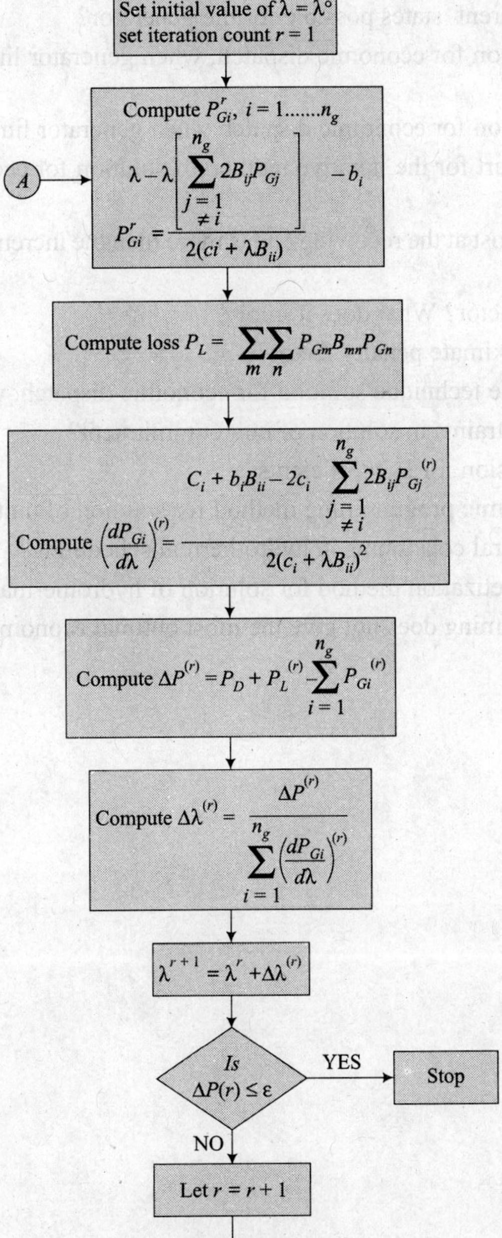

REVIEW QUESTIONS

1. Distinguish between the problem of unit commitment an economic dispatch.
2. What are the different states possible for the generator?
3. Derive the condition for economic dispatch, when generator limits and network losses are neglected.
4. Derive the condition for economic dispatch when generator limits are included.
5. Draw the flow chart for the iterative method of solution for economic dispatch of generation.
6. The incremental cost at the receiving end is more than the incremental cost at the generation end. Comment.
7. What is penalty factor? What does it imply?
8. What is the approximate penalty factor?
9. Derive the iterative technique solution for economic dispatch, with losses.
10. What are the constraints in solution of unit commitment?
11. Derive the expression for B-coefficients.
12. Describe the dynamic programming method for solution of unit commitment problem.
13. What are the general constraints in hydrothermal scheduling?
14. Describe the discretization method for solution of hydrothermal scheduling.
15. Dynamic programming does not give the most optimal economic schedule. Comment.

Chapter 9

Power System Stability

9.1 INTRODUCTION

Power system stability of modern large interconnected systems is a major problem for secure operation of the system. Recent major black-outs across the globe caused by system instability, even in very sophisticated and secure systems, illustrate the problems facing secure operation of power systems. Earlier, stability was defined as the ability of a system to return to normal or stable operation after having been subjected to some form of disturbance. This fundamentally refers to the ability of the system to remain in synchronism. However, modern power systems operate under complex interconnections, controls and extremely stressed conditions. Further, with increased automation and use of electronic equipment, the quality of power has gained utmost importance, shifting focus on to concepts of voltage stability, frequency stability, inter-area oscillations, etc.

The IEEE/CIGRE Joint Task Force on stability terms and conditions have proposed the following definition in 2004: "*Power system stability is the ability of an electric power system, for a given initial operating condition, to regain a state of operating equilibrium after being subjected to a physical disturbance, with most system variables bounded, so that practically the entire system remains intact*".

A power system is an extremely non-linear and dynamic system, with operating parameters continuously varying. Stability is hence, a function of the initial operating condition and the nature of the disturbance. Power systems are continually subjected to small disturbances in the form of load changes. The system must be in a position to be able to adjust to the changing conditions and operate satisfactorily. The system must also withstand large disturbances, which may even cause structural changes due to isolation of some faulted elements.

A power system may be stable for a particular (large) disturbance and unstable for another disturbance. It is impossible to design a system which is stable under all disturbances. The power system is generally designed to be stable under those disturbances which have a high degree of occurrence. The response to a disturbance is extremely complex and involves practically all the equipment of the power system. For example, a short circuit leading to a line isolation by circuit breakers will cause variations in the power flows, network bus voltages and generators rotor speeds. The voltage variations will actuate the voltage regulators in the system and generator speed variations will actuate the prime mover governors; voltage and frequency variations will affect the system loads. In stable systems, practically all generators and loads remain connected, even though parts of the system may be isolated to preserve bulk operations. On the other hand, an unstable system condition could lead to cascading outages and a shutdown of a major portion of the power system.

9.2 CLASSIFICATION OF POWER SYSTEM STABILITY

The high complexity of stability problems has led to a meaningful classification of the power system stability into various categories as shown in Fig. 9.1.

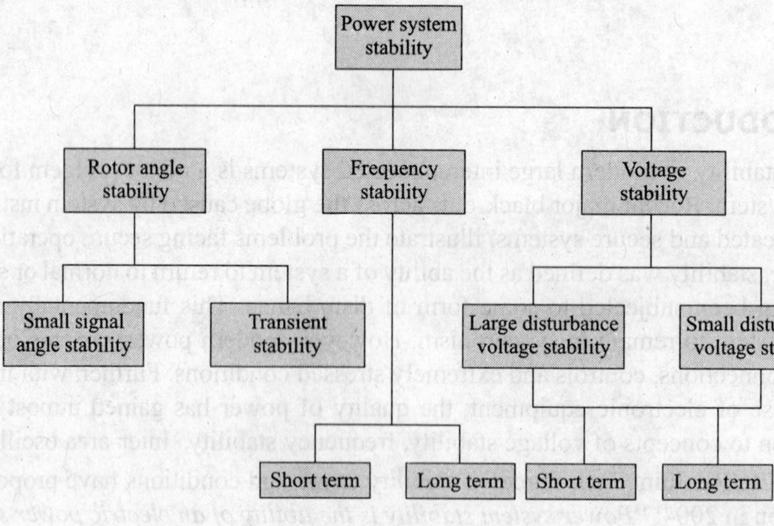

Fig. 9.1 Classification of power system stability.

The classification takes into account

- The main system variable in which instability can be observed.
- The size of the disturbance.
- The time span to be considered for assessing stability.

9.2.1 Rotor Angle Stability

Rotor angle stability refers to the ability of the synchronous machines of an interconnected power system to remain in synchronism after being subjected to a disturbance. Instability results in some generators accelerating (decelerating) and losing synchronism with other generators. Rotor angle stability depends on the ability of each synchronous machine to maintain equilibrium between electromagnetic torque and mechanical torque. Under steady state, there is equilibrium between the input mechanical torque and output electromagnetic torque of each generator, and its speed remains a constant. Under a disturbance, this equilibrium is upset and the generators accelerate/decelerate according to the mechanics of a rotating body. Rotor angle stability is further categorized as follows:

Small Signal (or Small Disturbance) Rotor Angle Stability

It is the ability of the power system to maintain synchronism under small disturbances. In this case, the system equation can be linearized around the initial operating point and the stability depends only on the operating point and not on the disturbance. Instability may result in

(i) A non-oscillatory or a periodic increase of rotor angle.

(ii) Increasing amplitude of rotor oscillations due to insufficient damping.

The first form of instability is largely eliminated by modern fast acting voltage regulators and the second form of instability is more common. The time frame of small signal stability is of the order of 10-20 seconds after a disturbance.

Large-signal Rotor Angle Stability or Transient Stability

This refers to the ability of the power system to maintain synchronism under large disturbances, such as short circuit, line outages, etc. The system response involves large excursions of the generator rotor angles. Transient stability depends on both the initial operating point and the disturbance parameters like location, type, magnitude, etc. Instability is normally in the form of a periodic angular separation. The time frame of interest is 3-5 seconds after disturbance.

The term dynamic stability was earlier used to denote the steady-state stability in the presence of automatic controls (especially excitation controls) as opposed to manual controls. Since, all generators are equipped with automatic controllers today, dynamic stability has lost relevance and the Task Force has recommended against its usage.

9.2.2 Voltage Stability

Voltage stability refers to the ability of a power system to maintain steady voltages at all buses in the system after being subjected to a disturbance. It depends on the ability of the system to maintain equilibrium between load demand and load supply. Instability results in a progressive fall or rise of voltages of some buses, which could lead to loss of load in an area or tripping of transmission lines, leading to cascading outages. This may eventually lead to loss of synchronism of some generators.

The cause of voltage instability is usually the loads. A run-down situation causing voltage instability occurs when load dynamics attempt to restore power consumption beyond the capability of the transmission network. Voltage stability is also threatened when a disturbance increases the reactive power demand beyond the sustainable capacity of the available reactive power resources. Voltage stability is categorized into the following sub-categories:

Small disturbance voltage stability

It refers to the system's ability to maintain steady voltages when subjected to small perturbations such as incremental changes in load. This is primarily influenced by the load characteristics and the controls at a given point of time.

Large disturbance voltage stability

It refers to the system's ability to maintain steady voltages following large disturbances. It requires computation of the non-linear response of the power system to include interaction between various devices like motors, transformer tap changers and field current limiters. Short-term voltage stability involves dynamics of fast acting load components and period of interest is in the order of several seconds. Long-term voltage stability involves slower acting equipment like tap-changing transformers and generator current limiters. Instability is due to loss of long-term equilibrium.

9.2.3 Frequency Stability

Frequency stability refers to the ability of a power system to maintain steady frequency following a severe disturbance, causing considerable imbalance between generation and load. Instability occurs in the form of sustained frequency swings leading to tripping of generating units or loads. During frequency swings, devices such as under frequency load shedding, generator controls and protection equipment get activated in a few seconds. However, devices such as prime mover energy supply systems and load voltage regulators respond after a few minutes. Hence, frequency stability could be a short-term or a long-term phenomenon.

9.3 MECHANICS OF ROTATORY MOTION

Since a synchronous machine is a rotating body, the laws of mechanics of rotating bodies are applicable to it. In rotation we first define the fundamental quantities. The angle θ_m is defined, with respect to a circular arc with its centre at the vertex of the angle, as the ratio of the arc length s to radius r.

$$\theta_m = \frac{s}{r} \qquad \text{...(9.1)}$$

The unit is radian. Angular velocity ω_m is defined as

$$\omega_m = \frac{d\theta_m}{dt} \text{ rad/s} \qquad \text{...(9.2)}$$

and angular acceleration as

$$\alpha = \frac{d\omega_m}{dt} = \frac{d^2\theta_m}{dt^2} \text{ rad/s}^2 \qquad \text{...(9.3)}$$

The torque on a body due to a tangential force F at a distance r from axis of rotation is given by

$$T = rF \qquad \text{...(9.4)}$$

The total torque is the summation of infinitesimal forces, given by

$$T = \int r\,dF \qquad \text{...(9.5)}$$

The unit of torque is N-m. When torque is applied to a body, the body experiences angular acceleration. Each particle experiences a tangential acceleration $a = r\alpha$, where r is the distance of the particle from axis of rotation. The tangential force required to accelerate a particle of mass dm is

$$dF = a\,dm = r\,\alpha\,dm \qquad \text{...(9.6)}$$

The torque required for the particle is

$$dT = rdF = r^2 \alpha \, dm \qquad ...(9.7)$$

and that required for the whole body is given by

$$T = \alpha \int r^2 dm = I\alpha \qquad ...(9.8)$$

Here,

$$I = \int r^2 dm \qquad ...(9.9)$$

is called the *moment of inertia* of the body. The unit is kg-m^2. If the torque is assumed to be the result of a number of tangential forces F, which act at different points of the body

$$T = \sum rF$$

Now each force acts through a distance

$$ds = r\, d\theta_m$$

The work done is $\sum F.ds$

$$dW = \sum Fr\, d\theta_m = d\theta_m T$$

$$W = \int T d\theta_m \qquad ...(9.10)$$

and,

$$T = \frac{dW}{d\theta_m} \qquad ...(9.11)$$

Thus, the unit of torque may also be joule per radian.

The power is defined as rate of doing work. Using (9.11)

$$P = \frac{dW}{dt} = \frac{Td\theta_m}{dt} = T\omega_m \qquad ...(9.12)$$

The angular momentum M is defined as

$$M = I\omega_m \qquad ...(9.13)$$

and the kinetic energy is given by

$$KE = \frac{1}{2} I\omega_m^2 = \frac{1}{2} M\omega_m \qquad ...(9.14)$$

From (9.14) we can see that the unit of M is J-sec/rad.

9.4 SWING EQUATION

The laws of rotation developed in section 9.3 are applicable to the synchronous machine. From (9.8)

$$I\alpha = T$$

or, $$\frac{Id^2\theta_m}{dt^2} = T \qquad ...(9.15)$$

Here T is the net torque of all torques acting on the machine, which includes the shaft torque (due to prime mover of a generator or load on a motor), torque due to rotational losses (friction, windage and core loss) and electromagnetic torque.

Let T_m = shaft torque or mechanical torque corrected for rotational losses.

T_e = electromagnetic or electrical torque.

For a generator T_m tends to accelerate the rotor in positive direction of rotation as shown in Fig. 9.2(a). Fig. 9.2(b) shows the corresponding torque for a motor with respect to the direction of rotation.

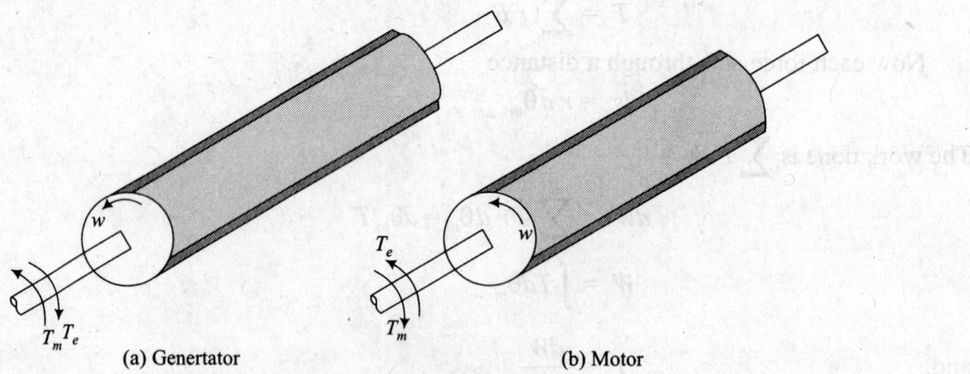

Fig. 9.2 Torque acting on a synchronous machine.

The accelerating torque for a generator

$$T_a = T_m - T_e \qquad ...(9.16)$$

Under steady-state operation of the generator, T_m is equal to T_e and the accelerating torque is zero. There is no acceleration or deceleration of the rotor masses and the machines run at a constant synchronous speed. In the stability analysis in the following sections, T_m is assumed to be a constant, since the prime movers (steam turbines or hydro turbines) do not act during the short time period in which rotor dynamics are of interest in the stability studies.

Now (9.15) has to be solved to determine θ_m as a function of time. Since, θ_m is measured with respect to a stationary reference axis on the stator, it is the measure of the absolute rotor angle and increases continuously with time even at constant synchronous speed. Since, machine acceleration/deceleration is always measured relative to synchronous speed, the rotor angle is measured with respect to a synchronously rotating reference axis. Let

$$\delta_m = \theta_m - \omega_{sm}t \qquad ...(9.17)$$

where, ω_{sm} is the synchronous speed in mechanical rad/s and δ_m is the angular displacement in mechanical radians.

Taking the derivative of (9.17), we get

$$\frac{d\delta_m}{dt} = \frac{d\theta_m}{dt} - \omega_{sm}$$

$$\frac{d^2\delta_m}{dt^2} = \frac{d^2\theta_m}{dt^2} \qquad \text{...(9.18)}$$

Substituting (9.18) in (9.15), we get

$$I\frac{d^2\delta_m}{dt^2} = T_a = T_m - T_e \text{ N-m} \qquad \text{...(9.19)}$$

Multiplying by ω_m on both sides, we get

$$\omega_m I\frac{d^2\delta_m}{dt^2} = \omega_m(T_m - T_e) \text{ N-m} \qquad \text{...(9.20)}$$

From (9.12) and (9.13), we can write

$$M\frac{d^2\delta_m}{dt^2} = P_m - P_e \text{ watts} \qquad \text{...(9.21)}$$

where, M is the angular momentum, also called inertia constant.

P_m = shaft power input less rotational losses.

P_e = electrical power output corrected for losses.

P_a = acceleration power.

M depends on the angular velocity ω_m, and hence is strictly not a constant, because ω_m deviates from the synchronous speed during and after a disturbance. However, under stable conditions ω_m does not vary considerably and M can be treated as a constant. (9.21) is called the *"swing equation"*. The constant M depends on the rating of the machine and varies widely with the size and type of the machine. Another constant called H constant (also referred to as inertia constant) is defined as

$$H = \frac{\text{Stored kinetic energy in mega joules at sychronous speed}}{\text{Machine rating in MVA}} \text{ MJ/MVA} \qquad \text{...(9.22)}$$

H falls within a narrow range and typical values are given in Table 9.1.

If the rating of the machine is G MVA, from (9.22) the stored kinetic energy is GH Mega Joules. From (9.14)

$$GH = \frac{1}{2}M\text{w}_{sm} \text{ MJ} \qquad \text{...(9.23)}$$

or,

$$M = \frac{2GH}{\omega_{sm}} \text{ MJ-s/mech rad} \qquad \text{...(9.24)}$$

The swing equation (9.21) is written as

$$\frac{2H}{\omega_{sm}}\frac{d^2\delta_m}{dt^2} = \frac{P_a}{G} = \frac{P_m - P_e}{G} \qquad \text{...(9.25)}$$

In (9.25) δ_m is expressed in mechanical radians and ω_{sm} in mechanical radians per second (the subscript m indicates mechanical units). If δ and ω have consistent units (both are mechanical or electrical units). (9.25) can be written as

$$\frac{2H}{\omega_s}\frac{d^2\delta}{dt^2} = P_a = P_m - P_e \text{ pu} \qquad ...(9.26)$$

Here, ω_s is the synchronous speed in electrical rad/s $\left(\omega_s = \left(\frac{p}{2}\right)\omega_{sm}\right)$ and P_a is acceleration power in per unit on same base as H. For a system with an electrical frequency f Hz, (9.26) becomes

$$\frac{H}{\pi f}\frac{d^2\delta}{dt^2} = P_a = P_m - P_e \text{ pu} \qquad ...(9.27)$$

when δ is in electrical radians and

$$\frac{H}{180f}\frac{d^2\delta}{dt^2} = P_a = P_m - P_e \text{ pu} \qquad ...(9.28)$$

when δ is in electrical degrees.

(9.27) and (9.28) also represent the swing equation. It can be seen that the swing equation is a second order differential equation which can be written as two first order differential equations:

$$\frac{2H}{\omega_s}\frac{d\omega}{dt} = P_m - P_e \text{pu} \qquad ...(9.29)$$

$$\frac{d\delta}{dt} = \omega - \omega_s \qquad ...(9.30)$$

in which ω, ω_s and δ are in electrical units. In deriving the swing equation, damping has been neglected.

Table 9.1 *H* constants of synchronous machines

Type of machine		*H(MJ/MVA)*
Turbine generator condensing	1800 rpm	9 – 6
	3600 rpm	7 – 4
Non-condensing	3600 rpm	4 – 3
Water wheel generator		
	Slow speed < 200 rpm	2 – 3
	High speed > 200 rpm	2 – 4
Synchronous condenser		
	Large	1.25 } 25% less for hydrogen cooled
	Small	1.0 } 25% less for hydrogen cooled
Synchronous motor with load varying from 1.0 to 5.0		2.0

In defining the inertia constant H, the MVA base used is the rating of the machine. In a multi machine system, swing equation has to be solved for each machine, in which case, a common MVA base for the system has to be chosen. The constant H of each machine must be consistent with the system base.

Let

G_{mach} = Machine MVA rating (base)

G_{system} = System MVA base

In (9.25), H is computed on the machine rating $G = G_{mach}$.

Multiplying (9.25) by $\frac{G_{mach}}{G_{system}}$ on both sides, we get

$$\left(\frac{G_{mach}}{G_{system}}\right)\frac{2H}{\omega_{sm}}\frac{d^2\delta_m}{dt^2} = \frac{P_m - P_e}{G_{mach}}\left(\frac{G_{mach}}{G_{system}}\right) \quad ...(9.31)$$

$$\frac{2H_{system}}{\omega_{sm}}\frac{d^2\delta_m}{dt^2} = P_m - P_e \text{ pu (on system base)}$$

where,

$$H_{system} = H\frac{G_{mach}}{G_{system}} \quad ...(9.32)$$

In the stability analysis of a multi machine system, computation is considerably reduced if the number of swing equations to be solved is reduced. Machines within a plant normally swing together after a disturbance. Such machines are called *coherent machines* and can be replaced by a single equivalent machine, whose dynamics reflects the dynamics of the plant. The concept is best understood by considering a two machine system.

9.4.1 Swing Equation of Two Coherent Machines

The swing equations for two machines on a common system base are:

$$\frac{2H_1}{\omega_s}\frac{d^2\delta_1}{dt^2} = P_{m1} - P_{e1} \text{ pu} \quad ...(9.33)$$

$$\frac{2H_2}{\omega_s}\frac{d^2\delta_2}{dt^2} = P_{m2} - P_{e2} \text{ pu} \quad ...(9.34)$$

Now $\delta_1 = \delta_2 = \delta$ (since they swing together). Adding (9.33) and (9.34), we get

$$\frac{2H_{eq}}{\omega_s}\frac{d^2\delta}{dt^2} = P_m - P_e \text{ pu} \quad ...(9.35)$$

where,

$$H_{eq} = H_1 + H_2$$

$$P_m = P_{m1} + P_{m2}$$

$$P_e = P_{e1} + P_{e2}$$

(9.35) represents the dynamics of the single equivalent machine.

9.4.2 Swing Equation of Two Non-coherent Machines

For any two non-coherent machines also (9.33) and (9.34) are valid. Subtracting (9.34) from (9.33), we get

$$\frac{2}{\omega_s}\frac{d^2\delta_1}{dt^2} - \frac{2}{\omega_s}\frac{d^2\delta_2}{dt^2} = \frac{P_{m1}-P_{e1}}{H_1} - \frac{P_{m2}-P_{e2}}{H_2} \quad \text{...(9.36)}$$

Multiplying both sides by $\dfrac{H_1 H_2}{H_1 + H_2}$, we get

$$\frac{2}{\omega_s}\left(\frac{H_1H_2}{H_1+H_2}\right)\frac{d^2(\delta_1-\delta_2)}{dt^2} = \frac{P_{m1}H_2 - P_{m2}H_1}{H_1+H_2} - \frac{P_{e1}H_2 - P_{e2}H_1}{H_1+H_2}$$

i.e.,

$$\frac{2}{\omega_s}H_{12}\frac{d^2\delta_{12}}{dt^2} = P_{m12} - P_{e12} \quad \text{...(9.37)}$$

where, $\delta_{12} = \delta_1 - \delta_2$, the relative angle of the two machines

$$H_{12} = \frac{H_1H_2}{H_1+H_2}$$

$$P_{m12} = \frac{p_{m1}\,H_2 - p_{m2}\,H_1}{H_1+H_2}$$

$$P_{e12} = \frac{p_{e1}\,H_2 - p_{e2}\,H_1}{H_1+H_2}$$

From (9.37) it is obvious that the swing of a machine is associated with dynamics of other machines in the system. To be stable, the angular differences between all the machines must decrease after the disturbance. In many cases, when the system loses stability, the machines split into two coherent groups, swinging against each other. Each coherent group of machines can be replaced by a single equivalent machine and the relative swing of the two equivalent machines solved using an equation similar to (9.37), from which stability can be assessed.

The acceleration power is given by $P_a = P_m - P_e$. Hence, under the condition that P_m is a constant, an accelerating machine should have a power characteristic which would increase P_e as δ increases. This would reduce P_a and hence the acceleration and help in maintaining stability. If on the other hand, P_e decreases when δ increases, P_a would further increase which is detrimental to stability. Therefore, $\dfrac{\partial P}{\partial \delta}$ must be positive for a stable system. Thus, the power-angle relationship plays a crucial role in stability.

Example 9.1: A 50 Hz, 4 pole turbo alternator rated 150 MVA, 11 kV has an inertia constant of 9 MJ/MVA. Find the (a) stored energy at synchronous speed (b) the rotor acceleration if the input mechanical power is raised to 100 MW when the electrical load is 75 MW, (c) the speed at the end of 10 cycles if acceleration is assumed constant at the initial value.

Solution:

(a) Stored energy = GH = 150 × 9 = 1350 MJ

(b) $P_a = P_m - P_e = 100 - 75 = 25$ MW

$$M = \frac{GH}{180f} = \frac{1350}{180 \times 50} = 0.15 \text{ MJ} - \text{s/°e}$$

$$0.15\frac{d^2\delta}{dt^2} = 25$$

Acceleration
$$\alpha = \frac{d^2\delta}{dt^2} = \frac{25}{0.15} = 166.6 \text{ °e/s}^2$$

$$= 166.6 \times \frac{2}{P} \text{ °m/s}^2$$

$$= 166.6 \times \frac{2}{P} \times \frac{1}{360} \text{ rpm/s}$$

$$= 166.6 \times \frac{2}{P} \times \frac{1}{360} \times 60 \text{ rpm/s}$$

$$= 13.88 \text{ rpm/s}$$

*Note °e = electrical degree; °m = mechanical degree; P = number of poles.

(c) 10 cycles = $\frac{10}{50}$ = 0.2 s

$$N_S = \text{Synchronous speed} = \frac{120 \times 50}{4} = 1500 \text{ rpm}$$

Rotor speed at end of 10 cycles = $N_S + \alpha \times 0.2$

$$= 1500 + 13.88 \times 0.2 = 1502.776 \text{ rpm}$$

Example 9.2: Two 50 Hz generating units operate in parallel within the same plant, with the following ratings:

Unit 1: 500 MVA, 0.8 pf, 13.2 kV, 3600 rpm: H = 4 MJ/MVA

Unit 2: 1000 MVA, 0.9 pf, 13.8 kV, 1800 rpm: H = 5 MJ/MVA

Calculate the equivalent H constant on a base of 100 MVA.

Solution:
$$H_{1\text{system}} = H_{1\text{mach}} \times \frac{G_{1\text{mach}}}{G_{\text{system}}}$$
$$= 4 \times \frac{500}{100} = 20 \text{ MJ/MVA}$$
$$H_{2\text{system}} = H_{2\text{ mach}} \times \frac{G_{2\text{mach}}}{G_{\text{system}}}$$
$$= 5 \times \frac{1000}{100} = 50 \text{ MJ/MVA}$$
$$H_{eq} = H_1 + H_2 = 20 + 50 = 70 \text{ MJ/MVA}$$

This is the equivalent inertia constant on a base of 100 MVA and can be used when the two machines swing coherently.

9.5 POWER-ANGLE EQUATION

In solving the swing equation, certain assumptions are normally made

(i) Mechanical power input P_m is a constant during the period of interest, immediately after the disturbance.

(ii) Rotor speed changes are insignificant.

(iii) Effect of voltage regulating loop during the transient is neglected, that is, the excitation is assumed to be a constant.

As discussed in section 9.4, the power-angle relationship plays a vital role in the solution of the swing equation.

9.5.1 Power-angle Equation for a Non-salient Pole Machine (classical model)

The simplest model for the synchronous generator is that of a constant voltage behind an impedance. This model is called the *classical model* and can be used for cylindrical rotor (non-salient pole) machines. Practically, all high-speed turbo alternators are of cylindrical rotor construction, where the physical air gap around the periphery of the rotor is uniform. This type of generator has approximately equal magnetic reluctance, regardless of the angular position of the rotor, with respect to the armature mmf. The phasor diagram of the voltages and currents at constant speed and excitation is shown in Fig. 9.3.

E_g = Generator internal emf

V_t = Terminal voltage

θ = Power factor angle

I_a = Armature current

R_a = Armature resistance

x_d = Direct axis reactance

The power output of the generator is given by the real part of $E_g I_a^*$.

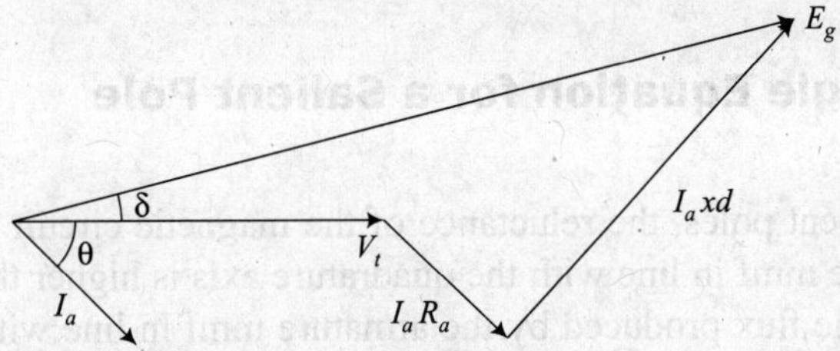

Fig. 9.3 Phasor diagram of a non-salient pole synchronous generator.

$$I_a = \frac{E_g \angle\delta - V_t \angle 0°}{R_a + jx_d} \qquad ...(9.38)$$

Neglecting R_a,
$$I_a = \frac{E_g \angle\delta - V_t \angle 0°}{jx_d}$$

$$P = \mathcal{R}\left\{(E_g \angle\, \delta)\left(\frac{E_g \angle \delta - 90°}{xd} - \frac{V_t \angle -90°}{xd}\right)\right\}$$

$$= \frac{E_g^{\,2} \cos 90°}{x_d} - \frac{E_g V_t \cos(90° + \delta)}{x_d}$$

$$= \frac{E_g V_t \sin\delta}{x_d} \qquad ...(9.39)$$

(Note-$\mathcal{R}$ stands for real part of)

The graphical representation of (9.39) is called the power angle curve and is shown in Fig. 9.4.

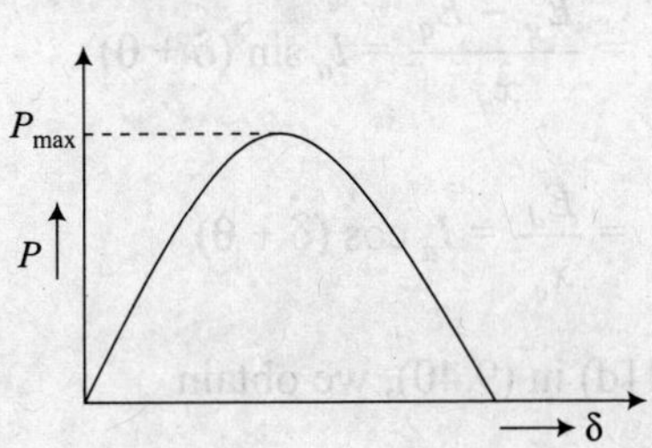

Fig. 9.4 Power angle curve of a non-salient pole machine.

The maximum power that can be transferred for a particular excitation is given by $\frac{E_g V_t}{x_d}$ at $\delta = 90°$. This is also called the steady state stability limit.

9.5.2 Power Angle Equation for a Salient Pole Machine

Here because of the salient poles, the reluctance of the magnetic circuit in which flows the flux produced by an armature mmf in line with the quadrature axis is higher than that of the magnetic circuit in which flows the flux produced by the armature mmf in line with the direct axis. These two components of armature mmf are proportional to the corresponding components of armature current. The component of armature current producing an mmf acting in line with direct axis is called the direct component, I_d. The component of armature current producing an mmf acting in line with the quadrature axis is called the quadrature axis component, I_q. The phasor diagram is shown in Fig. 9.5, with same terminology as Fig. 9.3 and R_a neglected.

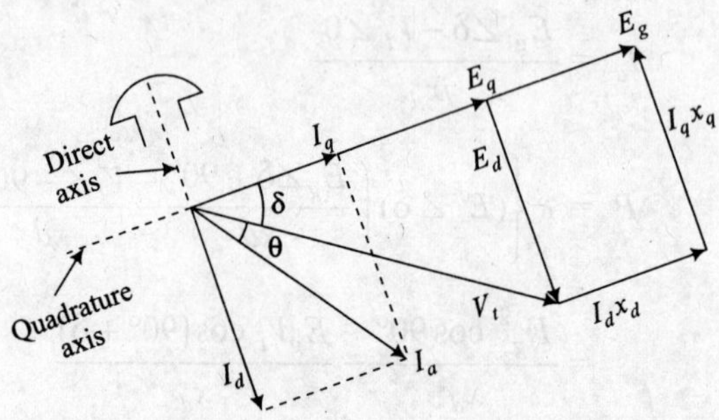

Fig. 9.5 Phasor diagram of a salient pole machine.

Power output $\quad P = V_t I_a \cos\theta$

$$= E_d I_d + E_q I_q \qquad \text{...(9.40)}$$

From Fig 9.5: $\quad E_d = V_t \sin\delta \qquad$ ...(9.41a)

$$E_q = V_t \cos\delta \qquad \text{...(9.41b)}$$

$$I_d = \frac{E_g - E_q}{x_d} = I_a \sin(\delta + \theta) \qquad \text{...(9.41c)}$$

$$I_q = \frac{E_d}{x_q} = I_a \cos(\delta + \theta) \qquad \text{...(9.41d)}$$

Substituting (9.41c) and (9.41d) in (9.40), we obtain

$$P = \frac{E_g V_t \sin\delta}{x_d} + \frac{V_t^2 (x_d - x_q)\sin 2\delta}{2 x_d x_q} \qquad \text{...(9.42)}$$

(9.42) gives the steady state power angle relationship for a salient pole machine. The second term does not depend on the excitation and is called the reluctance power component. This component makes the maximum power greater than in the classical model. However, the angle at which the maximum power occurs is less than 90°.

Example 9.3: Obtain the power angle relationship and the generator internal emf for (i) classical model (ii) salient pole model with following data:

$$x_d = 1.0 \text{ pu};\ x_q = 0.6 \text{ pu};\ V_t = 1.0 \text{ pu};\ I_a = 1.0 \text{ pu at upf}$$

Solution:

(i) **Classical model:** The phasor diagram is shown in Fig. 9.6.

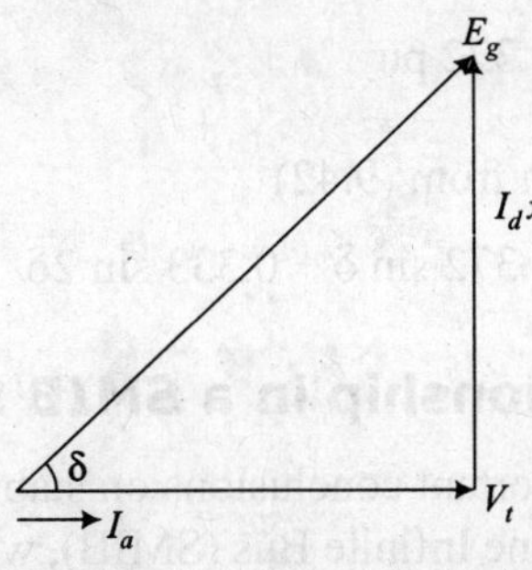

Fig. 9.6 Example 9.3, case(i).

$$|E_g| = \sqrt{V_t^2 + (I_a x_d)^2} = \sqrt{(1.0)^2 + (1.0 \times 1.0)^2} = 1.414$$

$$\delta = \tan^{-1}\frac{I_a x_d}{V_t} = \tan^{-1}\frac{1.0}{1.0} = 45°$$

$$\therefore \quad E_g = 1.414 \angle 45°.$$

If the excitation is held constant so that $|E_g| = 1.414$, then power output

$$P = \frac{1.414 \times 1.0 \sin\delta}{1.0} = 1.414 \sin\delta$$

(ii) **Salient pole:** From Fig. (9.5), we get using (9.41a) to (9.41d)

$$E_g = E_q + I_d x_d = V_t \cos\delta + I_d x_d$$

$$= V_t \cos\delta + I_a \sin\delta\, x_d$$

(*θ = 0°, since we are considering *upf*)

Substituting given values, we get

$$E_g = \cos\delta + \sin\delta.$$

Again from Fig. (9.5), we have

$$E_d = V_t \sin\delta = I_q x_q$$

$$\therefore \quad V_t \sin\delta - I_q x_q = 0$$

$$V_t \sin\delta - I_a \cos\delta \, x_q = 0$$

Substituting the given values, we get

$$0 = \sin\delta - 0.6\cos\delta$$

We thus have two simultaneous equations.

$$E_g = \cos\delta + \sin\delta$$

$$0 = \sin\delta - 0.6\cos\delta$$

Solving, we get $\delta = 30.96^\circ$

$$E_g = 1.372 \text{ pu}$$

If the excitation is held constant, then from (9.42)

$$P = 1.372\sin\delta + 0.333\sin 2\delta$$

9.5.3 Power Angle Relationship in a SMIB System

Without loss of generality, many important conclusions on stability can be arrived at by considering the simple case of a Single Machine Infinite Bus (SMIB), where a generator supplies power to an infinite bus. The concept of an infinite bus arises from the fact that if we connect a generator to a much larger power system, it is reasonable to assume that the voltage and frequency of the larger system will not be affected by control of the generator parameters. Hence, the external system can be approximated by an infinite bus, which is equivalent to an ideal voltage source, whose voltage and frequency are constant. The one line diagram is shown in Fig. 9.7.

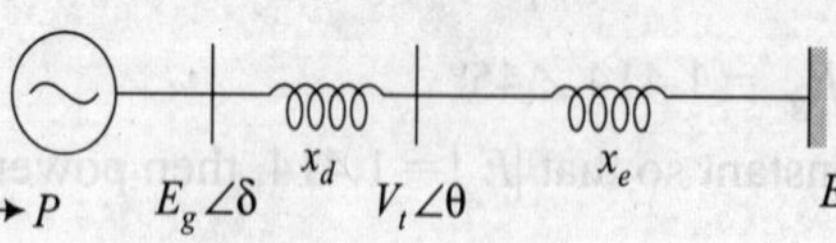

Fig. 9.7 SMIB system.

In Fig. 9.7, the infinite bus voltage is taken as reference and δ is the angle between E_g and E_b. The generator is assumed to be connected to the infinite bus through a lossless line of reactance x_e. The power transferred (using classical model) is given by

$$P = \frac{E_g E_b}{x_d + x_e}\sin\delta \qquad \text{...(9.43)}$$

and using salient pole model,

$$P = \frac{E_g E_b}{x_d + x_e}\sin\delta + \frac{E_b^2 (x_d - x_q)}{2(x_d + x_e)(x_q + x_e)}\sin 2\delta \qquad \text{...(9.44)}$$

An important measure of performance is the *steady state stability limit*, which is defined as *the maximum power that can be transmitted in steady state without loss of synchronism*, to the receiving end. If transient analysis is required, respective transient quantities, namely, Eg', xd', and x_q' are used in (9.43) and (9.44) to calculate the power output.

Example 9.4: Determine the steady state stability limit of the system shown in Fig. 9.8, if V_t = 1.0 pu and the reactances are in pu.

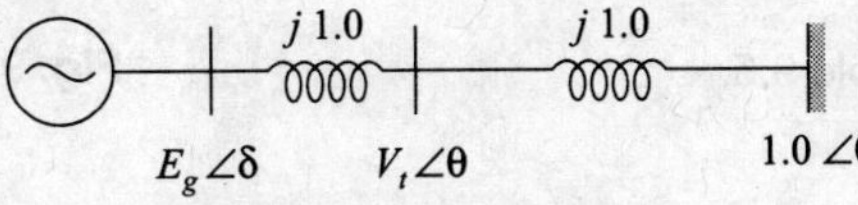

Fig. 9.8 Example 9.4.

Solution:

$$\text{Current } I = \frac{V_t \angle\theta - 1.0 \angle 0°}{j1.0} = \frac{1.0 \angle\theta - 1.0 \angle 0°}{j1.0}$$

$$E_g \angle\delta = V_t \angle\theta + j1.0\,(I)$$

$$= 1\angle\theta + \frac{j1.0\,(1.0 \angle\theta - 1.0 \angle 0°)}{j1.0}$$

$$= \cos\theta + j\sin\theta + \cos\theta + j\sin\theta - 1.0$$

$$= 2\cos\theta - 1 + j\,2\sin\theta$$

When maximum power is transferred $\delta = 90°$; which means real part of $E_g = 0$

$$\therefore \quad 2\cos\theta - 1 = 0$$

$$\theta = \cos^{-1} 0.5 = 60°$$

$$|E_g| = 2 \times \sin 60° = 1.732$$

$$E_g = 1.732 \angle 90° \text{ (for maximum power)}$$

$$\text{Steady state stability limit} = \frac{1.732 \times 1.0}{1.0 + 1.0} = 0.866 \text{ pu}$$

Example 9.5: Derive expression for the sending and receiving end power of a SMIB system, whose line impedance is $Z \angle\theta$.

Solution: The single line diagram is shown in Fig. 9.9a.

$$Z \angle\theta = R + jX$$

$$I = \frac{E_g \angle\delta - E_b \angle 0°}{Z \angle\theta} = \frac{E_g \angle\delta - \theta}{Z} - \frac{E_b \angle -\theta}{Z}$$

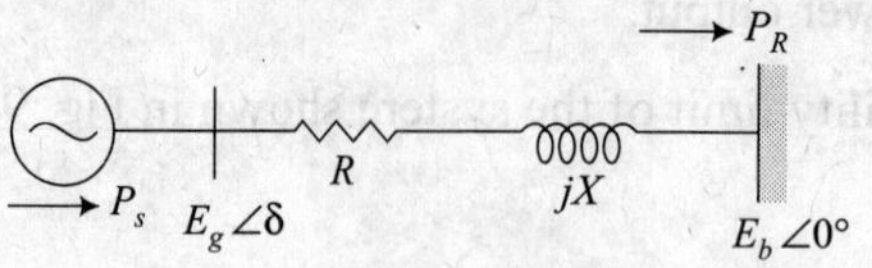

Fig. 9.9a Example 9.5.

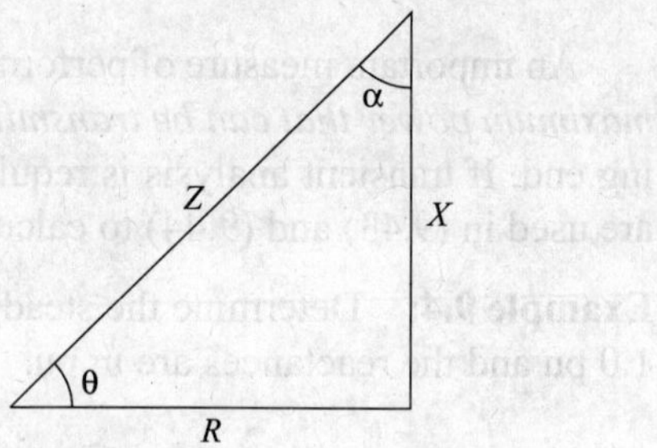

Fig. 9.9b Impedance diagram.

$$I^* = \frac{E_g \angle\theta - \delta}{Z} - \frac{E_b \angle\theta}{Z}$$

Sending end power $P_s = \mathcal{R}(E_g I^*)$

$$= \mathcal{R} E_g \angle\delta \left[\frac{E_g \angle\theta - \delta}{Z} - \frac{E_b \angle\theta}{Z}\right]$$

*$\mathcal{R}$ stands for real part of

Let $\alpha = 90 - \theta$ (Refer Fig. 9.9b); $\sin\alpha = R/Z$;

Then,
$$P_s = \frac{E_g^2 \sin\alpha}{Z} + \frac{E_g E_b \sin(\delta - \alpha)}{Z}$$

$$= \frac{E_g^2 R}{Z^2} + \frac{E_g E_b \sin(\delta - \alpha)}{Z}$$

Receiving end power

$$P_R = \mathcal{R}(E_b I^*) = \mathcal{R} E_b \angle 0° \left[\frac{E_g \angle\theta - \delta}{Z} - \frac{E_b \angle\theta}{Z}\right]$$

$$= \frac{E_g E_b \sin(\delta + \alpha)}{Z} - \frac{E_b^2 R}{Z^2}$$

9.6 SMALL SIGNAL STABILITY

Small signal stability (also called small-disturbance stability or steady state stability) is the ability of the system to maintain stability under minor disturbances. After the disturbance disappears, stability is assured if the system returns to its *original state*. Since, the disturbance is minor, the disturbance can be linearized about its initial operating point. Consider the SMIB system of Fig. 9.7 where power transferred is given by (9.43).

$$P_e = \frac{E_g E_b \sin\delta}{x_d + x_e} = P_{max} \sin\delta$$

Let the generator be operating initially at a steady state angle δ_o, supplying power P_{eo}. Since, under steady state $P_m = P_e$, $P_m = P_{eo}$. If now the electrical power changes by a small amount ΔP, the torque angle also changes by a small amount $\Delta\delta$ as shown in Fig. 9.10 (assuming P_m is a constant).

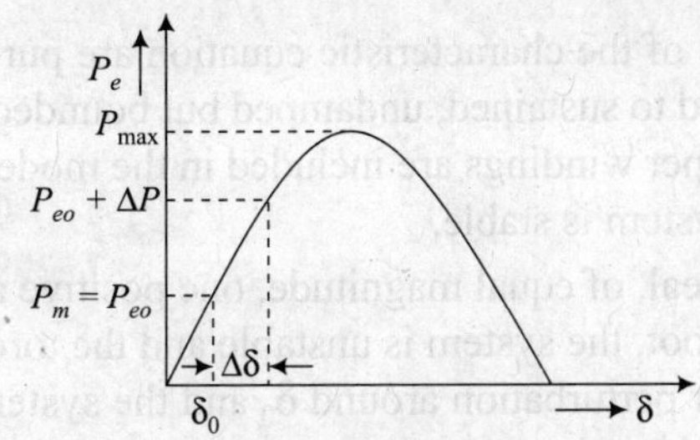

Fig. 9.10 Power angle curve.

Linearizing about the initial operating point (P_{eo}, δ_o), we get

$$\Delta P_e = \left(\frac{\partial P_e}{\partial \delta}\right)_O \Delta\delta \quad ...(9.45)$$

where, $\left(\frac{\partial P_e}{\partial \delta}\right)_O = \frac{\partial P_e}{\partial \delta}$ evaluated at $\delta = \delta_o$

$$= P_{max} \cos \delta_o \quad ...(9.46)$$

The changes in $\Delta\delta$ are described by the linearized swing equation.

$$M\frac{d^2\Delta\delta}{dt^2} = P_m - (P_{eo} + \Delta P_e) = -\Delta P_e \quad ...(9.47)$$

(as $p_m = p_{eo}$)

or,
$$M\frac{d^2\Delta\delta}{dt^2} + \left(\frac{\partial P_e}{\partial \delta}\right)_O \Delta\delta = 0 \quad ...(9.48)$$

The characteristic equation of the above second-order differential equation in $\Delta\delta$, is given by

$$Ms^2 + \left(\frac{\partial P_e}{\partial \delta}\right)_O = 0 \quad ...(9.49)$$

The roots of the characteristic equation are

$$s = \pm \frac{\left(-\left(\frac{\partial P_e}{\partial \delta}\right)_O\right)^{\frac{1}{2}}}{M} \quad ...(9.50)$$

$\left(\frac{\partial P_e}{\partial \delta}\right)_O$ is called the synchronizing power coefficient. Let $\left(\frac{\partial P_e}{\partial \delta}\right)_O = P_S$. We have the following possible cases:

(i) If $P_S > 0$, then the roots of the characteristic equation are purely imaginary and conjugate. The response would lead to sustained, undamped but bounded oscillations. However, if the line resistance and damper windings are included in the model, the system oscillations will decay. Therefore, the system is stable.

(ii) If $P_S < 0$, the roots are real, of equal magnitude, one positive and the other negative. Since, there is a positive real root, the system is unstable and the torque angle δ increases without bound, even for a minor perturbation around δ_0 and the system is unstable.

Thus, for stability under minor disturbances, the synchronizing power coefficient P_S should be greater than zero. From (9.46) it follows that $\cos \delta_0 > 0$ or $\delta_0 < 90°$. The maximum power that can be transferred under steady state i.e., the *steady state stability limit*, occurs when $\delta_0 = 90°$ and is given by

$$P_{max} = \frac{E_g E_b}{x_d + x_e} \quad ...(9.51)$$

In the derivation of the swing equation damping has been neglected. However, whenever there is a difference in the angular velocity between the rotor and the rotating air gap flux, induction motor action takes place. This sets up a torque, called the damping torque, which tends to minimize the difference between the two angular velocities. The damping power P_D is proportional to the speed deviation.

$$P_D = D\frac{d\delta}{dt} \quad ...(9.52)$$

When P_S is positive, the damping power aids in eventually damping out the oscillations and restoring equilibrium. With damping, the linearized swing equation is written as,

$$M\frac{d^2\Delta\delta}{dt^2} + D\frac{d\Delta\delta}{dt} + P_s\Delta\delta = 0$$

or,

$$\frac{d^2\Delta\delta}{dt^2} + \frac{D}{M}\frac{d\Delta\delta}{dt} + \frac{P_S}{M}\Delta\delta = 0 \quad ...(9.53)$$

(9.53) can be written in the standard second order differential equation form as

$$\frac{d^2\Delta\delta}{dt^2} + 2\,\xi\,\omega_n\frac{d\Delta\delta}{dt} + \omega_n^{\,2}\,\Delta\delta = 0 \quad ...(9.54)$$

where, ω_n = undamped (natural) frequency of oscillation.

$$= \sqrt{\frac{P_S}{M}} \quad ...(9.55)$$

$$\xi = \text{damping ratio}$$

$$= \frac{D}{2}\sqrt{\frac{1}{MP_S}} \qquad ...(9.56)$$

The characteristic equation of (9.54) is given by

$$s^2 + 2\xi\,\omega_n\,s + \omega_n^{\,2} = 0$$

The roots are given by

$$s_1, s_2 = -\xi\,\omega_n \pm j\,\omega_d \qquad ...(9.57)$$

where, ω_d = damped frequency of oscillation.

$$= \omega_n\sqrt{1-\xi^2} \qquad ...(9.58)$$

The roots have negative real parts provided $P_S > 0$ and the system is stable. If the rotor is perturbed by a small angle $\Delta\delta_0$, then the solution of (9.54) yields.

$$\Delta\delta = \frac{\Delta\delta_0}{\sqrt{1-\xi^2}}\,e^{-\xi\omega_n t}\,\sin(\omega_d t + \theta) \qquad ...(9.59)$$

where, $\theta = \cos^{-1}\xi$

The torque angle with respect to the synchronously rotating field is given by

$$\delta = \delta_0 + \Delta\delta$$

$$= \delta_0 + \frac{\Delta\delta_0}{\sqrt{1-\xi^2}}\,e^{-\xi\omega_n t}\sin(\omega_d t + \theta) \qquad ...(9.60)$$

The response time constant

$$\tau = \frac{1}{\xi\omega_n} = \frac{2M}{D} \qquad ...(9.61)$$

with a settling time approximately of 4τ. It can be observed from (9.55) and (9.56) that as H increases, the natural frequency and damping ratio decrease, resulting in a longer settling time. An increase in the synchronizing power coefficient results in an increase in the natural frequency and damping ratio.

In complex systems with excitation control, governor control, etc., small signal stability analysis is carried out in the same manner. The describing differential equation is linearized about the initial operating point. The roots of the characteristic equation indicate the condition for stability. In practical systems the excitation is not held constant, since constant excitation would lead to large dips in the terminal voltage under heavy loading. Instead, the terminal voltage is held a constant and the excitation adjusted for every load increase to keep $|V_t|$ constant.

9.7 STEADY STATE STABILITY LIMIT WITH RESISTANCE INCLUDED

Example 9.6: Derive expression for steady state stability limit if the resistance and shunt capacitance of the transmission line are considered.

Solution: The receiving end power is given by

$$P_R = \frac{|V_S||V_R|}{|B|}\cos(\beta-\delta) - \left|\frac{A}{B}\right| |V_R|^2 \cos(\beta-\alpha)$$

where, $A = |A| \angle\alpha$ and $B = |B| \angle\beta$ are the generalized circuit constants.

The synchronizing coefficient $P_s = \frac{\partial P_R}{\partial \delta} = \frac{|V_S||V_R|}{|B|}\sin(\beta-\delta)$.

For P_S to be positive $\delta < \beta$. Maximum power limit is obtained when $\beta = \delta$ and is given by

$$P_{\max} = \frac{|V_S||V_R|}{|B|} - \frac{|A||V_R|^2}{|B|}\cos(\beta-\alpha)$$

In nominal Π model of the transmission line, $A = 1 + \frac{1}{2}YZ$ and $B = Z$. If shunt capacitance is neglected, but resistance is considered $A = 1\angle 0°$ and $B = R + jX$.

$\therefore \qquad |B| > X$ and $\beta < 90°$

$$P_{\max} = \frac{|V_S||V_R|}{|B|} - \frac{|A||V_R|^2}{|B|}\cos\beta.$$

This value is less than the value if capacitance were to be considered. Therefore, neglecting capacitance gives pessimistic results.

If the resistance is also neglected, $A = 1\ \angle 0°$ and $B = X\angle 90°$.

$$P_{\max} = \frac{|V_S||V_R|}{X}\sin\delta$$

This value is greater than the value if resistance were considred. Hence, neglecting resistance gives optimistic results.

Example 9.7: A 50 Hz synchronous generator having an internal voltage 1.2 pu, $H = 5.2$ MJ/MVA and a reactance of 0.4 pu is connected to an infinite bus through a double circuit line, each line of reactance 0.35 pu. The generator is delivering 0.8pu power and the infinite bus voltage is 1.0 pu.

Determine

(a) Maximum power transfer.

(b) Steady state operating angle.

(c) Natural frequency of oscillation if damping is neglected.

Solution: The one line diagram is shown in Fig. 9.11.

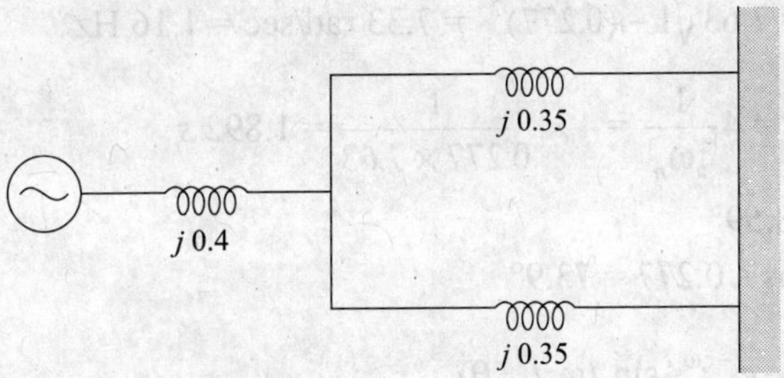

Fig. 9.11 Example 9.6.

(a) $X = 0.4 + \dfrac{0.35}{2} = 0.575$ pu

$$P_{max} = \frac{E_g E_b}{X} = \frac{1.2 \times 1.0}{0.575} = 2.087 \text{ pu}$$

(b) $P_e = P_{max} \sin \delta_0$

$$\therefore \quad \delta_0 = \sin^{-1}\frac{P_e}{P_{max}} = \sin^{-1}\left(\frac{0.8}{2.087}\right) = 22.54°.$$

(c) $P_s = P_{max} \cos \delta_0 = 2.087 \cos (22.54°)$

$= 1.927$ MW (pu)/elec rad.

$$M(\text{pu}) = \frac{H}{\Pi f} = \frac{5.2}{\Pi \times 50} = 0.0331 \text{ s}^2\text{/elec rad}$$

Without damping $s = \pm j\sqrt{\dfrac{P_S}{M}} = \pm j\sqrt{\dfrac{1.927}{0.0331}}$

$= \pm j7.63$ rad/sec

$= 1.21$ Hz

Natural frequency of oscillation $\omega_n = 1.21$ Hz.

Example 9.8: In example 9.6, if the damping is 0.14 and there is a minor disturbance of $\Delta\delta = 0.15$ rad from the initial operating point, determine

(a) ω_n (b) ξ (c) ω_d (d) settling (e) expression for δ.

Solution:

(a) $\omega_n = \sqrt{\dfrac{P_S}{M}} = \sqrt{\dfrac{1.927}{0.0331}} = 7.63$ rad/sec $= 1.21$ Hz

(b) $\xi = \dfrac{D}{2}\sqrt{\dfrac{1}{M P_S}} = \dfrac{0.14}{2}\sqrt{\dfrac{1}{0.0331 \times 1.927}} = 0.277$

(c) $\omega_d = \omega_n\sqrt{1-\xi^2} = 7.63\sqrt{1-(0.277)^2} = 7.33$ rad/sec = 1.16 Hz

(d) Settling time $= 4\tau = 4\dfrac{1}{\xi\omega_n} = 4 \times \dfrac{1}{0.277 \times 7.63} = 1.892\ s$

(e) $\Delta\delta_0 = 0.15$ rad $= 8.59°$

$\theta = \cos^{-1}\xi = \cos^{-1} 0.277 = 73.9°$

$$\delta = \delta_0 + \frac{\Delta\delta_0}{\sqrt{1-\xi^2}} e^{-\xi\omega_n t} \sin(\omega_d t + \theta)$$

$$= 22.54° + \frac{8.59}{\sqrt{1-0.277^2}} e^{-0.277\times 7.6t} \sin(7.33t + 73.9°)$$

$$= 22.54° + 8.94\, e^{-2.11t} \sin(7.33t + 73.9°)$$

The variation of delta with respect to time is shown below. It can be observed that the angle reaches the steady state value of 22.54° after the initial transient. It should be noted that the magnitudes of the swings decrease in a stable system with damping.

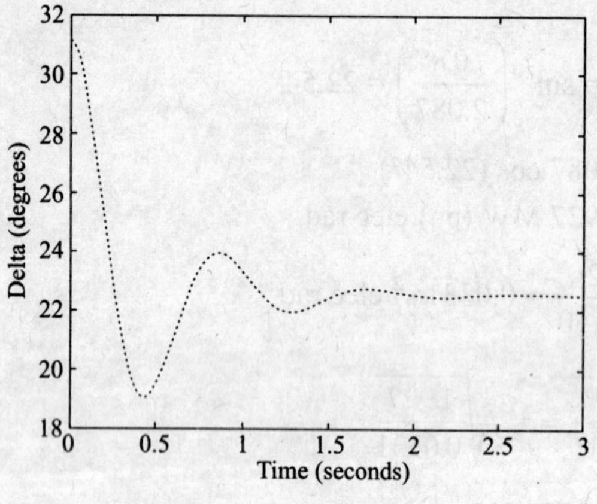

Swing curve for example 9.7

Example 9.9: In example 9.6, find the power angle relationship

(i) For the given network.

(ii) If a short circuit occurs in the middle of a line.

(iii) If fault is cleared by line outage.

Assume the generator to be supplying 1.0 pu power initially.

Solution:

(i) From example 9.6, $P_{max} = 2.087$, $P_e = 2.087 \sin\delta$.

(ii) If a short circuit occurs in the middle of the line, the network equivalent can be drawn as shown in Fig. 9.12a.

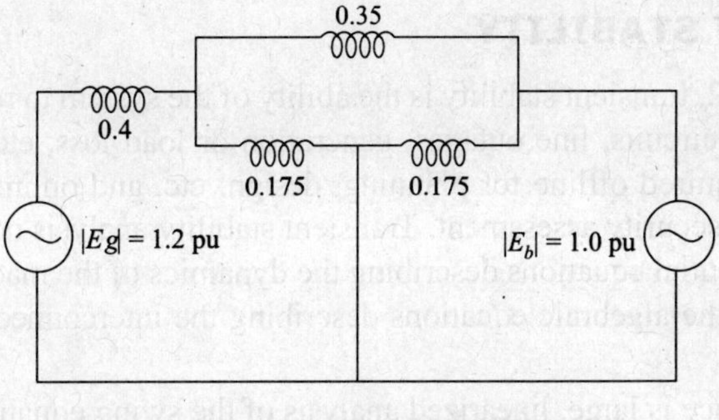

Fig. 9.12a Short circuit in middle of line.

The network is reduced by converting the delta to star and again the resulting star to delta as shown in Fig. 9.12b and 9.12c.

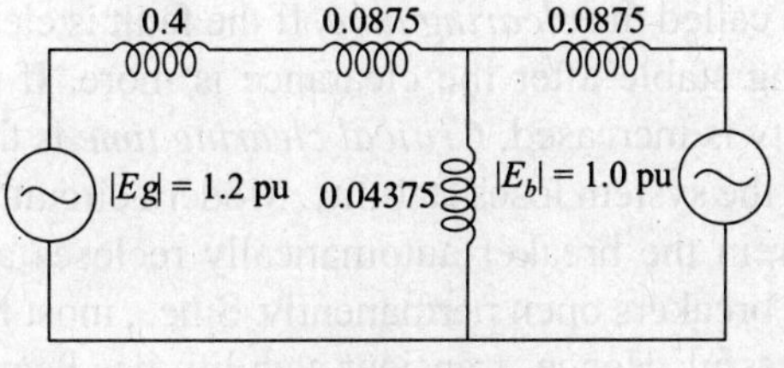

Fig. 9.12b

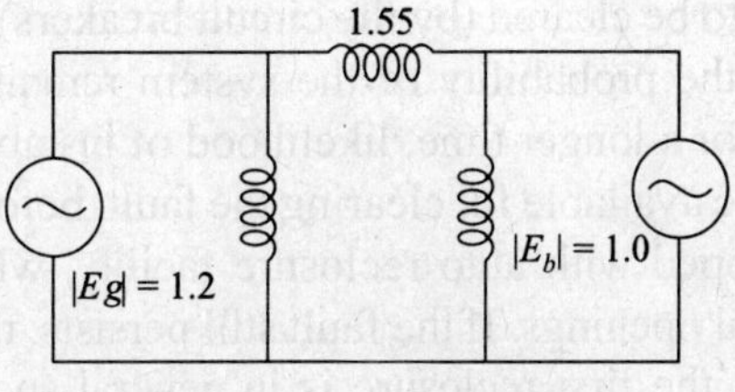

Fig. 9.12c

The transfer reactance is 1.55 pu. Hence,

$$P_{\max} = \frac{1.2 \times 1.0}{1.55} = 0.744$$

$$P_e = 0.744 \sin \delta$$

(iii) When there is a line outage

$$X = 0.4 + 0.35 = 0.75$$

$$P_{\max} = \frac{1.2 \times 1.0}{0.75} = 1.6$$

$$P_e = 1.6 \sin \delta$$

The system cannot be operated above its steady state stability limit, even though it can be operated above its transient stability limit. Steady state stability limit can be raised by decreasing reactance of the line. This can be achieved by using two parallel lines or by using a series capacitor to reduce the reactance of the line. Higher excitation voltages and fast acting excitation systems also improve the steady state stability limit.

9.8 TRANSIENT STABILITY

As defined in section 9.2, transient stability is the ability of the system to remain stable under large disturbances like short circuits, line outages, generation or load loss, etc. The evaluation of the transient stability is required offline for planning, design, etc. and online for load management, emergency control and security assessment. Transient stability analysis deals with actual solution of the non-linear differential equations describing the dynamics of the machines and their controls and interfacing it with the algebraic equations describing the interconnections through the transmission network.

Since, the disturbance is large, linearized analysis of the swing equation (which describes the rotor dynamics) is not possible. Further, the fault may cause structural changes in the network, because of which the power angle curve prior to fault, during the fault and post fault may be different (See example 9.8). Due to these reasons, a general stability criteria for transient stability cannot be established, as was done in the case of steady state stability (namely, $P_S > 0$). Stability can be established, for a given fault, by actual solution of the swing equation. The time taken for the fault to be cleared (by the circuit breakers) is called the *clearing time*. If the fault is cleared fast enough, the probability of the system remaining stable after the clearance is more. If the fault persists for a longer time, likelihood of instability is increased. *Critical clearing time* is the maximum time available for clearing the fault, before the system loses stability. Modern circuit breakers are equipped with auto reclosure facility, wherein the breaker automatically recloses after two sequential openings. If the fault still persists, the breakers open permanently. Since, most faults are transient, the first reclosure is in general successful. Hence, transient stability has been greatly enhanced by auto closure breakers.

Some common assumptions made during transient stability studies are as follows:

1. Transmission line and synchronous machine resistances are neglected. Since, resistance introduces a damping term in the swing equation, this gives pessimistic results.
2. Effect of damper windings is neglected which again gives pessimistic results.
3. Variations in rotor speed are neglected.
4. Mechanical input to the generator is assumed constant. The governor control loop is neglected. This also leads to pessimistic results.
5. The generator is modeled as a constant voltage source behind a transient reactance, neglecting the voltage regulator action.
6. Loads are modeled as constant admittances and absorbed into the bus admittance matrix.

The above assumptions, vastly simplify the equations. A digital computer program for transient stability analysis can easily include more detailed generator models and effect of controls, the discussion of which is beyond the scope of present treatment. Studies on the transient stability of an SMIB system, can shed light on some important aspects of stability of larger systems. One of the important methods for studying the transient stability of an SMIB system is the application of equal-area criterion.

9.9 EQUAL-AREA CRITERION

Transient stability assessment of an SMIB system is possible without resorting to actual solution of the swing equation, by a method known as equal-area criterion. In a SMIB system, if the system is unstable after a fault is cleared, $\delta(t)$ increases indefinitely with time, till the machine loses synchronism. In contrast, in a stable system, $\delta(t)$ reaches a maximum and then starts reducing as shown in Fig. 9.13.

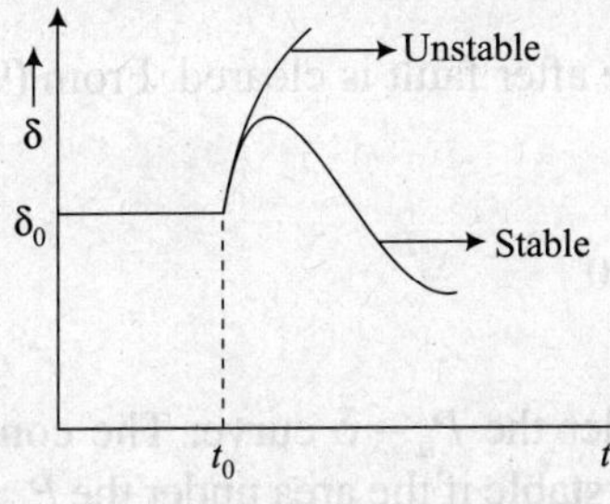

Fig. 9.13 $\delta V_S t$, for stable and unstable system.

Mathematically stated,

$$\frac{d\delta(t)}{dt} = 0$$

some time after the fault is cleared in a stable system and $\frac{d\delta}{dt} > 0$, for a long time after the fault is cleared in an unstable system.

Consider the swing equation (9.21)

$$M\frac{d^2\delta}{dt^2} = P_m - P_e = P_a$$

$$\frac{d^2\delta}{dt^2} = \frac{P_a}{M}$$

Multiplying both sides by $2\,\frac{d\delta}{dt}$, we get

$$2\frac{d\delta}{dt}\frac{d^2\delta}{dt^2} = 2\frac{d\delta}{dt}\frac{P_a}{M}$$

This may be written as

$$\frac{d}{dt}\left[\left(\frac{d\delta}{dt}\right)^2\right] = 2\frac{d\delta}{dt}\frac{P_a}{M}$$

Integrating both sides, we get

$$\left(\frac{d\delta}{dt}\right)^2 = \frac{2}{M}\int_{\delta_o}^{\delta} P_a d\delta$$

or, $$\frac{d\delta}{dt} = \sqrt{\frac{2}{M}\int_{\delta_o}^{\delta} P_a d\delta} \qquad ...(9.62)$$

For stability $\frac{d\delta}{dt} = 0$, some time after fault is cleared. From (9.62), this means

$$\int_{\delta_o}^{\delta} P_a d\delta = 0 \qquad ...(9.63)$$

The integral gives the area under the $P_a - \delta$ curve. The condition for stability can be, thus stated as follows: A SMIB system is stable if the area under the $P_a - \delta$ curve, becomes zero at some value of δ. This means that the accelerating (positive) area under $P_a - \delta$ curve, must equal the decelerating (negative) area under $P_a - \delta$ curve. Application of equal area criterion for several disturbances is discussed next.

9.9.1 Sudden Change in Mechanical Input

Consider the SMIB system shown in Fig. 9.14.

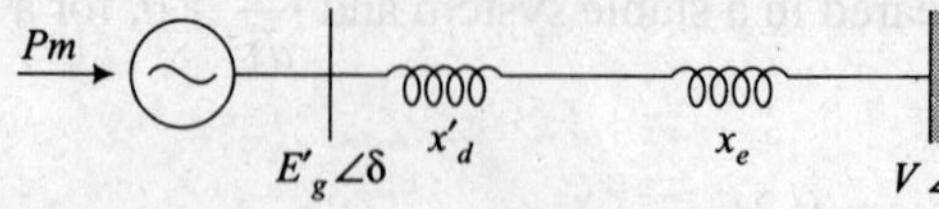

Fig. 9.14 SMIB system.

The electrical power transferred is given by

$$P_e = P_{max} \sin \delta$$

$$P_{max} = \frac{E'_g V}{x'_d + x_e}$$

Here we use E'_g as we are considering the transient period.

Under steady state $P_m = P_e$. Let the machine be initially operating at a steady state angle δ_0, at synchronous speed ω_s, with a mechanical input P_{m0}, as shown in Fig. 9.15 (point *a*).

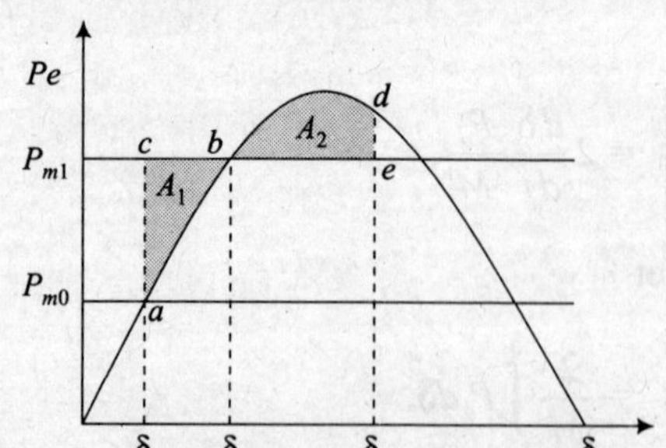

Fig. 9.15 Equal area criterion-sudden change in mechanical input.

If there is a sudden step increase in input power to P_{m1} the accelerating power is positive (since, $P_{m1} > P_{m0}$) and power angle δ increases. With increase in δ, the electrical power P_e increases, the accelerating power decreases, till at $\delta = \delta_1$, the electrical power matches the new input P_{m1}. The area A_1, during acceleration is given by

$$A_1 = \int_{\delta_0}^{\delta_1} (P_{m1} - P_e)\, d\delta$$

$$= P_{m1}(\delta_1 - \delta_0) - P_{max}(\cos\delta_0 - \cos\delta_1) \quad \text{...(9.64)}$$

At b, even though the accelerating power is zero, the rotor is running above synchronous speed. Hence, δ and P_e increase beyond b, wherein $P_e < P_{m1}$ and the rotor is subjected to deceleration. The rotor decelerates and speed starts dropping, till at point d, the machine reaches synchronous speed and $\delta = \delta_{max}$. The area A_2, during deceleration is given by

$$A_2 = \int_{\delta_1}^{\delta_{max}} (P_e - P_{m1})\, d\delta = P_{max}(\cos\delta_1 - \cos\delta_{max}) - P_{m1}(\delta_{max} - \delta_1) \quad \text{...(9.65)}$$

By equal area criterion $A_1 = A_2$. The rotor would then oscillate between δ_0 and δ_{max} at its natural frequency. However, damping forces will reduce subsequent swings and the machine finally settles down to the new steady state value δ_1 (at point b). Stability can be maintained only if area A_2 at least equal to A_1, can be located above P_{m1}. The limiting case is shown in Fig. 9.16, where A_2 is just equal to A_1.

Here δ_{max} is at the intersection of P_e and P_{m1}. If the machine does not reach synchronous speed at d, then beyond d, P_e decreases with increase in δ, causing δ to increase indefinitely. Applying equal area criterion to Fig. 9.16, we get

$$A_1 = A_2.$$

From (9.64) and (9.65), we get

$$P_{m1}(\delta_{max} - \delta_0) = P_{max}(\cos\delta_0 - \cos\delta_{max})$$

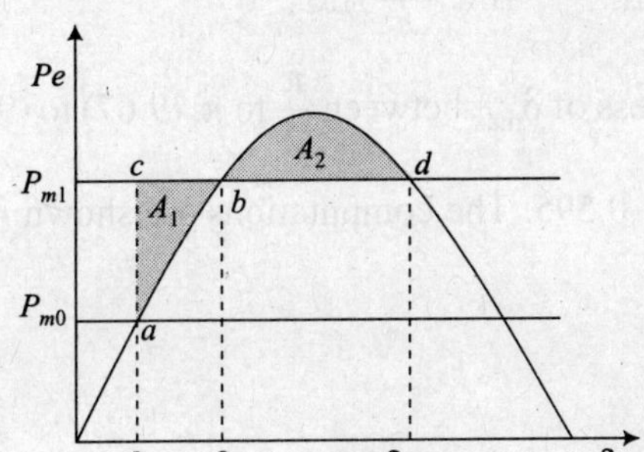

Fig. 9.16 Maximum increase in mechanical power.

Substituting $P_{m1} = P_{max} \sin \delta_{max}$, we get

$$(\delta_{max} - \delta_0) \sin \delta_{max} + \cos \delta_{max} = \cos \delta_0 \qquad ...(9.66)$$

(9.66) is a non-linear equation in δ_{max} and can be solved by trial and error or by using any numerical method for solution of non-linear algebraic equation (like Newton-Raphson, bisection, etc). From solution of δ_{max}, P_{m1} can be calculated. $P_{m1} - P_{m0}$ will give the maximum possible increase in mechanical input before the machine looses stability.

Example 9.10 : A generator supplies active power of 1.0 pu to an infinite bus, through a lossless line of reactance $x_e = 0.6$ pu. The reactance of the generator and the connecting transformer is 0.3 pu. The transient internal voltage of the generator is 1.12 pu and infinite bus voltage is 1.0 pu. Find the maximum increase in mechanical power that will not cause instability.

Solution:

$$P_{max} = \frac{1.12 \times 1.0}{0.9} = 1.244 \text{ pu}$$

$$P_{m0} = P_{e0} = 1.0 = P_{max} \sin \delta_0 = 1.244 \sin \delta_0$$

$$\therefore \quad \delta_0 = \sin^{-1} \frac{1.0}{1.244} = 53.47° = 0.933 \text{ rad.}$$

(9.66) can be solved by N-R method, since it is of the form $f(d_{max}) = K$. Applying N-R method, at any iteration 'r', we get

$$\Delta\delta_{max}^{(r)} = \frac{K - f\left(\delta_{max}^{(r)}\right)}{\dfrac{df}{d\delta_{max}^{(r)}}} \qquad ...(9.67)$$

$$\frac{df}{d\delta_{max}^{(r)}} = (\delta_{max}^{(r)} - d_0)\cos \delta_{max}^{(r)} \qquad ...(9.68)$$

(This is the derivative evaluated at a value of $\delta = \delta_{max}^{(r)}$)

$$\delta_{max}^{(r+1)} = \delta_{max}^{(r)} + \Delta\delta_{max}^{(r)} \qquad ...(9.69)$$

Starting from an initial guess of δ_{max} between $\frac{\pi}{2}$ to π, (9.67) to (9.69) are solved iteratively till $\Delta\delta_{max}^{(r)} \leq \epsilon$. Here $K = \cos \delta_0 = 0.595$. The computations are shown in Table 9.2, starting from an initial guess $\delta_{max}^{(1)} = 1.745$ rad.

Table 9.2 Example 9.9

Interaction r	$\delta_{max}^{(r)}$	$\frac{df}{d\delta_{max}^{(r)}}$	$f(\delta_{max}^{(r)})$	$\Delta\delta_{max}^{(r)}$	$\delta_{max}^{(r+1)}$
1	1.745	– 0.1407	0.626	0.22	1.965
2	1.965	– 0.396	0.568	– 0.068	1.897
3	1.897	– 0.309	0.592	– 0.0097	1.887
4	1.887	– 0.2963	0.596	– 0.0033	1.883

Since, $\Delta\delta_{max}^{(r)}$ is sufficient by small, we can take

$$\delta_{max} = 1.883 \text{ rad} = 107.88°$$

$$\delta_1 = 180 - \delta_{max} = 72.1°$$

$$P_{m1} = P_{max} \sin \delta_{max} = 1.183$$

Maximum step increase permissible is $P_{m1} - P_{m0}$, = 1.183 – 1.0 = 0.183 pu

Example 9.11: Transform a two machine system to an equivalent SMIB system and show how equal area criterion is applicable to it.

Solution: Consider the two machine system show in Fig. 9.17.

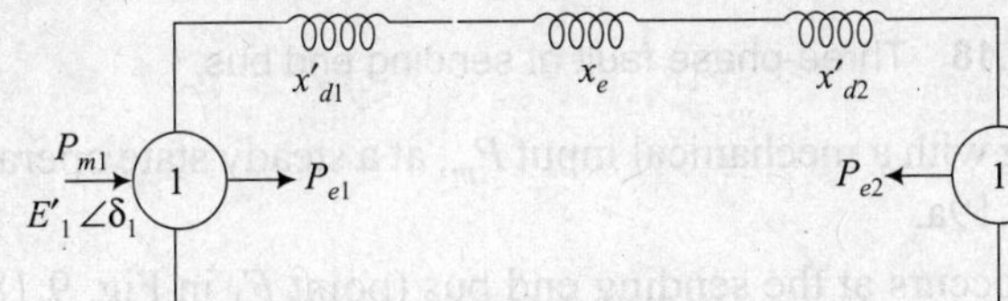

Fig. 9.17 Two machine system under steady state (neglecting losses in the machine).

$$P_{m1} = -P_{m2} = P_m$$

$$P_{e1} = -P_{e2} = P_e$$

The swing equations are

$$\frac{d^2\delta_1}{dt^2} = \frac{P_{m1} - P_{e1}}{M_1} = \frac{P_m - P_e}{M_1} \quad ...(9.70)$$

$$\frac{d^2\delta_2}{dt^2} = \frac{P_{m2} - P_{e2}}{M_2} = \frac{P_e - P_m}{M_2} \quad ...(9.71)$$

Subtracting (9.71) from (9.70), we get

$$\frac{d^2(\delta_1 - \delta_2)}{dt^2} = \frac{M_1 + M_2}{M_1\ M_2}(P_m - P_e)$$

or,

$$M_{eq}\frac{d^2\delta}{dt^2} = P_m - P_e \quad ...(9.72)$$

where
$$M_{eq} = \frac{M_1 M_2}{M_1 + M_2}$$
$$\delta = \delta_1 + \delta_2$$
$$P_e = \frac{E_1' E_2'}{x_{d1}' + x_e + x_{d2}'} \sin \delta$$

Equation (9.72) is identical to that of an SMIB system in form and can be used to determine the relative swing $(\delta_1 - \delta_2)$ between the two machines to assess the stability.

9.9.2 Three-phase Fault at Sending End Bus

Consider a generator supplying power to an infinite bus through a double circuit line as shown in Fig. 9.18.

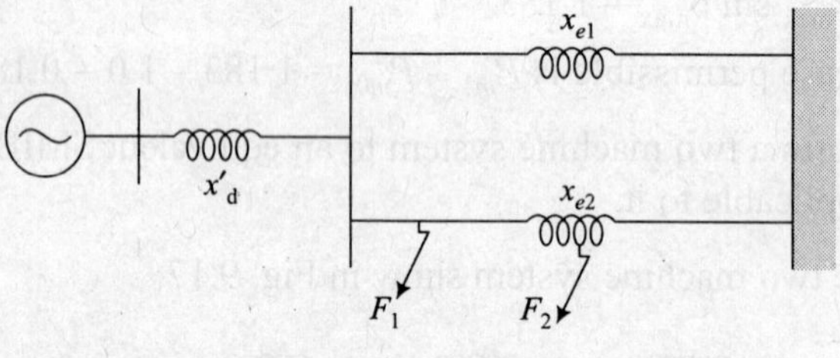

Fig. 9.18 Three-phase fault at sending end bus.

Let the system be operating with a mechanical input P_m, at a steady state operating angle δ_0, as indicated by point 'a' on Fig. 9.19a.

If now a three phase fault occurs at the sending end bus (point F_1 in Fig. 9.18), the electrical output of the generator reduces to zero and the operating point shifts to b. (Fig 9.19a). The generator is subject to acceleration and rotor angle moves along bc. Let the fault be cleared at point c, at a time t_c called clearing time, corresponding to an angle δ_1 called the clearing angle. The power angle curve is now same as the pre-fault curve and the operating point shifts to e (Fig. 9.19a). The rotor is now subject to deceleration and rotor angle moves along ef till at f, when $A_2 = A_1$, the rotor runs at synchronous speed and rotor angle is δ_{max}. The rotor now again moves along fe in reverse

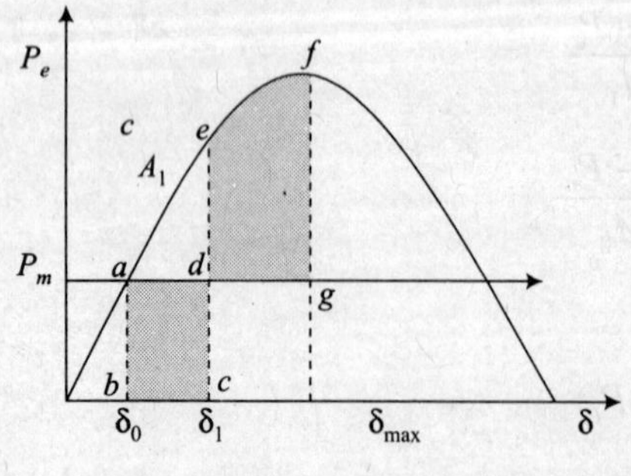

Fig. 9.19a Equal area criterion for a three phase fault at sending end.

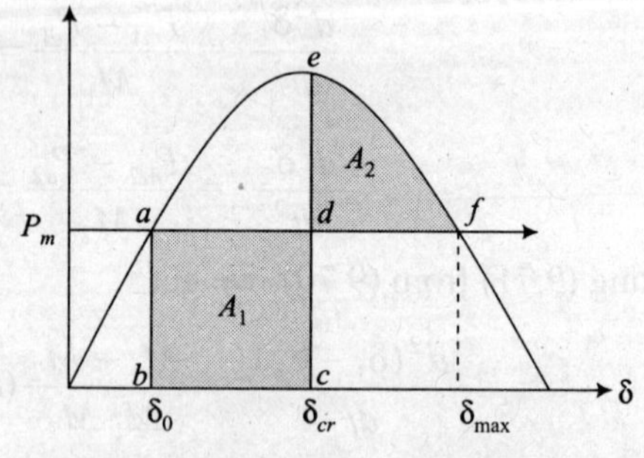

Fig. 9.19b Equal area criterion for critical clearing angle.

direction, since it is still subject to deceleration. Due to damping, the machine oscillates around δ_0 for a few swings and then settles down at point *a*.

As the clearing time t_c increases, so does δ_1, causing an increase in area A_1, till a critical case is reached beyond which $A_2 < A_1$. The critical case is shown in Fig. 9.19b. Here, the rotor should reach synchronous speed at *f*. Else, if it moves beyond *f*, with a positive velocity deviation, the acceleration will cause the velocity deviation to increase further and stability is lost. This maximum clearing time is called the *critical clearing time* (t_{cr}) and the angle at the time of clearing is called *critical clearing angle* (δ_{cr}). If the fault persists longer than t_{cr}, the system loses stability.

From Fig. 9.19b,

$$\delta_{max} = \pi - \delta_0 \quad ...(9.73)$$

$$P_m = P_{max} \sin \delta_0 = P_{max} \sin \delta_{max} \quad ...(9.74)$$

$$A_1 = P_m (\delta_{cr} - \delta_0)$$

$$A_2 = \int_{\delta_{cr}}^{\delta_{max}} (P_{max} \sin \delta - P_m)\, d\delta = P_{max}(\cos \delta_{cr} - \cos \delta_{amx}) - P_m(\delta_{max} - \delta_{cr})$$

From equal area criterion $A_1 = A_2$,

$$P_m (\delta_{cr} - \delta_0) = P_{max}(\cos \delta_{cr} - \cos \delta_{max}) - P_m(\delta_{max} - \delta_{cr})$$

$$\cos \delta_{cr} = \frac{P_m}{P_{max}}(\delta_{max} - \delta_0) + \cos \delta_{max} \quad ...(9.75)$$

Substituting (9.73) and (9.74) in (9.75), the critical clearing angle δ_{cr} is given by

$$\delta_{cr} = \cos^{-1} [(\pi - 2\,\delta_0) \sin \delta_0 - \cos \delta_0] \quad ...(9.76)$$

During the fault $P_e = 0$, hence $P_a = P_m$. From swing equation

$$\frac{d^2\delta}{dt^2} = \frac{\pi f}{H} P_m$$

Integrating twice,

$$\delta = \frac{\pi f}{2H} P_m t^2 + \delta_0$$

or,

$$\delta_{cr} = \frac{\pi f}{2H} P_m t^2_{cr} + \delta_0$$

or,

$$t_{cr} = \sqrt{\frac{2H}{\pi f} \frac{(\delta_{cr} - \delta_o)}{P_m}} \quad ...(9.77)$$

An explicit expression for critical clearing time is generally not possible, since the swing equation does not yield a closed form solution, expect when $P_e = 0$.

9.9.3 Three-phase Fault Away from Sending End

Consider a fault at location F_2 in Fig. 9.18. Let the machine be initially operating with a mechanical input P_m, at an operating angle δ_0 as shown in Fig 9.20a. The power angle curve is given by curve I in Fig. 9.20a.

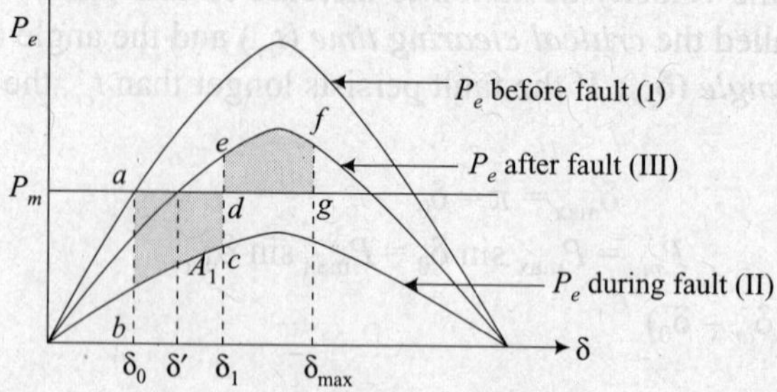

9.20a Equal area criterion for a three-phase fault away from sending end.

During the fault, P_e is not reduced to zero. The transfer reactance between sending end and receiving end increases (refer example 9.8), reducing the power transfer capability as shown by curve II. The generator now accelerates along *bc*. Let the fault be cleared at an angle δ_1, with outage of the faulted line. The power angle curve is now given by curve III. The generator now decelerates along *ef*, till $A_2 = A_1$. Since, $P_e > P_m$ the generator retraces path along *fe* and rotor oscillates around the final operating point δ' at its natural frequency. Eventually because of damping, oscillations reduce in subsequent swings and rotor operates at an angle δ' where P_m is equal to P_e(on the post fault curve).

The critical case is when any further increase in δ_1, causes A_2, the decelerating energy to become less than the area representing the accelerating energy A_1. This happens when δ_{max} lies on the intersection of P_m and post fault power angle curve as shown in Fig. 9.20b.

Any further increase in clearing time, makes the system unstable. From Fig. 9.20b,

$$\delta_{max} = \pi - \sin^{-1}\left(\frac{P_m}{P_{max\,III}}\right) \quad \text{...(9.78)}$$

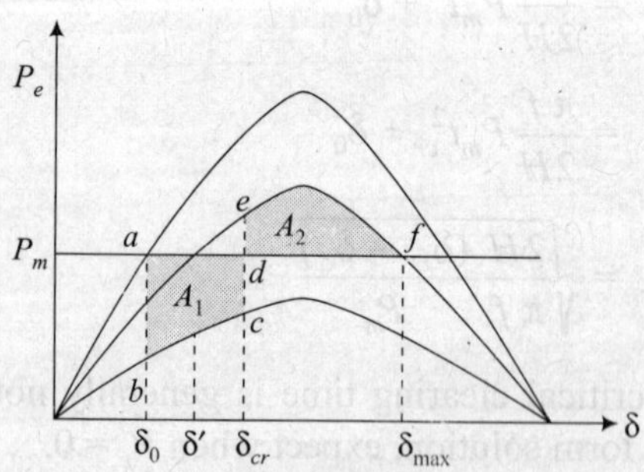

Fig. 9.20b Critical clearing angle.

$$\delta' = \sin^{-1}\left(\frac{P_m}{P_{\max III}}\right)$$

$$A_1 = \int_{\delta_o}^{\delta_{cr}} (P_m - P_{\max II} \sin \delta)\, d\delta = P_m(\delta_{cr} - \delta_0) + P_{\max II}(\cos \delta_{cr} - \cos \delta_0)$$

$$A_2 = \int_{\delta_{cr}}^{\delta_{\max}} (P_{\max III} \sin d - P_m)\, d\delta = P_{\max III} (\cos \delta_{cr} - \cos \delta_{\max}) - P_m(\delta_{\max} - \delta_{cr})$$

Applying equal area criterion $A_1 = A_2$,

$$P_m(\delta_{cr} - \delta_0) + P_{\max II}(\cos \delta_{cr} - \cos \delta_0) = P_{\max III}(\cos \delta_{cr} - \cos \delta_{\max}) - P_m(\delta_{\max} - \delta_{cr})$$

Rearranging terms,

$$\cos d_{cr} = \frac{P_m(\delta_{\max} - \delta_o) - P_{\max II} \cos\delta_o + P_{\max III} \cos\delta_{\max}}{P_{\max III} - P_{\max II}} \qquad ...(9.79)$$

In (9.79) the angles are in radians. To obtain the critical clearing time, the swing equation has to be solved. The time at which δ reaches δ_{cr} is the critical clearing time. This is illustrated in section 9.9.

Example 9.12: *A* 50 Hz, synchronous generator having inertia constant $H = 5.2$ MJ/MVA and $x_d' = 0.3$ pu is connected to an infinite bus through a double circuit line as shown in Fig. 9.21. The reactance of the connecting *HT* transformer is 0.2 pu and reactance of each line is 0.4 pu. $|E_g| = 1.2$ pu and $|V| = 1.0$ pu and $P_e = 0.8$ pu. Determine the critical clearing angle and the post fault steady state operating angle in the following cases:

(i) A three phase fault occurs at the *HT* terminals of the transformer and is cleared with both lines intact.

(ii) A three phase fault occurs at the middle of one of the transmission lines and is cleared without line outage.

(iii) A three phase fault occurs at the middle of one of the transmission lines and is cleared by isolating the faulted line.

Solution: Before fault transfer reactance between generator and infinite bus

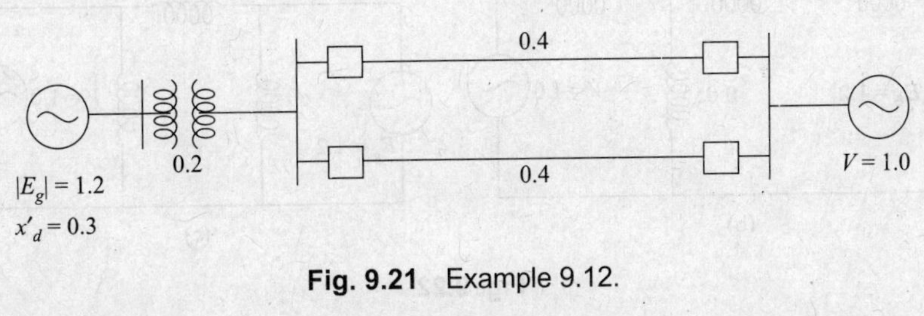

Fig. 9.21 Example 9.12.

$$X_I = 0.3 + 0.2 + \frac{0.4}{2} = 0.7 \text{ pu}$$

$$P_{\max I} = \frac{1.2 \times 1.0}{0.7} = 1.714 \text{ pu.}$$

Initial $\quad P_e = 0.8 \text{ pu} = P_m$

Initial operating angle $\quad \delta_0 = \sin^{-1} \frac{0.8}{1.714} = 27.82° = 0.485$ rad.

Case (i). $\delta_{max} = \pi - \delta_0 = 2.656$ rad. (refer Fig. 9.19b).

From (9.76),

$$\delta_{cr} = \cos^{-1} [(\pi - 2\,\delta_0) \sin \delta_0 - \cos \delta_0]$$
$$= \cos^{-1} [(\pi - 2 \times 0.485) \sin (0.485) - \cos (0.485)]$$
$$= 1.442 \text{ rad} = 82.5°$$

For this case critical clearing time can be calculated

$$t_{cr} = \sqrt{\frac{2H(\delta_{cr} - \delta_o)}{\pi f P_m}} = \sqrt{\frac{2 \times 5.2\,(1.442 - 0.485)}{\pi \times 50 \times 0.8}} = 0.28 \text{ sec}$$

Post fault steady state operating angle = δ_0 = 27.82°

Case (ii). When fault occurs at middle of one of the transmission lines, the network and its reduction is as shown in Fig. 9.22a to Fig. 9.22c.

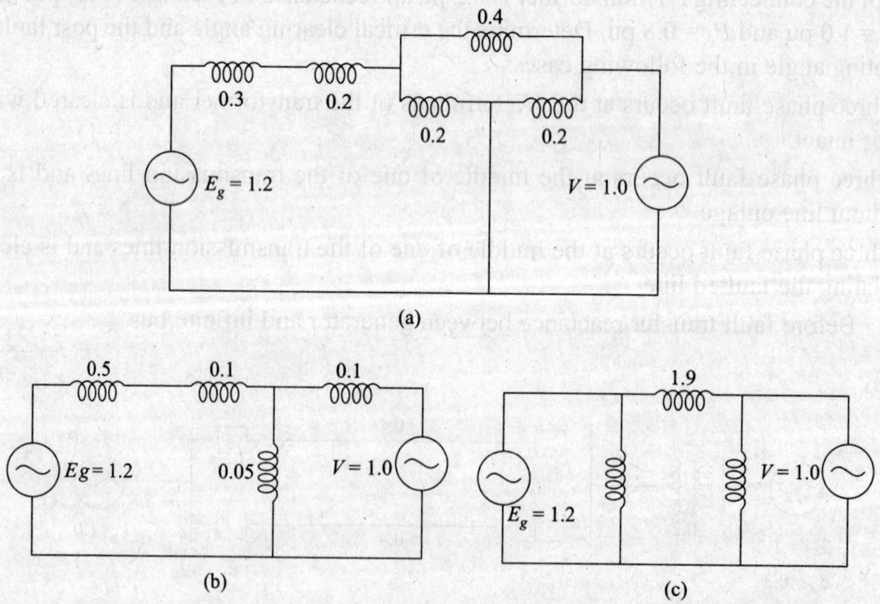

Fig. 9.22

The transfer reactance is 1.9 pu.

$$P_{\max II} = \frac{1.2 \times 1.0}{1.9} = 0.63 \text{ pu}$$

Since, there is no outage, $P_{\max III} = P_{\max I} = 1.714$

$$\delta_{\max} = \pi - \sin^{-1}\left(\frac{P_m}{P_{\max III}}\right) = \pi - \sin^{-1}\left(\frac{0.8}{1.714}\right) = 2.656 \text{ rad}$$

$$\cos \delta_{cr} = \frac{P_m(\delta_{\max} - \delta_o) - P_{\max II}\cos\delta_o + P_{\max III}\cos\delta_{\max}}{P_{\max III} - P_{\max II}}$$

$$= \frac{0.8\,(2.656 - 0.485) - 0.63\cos(0.485) + 1.714\cos(2.656)}{1.714 - 0.63}$$

$$= \frac{1.7368 - 0.5573 - 1.5158}{1.084} = -0.3102$$

$$\delta_{cr} = \cos^{-1}(-0.3102) = 1.886 \text{ rad} = 108.07°$$

Final steady state operating angle = δ_0 (since, there is no line outage)

Case (iii). With line outage

$$X_{III} = 0.3 + 0.2 + 0.4 = 0.9 \text{ pu}$$

$$P_{\max III} = \frac{1.2 \times 1.0}{0.9} = 1.333 \text{ pu}$$

$$\delta_{\max} = \pi - \sin^{-1}\frac{0.8}{1.333} = 2.498 \text{ rad}$$

$$\text{Cos}\,\delta_{cr} = \frac{0.8\,(2.498 - 0.485) - 0.63\cos(0.485) + 1.333\cos(2.498)}{1.333 - 0.63}$$

$$= \frac{1.61 - 0.5573 - 1.066}{0.703} = -0.012$$

$$\delta_{cr} = \cos^{-1}(-0.012) = 1.59 \text{ rad} = 91.13°$$

Post fault steady state operating angle $= \sin^{-1}\dfrac{0.8}{1.333} = 36.8°$

9.10 NUMERICAL SOLUTION OF SWING EQUATION

A closed form solution for the swing equation is not possible except in the simplest case presented in section 9.8.1. In a multi-machine system, the swing equation has to be solved for each machine, by use of a numerical method. Even in an SMIB system, we have to resort to a numerical technique to evaluate the variation of δ with time and to determine the critical clearing time. All numerical methods use the concept of discretization of the variables, over suitable time intervals. Many numerical techniques for solution of differential equations are available. In this section four methods are presented for solution of the swing equation.

1. Point-by-point method
2. Modified Euler's method
3. Runge-Kutta method
4. Milne's method

9.10.1 Point-by-point Method (Step-by-Step Method)

Consider the swing equation

$$M\frac{d^2\delta}{dt^2} = P_m - P_{\max}\sin\delta = P_a$$

or,

$$\frac{d^2\delta}{dt^2} = \frac{P_a}{M}$$

$$\left(\text{In pu } M = \frac{H}{180f}\,\text{s}^2/^\circ\text{ elec or } \frac{H}{\pi f}\,\text{s}^2/^\circ\text{rad}\right)$$

The solution of $\delta(t)$ is obtained in discrete intervals of time, with a uniform interval of Δt. Discretization is done as follows:

(1) The acceleration calculated at the beginning of an interval is assumed to remain constant from the middle of the preceding interval to the middle of the interval being considered. This assumption also implies that the accelerating power P_a computed at the beginning of an interval is assumed constant from the middle of the preceding interval to the middle of the interval being considered.

(2) The speed deviation $\omega = \frac{d\delta}{dt}$ is assumed constant over the entire interval computed at the middle of the interval. Note that here ω is the speed above the synchronous speed, ω_s.

These are graphically represented in Fig. 9.23. The numbers on $t/\Delta t$ axis refers to the end of the interval. Consider the nth interval, beginning at $t = (n - 1)\Delta t$. The angular position is δ_{n-1}. The acceleration $\alpha_{(n-1)}$ as calculated at this interval is assumed constant from

$t = \left(n - \frac{3}{2}\right)\Delta t$ to $\left(n - \frac{1}{2}\right)\Delta t$. The change in speed during this time is given by

$$\Delta\omega_{\left(n-\frac{1}{2}\right)} = \Delta t\,\alpha_{(n-1)} = \Delta t\frac{P_{a(n-1)}}{M} \qquad \text{...(9.80)}$$

where, $P_{a(n-1)} = P_m - P_{\max}\sin(\delta_{(n-1)})$

The speed at the end of this interval is

$$\omega_{n-\frac{1}{2}} = \omega_{n-\frac{3}{2}} + \Delta\omega_{n-\frac{1}{2}} \qquad ...(9.81)$$

From our assumption, the speed is constant throughout the nth interval. The change in angular position during the nth interval is

$$\Delta\delta_n = \Delta t\left[\omega_{n-\frac{1}{2}}\right] = \Delta t\left[\omega_{n-\frac{3}{2}} + \Delta\omega_{n-\frac{1}{2}}\right]$$

$$= \Delta t\omega_{n-\frac{3}{2}} + \Delta t^2 \frac{P_{a(n-1)}}{M} \qquad ...(9.82)$$

Similarly, we can write

$$\Delta\delta_{n-1} = \Delta t\omega_{\left(n-\frac{3}{2}\right)} \qquad ...(9.83)$$

Subtracting (9.83) from (9.82), we get

$$\Delta\delta_n = \Delta\delta_{n-1} + \Delta t^2 \frac{P_{a(n-1)}}{M} \qquad ...(9.84)$$

The angular position at the end of the nth interval is given by

$$\delta_n = \delta_{n-1} + \Delta\delta_n \qquad ...(9.85)$$

The process is repeated to obtain $P_{a(n)}$, $\Delta\delta_{n+1}$ and δ_{n+1}. The time solution is carried out till the time of interest. Greater accuracy can be obtained by considering smaller time intervals.

Any switching event (occurrence of a fault, clearance of a fault, line opening, etc.) causes discontinuity in the accelerating power. If the discontinuity occurs at the beginning of an interval 'n', the average of P_a, before ($P_{a(n-)}$) and after ($P_{a(n+)}$) the discontinuity must be used. If discontinuity occurs at the middle of an interval, no adjustment is required and the increment in angle is calculated from the value of P_a at the beginning of the interval. The following example illustrates the method.

Example 9.13: For the network of example 9.11, case (iii) evaluate the following using point-by-point method.

(i) $\delta(t)$ for a sustained fault and the critical clearing time.

(ii) $\delta(t)$ when fault is cleared in 2.5 cycles.

(iii) $\delta(t)$ when fault is cleared in 6.25 cycles.

Solution: (Refer example 9.11)

$$\delta_0 = 27.82°$$

Prefault maximum power $P_{\max I} = 1.714$ pu.

Fault on (during fault) maximum power $P_{\max II} = 0.63$ pu.

Post fault maximum power $P_{\max III} = 1.333$ pu.

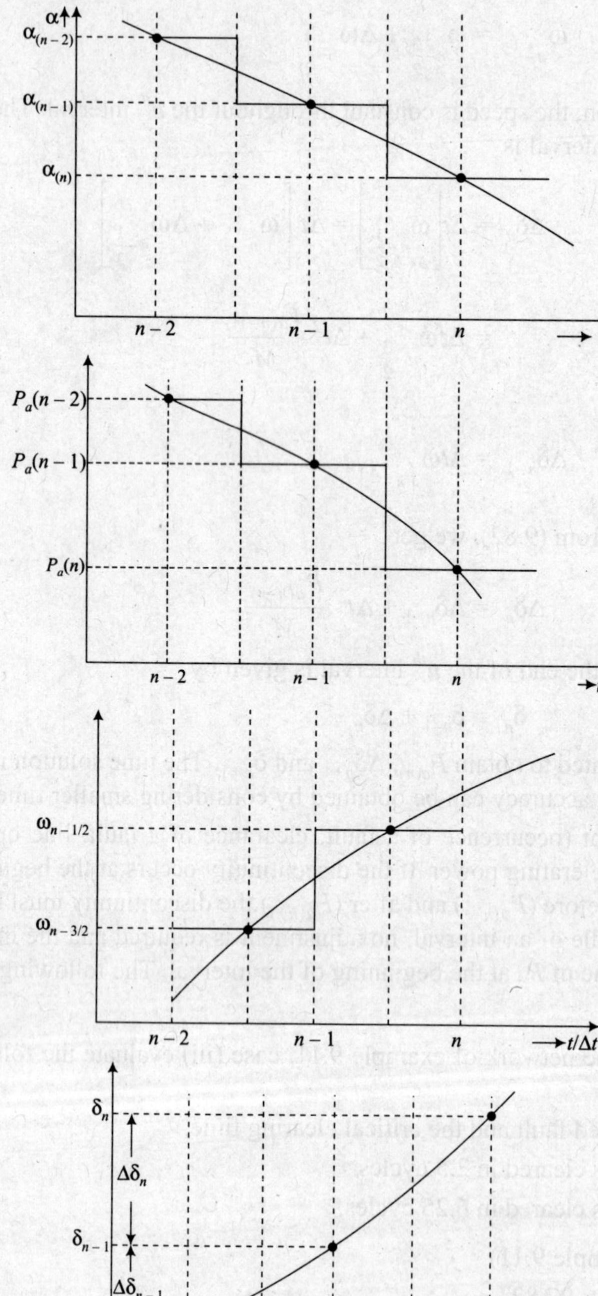

Fig. 9.23 Discretization of parameters for point-by-point method.

$$\delta_{cr} = 91.13°$$

$$P_m = 0.8 \text{ pu.}$$

$$M(\text{pu}) = \frac{H}{180f} = \frac{5.2}{180 \times 50} = 5.77 \times 10^{-4} \text{ s}^2/°\text{elec}$$

Let $\Delta t = 0.05\ s$

$$\frac{\Delta t^2}{M} P_a = \frac{(0.05)^2}{5.77 \times 10^{-4}} \times P_a = 4.32\ P_a$$

Case (i). Sustained fault.

$$P_e = 1.714 \sin \delta \text{ (before fault)}$$

$$= 0.63 \sin \delta \text{ (during fault)}$$

Calculations use the following formulae:

$$P_{a(n-1)} = P_m - P_{\max} \sin \delta_{n-1}$$

$$\Delta\delta_n = \Delta\delta_{n-1} + \frac{(\Delta t)^2}{M} P_{a(n-1)}$$

$$\delta_n = \delta_{n-1} + \Delta\delta_n$$

Assuming short circuit occurs at $t = 0$,

At $(0^-)\ P_e = 1.714 \sin \delta_0 = 0.8$

$P_a(0^-) = 0.8 - 0.8 = 0.0$ (steady state)

At $(0^+)\ P_e = 0.63 \sin \delta_0 = 0.293$

$P_a(0^+) = 0.8 - 0.293 = 0.507$

$$P_{av}(0) = \frac{P_a(0^-) + P_a(0^+)}{2} = \frac{0.0 + 0.507}{2} = 0.2535 \text{ pu.}$$

The calculations are shown in Table 9.3. An illustrative calculation is as follows:

At $t = 0.15$, δ_{n-1} (at the beginning of the interval) = 37.43°

$$P_e = 0.63 \sin (37.43°) = 0.383 \text{ pu}$$

$$P_{a(n-1)} = 0.8 - 0.383 = 0.417$$

$$\frac{\Delta t^2}{M} P_a = 4.32 \times 0.417$$

$$= 1.801.$$

$$\Delta\delta(t = 0.15\ s) = \Delta\delta(t = 0.1) + \frac{\Delta t^2}{M} P_a(t = 0.15\ s)$$

$$= 5.237 + 1.801 = 7.038°$$

$$\delta(t = 0.2\ s) = \delta(t = 0.15\ s) + \Delta\delta(t = 0.15s)$$
$$= 37.43° + 7.038° = 44.46°$$

(This is angle at the beginning of $t = 0.2$ sec)

Table 9.3 Computation of swing curve for sustained fault

t	P_{max}	$\delta_{(deg)}$	$Sin\ \delta$	$P_e = P_{max}$ $sin\ \delta$	$P_a =$ $0.8 - P_e$	$\frac{\Delta t^2}{M} P_a$ $4.32\ P_a$	$\Delta\delta$
0^-	1.714	27.820	0.466	0.800	0.000	–	–
0^+	0.630	27.820	0.466	0.293	0.507	–	–
0_{av}	–	27.820	–	–	0.253	1.095	1.095
0.05	0.630	28.960	0.484	0.305	0.495	2.138	3.233
0.10	0.630	32.190	0.533	0.336	0.464	2.004	5.237
0.15	0.630	37.430	0.608	0.383	0.417	1.801	7.038
0.20	0.630	44.460	0.700	0.441	0.359	1.550	8.588
0.25	0.630	53.050	0.800	0.503	0.296	1.281	9.869
0.30	0.630	62.920	0.890	0.561	0.239	1.032	10.901
0.35	0.630	73.820	0.960	0.605	0.195	0.842	11.743
0.40	0.630	85.560	0.997	0.628	0.172	0.742	12.485
0.45	0.630	98.050	0.990	0.623	0.176	0.761	13.246
0.50	0.630	111.29	0.930	0.587	0.213	0.920	14.16
0.55	0.630	125.46	–	–	–	–	–

From the swing curve, the time at which $\delta(t) = \delta_{cr}$, is determined to obtain the critical clearing time. This is shown in Fig. 9.24. The critical clearing time is found to be 0.42 seconds.

Case (ii). Fault cleared in 2.5 cycles:

$$2.5 \text{ cycles } = \frac{2.5}{50} = 0.05\ s$$

Since, the discontinuity occurs at the beginning of an interval $P_{a(av)}$ value is calculated from

$\frac{P_a(t = 0.05-) + P_a(t = 0.05+)}{2}$. The rest of computations are same as before. Computations are shown in Table 9.4.

Table 9.4 Computation of swing curve for fault cleared in 2.5 cycles

t	P_{max}	$\delta_{(deg)}$	$Sin\ \delta$	$P_e = P_{max}\ \sin\delta$	$P_a = 0.8 - P_e$	$\frac{\Delta t^2}{M} P_a$ 4.32 P_a	$\Delta\delta$
0^-	1.714	27.820	0.466	0.800	0.000	–	–
0^+	0.630	27.820	0.466	0.293	0.507	–	–
0_{av}	–	27.820	–	–	0.253	1.095	1.095
$0.05_{(-)}$	0.630	28.960	0.484	0.305	0.495	–	–
$0.05_{(+)}$	1.333	28.960	0.484	0.645	0.155	–	–
0.05_{av}	–	28.960	0.484	–	0.325	1.404	2.500
0.10	1.333	31.460	0.522	0.695	0.105	0.454	2.953
0.15	1.333	34.410	0.565	0.753	0.047	0.203	3.156
0.20	1.333	37.570	0.609	0.813	–0.013	–0.056	3.100
0.25	1.333	40.610	0.651	0.869	–0.069	–0.298	2.802
0.30	1.333	43.490	0.688	0.917	–0.117	–0.505	2.297
0.35	1.333	45.780	0.716	0.954	–0.154	–0.665	1.631
0.40	1.333	47.410	0.736	0.981	–0.181	–0.782	0.849
0.45	1.333	48.260	0.746	0.994	–0.194	–0.838	0.010
0.50	1.333	48.270	0.746	0.994	–0.194	–0.838	–0.828
0.55	1.333	47.440	–	–	–	–	

Case (iii). Fault cleared in 6.25 cycles

$$6.25 \text{ cycles} = \frac{6.25}{50} = 0.125\ s$$

Since, discontinuity occurs at the middle of an interval no special method is needed. The computations are shown in Table 9.5.

Table 9.5 Computation of swing curve for fault cleared in 6.25 cycles

t	P_{max}	$\delta_{(deg)}$	$Sin\ \delta$	$P_e = P_{max}\ \sin\delta$	$P_a = 0.8 - P_e$	$\frac{\Delta t^2}{M} P_a$ 4.32 P_a	$\Delta\delta$
$0_{a\theta}$	–	27.820	–	–	0.253	1.095	1.095
0.05	0.630	28.960	0.484	0.305	0.495	2.138	3.233
0.10	0.630	32.190	0.533	0.336	0.464	2.004	5.237

0.15	1.333	37.430	0.608	0.810	–0.010	–0.043	5.194
0.20	1.333	42.620	0.677	0.902	–0.102	–0.440	4.754
0.25	1.333	47.370	0.736	0.981	–0.181	–0.782	3.972
0.30	1.333	51.350	0.781	1.041	–0.241	–1.041	2.931
0.35	1.333	54.280	0.812	1.079	–0.279	–1.205	1.726
0.40	1.333	56.000	0.829	1.105	–0.305	–1.317	0.409
0.45	1.333	56.410	0.839	1.106	–0.306	–1.322	–0.913
0.50	1.333	55.490	0.824	1.098	–0.298	–1.287	–2.200
0.55	1.333	53.300	–	–	–	–	–

The swing curves for all the three cases are shown in Fig. 9.24. It can be seen that the swing is greater when the fault clearing time is increased.

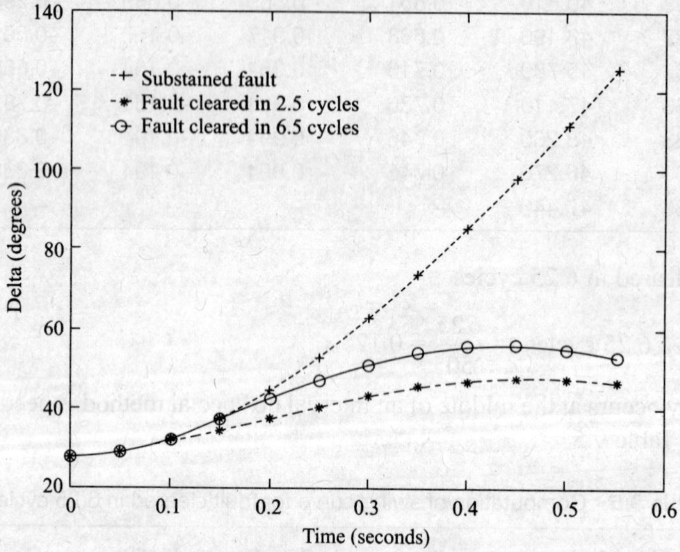

Fig. 9.24 Swing curves for example 9.12.

9.10.2 Modified Euler's method

Euler's method is one of the easiest methods to program for solution of differential equations using a digital computer. It uses the Taylor's series expansion, discarding all second-order and higher-order terms. Modified Euler's algorithm uses the derivatives at the beginning of a time step, to predict the values of the dependent variables at the end of the step ($t_1 = t_0 + \Delta t$). Using the predicted values, the derivatives at the end of the interval are computed. The average of the two derivatives is used in updating the variables. Consider two simultaneous differential equations:

$$\frac{dx}{dt} = f_y(x, y, t)$$

$$\frac{dy}{dt} = f_y(x, y, t)$$

Starting from initial values $x_{0,}\, y_0,\, t_0$ at the beginning of a time step and a step size Δt we solve as follows:

Let

$$D_x = f_x(x_0, y_0, t_0) = \left.\frac{dx}{dt}\right|_0$$

$$D_y = f_y(x_0, y_0, t_0) = \left.\frac{dy}{dt}\right|_0$$

$$x^p = x_o + D_x \Delta t$$

$$y^p = y_o + D_y\, \Delta t$$

$$t_1 = t_o + \Delta t = t_o + h \quad (h \text{ is normally used for step size})$$

$$D_{xP} = \left.\frac{dx}{dt}\right|_P = f_x(x^P, y^P, t_1)$$

$$D_{yP} = \left.\frac{dy}{dt}\right|_P = f_y(x^P, y^P, t_1)$$

$$x_1 = x_0 + \left(\frac{D_x + D_{xp}}{2}\right)\Delta t$$

$$y_1 = y_0 + \left(\frac{D_y + D_{yp}}{2}\right)\Delta t.$$

x_1 and y_1 are used as initial values to compute x_2 and y_2 at end of $t_2 = t_1 + \Delta t$ and so on. In general at any time step $(t + 1)$ we compute the values of x_{t+1} and y_{t+1} as follows.

$$D_x = f_x(x_t, y_t, t) = \left.\frac{dx}{dt}\right|_t$$

$$D_y = f_y(x_t, y_t, t) = \left.\frac{dy}{dt}\right|_t$$

$$x^p = x_t + D_x\, \Delta t$$

$$y^p = y_t + D_y\, \Delta t$$

$$D_{xp} = f_x(x^P, y^P, t+1) = \left.\frac{dx}{dt}\right|_p$$

$$D_{yp} = f_y(x^P, y^P, t+1) = \left.\frac{dx}{dt}\right|_p$$

$$x_{t+1} = x_t + \left(\frac{D_x + D_{xp}}{2}\right)\Delta t$$

$$Y_{t+1} = y_t + \left(\frac{D_y + D_{yp}}{2}\right)\Delta t$$

To solve the swing equation by Modified Euler's method, it is written as two first order differential equations:

$$\frac{d\delta}{dt} = \omega$$

$$\frac{d\omega}{dt} = \frac{P_a}{M} = \frac{P_m - P_{\max}\sin\delta}{M}$$

Starting from an initial value δ_0, ω_0 at the beginning of any time step, and choosing a step size Δt s, the equations to be solved in modified Euler's method are as follows:

$$\left.\frac{d\delta}{dt}\right|_0 = D_1 = \omega_0$$

$$\left.\frac{d\omega}{dt}\right|_0 = D_2 = \frac{P_m - P_{\max}\sin\delta_0}{M}$$

$$\delta^P = \delta_0 + D_1\,\Delta t$$

$$\omega^P = \omega_0 + D_2\,\Delta t$$

$$\left.\frac{d\delta}{dt}\right|_P = D_{1P} = \omega^P$$

$$\left.\frac{d\omega}{dt}\right|_P = D_{2P} = \frac{P_m - P_{\max}\sin\delta^P}{M}$$

$$\delta_1 = \delta_0 + \left(\frac{D_1 + D_{1P}}{2}\right)\Delta t$$

$$\omega_1 = \omega_0 + \left(\frac{D_2 + D_{2P}}{2}\right)\Delta t$$

δ_1 and ω_1 are used as initial values for the successive time step.

In any time step $t + 1$,

$$D_1 = \left.\frac{d\delta}{dt}\right|_t = \omega_t$$

$$D_2 = \left.\frac{d\omega}{dt}\right|_t = \frac{P_m - P_{\max} \sin\delta_t}{M}$$

$$\delta^p = \delta_t + D_1 \, \Delta t$$
$$\omega^p = \omega_t + D_2 \, \Delta t$$

$$D_{1p} = \frac{d\delta}{dt}|_p = \omega^p$$

$$D_{2p} = \frac{d\omega}{dt}|_p = \frac{P_m - P_{\max} \sin\delta^P}{M}$$

$$\delta_{t+1} = \delta_t + \left(\frac{D_1 + D_{1P}}{2}\right) \Delta t$$

$$\omega_{t+1} = \omega_t + \left(\frac{D_2 + D_{2P}}{2}\right) \Delta t$$

The value of P_{max} to be used will depend on the time of computation. We use the appropriate value of P_{max}, before fault, during fault and post-fault.Numerical errors are introduced because of discarding higher-order terms in Taylor's expansion. Errors can be decreased by choosing smaller values of step size. Too small a step size, will increase computation, which can lead to large errors due to rounding off. The Runge-Kutta method which uses higher-order terms is more popular.

Example 9.14: Obtain the swing curve for example 9.12, case (iii) using modified Euler's method.

Solution:

$$\delta_0 = 27.82° = 0.485 \text{ rad}$$

$$\omega_0 = 0.0 \text{ rad/sec (at } t = 0°)$$

Choosing a step size of 0.05 s, the swing is computed. Table 9.6a gives the values of the derivatives and predicted values. Table 9.6b gives the initial values δ_t, ω_t and the values at the end of the interval δ_{t+1}, ω_{t+1}. Calculations are illustrated for the time step $t = 0.2$ s.

$$\delta_0 = 0.761$$

$$\omega_0 = 2.072$$

$$P_m = 0.8$$

$$M = \left[\frac{5.2}{\pi \times 50}\right] = 0.0331 \text{ s}^2/\text{rad}$$

P_{max} (after fault clearance) = 1.333 pu

$$D_1 = 2.072$$

$$D_2 = \frac{0.8 - 1.333\sin(0.761)}{0.0331} = -3.604$$

$$\delta^P = 0.761 + (2.072 \times 0.05) = 0.865$$

$$\omega^P = 2.072 + (-3.604 \times 0.05) = 1.892$$

$$D_{1P} = 1.892$$

$$D_{2P} = \frac{0.8 - 1.333\sin(0.865)}{0.0331} = -6.482$$

$$\delta = 0.761 + \left(\frac{2.072 + 1.892}{2}\right)0.05 = 0.860$$

$$\omega_1 = 2.072 + \left(\frac{-3.604 - 6.482}{2}\right)0.05 = 1.82$$

δ_1, ω_1 are used as initial values in next time step.

Table 9.6a Calculation of derivatives in modified Euler's method

t	D_1	D_2	δ^P	ω^P	D_{1P}	D_{2P}
0+	0.0	15.296	0.485	0.765	0.765	15.296
0.05	0.765	14.977	0.542	1.514	1.514	14.350
0.10	1.498	14.043	0.636	2.200	2.200	12.860
0.15	2.17	–0.299	0.761	2.155	2.155	–3.600
0.20	2.072	–3.604	0.865	1.892	1.892	–6.482
0.25	1.820	–6.350	0.951	1.502	1.502	–8.612
0.30	1.446	–8.424	1.015	1.025	1.025	–10.041
0.35	0.984	–9.827	1.054	0.493	0.493	–10.843
0.40	0.467	–10.602	1.065	–0.063	–0.063	–11.060
0.45	–0.074	–10.803	1.048	–0.614	–0.614	–10.720
0.50	–0.612	–10.46	1.004	–1.135	–1.135	–9.800

Table 9.6b Calculations of δ_t, ω_t and δ_{t+1}, ω_{t+1} in modified Euler's method

t	P_{max} (pu)	δ_t rad	ω_t rad/sec	δ_{t+1} rad	ω_{t+1} rad/sec	δ_{t+1} deg
0–	1.714	0.485	0.0	–	–	–
0+	0.630	0.485	0.0	0.504	0.765	28.87
0.05	0.630	0.504	0.765	0.561	1.498	32.14
0.10	0.630	0.561	1.498	0.653	2.170	37.41
0.15	1.333	0.653	2.170	0.761	2.072	43.60
0.20	1.333	0.761	2.072	0.860	1.820	49.27

0.25	1.333	0.860	1.820	0.943	1.446	54.03
0.30	1.333	0.943	1.446	1.005	0.984	57.58
0.35	1.333	1.005	0.984	1.042	0.467	59.70
0.40	1.333	1.042	0.467	1.052	−0.074	60.27
0.45	1.333	1.052	−0.074	1.035	−0.612	59.30
0.50	1.333	1.035	−0.612	0.991	−1.118	56.78

9.10.3 Runge-Kutta Method

In Runge-Kutta method, the changes in dependent variables are calculated from a given set of formulae, derived by using an approximation, to replace a truncated Taylor's series expansion. The formulae for the Runge-Kutta fourth order approximation, for solution of two simultaneous differential equations are given below:

Given
$$\frac{dx}{dt} = f_x(x, y, t)$$
$$\frac{dy}{dt} = f_y(x, y, t)$$

Starting from initial values x_0, y_0, t_0 and step size h, the updated values are:

$$x_1 = x_0 + \frac{1}{6}(k_1 + 2k_2 + 2k_3 + k_4)$$
$$y_1 = y_0 + \frac{1}{6}(l_1 + 2l_2 + 2l_3 + l_4)$$

where,
$$k_1 = f_x(x_0, y_0, t_0)h$$
$$k_2 = f_x\left(x_0 + \frac{k_1}{2}, y_0 + \frac{l_1}{2}, t_0 + \frac{h}{2}\right)h$$
$$k_3 = f_x\left(x_0 + \frac{k_2}{2}, y_0 + \frac{l_1}{2}, t_0 + \frac{h}{2}\right)h$$
$$k_4 = f_x(x_0 + k_3, y_0 + l_3, t_0 + h)h$$
$$l_1 = f_y(x_0, y_0, t_0)h$$
$$l_2 = f_y\left(x_0 + \frac{k_1}{2}, y_0 + \frac{l_1}{2}, t_0 + \frac{h}{2}\right)h$$
$$l_3 = f_y\left(x_0 + \frac{k_2}{2}, y_0 + \frac{l_1}{2}, t_0 + \frac{h}{2}\right)h$$
$$l_4 = f_y(x_0 + k_3, y_0 + l_3, t_0 + h)h$$

In the above equations x_0, y_0, t_0 are the initial values in any time step and x_1, y_1, t_1 are the values at the end of the time step. In the subsequent step, x_1, y_1, t_1 are used as initial values to compute x_2, y_2, t_2 and so on.

The two first order differential equations to be solved to obtain solution for the swing equation are:

$$\frac{d\delta}{dt} = \omega$$

$$\frac{d\omega}{dt} = \frac{P_a}{M} = \frac{P_m - P_{max} \sin\delta}{M}$$

(Comparing with the general equations, $x = \delta$ and $y = \omega$)

Starting from initial value δ_0, ω_0, t_0 and a step size of Δt the formulae are as follows:

$$k_1 = \omega_0 \Delta t$$

$$l_1 = \left[\frac{P_m - P_{max} \sin\delta_0}{M}\right]\Delta t$$

$$k_2 = \left(\omega_0 + \frac{l_1}{2}\right)\Delta t$$

$$l_2 = \left[\frac{P_m - P_{max} \sin\left(\delta_0 + \frac{k_1}{2}\right)}{M}\right]\Delta t$$

$$k_3 = \left(\omega_0 + \frac{l_2}{2}\right)\Delta t$$

$$l_3 = \left[\frac{P_m - P_{max} \sin\left(\delta_0 + \frac{k_2}{2}\right)}{M}\right]\Delta t$$

$$k_4 = (\omega_0 + l_3)\Delta t$$

$$l_4 = \left[\frac{P_m - P_{max} \sin(\delta_0 + k_3)}{M}\right]\Delta t$$

$$\delta_1 = \delta_0 + \frac{1}{6}[k_1 + 2k_2 + 2k_3 + k_4]$$

$$\omega_1 = \omega_0 + \frac{1}{6}[l_1 + 2l_2 + 2l_3 + l_4]$$

Use δ_1 , ω_1 as initial values in the next time step and update the variables as before. Appropriate value of P_{max} must be used in the equations.

Example 9.15: Obtain the swing curve for example 9.12, for case (iii) using Runge-Kutta method.

Solution:

$$\delta_0 = 27.82° = 0.485 \text{ rad.}$$

$$\omega_0 = 0.0 \text{ rad/sec. (at } t = 0^-)$$

Choosing a step size of 0.05 s, the coefficient k_1, k_2, k_3, k_4 and l_1, l_2, l_3, and l_4 are calculated for each time step. The values of δ and ω are then updated. Table 9.7a gives the coefficient for different time steps. Table 9.7b gives the starting values δ_0, ω_0 for a time step and the updated values δ_1, ω_1 obtained by Runge-Kutta method. The updated values are used as initial values for the next time step and process continued. Calculations are illustrated for the time step $t = 0.2$ s.

$$\delta_0 = 0.756$$

$$M = 0.0331 \text{ s}^2/\text{rad}$$

$$\omega_0 = 2.067$$

$$P_m = 0.8$$

$$P_{max} = 1.333 \text{ (after fault is cleared)}$$

$$k_1 = 2.067 \times 0.05 = 0.103$$

$$l_1 = \left[\frac{0.8 - 1.333 \sin(0.756)}{0.0331}\right] \times 0.05 = -0.173$$

$$k_2 = \left[2.067 - \frac{0.173}{2}\right] 0.05 = 0.099$$

$$l_2 = \left[\frac{0.8 - 1.333 \sin\left(0.756 + \frac{0.103}{2}\right)}{0.0331}\right] \times 0.05 = -0.246$$

$$k_3 = \left[2.067 - \frac{0.246}{2}\right] 0.05 = 0.097$$

$$l_3 = \left[\frac{0.8 - 1.333 \sin\left(0.756 + \frac{0.099}{2}\right)}{0.0331}\right] \times 0.05 = -0.244$$

$$k_4 = (2.067 - 0.244)\, 0.05 = 0.091$$

$$l_4 = \left[\frac{0.8 - 1.333 \sin (0.756 + 0.097)}{0.0331}\right] \times 0.05 = -0.308$$

$$\delta_1 = 0.756 + \frac{1}{6}[0.103 + 2 \times 0.099 + 2 \times 0.097 + 0.091] = 0.854$$

$$\omega_1 = 2.067 + \frac{1}{6}[-0.173 + 2 \times -0.246 + 2 \times -0.244 - 0.308] = 1.823$$

Now $\delta = 0.854$ and $\omega = 1.823$ are used as initial values for the next time step. The computations have been rounded off to three digits. Greater accuracy is obtained by reducing the step size.

Table 9.7a Coefficients in Runge-Kutta method

T	k_1	l_1	k_2	l_2	k_3	l_3	K_4	l_4
0.0	0.0	0.764	0.019	0.764	0.019	0.757	0.038	0.749
0.05	0.031	0.749	0.056	0.736	0.056	0.736	0.075	0.703
0.10	0.075	0.704	0.092	0.674	0.091	0.667	0.108	0.632
0.15	0.108	– 0.010	0.108	– 0.094	0.106	– 0.095	0.103	– 0.173
0.20	0.103	– 0.173	0.099	– 0.246	0.097	– 0.244	0.091	– 0.308
8.25	0.091	– 0.309	0.083	– 0.368	0.082	– 0.363	0.073	– 0.413
0.30	0.073	– 0.413	0.063	– 0.455	0.061	– 0.450	0.050	– 0.480
0.35	0.050	– 0.483	0.038	– 0.510	0.037	– 0.504	0.025	– 0.523
0.40	0.025	– 0.523	0.012	– 0.536	0.011	– 0.529	– 0.001	– 0.534
0.45	– 0. 001	– 0.534	– 0.015	– 0.533	– 0.015	– 0.526	– 0.027	– 0.519
0.50	– 0.028	– 0.519	– 0.040	– 0.504	– 0.040	– 0.498	– 0.053	– 0.476

Table 9.7b δ, ω computations by Runge-Kutta method

t (sec)	P_{max} (pu)	δ_t (rad)	ω_t rad/sec	δ_{t+1} rad	ω_{t+1} rad/sec	δ_{t+1} deg
0^-	1.714	0.485	0.0			
0^+	0.630	0.485	0.0	0.504	0.759	28.87
0.05	0.630	0.504	0.756	0.559	1.492	32.03
0.10	0.630	0.559	1.492	0.650	2.161	37.24
0.15	1.333	0.650	2.161	0.756	2.067	43.32
0.20	1.333	0.756	2.067	0.854	1.823	48.93
0.25	1.333	0.854	1.823	0.936	1.459	53.63
0.30	1.333	0.936	1.459	0.998	1.008	57.18
0.35	1.333	0.998	1.008	1.035	0.502	59.30
0.40	1.333	1.035	0.502	1.046	– 0.029	59.93
0.45	1.333	1.046	– 0.029	1.031	– 0.557	59.07
0.50	1.333	1.031	– 0.557	0.990	– 1.057	56.72

Note: δ_0, ω_0 indicate values at beginning of interval and δ_1, ω_1 at end of interval. The fault is cleared at 0.125 seconds. $\therefore$ $P_{max} = 0.63$ at $t = 0.1$ sec and $P_{max} = 1.333$ at $t = 0.15$ sec, since fault is already cleared at that time. The swing curves obtained from modified Euler's method and Runge-Kutta method are shown in Fig. 9.25. It can be seen that the two methods yield very close results.

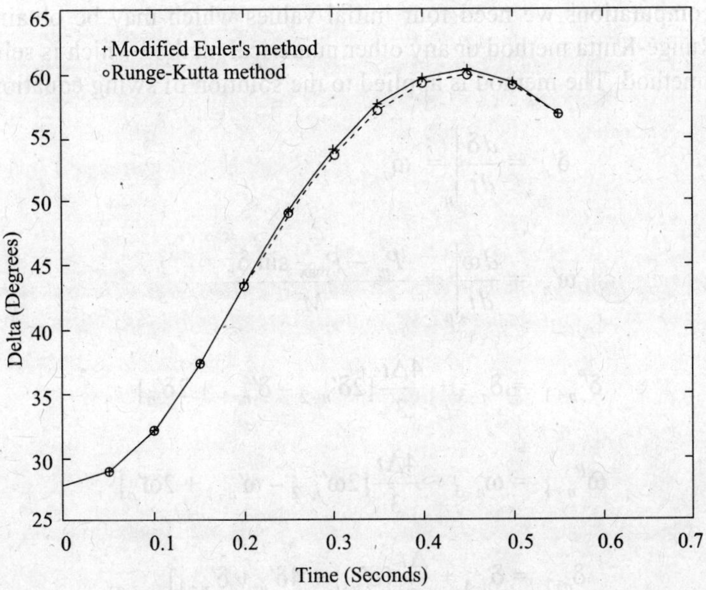

Fig. 9.25 Swing curves with modified Euler and Runge-Kutta methods.

9.10.4 Milne's Predictor Corrector Method

The Milne's formulae for solving two simultaneous differential equations are given below.

Consider
$$\frac{dx}{dt} = f_x(x, y, t)$$

$$\frac{dy}{dt} = f_y(x, y, t)$$

With values of x and y known for four consecutive previous times, the predicted value for $n + 1^{th}$ time step is given by

$$x^P_{n+1} = x_{n-3} + \frac{4h}{3}[2x'_{n-2} - x'_{n-1} + 2x'_n]$$

$$y^P_{n+1} = y_{n-3} + \frac{4h}{3}[2y'_{n-2} - y'_{n-1} + 2y'_n]$$

Where, x' and y' are derivatives at the corresponding time step. The corrected values are

$$x_{n+1} = x_{n-1} + \frac{h}{3}[x'_{n-1} + 4x'_n + x'_{n+1}]$$

$$y_{n+1} = y_{n-1} + \frac{h}{3}[y'_{n-1} + 4y'_n + y'_{n+1}]$$

where,
$$x'_{n+1} = f_x[x^P_{n-1}, y^P_{n+1}, t_{n+1}]$$
$$y'_{n+1} = f_y[x^P_{n+1}, y^P_{n+1}, t_{n+1}]$$

To start the computations we need four initial values which may be obtained by modified Euler's method, Runge-Kutta method or any other numerical method which is self starting, before applying Milne's method. The method is applied to the solution of swing equation as follows:

Define
$$\delta'_n = \left.\frac{d\delta}{dt}\right|_n = \omega_n$$

$$\omega'_n = \left.\frac{d\omega}{dt}\right|_n = \frac{P_m - P_{\max} \sin \delta_n}{M}$$

$$\delta^P_{n+1} = \delta_{n-3} + \frac{4\Delta t}{3}[2\delta'_{n-2} - \delta'_{n-1} + 2\delta'_n]$$

$$\omega^P_{n+1} = \omega_{n-3} + \frac{4\Delta t}{3}[2\omega'_{n-2} - \omega'_{n-1} + 2\omega'_n]$$

$$\delta_{n+1} = \delta_{n-1} + \frac{\Delta t}{3}[\delta'_{n-1} + 4\delta'_n + \delta'_{n+1}]$$

$$\omega_{n+1} = \omega_{n-1} + \frac{\Delta t}{3}[\omega'_{n-1} + 4\omega'_n + \omega'_{n+1}]$$

where,
$$\delta'_{n+1} = \omega^P_{n+1}$$

$$\omega'_{n+1} = \frac{P_m - P_{\max} \sin \delta^P_{n+1}}{M}$$

Example 9.16: Solve example 9.14 using Milne's method.

Solution: To start the process, we take the first four computations from Runge-Kutta method (Refer Table 9.7b).

$t = 0.0$ s $\delta_1 = 0.504$ $\omega_1 = 0.759$

$t = 0.05$ s $\delta_2 = 0.559$ $\omega_2 = 1.492$

$t = 0.10$ s $\delta_3 = 0.650$ $\omega_3 = 2.161$

$t = 0.15$ s $\delta_4 = 0.756$ $\omega_4 = 2.067$

The corresponding derivatives are calculated using the formulae for δ'_n and ω'_n. We get

$\delta'_1 = 0.759$ $\omega'_1 = 14.97$

$$\delta'_2 = 1.492 \quad \omega'_2 = 14.075$$
$$\delta'_3 = 2.161 \quad \omega'_3 = 12.65$$
$$\delta'_4 = 2.067 \quad \omega'_4 = -3.46$$

We now compute δ_5 and ω_5, at the next time step i.e., $t = 0.2$ s.

$$\delta^P_5 = \delta_1 + \frac{4\Delta t}{3}[2\delta'_2 - \delta'_3 + 2\delta'_4]$$

$$= 0.504 + \frac{4 \times 0.05}{3}[2 \times 1.492 - 2.161 + 2 \times 2.067] = 0.834$$

$$\omega^P_5 = \omega_1 + \frac{4\Delta t}{3}[2\omega'_2 - \omega'_3 + 2\omega'_4]$$

$$= 0.759 + \frac{4 \times 0.05}{3}[2 \times 14.075 - 12.65 + 2 \times (-3.46)] = 1.331$$

$$\delta'_5 = 1.331$$

$$\omega'_5 = \frac{0.8 - 1.333 \sin(0.834)}{0.0331} = -5.657$$

$$\delta_5 = \delta_3 + \frac{\Delta t}{3}[\delta'_3 + 4\delta'_4 + \delta'_5]$$

$$= 0.65 + \frac{0.05}{3}[2.161 + 4 \times 2.067 + 1.331] = 0.846$$

$$\omega_5 = \omega_3 + \frac{\Delta t}{3}[\omega'_3 + 4\omega'_4 + \omega'_5]$$

$$= 2.161 + \frac{0.05}{3}[12.65 - 4 \times 3.46 - 5.657] = 2.047$$

$$\delta'_5 = \omega_5 = 2.047$$

$$\omega'_5 = \frac{0.8 - 1.333 \sin(0.846)}{0.0331} = -5.98$$

The computations are continued for the next time step in a similar manner.

9.11 MULTI MACHINE TRANSIENT STABILITY ANALYSIS

A typical modern power system consists of a few thousands of nodes with heavy interconnections. Computation simplification and memory reduction have been two major issues in the development of mathematical models and algorithms for digital computation of transient stability. In its simplest form, the problem of a multi machine power system undergoing a disturbance can be mathematically stated as follows:

$$\dot{x}(t) = f_{\text{I}}(x(t)) \qquad -\infty \le t \le 0 \qquad \text{...(9.86a)}$$

$$\dot{x}(t) = f_{\text{II}}(x(t)) \qquad -0 < t \le t_{ce} \qquad \text{...(9.86b)}$$

$$\dot{x}(t) = f_{\text{III}}(x(t)) \qquad t_{ce} < t < \infty \qquad \text{...(9.86c)}$$

$x(t)$ is the vector of state variables to describe the differential equations governing the generator rotor dynamics, dynamics of flux decay and associated generator controller dynamics (like excitation control, PSS, governor control, etc). The function f_{I} describes the dynamics prior to the fault. Since, the system is assumed to be in steady state, all the state variable are constant. If the fault occurs at $t = 0$, f_{II} describes the dynamics during fault, till the fault is cleared at time t_{cl}. The post–fault dynamics is governed by f_{III}. The state of the system x_{cl} at the end of the fault-on period (at $t = t_{cl}$) provides the initial condition for the post-fault network described by (9.86c), which determines whether a system is stable or not after the fault is cleared. Some methods are presented in the following sections to evaluate multi machine transient stability. However, a detailed exposition is beyond the scope of the present book.

9.11.1 Reduced Order Model

This is the simplest model used in stability analysis and requires minimum data. The following assumptions are made:

- Mechanical power input to each synchronous machine is assumed to be constant.
- Damping is neglected.
- Synchronous machines are modeled as constant voltage sources behind transient reactance.
- Loads are represented as constant impedances.

With these assumptions, the multi machine system is represented as in Fig. 9.26.

Nodes 1, 2 ... n are introduced in the model and are called internal nodes (the terminal node is

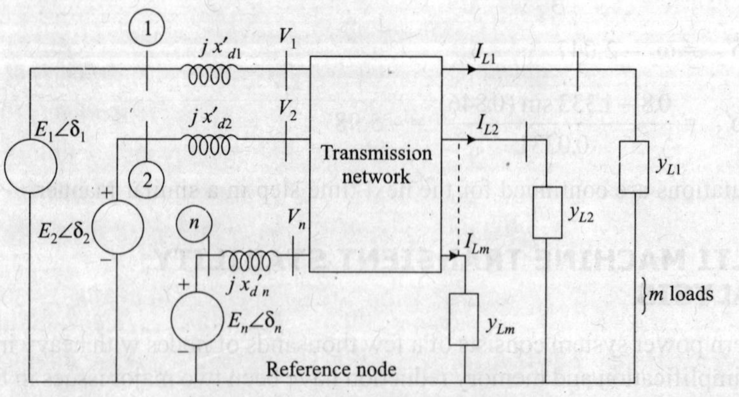

Fig. 9.26 Multi machine system.

the external node connected to the transmission network). The swing equations are formed for the various generators using the following steps:

Step 1. All system data is converted to a common base.

Step 2. A prefault load flow is performed, to determine the prefault steady state voltages, at all the external buses. Using the prefault voltages, the loads are converted into equivalent shunt admittance, connected between the respective bus and the reference node. If the complex load at bus i is given by

$$S_i = P_{Li} + Q_{Li}$$

the equivalent admittance is given by

$$Y_{Li} = \frac{S^*_{Li}}{|V_{Li}|^2} = \frac{P_{Li} - jQ_{Li}}{|V_{Li}|^2} \qquad ...(9.87)$$

Step 3. The internal voltages are calculated from the terminal voltages, using

$$|E_i| \angle\delta'_i = |V_i| + jx'_{di} I_i$$

$$= |V_i| + jx'_{di} \frac{S^*_{Gi}}{|V_i|}$$

$$= |V_i| + j\,x'_{di} \frac{(P_{Gi} - jQ_{Gi})}{|V_i|} \qquad ...(9.88)$$

In (9.88), δ'_i is the angle of E_i with respect to V_i. If the angle of V_i is β_i, then the angle of E_i, with respect to common reference is given by $\delta_i = \delta'_i + \beta_i$. P_{Gi} and Q_{Gi} are obtained from load flow solution.

Step 4. The bus admittance matrix Y_{bus} formed to run the load flow is modified to include the following.

(i) The equivalent shunt load admittance given by (9.87), connected between the respective load bus and the reference node.

(ii) Additional nodes are introduced to represent the generator internal nodes. Appropriate values of admittances corresponding to x'_d, connected between the internal nodes and terminal nodes are used to update the Y_{bus}.

(iii) Y_{bus} corresponding to the faulted network is formed. Generally, transient stability analysis is performed, considering three phase faults, since they are the most severe. The Y_{bus} during the fault is obtained by setting the elements of the row and column corresponding to the faulted bus to zero.

(iv) Y_{bus} corresponding to the post-fault network is obtained, taking into account line outages if any. If the structure of the network does not change, the Y_{bus} of the post-fault network is same as the prefault network.

Step 5. The admittance form of the network equations is

$$I = Y_{\text{bus}}\ V \qquad \text{...(9.89)}$$

Since, loads are all converted into passive admittances, current injections are present only at the n generator internal nodes. The injections at all other nodes are zero. Therefore, the current vector I can be partitioned as

$$I = \left[\frac{I_n}{0}\right] \qquad \text{...(9.90)}$$

where, I_n is the vector of current injections corresponding to the n generator internal nodes. Y_{bus} and V are also partitioned appropriately, so that

$$\left[\frac{I_n}{0}\right] = \left[\begin{array}{c|c} Y_1 & Y_2 \\ \hline Y_3 & Y_4 \end{array}\right]\left[\frac{E_n}{V_t}\right] \qquad \text{...(9.91)}$$

where, E_n is the vector of internal emfs of the generators and V_t is the vector of external bus voltages. From (9.91), we can write

$$I_n = Y_1\, E_n + Y_2\, V_{\text{t}} \qquad \text{...(9.92a)}$$

$$0 = Y_3\, E_n + Y_4\, V_t \qquad \text{...(9.92b)}$$

From (9.92b), we get

$$V_t = -Y^{-1}{}_4 Y_3 E_n \qquad \text{...(9.93)}$$

Substituting (9.93) in (9.92a) we get

$$I_n = (Y_1 - Y_2\, Y^{-1}{}_4 Y_3)E_n = \hat{Y}\, E_n \qquad \text{...(9.94)}$$

where, $\hat{Y} = Y_1 - Y_2\, Y^{-1}{}_4 Y_3$ is called the reduced admittance matrix and has dimension $n \times n$. $\hat{Y}$ gives the relationship between the injected currents and the internal generator voltages. It is to be noted that (9.94) has eliminated all nodes except the n internal nodes.

Step 6. The electric power output of the generators are given by

$$P_{Gi} = \mathcal{R}[E_i\, I^*_{i}.]$$

Substituting for I_i from (9.94), we get

$$P_{Gi} = |E_i|^2\, \hat{G}_{ii} + \sum_{\substack{j=1 \\ \neq i}}^{n} |E_i|\,|E_j|\,(\hat{B}_{ij} \sin(\delta_i - \delta_j) + \hat{G}_{ij} \cos(\delta_i - \delta_j)) \qquad \text{...(9.95)}$$

(This equation is derived in chapter on load flows)

Step 7. The rotor dynamics representing the swing is now given by

$$M_i \frac{d^2\delta_i}{dt^2} = P_{Mi} - P_{Gi} \quad i = 1 \ldots n \qquad \text{...(9.96)}$$

The mechanical power P_{Mi} is equal to the pre-fault electrical power output, obtained from pre-fault load flow solution.

Step 8. The n second order differential equations given by (9.96) can be decomposed into $2n$ first order differential equations which can be solved by any numerical method as illustrated for the SMIB case in example 9.12, 9.13 and 9.14. During the solution, appropriate Y_{bus} for prefault, fault-on (during fault) and post fault periods are used in (9.95) to obtain the generator output.

Though reduced order models, also called classical models, require less computation and memory, their results are not reliable. Further, the interconnections of the physical network of the system is lost. This led to development of structure preserving models.

9.11.2 Structure Preserving Model

The physical structure of the network is preserved and no reduction to internal nodes is done. Here, the dynamics of the generator and the controllers are represented by a set of differential equations given by

$$\dot{x} = f(x, y) \quad \text{...(9.97)}$$

where, x is the vector of state variables and y is the vector of bus voltages (including generator terminal voltages). The power balance equations of the network constitute a set of algebraic equations given by

$$0 = g(x, y, p) \quad \text{...(9.98)}$$

where, p is the set of parameters like load powers, etc. which affect the power balance. (9.98) is typically the power-flow (load-flow) equations of the system. Equations (9.97) and (9.98) together are called as Differential-Algebraic Equations (DAE). Changes in the network occurring due to faults or fault clearing, cause the structure of (9.98) to change. Initial values of y_0 are obtained from the steady state load flow solution. The initial values of x_0 are computed from y_0 and the power outputs of the generators.

Both (9.97) and (9.98) are non-linear and have to be solved using numerical methods. The solution method must have the following desirable features.

- Be sufficiently accurate. The level of accuracy desired, will depend on the reason for the stability analysis being carried out.
- Be reliable and free from convergence problems.
- Use computer memory effectively.
- Be easy to program, update and flexible to incorporate any changes in data, models, etc.

The total computing time depends on the computation at each time step (which is dependent on the choice of the numerical method) and the time step used. In general, smaller step sizes improve accuracy. The solution of DAE is classified into

(i) Partitioned or alternating method.
(ii) Simultaneous or combined method.

(i) Partitioned or alternating method

In the partitioned method, (9.97) is solved for x by any numerical integration method. (9.98) is solved separately for y, using any numerical method for solution of non-linear algebraic equations.

In solving (9.97) generally a constant time step $h = \Delta t$ s, is assumed so that

$$t_k = t_0 + hk, \quad k = 0, 1, 2...$$

The numerical method generates value of x_{k+1} at end of kth time interval, given x_k at the beginning of the time step. The numerical method is of two types:

(i) **Explicit:** where the algorithm gives an explicit expression to compute x_{k+1}, given x_k. An example is Forward Euler's method where if, $\dot{x} = f(x, t)$ then the value of x in the $k + 1^{th}$ interval is given by

$$x_{k+1} = x_k + f(x_k, t_k) \quad ...(9.99)$$

The computation of x_{k+1} from x_k is straightforward.

(ii) **Implicit:** where the algorithm gives an algebraic equation in x_{k+1}, which has to be solved using a suitable numerical technique. An example is the Backward Euler's method, where

$$x_{k+1} = x_k + h\, f(x_{k+1}, t_{k+1}) \quad ...(9.100)$$

Solution of (9.100) requires an iterative technique like Newton's method.

In the partitioned method either Explicit or Implicit numerical integration methods can be used. The solution procedure in a given time step, is as follows:

(i) Solve (9.97) using x_k, y_k.

(ii) Solve (9.98) using y_k and value of x_k obtained in previous step.

(iii) Repeat steps (i) and (ii) till desired accuracy is reached, before proceeding to next time step or next iteration in the same time step. The values obtained in step (i) are not accurate and are used in step (ii). This causes an error called interface error. Step (iii) is needed to reduce the interface error.

Since, (9.97) and (9.98) are solved one after the other, the method is also called alternating method.

(ii) Simultaneous method

This method is devised to eliminate the interface error present in the partitioned approach. Here the differential equation set (9.97) is converted into algebraic equations using any Implicit integration method. For example, using the Trapezoidal method of integration, we can write

$$x_{k+1} = x_k + \frac{h}{2}\,[f(x_k, y_k, t_k) + f(x_{k+1}, y_{k+1}, t_{k+1})] \quad ...(9.101)$$

These algebraic equations are lumped together with (9.98) and solved simultaneously for x_{k+1} and y_{k+1}, using any numerical method. Newton's method or VDHN (Very Dishonest Newton) method are commonly used. Since, x_{k+1} and y_{k+1} are solved simultaneously, interface errors are avoided.

Transient stability analysis programs can be written using standard software packages like EMTP, MATLAB, etc. or using higher level languages like Fortran, C and C++.

9.12 FACTORS AFFECTING TRANSIENT STABILITY

The relative swing of a machine and the critical clearing time are a measure of the stability of a generating unit. From the swing equation, it is obvious that the generating units with smaller H, have larger angular swings at any time interval. The maximum power transfer $P_{max} = \frac{E_g V}{x'_d}$, where V is the terminal voltage of the generators. Therefore, an increase in x'_d, would reduce P_{max}. Hence, to transfer a given power P_e, the angle δ would increase since $P_e = P_{max} \sin \delta$, for a machine with larger x'_d. This would reduce the critical clearing time, thus, increasing the probability of losing stability.

Generating units of present day have lower values of H, due to advanced cooling techniques, which have made it possible to increase the rating of the machines without significant increase in the size. Modern control schemes like generator excitation control, Turbine valve control, single-pole operation of circuit breakers and fast-acting circuit breakers with auto reclosure facility have helped in enhancing overall system stability. Factors which can improve transient stability are:

(i) Reduction of transfer reactance by using parallel lines.

(ii) Reducing transmission line reactance by reducing conductor spacing and increasing conductor diameter, by using hollow cores.

(iii) Use of bundled conductors.

(iv) Series compensation of the transmission lines with series capacitors. This also increases the steady state stability limit. However, it can lead to problem of sub-synchronous resonance.

(v) Since, most faults are transient, fast acting circuit breakers with rapid re-closure facility can aid stability.

(vi) The most common type of fault being the single line-to-ground fault, selective single pole opening and reclosing can improve stability.

(vii) Use of braking resistors at generator buses. During a fault, there is a sudden decrease in electric power output of generator. A large resistor, connected at the generator bus, would partially compensate for the load loss and help in decreasing the acceleration of the generator. The braking resistors are switched during a fault through circuit breakers and remain for a few cycles after fault is cleared till system voltage is restored.

(viii) Short circuit current limiters, which can be used to increase transfer impedance during fault, thereby reducing short circuit currents.

(ix) A recent method is fast valving of the turbine wherein the mechanical power is lowered quickly during the fault and restored once fault is cleared.

9.13 SUMMARY

This chapter presents the important basic aspects of stability. The classification of stability into rotor angle stability, voltage stability and frequency stability is discussed. The derivation of the swing equation is obtained, from the mechanics of rotary motion. Two or more machines swinging coherently can be reduced to a single equivalent machine. Under minor deviations, the system is stable if the synchronizing coefficient.

$$P_S = \frac{\partial P}{\partial \delta} > 0.$$

The steady state stability limit (the maximum power that can be transferred under steady state) is reached when the torque angle $\delta = 90°$.

Equal area criterion can be applied, to assess the critical clearing angle, in case of SMIB systems. To evaluate the critical clearing time, the swing equation has to be numerically solved. The numerical technique will be more accurate if smaller step sizes are used. In case of multi-machine systems, reduced models are used to reduce computation involved in stability analysis. Structure preserving models are possible, wherein partitioned or simultaneous approach can be used for solution of the DAE. Stability of power systems can be greatly improved by using fast excitation systems, parallel lines, fast valving of turbines, braking resistors, series capacitors, bundled conductors and fast acting circuit breakers with auto re-closure facility.

9.14 SUPPLEMENTARY PROBLEMS

(1) A two pole, 60 Hz, 13.2 kV turbo alternator has a rating of 100 MW, pf 0.85 lag. The rotor movement of inertia is 10200 kg-m^2, Calculate M and H.

[Hint: $\text{KE} = \frac{1}{2} I\omega_{sm}^2\ 10^{-6}\ \text{MJ} \frac{1}{2}\left[I\left(\frac{2}{P}\right)^2 \omega_s \times 10^{-6}\right]\omega_s\ \text{MJ}$

$\therefore M = I\left(\frac{2}{P}\right)^2 \omega_s\ 10^{-6}$ MJ–s/elec rad].

(2) A power station has four identical generator sets, each rated at 90 MVA, having an inertia constant 6.5 MJ/MVA; while station 2 has four generators rated 150 MVA, 4 MJ/MVA. If the stations are close by and have to be replaced by a single equivalent machine, find the inertia constant of the single machine on a 100 MVA base.

(3) Area 1 supplies 60 MW over a tie-line to area 2. The maximum power transfer capability of area 1 is 110 MW. Find the maximum allowable load that can be switched on, without loss of stability.

(4) An SMIB system is shown in Fig. P9.4.

(i) What is the maximum power that can be transferred?

(ii) If the generator is loaded to 80% of its maximum limit, what are the operating angles possible? Which of them is stable?

(iii) In (ii) above, if $H = 5$ MJ/MVA and $f = 50$ Hz, determine the natural frequency of

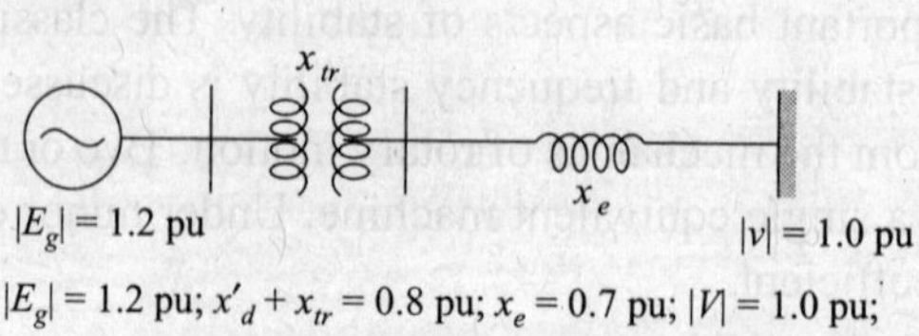

$|E_g| = 1.2$ pu; $x'_d + x_{tr} = 0.8$ pu; $x_e = 0.7$ pu; $|V| = 1.0$ pu;

Fig. P9.4

oscillation.

(5) In Fig. 9.26, if a shunt capacitor is connected in the middle of the line x_e, and the capacitive reactance $x_c = 0.8$ pu, what is the steady state power limit? What is the limit, if a shunt reactor of value $x_L = 0.8$ pu is connected instead of the capacitor.

(6) A 50 Hz tie-line, 400 km long has following parameters.

$$R = 0.1\ \Omega/\text{km} \qquad C = 0.008\mu\ F/\text{km}$$
$$L = 1.39\ \text{mH/km} \quad G = 0$$

If $|V_s| = |V_R| = 220$ kV, what is the steady state stability limit? What is the limit if capacitance is neglected? What if resistance is also neglected?

(7) A generator operates at 50 Hz, delivering a power of 1 pu to an infinite bus. The maximum power $P_{\max I} = 2.0$ pu before the fault, $P_{\max II} = 0.5$ pu during the fault and $P_{\max III} = 1.5$ pu after fault is cleared. Determine the critical clearing angle. If $H = 5.2$, find the critical clearing time.

(8) In problem 9.7 plot the swing curve, if the fault is cleared in 6.25 cycles using
   (i) step by step method
   (ii) Modified Euler's method
   (iii) Runge-Kutta method.

(9) A 60 Hz generator is supplying 50% of $P_{\max}$ to an infinite bus through a reactive network. A fault occurs which increases the reactance of the network between the generator internal bus and the infinite bus by 350%. When fault is cleared, the maximum power that can be delivered is reduced to 75% of the original value. Determine δ_{cr}. If H = 4.8 MJ/MVA and $P_m = 0.9$ pu, find t_{cr}.

(10) Solve problem 9 for the swing curve, if the fault is cleared in 3 cycles.

REVIEW QUESTIONS

1. Define stability.
2. The rotor angle stability depends primarily on the generation. Comment.
3. The voltage stability depends on the load characteristics. Comment.
4. How is frequency stability different from voltage stability?
5. Any instability will ultimately result in rotor angle instability? Justify.
6. Derive expressions for the power of a salient and non-salient pole machine. Draw the power angle characteristics.
7. Under steady state conditions the machine angle cannot exceed 90°. Justify.
8. Derive the condition for small signal stability of a generator connected to an infinite bus (i) when damping is neglected (ii) When damping is considered.
9. A system can be steady state stable, but need not be transiently stable. Discuss the implication of the above statement.
10. What is equal area criterion? How is it useful in determining stability?
11. What is critical clearing time? Can it be defined for a given system?

12. How is the critical clearing time determined?
13. Steady state stability is a function of the network interconnections, while transient stability depends on many other factors. Justify.
14. Derive the expression for the maximum increase in mechanical power possible, without the system losing stability.
15. Derive expression for the critical clearing angle for an SMIB system, which supplies a maximum power Pmax1 pre-fault, Pmax2 during fault, and Pmax3 post fault. Can the critical clearing time be determined from this?
16. Discuss the methods for transient stability improvement.
17. Adding parallel line, in an SMIB system, improves the steady state stability. Justify.
18. What is the effect of resistance and damping on stability?
19. Fast acting circuit breakers have improved stability of modern power systems. Comment on the statement.
20. Interconnections can help in improving stability, but at times, it is necessary to isolate regions of the grid, to keep the interconnected system stable. Discuss the statement.

Chapter 10

Modeling of Synchronous Machine

Synchronous machines are the most widely used machines in the power system. Modeling of synchronous machines is a challenge and has been widely addressed. This chapter deals with the development of detailed mathematical model for the synchronous machine.

10.1 STRUCTURE OF THE MACHINE

The cross section of the synchronous machine is shown in Fig. 10.1. Two poles are considered on the rotor.

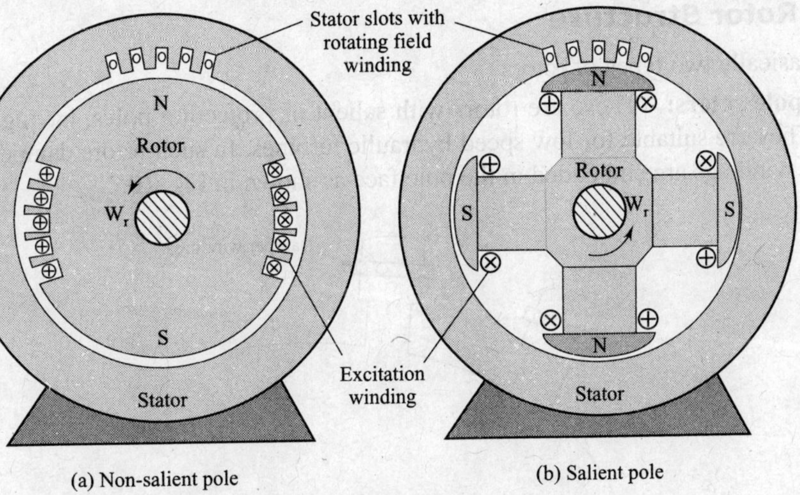

Fig. 10.1 3 – ϕ synchronous machine.

Note the following with respect to the schematic,

- ω_r is the rotor speed
- Field winding is on the rotor
- The d-axis (direct axis) is along the field axis.
- The q-axis is in quadrature and leads the d-axis

10.1.1 Description

The synchronous machine essentially consists of the field and the armatures .The field is normally the rotor. The field winding carries dc current supplied by the excitation system. It produces a magnetic field which induces alternating voltage in the armature windings.

The armature is normally the stator and consists of three-phase windings distributed 120° apart in space. When a unified magnetic field is present, three-phase voltages displaced in time by 120° are induced in them. Since the armature is subjected to varying magnetic fields, the stator is made up of thin laminations to reduce eddy current losses.

10.1.2 Torque Production

With balanced three-phase currents in the armature, a rotating magnetic field at synchronous speed is produced in the air gap. The synchronous speed is given by

$$N = \frac{120f}{P} \text{ rpm} \tag{10.1}$$

where P is the number of field poles.

The field produced by the dc field winding rotates with the rotor. For the production of a steady torque, the magnetic fields produced by the rotor and stator must rotate at the same speed. Therefore, the rotor must be run at synchronous speed.

10.1.3 Rotor Structure

There are basically two types of rotors.

(i) Salient pole rotors: These are rotors with salient or projecting poles, having concentrated windings. They are suitable for low speed hydraulic turbines. In such rotors damper windings or amortisseur windings are embedded in the pole face as shown in Fig. 10.2.

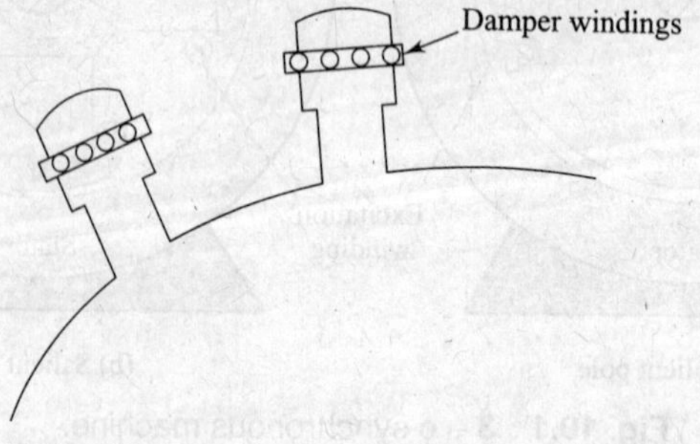

Fig 10.2 Damper windings.

The damper windings damp out oscillations in the speed. The pole faces are usually laminated.

(ii) Non-salient pole rotor: Here the rotors are cylindrical and made up of solid steel forgings. They have normally two or four poles formed by distributed windings placed in slots around the rotor. The windings are held in place by solid steel wedges. Though they do not have separate damper windings, the solid steel rotor provides paths for eddy currents which damp speed oscillations.

If there are more than one pair of field poles, then the electrical angle θ is given by

$$\theta = \frac{P}{2}\,\theta_m \tag{10.2}$$

where θ_m is the mechanical angle.

10.1.4 MMF

The MMF in the air gap due to the stator currents has a sinusoidal spatial distribution. It has a constant amplitude and moves at a constant angular velocity, ω_s elec. rad/s. The speed of rotation of the stator field is

$$\omega_{sm} = \frac{2}{P}\,\omega_s \text{ mech rad /s}$$

$$N_s = \frac{60\omega_{sm}}{2\pi} = \frac{120f}{P} \text{ rpm}$$

This is same as the synchronous speed. Therefore for balanced stator currents, the mmf due to stator currents is stationary with respect to rotor. The stator mmf wave magnitude and phase with respect to rotor mmf wave depends on the synchronous machine output or load on the machine. The electromagnetic torque acts to bring the magnetic fields into alignment. When the rotor mmf leads the almature mmf, the torque acts opposite to the rotor direction and the machine acts as a generator. When the rotor mmf lags the armature mmf, the torque acts in the direction of rotation, and the machine acts as a motor. For generator action, the rotor mmf is moved ahead of the armature field by the prime mover while in a motor the rotor field lags because of the opposing or retarding torque of the mechanical load.

10.2 DIRECT AXIS AND QUADRATURE AXIS

All the magnetic circuits and rotor windings are symmetrical with respect to field polar axis and inter-polar axis. We define too axes as follows:

- The d-axis or direct axis, centred magnetically in the center of the north pole.
- The q-axis which is in quadrature with d-axis. It can be taken to lead or lag the d-axis. IEEE standard -100 (1977) definition recommends selection of q-axis to lead the d-axis.

Figure 10.3 shows the stator and rotor circuits of a synchronous machine.

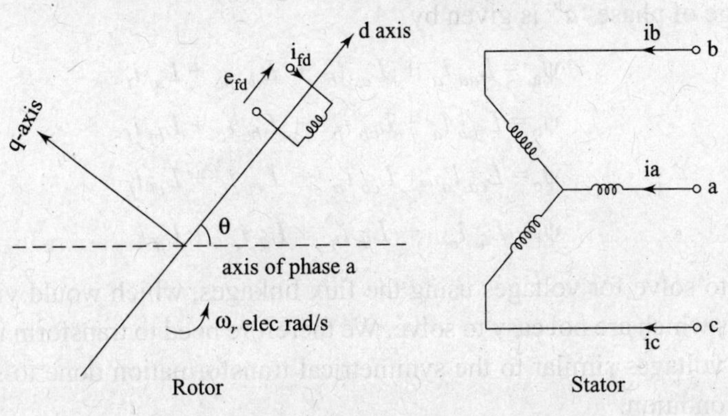

Fig. 10.3 Circuits on rotor and stator.

The figure does not show damper windings. The rotor has a field winding along the direct-axis and the stator comprises three windings.

θ is the angle of the d-axis with the phase 'a' winding axis, in the direction of rotation.

$$\theta = \omega_r t + \theta_o$$

where θ_o is the initial angle.

10.3 BASIC MATHEMATICAL EQUATIONS

The mutual and self-inductances of the stator circuits vary with rotor position because of variation in the permeance of the magnetic flow due to the non-uniform air gap. The effect is more prominent in salient pole machines but not insignificant in non-salient pole machines.

All the self and mutual inductance terms contain a constant term plus a second harmonic term. The self inductances of the stator windings are given by

$$L_{aa} = L_s + L_m \cos 2\theta \,;\, L_s > L_m \tag{10.3a}$$

$$L_{bb} = L_s + L_m \cos 2\,(\theta - 120°)\,; \tag{10.3b}$$

$$L_{cc} = L_s + L_m \cos 2\,(\theta + 120°)\,; \tag{10.3c}$$

We can observe that L_{aa} has a maximum value when $\theta = 0°$ or $180°$ which means when the phase *'a'* axis coincides with the direct axis of the rotor. Similarly, the inductance of coils *b* and *c* is greatest when their respective coil axes coincide with the direct axis. The mutual inductance between two stator windings also varies with the rotor position. They are given by

$$L_{ab} = L_{ba} = -M_s + L_m \cos 2\,(\theta - 30°) \tag{10.4a}$$

$$L_{bc} = L_{cb} = -M_s + L_m \cos 2\theta \tag{10.4b}$$

$$L_{ca} = L_{ac} = -M_s + L_m \cos 2\,(\theta + 150°) \tag{10.4c}$$

In case of mutual inductance between the rotor windings and stator windings it is maximum when the direct axis coincides with the axis of that phase. Here, we assume only one field winding on the direct axis.

$$L_{af} = M_f \cos \theta = L_{fa} \tag{10.5a}$$

$$L_{bf} = M_f \cos (\theta - 120°) = L_{fb} \tag{10.5b}$$

$$L_{cf} = M_f \cos (\theta + 120°) = L_{fc} \tag{10.5c}$$

The self inductance of the field windings is a constant, L_{ff}, and independent of rotor position. Now the flux linkage of phase *'a'* is given by

$$\psi_a = L_{aa}\, i_a + L_{as}\, i_b + L_{ac\,pe} + L_{af}\, i_f \tag{10.6a}$$

$$\psi_b = L_{ba}\, i_a + L_{bb}\, i_b + L_{bc}\, i_c + L_{bf}\, i_f \tag{10.6b}$$

$$\psi_c = L_{ca}\, i_a + L_{cb}\, i_b + L_{cc}\, i_c + L_{cf}\, i_f \tag{10.6c}$$

$$\psi_f = L_{fa}\, i_a + L_{fb}\, i_b + L_{fc}\, i_c + L_{ff}\, i \tag{10.6d}$$

We next have to solve for voltages using the flux linkages, which would yield time-varying diferential equations which are not easy to solve. We therefore need to transform into a meaningful set of currents and voltages similar to the symmetrical transformation done to simplify analysis under unbalanced condition.

10.3.1 dqo Transformation

The stator currents are transformed into a set of three currents in the 'dqo' frame as follows:

$$i_d = K_d\,[\,i_a \cos\theta + i_b \cos(\theta - 120°) + i_c \cos(\theta + 120°)] \tag{10.7a}$$

$$i_q = -K_q\,[\,i_a \sin\theta + i_b \sin(\theta - 120°) + i_c \sin(\theta + 120°)] \tag{10.7b}$$

$$i_o = K_o\,[i_a + i_b + i_c]$$

The choices of K_d, K_q and K_o are arbitrary and are generally chosen to simplify the mathematical computations with the transformed variables. Two commonly used transformation coefficients are

(i) $K_d = K_q = 2/3\ ;\ K_o = 1/3$

(ii) $k_d = k_q = \sqrt{\dfrac{2}{3}}\ ;\ K_o = 1/3$

We choose the first set wherein the peak value of the stator current is equal to the peak value of i_d and i_q. The second set results in a power invariant transformation. We have

$$\begin{bmatrix} i_d \\ i_q \\ i_o \end{bmatrix} = \frac{2}{3}\begin{bmatrix} \cos\theta & \cos(\theta - 120°) & \cos(\theta + 120°) \\ -\sin\theta & -\sin(\theta - 120°) & -\sin(\theta + 120°) \\ \dfrac{1}{2} & \dfrac{1}{2} & \dfrac{1}{2} \end{bmatrix}\begin{bmatrix} i_a \\ i_b \\ i_c \end{bmatrix} \tag{10.8}$$

The field current, i_f, will not be transformed. The inverse transformation is given by

$$\begin{bmatrix} i_a \\ i_b \\ i_c \end{bmatrix} = \begin{bmatrix} \cos\theta & -\sin\theta & 1 \\ \cos(\theta - 120°) & -\sin(\theta - 120°) & 1 \\ \cos(\theta + 120°) & -\sin(\theta + 120°) & 1 \end{bmatrix} = \begin{bmatrix} i_d \\ i_q \\ i_o \end{bmatrix} \tag{10.9}$$

The above transformations are also applicable to stator flux linkages and voltages.

We now apply the transformation to (10.6) to get ψ_d, ψ_q and i_o. We next substitute for i_a, i_b, i_c from (10.9) to get

$$\psi_d = (L_s + M_s + \frac{3}{2} L_m)\, i_d + M_f i_f \tag{10.10a}$$

$$\psi_q = (L_s + M_s - \frac{3}{2} L_m)\, i_q \tag{10.10b}$$

$$\psi_o = (L_s - 2M_s)\, i_o \tag{10.10c}$$

$$\psi_f = \frac{3}{2} M_f i_d + L_{ff} i_f \tag{10.10d}$$

In the above equations, the coefficients are all constants and independent of the rotor position, θ. Further, the equations are almost completely decoupled, we define,

$$L_d = L_s + M_s + \frac{3}{2} L_m = \text{direct-axis synchronous inductance} \quad (10.11)$$

$$L_q = L_s + M_s - \frac{3}{2} L_m = \text{quadrature axis synchronous inductance} \quad (10.12)$$

$$L_o = L_s - 2 M_s = \text{zero sequence inductance} \quad (10.13)$$

With this (10.10) can be written as

$$\psi_d = L_d i_d + M_f i_f \quad (10.14a)$$

$$\psi_q = L_q i_q \quad (10.14b)$$

$$\psi_o = L_o i_o \quad (10.14c)$$

$$\psi_f = \frac{3}{2} M_f i_d + L_{ff} i_f \quad (10.14d)$$

A physical interpretation has been given to the Park's transformed variables. The current i_d may be interpreted as the instantaneous current in a fictitious armature winding, which rotates at same speed as the field winding and remains with its axis coinciding with the direct axis of the field so that it gives the same mmf on this axis as the instantaneous armature currents flowing in the stator windings. 'i_q' can be similarly interpreted, except that it acts in the quadrant axis instead of in the direct axis. i_o is identical to the zero-sequence current except that it is defined for instantaneous phase currents. It does not give space fundamental air-gap flax.

10.3.2 Voltage Equations in Terms of iqo Variables

If we consider a coil as shown in Fig 10.4, we can write for a *motor* its

$$v = ri + \frac{d\psi}{dt} \quad (10.15)$$

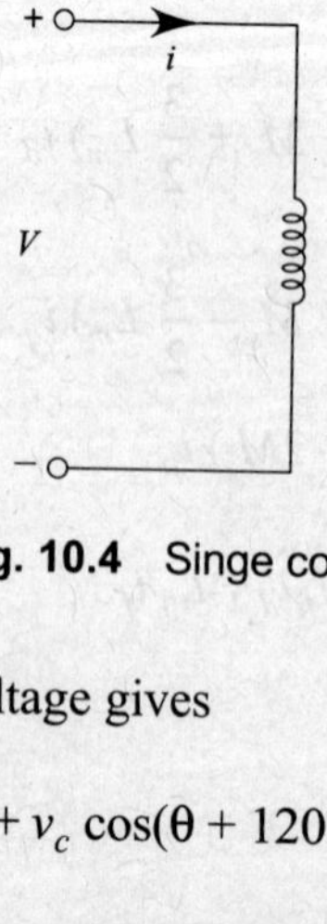

Fig. 10.4 Singe coil.

where ψ is the total flux linkage

The Park's transformation for the voltage gives

$$v_d = \frac{2}{3} [v_a \cos\theta + v_b \cos(\theta - 120°) + v_c \cos(\theta + 120°) \quad (10.16)$$

Substituting (10.15) in (10.16), we can write

$$v_d = \left[\left(ri_a + \frac{d\psi_a}{dt}\right)\cos\theta + \left(ri_b + \frac{d\psi_b}{dt}\right)\cos(\theta - 120°) + \left(ri_c + \frac{d\psi_c}{dt}\right)\cos(\theta + 120°)\right]$$

$$= \frac{2}{3} r\,[i_a \cos\theta + i_b \cos(\theta - 120°) + i_c \cos(\theta + 120°)]$$

$$+ \frac{2}{3}\left[\frac{d\psi_a}{dt}\cos\theta + \frac{d\psi_b}{dt}\cos(\theta - 120°) + \frac{d\psi_c}{dt}\cos(\theta + 120°)\right]$$

$$= r\, i_d + \frac{d\psi_d}{dt} - \omega\,\psi_q \tag{10.17}$$

[* Note (10.17) can be derived by noting $\theta = \omega t + \theta_o$]

Similarly,

$$v_q = ri_q + \frac{d\psi_q}{dt} + \omega\psi_d \tag{10.18}$$

$$v_o = ri_o + \frac{d\psi_o}{dt} \tag{10.19}$$

Substituting for ψ_d, ψ_q and ψ_o from (10.14), we get

$$v_d = ri_d + L_d\frac{di_d}{dt} - \omega\, L_q\, i_q + M_f\frac{di_f}{dt} \tag{10.20a}$$

$$v_q = ri_q + L_q\frac{di_q}{dt} + \omega\,(L_d\, i_d + M_f i_f) \tag{10.20b}$$

$$v_o = ri_o + L_o\frac{di_o}{dt} \tag{10.20c}$$

$$v_f = r_f i_f + L_{ff}\frac{di_f}{dt} + \frac{3}{2}M_f\frac{di_d}{dt} \tag{10.20d}$$

These equations too have constant coefficients and hence are simpler to handle.

The direct axis transient inductance $L_d' < L_d$ if there are no damper windings then the sub-transient direct axis reactance is equal to L_d'. In the presence of damper windings $L_d'' < L_d'$. Unless there is a damper winding on the quadrature axis, $L_q'' < L_q' < L_q$.

If an additional damper winding is present on the quadrature axis, $L_q' = L_q$ and $L_q'' < L_q'$.

The negative sequence inductance $L_2 = \dfrac{L_d' + L_q}{2}$ if there are no other rotor circuits, except the field winding. If damper windings are present in both the axes, then $L_2 = \dfrac{L_d'' + L_q''}{2}$.

10.4 VECTOR DIAGRAM

Consider the following instantaneous current expressions.

$$i_a = I_m \cos(\omega t + \phi) \quad (10.21a)$$

$$i_b = I_m \cos(\omega t + \phi - 120°) \quad (10.21b)$$

$$i_c = I_m \cos(\omega t + \phi + 120°) \quad (10.21c)$$

$$i_f = I_f \text{(constant)} \quad (10.21d)$$

when substituted in (10.8) we get

$$i_d = I_m \cos\phi = I_d \text{ (a constant value)} \quad (10.22a)$$

$$i_q = I_m \sin\phi = I_q \text{ (a constant value)} \quad (10.22b)$$

$$i_o = 0 \quad (10.22c)$$

The fluxes are given by

$$\psi_d = L_d I_d + M_f I_f \quad (10.23a)$$

$$\psi_q = L_q I_q \quad (10.23b)$$

$$\psi_o = 0 \quad (10.23c)$$

$$\psi_f = \frac{3}{2} M_f I_d + L_{ff} I_f \quad (10.23d)$$

It can be observed that all the fluxes are constant and hence their derivatives are zero. Substituting in (10.17) we get

$$\begin{aligned} v_d &= r I_d - \omega \psi_q \\ &= r I_d - \omega L_q I_q = r I_d - x_q I_q \\ &= V_d \text{ (a constant)} \end{aligned} \quad (10.24)$$

$$\begin{aligned} v_q &= r I_q + \omega \psi_d \\ &= r I_q + \omega L_d I_d + \omega M_f I_f \\ &= r I_q + x_d I_d + E_q \\ &= V_q \text{ (a constant)} \end{aligned} \quad (10.25)$$

where $$E_q = \omega M_f I_f \quad (10.26)$$

E_q. is the excitation voltage or the steady state internal voltage. It is directly proportional to the field current. The positive sequence current is given by

$$\boldsymbol{I} = I_d + j I_q \quad (10.27)$$

Similarly $$\mathbf{V} = V_d + j V_q \quad (10.28)$$

and $$\boldsymbol{E} = 0 + j E_q = j \omega M_f I_f \quad (10.29)$$

Substituting from (10.24) and (10.25) into (10.28), we get

$$\boldsymbol{V} = \boldsymbol{E} + r\boldsymbol{I} + jx_d I_d - x_q I_q \tag{10.30}$$

The vector diagram can be drawn as shown in Fig 10.5.

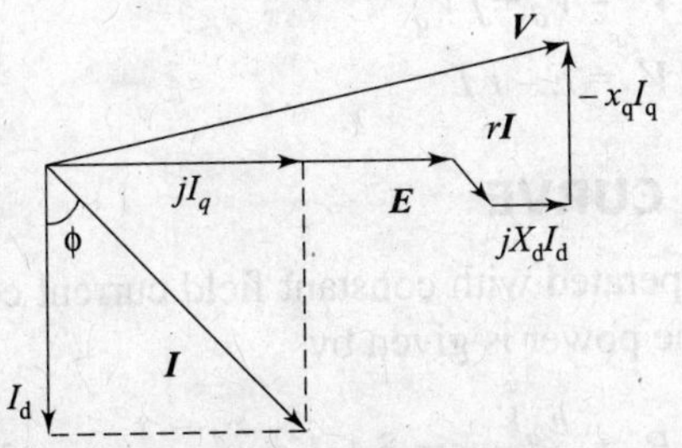

Fig. 10.5 Steady state vector diagram of a synchronous motor.

The above vector diagram corresponds to a motor with lagging current corresponding to positive I_d.

If we write the equations for a generator, we get

$$\mathbf{V} = \mathbf{E} - \mathbf{ZI} \tag{10.31}$$

Here positive I_d is demagnetizing. The complete vector diagram for the generator is shown in Fig 10.6.

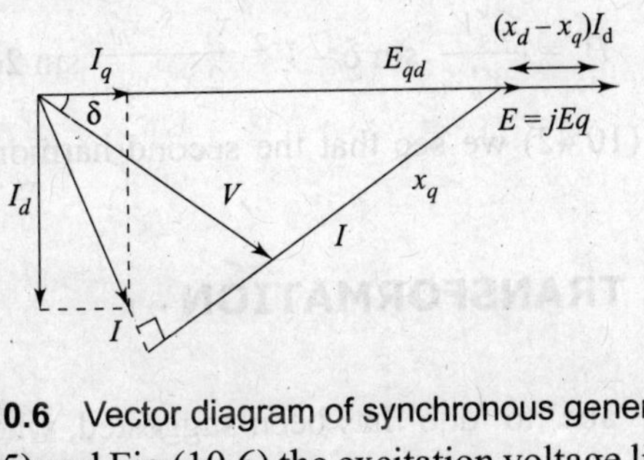

Fig. 10.6 Vector diagram of synchronous generator.

Note that in both Fig (10.5) and Fig (10.6) the excitation voltage lies along the quarantine axis. The complete set of equations for the generator is given below. The algebraic signs of i_d and i_q are reversed.

$$\psi_d = M_f i_f - L_d i_d \tag{10.32a}$$

$$\psi_q = L_q i_q \tag{10.32b}$$

$$\psi_f = L_{ff} i_f - \frac{3}{2} M_f i_d \tag{10.33}$$

$$v_d = -r i_d + \frac{d\psi_d}{dt} - \omega \psi_q \tag{10.34a}$$

$$v_q = -r i_q + \frac{d\psi_d}{dt} + \omega \psi_d \tag{10.34b}$$

$$V_d = -r I_d + x_q I_q \tag{10.35a}$$

$$V_q = -r I_q - x_d I_d + E_q \tag{10.35b}$$

$$E_q = \omega M_f I_f \tag{10.36}$$

$$\boldsymbol{I} = I_d + j I_q \tag{10.37}$$

$$V = V_d + j V_q \tag{10.38}$$

$$V = E - r I \tag{10.39}$$

10.5 POWER ANGLE CURVE

If a synchronous machine is operated with constant field current connected to an infinite bus, E_q and V are independent of δ. The power is given by

$$P = \frac{E_q V}{x_d} \sin \delta + V^2 \frac{x_d - x_q}{2 x_d x_q} \sin 2\delta \tag{10.40}$$

Under transient condition the armature current changes quickly. Such a change could be due to a fault or a power swing. During such a change, a new fictitious armature voltage is defined as

$$E'_q = \frac{\omega M_f}{L_{ff}} \psi_f \tag{10.41}$$

The power-angle relationship is then given by

$$P = \frac{E'_q V}{x'_d} \sin \delta - V^2 \frac{x_q - x'_d}{2 x'_d x_q} \sin 2\delta \tag{10.42}$$

Comparing (10.40) and (10.42) we see that the second harmonic term is reversed. This is because $x_q > x'_d$.

10.6 ALTERNATIVE TRANSFORMATION

Another transformation from 'abc' to 'dqo' has been suggested, where $K_d = K_q = \sqrt{\frac{2}{3}}$ and $K_o = 1/\sqrt{2}$ therefore

$$\begin{bmatrix} i_d \\ i_q \\ i_o \end{bmatrix} = \sqrt{\frac{2}{3}} \begin{bmatrix} \cos\theta & \cos(\theta - 120°) & \cos(\theta + 120°) \\ -\sin\theta & -\sin(\theta - 120°) & -\sin(\theta + 120°) \\ \sqrt{1/2} & \sqrt{1/2} & \sqrt{1/2} \end{bmatrix} \begin{bmatrix} i_a \\ i_b \\ i_c \end{bmatrix} \tag{10.43}$$

This transformation is orthogonal. The inverse matrix is transpose of the transformation matrix. This means the transformation is power invariant.

$$\begin{aligned} P &= e_a i_a + e_b i_b + e_c i_c \\ &= e_d i_d + e_q i_q + e_o i_o \end{aligned} \tag{10.44}$$

Further, with this transformation all mutual inductances are reciprocal. However, it is difficult to give a physical interpretation to i_d and i_q as was done with the previous transformation.

REVIEW QUESTIONS

1. Explain the model for cylindrical and salient pole synchronous machine.
2. Discuss the need for Park's transformation.
3. Briefly explain the Park's transformation.
4. Prove that the equations are decoupled in Park's transformation.
5. Give the physical interpretation of Park's transformation.
6. Compare Park's transformation and symmetrical component transformation.

Chapter 11

Modeling of Excitation Systems

11.1 INTRODUCTION

The field winding of the AC generators needs a DC excitation. The excitation system consists of the device supplying the DC, called the exciter and a voltage regulator which controls the output of the exciter so that the required voltage is generated.

The voltage regulator is called automatic voltage regulator (AVR). The functions of the AVR are:

- To control the system voltage to maintain within prescribed limits. This permits operation of the machine nearer to the steady state limit.
- To maintain system voltage even under fault to ensure operation of protective switchgear.
- To regulate sharing of the reactive load between the machines operating in parallel. When the load changes, suddenly, the excitation should give the same voltage under new load conditions.

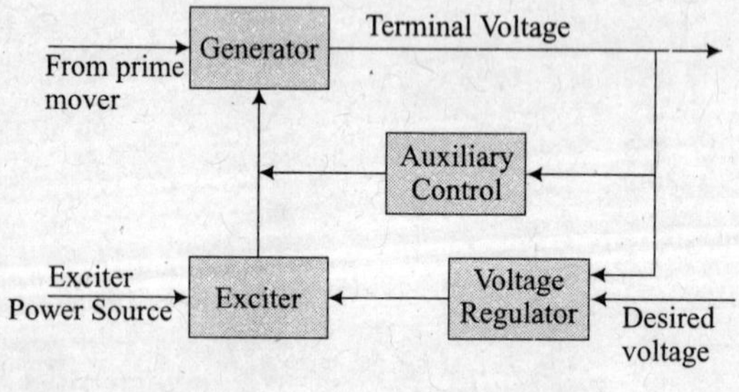

Fig. 11.1 Excitation system.

The exciters can be DC exciters (using DC generators), AC exciters (using AC generators in conjunction with rectifiers) or static exciters (without any rotating elements). The voltage regulator is normally a continuously acting system, which takes corrective action for any deviation in the AC terminal voltage. With the large-scale use of computers, the modeling of the excitation system has gained importance. IEEE committee has suggested standard computer models for different types of excitation systems, which are adequate to represent all modern systems.

The general arrangement of excitation system components are shown in Fig. 11.1.

The voltage regulator controls the exciter output, such that the terminal voltage of the AC generator equals the desired voltage, often called the reference voltage. The auxiliary controls are desirable for feedback of speed, frequency, acceleration and other signals to improve system conditions like stability, damping overshoot, etc. Some standard definitions used in excitation systems are given below.

11.2 DEFINITIONS

(a) **Exciter:** The source of field current for the excitation of the generator.

(b) **DC generator-Commutator exciter:** An exciter whose energy is derived from a DC generator. The exciter includes a DC generator and its commutator. The exciter may be driven by a motor, prime mover or the shaft of the synchronous machine.

(c) **Alternator-Rectifier exciter:** An exciter whose energy is derived from an alternator and converted to DC by rectifiers. The exciter includes an alternator and power rectifiers, which may be either controlled or uncontrolled, including the gate circuitry. The alternator may be driven by a motor, prime mover or by the shaft of the synchronous machine. The rectifiers may be stationary or rotating with the alternator shaft.

(d) **Compound-Rectifier exciter:** An exciter whose energy is derived from the currents and potentials of the AC terminals of the synchronous machine and converted to DC by rectifiers. The exciter includes the power transformers, power reactors, power rectifiers including the gate circuitry.

(e) **Potential source rectifier exciter:** An exciter whose energy is derived from a stationary AC potential source and converted to DC by rectifiers.

(f) **Pilot exciter:** The source of field current for the excitation of another exciter.

(g) **Primary detecting element:** The portion of the feedback elements which first utilizes or transforms energy from the controlled medium, to produce a signal which is a function of the value of the directly controlled variable.

(h) **Compensator:** A feedback element of the regulator, which acts to compensate for the effect of a variable, by modifying the function of the primary detecting element.

(i) **Limiter:** A feedback element of the excitation system, which acts to limit a variable, by modifying or replacing the function of the primary detector element, when predetermined conditions are reached.

(j) **Continuously acting regulator:** A regulator that initiates corrective action for a sustained infinitesimal change in the controlled variable.

(k) **Non-continuously acting regulator:** A regulator that initiates corrective action for a finite change in the controlled variable.

(l) **Air gapline:** The extended linear part of the no-load saturation curve.

(m) **Exciter ceiling voltage:** The maximum voltage that may be attained by an exciter under specified conditions.

(n) **Excitation system voltage-time response:** The excitation system output voltage expressed as a function of time, under specified conditions.

(o) **Excitation system voltage response time:** The time in seconds for the excitation voltage to reach 95% of the ceiling voltage.

11.3 EXCITERS

The exciter is the device which feeds DC supply to the main generator field. Some standard exciters are discussed below.

11.3.1 DC Excitation Systems

These utilize a DC generator for the DC supply to the main generator field. Different configurations are possible, which are discussed briefly below.

(i) **Main exciter only:** The arrangement of an excitation system with a main exciter is shown in Fig. 11.2. Here the exciter is a DC shunt generator.

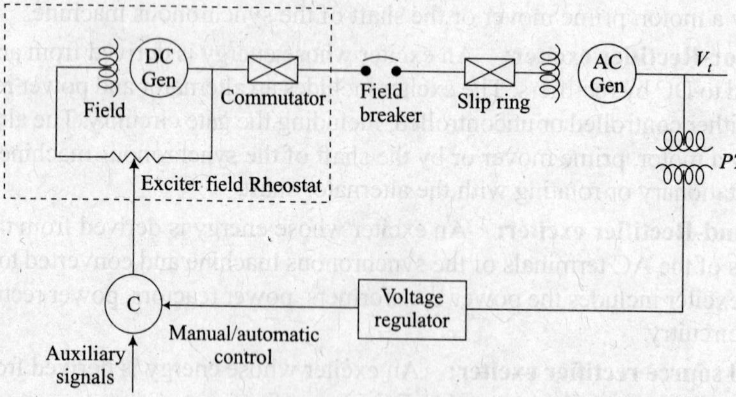

Fig. 11.2 Main DC exciter.

The output of the DC generator is connected to the alternator field through slip rings.

The output of the exciter is regulated by the control of the field rheostat. The voltage regulator detects changes in voltage level and includes a mechanical device to change the exciter field rheostat. This system is commonly used with small AC generators. The response of this system is slow.

(ii) **Main and pilot exciters:** Here the main exciter is a separately excited DC generator. The field of the main exciter is fed by another DC generator called the pilot exciter as shown in Fig.11.3.

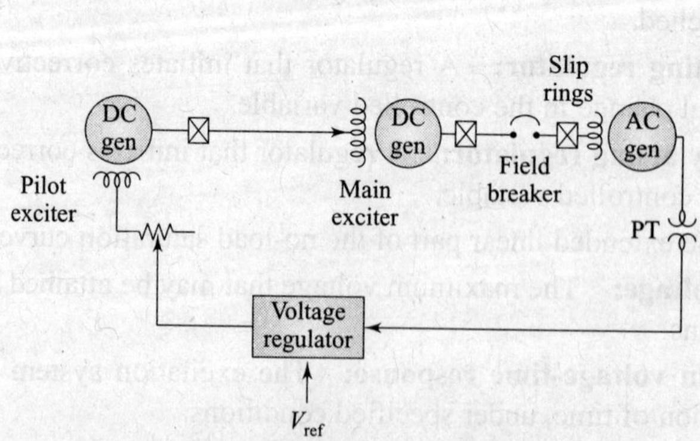

Fig. 11.3 Excitation system with pilot exciter.

Since, the main exciter is separately excited, this system has a faster response, as the exciter field control is independent of the exciter output voltage.

The pilot exciter may be replaced by a solid state rectifier. The AC supply for this rectifier may be either the main power system or a permanent magnet generator directly connected to the main machine.

11.3.2 AC Exciter

With the advent of solid state technology and availability of reliable high-current rectifiers, an AC generator can be used for excitation. The output is rectified to provide DC current. One such arrangement is shown in Fig. 11.4.

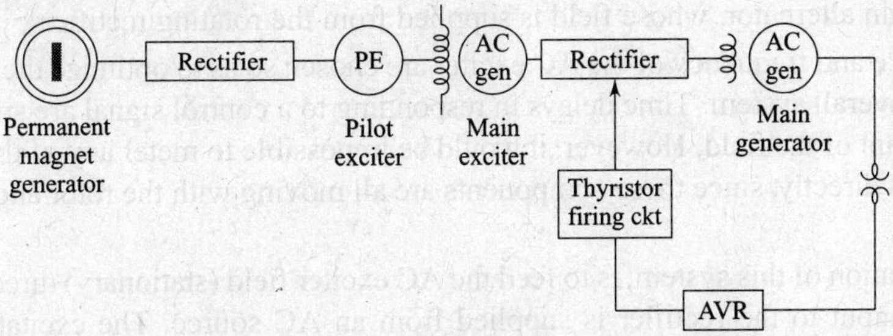

Fig. 11.4 AC exciter.

Here, the AC exciter machine is mounted on the main shaft and its field is excited by the pilot exciter. The field of the pilot exciter, in turn, is derived from a permanent magnet generator. The output of the AC exciter is fed to a floor standing rectifier. The rectified DC output is fed directly to the field of the main generator through conventional slip rings. A fully controlled three-phase bridge rectifier is normally used because of lower voltage stresses on the semiconductors and greater utilization of the transformer capacity.

11.3.3 Brushless Excitation Systems

With increase in ratings of generators, large DC field excitation is required. Cooling and maintenance problems are associated with slip rings, commutators and brushes. These problems are minimized by reducing the use of sliding contacts, slip rings and brushes, giving rise to a class of excitation systems called brushless excitation system. One such arrangement is shown in Fig.11.5.

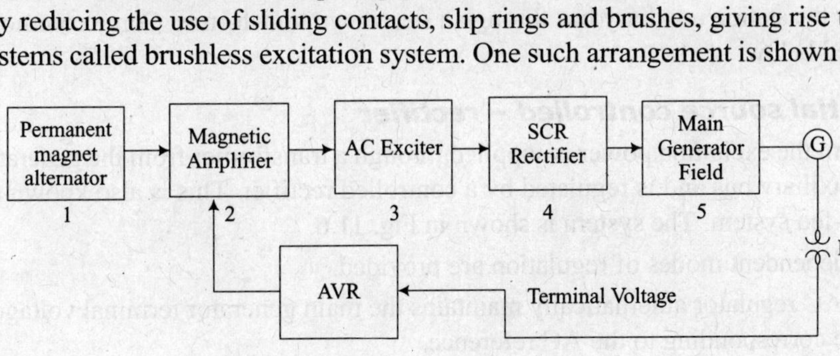

Fig. 11.5 Brushless excitation system.

The components are:

1. The permanent magnet alternator, which has a rotating field and stationary armature. This is driven from the main shaft itself.
2. The magnetic amplifier, whose output is controlled, depending on the terminal voltage and is fed to the field of the AC exciter.
3. The main AC exciter, which has a rotating armature and a stationary field.
4. The rotating armature of the exciter feeds power along the shaft to the rectifier, which is also mounted on the same shaft as the generator field. DC excitation to the main generator field is fed from these rotating rectifiers. The coupling between stationary and rotating elements is electromagnetic, eliminating commutators, slip rings and brushes.
5. The main alternator, whose field is supplied from the rotating rectifiers.

The voltage and frequency of the AC exciter are chosen so as to optimize the performance and design of the overall system. Time delays in responding to a control signal are small, compared to the time constant of the field. However, it would be impossible to meter any of the main generator field quantities directly, since these components are all moving with the rotor and no slip rings are used.

A modification of this system, is to feed the AC exciter field (stationary) directly from an SCR rectifier. The input to the rectifier is supplied from an AC source. The excitation level can be changed by simply changing the firing angle of the SCR's. This is very fast acting. However, this would require a larger AC-exciter than would otherwise be needed, since it must have a rating capable of continuous operation at ceiling voltage. In conventional exciters the ceiling voltage is reached after a delay and sustained operation at that level is unlikely. Different configurations are available by different vendors.

11.3.4 Static Exciter

These exciters are entirely static. Here the main generator is self-excited. The generator output is rectified by means of SCR-rectifier units. Static voltage regulators generate a control signal, by which the SCR gating may be controlled. This type of control is very fast, since there is negligible time delay in shifting the firing angle of the SCR's. The rectified output is fed directly to the field of the main generator through slip rings. The supply of power to the rectifiers is from the main generator or the station auxiliary bus through a step down transformer. Some common forms used are discussed below.

(i) Potential source controlled – rectifier

In this system the excitation power is supplied through a transformer from the generator terminals, or station auxiliary bus and is regulated by a controlled rectifier. This is also known as bus-fed or transformer-fed system. The system is shown in Fig. 11.6.

Two independent modes of regulation are provided.

(a) The AC regulator automatically maintains the main generator terminal voltage at a desired value corresponding to the AC reference.

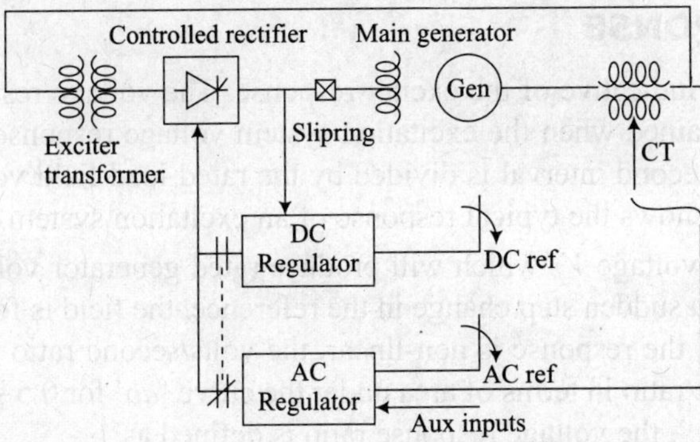

Fig. 11.6 Potential source controlled rectifier.

(b) The DC regulator maintains a constant generator field voltage determined by the DC reference. It is also called the manual control mode and is used when the AC regulator is not operating. This excitation system has a small time constant. The ceiling voltage is dependent on the input AC voltage. The major disadvantage of the scheme is that during faults, because of drop in terminal voltage, the AC input to exciter is also reduced.

(ii) Compound source rectifier systems

Here the power to the excitation system is formed by utilizing the current as well as voltage of the main generator. This is shown in Fig.11.7.

The regulator controls the exciter output through controlled saturation of excitation transformer. On no-load, since current is zero, the entire excitation power is supplied by the Power Potential Transformer (PPT). During faults when voltage dips, the current input provides sufficient power to the exciter.

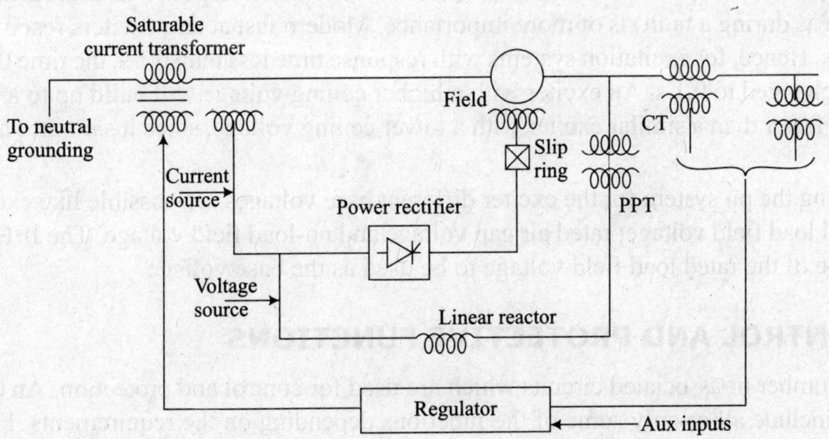

Fig. 11.7 Compound source rectifier systems.

11.4 EXCITER RESPONSE

The voltage response ratio is indicative of the exciter response. The voltage response ratio is the numerical value which is obtained when the excitation system voltage response in volts/second, measured over the first half-second interval is divided by the rated-load field voltage of the synchronous machine. Fig. 11.8 shows the typical response of an excitation system.

'*oa*' is the value of field voltage V_F which will produce rated generator voltage under name plate loading. Responding to a sudden step change in the reference, the field is forced along curve '*ab*' to ceiling voltage. Since, the response is non-linear, the volts/second ratio is not a constant. Therefore, we define response ratio in terms of area under the curve '*ab*' for 0.5 s. Approximating the curve by a straight line '*ac*', the voltage response ratio is defined as

$$\frac{Cd}{(Oa)\,(0.5)} \text{ pu voltage/second}$$

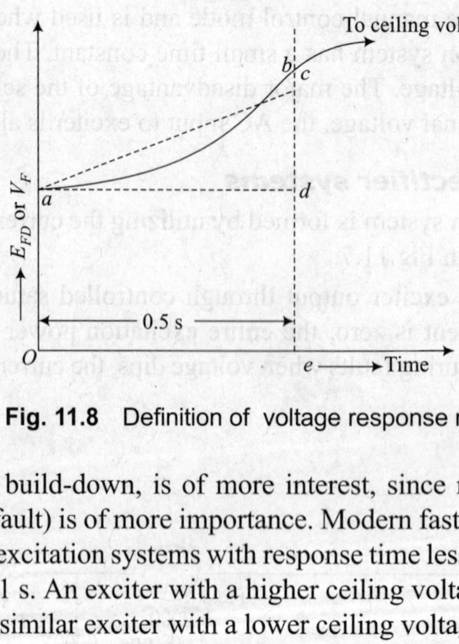

Fig. 11.8 Definition of voltage response ratio.

Build-up, rather than build-down, is of more interest, since response to a drop in terminal voltage (such as during a fault) is of more importance. Modern fast acting exciters reach ceiling in less than 0.5 s. Hence, for excitation systems with response time less than 0.1 s, the time 0.5 s in the definition is changed to 0.1 s. An exciter with a higher ceiling voltage will build up to a particular voltage level faster than a similar exciter with a lower ceiling voltage, since it saturates at a higher value.

In adopting the pu system for the exciter different base voltages are possible like exciter rated voltage; rated load field voltage; rated air gap voltage and no-load field voltage. The IEEE recommends the use of the rated load field voltage to be used as the base voltage.

11.5 CONTROL AND PROTECTIVE FUNCTIONS

There are a number of associated circuits which are used for control and protection. An excitation system may include all or only some of the functions depending on the requirements. Frequently used controls and protective functions and the associated components are briefly discussed.

1.5.1 Voltage Regulators

he voltage regulator is the device that senses changes in the terminal voltage or current and causes orrective action to take place. The voltage regulator in other words, drives the exciter to change its utput. The performance of the excitation system thus depends on the performance of the regulator. he regulator has to be a continuously acting type so that corrective action is taken proportional to eviations in AC terminal voltage, irrespective of how small the deviation is. Dead bands cannot be lerated. A brief explanation of regulators is given.

Earlier vibrating type or tirril regulators had contacts which open and close continuously sev-ral times a second. Each time a contact closes, it short circuits the regulating field rheostat and hen it opens the field rheostat is included in the field circuit. The average value of field voltage nd current depends on the time of engagement of the contacts. A change in this affects the average alue of the field rheostat.

In rheostatic voltage regulators, a regulating resistance is continuously varied. In direct acting pe, the voltage sensitive element of the regulator (like an electromagnet) controls the rheostat rough direct mechanical connection. In indirect acting type, the voltage sensitive element oper-tes contacts which, in turn, control a motor to drive the rheostat.

Modern excitation systems use static regulators which employ electronic circuits for compari-on and regulation. These are fast acting compared to older electromechanical regulators.

1.5.2 Excitation System Stabilizers (ESS)

xcitation systems, especially using DC and AC exciters, comprise elements with significant time elays giving a poor dynamic response. When the generator is operating on no-load (where varia-ons in terminal voltage are not significant) a high gain in regulator can lead to an unstable xcitation control. The ESS is used to improve the dynamic performance through a feedback com-ensation which is normally a derivative feedback as shown in Fig. 11.9a.

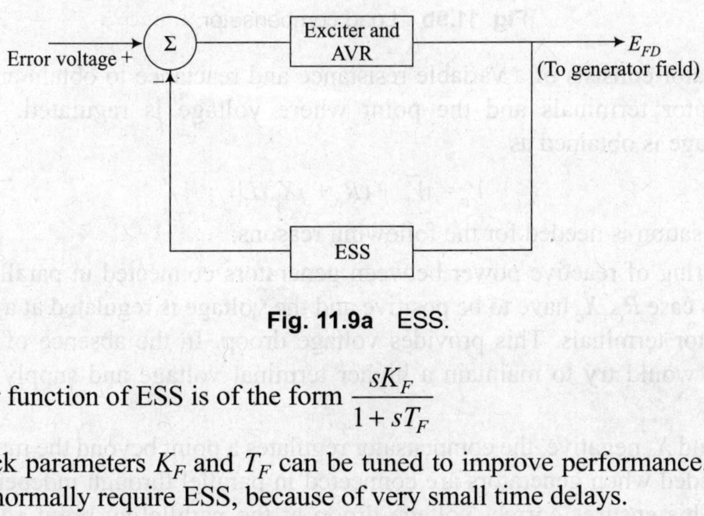

Fig. 11.9a ESS.

The transfer function of ESS is of the form $\dfrac{sK_F}{1+sT_F}$

The feedback parameters K_F and T_F can be tuned to improve performance. Static excitation ystems do not normally require ESS, because of very small time delays.

11.5.3 Power System Stabilizer

It has been established that fast acting exciters with high gain AVR's can contribute to oscillatory instability. The frequency range is between 0.2 – 2.0 Hz. This can be a threat to stable and secure operation of the power system. Power system stabilizers are supplementary controls used to provide damping of these low frequency oscillations under transient state. They have a designed wash-out circuit which prevents operation under steady state. It has become a standard control with modern fast acting static exciters. Commonly used signals for the PSS are the acceleration power, frequency and speed deviation. A detailed exposition is beyond the scope of the book.

11.5.4 Load Compensator

The AVR is normally used to control the generator stator terminal voltage. Sometimes, it may be necessary to regulate voltage at a point within or external to the generator terminals (like the HT bus of the transformer), which requires additional circuits as shown in Fig. 11.9b.

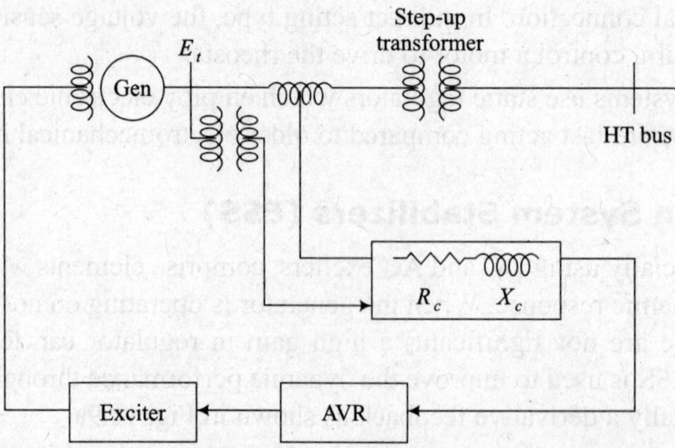

Fig. 11.9b Load compensator.

The compensator consists of a variable resistance and reactance to obtain an impedance between the generator terminals and the point where voltage is regulated. Using this, the compensated voltage is obtained as

$$V_c = |\overline{V}_t + (R_c + jX_c)I_t|$$

Load compensation is needed for the following reasons:

(a) Proper sharing of reactive power between generators connected in parallel to a common bus. In this case R_c, X_c have to be positive and the voltage is regulated at a point internal to the generator terminals. This provides voltage droop. In the absence of this, one of the generators would try to maintain a higher terminal voltage and supply all the reactive power.

(b) With R_c, and X_c negative, the compensator regulates a point beyond the machine terminals. This is needed when generators are connected in parallel through independent unit transformers. This ensures correct voltage droop at the paralleling point so that generators operate satisfactorily.

11.5.5 Limiters

Under excitation limiter is provided to prevent reduction of generator excitation to a level where small signal stability is a threat. Over excitation limiters are provided to protect the generator from over heating due to prolonged field over current. When the excitation exceeds a preset limit, the control changes to DC regulator from AC regulator, such that the exciter is set to a predetermined value.

11.5.6 Field Shorting Circuits

Rectifiers cannot conduct current in reverse directions. During system short circuits, the induced current in the generator field winding may be negative. If path is not provided for this negative current, high voltages are developed across the field circuit. The field shorting circuits, bypass the exciter to allow negative field current to flow. One way is to connect a resistor across the field. The resistor is a non-linear resistor. Under normal exciter voltage, it has high resistance and negligible current. When the voltage crosses a threshold, its resistance falls and current through it increases rapidly.

11.6 COMPUTER MODELS OF EXCITATION SYSTEMS

IEEE has proposed standard models for representation of excitation systems for system studies. In the development of the block diagrams the pu system is used. Options for choosing the base voltage are:

(a) Exciter rated voltage – This gives very low pu values under normal operating conditions.

(b) Rated load field voltage – The field voltage necessary to produce rated voltage on rated load.

(c) Rated air gap voltage – The voltage necessary to produce rated voltage on the air gap line of the main machine.

In the models discussed, the rated air gap voltage is taken as base for the generator field. One per unit exciter output voltage is that voltage required to produce rated generator voltage on the generator air gap. This means that at no-load and neglecting saturation, $E_{FD} = 1.0$ pu gives exactly $V_t = 1.0$ pu.

11.6.1 Inclusion of Limits

In the computer representation various transfer function block have limiters. The limiters are basically.

(a) Windup limiter (soft limiter)

The limiter action is as follows:

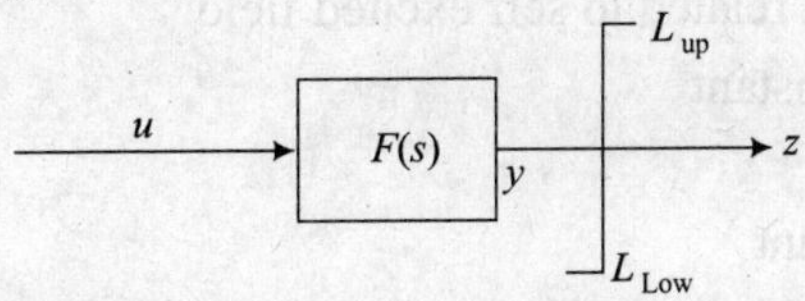

Fig. 11.10 Soft limiter.

If $L_{up} \leq y \leq L_{Low}$, then $z = y$

If $y \leq L_{Low}$, then $z = L_{Low}$

If $y \geq L_{up}$, then $z = L_{up}$

Here there is no limit on y. Hence, the output variable z cannot come off a limit unless y come within the limits.

(b) Non-winding limiter (hard limiter)

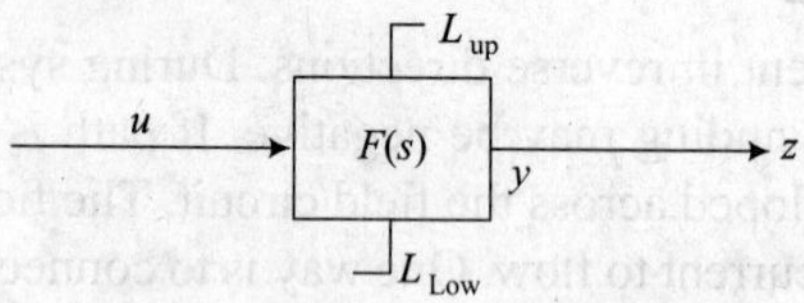

Fig. 11.11 Hard limiter.

The limiter action is as follows:

$$\text{If } L_{up} \leq y \leq L_{Low}, \text{ then } \frac{dy}{dt} = f$$

$$y \leq L_{Low}, \frac{dy}{dt} = 0; y = L_{Low}$$

$$y \geq L_{up}, \frac{dy}{dt} = 0; y = L_{up}$$

Here the output variable y is limited and comes off limit depending on the input.

The computer models developed are meant to describe the relationship between various parameters. State equations are derived and integrated with the system equations for dynamic simulation.

11.6.2 Symbols used in Models

E_{FD} - Exciter output voltage

V_t - Generator terminal voltage

I_F - Generator field current

I_T - Generator terminal current

K_A - Regulator gain

T_A - Regulator amplifier time constant

K_E - Exciter constant related to self excited field

T_E - Exciter time constant

K_F - ESS gain

K_F - ESS time constant

K_I - Current circuit gain (type 3)

K_P - Potential circuit gain (type 3)
S_E - Exciter saturation function
T_R - Regulator input filter time constant
T_{RH} - Rheostat time constant (type 4)
V_R - Regulator output voltage
V_R min - Limiting values of V_R
VR max,
V_{REF} - Regulator reference voltage setting
V_S - Auxiliary signals

11.6.3 Type 1 System–Continuously Acting Exciter and Regulator

The block diagram of TYPE 1 system is shown in Fig. 11.12. It is representative of a majority of modern systems and includes most continuously acting systems with rotating amplifiers like Allis Chalmers-Regulex regulator, *GE*-amplidyne regulator, Westing house-Rototrol, Silverstat, etc. Fig. 11.12 shows the significant transfer functions to be included in computer models. In the representation, provision is made for first order smoothing or filtering of the terminal voltage V_t with a filter time constant T_R. Usually this is very small. The main regulator has a time constant T_A and gain K_A and its output is limited by $V_{R\max}$ and $V_{R\min}$. If ESS is not present, K_F can be taken as zero, since there is no feedback.

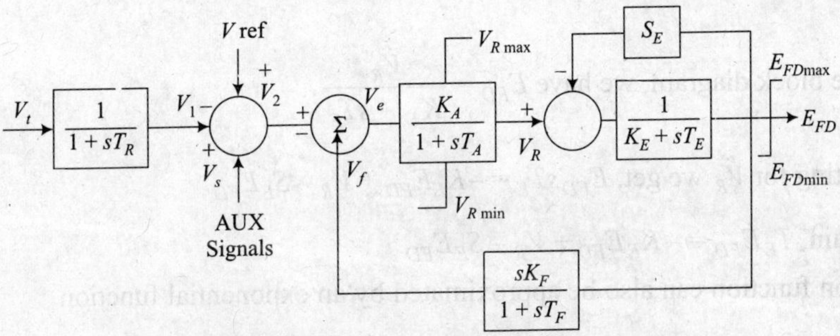

Fig. 11.12 Type 1 Excitation system model.

The first summing point compares the regulator reference with the output of the input filter, to obtain the error voltage input to the regulator amplifier. The input to the amplifier is the error voltage.

$$V_e = V_{ref} - V_1 + V_s - V_F \quad ...(11.1)$$

The exciter is represented as a first order linear system with time constant T_E and gain K_E. S_E is the saturation function to include the effect of saturation in exciter. It is defined as shown in Fig. 11.13 by the relation.

$$S_E = \frac{(A-B)}{B} = f(E_{FD}) \quad ...(11.2)$$

Thus, S_E is a non-linear function of E_{FD}. The amplifier voltage V_R is modified by an amount $S_E E_{FD}$ to give the effective value $\hat{V}_R$,

$$\hat{V}_R = V_R - S_E E_{FD} \quad ...(11.3)$$

This is operated upon linearly by the exciter transfer function. In the linear region $S_E = 0$. The exciter transfer function is given by

$$G_E(s) = \frac{1}{K_E + sT_E} \quad ...(11.4)$$

When a self-excited shunt field is used, K_E represents setting of the shunt field rheostat and provides a positive feedback of exciter output.

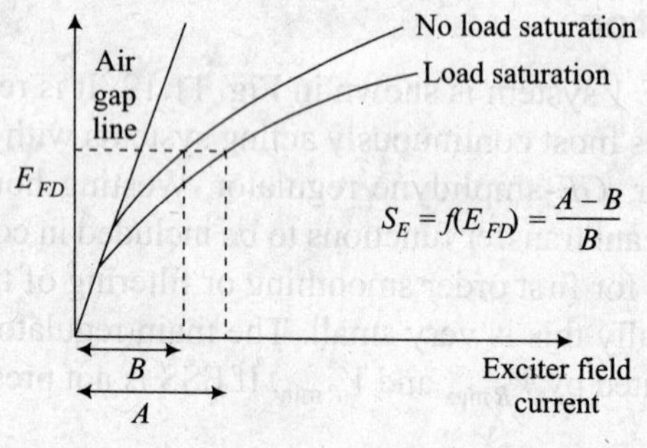

Fig. 11.13 Saturation function.

From the block diagram, we have $E_{FD} = \dfrac{\hat{V}_R}{K_E + sT_E}$. ...(11.5)

Substituting for $\hat{V}_R$ we get, $E_{FD} sT_E = -K_E E_{FD} + V_R - S_E E_{FD}$. ...(11.6)

In time domain, $T_E \dot{E}_{FD} = -K_E E_{FD} + V_R - S_E E_{FD}$ ...(11.7)

The saturation function can also be approximated by an exponential function

$$S_E = f(E_{FD}) = A_{EX} e^{(B_{EX} E_{FD})} \quad ...(11.8)$$

The coefficients A_{EX} and B_{EX} are computed from saturation data where S_E and E_{FD} are specified at two points, usually the exciter ceiling voltage and 75% of ceiling. Major loop damping is provided by the ESS whose transfer function is given by

$$G_{ESS}(S) = \frac{sK_F}{1 + sT_F} \quad ...(11.9)$$

This introduces a zero on the negative real axis. It introduces both a derivative feedback and a first-order lag.

To set initial conditions, V_{REF} is so chosen such that V_T is at the correct value and error is zero. In shunt machines, K_E is chosen such that it is equal in magnitude to the saturation function at the initial value of E_{FD}. At this value, the shunt field exactly compensates for exciter saturation and no regulator output is required to establish initial value of E_{FD}. In separately-excited systems, regulator output is required to supply exciter field and establish initial E_{FD}.

The interrelation between the exciter ceiling $E_{FD\max}$ and regulator ceiling $V_{R\max}$, the saturation function S_E and the gain K_E must be noted. Under steady state (refer 11.7) $\dot{E}_{FD} = 0$, giving

$$V_R - (K_E + S_E)E_{FD} = 0, \quad E_{FD}\min \le E_{FD} \le E_{FD}\max$$

At ceiling $E_{FD} = E_{FD\max}$, $V_{R\max} - (K_E + S_E\max)E_{FD\max} = 0.$ ...(11.10)

K_E is always specified. Specification of any two out of the three constants $V_{R\max}$, $S_{E\max}$ and $E_{FD\max}$ establishes the third from (11.11).

A special case of type 1 is the power source static exciter employing an excitation source from terminal voltage only. This responds quickly but regulator ceiling is proportional to V_T.

$$V_{R\max} = K_R V_T$$

Further, in such systems $K_E = 1$, $T_E = 0$, $S_E = 0$

The state equations of the type 1 system is shown in Fig. 11.12 are given by

$$\dot{E}_{FD} = \left(\frac{1}{T_E}\right)V_R - \left(\frac{K_E + S_E}{T_E}\right)E_{FD} \qquad \text{...(11.11a)}$$

$$\dot{V}_R = \left(\frac{K_A}{T_A}\right)V_e - \left(\frac{1}{T_A}\right)V_R \qquad \text{...(11.11b)}$$

$$\dot{V}_1 = \left(\frac{1}{T_R}\right)V_T - \left(\frac{1}{T_R}\right)V_R \qquad \text{...(11.11c)}$$

$$\dot{V}_F = \frac{1}{T_F}(K_F\dot{E}_{FD} - V_F) = \left(\frac{K_F}{T_F T_E}\right)V_R - K_F\left(\frac{K_E + S_E}{T_F T_E}\right)E_{FD} - \frac{V_F}{T_F} \qquad \text{...(11.11d)}$$

$$V_e = V_{\text{ref}} + V_S - V_1 - V_F \qquad \text{...(11.11e)}$$

11.6.4 Type 2 Excitation System-Rotating Rectifier System

The rotating rectifier systems incorporate damping loops that originate from the regulator output rather than from the excitation voltage. This is because being brushless, the excitation voltage is not available for feedback. The block diagram is shown in Fig. 11.14.

Note that this damping loop has two time constants, T_{F1} and T_{F2}. One of them approximates the exciter time delay and is referred to as 'major damping'. The second 'minor damping' is used to damp higher frequencies.

The state equations of the system are

$$\dot{V}_1 = \left(\frac{K_R}{T_R}\right)V_T - \left(\frac{1}{T_R}\right)V_1 \qquad \text{...(11.12a)}$$

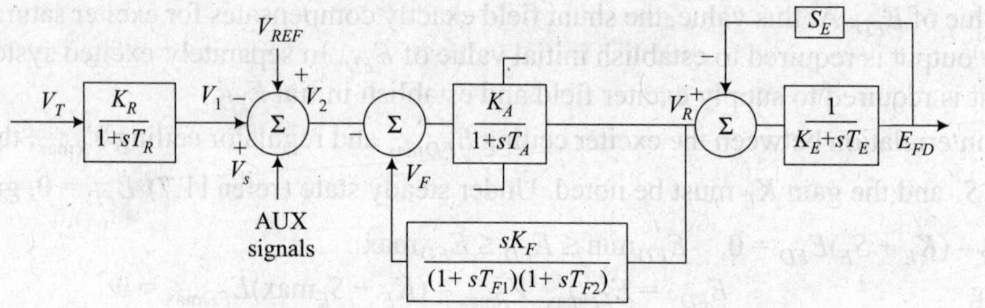

Fig. 11.14 Type 2 excitation system model.

$$\dot{V}_R = \left(\frac{K_A}{T_A}\right)V_e - \left(\frac{1}{T_A}\right)V_R \quad \text{...(11.12b)}$$

$$\dot{E}_{FD} = \left(\frac{1}{T_E}\right)V_R - \left(\frac{K_E + S_E}{T_E}\right)E_{FD} \quad \text{...(11.12c)}$$

$$\ddot{V}_F = \frac{K_F\dot{V}_R}{T_{F1}T_{F2}} - \left(\frac{T_{F1}+T_{F2}}{T_{F1}T_{F2}}\right)\dot{V}_F - \frac{1}{T_{F1}T_{F2}}V_F \quad \text{...(11.12d)}$$

$$= \left(\frac{K_F K_A}{T_A T_{F1} T_{F2}}\right)Ve - \left(\frac{K_F}{T_A T_{F1} T_{F2}}\right)V_R - \left(\frac{T_{F1}+T_{F2}}{T_{F1}T_{F2}}\right)\dot{V}_F - \frac{1}{T_{F1}T_{F2}}V_F$$

$$V_e = V_{\text{REF}} + V_S - V_1 - V_F \quad \text{...(10.12e)}$$

11.6.5 Type 3 Excitation System–Static with Terminal Current and Potential Sources

As mentioned in the section under static exciters, some static systems use generator terminal current with potential as the excitation source. The representation is given in Fig. 11.15.

The transfer function of the regulator is similar to type 1. The next summing point combines regulator output V_R with a signal representing the self-excitation from the generator terminals. K_P is the coefficient of shunt excitation proportional to the terminal voltage and K_I is the coefficient of the supply obtained from the terminal saturable CT's. The multiplier accounts for variation of self-excitation with change in the angular position of field current I_{FD} and self-excitation voltage V_{th}. The transfer function $\frac{1}{(K_E + sT_E)}$ represents the constants of the excitation transformer. $\frac{sK_F}{1+sT_F}$ represents damping.

Since, the system is non-linear, we can approximate it to formulate a linearized state space representation. The self-excitation voltage can be written as

$$V_C = K_1 V_T + K_2 I_T + K_3 I_{FD} \quad \text{...(11.13)}$$

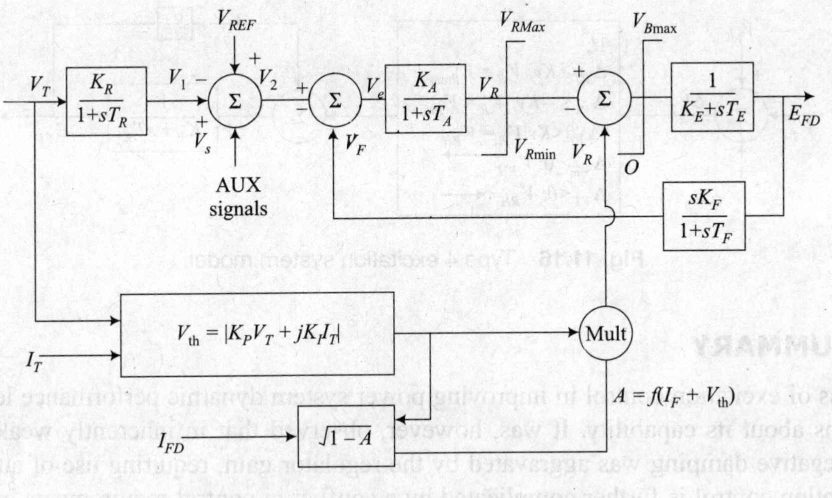

Fig. 11.15 Type 3 excitation model.

The state equations are given by

$$\dot{V}_1 = \left(\frac{K_R}{T_R}\right)V_T - \left(\frac{1}{T_R}\right)V_1 \quad ...(11.14a)$$

$$\dot{V}_R = \left(\frac{K_A}{T_A}\right)V_e - \left(\frac{1}{T_A}\right)V_R \quad ...(11.14b)$$

$$\dot{E}_{FD} = \left(\frac{1}{T_E}\right)V_B - \left(\frac{K_E}{T_E}\right)E_{FD} \quad ...(11.14c)$$

$$\dot{V}_F = \left(\frac{K_F}{T_F}\right)\dot{E}_{FD} - \frac{1}{T_F}V_F \quad ...(11.14d)$$

$$V_B = V_R + V_C \quad ...(11.14e)$$

$$V_e = V_{REF} + V_s - V_1 - V_F \quad ...(11.14f)$$

11.6.6 Type 4 Excitation System–Non–Continuous Acting

Type 1, 2, 3 are all representations of fast-acting systems. Some earlier systems have a dead zone in which system operates in open loop mode. The representation is shown in Fig. 11.16.

V_{RH} is rheostat setting. K_V is the raise-lower contact setting that controls the rheostat setting. Depending on magnitude of error voltage ΔV_T, different regulator modes are specified. If error voltage is greater than the fast raise/lower contact setting K_V, $V_{R\,max}$ or $V_{R\,min}$ is applied to the exciter, depending on the sign of the error voltage. For an error voltage less than K_V, the exciter input equals the rheostat setting V_{RH}. The time constant representing the slow adjustment of exciter field voltage is T_{RH}. Since, the system is highly non-linear, state equation representation is not possible.

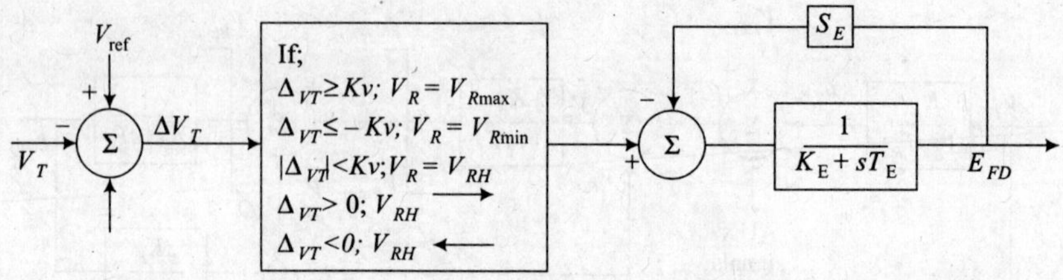

Fig. 11.16 Type 4 excitation system model.

11.7 SUMMARY

The success of excitation control in improving power system dynamic performance lead to great expectations about its capability. It was, however, observed that in inherently weakly coupled systems, negative damping was aggravated by the regulator gain, requiring use of auxiliary signals. Excitation control is further complicated by a conflict in control requirements in the period following the initiation of a transient. The control immediately after the transient may be very different from the requirement after a few seconds. In general, fast acting excitation systems increase the steady state stability limit. The effect on first swing transient is marginal for less severe transients. Voltage regulators improve synchronizing torque. The effect of the excitation system on subsequent swings in case of more severe transients (faults of longer duration) is more pronounced. The recovery of the terminal voltage is faster, with fast acting exciters. Standard computer models are used to represent most of the excitation systems. By suitable choice of the parameters, the models can be modified if necessary to suit specific requirements. The mathematical representation of the excitation system equations in the state variable form, facilitates the differential equations to be solved along with the system dynamic equations. A more detailed discussion is beyond the scope of the book.

REVIEW QUESTIONS

1. With a neat block diagram, explain the excitation system and its main components.
2. Define ceiling voltage and excitation system voltage-time response.
3. DC exciters are simpler than AC exciters. Comment.
4. What is the main drawback of potential source controlled rectifier? How is it corrected?
5. Describe a scheme for brushless excitation.
6. What is a pilot exciter? What is the advantage of having one?
7. What is a power system stabilizer? Why is it needed?
8. Describe the working of a hard and soft limiter. What is the functional difference between the two?
9. With a block diagram explain IEEE type 1 model for excitation systems.
10. A simple first order transfer function is sufficient to model modern exciters. Comment.
11. The exciter gain has an effect on the stability of the system. Comment.
12. With a block diagram explain IEEE type 2 and 3 models for exciters.

Chapter 12

Modeling of Prime Movers and Loads

12.1 INTRODUCTION

Steam turbines and hydro-turbines are the most widely used as prime movers for generators, followed by pressurized water reactor nuclear units. The regulation of the frequency in the power system, requires the speed control of the prime mover, using the governor. This would require sensing of the speed and translating it to suitable control action. The prime mover control must have a drooping characteristic to ensure proper division of load, when generators are operating in parallel. It is also necessary for the prime mover control to adjust the generation according to economic dispatch schedule. There are two types of speed governing system.

- Mechanical-hydraulic
- Electro-hydraulic

In both types, hydraulic motors are used for positioning the valve or gate, controlling the steam or water flow, respectively. The speed sensing and conditioning is done by electronic circuits in electro-hydraulic systems and by mechanical components in mechanical-hydraulic systems. The functional block diagram of the power system, including the prime mover system, is shown in Fig. 12.1. Suitable mathematical models have to be developed, to integrate the prime mover and governor dynamics along with generators and the network. These are presented in the following sections, based on *IEEE* proposed standards.

12.2 SPEED GOVERNING SYSTEMS FOR STEAM TURBINES

The models presented are applicable to conventional fossil-fired units or turbines supplied by pressurized water reactors. The models do not include governor dead band, contingency valving or fast valving.

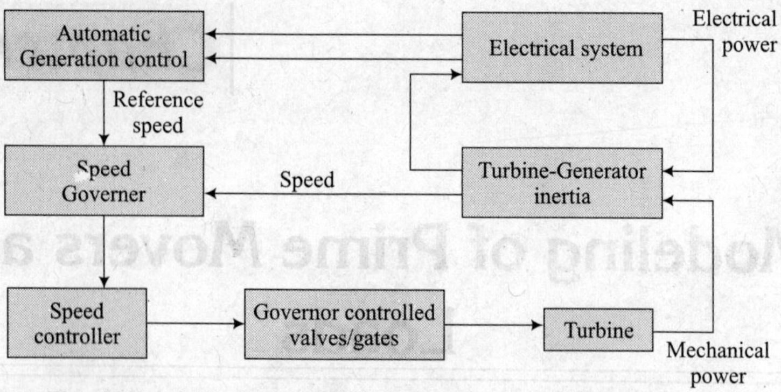

Fig. 12.1 Speed-governing systems for steam turbines.

12.2.1 Mechanical-hydraulic System

The block diagram of the mechanical-hydraulic system for steam turbines is shown in Fig. 12.2.

The speed signal is transformed by the speed governor into a position signal, which is sub-

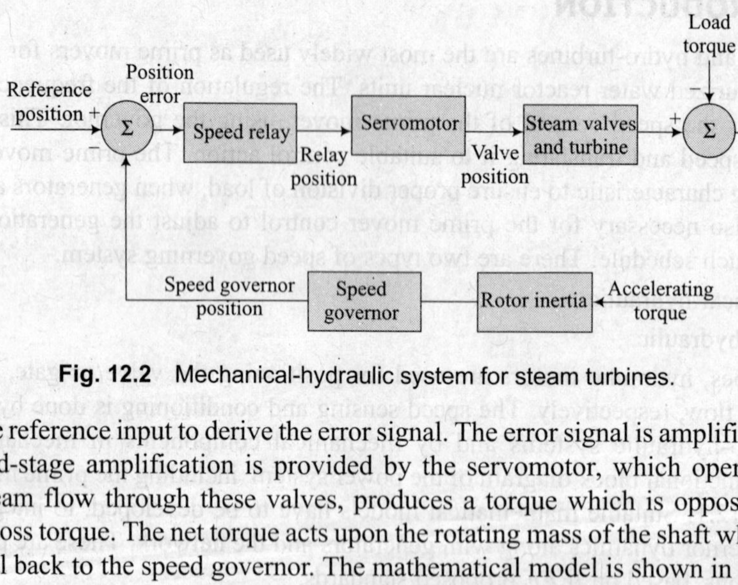

Fig. 12.2 Mechanical-hydraulic system for steam turbines.

tracted from the reference input to derive the error signal. The error signal is amplified by the speed relay. A second-stage amplification is provided by the servomotor, which operates the steam valves. The steam flow through these valves, produces a torque which is opposed by the load torque and the loss torque. The net torque acts upon the rotating mass of the shaft which feeds back the speed signal back to the speed governor. The mathematical model is shown in Fig. 12.3.

The speed governor is represented by a gain K_G, which is reciprocal of regulation or droop. S_R is the reference signal, determined by the automatic generation control. It is generally kept constant over the interval of a stability study. The speed relay is represented as an integrator with a time constant T_{SR} and direct feed back. The servomotor is represented by an integrator with a time constant T_{SM}. It physically moves the valves and is generally large. Rate limits ($\dot{C}_{V\,\mathrm{OPEN}}$ and $\dot{C}_{V\,\mathrm{CLOSE}}$) are shown as input to the integrator and position limits ($\dot{C}_{V\,\max}$ and $C_{V\,\min}$) are shown at the integrator output. Typic parameters are

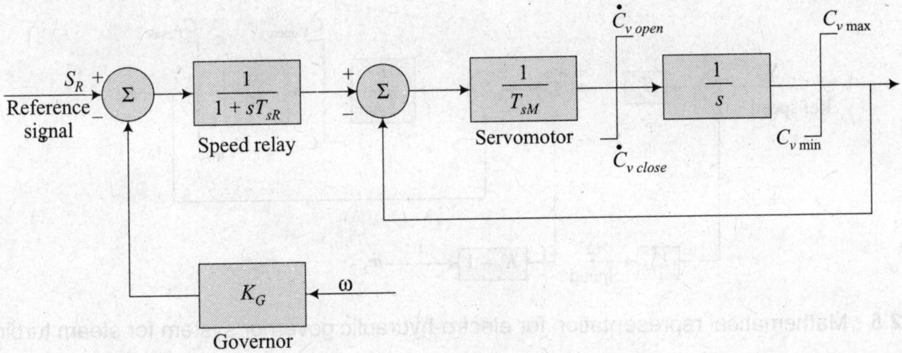

Fig. 12.3 Mathematical representation of mechanical hydraulic governor for steam turbine. (C_v is effective valve position).

$K_G = 20.0$; $T_{SM} = 0.2$ to 0.3s $\dot{C}_{v\ OPPEN} = 0.1$ pu per sec per valve

$T_{SR} = 0.1$ sec $C_{v\ CLOSE} = 1.0$ pu per sec per valve

12.2.2 Electro-hydraulic System

Electronic circuits are used in low-power portions, leading to more flexibility. The block diagram of Fig. 12.4 shows a typical configuration.

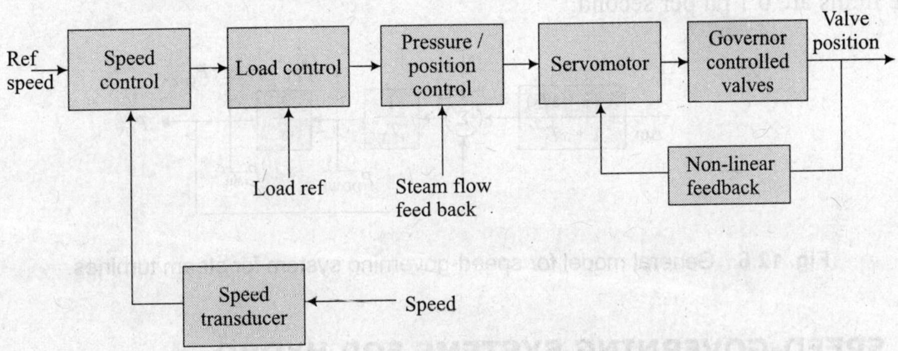

Fig. 12.4 Electro-hydraulic governor system for steam turbines.

The steam flow feedback and servo motor feedback provide improved linearity over the mechanical-hydraulic system. The mathematical model is shown in Fig. 12.5 for a typical control system.

($\dot{m}_{HP}$ is high pressure turbine steam flow)

Typical values are $K_G = 20.0$; $K_P = 3.0$ with steam flow feedback

$K_P = 1.0$ without steam flow feedback

$T_{SM} = 0.1$ seconds

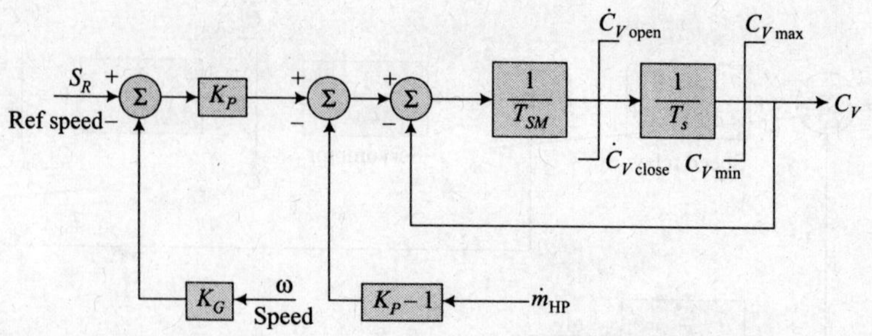

Fig. 12.5 Mathematical representation for electro-hydraulic governor system for steam turbines.

A general speed-governing model is shown in Fig. 12.6, which can be used for both electro-hydraulic and mechanical-hydraulic systems, by suitable choice of parameters. The load reference is shown as an initial power P_o, which is combined with the increments due to speed deviation to obtain the total power P_{GV}, subjected to the time lag T_3, introduced by the servomotor.

Typical values of time constants (seconds) are as follows.

Mechanical-hydraulic governor: $T_1 = 0.2 - 0.3$ s; $T_2 = 0$; $T_3 = 0.1$ s

$$\dot{P}_{UP} = 0.1 \text{ pu per second}$$

$$\dot{P}_{DOWN} = 1.0 \text{ pu per second}$$

Electro-hydraulic governor: $T_1 = T_2$, $T_3 = 0.025 - 0.15$ s.

Rate limits are 0.1 pu per second.

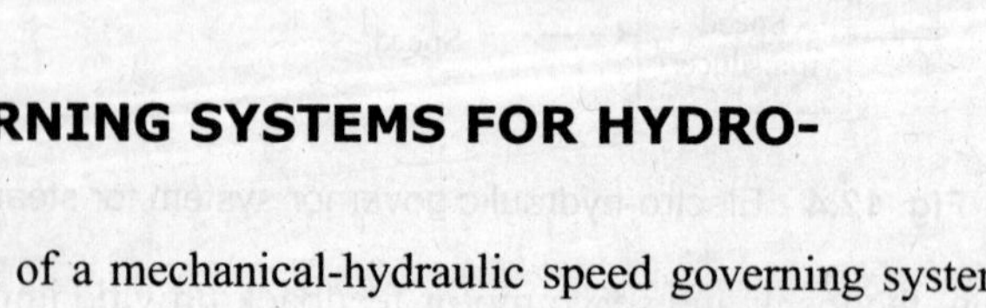

Fig. 12.6 General model for speed-governing system for steam turbines.

12.3 SPEED-GOVERNING SYSTEMS FOR HYDRO-TURBINES

A functional block diagram of a mechanical-hydraulic speed governing system, for hydro-turbines, is shown in Fig. 12.7.

The mathematical model is shown in Fig. 12.8.

The rate limits may be present for large rapid speed excursions. Position limits exist corresponding to the extremes of gate opening. δ is the transient speed droop coefficient. T_P and T_G are the pilot valve time constant and gate servomotor time constant, respectively. T_R is the dash pot time constant. Typical values are

$$T_R = 5.0 \text{ s}; T_G = 0.2 \text{ s}; T_P = 0.04 \text{ s}; d = 0.3 \text{ and } \sigma = 0.05$$

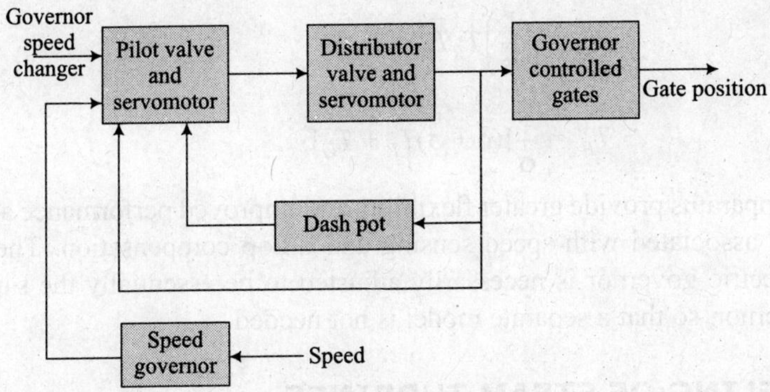

Fig. 12.7 Functional diagram of mechanical-hydraulic governing system.

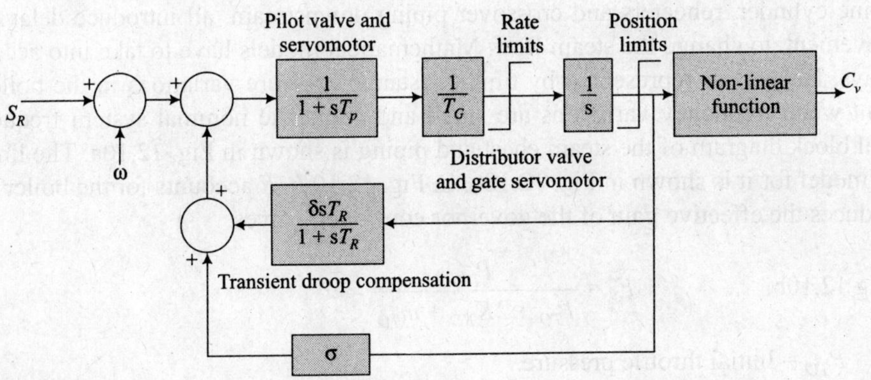

Fig. 12.8 Mathematical model of mechanical-hydraulic speed governing system.

A simplified model is shown in Fig. 12.9.

Here, T_P is neglected and the gate servomotor limits are also neglected. The constants, of Fig. 12.9 are related to Fig. 12.8 as follows:

$$K = \frac{1}{\sigma} \qquad ...(12.1)$$

$$T_1, T_3 = \frac{T_B}{2} \pm \sqrt{\frac{T_B^2}{4} - T_{A_{P_o}}} \qquad ...(12.2)$$

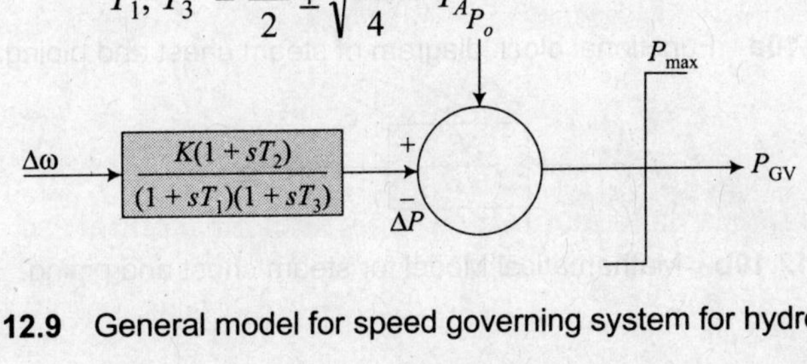

Fig. 12.9 General model for speed governing system for hydro-turbines.

where,
$$T_A = \left(\frac{1}{\sigma}\right) T_R T_G \quad ...(12.3)$$

$$T_B = \frac{1}{\sigma}[(\delta + \sigma)T_R + T_G] \quad ...(12.4)$$

Electronic apparatus provide greater flexibility and improved performance and are used in low power functions associated with speed sensing and droop compensation. The dynamic performance of the electric governor is necessarily adjusted to be essentially the same as that for the mechanical governor, so that a separate model is not needed.

12.4 MODELING OF STEAM TURBINES

Steam turbine systems are compound, with more than one pressure. Governor-controlled valves, control steam flow at the inlet to the high pressure turbine. The steam chest and inlet piping to the first turbine cylinder, reheaters and crossover piping downstream, all introduce delays after the valve movement, to change the steam flow. Mathematical models have to take into account these time delays. Delays are represented by time constants. Pressure variations in the boiler are not significant when frequency variations are small and about the nominal system frequency. The functional block diagram of the steam chest and piping is shown in Fig. 12.10a. The linear mathematical model for it is shown in Fig. 12.10b. In Fig. 12.10 *b*, *F* accounts for the boiler tube drop which reduces the effective gain of the governor controlled valves.

In Fig 12.10b,
$$F = \frac{P_{TO}^2}{P_{TO} + 2K_{PO} + \dot{m}_{TO}^2} \quad ...(12.5)$$

where, P_{TO} = Initial throttle pressure.

K_{PO} = Steam lead pressure drop coefficient.

$\dot{m}_{TO}$ = Initial throttle valve steam flow.

T_{CH} = Steam chest time constant.

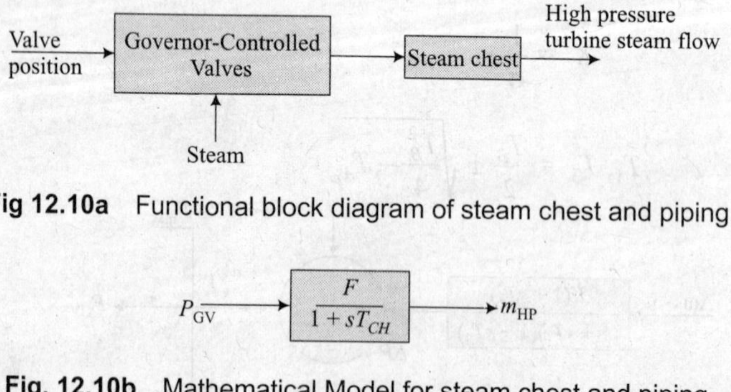

Fig 12.10a Functional block diagram of steam chest and piping.

Fig. 12.10b Mathematical Model for steam chest and piping.

12.4.1 Common Steam System Configurations

Six common steam systems are used. In the models, the steam chest and inlet piping, reheaters and crossover piping are represented by time constants T_{CH}, T_{RH} and T_{CO}, respectively. F_{VHP}, F_{HP}, F_{IP}, and F_{LP} represent fractions of the total power developed in various cylinders (very high pressure, high pressure, intermediate pressure, low pressure). The configurations and their corresponding models are shown in Fig. 12.11a to 12.11f, along with the typical parameter values for each system.

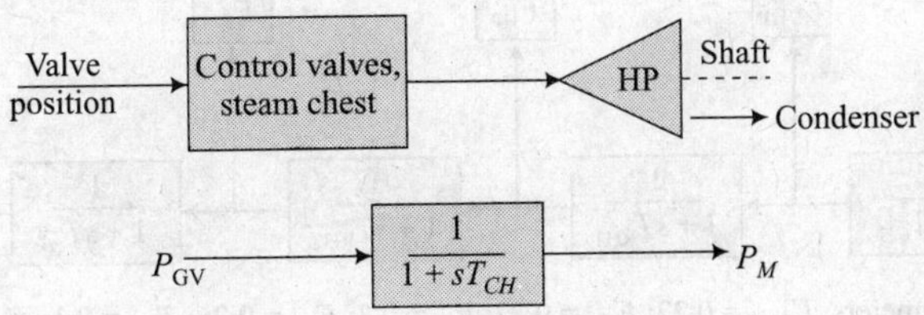

Typical parameters: $T_{CH} = 0.2 - 0.5$ s;

Fig. 12.11a Non-reheat system.

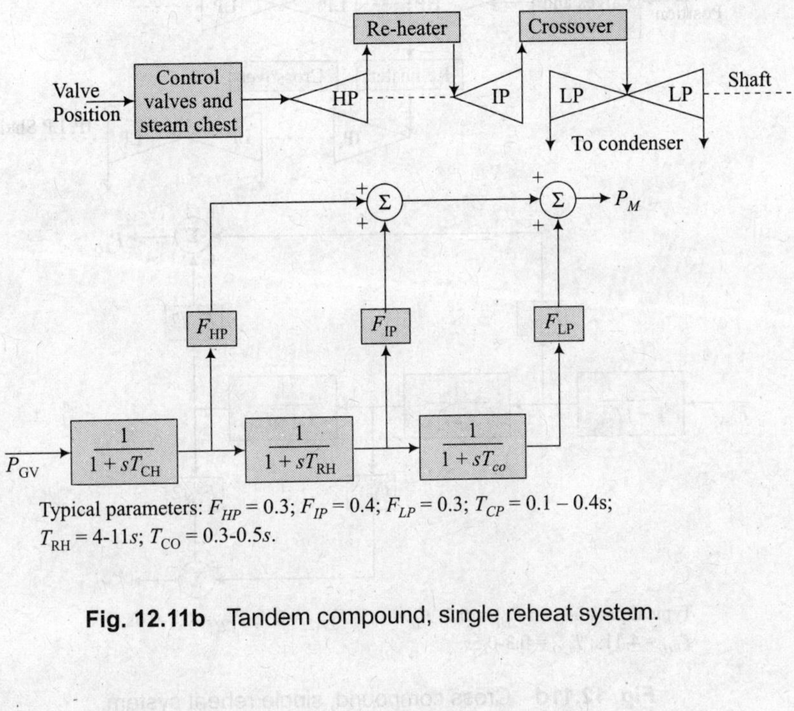

Typical parameters: $F_{HP} = 0.3$; $F_{IP} = 0.4$; $F_{LP} = 0.3$; $T_{CP} = 0.1 - 0.4$s; $T_{RH} = 4\text{-}11s$; $T_{CO} = 0.3\text{-}0.5s$.

Fig. 12.11b Tandem compound, single reheat system.

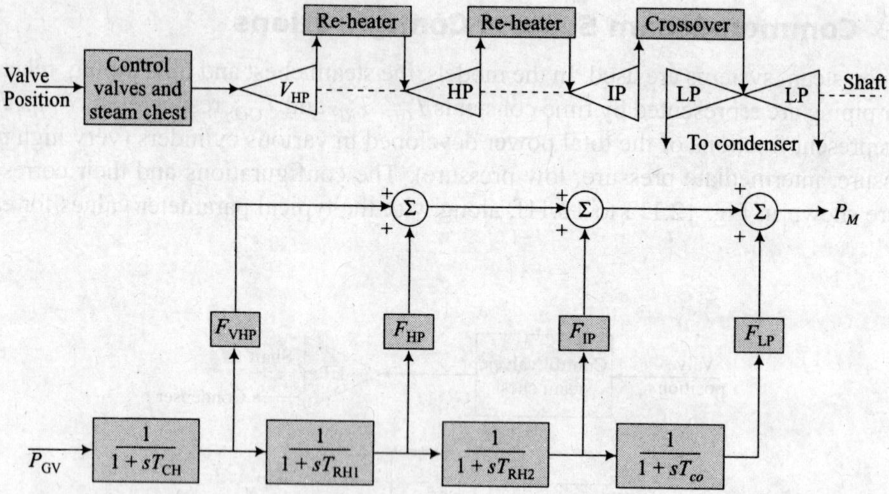

Typical parameters: $F_{VHP} = 0.22$; $F_{HP} = 0.22$; $F_{IP} = 0.3$; $F_{LP} = 0.26$; $T_{CH} = 0.1 - 0.4$s; $T_{RH1} = 4$-11s; $T_{CO} = 0.3$-0.5s.

Fig. 12.11c Tandem compound, double reheat system.

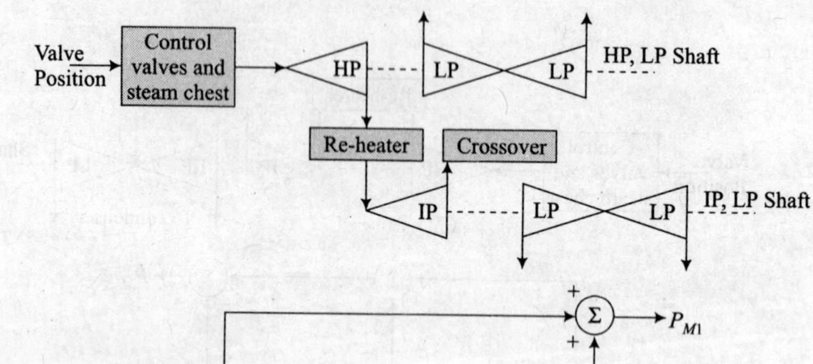

Typical parameters: $F_{HP} = 0.3$; $F_{IP} = 0.3$; $F_{LP/2} = 0.4$; $T_{CH} = 0.1$-0.4s; $T_{RH} = 4$-11s; $T_{CO} = 0.3$-$0.5s$.

Fig. 12.11d Cross compound, single reheat system.

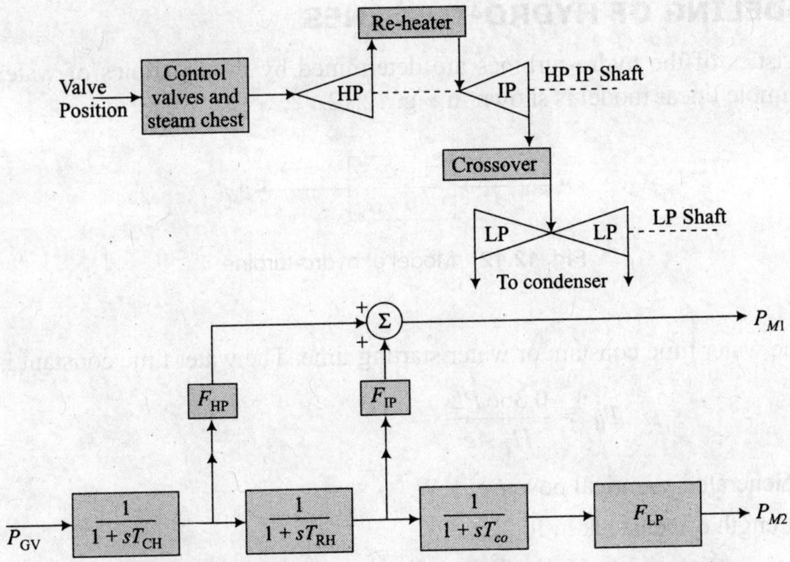

Typical parameters: $F_{HP} = 0.25$; $F_{IP} = 0.25$; $F_{LP} = 0.5$; $T_{CH} = 0.1\text{-}0.4$s; $T_{RH} = 4\text{-}11$s; $T_{CO} = 0.3\text{-}0.5$s.

Fig. 12.11e Cross compound, single reheat system (second model).

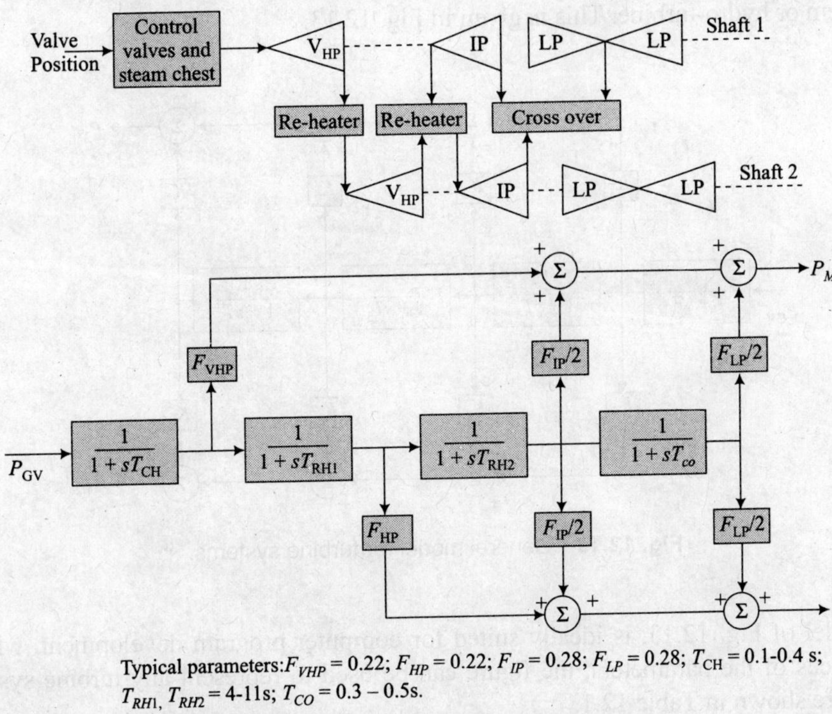

Typical parameters: $F_{VHP} = 0.22$; $F_{HP} = 0.22$; $F_{IP} = 0.28$; $F_{LP} = 0.28$; $T_{CH} = 0.1\text{-}0.4$ s; T_{RH1}, $T_{RH2} = 4\text{-}11$s; $T_{CO} = 0.3 - 0.5$s.

Fig. 12.11f Cross compound, double reheat system.

12.5 MODELING OF HYDRO-TURBINES

The characteristics of the hydro-turbines are determined by the dynamics of water flow in the penstock. A simple linear model is shown in Fig. 12.12.

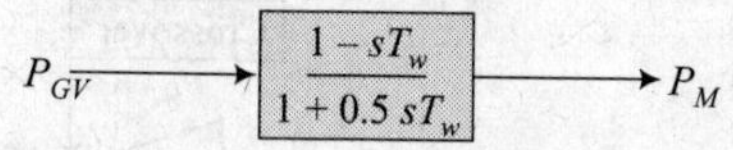

Fig. 12.12 Model of hydro-turbine.

T_W is called the water time constant or water starting time. The water time constant is given by

$$T_W = \frac{0.366\,PL}{H_T^2\,Ae} \quad ...(12.6)$$

where, P = Generated electrical power in KW.

L = Length of penstock in ft.

A = average penstock area in ft^2.

H_T = Total head in ft.

e = product of turbine efficiency and generator efficiency.

A general mathematical model can be given to represent any of the six configurations of steam turbine system or hydro-turbine. This is given in Fig. 12.13.

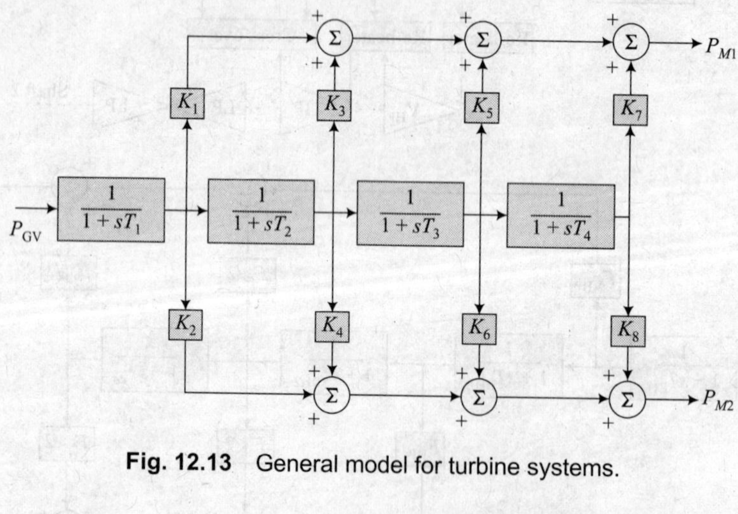

Fig. 12.13 General model for turbine systems.

The model of Fig. 12.13, is ideally suited for computer program development, wherein by suitable choices of the parameters, the figure can be used to represent any turbine system. The parameters are shown in Table 12.1.

Table 12.1 Parameters for general model of turbines

System	*Time constants*											
	T_1	T_2	T_3	T_4	K_1	K_2	K_3	K_4	K_5	K_6	K_7	K_8
Fig 12.11a	T_{CH}	-	-	-	1	0	0	0	0	0	0	0
Fig 12.11b	T_{CH}	T_{RH}	T_{CO}	-	F_{HP}	0	F_{IP}	0	F_{LP}	0	0	0
Fig 12.11c	T_{CH}	T_{RH1}	T_{RH2}	T_{CO}	F_{VHP}	0	F_{HP}	0	F_{IP}	0	F_{LP}	0
Fig 12.11d	T_{CH}	T_{RH}	T_{CO}	-	F_{HP}	0	0	F_{IP}	$F_{LP}/2$	$F_{LP}/2$	0	0
Fig 12.11e	T_{CH}	T_{RH}	T_{CO}	-	F_{HP}	0	F_{IP}	0	0	F_{LP}	0	0
Fig 12.11f	T_{CH}	T_{RH1}	T_{RH2}	T_{CO}	F_{VHP}	0	0	F_{HP}	$F_{IP}/2$	$F_{IP}/2$	$F_{LP}/2$	$F_{LP}/2$
Fig 12.12	0	$T_w/2$	-	-	-2	0	3	0	0	0	0	0

The models presented above are sufficient to develop models to be used in computer programs for transient stability analysis. The models can be extended to include fast valving and introduction of boiler pressure at the governing valve.

12.6 LOAD MODELING

Decisions of the power system planner, concerning system reinforcements and system performances, are largely based on results of power flow and stability studies. Improper or inadequate modeling of loads can lead to incorrect results, which can prove costly, since load representation can have a significant impact on the analysis of the results. The modeling of loads is complicated because the load at a bus is made of diverse devices such as heaters, furnaces, lighting loads, air-conditioners, large induction motors, compressors, etc. The exact composition is difficult to predict or estimate. It is impractical to model each load and necessary to develop composite load models at the buses. In this section, static load models, dynamic load models and models of induction motors, which constitute a major portion of the load are developed.

12.6.1 Static Load Models

These models express the active and reactive powers, at any instant of time, as an algebraic function of the bus voltage magnitude and frequency. Different models have been proposed and are commonly used.

(a) Exponential Models: Here P and Q are expressed as some exponential components of their values at initial operating conditions.

$$P_L = P_o\left(\frac{V}{V_o}\right)^{np} \quad ...(12.7)$$

$$Q_L = Q_o\left(\frac{V}{V_o}\right)^{nq} \quad ...(12.8)$$

where, P_L = Active load power

Q_L = Reactive load power

P_o = Active load power at rated voltage, V_o

Q_o = Reactive load power at rated voltage, V_o

V_o = Rated voltage

V = Supply voltage (or bus voltage)

np = Active power exponent

nq = Reactive power exponent

(b) Constant impedance model: Here the power is assumed to vary directly with square of the voltage magnitude, so that the impedance of the load is a constant. It is also called constant admittance model. This model has been widely used for real and reactive components of loads in stability studies, to avoid complication of non-linear elements in network equations.

(c) Constant current model: Here the power is assumed to vary directly with voltage magnitude, so that the current is a constant.

(d) Constant power model: The power does not vary with changes in voltage magnitude. It is also called constant MVA model.

The relationships in the three models are shown in Fig. 12.14a and 12.14b.

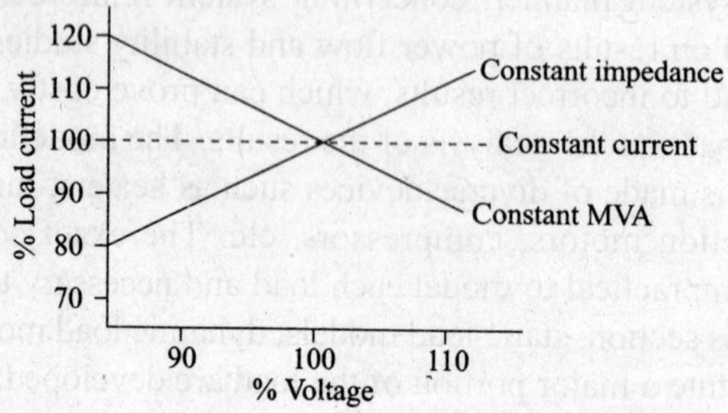

Fig. 12.14a Load current *Vs* bus voltage.

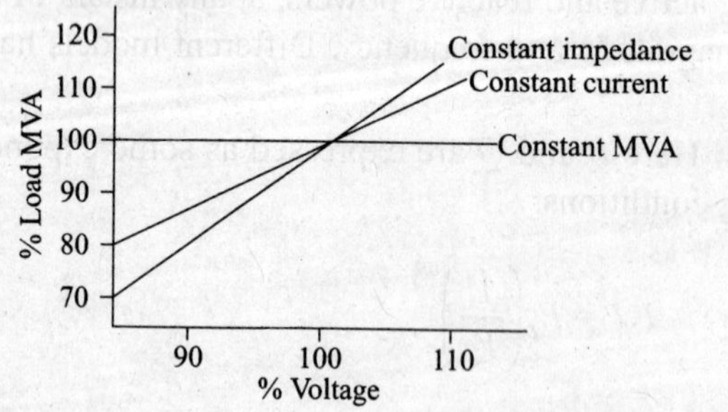

Fig. 12.14b Load MVA vs bus voltage.

The response of nearly all loads to voltage changes can be represented by some combination of constant impedance, constant current, and constant MVA devices. The constant MVA type load representation is the most severe representation from stability point of view, since it amplifies voltage oscillations. For example, a drop in voltage causes an increase in load current (to keep MVA constant) resulting in a further voltage drop. Hence, constant power load models can lead to convergence problems in transient stability programs. In contrast, constant impedance loads have a damping effect on voltage oscillations. Generally, heating and lighting equipment exhibit a constant, near unity power factor. Motors will normally be constant MVA for moderate voltage changes. Agriculture and industrial loads are mostly motor loads and have more constant MVA load than commercial and residential loads. Behaviour of loads for changing frequency is more uncertain. Table 12.2 gives the voltage and frequency dependency of various loads and Table 12.3 gives the typical characteristics of different load classes. Different characteristics of static loads tend to compensate each other, resulting in a composite effect of constant impedance.

Table 12.2 Voltage and frequency dependence of loads

Component	*Power factor*	$\frac{\partial P}{\partial V}$	$\frac{\partial Q}{\partial V}$	$\frac{\partial P}{\partial f}$	$\frac{\partial Q}{\partial f}$
Air-conditioner					
3-ø central	0.9	0.088	2.5	0.98	−1.3
window type	0.82	0.468	2.5	0.56	−2.8
Heater, ovens	1.0	2.0	0	0	0
Refrigerator	0.8	0.77	2.5	0.53	-1.5
Television	0.8	2.0	5.1	0	-4.5
Incandescent light	1.0	1.55	0	0	0
Fluorescent light	0.9	0.96	7.4	1.0	-2.8
Industrial motors	0.88	0.07	0.5	2.5	1.2
Fan motors	0.87	0.08	1.6	2.9	1.7
Agriculture pumps	0.85	1.4	1.4	5.0	4.0
Arc furnace	0.7	2.3	1.6	-1.0	-1.0
Washing machine	0.65	0.08	1.6	3.0	1.8

Table 12.3 Typical characteristics of different load classes

Component	*Power factor*	$\frac{\partial P}{\partial V}$	$\frac{\partial Q}{\partial V}$	$\frac{\partial P}{\partial f}$	$\frac{\partial Q}{\partial f}$
Residential					
Summer	0.9	1.2	2.9	0.8	−2.2
Winter	0.99	1.5	3.2	1.0	−1.5
Commercial					
Summer	0.85	0.99	3.5	1.2	-1.6
Winter	0.9	1.3	3.1	1.5	-1.1
Industrial	0.85	0.18	6.0	2.6	1.6

(e) Polynomial model: It represents the load power relationship to voltage magnitude as a polynomial equation.

$$P_L = P_o\left[a_1\left(\frac{V}{V_o}\right)^2 + a_2\left(\frac{V}{V_o}\right) + a_3\right] \quad ...(12.9)$$

$$Q_L = Q_o\left[b_1\left(\frac{V}{V_o}\right)^2 + b_2\left(\frac{V}{V_o}\right) + b_3\right] \quad ...(12.10)$$

where a_1, b_1 = Fractions of constant impedance component.

a_2, b_2 = Fractions of constant current component.

a_3, b_3 = Fractions of constant power component.

V_o = rated voltage.

P_o, Q_o = Power consumed at rated voltage.

(f) Frequency dependent model: It is obtained by multiplying either a polynomial or exponential load model by a factor.

$$a = 1 + a_f(f - f_o) \quad ...(12.11)$$

where, f = frequency of bus voltage.

a_f = frequency sensitivity parameter.

f_o = rated frequency.

The bus voltage frequency can be taken to be the numerical derivative of the bus-voltage angle.

12.6.2 Dynamic Load Model

The response of most composite loads to voltage and frequency changes is fast. Hence, static load models can be used. However, studies on interarea oscillations, voltage stability, and long-term stability require the modeling of the load dynamics. Dynamics have to be included in

- Large motors, which constitute 60 – 70% of total loads.
- Extinction of discharge lamps below a certain voltage and their restart after voltage recovery.
- Representation of effect of ULTCs connected to distribution transformer, voltage regulators and voltage–controlled capacitor banks.
- Operation of thermal or over current protective delays which operate contactors of large motors if voltage drops below 0.6–0.75 pu.

Thermostatically controlled load

A realistic dynamic model, proposed for thermostatically controlled loads is shown in Fig. 12.15.

Here, τ_{ref} : Reference temperature

τ_H : Temperature of heated area

τ_A : Ambient temperature

K_P : Gain of proportional controller

K_I : Gain of integral controller

G : Load conductance

K_1 : Gain associated with load model

T_C : Time constant of integral controller

P_H : Power from heater = $K_H GV^2$

P_L : Heat lost to ambience = $K_A(\tau_H - \tau_A)$

The dynamic equation is given by

$$K\frac{d\tau_H}{dt} = P_H - P_L \qquad ...(12.12)$$

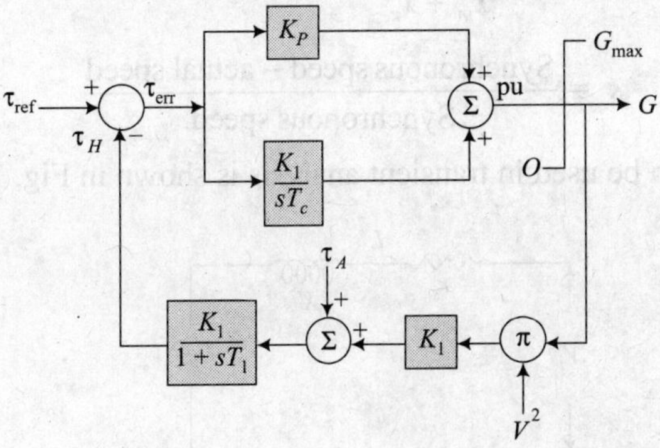

Fig. 12.15 Thermostatically controlled load.

The temperature τ_H is compared with the reference temperature, and the error signal controls the load conductance through a proportional plus integral controller. When all the thermostatically controlled loads are on, G reaches maximum value of G_{max}.

12.6.3 Induction Motor Model

In some studies, induction motors can be represented by shunt impedances. However, a more detailed model is necessary for large induction motors. Since, there is no excitation source applied to rotor windings, the dynamics of the rotor is determined by slip, rather than excitation. The single phase equivalent circuit of an induction motor in steady state is shown in Fig. 12.16.

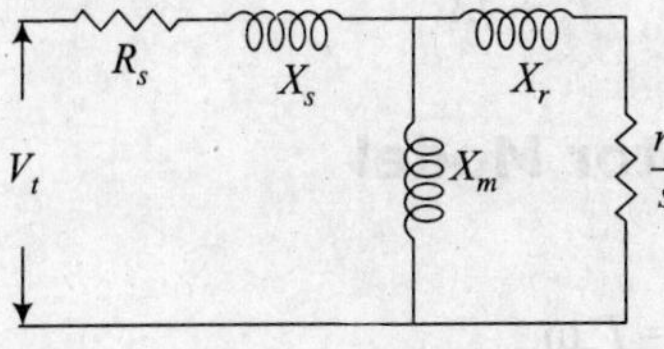

Fig. 12.16 Steady state equivalent circuit of induction motor.

In Fig. 12.16 r_s : Stator resistance in pu.

x_s : Stator reactance in pu.

x_m : Magnetizing reactance in pu.

x_r : Rotor reactance in pu.

s : Slip

r_r : Rotor resistance in pu.

Since, r_r is small, it can be neglected. The open circuit reactance is approximately given by

$$X_o = x_s + x_m \quad ...(12.13)$$

And blocked rotor reactance

$$X^1 = x_s + \frac{x_m x_r}{x_m + x_r} \quad ...(12.14)$$

$$s = \frac{\text{Synchronous speed} - \text{actual speed}}{\text{Synchronous speed.}} \quad ...(12.15)$$

The equivalent circuit to be used in transient analysis is shown in Fig. 12.17.

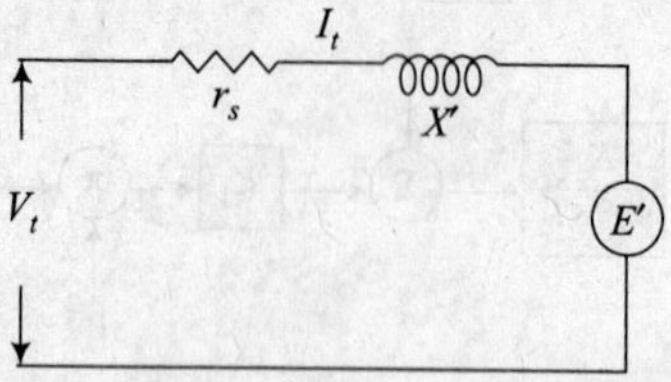

Fig. 12.17 Equivalent circuit of induction motor for transient analysis.

The differential equation describing rate of change of voltage behind transient reactance is given by

$$\frac{dE'}{dt} = -j\omega sE' - \frac{1}{T_o}[E' - j(X_o - X')I_t] \quad ...(12.16)$$

where, T_o = rotor open circuit time constant = $\dfrac{x_r + x_m}{2\pi f r_r}$.

and,

$$I_t = \frac{V_t - E'}{r_s + jX'} \quad ...(12.17)$$

12.6.4 Sychronous Motor Model

The load torque is modeled as

$$T_m = T_o \omega_m \quad ...(12.18)$$

and the rotor acceleration is given by

$$\frac{d\omega}{dt} = \frac{1}{2H}(T_e - T_m) \qquad ...(12.19)$$

In (12.19) H is the combined inertia constant of the motor and load.

12.7 LOAD MODELING TECHNIQUES

There are two approaches to modeling the load at a feeder/bus.

(a) Component-based approach wherein, the load composition on a feeder is first estimated. Depending on the characteristic of the loads, (refer Tables 12.2 and 12.3) an aggregate static load model is developed. This model can be used for any of the power system analysis.

(b) Measurement-based approach: Here, measurement and data acquisition devices are installed at points where bus loads are to be represented. Voltage, frequency, active power and reactive power are measured at the bus and the mathematical relationship between load powers (active and reactive) and voltage/frequency are obtained by some statistical or curve fitting technique.

12.8 ISSUES IN LOAD-MODELING

Accurate modeling of load is a difficult task due to

- Large number of load components and their characteristics.
- Ownership and location of load devices in customer facilities not directly accessible to the utilities.
- Changing load composition with days, seasons, etc.
- Lack of precise information on composition of the load.
- Uncertain load behaviour with large frequency and voltage deviations.

If the load representation is presently overly pessimistic, then better and more accurate modeling will help in deferring or avoiding the expense of system modification and equipment additions. If the load representation is presently overly optimistic, accurate modeling will help in preventing system inadequacies that may occur during operation.

The range of voltage and frequency excursions of interest depends on the nature of disturbances in the system. While frequency excursions are limited to less than ± 5%, the voltage excursions are much higher . Accidental loss of load often leads to high frequency and high voltage conditions; loss of generation and system collapse due to excessive rise of load usually lead to low frequency and low voltage conditions; a short circuit fault and clearance often leads to severe voltage depressions followed by oscillations in voltage and frequency. Customarily, the load is converted to a constant impedance model, under very low voltages.

During short circuits, the load characteristics during a fault require careful attention. The bus voltages are depressed under short circuit and the generators start accelerating. The load characteristics can have a significant effect on the acceleration or deceleration of the individual generators during the fault. The acceleration is less if local constant-power loads are used than if constant-impedance loads are used. The frequency dependency of the loads is an important contributor to system damping. The total active power drawn by the motors depends not only on the system frequency, but also on the rate of the frequency change. For example, all the motors try to run faster when the system frequency is increasing and consequently draw more active power to increase the kinetic energy associated with the inertia, as well as to compensate the increased losses and shaft loads. Thus, load modeling is crucial and has wide impact on a variety of system operating parameters.

12.9 SUMMARY

It is imperative to develop standard mathematical models for the prime mover system, which includes the governing system and the turbine. Speed governing system is either mechanical hydraulic or electro hydraulic. Simplified mathematical models are derived for the governing system of both steam and hydro-turbines. Steam turbines are broadly classified into six categories, while hydro-turbines can be represented by a single model. A general mathematical model can be given to represent any steam turbine configuration or hydro-turbine, by suitably modifying the parameters.

Since, load models play a decisive role in planning and in stability, analysis, and accurate load model is to be developed. Static load models essentially model the load as constant impedance, constant current, constant power or a combination of both. Different load classes differ widely in their voltage and frequency dependence. The choice of model depends also on the application of the model for a particular study. For example, we have seen that in load flow studies, it is convenient to model the load as constant power loads. At the same time, in stability studies it is customary to use the constant impedance model for the loads. Dynamic models represent the dynamic variation of load with time. Induction motors, which constitute about 60–70% of the load, have to be modeled in greater detail instead of as a constant impedance. The diverse nature of the loads and their characteristics make it difficult to model them accurately.

REVIEW QUESTIONS

1. With a block diagram explain the speed-governing systems for steam turbines.
2. Explain the hydraulic governor system for steam turbines.
3. Draw the general transfer function blocks, explaining each component, for a speed-governing system.
4. Discuss the various steam system configurations.
5. Static load models can approximately represent the characteristics of loads. Justify.
6. How are thermostatically controlled loads modeled?
7. Derive the induction motor model.
8. Discuss the two techniques used in load modeling.
9. What are the major issues in load modeling?
10. What are composite loads?

Chapter 13

Load Frequency Control

13.1 INTRODUCTION

The preceding chapters were devoted to problems associated with the selection of a normal operating state for the power system and optimum scheduling of generation. The present chapter deals with the continuous control of active and reactive power in order to keep the system in steady state. The power system being dynamic, the demand continuously deviates from its normal value. This leads to a small change in the state of the system. The automatic control should act in a closed loop manner, to detect these changes and initiate actions to eliminate the deviations. Briefly stated, the control strategy should be designed to deliver power to an interconnected system economically and reliably, while maintaining the voltage and frequency within the permissible limits.

Changes in real power mainly affect the system frequency and changes in reactive power mainly depend on changes in voltage magnitude and are relatively less sensitive to changes in frequency. Thus, real and reactive powers can be controlled separately. The Automatic Load Frequency Control (ALFC) controls the real power and the Automatic Voltage Regulator (AVR) regulates the voltage magnitude and hence the reactive power. The two controls, along with the generator and prime mover are shown in Fig. 13.1. Unlike the AVR, ALFC is not a single loop. A fast primary loop responds to the frequency changes and regulates the steam (water) flow via the speed governor and control valves to match the active power output with that of the load. The time period here is a few seconds. The frequency is controlled via control of the active power.

A slower secondary loop maintains fine frequency adjustment to maintain proper active power exchange with other interconnected networks via tie-lines. This loop does not respond to fast load changes but instead focuses on changes, which lead to frequency drifting over several minutes.

Since, the AVR loop is much faster than the ALFC loop, the AVR dynamics settle down before they affect the ALFC control loop. Hence, cross-coupling between the controls can be neglected. With the growth of large interconnected systems, ALFC has gained importance in recent times. This chapter presents an introduction to power system controls.

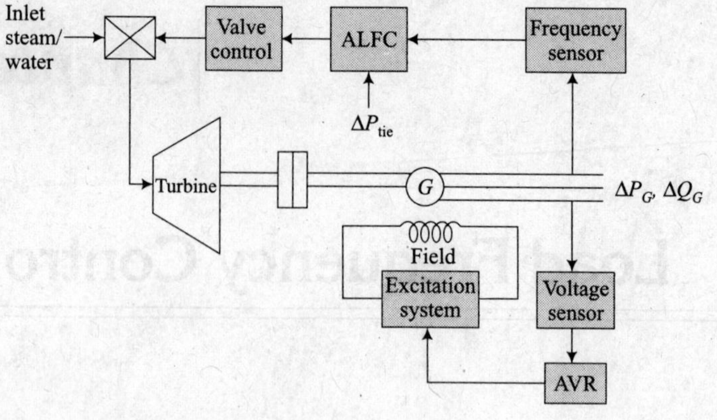

Fig. 13.1 ALFC and AVR.

13.2 AUTOMATIC LOAD FREQUENCY CONTROL

The functions of the ALFC are to maintain steady frequency, control tie-line power exchange and divide the load between the generators. The tie-line power deviation is given by ΔP_{tie} and the change in frequency Δf, is measured by $\Delta\delta$, the change in the rotor angle δ. The error signals Δf and ΔP_{tie} are amplified, mixed and transformed to a real power signal, which controls the valve position to generate a command signal ΔP_v. ΔP_v is sent to the prime mover to initiate change in its torque. The prime mover changes the generator output by ΔP_G, so as to bring Δf and ΔP_{tie} within acceptable limits. The next step in the analysis is to build the mathematical model for the ALFC.

13.2.1 Generator Model

We can apply the swing equation to a small perturbation (refer chapter 9) to obtain the linearized equation.

$$\frac{2H}{\omega_s}\frac{d^2\Delta\delta}{dt^2} = \Delta P_m - \Delta P_e \qquad ...(13.1)$$

Expressing speed deviation in pu, (13.1) can be written as

$$\frac{d\Delta\omega}{dt} = \frac{1}{2H}(\Delta P_m - \Delta P_e) \qquad ...(13.2)$$

where $\Delta\omega = \dfrac{d\Delta\delta}{dt}$

Taking the Laplace transform of (13.2), we get

$$\Delta\omega(s) = \frac{1}{2Hs}(\Delta P_m(s) - \Delta P_e(s)) \qquad ...(13.3)$$

The above equation can be represented as a block diagram shown in Fig. 13.2.

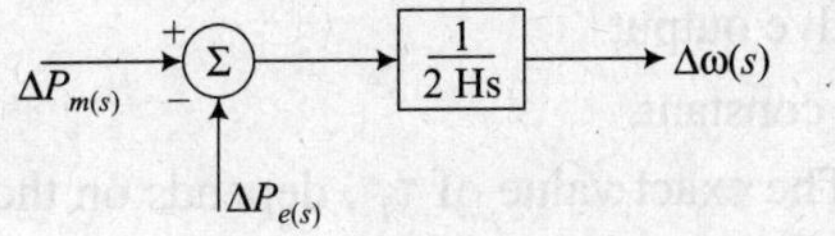

Fig.13.2 Generator model.

13.2.2 Load Model

The details of load modeling are covered in chapter 12. In general, the loads are composite. Resistive loads such as lighting and heating loads are independent of frequency. However, in case of electric motors, the power is dependent on frequency. We can arrive at a composite frequency dependent load characteristic given by

$$\Delta P_e = \Delta P_L + D\Delta\omega \qquad ...(13.4)$$

where, ΔP_L = Non-frequency sensitive load change

D = Load damping constant

$D\Delta\omega$ = Frequency sensitive load change

The damping constant is expressed as per cent change in load for one per cent change in frequency. A value of $D = 1.2$ means a change in frequency by 1% causes the load to change by 1.2%. The block diagram of Fig. 13.2 is modified as shown in Fig. 13.3 to include the frequency dependence of the load. Fig. 13.3 can be reduced to Fig. 13.4.

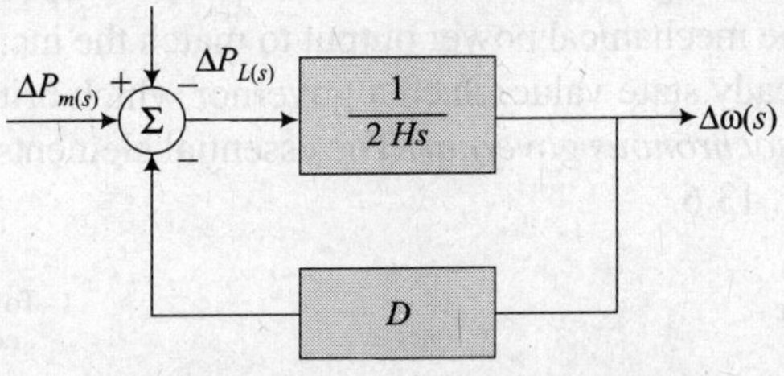

Fig. 13.3 Generator and load model.

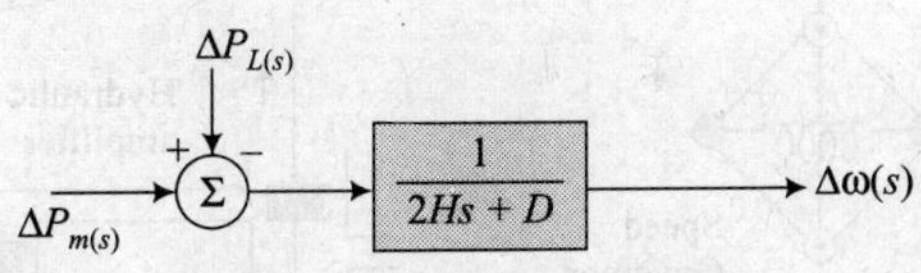

Fig. 13.4 Reduced model.

13.2.3 Turbine Model

The prime mover for the generator is the turbine, which is mostly a steam turbine or a hydro-turbine. The detailed models are presented in chapter 12. In the simplified model, the turbine can be represented by a first order, single time constant transfer function given by

$$G_T(s) = \frac{\Delta P_m(s)}{\Delta P_V(s)} = \frac{1}{1+\tau_T s} \qquad ...(13.5)$$

where, ΔP_V = Change in valve output.

τ_T = Turbine time constant.

τ_T varies from 0.2 – 2 secs. The exact value of τ_T, depends on the type of turbine. (13.5) can be represented in block diagram form as shown in Fig. 13.5.

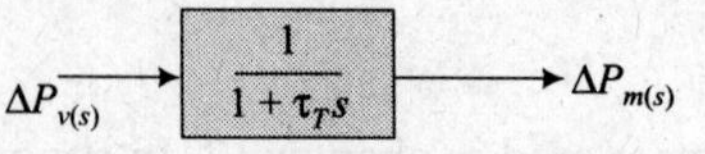

Fig. 13.5 Turbine model.

13.2.4 Governor Model

During a transient, the shaft of the turbine-generator unit accelerates (or decelerates) and the kinetic energy stored in the rotating inertia changes. We have,

$$P_m = P_e + \frac{d}{dt}(W_{\text{kinetic}})$$

If the electrical load on the generator suddenly increases with mechanical power being constant the output electrical power exceeds the input mechanical power. The difference is supplied by the kinetic energy stored in the system. The reduction in the kinetic energy causes the turbine speed and frequency to fall. The turbine governor reacts to this change in speed, and adjusts the turbine input valve/gate to change the mechanical power output to match the increased power demand and bring the frequency to its steady state value. Such a governor which brings back the frequency to its nominal value is called *isochronous governor*. The essential elements of a conventional governor system are shown in Fig. 13.6.

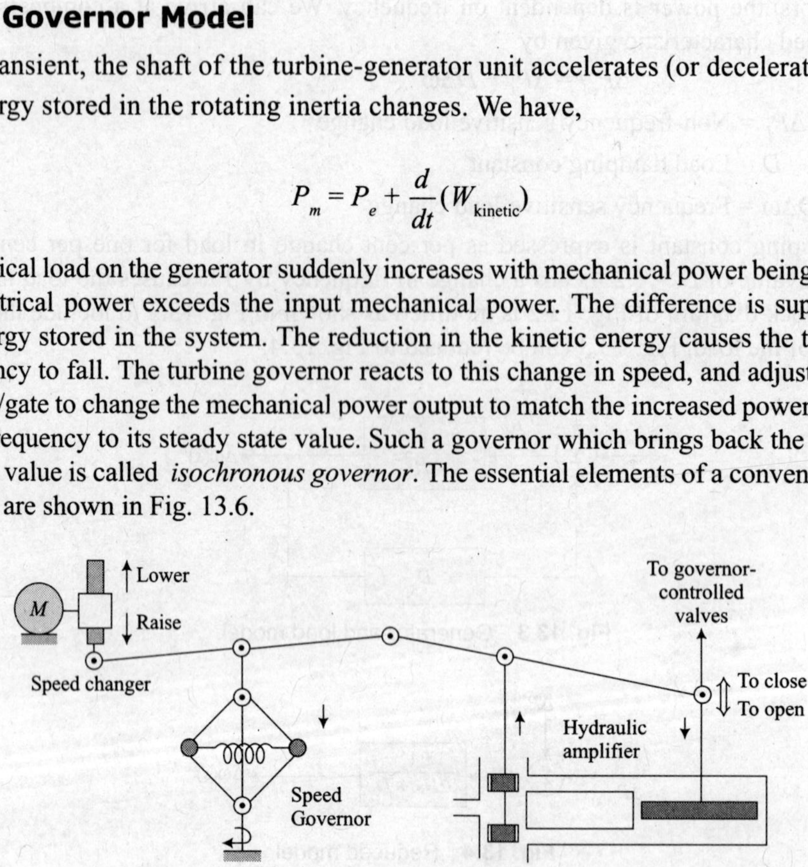

Fig. 13.6 Conventional governor.

The major parts are:

(i) **Speed Governor:** This consists of centrifugal flyballs driven directly or through gears by the turbine shaft, to provide upward and downward vertical movements proportional to the change in speed.

(ii) **Linkage mechanism:** This transforms the flyball movement to the turbine valve through a hydraulic amplifier and provides a feed back from turbine valve movement.

(iii) **Hydraulic amplifiers:** These transform the governor movements into high power forces via several stages of hydraulic amplifiers to build mechanical forces large enough to operate the steam valves or water gates.

(iv) **Speed changer:** This consists of a servomotor which is used to schedule the load at nominal frequency. By adjusting its set point, a desired load dispatch can be scheduled.

An isochronous governor works satisfactorily only when a generator is supplying an isolated load, or when only one generator is required to respond to change in load in a multigenerator system. For proper power sharing between a number of generators connected to the system, the governors are designed to permit the speed to drop as the load is increased. This provides the speed-output characteristic a droop as shown in Fig. 13.7. The speed regulation R is given by the slope of the speed-output characteristic.

$$R = \frac{\Delta\omega}{\Delta P} \quad \text{...(13.6)}$$

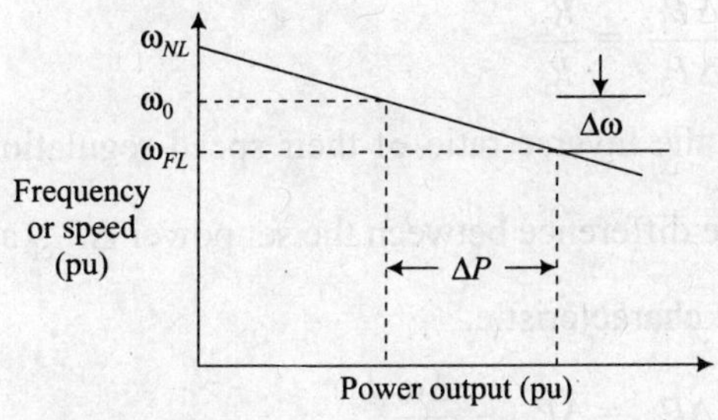

Fig. 13.7 Governor speed characteristic.

Governor % speed regulation is defined as

$$\%R = \left(\frac{\omega_{NL} - \omega_{FL}}{\omega_o}\right) \times 100 \quad \text{...(13.7)}$$

where, ω_{NL} = No-load speed

ω_{FL} = Full load speed

ω_O = Nominal speed

Note that in pu, $\Delta\omega = \Delta f$

To illustrate how load is shared between two generators, consider two generators with droop characteristics as shown in Fig. 13.8.

Let the initial frequency be f_0 and the outputs of the two generators be P_{10} and P_{20}, respectively. If now the load increases by an amount ΔP_L, the units slow down and the governors increase the output until a common operating frequency f_1 is reached. The amount of load picked up by each generator to meet the increased demand ΔP_L depends on the value of the regulation.

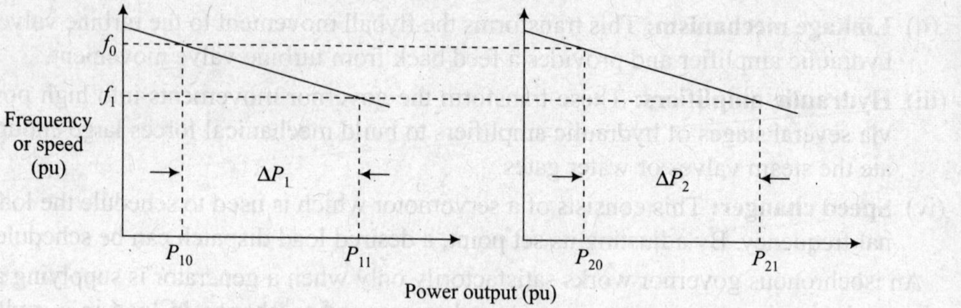

Fig. 13.8 Load sharing.

$$\Delta P_1 = \frac{\Delta f}{R_1} \quad \text{...(13.8)}$$

$$\Delta P_2 = \frac{\Delta f}{R_2} \quad \text{...(13.9)}$$

$$\frac{\Delta P_1}{\Delta P_2} = \frac{R_2}{R_1} \quad \text{...(13.10)}$$

The output is shared in the inverse ratio of their speed regulation. The output of the speed governor is ΔP_g, which is the difference between the set power ΔP_{ref} and the power $\frac{\Delta\omega}{R}$ which is given by the governor speed characteristic.

$$\Delta P_g = \Delta P_{\text{ref}} - \frac{\Delta\omega}{R} \quad \text{...(13.11)}$$

$$\Delta P_g(s) = \Delta P_{\text{ref}}(s) - \frac{\Delta\omega(s)}{R} \quad \text{...(13.12)}$$

The hydraulic amplifier transforms the command into valve/gate position ΔP_V. Assuming a time constant τ_g for the governor,

$$\Delta P_V(s) = \frac{1}{1+\tau_g s}\Delta P_g(s) \quad \text{...(13.13)}$$

(13.13) can be represented by the block diagram of Fig. 13.9.

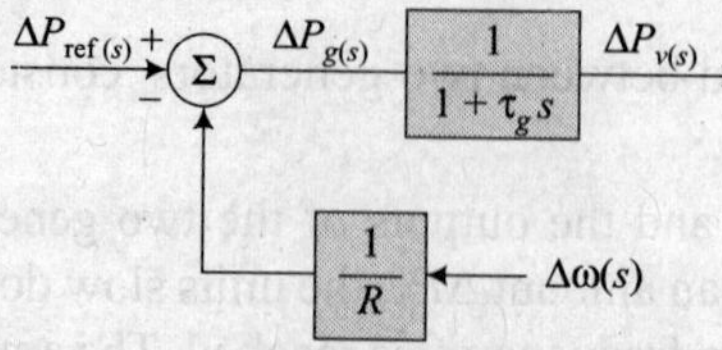

Fig. 13.9 Governor model.

13.2.5 Complete ALFC Block Diagram

The block diagrams of Fig. 13.4, Fig. 13.5 and Fig. 13.9 can be combined as shown in Fig. 13.10.

Figure 13.10 shows the complete load frequency control for an isolated generator supplying a load. Since, we are interested in the change in speed for change in load, we can obtain the transfer function $\dfrac{\Delta\omega(s)}{-\Delta P_L(s)}$ from Fig. 13.10, by reducing it as shown in Fig. 13.11.

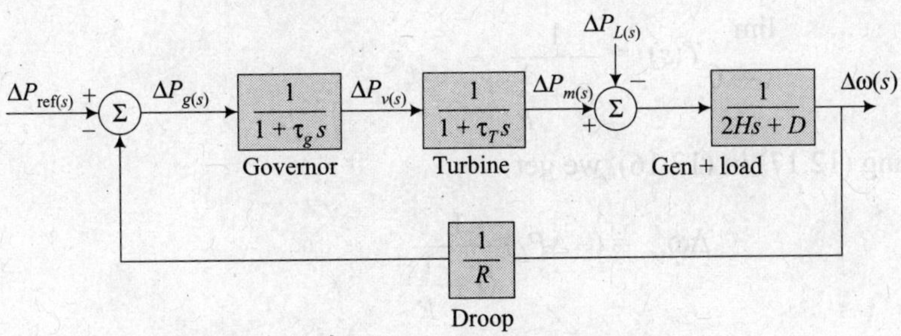

Fig. 13.10 Block diagram of complete governor system.

The closed loop transfer function is obtained from Fig. 13.11.

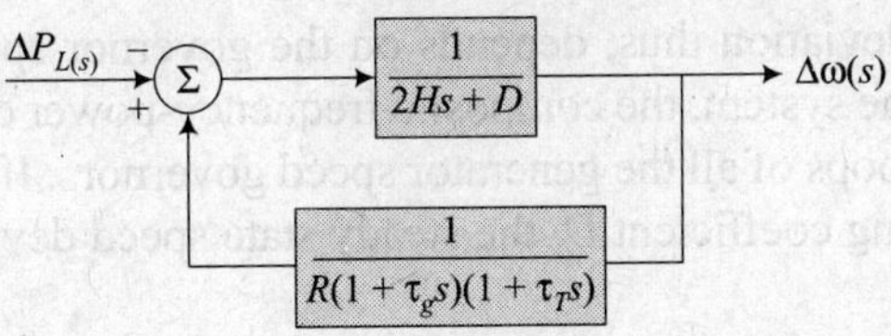

Fig. 13.11 Reduced block diagram.

$$\frac{\Delta\omega(s)}{-\Delta P_L(s)} = \frac{\left(\dfrac{1}{2Hs+D}\right)}{1+\left(\dfrac{1}{2Hs+D}\right)\left(\dfrac{1}{R(1+\tau_g s)(1+\tau_T s)}\right)}$$

$$= \frac{(1+\tau_g s)(1+\tau_T s)}{(2Hs+D)(1+\tau_g s)(1+\tau_T s)+\dfrac{1}{R}} \qquad \text{...(13.14)}$$

From (13.14), we can write

$$\Delta\omega(s) = -\Delta P_L(s)\; T(s) \qquad \text{...(13.15)}$$

If we consider a step change ΔP_L in the load,

$$\Delta P_L(s) = \frac{\Delta P_L}{s}$$

The steady state frequency deviation $\Delta\omega_{ss}$ is given by the limit of $\Delta\omega$ as $t \to \infty$ This can be obtained from (13.15) by application of the final value thereon

$$\Delta\omega_{ss} = \lim_{t\to\infty} \Delta\omega = \lim_{s\to 0} s\Delta\omega(s)$$

$$= \lim_{s\to 0} s\left(\frac{-\Delta P_L}{s} T(s)\right) \qquad \text{...(13.16)}$$

Now,
$$\lim_{s\to 0} T(s) = \frac{1}{D + \frac{1}{R}} \qquad \text{...(13.17)}$$

Substituting (12.17) in (12.16), we get

$$\Delta\omega_{ss} = (-\Delta P_L)\frac{1}{D + \frac{1}{R}} \qquad \text{...(13.18)}$$

If there are no frequency sensitive loads, $D = 0$; in which case

$$\Delta\omega_{ss} = \frac{-\Delta P_L}{\frac{1}{R}} \qquad \text{...(13.19)}$$

The steady state speed deviation thus, depends on the governor speed regulation. If several generators are connected to the system, the composite frequency-power characteristics depends on the combined effect of the droops of all the generator speed governors. If we consider n generators with a composite load damping coefficient D, the steady state speed deviation after a load change ΔP_L is given by

$$\Delta\omega_{ss} = \frac{-\Delta P_L}{D + \left(\frac{1}{R_1} + \frac{1}{R_2} + \ldots \frac{1}{R_n}\right)} = \frac{-\Delta P_L}{D + \frac{1}{R_{eq}}} \qquad \text{...(13.20)}$$

where,
$$\frac{1}{R_{eq}} = \frac{1}{R_1} + \frac{1}{R_2} + \ldots \frac{1}{R_n} \qquad \text{...(13.21)}$$

The *stiffness* of the system, β is given by

$$\beta = \frac{-\Delta P_L}{\Delta\omega_{ss}} = \frac{1}{R_{eq}} + D \text{ MW/Hz} \qquad \text{...(13.22)}$$

β is also called the *frequency bias factor* and is indicative of the change in frequency which would occur for a change in the load.

(* The pu speed deviation $\Delta\omega$ is same as pu frequency deviation Δf)

Example 13.1: A system consists of 4 identical 250 MVA generators feeding a load of 510 MW. The inertia constant H of each unit is 2.5 on the machine base. The total load varies by 1.4% for a

1% change in frequency. If there is a drop in load of 10 MW, determine the system block diagram expressing H and D on a base of 1000 MVA. Give expression for the speed deviation, assuming there is no speed governor.

Solution: H for 4 units on 1000 MVA base $= 4 \times 2.5 \times \dfrac{250}{1000} = 2.5$

Load after drop of 10 MW = 510 – 10 = 500 MW

D for load on base of 1000 MVA is given by

$$D = 1.4 \times \frac{500}{1000} = 0.7\%$$

[Note that a change of load of 1.4% on base 500 MW corresponds to 0.7% on base of 1000 MVA]

Block diagram is shown in Fig.13.12a and the block diagram in the standard first order form with transfer function given by $\dfrac{K}{1+sT}$, is shown in Fig. 13.12b.

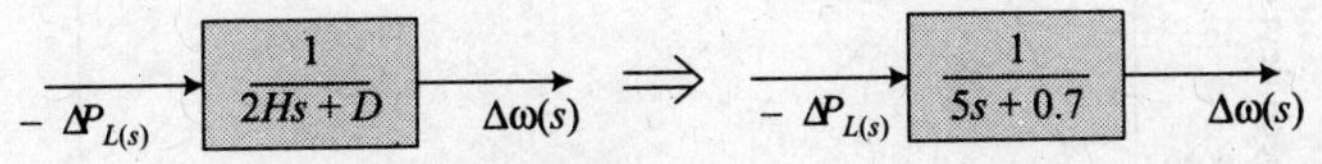

Fig. 13.12a Block diagram of example 13.1.

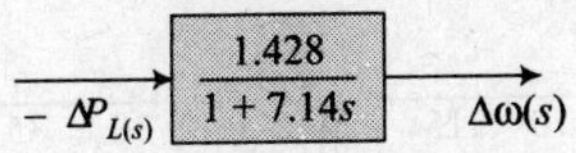

Fig. 13.12b First order transfer function.

The gain = 1.428 and time constant = 7.14 sec

$$\Delta P_L = -10 \text{ MW} = \frac{-10}{1000} = -0.01 \text{ pu.}$$

$$\Delta P_L(s) = \frac{-0.01}{s} \text{ pu}$$

From block diagram

$$\Delta\omega(s) = -\left(\frac{-0.01}{s}\right)\left(\frac{1.428}{1+7.14s}\right)$$

$$= \left(\frac{0.01}{s}\right)\left(\frac{1.428}{1+7.14s}\right) = \frac{0.01428}{s} - \frac{0.10196}{1+7.14s}$$

$$= \frac{0.01428}{s} - \frac{0.01428}{s+0.14}$$

Taking inverse Laplace transform

$$\Delta\omega(t) = 0.01428\ (1 - e^{-0.14t})$$

The pu speed deviation as function of time is shown in Fig. 13.13.

The steady state speed deviation is 0.01428 pu. If frequency is 50 Hz, steady state frequency deviation = 50 × 0.01428 = 0.714 Hz. The frequency deviation is positive since, a decrease in load leads to increase in frequency. The steady state frequency is 50.714 Hz.

Example 13.2: A power system has a load of 1250 MW at 50 Hz. $D = 1.5$. If 50 MW is tripped, find the steady state frequency deviation when (a) there is no speed control (b) the system has a

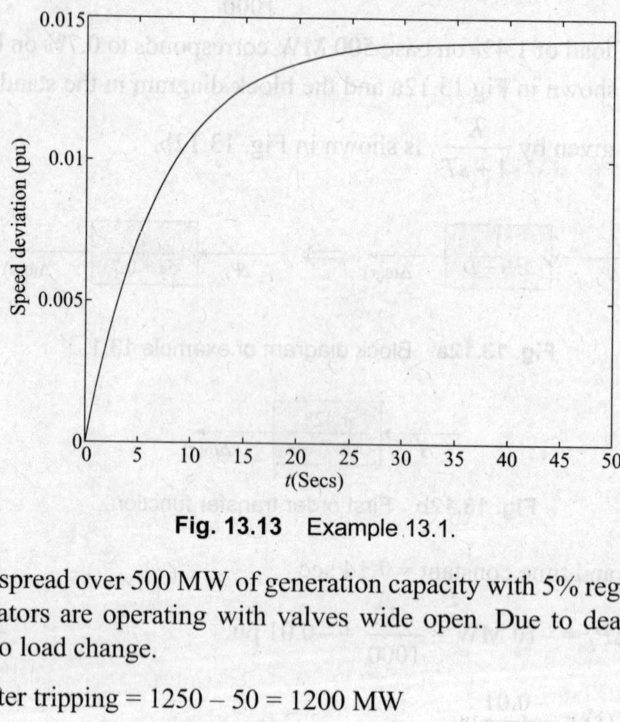

Fig. 13.13 Example 13.1.

reserve of 200 MW spread over 500 MW of generation capacity with 5% regulation on this capacity. All other generators are operating with valves wide open. Due to dead band only 80% of governors respond to load change.

Solution: Load after tripping = 1250 – 50 = 1200 MW

$D = 1.5\%$; In terms of MW/Hz

$$D = \frac{\left(\dfrac{1.5}{100} \times 1200\right)}{\left(50 \times \dfrac{1}{100}\right)} = 36 \text{ MW/Hz}$$

(a) With no speed control steady state frequency deviation

$$\Delta f = \frac{-\Delta P_L}{D} = \frac{-(-50)}{36} = 1.39 \text{ Hz}$$

Thus, frequency tends to increase.

(b) Since, the frequency increases, all generators respond.

Total spinning generator capacity is equal to load + reserve = 1250 + 200 = 1450 MW.

80% of this contribute for regulation = 0.8 × 1450 = 1160 MW

5% regulation means a change of 5% in frequency causes a 100% change in power generation (refer eq. 13.6). Therefore,

$$\frac{1}{R} = \frac{1160}{\left(\frac{5}{100}\right) \times 50} = 464 \text{ MW/Hz}$$

Steady state frequency deviation

$$\Delta f = \frac{-\Delta P_L}{\frac{1}{R} + D} = \frac{-(-50)}{464 + 36} = 0.1 \text{ Hz.}$$

Thus, it can be seen that the frequency deviation reduces if governor control is present.

Example 13.3: An isolated generator and its control have following parameters:

Generator inertia constant = 5 sec

Governor time constant $\tau_g = 0.25$ sec

Turbine time constant $\tau_T = 0.6$ sec

Governor speed regulation = 0.05 pu

$D = 0.8$

The turbine rated output is 200 MW at 50 Hz. The load suddenly increases by 50 MW. Find the steady state frequency deviation. Plot the frequency deviation as a function of time.

Solution: From (13.15) the transfer function is given by

$$\frac{\Delta\omega(s)}{-\Delta P_L(s)} = T(s) = \frac{(1 + 0.25\,s)(1 + 0.6s)}{(10\,s + 0.8)(1 + 0.25\,s)\,(1 + 0.6\,s) + \frac{1}{0.05}}$$

$$= \frac{0.15s^2 + 0.85\,s + 1}{1.5\,s^3 + 8.62\,s^2 + 10.68\,s + 20.8}$$

$$\Delta P_L = \frac{50}{200} = 0.25 \text{ pu}$$

From (12.18)

$$\Delta\omega_{ss} = \frac{-\Delta P_L}{D + \frac{1}{R}} = \frac{-0.25}{0.8 + \frac{1}{0.05}} = -0.01202 \text{ pu.}$$

Steady state frequency deviation

$$\Delta f_{ss} = -0.01202 \times 50 = -0.601 \text{ Hz.}$$

The frequency decreases since the load has increased. The time response to a step deviation of 0.25 is shown Fig. 13.14. The plot has been obtained from MATLAB as follows.

```
Num = [0.15  0.85  1]
Den  = [1.5 8.62 10.68 20.8]
Sys   = TF (Num, Den)
Step (sys * – 0.25)
```

The command sys = TF (Num Den) creates a system with the transfer function defined.The step command in MATLAB directly plots step response.

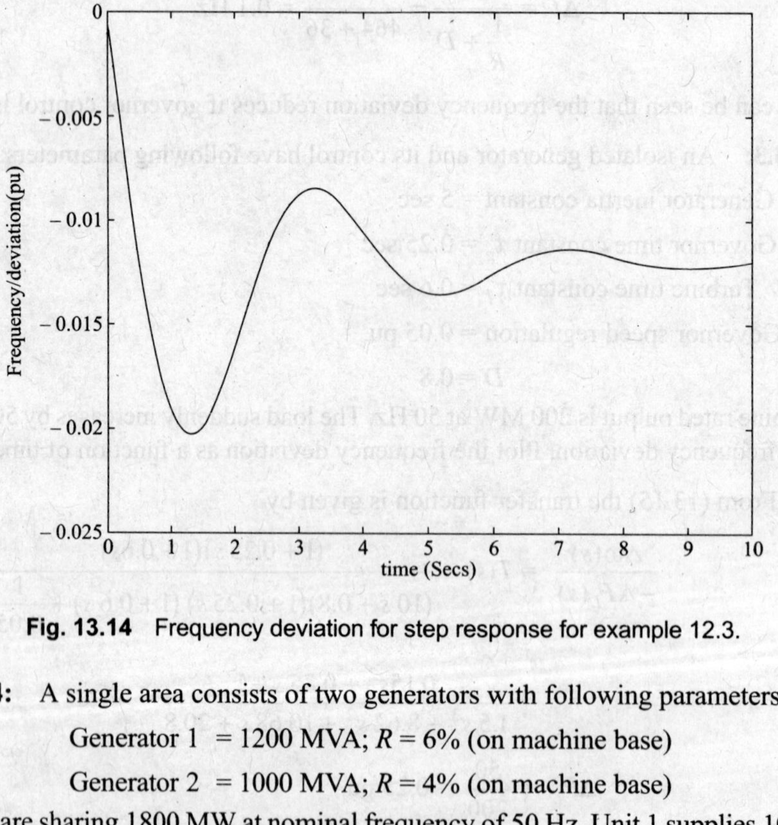

Fig. 13.14 Frequency deviation for step response for example 12.3.

Example 13.4: A single area consists of two generators with following parameters

Generator 1 = 1200 MVA; R = 6% (on machine base)

Generator 2 = 1000 MVA; R = 4% (on machine base)

The units are sharing 1800 MW at nominal frequency of 50 Hz. Unit 1 supplies 1000 MW and unit 2 supplies 800 MW. The load is now increased by 200 MW.

(a) Find steady state frequency and generation of each unit if $D = 0$.

(b) Repeat (a) if $D = 1.5$.

Solution: Choose a common base of 2000 MVA. The speed regulation is now expressed on the common MVA.

$$R_1 = \frac{2000}{1200} \times 0.06 = 0.1 \text{ pu}$$

$$R_2 = \frac{2000}{1000} \times 0.04 = 0.08 \text{ pu}$$

$$\Delta P_L = \frac{200}{2000} = 0.1 \text{ pu}$$

(a) If $D = 0$, from (13.19)

$$\Delta\omega_{ss} = \frac{-\Delta P_L}{\dfrac{1}{R_1} + \dfrac{1}{R_2}} = \frac{-0.1}{\dfrac{1}{0.1} + \dfrac{1}{0.08}} = -0.0044 \text{ pu}$$

Steady state frequency deviation

$$\Delta f = -0.0044 \times 50 = -0.22 \text{ Hz.}$$

New steady state frequency after load increases

$$f_1 = f_0 + \Delta f = 50 - 0.22 = 49.78 \text{ Hz}$$

The change in generations are given by

$$\Delta P_1 = \frac{-\Delta f}{R_1} = \frac{0.0044}{0.1} = 0.044 \text{ pu} = 88.88 \text{ MW}$$

$$\Delta P_2 = \frac{-\Delta f}{R_2} = \frac{0.0044}{0.08} = 0.055 \text{ pu} = 111.11 \text{ MW}$$

Thus, unit 1 supplies (1000 + 88.88) 1088.88 MW and unit 2 supplies (800 + 111.11) 911.11 MW at a frequency of 49.78 Hz.

(b) For $D = 1.5$, we use (12.20)

$$\Delta\omega_{ss} = \frac{-0.1}{\dfrac{1}{0.1} + \dfrac{1}{0.08} + 1.5} = -4.166 \times 10^{-3} \text{ pu}$$

Steady state frequency deviation Δf is given by

$$\Delta f = 4.166 \times 10^{-3} \times 50 = 0.2083 \text{ Hz}$$

The new operating frequency is

$$50 - 0.2083 = 49.79 \text{ Hz.}$$

$$\Delta P_1 = -\left(\frac{-4.166 \times 10^{-3}}{0.1}\right) \times 2000 = 83.33 \text{ MW}$$

$$\Delta P_2 = -\left(\frac{-4.166 \times 10^{-3}}{0.08}\right) \times 2000 = 104.16 \text{ MW}$$

Unit 1 supplies, 1000 + 83.33 = 1083.33 MW

Unit 2 supplies, 800 + 104.16 = 904.16 MW

Total generation = 1083.33 + 904.16 = 1987.49 MW, which is less than the load of 2000 MW. This is because the load decreases due to frequency drop by an amount.

$$D\Delta\omega = 1.5 \times (-4.166 \times 10^{-3}) \times 2000 = -12.5 \text{ MW}.$$

[This is exactly difference between generation and load]

Example 13.5: Determine the primary ALFC loop parameters for a control area with following data:

Total generation capacity = 2500 MW

Normal operating load = 1500 MW

$H = 5.0$ s; $D = 1\%$; and $f = 50$ Hz

$R = 2.5$ Hz/pu MW

Solution: $D = 1\%$ means the load increases by 1% for an increase in frequency of 1%.

1% load = 15 MW;

1% frequency = 0.5 Hz,

$$D = \frac{15}{0.5} = 30 \text{ MW/Hz} = 30/2500 = 0.012 \text{ pu MW/Hz}$$

Since, the load is a constant, the transfer function is given by

$$T(s) = \frac{1}{2Hs + D} = \frac{\dfrac{1}{D}}{\dfrac{2Hs}{D} + 1} = \frac{K}{sT + 1}$$

The gain $$K = \frac{1}{D} = \frac{1}{0.012} = 83.33 \text{ Hz/pu MW}$$

Time constant $$T = \frac{2H}{D} = \frac{2 \times 5}{0.012} = 833.3 \text{ pu}$$

$$= \frac{833.3}{50} = 16.66 \text{ sec}$$

Example 13.6: In example 13.5 if there is a 1.5% increase in load find frequency drop

(a) without governor control

(b) with governor control

Solution: (a) without governor control $\Delta f = \dfrac{-\Delta P_L}{D}$

$$\Delta P_L = 1.5\% = 0.015 \text{ pu}$$

$$\Delta f = \frac{-0.015}{0.012} = -1.25 \text{ Hz}$$

(b) with governor control $\Delta f = \dfrac{-\Delta P_L}{D + \dfrac{1}{R}} = \dfrac{-0.015}{0.012 + \dfrac{1}{2.5}} = -0.0364$ Hz

Example 13.7: A 500 MVA synchronous generator operates on full load at a frequency of 50 Hz. The load is reduced to 400 MW. The steam valve begins to operate with a time lag of 0.5 seconds. Determine the change in frequency if $H = 5$ kW – sec/kVA.

Solution: Kinetic energy stored in rotating mass (generator and turbine)

$$= GH = 5 \times 500 \times 1000 = 25 \times 10^5 \text{ kW-sec.}$$

If load drops to 400 MW, excess mechanical input before steam valve operates

$$= 100 \text{ MW.}$$

Excess energy input in 0.5 seconds to the rotating parts

$$= 100 \times 1000 \times 0.5 = 0.5 \times 10^5 = \text{kW-sec}$$

Since, kinetic energy is proportional to ω^2 or (frequency)^2, frequency is proportional to $\sqrt{KE}$

∴ frequency at the end of 0.5 sec

$$= 50 \times \left(\frac{25 \times 10^5 + 0.5 \times 10^5}{25 \times 10^5} \right)^{\frac{1}{2}} = 50.5 \text{ Hz.}$$

Example 13.8: Two generators rated 250 MW and 500 MW are operating in parallel. The droop characteristics of their governors are 4% and 5% from no load to full load. If the nominal frequency is 50 Hz at no load, how would a load of 750 MW be shared between them? What is system frequency?

Solution: Since, the two generators are in parallel they operate at same frequency under steady state.

Unit 1. Let load be x MW. The generator output changes from zero at no load to x MW.

$$R_1 = \frac{\Delta f}{\Delta P_1} = \frac{\Delta f}{x}$$

Given $\quad R_1 = 4\% = \dfrac{0.04 \times 50}{250} = 0.008$ Hz/MW

Unit 2. Load is $750 - x$

$$R_2 = \frac{\Delta f}{\Delta P_2} = \frac{\Delta f}{750 - x}$$

Given $$R_2 = 5\% = \frac{0.05 \times 50}{500} = 0.005 \text{ Hz/MW}$$

We can equate Δf obtained from the two equations

$$x \times 0.008 = (750 - x) \times 0.005$$

$$x = 288.46 \text{ MW (load on unit 1)}$$

$$750 - x = 461.54 \text{ MW (load on unit 2)}$$

$$\Delta f = 288.46 \times 0.008 = 2.3 \text{ Hz}$$

$$f = 50 - 2.3 = 47.7 \text{ Hz}$$

Due to the difference in the droop characteristics of the governors, unit 1 gets over loaded and unit 2 is under loaded. If the droops are identical, they will share the load corresponding to their ratings.

13.2.6 Actual Speed-Droop Characteristic

The linear speed-droop characteristic represents an ideal condition. In practical systems the speed droop characteristic is as shown in Fig. 13.15.

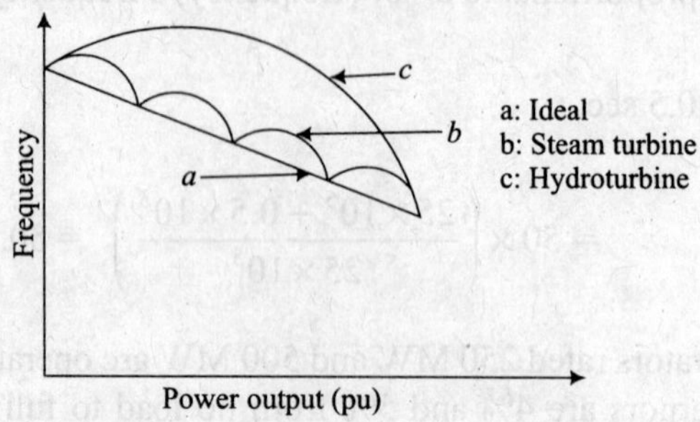

Fig. 13.15 Speed-droop characteristic.

Steam turbines have a number of control valves with non-linear flow area versus position characteristic. Hence, they have a characteristic as shown in curve *b*. Each section represents one control valve. Thus, the regulation may vary incrementally from 2% to 12% depending on the unit output. Hydro-turbines, which have a single gate have a characteristic as shown in curve *c*.

13.3 CONCEPT OF AUTOMATIC GENERATION CONTROL (AGC)

With the primary ALFC, it was seen that a change in the system load results in a steady state frequency deviation, depending on the regulation and frequency sensitivity (as indicated by *D*) of the load. All the connected generator units of the system contribute to the overall change in generation, irrespective of the location of the load change. Thus, restoration to nominal system frequency requires additional control action which changes the load reference set point to match the variations in the system load. This control scheme is called the Automatic Generation Control (AGC).

The main objectives of the AGC are to regulate the system frequency and maintain the scheduled power inter changes, between the interconnected areas, via the tie-lines. A secondary objective of the AGC is to distribute the required change in generation among the various units to obtain least operating costs. During large transient disturbances and emergencies, AGC is bypassed and other emergency controls act.

13.3.1 AGC in a Single Area

In a single area system, since there is no tie-line schedule to be maintained, the function of AGC is only to bring the frequency to the nominal value. This is achieved by introducing an integral controller to change the load reference setting so as to change the speed set point. The integral controller forces the steady state speed deviation to zero. The ALFC with the integral controller is shown in Fig. 13.16.

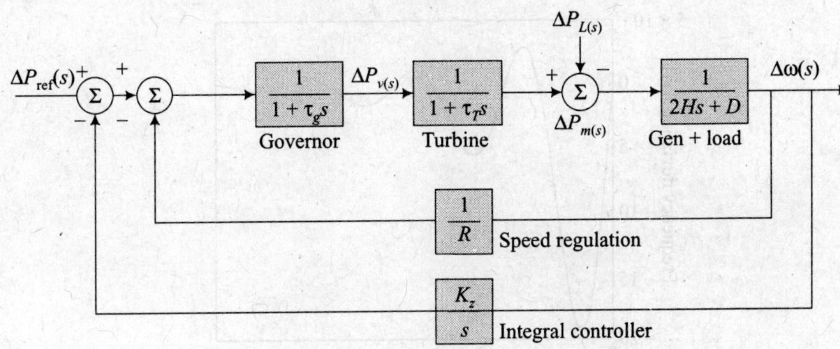

Fig. 13.16 AGC of a single area system.

The gain K_I of the integral controller needs to be adjusted for satisfactory response in terms of over shoot, setting time, etc. The equivalent block diagram is shown in Fig. 13.17.

The closed loop transfer function for Fig. 13.17 is given by

$$T(s) = \frac{\Delta\omega(s)}{-\Delta P_L(s)} = \frac{s(1+\tau_g s)(1+\tau_T s)}{s(2Hs+D)(1+\tau_g s)(1+\tau_T s)+K_I+\dfrac{s}{R}} \quad \text{...(13.23)}$$

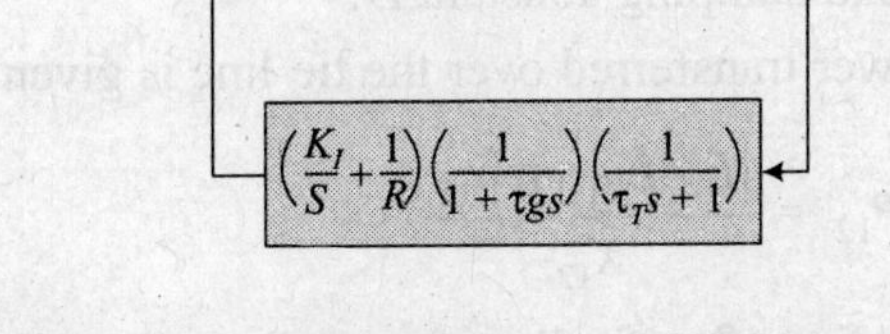

Fig. 13.17 Reduced block diagram of single area AGC.

Example 13.9: In the example of 13.3, an integral controller with gain $K_I = 6$ is added as in Fig. 13.16. Obtain the dynamic response, if all other conditions are same.

Solution: From (13.23) the transfer function

$$\frac{\Delta\omega(s)}{-\Delta P_L(s)} = \frac{0.15\,s^3 + 0.85\,s^2 + s}{1.5\,s^4 + 8.62\,s^3 + 10.68\,s^2 + 20.8\,s + 6}$$

The change in P_L, $\Delta P_L = 0.25$ pu.

The response of the above function for a step change of 0.25 pu is plotted using MATLAB. The response is shown if Fig. 13.18.

Comparing with Fig. 13.14, it can be seen that the steady state frequency deviation is now zero. However, the overshoot and settling time are more.

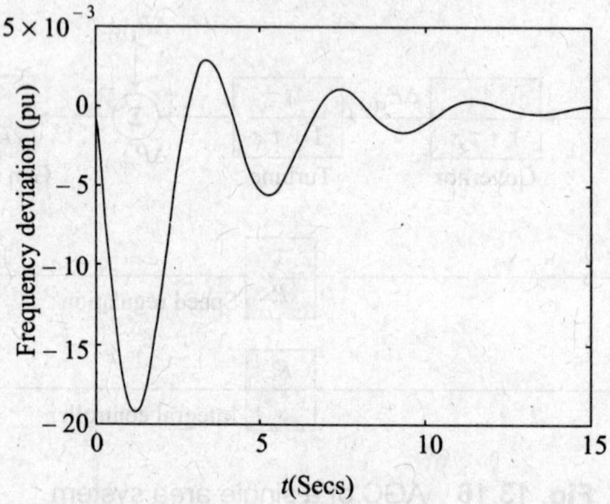

Fig. 13.18 Dynamic response of example 13.9.

13.3.2 AGC in Multi Area Systems

In interconnected systems, a group of generators are closely coupled internally and swing in phase. Such a group is called coherent group. The ALFC loop can be represented for the whole area, referred to as control area.

Consider two areas interconnected by lossless tie-line of reactance X_{tie}, with a power flow P_{12} from area 1 to area 2 as shown in Fig. 13.19a. Let the generators be represented by a single equivalent generator for area 1 and area 2. The generators are modeled simply as constant voltage sources behind reactance as shown in Fig. 13.19b. We first consider only primary ALFC loop as shown in Fig. 13.19c. H is the equivalent inertia constant of each area. The turbines are represented by the effective speed droop R and load damping constant D.

Under steady state the power transferred over the tie-line is given by

$$P_{12} = \frac{|E_1||E_2|\sin\delta_{12}}{X_{12}} \qquad \text{...(13.24)}$$

where, $X_{12} = X_1 + X_{tie} + X_2$ and $\delta_{12} = \delta_1 - \delta_2$. For a small deviation ΔP_{12} of the tie-line power flow, we can write

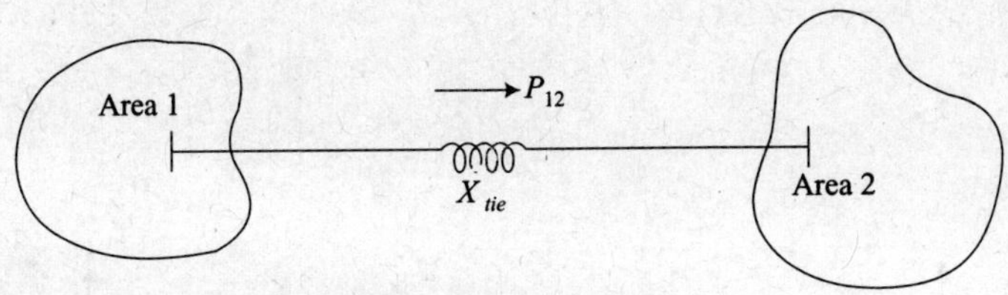

Fig. 13.19a Two area system.

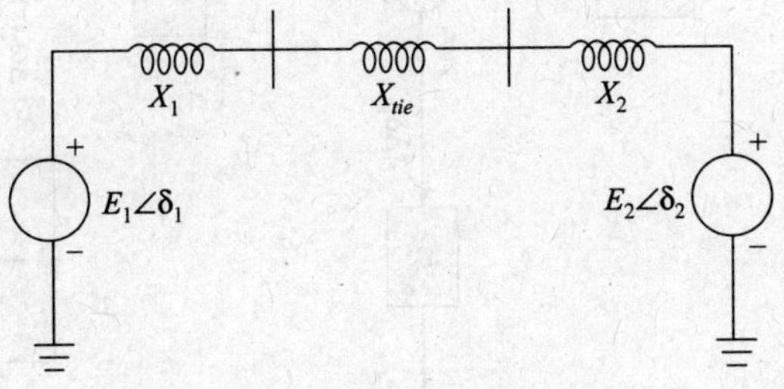

Fig. 13.19b Electrical equivalent.

$$\Delta P_{12} = \left.\frac{\partial P_{12}}{\partial \delta_{12}}\right|_{\delta_{120}} \Delta\delta_{12} = P_S\,\Delta\delta_{12} \qquad \text{...(13.25)}$$

where, $\left.\frac{\partial P_{12}}{\partial \delta_{12}}\right|_{\delta 120} = P_S$, is the slope of the power angle curve evaluated at the initial operating point ($\delta_{120} = \delta_{10} - \delta_{20}$) and is the synchronizing power coefficient as defined in chapter 9.

$$P_S = \left.\frac{\partial P_{12}}{\partial \delta_{12}}\right|_{\delta_{120}} = \frac{|E_1||E_2|}{X_{12}} \cos \delta_{120} \qquad \text{...(13.26)}$$

(12.25) can be written as

$$\Delta P_{12} = P_S(\Delta\delta_1 - \Delta\delta_2) \qquad \text{...(13.27)}$$

A positive ΔP_{12} occurs when $\Delta\delta_1 > \Delta\delta_2$ and indicates a flow of real power from area 1 to area 2. This has the effect of increasing load on area 1 and decreasing load on area 2. Hence, ΔP_{12} has negative sign for area 1 and positive sign for area 2 in Fig. 13.19c.

Consider a change in load ΔP_{L1} in area 1. The steady state frequency deviation Δf is same for both the areas. Hence,

$$\Delta f = \Delta f_1 = \Delta f_2.$$

For area 1,

$$\Delta P_{m1} - \Delta P_{12} - \Delta P_{L1} = \Delta f D_1 \qquad \text{...(13.28)}$$

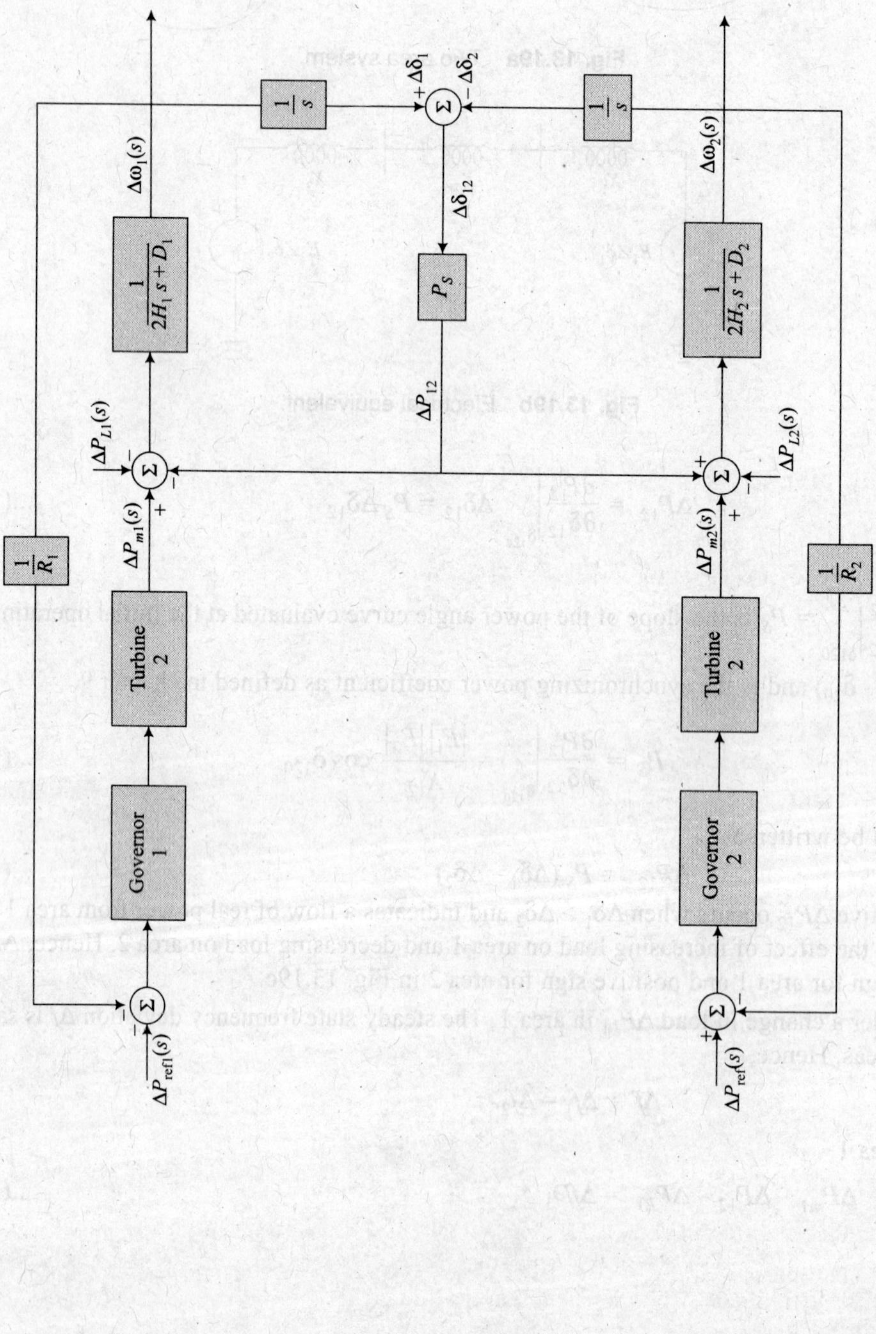

Fig. 13.19c Two area system with primary ALFC.

For area 2,

$$\Delta P_{m2} + \Delta P_{12} = \Delta f D_2 \qquad ...(13.29)$$

The change in mechanical powers depends on the respective regulations.

$$\Delta P_{m1} = \frac{-\Delta f}{R_1} \qquad ...(13.30)$$

$$\Delta P_{m2} = \frac{-\Delta f}{R_2} \qquad ...(13.31)$$

Substituting (13.30) in (13.28), we get

$$\Delta f\left(\frac{1}{R_1} + D_1\right) = -\Delta P_{12} - \Delta P_{L1} \qquad ...(13.32)$$

$$\frac{1}{R_1} + D_1 = \beta_1, \text{ the frequency bias factor for area 1.}$$

Hence, (13.32) can be written as

$$\Delta f\,\beta_1 = -\Delta P_{12} - \Delta P_{L1} \qquad ...(13.33)$$

Similarly, substituting (13.31) in (13.29), we get

$$\Delta f\,\beta_2 = \Delta P_{12} \qquad ...(13.34)$$

Solving for Δf from (13.33) and (13.34), we get

$$\Delta f = \frac{-\Delta P_{L1}}{\beta_1 + \beta_2} \qquad ...(13.35)$$

From (13.34), $$\Delta P_{12} = \frac{-\Delta P_{L1}\,\beta_2}{\beta_1 + \beta_2} \qquad ...(13.36)$$

Thus, an increase of load in area 1 reduces frequency in both areas. Similarly, for a change in load ΔP_{L2} in area 2,

$$\Delta f = \frac{-\Delta P_{L2}}{\beta_1 + \beta_2} \qquad ...(13.37)$$

and, $$\Delta P_{12} = -\Delta P_{21} = \frac{-\Delta P_{L2}\,\beta_1}{\beta_1 + \beta_2} \qquad ...(13.38)$$

Example 13.10: A two area system connected by a tie-line has following parameters on 1000 MVA base;

$$R_1 = 4.5\%;\ D_1 = 0.6;\ H_1 = 4.5;$$

$$R_2 = 6\%;\ D_2 = 0.85;\ H_2 = 5.0;$$

The units are running in parallel at a frequency of 50 Hz. The synchronizing power coefficient is 1.9 pu at the initial operating angle. A load increase of 150 MW occurs in area 1. Determine the new steady state frequency and the change in tie-line power flow.

Solution: $\Delta P_{L1} = \dfrac{150}{1000} = 0.15$ pu.

Steady state frequency deviation is

$$\Delta f = \frac{-\Delta P_{L1}}{\beta_1 + \beta_2}$$

$$\beta_1 = \frac{1}{R_1} + D_1 = \frac{1}{0.045} + 0.6 = 22.822$$

$$\beta_2 = \frac{1}{R_2} + D_2 = \frac{1}{0.06} + 0.85 = 17.516$$

$$\Delta f = \frac{-0.15}{22.822 + 17.516} = -0.0037 \text{ pu}$$

Steady state frequency = 50 – (0.0037 × 50) = 49.815 Hz.

$$\Delta P_{12} = \Delta f \beta_2 = -0.0037 \times 17.516 = -0.0648 \text{ pu} = -64.8 \text{ MW.}$$

Since, ΔP_{12} is negative, it implies that 64.8 MW flows from area 2 to area 1.

Change in mechanical powers is given by

$$\Delta P_{m1} = \frac{-\Delta f}{R_1} = -\left(\frac{-0.0037}{0.045}\right) = 0.082 \text{ pu} = 82 \text{ MW}$$

$$\Delta P_{m2} = \frac{-\Delta f}{R_2} = -\left(\frac{-0.0037}{0.06}\right) = 0.0617 \text{ pu} = 61.7 \text{ MW}$$

Change in load in area 1 due to frequency sensitive loads is $\Delta f D_1 = (-0.0037)(0.6) = -0.0022$ pu = – 2.2 MW. Similarly, for area 2 $\Delta f D_2 = (-0.0037)(0.85) = -0.0031$ pu = – 3.1 MW. Total change in load is –5.3 MW. The power flow of 64.8 MW from area 2 to area 1 is contributed by an increase in generation of area 2 by 61.7 MW and reduction in load of area 2 by 3.1 MW.

13.3.3 Tie-Line Bias Control

From discussion in section 13.3.2, it can be seen that, if the areas are equipped only with primary control of the ALFC, a change in load in one area met is with change in generation in both areas, change in tie-line power and a change in the frequency. Hence, a supplementary control is necessary to maintain

- Frequency at the nominal value.
- Maintain net interchange power with other areas at the scheduled values.
- Let each area absorb its own load.

Hence, the supplementary control should act only for the areas where there is a change in load. To achieve this, the control signal should be made up of the tie-line flow deviation plus a signal proportional to the frequency deviation. A suitable proportional weight for the frequency deviation is the frequency-response characteristic β. This is the reason why β is also called the frequency bias factor. This control signal is called the area control error (ACE). In a two area system

$$\text{ACE}_1 = \Delta P_{12} + B_1\,\Delta f;\ B_1 = \beta_1 \quad \text{...(13.39)}$$

$$\text{ACE}_2 = \Delta P_{21} + B_2\,\Delta f;\ B_2 = \beta_2 \quad \text{...(13.40)}$$

The ACE represents the required change in area generation and its unit is MW. ACEs are used as control signals to activate changes in the reference set points. Under steady state, ΔP_{12} and Δf will be zero. The block diagram with the supplementary control is shown in Fig. 13.20. It is applied to selected units in each area.

The operation of the ACE can be explained as follows. Consider an increase in load of area 1, which leads to a decrease in system frequency. The primary ALFC loop limits the frequency deviation to

$$\Delta f = \frac{-\Delta P_{L1}}{\beta_1 + \beta_2}$$

The tie-line power has a deviation $\Delta P_{12} = \beta_2 \Delta f$. The slower acting supplementary control, starts responding now.

$$\text{ACE}_1 = \Delta P_{12} + B_1\Delta f = \frac{(\beta_1 + \beta_2)}{(\beta_1 + \beta_2)}(-\Delta P_{L1}) = -\Delta P_{L1} \quad \text{...(13.41)}$$

and,

$$\text{ACE}_2 = -\Delta P_{12} + B_2\Delta f = \frac{(-\Delta P_{L1})}{(\beta_1 + \beta_2)}(-\beta_2 + \beta_2) = 0, \quad \text{...(13.42)}$$

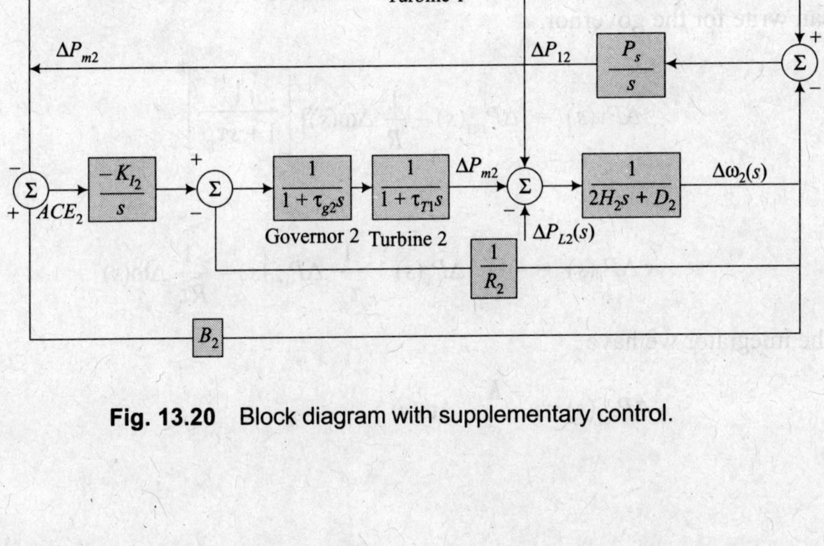

Fig. 13.20 Block diagram with supplementary control.

Thus, only supplementary control of area 1 responds to ΔP_{L1} and the generation changed so that ACE_1 becomes zero.

13.3.4 Performance of AGC Under Abnormal Conditions

Under normal conditions, AGC helps to maintain system frequency at nominal value and maintain the inter area power transfers at scheduled values. Under abnormal conditions, a few areas may not be able to increase their generation due to lack of sufficient reserve. In such cases, other areas assist by allowing the tie-line powers to deviate from scheduled values. The frequency also deviates from the nominal value. The areas participate in frequency regulation, in proportion to their available regulating capacity.

13.3.5 State Space Model

We have the following equations derived.

$$\Delta\omega(s) = \frac{1}{(2Hs + D)}\,[\Delta P_m(s) - \Delta P_L(s)]$$

From which we can write,

$$s\,\Delta\omega(s) = \frac{-D}{2H}\Delta\omega(s) + \frac{1}{2H}\Delta P_m(s)\,\frac{1}{2H}\Delta P_L(s) \qquad \text{(i)}$$

For the turbine we have (refer Fig. 13.16)

$$\Delta P_m(s) = \Delta P_v(s)\left[\frac{1}{1+\tau_T s}\right]$$

or

$$s\,\Delta P_m(s) = -\frac{1}{\tau_T}\Delta P_m(s) + \frac{1}{\tau_T}\Delta P_v(s) \qquad \text{(ii)}$$

We can write for the governor,

$$\Delta Pv(s) = [\Delta P_{\text{ref}}(s) - \frac{1}{R}\Delta\omega(s)]\left[\frac{1}{1+s\tau_g}\right]$$

or

$$s\,\Delta P_v(s) = -\frac{1}{\tau_g}\Delta P_v(s) + \frac{1}{\tau_g}\Delta P_{\text{ref}}(s) - \frac{1}{R\tau_g}\Delta\omega(s) \qquad \text{(iii)}$$

For the integrator we have

$$\Delta P_{\text{ref}}(s) = \frac{-K_I}{s}\Delta\omega(s)$$

or $$s\,\Delta P_{\text{ref}}(s) = -K_I\,\Delta\omega(s) \tag{iv}$$

Equation (i) – (iv) are the state equations of the single area system. The state vector is defined as

$$X = [\Delta\omega\ \Delta P_m\ \Delta P_v\ \Delta P_{\text{ref}}]^T$$

The external input is ΔP_L. Changing (i) to (iv) to time domain state equations, we get

$$\begin{bmatrix} \Delta\dot{\omega} \\ \Delta\dot{P}_m \\ \Delta\dot{P}_v \\ \Delta\dot{P}_{ref} \end{bmatrix} = \begin{bmatrix} \frac{-D}{2H} & \frac{1}{2H} & 0 & 0 \\ 0 & \frac{-1}{\tau_T} & \frac{1}{\tau_T} & 0 \\ \frac{-1}{R\tau_g} & 0 & \frac{-1}{\tau_g} & \frac{1}{\tau_g} \\ -K_I & 0 & 0 & 0 \end{bmatrix} \begin{bmatrix} \Delta\omega \\ \Delta P_m \\ \Delta P_v \\ \Delta P_{ref} \end{bmatrix} + \begin{bmatrix} \frac{-1}{2H} \\ 0 \\ 0 \\ 0 \end{bmatrix} \Delta P_L$$

Which is in standard $\dot{X} = AX + Bu$ form. The output,

$$y = \Delta\omega$$

The state space model is useful in finding the response to step changes in load.

Example 13.11: Two areas are connected via an inter tie. The load at 50 Hz, is 15,000 MW in area 1 and 35,000 MW in area 2. Area 1 is importing 1500 MW from area 2. The load damping constant in each area is $D = 1.0$ and the regulation $R = 6\%$ for all units. Area 1 has a spinning reserve of 800 MW spread over 4000 MW of generation capacity and area 2 has a spinning reserve of 1000 MW spread over 10,000 MW generation. Determine the steady state frequency, generation and load of each area and tie-line power for

(a) Loss of 1000 MW in area 1, with no supplementary control.

(b) Loss of 1000 MW in area 1, with supplementary controls provided on generators with reserve.

$\beta_1 = 250$ MW/0.1 Hz and $\beta_2 = 400$ MW/0.1 Hz

Solution: (a) Assume a lossless system

Area 1: Load = 15,000 MW

Power import from area 2 = 1,500 MW

Generation = 15,000 – 1500 = 13,500 MW

Reserve = 800 MW

Total generation capacity = 13,500 + 800 = 14,300 MW

Area 2: Load = 35,000 MW

Export to area 1 = 1,500 MW

Generation = 35,000 + 1,500 = 36,500 MW

Reserve = 1000 MW

Total generation capacity = 36,500 + 1000 = 37,500 MW

Regulation of 6% on a generation capacity of 14,300 MW (including reserve) corresponds to

$$R_1 = \frac{0.06 \times 50}{14300}$$

$$\frac{1}{R_1} = \frac{1}{0.06} \times \frac{14300}{50} = 4{,}766.67 \text{ MW/Hz}$$

Similarly,

$$\frac{1}{R_2} = \frac{1}{0.06} \times \frac{37500}{50} = 12{,}500 \text{ MW/Hz}$$

$$\frac{1}{R_{eq}} = \frac{1}{R_1} + \frac{1}{R_2} = 17{,}266.6 \text{ MW/Hz.}$$

Load damping $D = 1.0$, which means a 1% change in load occurs for a 1% change in frequency. D_1 is computed for area 1 on a total load of 14000 MW, considering the loss of 1000 MW.

$$D_1 = 1 \times \frac{14000}{100} \times \frac{100}{50} = 280 \text{ MW/Hz.}$$

Similarly, D_2 is calculated on a load of 35,000 MW, since there is no change in load of area 2.

$$D_2 = 1 \times \frac{35000}{100} \times \frac{100}{50} = 700 \text{ MW/Hz.}$$

$$D_{eq.} = D_1 + D_2 = 980 \text{ MW/Hz.}$$

The change in system frequency is given by

$$\Delta f = \frac{-\Delta P_L}{D_{eq} + \dfrac{1}{R_{eq}}} = \frac{-(-1000)}{17{,}266.6 + 980} = 0.0548 \text{ Hz.}$$

Load changes due to load damping are,

$$\Delta P_{D1} = D_1 \, \Delta f = 280 \times 0.0548 = 15.344 \text{ MW}$$

$$\Delta P_{D2} = D_2 \, \Delta f = 700 \times 0.0548 = 38.36 \text{ MW}$$

Change in generations are

$$\Delta P_{G1} = \frac{-\Delta f}{R_1} = -4766.6 \times 0.0548 = -261.2 \text{ MW}$$

$$\Delta P_{G2} = \frac{-\Delta f}{R_2} = -12500 \times 0.0548 = -685 \text{ MW}$$

We recalculate powers as follows:

Area 1: New load = 15,000 – 1,000.00 + 15.344 = 14,015.344 MW

New generation = 13,500 – 261.2 = 13,238.8 MW

Deficit = 14,015.34 – 13,238.8 = 776 MW

Area 2: New load $= 35{,}000 + 38.36 = 35{,}038.36$ MW

New generation $= 36{,}500 - 685 = 35{,}815$ MW

Excess $= 35{,}815 - 35{,}038.36 = 776$ MW

Thus, tie-line power is 776 MW and flows from area 2 to area 1.

Steady state frequency $= 50 + 0.0548 = 50.0548$ Hz

(b) With supplementary control and $\beta_1 = 250$ MW/0.1 Hz and $\beta_2 = 400$ MW/0.1 Hz

Generating capacity with supplementary control in area 1 is 4000 MW (on reserve) and in area 2 it is 1000 MW. These supplementary controls will keep ACE_1 and ACE_2 at zero.

$$ACE_1 = \beta_1 \Delta f + \Delta P_{12} = 0$$

$$ACE_2 = \beta_2 \Delta f + \Delta P_{21} = \beta_2 \Delta f - \Delta P_{12} = 0$$

This means $\Delta P_{12} = 0$ and $\Delta f = 0$.

Thus, the load and generation in area 1 are reduced by 1000 MW. There is no steady state deviation of tie-line power flow and frequency. The generation and load of area 2 also do not change.

Example 13.12: In example 13.11, with supplementary controls as mentioned, find steady state frequency, generation and load of each area and tie–line power for following cases:

(i) Loss of 500 MW generation in area 1, carrying part of spinning reserve.

(ii) Loss of 1500 MW generation in area 1, not carrying spinning reserve.

(iii) Tripping of tie-line, assuming no change to inter change schedule of the supplementary control.

(iv) Tripping of tie-line, and changing schedule of supplementary control to zero.

Solution: (i) Loss of 500 MW in area 1, carrying part of spinning reserve in area 1.

The supplementary reserve in area 1 is 800 MW spread over 4000 MW of capacity. 4000 MW, therefore, includes 3200 MW of generation and 800 MW of reserve. With generation loss of 500 MW, spinning reserve is lost proportionally. Spinning reserve lost is

$$\frac{500}{3200} \times 800 = 125 \text{ MW}.$$

Remaining spinning reserve is $800 - 125 = 675$ MW. This reserve can makeup for the loss of 500 MW generation. Therefore, the generation and load of the two areas are restored to their pre-disturbance values. There is also no frequency and tie-line flow deviation.

(ii) Loss of 1500 MW generation in area 1 and not carrying spinning reserve.

800 MW of generation loss is made up of by the spinning reserve of area 1. After this, ACE_1 can no longer act. Supplementary control in area 2 can act.

$$ACE_2 = \beta_2 \Delta f - \Delta P_{12} = 0$$

$$\therefore \quad \Delta P_{12} = \beta_2 \Delta f;$$

$$\beta_2 = 400 \text{ MW/0.1 Hz} = 4000 \text{ MW/Hz}.$$

$$\Delta P_{12} = 4000\, \Delta f$$

There is net reduction in system frequency which causes a reduction in load. The balance generation loss of 700 MW, is made up by reduction in load in area 1 and tie-line flow increase from area 2.

$$\text{Area 1 load damping } D_1 = 1 \times \frac{15000}{100} \times \frac{100}{5} = 300 \text{ MW/Hz}$$

$$-700 = D_1 \Delta f + \Delta P_{12} = 300\,\Delta f + 4000\,\Delta f$$

$$\Delta f = \frac{-700}{4300} = -0.1628 \text{ Hz.}$$

Change in area 1 load is

$$\Delta P_{D1} = D_1 \Delta f = 300 \times (-0.1628) = -48.84 \text{ MW.}$$

$$\Delta P_{12} = 4000 \times (-0.1628) = -651.2 \text{ MW.}$$

[ΔP_{12} negative means there is increase in flow from area 2 to area 1]

Change in load of area 2

$$\Delta P_{D2} = D_2 \Delta f = 700 \times (-0.1628) = -113.96 \text{ MW.}$$

The load and generation is now calculated as follows.

Area 1. Load = 15,000 – 48.84 = 14,951.16 MW

Generation = 13,500 – 700 = 12,800 MW

Area 2. Load = 35,000 – 113.96 = 34,886.04 MW

Generation = 36,500 – 113.96 + 651.2 = 37,037.24 MW

Steady state power flow over tie–line from area 2 to area 1 is 2151 MW [an increase of 651 MW]. Steady state frequency is 50.0 – 0.1628 = 49.8372 Hz.

(iii) Tripping of tie-line, with no change in inter change schedule.

The supplementary control in area 1 attempts to maintain the inter change schedule at 155 MW. Therefore,

$$ACE_1 = \Delta P_{12} + \beta_1 \Delta f_1 = 1500 + 2500\,\Delta f_1$$

Solving $\Delta f_1 = -0.6$ Hz

Change in load due to load damping in area 1 is

$$D_1 \Delta f_1 = 300 \times (-0.6) = -180 \text{ MW}$$

Similarly, $$\Delta f_2 = \frac{1500}{4000} = 0.375 \text{ Hz}$$

$$D_2 \Delta f_2 = 700 \times (0.375) = 262.5 \text{ MW.}$$

Thegeneration and loads are as follows.

Area 1. Load = 15000 – 180 = 14,820 MW

Generation = 14,820 MW

Since, the generation required is more than the capacity of 14,300 MW (including reserve) area 1 cannot meet the load independently. Hence, load shedding has to be done to the tune of 520 MW to prevent isolated running of the area at under frequency.

Area 2. Load $= 35{,}000 + 262.5 = 35{,}262.5$ MW

Generation $= 35{,}262.5$ MW

$f_2 = 50 + 0.375 = 50.375$ Hz.

There is no tie-line flow.

(iv) Tripping of the tie-line, with inter change schedule switched to zero.

In this case area 1 supplementary control tries to pick up 1500 MW to make up for loss of import from area 2. However, the spinning reserve is 800 MW. Hence, load shedding has to be done if frequency has to be maintained at nominal value. The supplementary control of area 2, reduces its generation by 1500 MW to make up for loss of export to area 1. Its frequency is maintained at 50 Hz.

Example 13.13: If $R = 5\%$, and frequency changes from 60 Hz to 59 Hz, find (i) the increase in generation

(ii) what change in frequency would cause generation to change from 0 to 1 pu?

Solution:

(i)
$$\Delta P_G = \frac{\Delta\omega}{R} = \frac{\frac{60-59}{60}}{0.05} = 0.333 \text{ pu.}$$

If the machine were rated at say 1000 MW, then a drop of 1Hz would cause generation to increase by 333 MW.

(ii) Since $R = 5\%$, we need a change in frequency of 5% to change generation from 0 (no load) to 1Pu (full load).

$$5\% = 0.05 \times 60 = 3\text{Hz}.$$

So if generator was running at nominal frequency on no-load, the frequency drops to 47 Hz on full load.

13.4 SUPPLEMENTARY ISSUES IN LOAD FREQUENCY CONTROL

13.4.1 Load Frequency Control with Generation Rate Constraints (GRC)

In section 13.3, we have not used any limits on the rate of change of power generation. In thermal plants, the generation can change only at a specified maximum rate, due to safety considerations. The rate varies from 3% to 10% per minute. If these constraints are not taken into account, the controller tends to act for large momentary disturbances causing undue wear and tear. When GRC is considered, the model becomes non-linear when dynamic response is considered. Hence, when the state equations are solved, we have to check at each iteration for GRC violation. A second method is to add limits on governors in the two areas. With GRCs, R should be carefully selected. In hydrothermal systems, the GRC of hydro area is within safe limits and hence GRCs of hydro plants can be ignored.

13.4.2 Filtering of ACE

The generating units need not respond to fast random variations in load. Control action to these changes, do not reduce the area error but cause unnecessary wear and tear of the governor motors and turbine valves. Hence, the ACE signal is filtered to obtain a smoothed ACE to control the generation. A low order filter can be used to reduce the noise.

13.4.3 Speed Governor Dead-Band

The speed governor dead-band affects the response of the system. The dead-band is defined as the total magnitude of the change in steady state speed within which there is no resulting measurable change in the position of the governor controlled valves or gates. Dead-band is caused by Coloumb friction and valve overlaps in hydraulic relays. The dead-band introduces non-linearity in the governor action. The IEEE limits the maximum dead-band to 0.06%. One of the effects of the dead-band is to cause an apparent increase in the steady state speed regulation R. If the frequency deviation is small, it may lie within the dead-band. Hence, speed control will be inactive. Speed governor dead-band results in random frequency fluctuations in the range of 0.01 Hz.

13.4.4 AGC Tuning

The tuning of the AGC has a direct bearing on the performance of the power plant which they control. A smooth control is preferred to a rapid control that aims to bring the ACE to zero quickly. It is not possible to match load and generation continuously. It is found that execution of the AGC once every 2–4 seconds results in good performance. The stability of the control system is affected by phase lags caused by transducers, measurement and transmission of control signals.

If there is islanding in the system, AGC is suspended. If islanding leads to under generation in an area (see example 13.12) there is a decline in the frequency. To prevent operation at a lower frequency for a long time, we should resort to load shedding. Under frequency operation for a long time causes cumulative vibratory stress on the low-pressure turbine blades; reactor over-heat in nuclear plants and reduces plant capability due to reduced output of boiler feed pumps.

13.4.5 Digital Load Frequency Controllers

Digital computers are increasing becoming popular for implementing AGC control schemes, since digital control is more accurate, reliable and less sensitive to noise. The ACE is available in discrete form, since it is sampled every 2–4 seconds. In the discrete control mode, the control vector is constrained to remain constant between the sampling instants.

Digital control can be used to integrate the economic dispatch problem with ALFC. For optional economic dispatch, the units are required to be run at equal incremental costs. The details have been discussed in chapter 8. The AGC settings are compared with the optimal dispatch settings. If there is a deviation, the computer generates a signal to raise or lower the generation of the individual units. The generation schedule takes into account the interchange schedule between different areas.

13.5 REACTIVE POWER AND VOLTAGE CONTROL

Reactive power and voltage control is necessary for the following reasons:

- Voltages at the terminals of all equipment should be within the acceptable range, which depends on the voltage rating of the equipment. Operation of equipment beyond the range for a long time can damage it.
- Stability is enhanced by voltage control as seen in chapter 9.
- Control of reactive power enhances the efficiency of the transmission system.
- Different loads have different reactive power characteristics as mentioned in chapter 11. Moreover, reactive power cannot be transmitted over long distance and there is need for compensation throughout the system. This is unlike the frequency control which balances the active power of the entire system. The voltage is controlled at all levels in the system. Some of the devices used are:
  - Shunt capacitors
  - Shunt reactors
  - Synchronous condensers
  - Static var compensators
  - Series capacitors
  - Regulating transformers
  - Automatic voltage regulators

AVRs are discussed briefly in the next section.

13.5.1 Automatic Voltage Regulator (AVR)

The excitation system regulates the generator terminal voltage and hence controls the reactive power flow. The primary function of the AVR is to hold the terminal voltage magnitude at a specified value. The details of different excitation systems have been presented in chapter 10. The working of the AVR is briefly discussed below.

An increase in the reactive power demand is accompanied by a drop in terminal voltage magnitude of the generator. This is sensed by a potential transformer in one phase, rectified and compared with a DC reference set point to generate the error signal. This error signal is amplified and used to control the generator field current to increase the induced emf. The reactive power is increased, raising the terminal voltage to the reference value. The block diagram is shown in Fig. 13.21.

Simplified first order models are used for various components. The closed loop transfer function is given by

$$\frac{V_t(s)}{V_{\text{ref}}(s)} = \frac{K_A K_E K_G K_R (1+s\tau_R)}{(1+s\tau_A)(1+s\tau_E)(1+s\tau_G)(1+s\tau_R) + K_A K_E K_G K_R} \quad \text{...(13.43)}$$

$$V_t(s) = T_E(s)\, V_{\text{ref}}(s)$$

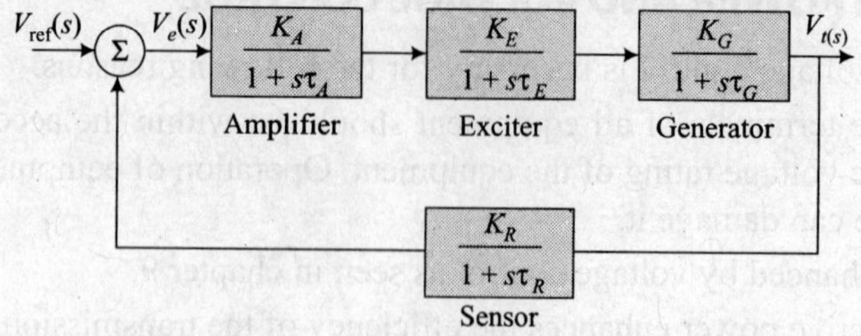

Fig. 13.21 Simplified block diagram of AVR.

If we consider a step change ΔV_{ref}, then

$$V_{ref}(s) = \frac{\Delta V_{ref}}{s} \text{ and } V_t(s) = T_E(s)\,\frac{\Delta V_{ref}}{s} \quad ...(13.44)$$

The steady state value of V_t is given by

$$V_{tss} = \lim_{s \to 0^s} V_t(s) = \frac{K_T}{1 + K_T}. \quad ...(13.45)$$

where, $K_T = K_A\, K_E\, K_G\, K_R$.

Example 13.14: The AVR system shown in Fig. 13.21 has the following parameters: $K_A = 8$; $\tau_A = 0.12$ sec; $K_E = 1$; $\tau_E = 0.4$ sec; $K_G = 1$; $\tau_G = 1.2$ sec; $K_R = 1$; $\tau_R = 60$ msec. Obtain the dynamic response for a unit step change in the reference.

Solution: $T_E(s)$ is obtained using (13.43).

$$T_E(s) = \frac{8(1 + s \times 0.06)}{(1 + s \times 0.12)\,(1 + s \times 0.4)\,(1 + s \times 1.2)\,(1 + s \times 0.06) + 8}$$

$$= \frac{8.48s + 8}{0.003456s^4 + 0.09792s^3 + 0.7752s^2 + 1.78s + 9}$$

For a unit step change in V_{ref}, $V_{ref}(s) = \frac{1}{s}$. The dynamic response is simulated and shown in Fig.13.22.

The steady state value of the terminal voltage is $\frac{8}{9} = 0.888$ pu. Thus, there is a steady state error, which can be decreased by increasing the gains. However, increase in the amplifier gain beyond a certain value, causes instability and results in unbounded response.

The stability is improved by adding a rate feedback to the control system as shown in Fig. 13.23.

By suitably adjusting values of K_F and τ_F we can obtain satisfactory response.

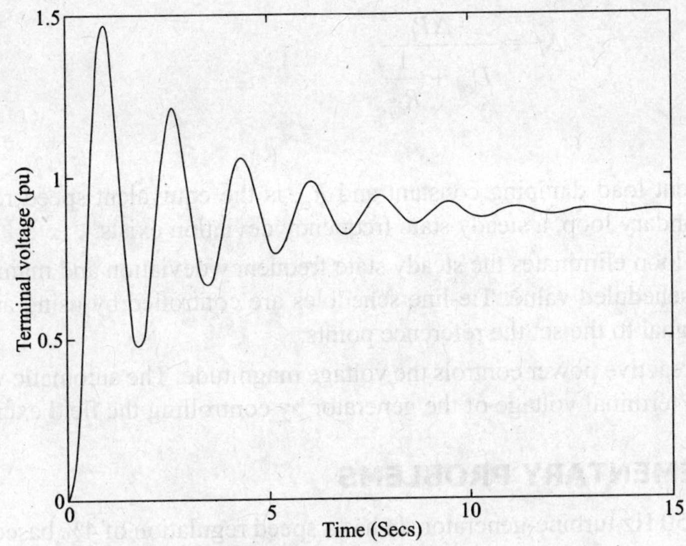

Fig. 13.22 Response of example 13.13.

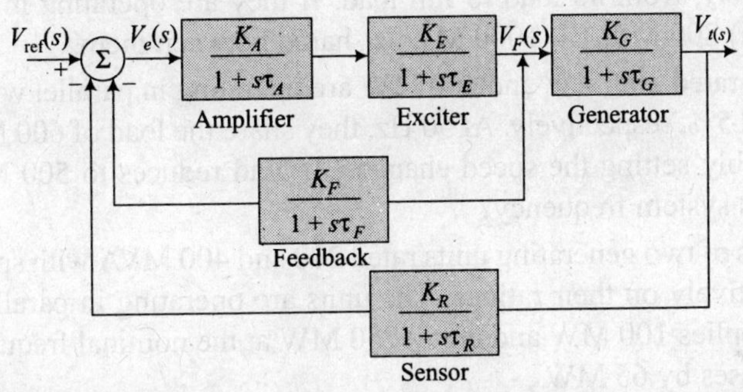

Fig. 13.23 AVR with feedback.

13.6 SUMMARY

The power system is extremely dynamic and there is a continuous deviation of the system state from its nominal state. The control strategy should be designed to deliver power to the interconnected system economically and reliably, at the same time maintaining the voltage and frequency within the acceptable limits.

Changes in real power directly affect the frequency of the system. The automatic load frequency control (ALFC) controls the real power flow in the system by regulating the output of the turbine. The ALFC consists of a primary loop which responds to fast changes in load and a secondary loop which maintains fine frequency adjustment to maintain proper power exchange with other networks via tie-lines. The response of the system to load changes depends on the value of the speed governor regulation (R) and the frequency dependency of the load. The change in the frequency Δf, for a change in the load by ΔP_L is given by

$$\Delta f = \frac{-\Delta P_L}{D_{eq} + \dfrac{1}{R_{eq}}}$$

where,

D_{eq} is the equivalent load damping constant and R_{eq} is the equivalent speed regulation. In the absence of the secondary loop, a steady state frequency deviation exists.

The secondary loop eliminates the steady state frequency deviation and maintains the tie-line power flow at the scheduled value. Tie-line schedules are controlled by using area control error (*ACE*) as control signal to the set the reference points.

The control of reactive power controls the voltage magnitude. The automatic voltage regulator (*AVR*) regulates the terminal voltage of the generator by controlling the field excitation.

13.7 SUPPLEMENTARY PROBLEMS

(1) A 500 MW, 50 Hz turbine-generator set has a speed regulation of 4% based on its rating. If frequency decreases to 49.6 Hz, find the decrease in the turbine power output.

(2) Two generating units rated 500 MW and 800 MW have a speed regulation of 6.0% and 6.5%, respectively, from no load to full load. If they are operating in parallel, determine how a load of (i) 800 MW (ii) 1000 MW is shared between them.

(3) Two generators rated 200 MW and 400 MW are operating in parallel with droop characteristics of 4% and 5%, respectively. At 50 Hz, they share the load of 600 MW in ratio of their ratings by suitably setting the speed changes. If load reduces to 500 MW, how will it be shared? What is system frequency?

(4) An area consists of two generating units rated 200 and 400 MVA with speed regulations 4% and 5%, respectively on their ratings. The units are operating in parallel and sharing 350 MW. Unit 1 supplies 100 MW and unit 2 250 MW at the nominal frequency of 50 Hz. The load now increases by 65 MW.

   (a) If $D = 0$, find the steady state frequency and the new generations.

   (c) If $D = 0.8$, find the steady state frequency deviation and the new generations.

(5) In example 12.3, determine steady state frequency deviation for an increase in load by 100 MW. Plot the dynamic response.

(6) Repeat problem 5 for a decrease in load by 100 MW.

(7) In example 12.9, determine the range of K_I for stable operation, using Routh-Hurwitz criterion.

(8) In example 12.9, plot the dynamic response for different values of K_I. Comment on the result.

(9) Repeat example 12.11 for a loss of 100 MW in area 2.

(10) In example 12.13, obtain the range of K_A for stable operation if the other gains are at 1. Plot dynamic response for different values of K_A.

REVIEW QUESTIONS

1. Draw the generator control loops with AVR and ALFC.
2. Derive the complete block diagram for the ALFC.
3. Derive expression for the steady state frequency deviation.
4. What is frequency bias factor?
5. Two generators are running in parallel. There is a sudden increase in load. How is it shared between the two?
6. Droop is necessary for the speed characteristic for proper sharing of load between alternators running in parallel.
7. What is AGC?
8. How does an integral controller in the feedback path of an ALFC help in improving stability?
9. Discuss the importance of AEC and tie-line power control.
10. Reactive power control is necessary for voltage control in the system. Comment.

Chapter 14

Power System Simulation with MATLAB

In this chapter, the MATLAB code for different power system simulation exercises are presented. The readers are urged to try out all of them and then do the exercises at the end of the chapter to get a good grip on programming. The comments are given in italics for readability of the programs. The program codes should be typed in MATLAB. All program codes should be typed in a .m file in MATLAB, preferably with meaningful file names, such as Pow_angle_char.m, ABCD_med_line.m, Powerflow_GS.m, etc. The function files should be saved in .m files with the same name as the function. For every program, a program code is given which is the main program, function files which could also be data files, sample data and results corresponding to the data.

PROGRAM 1

Power Angle Characteristics of Cylindrical Rotor Generator

This Program plots the power angle characteristics of a cylindrical (non salient) rotor alternator. (refer chapter 9).

Program Code

```
x=input ('Enter the value of the synchronous reactance\n'); %Get data from keyboard on execution
v=input ('Enter the value of the terminal voltage\n');
pow=input ('Enter the value of the active power\n');
pf=input('Enter the value of the power factor\n');

% Calculate the value of the excitation voltage.
ia=pow/(v*pf)
theta=acos(pf)
ia=complex(ia*cos(theta),-ia*sin(theta));
e=v+x*ia*i;                                % E=V+j*X*Ia
```

```
% Calculate the value of the power delivered by the synchronous machine.
pmax=abs(e)*abs(v)/x;                    % Maximum power is EV/X
del=0:0.1:pi; % Power angle varies from 0° to 180° in step of 0.1.
power= pmax*sin(del);                    % Power=pmax*sin(delta)
plot(del*180/pi, power,'r');             % Graph of Power delivered v/s the power angle.
title('Power angle characteristics of cylindrical rotor synchronous machine');
xlabel('Delta (deg)');
ylabel('Power ( pu) ');
legend('pmax*sin(delta)');

% To calculate the voltage regulation.
reg=(abs(e)-abs(v))/abs(v)*100;       %(E-V)/Vx100%

% Display the excitation voltage, peak power delivered and the voltage regulation values.
fprintf('\n the Excitation voltage=%fp.u\n',e);
fprintf('\n the Peak power=%f p.u\n',pmax);
fprintf('\n the Voltage regulation=%f%%\n',reg);
```

Data (entered via keyboard on prompting)

x=0.4

v=1.0

pow=0.8

pf=0.85

Result

The plot is shown in Fig. 14.1.

The Excitation voltage=1.198318p.u

The Peak power=3.100773 p.u

The Voltage regulation=24.030903%

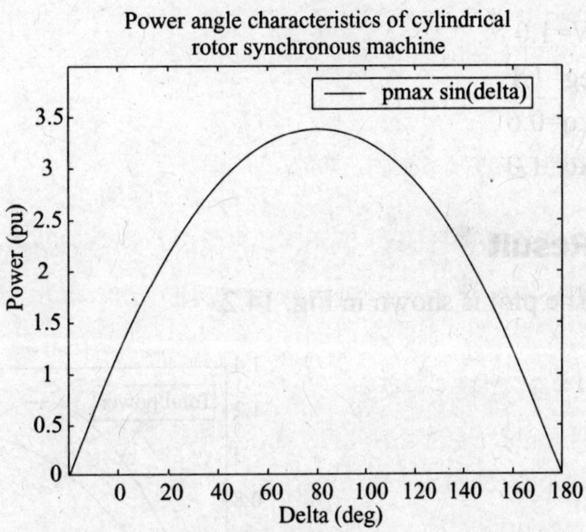

Fig. 14.1 Power angle characteristic of a non-salient pole machine.

PROGRAM 2

Power Angle Characteristic of a Salient Pole Machine

The power of a salient pole machine contains two terms. The first term is same as the non-salient pole term. The second term has double the system frequency and is due to saliency. This is called reluctance power. The values are entered in pu. (refer chapter 9)

Program Code

```
v= input('Enter the value of terminal voltage\n')  %Get inputs from keyboard on execution
eg=input('Enter the generator emf\n')
xq=input('Enter the quadrature axis reactance xq\n')
xd=input('Enter the direct axis reactance xd\n')
Pm1=abs(eg)*abs(v)/xd;% Maxi power of component similar to non-salient machine
Pm2=(abs(v)^2)*(xd-xq)/(2*xd*xq)%Maxi Reluctance power
del=linspace(0.0,pi,100);
y1=Pm1*sin(del);                    % Sinusoidal component of power
y2=Pm2*sin(2*del);                  % Reluctance power at double the frequency
del=del.*180/pi;                    % Convert angle to degree
plot(del,y1,'r')
hold on                             %This command in MATLAB holds the plot
plot(del,y2,'b')
hold on
y=y1+y2                             % Total power
plot(del,y,'m')
```

Data

V=1.0
eg=1.1
xq=0.6
xd=1.0

Result

The plot is shown in Fig. 14.2.

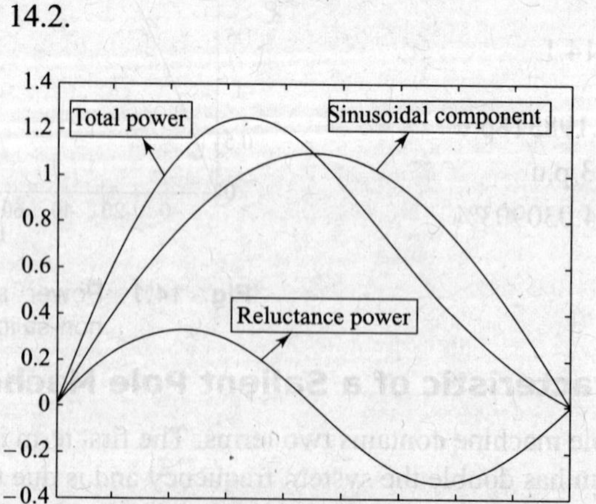

Fig. 14.2 Power anlge characteristic of a salient pole machine.

Exercise Modify the program to find the angle at which the maximum power occurs.

PROGRAM 3

Determination of ABCD Parameters of a Medium Line

Three phase transmission line can be represented on a single-phase basis, as a two port network. The input parameters are the sending end voltage and current and the output parameters are the receiving end voltage and current. The input and output parameters can related through the two port parameters, called the transmission parameters or the ABCD parameters. The relationship is given by,

$$V_s = AV_r + BI_r$$

$$I_s = CV_r + DI_r$$

The capacitance of the system is significant for transmission lines over 100 km. The R, L and C of the line are lumped, even though they are distributed. Depending on how they are modeled, there are two models for medium lines – T and π model.

Nominal T method In this method capacitance of each conductor is represented by an equal capacitance shunted between the conductor and neutral midway along the lines as shown in Fig. 14.3.

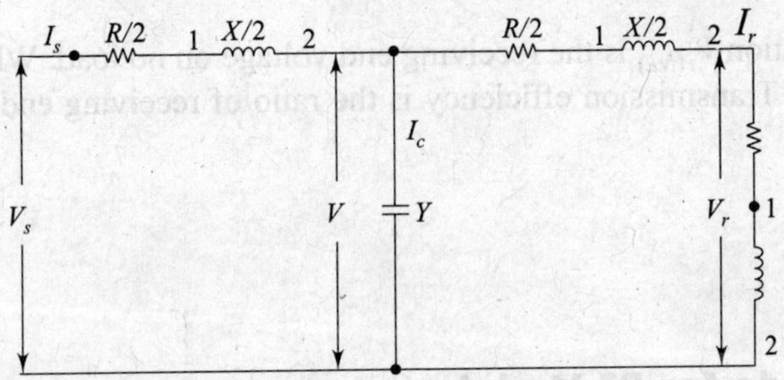

Fig. 14.3 T model of a medium line.

If $Z = R + jX$ and Y is the admittance of the shunt capacitance, then

$$A = 1 + \frac{YZ}{2}; B = Z\left(1 + \frac{YZ}{4}\right); C = Y; D = 1 + \frac{YZ}{2}$$

$$and\ \ AD - BC = \left(1 + \frac{YZ}{2}\right)^2 - YZ\left(1 + \frac{YZ}{4}\right) = 1$$

Nominal π method In π model, it is assumed that one half the total capacitance of each conductor is shunted at each end between the conductor and neutral as shown in Fig. 14.4.

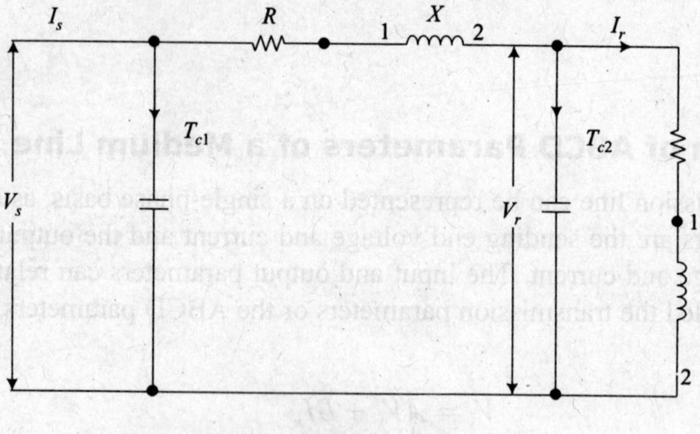

Fig. 14.4 π model of medium line.

A=D=1+YZ/2; B=Z; C =Y(1+YZ/4)

The voltage regulation is defined as percentage increase in voltage at the receiving end when full load is thrown off, the sending end voltage being kept unchanged. Percentage regulation is given by

$$\%regulation = \frac{V_{r(NL)} - V_r}{V_r} \times 100$$

In the above equation $V_{r(NL)}$ is the receiving end voltage on no load. When on no load, $Ir = 0$. Hence, $V_{r(NL)} = Vs/A$. Transmission efficiency is the ratio of receiving end power to the sending end power.

$$\%\eta = \frac{P_r}{P_s} \times 100$$

A - Program Code for PI Model

% Here the main program itself is written as a function. The inputs are read from different files.

```
function[a,b,c,d]=med_pi_abcd();  %Define the function to output ABCD
[r x y len]=medinput();           %Obtain line data
[ir vr pf]=medpow();              %Obtain receiving end data
z=len*complex(r,x)                % Length *(R+jX). Line data is normally per km
y=len*complex(0,y)
a=(1+(z*y)/2);
b=z;
c1=(1+(z*y)/4);
c=y*c1;
d=a;
```

```
Product=a*d-b*c
ABCD=[a b;c d]                          %Put ABCD in a matrix
vs_is=ABCD*[vr;ir]                      % Compute sending end values
vs_line=sqrt(3)*abs(vs_is(1))           %Sending end line voltage
is=1000*abs(vs_is(2))                   %Sending end current in amperes.
vr_line=vr*sqrt(3)                      %Receiving end line voltage
reg=((vs_line/abs(a)-vr_line)/vr_line)*100 % Vs_line is divided by A to get no-load receiving
                                        % end voltage

function[r,x,y,len]=medinput();
r=0.11;                                 % line resistance in ohms /km
x=0.3877;                               % line reactance in ohms/km
y=2.952e-6;                             % shunt admittance in mhos/km
len=250;                                % line length in kms

function [ir,vr,pf]=medpow();
sr=50;                                  % receiving  end MVA
vr=132;                                 % receiving end line volatge in kv
pf=0.85;                                 % power factor
theta=acos(pf);
pr=cos(theta);
qr=sin(theta);
sr1=(sr*complex(pr,-qr))/3;             % calculate P-jQ per phase
vr=complex(vr/sqrt(3),0);               %Receiving end L-N voltage
ir=(sr1)/conj(vr);                      %Ir =S*/V*, per phase
```

Result

A = 0.9642 + j0.0101=D

B= 27.5000 +j96.925

C=0.0000 + j0.0007

Vs = 89.7626 + j 15.6226 kV (Per phase)

Vs (Line –Line) magnitude = 157.8106 kV

Is= 180.1 - j54.0 A

Reg = 23.9811%

Another set of data with result

r=0.036 ohm/km

L=0.8 mH/km

C=0.0112iF/km

Length=130 km

Frequency = 50

Vr = 325kV

Sr=270 MVA

Pf=0.8

Result

A =D=0.9925 +j 0.0011

B= 4.6800 +j32.650

C=0.0000 + 0.0005i

Vs= 389.968 kV

Reg= 4.7735%

B - Program Code for T Model

```
function[a,b,c,d]=med_tee_abcd();
[r x y len]=medinput();
[ir vr pf]=medpow();
z=len*complex(r,x)
y=len*complex(0,y)
a=(1+(z*y)/2);
b1=(1+(z*y)/4);
b=z*b1;
c=y;
d=a;
x=a*d-b*c
ABCD=[a b;c d]
vs_is=ABCD*[vr;ir]                    % Compute sending end values
vs_line=sqrt(3)*abs(vs_is(1))         %Sending end line voltage
is=1000*abs(vs_is(2))                 %Sending end current in amperes.
vr_line=vr*sqrt(3)                    %Receiving end line voltage
reg=((vs_line/abs(a)-vr_line)/vr_line)*100
```

The medinput and medpow files remain the same.

Result

A= 0.9642 + j0101=D

B= 26.5165+j95.3313

C= 0 +j 0.0007i
Vs= 89.3962 +j15.4396 kV (per phase)
Is=180.4 -j53.0 A
Sending end line-line voltage magnitude =157.1311 kV
reg =23.4473 %

Exercise Modify the program to vary length from 200 km to 300 km, in steps of 10 km, and plot the regulation vs line length curve.

PROGRAM 4

ABCD Parameters of Long Transmission Line

Transmission lines above 260 km are long transmission lines. The ABCD parameters of long transmission lines are given by

$$A=D=\cosh(\gamma l);\ B=Z_c \sinh(\gamma l);\ C=\frac{1}{Z_c}\sinh(\gamma l)$$

AD-BC$=(\cosh(\gamma l))^2 - (\mathrm{Sinh}(\gamma l))^2 = 1$

where $\gamma=\sqrt{yz}$; y= admittance per km and z is impedance per km

$Z_c = \sqrt{Z/y}$; the characteristic impedance of the line

Program Code

```
function[a,b,c,d]=long_abcd();      %function definition
[z y len]=long_input();             % calling input parameters
[ir vr pf]=long_pow();              %calling function to read receiving end data
gamma=sqrt(z*y);
zc=sqrt(z/y);
a=cosh(gamma*len)
b=zc*sinh(gamma*len)
c=(1/zc)*sinh(gamma*len)
d=a
x=a*d-b*c
ABCD=[a b;c d]
vs_is=ABCD*[vr;ir]                  % Compute sending end values
vs_line=sqrt(3)*abs(vs_is(1))       %Sending end line voltage
is=1000*abs(vs_is(2))               %Sending end current in amperes.
vr_line=vr*sqrt(3)                  %Receiving end line voltage
```

```
reg=((vs_line/abs(a)-vr_line)/vr_line)*100

function[z,y,len]=long_input();
r=0.045; % line resistance in ohms /km
x=0.4;                          % line reactance in ohms/km
z=r+x*i
y=4e-6;                         % shunt admittance in mhos/km
len=250;                        % line length in kms

function [ir,vr,pf]=long_pow();
sr=270;                         % receiving  end three phase MVA
vr=325;                         % receiving end line volatge in kv
pf=0.8;                         % power factor
theta=acos(pf);
pr=cos(theta);
qr=sin(theta);
sr1=(sr*complex(pr,-qr))/3;     % calculate P-jQ per phase
vr=complex(vr/sqrt(3),0);       %Receiving end L-N voltage
ir=(sr1)/conj(vr);              %Ir =S*/V*, per phase
```

Result

A=D=10.01+j0.05

B=9.6+100.37

C=0

Vs= 391.0043 kV

Reg= 19.5368%

PROGRAM 5

Formation of Y_{bus} with Off-nominal Turns Ratio by Inspection (Direct) Method

This program forms the bus admittance matrix by method of inspection Line Data. This is stored as a function file, in a separate .m file, named linedatas.m. This file is called from main file to read the data.

p = start bus, q = end bus, z =line impedance, hlc=half line charging admittance, a=tap ratio(if it is a transformer). Transformer's hlc=0. Here data of Example 3.1 is used.

Program Code

```
z=linedatas( );                % calling the function linedatas.
sb=z(:,1);                     % First column of matrix z is stored in variable sb (i.e. start bus data).
eb=z(:,2);                     % Second column of matrix z is stored in variable eb (i.e. end bus data).
zline=z(:,3);                  % line impedance values are stored in variable Z.
hlc=z(:,4);                    % hlc admittance values are stored in variable hlc.
a=z(:,5);                      % tap ratio is stored in variable a.
y=1./zline;                    % Admittance of line = 1/ impedance of line.
nbus=max(max(sb),max(eb));
Ybus=zeros(nbus);              % creates a zero matrix of the size nbus*nbus.
nline=length(sb);
for k=1:nline,
    p=sb(k);q=eb(k);           % The start and end bus of line k are determined
    y(k)=y(k)/a(k);            % The arm of the π model of off-nominal ratio transformer
    % Diagonal elements of the Ybus.

  Ybus(p,p)=Ybus(p,p)+y(k)+hlc(k)+(1-a(k))/a(k)*y(k);
  Ybus(q,q)=Ybus(q,q)+y(k)+hlc(k)+(a(k)-1)*y(k);

  % Off-Diagonal elements of the Ybus.
  Ybus(p,q)=Ybus(p,q)-y(k);
  Ybus(q,p)=Ybus(p,q);
end
Ybus                           % Display Ybus.
```

Data File Example 3.1

```
function z = linedatas()
%    p   q   z              hlc       a
z = [1   4   0.08+0.37i     0.007i    1;
     1   6   0.123+0.518i   0.01i     1;
     2   3   0.723+1.05i    0.0       1;
     2   5   0.282+0.64i    0.0       1;
     4   3   0.0+0.133i     0.0       0.909;
     4   6   0.097+0.407i   0.0076i   1 ;
     6   5   0.0+0.3i       0.0       0.976  ];
end
```

**Check that the result matches with that given in Example 3.1

Exercise This program will give error if there is an element between a bus and the reference bus. These elements are mainly shunt inductors or capacitors. Modify the above program to read shunt data and add them to the diagonal elements of the respective buses.

PROGRAM 6

Formation of Y_{bus} Using Singular Transformation Method with Half Line Charging (Or Shunt Admittance)

This program forms the bus admittance matrix using singular transformation when there is no mutual coupling between the lines. The line data is stored in a function file name sing_linedatas.m This cannot be used when transformers with a non unity off-nominal ratio are present. Other transformers can simply be modeled as a line with no hlc.

Program Code

```
z=sing_linedatas();          % calling the function linedatas.
sb=z(:,1);                   % First column of matrix z is stored in variable sb (i.e. start bus data).
eb=z(:,2);                   % Second column of matrix z is stored in variable eb (i.e. end bus data).
zline=z(:,3);                % line impedance values are stored in variable zline.
hlc=z(:,4);                  % half line charging admittance values are stored in variable hlc.
y=1./zline;                  % Admittance of line = 1/ impedance of line.
nbus=max(max(sb),max(eb));           % Number of buses is maximum value of the bus numbers
Y=zeros(nbus);               % Create a null matrix of size nbus x nbus
nline=length(sb);            % Find number of lines
nlb=nline+nbus;              % Augment with number of buses to include elements to ground
A=zeros(nlb,nbus);           % create null matrix for incidence matrix
                             % Formation Bus incidence Matrix
for k=1: nbus
  A(k,k)=1;                  % Entry for the elements between a node and ground (or reference).
end

for k=1:nline
    A(nbus+k,sb(k))=1; % Entry for the elements between two nodes. 1 for start-bus(away from
    A(nbus+k,eb(k))=-1; % node) and -1 for end bus (into a node)
end
% Formation of Primitive admittance matrix
sh=zeros(nbus);              % create a null matrix of size nbus x nbus
for k=1:nline
```

```
    sh(sb(k))=sh(sb(k))+hlc(k); % Add half line charging admittance to the diagonal elements
    sh(eb(k))=sh(eb(k))+hlc(k);
end
ypr=zeros(nlb,nlb);          % create a null matrix to form the primitive admittance matrix
for k=1:nbus
    ypr(k,k)=sh(k);          % The shunt elements are updated
end
for k=1:nline
    ypr(nbus+k,nbus+k)=y(k);          % update primitive matrix with the line data
end

format short;                         % Scaled fixed point format with 5 digits.
Ybus=A'*ypr*A                         % Singular transformation Ybus= A^T * ypr *A
```

Data File 1

```
function z = sing_linedatas()
% p = from bus, q = to bus, z =line impedance, hlc=half line charging admittance.
%  p  q       z          hlc
z= [5  4    0.02+0.06i   0.03i
    5  1    0.08+0.24i   0.025i
    4  1    0.06+0.18i   0.02i
    4  2    0.06+0.18i   0.02i
    4  3    0.04+0.12i   0.015i
    1  2    0.01+0.03i   0.01i
    2  3    0.08+0.24i   0.025i ];
end
```

Data File 2

```
function z = sing_linedatas()

% p = from bus, q = to bus, z =line impedance, hlc=half line charging admittance.
z= [1  4    0.08+0.37i      0.007i ;
    1  6    0.123+0.518i    0.01i  ;
    2  3    0.723+1.05i     0.0    ;
    2  5    0.282+0.64i     0.0    ;
    4  3    0.0+0.133i      0.0    ;
    4  6    0.097+0.407i    0.0076i ;
    6  5    0.0+0.3i        0.0i  ];
end
```

Exercise Modify the above program to read shunt element data (reactors or capacitors connected in shunt at a bus) and add them to the diagonal elements of the respective buses.

PROGRAM 7

Formation of Ybus Using Singular Transformation Method with Mutual Coupling

This program is used to find the bus admittance matrix, when there is mutual coupling between the elements. In such a case, it is not possible to form the admittance matrix by direct method. Singular transformation is the only method in such cases. Half line charging has not been considered.

Program Code

```
z=mutual_linedatas ();          % calling the function mutual_linedatas.
sb=z(:,1);                      % First column of matrix z is stored in variable sb i.e. start bus
data.
eb=z(:,2);                      % Second column of matrix z is stored in variable eb i.e. to bus
                                data.
zline=z(:,3);                   % line impedance values are stored in variable zline.
num_mut=z(:,4);                 % the element number with which it has mutual coupling
z_mut=z(:,5);                   % Value of the mutual impedance
nbus=max(max(sb),max(eb));      % Number of buses
Y=zeros(nbus);                  % Initialize Ybus Matrix
nline=length(sb);
% Formation of Bus incidence Matrix
A=zeros(nline,nbus);
for k=1:nline,
  if(eb(k)==0)
     A(k,sb(k))=1;
  elseif(sb(k)==0)
     A(k,eb(k))=-1;
  end
if(sb(k)~=0 & eb(k)~=0)
A(k,sb(k))=1;
A(k,eb(k))=-1;
end

end
% Formation of Primitive admittance matrix
zpr=zeros(nline,nline);                % Initialize primitive impedance matrix.
 for k=1:nline
```

```
        zpr(k,k)=zline(k);            % Diagonal element is value of the line impedance
        if(num_mut(k))
            zpr(k,num_mut(k))=z_mut(k); % Off-diagonal elements are the mutual impedances
            zpr(num_mut(k),k)=z_mut(k);
        end
end
ypr=inv(zpr);z        % Determine Primitive Admittance matrix with mutual coupling. Ypr=Zpr^-1
format short;         % Scaled fixed point format with 5 digits.
Ybus=A'*ypr*A         % Ybus= A^T*Ypr*A
```

Data File

```
function z = mutual_linedatas ()
% p = from bus, q = to bus, z = line impedance, num_mut = mutual coupling element number.
% p    q       z        num_mut  z_mut
z= [0  1    0.1+0.6i       0      0
    0  2    0.05+0.5i      1      0.1i
    2  3    0.1+0.5i       0      0
    0  1    0.07+0.4i      1      0.2i
    1  3    0.06+0.2i      0      0    ];
end
```

Result

```
Ybus = [ 1.7539 - 7.5581i    -0.0369 + 0.2054i    -1.3761 + 4.5872i
        -0.0369 + 0.2054i     0.6166 - 3.9753i    -0.3846 + 1.9231i
        -1.3761 + 4.5872i    -0.3846 + 1.9231i     1.7608 - 6.5102i]
```

Exercise Modify the above program to include half-line charging admittance.

PROGRAM 8

Z-bus Building Algorithm

The bus impedance matrix of a system can be built step by step. This algorithm describes it. The algorithm starts with assuming a null matrix for Zbus and then building it as each element is added. We define a branch as an element between an existing bus in the partially built Zbus and a new bus. A link is defined as an element between two buses in the partial Zbus. With this the elements are classified into 4 types.

- Type 1 – A branch added between a New bus and Reference bus.
- Type 2: A branch added between a New bus and an Old bus.
- Type 3: A link added between an Old bus and reference bus.
- Type 4: A link added between two Old buses

In this program mutual coupling is not considered. The building algorithm has to start with a Type 1 element. Normally the reference bus which is the ground, is taken to be bus 0. The size of Ybus and Zbus is equal to nbus x nbus where nbus is the number of external buses excluding the ground. In a practical system, the half line charging admittances are typically treated as elements between buses and grounds. If the half line charging admittance is zero, then one of the buses is taken to be the reference bus. MATLAB does not accept zero as index to an array. Hence, bus no 1 is taken to be the reference node.

Program Code

```
linedata=linedatas( );                % calling the function linedatas.

zbus=[0];                             % Initialize Zbus matrix to zero.
zline=linedata(:,3);                  % Line impedance values are stored in variable zline.
type=linedata(:,4);                   % Type of connection is stored in variable type,
nl=length(zline);                     % Number of elements
for k=1:nl

[rows cols]=size(zbus);               % Returns number of rows and columns of matrix zbus.
sb=linedata(k,1);                     % Number of the start bus of the line.
eb=linedata(k,2);                     % Number of end bus of the line.

% type 1 modification or type 2 modification
% Addition of branch increase the size of zbus by one.

if (type(k)==2||type(k)==1)% Check if it is a branch
    nb=min(sb,eb);
    zbus=[[zbus  zbus(:,nb)]; [zbus(nb,:)  zline(k)+zbus(nb,nb)]] % Augment partial zbus
continue;
end

if (type(k)==3||type(k)==4)           % Check if it is a link
    sb=linedata(k,1)
    eb=linedata(k,2)
    zncol=zbus(:,sb)-zbus(:,eb)
    znrow=zbus(sb,:)-zbus(eb,:)
    zndia=zbus(sb,sb)+zbus(eb,eb)-2*zbus(sb,eb)+zline(k)
    zbus=[zbus]-[(zncol*znrow)/zndia]
 end
end
zbus=zbus(2:n,2:n)
ybus=inv(zbus)
```

Data (Example 3.3)

```
function z = zbusdata()

% p = start bus, q = end bus, z =line impedance, type= element type.
%          p    q    z       type
    z  =[  1    2    0.1i     1
           3    1    0.15i    1
           4    2    0.4i     2
           3    2    0.6i     4
           3    4    0.4i     4 ];
end
```

Result

```
zbus =
     0 + 0.0831i    0 + 0.0253i    0 + 0.0542i
     0 + 0.0253i    0 + 0.1120i    0 + 0.0687i
     0 + 0.0542i    0 + 0.0687i    0 + 0.2614i
```

Data (Example 3.7)

```
function z = zbusdata()
%           p     q    z       type
    z  =  [ 1     2    1.25i   1
            2     3    0.25i   2
            3     4    0.4i    2
            i     4    1.25i   3
            4     5    0.2i    2
            3     5    0.125i  4];

end
```

Result

```
zbus =
     0 + 0.7166i    0 + 0.6099i    0 + 0.5334i    0 + 0.5805i
     0 + 0.6099i    0 + 0.7319i    0 + 0.6401i    0 + 0.6966i
     0 + 0.5334i    0 + 0.6401i    0 + 0.7166i    0 + 0.6695i
     0 + 0.5805i    0 + 0.6966i    0 + 0.6695i    0 + 0.7631i
```

PROGRAM 9

Iterative Technique for Economic Dispatch Without Losses and Generator Limits

This program is an iterative technique for economic dispatch without losses. Generator limits are not included. The input is given through a file eco_wol_input. The cost function is given by $F_i=a_i+b_iP_{Gi}+c_iP_{Gi}^2$. Cost coeffecients are read in a matrix in the order c_i, b_i, a_i for each generator in a line, along with the demand PD and initial lamda. Note that multiple inputs are read from a single file.

Program Code

```
clear;
diary('wer.txt')
y1=[];
[cost_coeff,PD,Ini_lam]=eco_wol_input;          %Read input from file
iter=0;                                          %Iteration count = 0
Delp=100;
n=length(cost_coeff(:,1));                       %Get number of generators
lam=Ini_lam;                                     %Initial value of Lambda
while(abs(Delp)>0.001)
  iter=iter+1;                                   %Update iteration count
  for i=1:n
    P(i)=(lam-cost_coeff(i,2))/(2*cost_coeff(i,1));   %Compute P_Gi
  end
  Ptotal=sum(P);                                 %total generation
  Delp=PD-Ptotal;                                %Demand - Generation
  sigma_a=sum(1./(2.*cost_coeff(:,1)));
  lam=lam+Delp/sigma_a;                          %Update Lambda
  y=[iter P Ptotal Delp lam];
  y1=[y1;y];
end
disp(y1)
diary('wer.txt')
```

Example 8.1 Data File

```
function[cost_coeff,PD,Ini_lam]=eco_wol_input();
cost_coeff=[0.1 20 1.5;
            0.1 30 1.9];
```

```
PD=200;
Ini_lam=35;
```

Result

```
1  75  25 100 100  45
2 125  75 200   0  45
```

Exercise Try the example with different initial values of lambda and observe how quickly the solution converges, even though the initial value is far away from the final value.

PROGRAM 10

Economic Dispatch Without Losses and Generator Limits Included

This program is for economic dispatch without losses with generator limits included.The input is given through a file eco_wol_genlit_in. The cost function is given by $F_i = a_i + b_iP_{Gi} + c_iP_{Gi}^2$. Cost coeffecients are read in a matrix in the order c_i, b_i, a_i for each generator in a line.

Program Code

```
clear;
diary('wer.txt')
y1=[];
[cost_coeff,PD,Ini_lam,gen_lim]=eco_wol_genlit_in;       %Read input from file
iter=0;                                                  %Iteration count = 0
Delp=100;
n=length(cost_coeff(:,1));                               %Get number of generators
lam=Ini_lam;                                             %Initial value of Lambda
while(abs(Delp)>0.001)
  iter=iter+1;                                           %Update iteration count
  for i=1:n
    P(i)=(lam-cost_coeff(i,2))/(2*cost_coeff(i,1));      %Compute PGi
  end
 for i=1:n
  if (P(i)<gen_lim(i,1))                                 %Check for generator limit violations
      P(i)=gen_lim(i,1);
  end
  if (P(i)>gen_lim(i,2))
    P(i)=gen_lim(i,2);
  end
 end
  Ptotal=sum(P); %Total generation
```

```
    Delp=PD-Ptotal; %Demand - Generation
    sigma_a=sum(1./(2.*cost_coeff(:,1)));
    lam=lam+Delp/sigma_a; %Update Lambda
    y=[iter P Ptotal Delp lam];
    y1=[y1;y];
end
disp(y1)
diary('wer.txt')
```

Data file(Example 8.7). This is a function file where data is stored.

```
function[cost_coeff,PD,Ini_lam,gen_lim]=eco_wol_genlit_in();
cost_coeff=[0.004  7.2  350;
     0.0025 7.3  500
     0.003  6.74 600];
PD=450;
Ini_lam=6.0;
gen_lim=[125 300
         125 300
         125 300]
end
```

Result of the program for example 8.7

```
 1.0000 125.0000 125.0000 125.0000 375.0000 75.0000 6.1525
 2.0000 125.0000 125.0000 125.0000 375.0000 75.0000 6.3051
 3.0000 125.0000 125.0000 125.0000 375.0000 75.0000 6.4576
 4.0000 125.0000 125.0000 125.0000 375.0000 75.0000 6.6102
 5.0000 125.0000 125.0000 125.0000 375.0000 75.0000 6.7627
 6.0000 125.0000 125.0000 125.0000 375.0000 75.0000 6.9153
 7.0000 125.0000 125.0000 125.0000 375.0000 75.0000 7.0678
 8.0000 125.0000 125.0000 125.0000 375.0000 75.0000 7.2203
 9.0000 125.0000 125.0000 125.0000 375.0000 75.0000 7.3729
10.0000 125.0000 125.0000 125.0000 375.0000 75.0000 7.5254
11.0000 125.0000 125.0000 130.9040 380.9040 69.0960 7.6660
12.0000 125.0000 125.0000 154.3263 404.3263 45.6737 7.7589
13.0000 125.0000 125.0000 169.8089 419.8089 30.1911 7.8203
14.0000 125.0000 125.0000 180.0432 430.0432 19.9568 7.8608
15.0000 125.0000 125.0000 186.8082 436.8082 13.1918 7.8877
16.0000 125.0000 125.0000 191.2800 441.2800  8.7200 7.9054
```

17.0000	125.0000	125.0000	194.2359	444.2359	5.7641	7.9171
18.0000	125.0000	125.0000	196.1899	446.1899	3.8101	7.9249
19.0000	125.0000	125.0000	197.4814	447.4814	2.5186	7.9300
20.0000	125.0000	126.0022	198.3352	449.3374	0.6626	7.9314
21.0000	125.0000	126.2718	198.5598	449.8315	0.1685	7.9317
22.0000	125.0000	126.3403	198.6169	449.9572	0.0428	7.9318
23.0000	125.0000	126.3577	198.6314	449.9891	0.0109	7.9318
24.0000	125.0000	126.3621	198.6351	449.9972	0.0028	7.9318
25.0000	125.0000	126.3633	198.6360	449.9993	0.0007	7.9318

Exercise Run the program with different initial values of lambda and note the effect on number of iterations. Run the program for other demand values also.

PROGRAM 11

Economic Dispatch with Losses

This program is for economic dispatch with losses. The input is given through a file eco_wol_input. The cost function is given by $F_i = a_i + b_iP_{gi} + c_iP_{gi}^2$. Cost coefficients are read in a matrix with c_i, b_i, a_i for each generator in a line.

Program Code

```
clear;
diary('wer.txt')
y1=[];
[cost_coef,PD,b_coef,lam]=eco_wl_in;          %Read input from file
ci_coef=cost_coef(:,1);
bi_coef=cost_coef(:,2);
cdiag=diag(ci_coef);
iter=0;                                       %Iteration count = 0
Delp=100;
n=length(cost_coef(:,1));                     %Get number of generators
 b_coef=b_coef./100

while(abs(Delp)>0.001)
  iter=iter+1;                                %Update iteration count
  P1=lam-bi_coef;
  for i=1:n
    P2(i)=2*(ci_coef(i)+lam*b_coef(i,i));
    P(i)=P1(i)/P2(i);
  end
```

```
    Ploss =P*b_coef*P';
    for i=1:n
        m1(i)=ci_coef(i)+b_coef(i,i)*bi_coef(i);
        m2(i)=2*(ci_coef(i)+lam*b_coef(i,i))*(ci_coef(i)+lam*b_coef(i,i));
        m(i)=m1(i)/m2(i);
    end
    Delp=PD+Ploss-sum(P);
    Dellam=Delp/sum(m);
    lam=lam+Dellam;                    %Update Lambda
    Ptotal=sum(P);
    y=[iter P Ptotal Delp lam];
    y1=[y1;y];
end
disp(y1)
disp('Demand')
```

Data file

```
function[cost_coeff,PD,b_coef,lam]=eco_wl_in();
cost_coeff=[0.008 7.0 200;
            0.009 6.3 180
            0.007 6.8 140];
PD=500;
b_coef=[ 0.0218  0.0     0.0;
         0.0     0.0228  0.0;
         0.0     0.0     0.0179;]
lam=8;
```

Result

Iter	PG1	PG2	PG3	Ptotal	Delp	Lambda
1.0000	51.3136	78.5292	71.1575	201.0003	301.8861	9.9797
2.0000	146.4134	163.1734	180.9440	490.5308	26.0737	10.1656
3.0000	154.9304	170.7746	190.7997	516.5046	1.8939	10.1792
4.0000	155.5512	171.3287	191.5181	518.3980	0.1349	10.1801
5.0000	155.5954	171.3682	191.5693	518.5329	0.0096	10.1802
6.0000	155.5986	171.3710	191.5730	518.5425	0.0007	10.1802

Demand	= PD	= 500 MW
Total generation	= Ptotal	= 518.5425 MW
Losses	= Ploss	= 18.5432 MW

PROGRAM 12

Swing Curve of an SMIB System Using Step-by-Step Method for Sustained Fault

In step-by-step or point-to-point method, we use two equations.

$$\Delta \delta_n = \Delta \delta_{n-1} + \Delta t^2 \frac{P_{a(n-1)}}{M}$$

$$\delta_n = \delta_{n-1} + \Delta\delta_n$$

In the above equations n is the time step count (or iteration count). If we use the unit of M in pu $s^2/^o$elec, then the above equations can be used directly in degrees. However, the argument for trigonometric functions in MATLAB is radians. So care must be taken to use correct units. Conversion from one to the other is resorted to in the program, since angle in degrees is more convenient to use. A smaller time step will give better results, but will be computationally expensive in terms of time and memory. Here the program for sustained fault is given, along with computation of critical clearing angle and critical clearing time. Fault is assumed to occur at time =0 secs.

Program Code

```
% start reading data
clear
pmech=0.8;              %Initial power output is equal to mechanical power input
x1=0.7;
x2=1.9;
x3=0.9;
H=5.2;
freq=50;
eg=1.2;                 %Generator emf
vt=1.0;                 % Infinite bus voltage
delta_t=0.05;           %Time step

deg_rad=pi/180;         %conversion for degree to radian.
M=H/180/freq;           % M(pu) s^2/°elec
iter=1;                 % Start iteration count
del_delta=0;            %Δδ=0 at starting
term=delta_t^2/M;       % Δt^2/M
```

```
pmax1=eg*vt/x1;                     %Maximum power before , during and after fault.
pmax2=eg*vt/x2;
pmax3=eg*vt/x3;

t(iter)=0;                          % Initial time =0
delta0=asin(pmech/pmax1);           % δo=sin-1(pmech/pmax1)
delta(iter)=delta0;                 %Initial angle
pa01=0;                             % Pa(0-)=0.0
delmax=pi-asin(pmech/pmax3);        % δmax

%Find critical clearing angle
cos_delcr=(pmech*(delmax-delta0)-pmax2*cos(delta0)+pmax3*cos(delmax))/(pmax3-pmax2);
delcr=acos(cos_delcr)               %critical clearing angle in radians
pe=pmax2*sin(delta0);               %Pe at t=0+
pa02=(pmech-pe);                    %Pa(0+)
pa=(pa01+pa02)/2;                   %Average of pa(0-) and pa(0+)
delta(iter)=delta(iter)/deg_rad;    % convert radians to degree
while(t(iter)<=0.55)
   term1=term*pa;                   % (Δt^2/M)*pa

   del_delta=del_delta+term1;       % Δδn = Δ δn-1 + Δt^2 Pa(n-1)/M

   iter=iter+1;
   t(iter)=t(iter-1)+delta_t;       % Increase time
   delta(iter)=delta(iter-1)+del_delta; % δn = Δn-1 + Δδn
   ang=delta(iter)*deg_rad;         % Convert angle to radians
   pe=pmax2*sin(ang);               % Determine pe=pmax2*sin(ä)
   pa=pmech-pe;
end
plot(t,delta)
critical_clearing_angle=delcr/deg_rad % critical clearing angle in degree

% Find critical clearing time. It can also be found graphically from swing curve as the time at
which delta is equal %to critical clearing angle
iter=1
while(t(iter)<=1.0)
  if(delta(iter)>=delcr/deg_rad)
    break;
```

```
  end
    iter=iter+1;
end
critical_clearing_time=t(iter)
```

Result

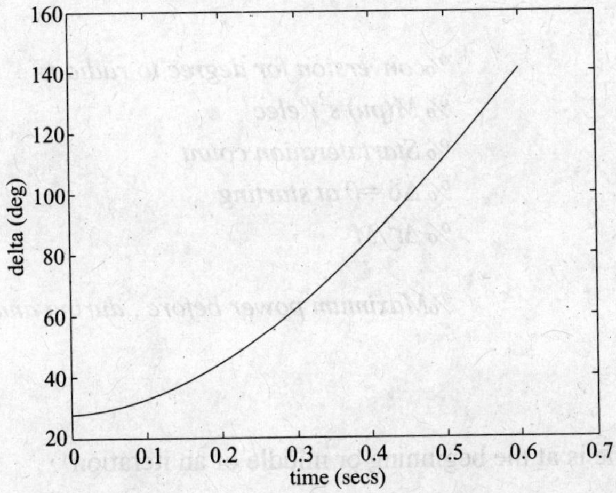

Fig. 14.5 Sustained fault using step-by-step procedure.

critical_clearing_angle = 91.2409°
critical_clearing_time = 0.45 secs

PROGRAM 13

Swing Curve by Step-by-Step Method

In this method if the fault is cleared in the beginning of any time step, the acceleration power to be used is the average of the accelerating power before and after fault clearing. If fault is cleared in middle of an interval, then we use corresponding accelerating power for time step before and after fault clearing.

Program Code

```
% start reading data
clear
pmech=0.8;                    %Initial power output is equal to mechanical power input
x1=0.7;
x2=1.9;
```

```
x3=0.9;
H=5.2;
freq=50;
eg=1.2;                              %Generator emf
vt=1.0;                              % Infinite bus voltage
delta_t=0.05;                        %Time step
tcr=input(‘enter clearing time’)
y=[];
deg_rad=pi/180;                      %conversion for degree to radian.
M=H/180/freq;                        % M(pu) s^2/°elec
iter=1;                              % Start iteration count
del_delta=0;                         % Δδ =0 at starting
term=delta_t^2/M;                    % Δt^2/M

pmax1=eg*vt/x1;                      %Maximum power before , during and after fault.
pmax2=eg*vt/x2;
pmax3=eg*vt/x3;

% Check if clearing time is at the beginning or middle of an iteration
interval=tcr/delta_t;
check=tcr-floor(interval)*delta_t;
if(check==0)
  fprintf(‘clearing time is at beginning of an interval\n’)
else
  fprintf(‘ clearing time is at the middle of an interval\n’)
end

t(iter)=0;                           % Initial time =0
delta0=asin(pmech/pmax1);            % äo=sin^-1 (pmech/pmax1)
delta(iter)=delta0;                  %Initial angle
pa01=0;                              % Pa(0-)=0.0
pmax=pmax2;
pe=pmax*sin(delta0);                 %Pe at t=0+
pa02=(pmech-pe);                     %Pa(0+)
pa=(pa01+pa02)/2;                    %Average of pa(0-) and pa(0+)
delta(iter)=delta(iter)/deg_rad;     % convert radians to degree

%start iterations using the same formula as in sustained fault
```

```
while(t(iter)<=0.55)
  term1=term*pa;                          % (Δt²/M)*pa
  del_delta=del_delta+term1;              % Δδn = Δδn-1 + Δt² Pa(n-1)/M
 y1=[t(iter) pmax delta(iter)];           %get values of time, pmax and delta into an array
y=[y;y1];                                 %Append the array to add new row of present values
  iter=iter+1;
  t(iter)=t(iter-1)+delta_t;              % Increase time
  delta(iter)=delta(iter-1)+del_delta; % δn = δn-1 + Δδn

 if(check==0)                             % If fault is at beginning of interval
    if(t(iter)<tcr)                       % Before clearing time
    pmax=pmax2;
   elseif(t(iter)==tcr)
    pmax=(pmax2+pmax3)/2;                 %At clearing time, pmax is average of pmax1 and pmax2
else
  pmax=pmax3;                             %After clearing pmax is pmax3
         end
end

if(check~=0)                              %If fault is cleared in middle of an interval
  if(t(iter)<tcr)
  pmax=pmax2;                             %Before clearing
  else
  pmax=pmax3;                             %After clearing
       end
end

ang=delta(iter)*deg_rad;                  % Convert angle to radians
    pe=pmax*sin(ang);                     % Determine pe=pmax2*sin(ä)
    pa=pmech-pe;
end
plot(t,delta)

%check for stability

stab=0;
  for k=2:iter
```

```
    if(delta(k)<delta(k-1))          %check if delta decreases at any time
    stab=1;
    end
  end
if(stab)
  display(' System is stable')
else
  display(' system is unstable')
end
fprintf(' time      Pmax       delta\n');
disp(y)
```

Result

Case (a) Clearing time is 0.1 secs which is beginning of an interval. The MATLAB results are as follows.

tcr =

0.1000

clearing time is at beginning of an interval

System is stable

time	Pmax	delta
0	0.6316	27.8181
0.0500	0.6316	28.9113
0.1000	0.9825	32.1447
0.1500	1.3333	36.5780
0.2000	1.3333	41.0347
0.2500	1.3333	45.1655
0.3000	1.3333	48.6665
0.3500	1.3333	51.2971
0.4000	1.3333	52.8869
0.4500	1.3333	53.3376
0.5000	1.3333	52.6220
0.5500	1.3333	50.7834

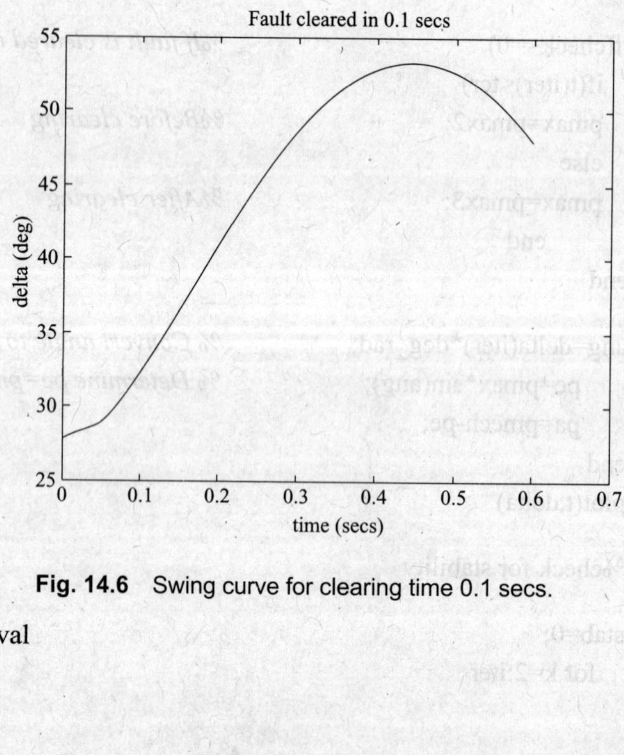

Fig. 14.6 Swing curve for clearing time 0.1 secs.

Case (b)

tcr =

0.1750

clearing time is at the middle of an interval

System is stable

time	Pmax	delta
0	0.6316	27.8181
0.0500	0.6316	28.9113
0.1000	0.6316	32.1447
0.1500	0.6316	37.3857
0.2000	1.3333	44.4290
0.2500	1.3333	50.8952
0.3000	1.3333	56.3460
0.3500	1.3333	60.4561
0.4000	1.3333	63.0086
0.4500	1.3333	63.8818
0.5000	1.3333	63.0365
0.5500	1.3333	60.5106

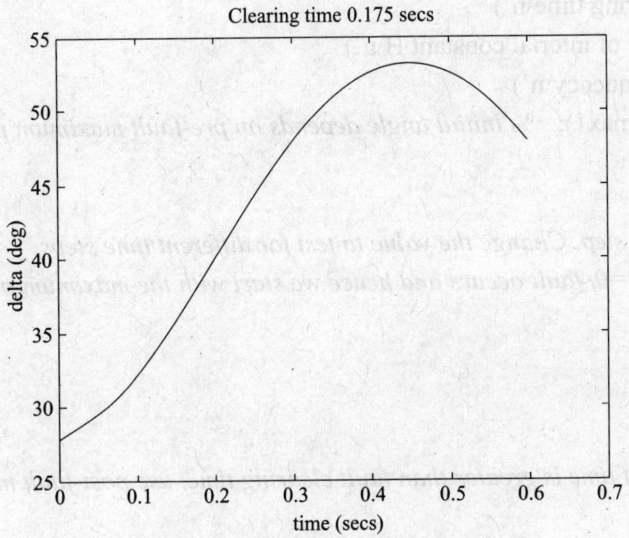

Fig. 14.7 Clearing time is 0.175 secs.

PROGRAM 14

Euler's Method for Solution of Swing Equation for an SMIB System

This program is to solve the swing equation using Euler's method. Detailed equations are given in chapter 9. In this program, we assume that the fault occurs at t=0 secs. The data consists of the maximum power transfers pre-fault, during fault and post-fault, along with system data. The time step plays a vital role in the solution. In an SMIB system, it may not drastically affect the conclusion on stability, but in multi-machine case, it will definitely affect the deduction of stability. **Sustained fault is simply simulated by giving a large fault clearing time.** It is not necessary to write a separate program. Values of power are in pu.

Program Code

```
% This program solves the swing equation by modified Euler's method.
clear
pmax1=input('enter the maximum power before fault\n');
pmax2=input('enter the maximum power during fault\n');
pmax3=input('enter the maximum power after fault\n');
power=input('enter the value of power transferred\n')
tcl=input('enter clearing time\n')
H=input('enter value of interial constant H\n')
freq=input('enter frequency\n')
delta0=asin(power/pmax1);  % Initial angle depends on pre-fault maximum power
omega0=0.0;
M=H/(pi*freq);
delta_t=0.01;% Time step. Change the value to test for different time steps
pmax=pmax2; % At t=0, fault occurs and hence we start with the maximum power during fault
y1=[];
z1=[];
for t=0.0:delta_t:0.5
if(t>tcl)
   pmax=pmax3; % If time is greater than fault clearing time, use post-fault maximum power
end
D1=omega0;
D2=(power-pmax*sin(delta0))/M;
delp=delta0+D1*delta_t; % Intermediate values of delta and omega
omegp=omega0+D2*delta_t;
D1p=omegp;
D2p=(power-pmax*sin(delp))/M;
delta0=delta0+(D1+D1p)/2*delta_t;
omega0=omega0+(D2+D2p)/2*delta_t;
delta=delta0*180/pi;  % Convert to delta to degrees
omega=omega0;
y=[t D1 D2 D1p D2p ]; % to display values of time and derivatives
z=[t pmax delta omega];
y1=[y1;y];
z1=[z1;z];
plot(t,delta,'r')
hold on
```

```
end
disp(y1)
disp(z1)
```

Data (Example 9.4)

Pmax1=1.714 pu
Pmax2=0.63 pu
Pmax3=1.333 pu
Power transferred =0.8 pu
Fault clearing time =0.125 secs
Frequency=50 Hz

Result

Three different cases are presented.

Case 1: Simulation time is 5.0 secs. Time step is 0.001 secs. Fault is cleared in 0.125 secs.

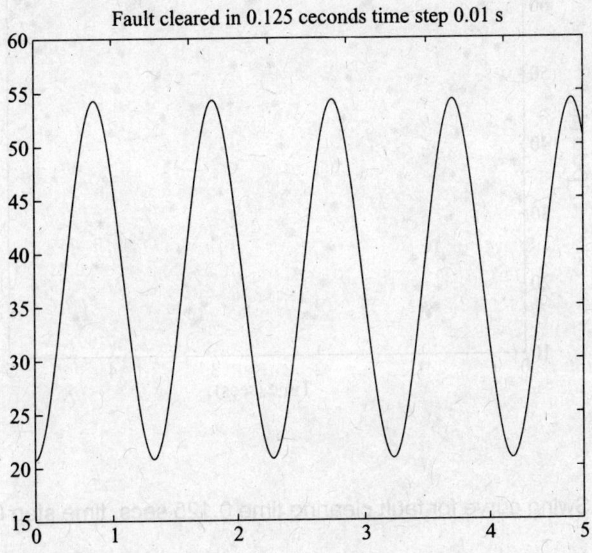

Fig. 14.8 Swing curve for fault clearing time 0.125 secs, time step 0.001 secs.

Note: The oscillations are sustained since damping has been neglected. System is stable.

Case 2: Simulation time is 5.0 secs. Time step is 0.05 secs. Fault is cleared in 0.125 secs.

The readings upto 0.05 seconds are given here. Compare it with what is given in chapter 9.

Time	Pmax	Delta	Omega
0.0000	0.6300	28.9178	0.7642
0.0500	0.6300	32.1787	1.4966
0.1000	0.6300	37.4711	2.1688
0.1500	1.3330	43.6605	2.0694
0.2000	1.3330	49.3286	1.8162
0.2500	1.3330	54.0751	1.4411
0.3000	1.3330	57.5990	0.9786
0.3500	1.3330	59.6985	0.4617
0.4000	1.3330	60.2620	-0.0798
0.4500	1.3330	59.2601	-0.6177
0.5000	1.3330	56.7427	-1.1235

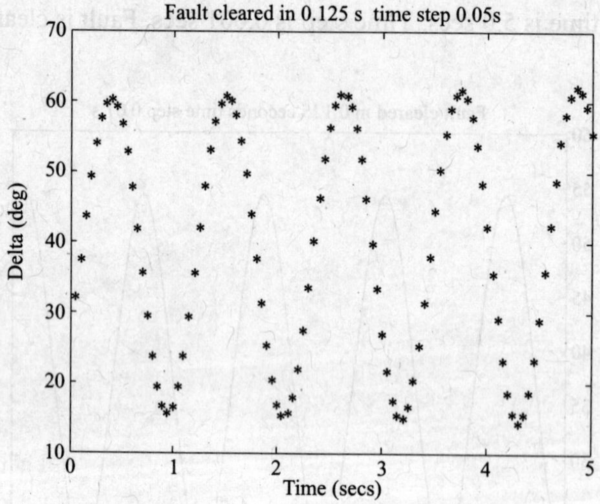

Fig. 14.9 Swing curve for fault clearing time 0.125 secs, time step 0.05 secs.

Case 3: Sustained fault

This is simulated by taking a large clearing time of 1 secs. The time step chosen is 0.01 secs.

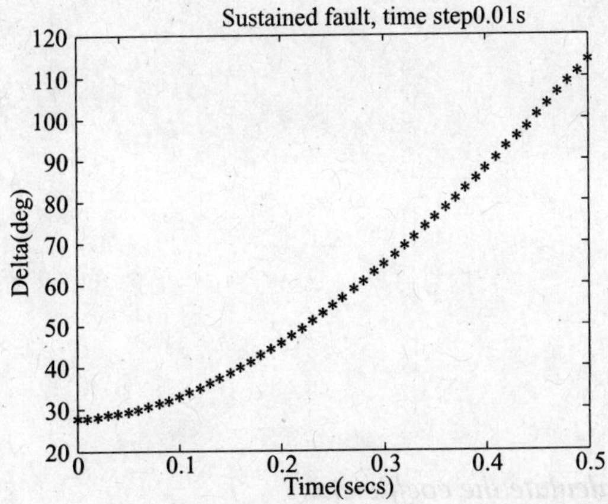

Fig. 14.10 Swing curve for sustained fault, time step 0.05 secs.

Note: Here delta increases continuously. The system is unstable.

Exercise (i) Study the effect of eliminating the intermediate updation of variables. Ii) Simulate the system is pmax3=pmax1 and compare with the results given iii) draw swing curve for pmax2=0.0 (dead short circuit at generator terminals.) iv) Determine the critical clearing time. (this is done by taking a small time step of 0.001 sec and slowly increasing the clearing time in steps of say 0.05 secs, until at one time it becomes unstable)

PROGRAM 15

Runge-Kutta Method for Solution of Swing Equation for an SMIB System

This program is to solve the swing equation using Range-Kutta method. Detailed equations are given in chapter 9. In this program, we assume that the fault occurs at t=0 secs. The data consists of the maximum power transfers pre-fault, during fault and post-fault, along with system data. **Sustained fault is simply simulated by giving a large fault clearing time.** Values of power are in pu.

Program Code

```
% This program is to solve the swing equation using Runge-Kutta method
clear
pmax1=input('enter the maximum power before fault\n');
pmax2=input('enter the maximum power during fault\n');
pmax3=input('enter the maximum power after fault\n');
power=input('enter the value of power transferred\n')
tcl=input('enter clearing time\n')
H=input('enter value of interial constant H\n')
freq=input('enter frequency\n')
delta0=asin(power/pmax1);
```

```
omega0=0.0;
M=H/(pi*freq);
delta_t=0.05;
pmax=pmax2;
y1=[];
z1=[];
for t=0.0:delta_t:0.5
if(t>tcl)
   pmax=pmax3;
end
k1=omega0*delta_t; % Calculate the coefficients
l1=(power-pmax*sin(delta0))/M*delta_t;
k2=(omega0+l1/2)*delta_t;
l2=(power-pmax*sin(delta0+k1/2))/M*delta_t;
k3=(omega0+l2/2)*delta_t;
l3=(power-pmax*sin(delta0+k2/2))/M*delta_t;
k4=(omega0+l3)*delta_t;
l4=(power-pmax*sin(delta0+k3))/M*delta_t;
delta0=delta0+(k1+2*k2+2*k3+k4)/6;
omega0=omega0+(l1+2*l2+2*l3+l4)/6;
delta=delta0*180/pi;
omega=omega0;
y=[t k1 l1 k2 l2 k3 l3 k4 l4 ];
z=[t pmax delta omega];
y1=[y1;y];
z1=[z1;z];
plot(t,delta,'*-r')
hold on
end
disp(y1)
disp(z1)
```

Data (Example 9.4)

Pmax1=1.714 pu

Pmax2=0.63 pu

Pmax3=1.333 pu

Power transferred =0.8 pu

Fault clearing time =0.125 secs

Frequency=50 Hz

Result

Three different cases are presented.

Case 1 Simulation time is 5.0 secs. Time step is 0.001 secs. Fault is cleared in 0.125 secs.

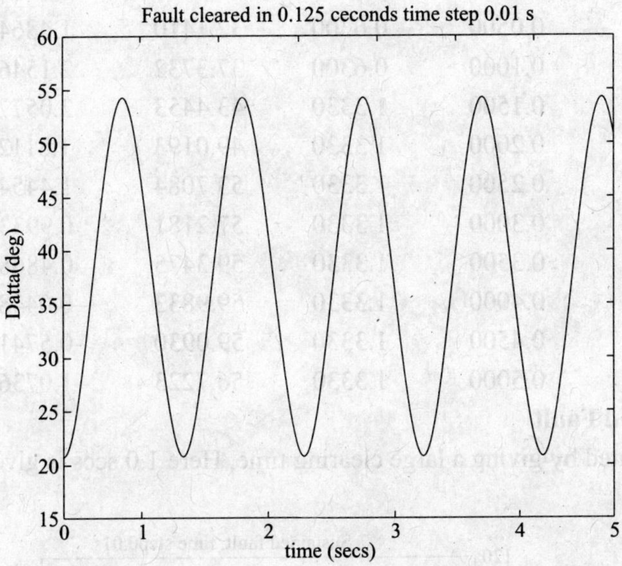

Fig. 14.11 Swing curve for fault clearing time 0.125 secs, time step 0.01 secs.

Case 2 Simulation time is 5.0 secs. Time step is 0.05 secs. Fault is cleared in 0.125 secs.

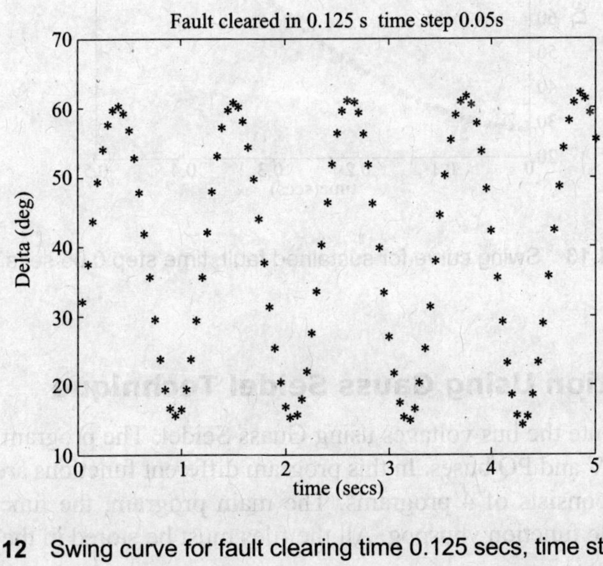

Fig. 14.12 Swing curve for fault clearing time 0.125 secs, time step 0.05 secs.

The actual values are given upto 0.5 seconds. The reader is urged to work out at least four-five iterations by hand, to understand how the algorithm works.

Time	Pmax	Delta	omega
0.0000	0.6300	28.9140	0.7588
0.0500	0.6300	32.1410	1.4864
0.1000	0.6300	37.3732	2.1546
0.1500	1.3330	43.4453	2.0577
0.2000	1.3330	49.0193	1.8112
0.2500	1.3330	53.7084	1.4454
0.3000	1.3330	57.2181	0.9933
0.3500	1.3330	59.3475	0.4868
0.4000	1.3330	59.9833	–0.0448
0.4500	1.3330	59.0930	–0.5741
0.5000	1.3330	56.7223	–1.0736

Case 3 : Sustained Fault

This is simulated by giving a large clearing time. Here 1.0 secs is given.

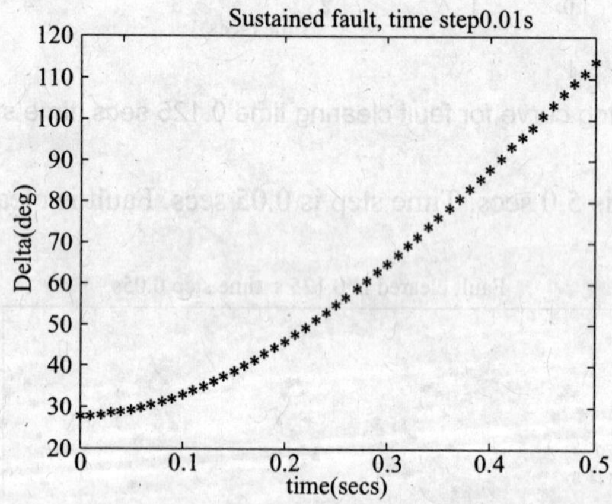

Fig. 14.13 Swing curve for sustained fault, time step 0.05 secs.

PROGRAM 16

Power Flow Solution Using Gauss Seidel Technique

This program is to compute the bus voltages using Guass Seidel. The program is general and can be used to handle both PV and PQ buses. In this program different functions are used to read data. The program basically consists of 4 programs. The main program, the function busdata6, the function linedata6 and the function ybusppg. All the files must be stored in the same directory.

Program Code

```
% Program for Gauss - Seidel Load Flow Analysis
% Bus 1 is considered as Slack bus.

ybus = ybusppg();          % Calling program "ybusppg.m" to get Y-Bus.
busdata = busdata6();      % Calling "busdata6.m" for bus data.

bus = busdata(:,1);        % Bus number.
type = busdata(:,2);       % Type of Bus 1-Slack, 2-PV, 3-PQ.
V = busdata(:,3);          % Initial Bus Voltages.
th = busdata(:,4);         % Initial Bus Voltage Angles. This is normally zero for all buses.
GenMW = busdata(:,5);      % PGi, Real Power injected into the buses by generator.
GenMVAR = busdata(:,6);    % QGi, Reactive Power injected into the buses.
LoadMW = busdata(:,7);     % PLi, Real Power Drawn from the buses.
LoadMVAR = busdata(:,8);   % QLi, Reactive Power Drawn from the buses.
Qmin = busdata(:,9);       % Minimum Reactive Power Limit
Qmax = busdata(:,10);      % Maximum Reactive Power Limit
nbus = max(bus);           % To get no. of buses
P = GenMW - LoadMW;        % Pi = PGi - PLi, Net real Power injection at the buses.
Q = GenMVAR - LoadMVAR;    % Qi = QGi - QLi, Net Reactive Power injection at the buses.
Vprev = V;
toler = 1;                 % Tolerence.
iteration = 1;             % iteration starting
while (toler > 0.001)      % Start of while loop
  for i = 2:nbus
    sumyv = 0;
    for k = 1:nbus
      if i ~= k
        sumyv = sumyv + ybus(i,k)* V(k);  %
      end
    end
    if type(i) == 2                                  % Computing Qi for PV bus
      Q(i) = -imag(conj(V(i))*(sumyv + ybus(i,i)*V(i)));
      if (Q(i) > Qmax(i)) || (Q(i) < Qmin(i))        % Checking for Qi Violation.
        if Q(i) < Qmin(i)                            % Whether violated the lower limit.
          Q(i) = Qmin(i);
        else                   % Upper limit is violated.
          Q(i) = Qmax(i);
        end
```

```
            type(i) = 3;  % If Violated, change PV bus to PQ bus.
        end
    end
    V(i) = (1/ybus(i,i))*((P(i)-j*Q(i))/conj(V(i)) - sumyv); % Compute Bus Voltages.
    if type(i) == 2 % For PV Buses, Voltage Magnitude remains same, but Angle changes.
        realv(i)=abs(Vprev(i))*cos(angle(V(i)));
        imagv(i)=abs(Vprev(i))*sin(angle(V(i)));
        V(i)=complex(realv(i),imagv(i))%
    end
  end
  iteration = iteration + 1;          % Increment iteration count.
  toler = max(abs(abs(V) - abs(Vprev))); % Calculate tolerance.
  Vprev = V; % Vprev is required for next iteration,
end    % End of while loop / Iteration

iteration                              % Total iterations.
V                                      % Bus Voltages in Complex form.
Vmag = abs(V)                          % Final Bus Voltage magnitudes.
Ang = 180/pi*angle(V)                  % Final Bus Voltage Angles in Degree.
```

Function to compute Ybus

```
% Program to form Admittance matrix. Compare this program with the program written earlier.
function ybus = ybusppg();  % Returns ybus

linedata = linedata6();           % Calling "linedata6.m" for Line Data
sb = linedata(:,1);               % Start bus number
eb = linedata(:,2);               % End bus number
r = linedata(:,3);                % Line Resistance
x = linedata(:,4);                % Line Reactance
b = linedata(:,5);                % Magnitude of Half line charging admittance
z = r + i*x;                      % Line impedance =R+jX
y = 1./z;                         % Line admittance is inverse of line impedance
b = i*b;                          % Half line charging admittance

nbus = max(max(sb),max(eb));% no. of buses.
nbranch = length(sb);             % no. of branches
ybus = zeros(nbus,nbus);% Initialise YBus elements to zero.
```

```
% Formation of the Off Diagonal Elements...
for k=1:nbranch
   ybus(sb(k),eb(k)) = -y(k);
   ybus(eb(k),sb(k)) = ybus(sb(k),eb(k));
end

% Formation of Diagonal Elements....
for m=1:nbus
   for n=1:nbranch
      if sb(n) == m | eb(n) == m
         ybus(m,m) = ybus(m,m) + y(n) + b(n); %Bc/2 is added to diagonal element.
      end
   end
end
ybus;                         % Bus Admittance Matrix
zbus = inv(ybus);             % Bus Impedance Matrix
```

Function to read busdata

```
% Bus data for Load Flow Analysis.

function busdata = busdata6()    % This function returns busdata.

        Bus no Type Vsp  theta  PG  QG  PL  QLi Qmin Qmax
busdata = [ 1   1   1.05   0    0.0  0   0    0    0    0;
            2   2   1.05   0    0.5  0   0    0   -0.5  1.0;
            3   2   1.07   0    0.6  0   0    0   -0.5  1.5;
            4   3   1.0    0    0.0  0   0.7  0.7  0    0;
            5   3   1.0    0    0.0  0   0.7  0.7  0    0;
            6   3   1.0    0    0.0  0   0.7  0.7  0    0 ];
```

Function to read line data

```
% Line Data for Y-Bus Formation.
function linedata = linedata6()    % Returns linedata.
%          | Start | End  |  R  |  X  | B/2 |
%          | Bus   | Bus  |     |     |     |
linedata = [1    2     0.10  0.20   0.02;
            1    4     0.05  0.20   0.02;
            1    5     0.08  0.30   0.03;
```

```
        2   3   0.05  0.25   0.03;
        2   4   0.05  0.10   0.01;
        2   5   0.10  0.30   0.02;
        2   6   0.07  0.20   0.025;
        3   5   0.12  0.26   0.025;
        3   6   0.02  0.10   0.01;
        4   5   0.20  0.40   0.04;
        5   6   0.10  0.30   0.03;];
```

Result of six bus data

Tolerance taken is 0.001 pu.

No of iterations taken for convergence is 4

Converged voltages in complex form.

```
1.0500 +0.0000i
1.0498 - 0.0178i
1.0699 - 0.0159i
0.9867 - 0.0400i
0.9840 - 0.0503i
1.0039 - 0.0513i
```

Voltage magnitudes	Voltage angles
1.0500	0
1.0500	-0.9723
1.0700	-0.8496
0.9876	-2.3209
0.9853	-2.9255
1.0052	-2.9279

Three bus data

```
% Bus data for Load Flow Analysis.
function busdata = busdata3()    % Returns busdata.
         Bus no Type Vsp  theta PG  QG  PL  QLi Qmin Qmax
busdata = [ 1    1   1.05   0    0   0   0   0    0    0 ;
            2    3   1.0    0    0   0   4   2.5  0    0;
            3    2   1.04   0   2.0  0   0   0   -0.5  2; ];
end

% Line Data for Y-Bus Formation.
```

```
function linedata = linedata3()    % Returns linedata.

%          sb   eb      R         X       B/2
linedata = [ 1    2    0.02      0.04    0.0;
             1    3    0.01      0.03    0.0;
             2    3    0.0125    0.025   0.0;];
```

Results

Complex bus voltages

V1 = 1.0500 + j 0.0000   0

-2.6422

-0.4591

V2 = 0.9707 - j0.0448

V3 = 1.0400 - j0.0083

Voltage magnitudes	Voltage Angles
1.0500	0.0
0.9718	-2.6422
1.0400	-0.4591

Exercise Modify the program to determine the losses in the system and also determine the slack bus power.

PROGRAM 17

Determination of Bus Currents, Line Flows and Line Losses

Once the voltages at the buses are solved using any power flow algorithm, it is important to determine the line flows to check if any line is overloaded. The line losses and the total system losses can also be determined. Consider the part of the network shown in Fig. 14.14.

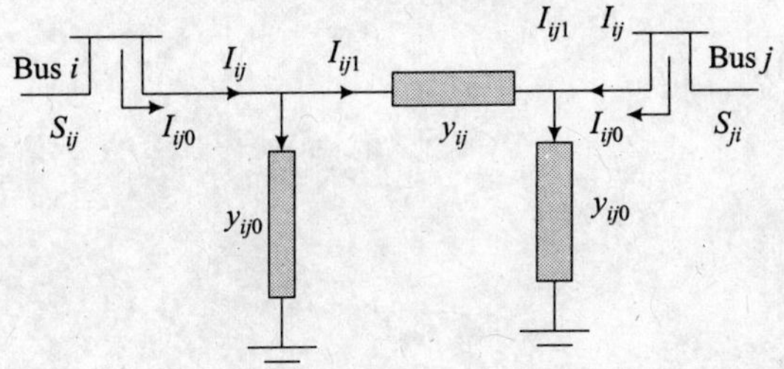

Fig. 14.14 Line between buses i and j.

In the figure the line is connected between buses i and j.

I_{ij} is current injected at bus i into the network.

I_{ji} is current injected at bus j into the network

y_{ijo} is shunt admittance at bus i

y_{jio} is shunt admittance at bus j

y_{ij} is admittance of line i-j

I_{ijo} is current drawn by y_{ijo}

I_{jio} is current drawn by y_{jio}

S_{ij} is complex power injection at bus i

S_{ji} is complex power injection at bus j

The equations from the figure are as follows:

$$I_{ij} = I_{ij1} + I_{ij0} = (V_i - V_j)y_{ij} + V_i y_{ij0}$$

$$I_{ji} = I_{ji1} + I_{ji0} = (V_j - V_i)y_{ij} + V_j y_{ji0}$$

$$S_{ij} = V_i\, I_{ij}^* = V_i\,(V_i^* - V_j^*)\,y_{ij}^* + V_i V_i^* y_{ij0}^*$$

$$S_{ji} = V_j\, I_{ji}^* = V_j\,(V_j^* - V_i^*)y_{ij}^* + V_j V_j^* y_{ji0}^*$$

The power loss in the (i-j)th line is the sum of S_{ij} and S_{ji}.The total transmission loss can be evaluated by summing all the line flows for all values of *i,j*. The slack bus power can also be found by summing the flows on the lines terminating at the slack bus.

Program Code

```
function[ybus,line flows]=line_losses();        %Definition of funciton
bmva=100;                                       %Base MVA
[z_data,vol_data]=pcal_input();                 %call input data
n=max(max(z_data(:,1)),max(z_data(:,2)))        % number of buses
r=[z_data(:,3)];                                %Formation of ybus
x=[z_data(:,4)];
z=complex(r,x);
y=1./z;
b=complex(0,z_data(:,5));
a=[z_data(:,6)];
nbr=length(r);
ybus=zeros(n,n);
lineflow=[];
for i=1:nbr
  sb=z_data(i,1);
  eb=z_data(i,2);
```

```
  ybus(sb,eb)=ybus(sb,eb)-y(i)/a(i);
  ybus(eb,sb)=ybus(sb,eb);
  ybus(sb,sb)=ybus(sb,sb)+(y(i)/(a(i)*a(i)))+b(i);
  ybus(eb,eb)=ybus(eb,eb)+y(i)+b(i);          % end of ybus computation
end
ybus
[vr,vi]=pol2cart(vol_data(:,3),vol_data(:,2)); % Convert voltages into Cartesian coordinates
v=complex(vr,vi)
i=ybus*v;
   % [I]=[Y][V]
s=v.*conj(i);
   % S=VI*
for k=1:nbr
  i=z_data(k,1);
  j=z_data(k,2);
  Iij=(v(i)-v(j))*(-ybus(i,j))+v(i)*b(k);              % Calculate Iij, Iji, Sij and Sji
  Sij=(v(i)*conj(Iij))*bmva;
  Iji=(v(j)-v(i))*(-ybus(j,i))+v(j)*b(k);
  Sji=(v(j)*conj(Iji))*bmva;
  loss(k)=Sij+Sji;
  y=[i j Sij Sji loss(k)];                              % Put line flows in an array
  lineflow=[lineflow;y];
end
display(lineflow)
Pl=sum(loss)
end

function[z_data,vol_data]=pcal_input();
%        Sb  eb   r       x      b    a
z_data= [1   2   0.02    0.04   0.0  1;
         2   3   0.0125  0.025  0.0  1;
         3   1   0.01    0.03   0.0  1;];

 %          Bus.No  V mag   Vang in radians
 vol_data=[  1      1.05     0;
             2      1.081   -0.0611;
             3      1.04    -0.0499;];

end
```

Result

i	j	Sij	Sji	Loss
1	2	199.5MW+j84.0Mvar	-191.0MW-j67.0Mvar	8.5MW+j17.0Mvar
2	3	-65.6MW-j43.2Mvar	66.4MW+j44.8Mvar	0.8MW+j1.6Mvar
3	1	-205.0MW-j90.0Mvar	210.0MW+j105.0Mvar	5.0MW+j15.0Mvar

Total loss = 14.300 MW + j33.60Mvar

PROGRAM 18

Jacobian Formation

The main equation to be solved in Newton Raphson method is given by

$$\begin{bmatrix} \Delta P \\ \Delta Q \end{bmatrix} = \begin{bmatrix} J_1 & J_2 \\ J_3 & J_4 \end{bmatrix} \begin{bmatrix} \Delta\delta \\ \Delta|V| \end{bmatrix}$$

The matrix $\begin{bmatrix} J_1 & J_2 \\ J_3 & J_4 \end{bmatrix}$ = [J] is called the Jacobian matrix. The submatrices are negated partial derivatives of ΔP and ΔQ with respect to corresponding δ and $|V|$. In this program the Ybus is read as data and the complex voltages at the buses is also read as data. These can be written as function files as in other examples. Here bus 1 is the slack bus, next we order the PV buses and then PQ buses. While ordering in sequence is not mandatory, it makes programming simpler. The data given in the program code is that of example 7.7

Program Code

```
% Enter ybus
ybus=[-15i 10i 5i;
   10i -15i 5i;
   5i 5i -10i]
 V=[ 1.0 1.1 1.0] % Enter voltage magnitudes. This is the initial voltage
 del=[0 0 0]% Enter initial voltage angles
 G=real(ybus)
 B=imag(ybus)
 nbus=3                              % Total number of buses
 npv=1                               % Number of PV buses
 npq=1                               % Number of PQ buses

% Formation of Jacobian
  % J1 - Derivative of Real Power Injections with Angles..
```

```
J1 = zeros(nbus-1,nbus-1); % Null matrix size if J1
for i = 1:(nbus-1)
   m = i+1;
   for k = 1:(nbus-1)
      n = k+1;
      if n == m
         for n = 1:nbus
               J1(i,k) = J1(i,k) + V(m)* V(n)*(-G(m,n)*sin(del(m)-del(n)) + B(m,n)*cos(del(m)-
del(n)));
         end
         J1(i,k) = J1(i,k) - V(m)^2*B(m,m);
      else
         J1(i,k) = V(m)* V(n)*(G(m,n)*sin(del(m)-del(n)) - B(m,n)*cos(del(m)-del(n)));
      end
   end
end

% J2 - Derivative of Real Power Injections with Voltage magnitudes
J2 = zeros(nbus-1,npq);
n1=2+npv;
for i = 1:(nbus-1)
   m = i+1;
   for k = n1:nbus
     n=k ;
      if k == m
         for n = 1:nbus
                     J2(i,k-n1+1) = J2(i,k-n1+1) + V(n)*(G(m,n)*cos(del(m)-del(n)) +
B(m,n)*sin(del(m)-del(n)));
         end
         J2(i,k-n1+1) = J2(i,k-n1+1) + 2*V(m)*G(m,m);
      else
         J2(i,k-n1+1) = V(m)*(G(m,n)*cos(del(m)-del(n)) + B(m,n)*sin(del(m)-del(n)));
      end
   end
end

% J3 - Derivative of Reactive Power Injections with Angles..
J3 = zeros(npq,nbus-1);
for i = n1:nbus
  m=i;
```

```
    for k = 1:(nbus-1)
       n = k+1;
       if n == m
          for n = 1:nbus
             J3(i-n1+1,k) = J3(i-n1+1,k) + V(m)* V(n)*(G(m,n)*cos(del(m)-del(n)) +
B(m,n)*sin(del(m)-del(n)));
          end
          J3(i-n1+1,k) = J3(i-n1+1,k) - V(m)^2*G(m,m);
       else
          J3(i-n1+1,k) = V(m)* V(n)*(-G(m,n)*cos(del(m)-del(n)) - B(m,n)*sin(del(m)-del(n)));
       end
    end
end

% J4 - Derivative of Reactive Power Injections with V.
J4 = zeros(npq,npq);
for i = 1:npq
    m=i+n1-1;
    for k = 1:npq
       n=k+n1-1;
       if n == m
          for n = 1:nbus
             J4(i,k) = J4(i,k) + V(n)*(G(m,n)*sin(del(m)-del(n)) - B(m,n)*cos(del(m)-del(n)));
          end
          J4(i,k) = J4(i,k) - V(m)^2*B(m,m);
       else
          J4(i,k) = V(m)*(G(m,n)*sin(del(m)-del(n)) - B(m,n)*cos(del(m)-del(n)));
       end
    end
end

J = [J1 J2; J3 J4]    % Jacobian Matrix..
```

Result

```
J =

    16.5000    -5.5000         0
    -5.5000    10.5000         0
     0          0              9.5000
```

Appendix A

MATRIX ALGEBRA

A.1 Basic Concepts

Definition of matrix: A matrix is a rectangular array of numbers, called elements arranged in m rows and n columns. The dimension of the matrix is specified as $m \times n$. The elements can be real or complex numbers. A matrix element is denoted by a_{ij}, where i stands for the row and j stands for the column.

Row matrix or row vector: It is a matrix with one row and more than one column. Dimension is $1 \times n$.

Column matrix or column vector: It is a matrix with one column and more than one row. Dimension is $m \times 1$.

Square matrix: A matrix where $m = n$ is called a square matrix. Elements a_{ij}, $i = j$, are called diagonal elements. Elements a_{ij}, $i \neq j$, are called off-diagonal elements.

Diagonal matrix: If all off-diagonal elements are zero, it is called a diagonal matrix.

Unit or Identity matrix: A diagonal matrix where all diagonal elements are 1, is called a unit or identity matrix. It is denoted by $[I]$ or $[U]$. Example,

$$[I] = \begin{bmatrix} 1 & 0 & 0 \\ 0 & 1 & 0 \\ 0 & 0 & 1 \end{bmatrix}$$

Null matrix: It is a matrix wherein all elements are zero.

Transpose of a matrix: If all rows and columns are interchanged, an $m \times n$ matrix results in an $n \times m$ matrix, called the transpose and denoted by A^T. Example,

$$[A] = \begin{bmatrix} 1 & 2 \\ 3 & 4 \\ 5 & 6 \end{bmatrix} \qquad [A^T] = \begin{bmatrix} 1 & 3 & 5 \\ 2 & 4 & 6 \end{bmatrix}$$

Symmetric matrix: In a square matrix if $a_{ij} = a_{ji}$, it is called a symmetric matrix. In a symmetric matrix $[A^T] = [A]$.

Skew-symmetric: A square matrix in which, $[A^T] = -[A]$ is called a skew symmetric matrix. Example,

$$[A] = \begin{bmatrix} 0 & -2 & 3 \\ 2 & 0 & 5 \\ -3 & -5 & 0 \end{bmatrix}$$

Orthogonal matrix: If $[A^T]\,[A] = [A]\,[A^T] = [I]$ for a square matrix with real elements, the matrix is said to be orthogonal.

Conjugate of a matrix: If all the elements of a matrix are replaced by their complex conjugates, the resultant matrix is called the conjugate and denoted by A^*. If

$$[A] = \begin{bmatrix} 3+j4 & 4+j7 \\ 6-j5 & 1+j0 \end{bmatrix} \quad [A^*] = \begin{bmatrix} 3-j4 & 4-j7 \\ 6+j5 & 1-j0 \end{bmatrix}$$

Hermatian matrix: If $[A] = [A^*]^T$ for a complex square matrix, the matrix is called a Hermatian matrix.

Skew-Hermatian matrix: If $[A] = -[A^*]^T$ for a complex square matrix, the matrix is called, skew-Hermitian matrix.

Unitary matrix: If $[A^*]^T\,[A] = [A]\,[A^*]^T = [I]$, the matrix is called a unitary matrix. A unitary matrix is orthogonal if all elements are real.

Determinant: The determinant is defined only for a square matrix.

$$[A] = \begin{bmatrix} a_{11} & a_{12} \\ a_{21} & a_{22} \end{bmatrix}$$

It is denoted by $|A| = (a_{11}, a_{22} - a_{21}a_{12})$. Determinants have following properties:

- The value of a determinant is zero,
  If all elements of a row or column are zero.
  The corresponding elements of two rows or two columns are zero.
  A row or column is a linear combination of one or more rows or columns.
- The value of a determinant is changed in sign only if two rows or columns are interchanged.
- $|A| = |A|^T$.
- The value of the determinant is unchanged it k times the elements of any row or column are added to the corresponding elements of another row (or column).
- If all the elements of a row or column are multiplied by a constant k, the value of the determinant is multiplied by k.
- In general, $|A + B|\ ^1\ |A| + |B|$
- $|ABC| = |A||B||C|$

Minors and cofactors: The determinant obtained by deleting the ith row and j^{th} column is the minor of the element $a_{ij.}$ The cofactor of an element a_{ij} is given by $A_{ij} = (-1)^{i+j}$ (minor of *aij*).

Adjoint: If each element of a square matrix is replaced by its cofactor and then the matrix is transposed, the resulting matrix is an adjoint and designated by A^+.

A.2 Matrix Operations

Equality: Two matrices are equal if all the corresponding elements are equal.

Addition and subtraction of matrices

Matrices of the same dimensions only can be added or subtracted. If $A \pm B = C$, then the elements of C are given by $c_{ij} = a_{ij} \pm b_{ij}$.

- $A + B = B + A$
- $A + B + C = (A + B) + C = A + (B + C)$

Multiplication by a scalar: When a matrix is multiplied by a scalar k, all the elements are multiplied by k.

- $kA = Ak$
- $k(A + B) = kA + kB$

Multiplication of two matrices: $AB = C$, is defined only if the number of columns of A equals the rows of B. Any element C_{ij} is the sum of the products of the corresponding elements of the ith row of A and j^{th} column of B.

- In general for a square matrix $AB \neq BA$
- $AB = 0$, does not imply $A = 0$ or $B = 0$.
- $CA = CB$, does not mean $A = B$.
- If $C = AB$, $C^T = B^T A^T$

Inverse of a matrix: If $AX = Y$ and $X = BY$, then B is called the inverse of A, denoted by A^{-1} and is given by $B = \dfrac{A^+}{|A|}$.

- The inverse does not exist if $|A|$ is zero.
- $AA^{-1} = I$
- $(AB)^{-1} = B^{-1}A^{-1}$

Rank of a matrix: The rank of an $m \times n$ matrix is equal to the maximum number of linearly independent columns of A (column rank) or the maximum number of linearly independent rows of A (row rank).

A.3 Partitioning of Matrices

A large matrix can be partitioned into smaller sub-matrices of smaller dimensions.

- $[A] = \begin{bmatrix} A_1 & A_2 \\ A_3 & A_4 \end{bmatrix}$

- $\begin{bmatrix} A_1 & A_2 \\ A_3 & A_4 \end{bmatrix} \pm \begin{bmatrix} B_1 & B_2 \\ B_3 & B_4 \end{bmatrix} = \begin{bmatrix} A_1 \pm B_1 & A_2 \pm B_2 \\ A_3 \pm B_3 & A_4 \pm B_4 \end{bmatrix}$
- $\begin{bmatrix} A_1 & A_2 \\ A_3 & A_4 \end{bmatrix} \begin{bmatrix} B_1 & B_2 \\ B_3 & B_4 \end{bmatrix} = \begin{bmatrix} C_1 & C_2 \\ C_3 & C_4 \end{bmatrix}$

where, $C_1 = A_1B_1 + A_2B_3$
$C_2 = A_1B_2 + A_2B_4$
$C_3 = A_3B_1 + A_4B_3$
$C_4 = A_3B_2 + A_4B_4$

The corresponding sub-matrices must be conformable.

- If $[A] = \begin{bmatrix} A_1 & A_2 \\ A_3 & A_4 \end{bmatrix}$ and $[A]^{-1} = \begin{bmatrix} B_1 & B_2 \\ B_3 & B_4 \end{bmatrix}$, then

$$B_1 = (A_1 - A_2A_4^{-1}A_3)^{-1}$$
$$B_2 = -B_1A_2A_4^{-1}$$
$$B_3 = -A_4^{-1}A_3B_1$$
$$B_4 = A_4^{-1} - A_4^{-1} - A_3B_2$$

And A_1 and A_4 must be square matrices.

$$(\bar{I}^*_{\text{Loop}})^t C^t \bar{e} = (\bar{i}^*)^t \bar{e}.$$

From the above we get $(\bar{i}^*)^t = (\bar{I}^*_{\text{Loop}})^t C^t$ or $\bar{i} = C^* \bar{I}_{\text{Loop}}$. Since, C is real $C^* = C$. We get, $\bar{i} = C\bar{I}_{\text{Loop}}$. Using this we can write,

$$C^t \bar{e} = C^t[z]C\bar{I}_{\text{Loop}} = \bar{E}_{\text{Loop}} \quad \text{...(B.10)}$$

From B.10, we get

$$Z_{\text{Loop}} = C^t[z]C \quad \text{...(B.11)}$$

Bibliography

Anderson, P.M., and Fouad, A.A., Power System Control and Stability, The Iowa State University Press, Ames, Iowa, 1977.

Arrillaga, J., Arnold, C.P., and Harker, B. J., Computer Modeling of Electrical Power Systems, John Wiley & Sons, Inc., New York, 1986.

Bergen, A.R., Power Systems Analysis, Prentice-Hall, Englewood, Cliffs, New Jersey, 1970.

Billinton, R., Ringlee, R., and Wood, A., Power System Reliability Calculations, MIT Press, Cambridge, Massachusetts, 1973.

Billinton, R., Power System Reliability Evaluations, Gordon and Breach, New York, 1970.

Byerly, R. T., and Kimbark, E. W., Stability of Large Electric Power Systems, IEEE Press, 1974.

Clarke, E., Circuit Analysis of AC Power Systems, John Wiley & Sons, Inc., New York, 1958.

Concordia, C., Synchronous Machine: Theory and Performance, John Wiley & Sons Inc., New York, 1951.

Crary, Power System Stability, Vol. I and II, John Wiley & Sons, Inc., New York, 1955.

Del Toro, V., Electric Power Systems, Prentice-Hall, Englewood Cliffs, New Jersey, 1992.

Dommel, H. W., and Sato, N., "Fast Transient Stability Solutions", IEEE Trans., Power Apparatus and Systems, PAS-91, pp. 1643-1650, October 1972.

Dommel, H.W., and Tinney, W.F., "Optimal Power Flow Solutions IEEE Trans.," Power Apparatus and Systems, PAS-87, pp. 1866-1876, October 1972.

Elgerd, O. I., Electric Energy Systems Theory, Second Edition, McGraw-Hill Book Company, New York, 1982.

Fink, D. G., and Beaty, H. W., Standard Handbook for Electrical Engineers, McGraw-Hill, New York, 1987.

Fitzgerald, A. E., Kingsley, C., and Umans, S., Electric Machinery, Fourth Edition, McGraw-Hill, New York, 1982.

Fouad, A. A. and Stanton, S.E., "Transient Stability of Multi-machine Power Systems, Part I and II", IEEE Trans., Vol. PAS-100, pp. 3408-3424, July 1981.

Fouad, A. A., Vijay, V.," Power System Transient Stability Analysis Using the Transient Energy Function Method", Prentice-Hall, Englewood. Cliffs, New Jersey, 1992.

Kimbark, E. W., Power System Stability, Vol. 1, Elements of Stability Calculations, John Wiley & Sons, Inc., New York, 1948.

Kimbark, E. W., Power System Stability, Vol. 2, Power Circuit Breakers and Protective Relays, John Wiley & Sons, Inc., New York, 1950.

Kimbark, E. W., Power System Stability, Vol. 3, Synchronous Machine, John Wiley & Sons, Inc., New York, 1956.

Kirchmayer, L.K., Economic Operation of Power Systems, John Wiley & Sons, Inc., New York, 1958.

Kron, G., "Tensorial Analysis of Integrated Transmission Systems, Part I, The Six Basic Reference Frames", AIEE Trans., Vol. 70, pp. 1239-1248, 1951.

Kundur, P., Power System Stability and Control, McGraw-Hill Book Company, New York, 1994.

Kusic, G. L., Computer Aided Power System Analysis, Prentice-Hall, Englewood Cliffs, New Jersey, 1986.

Nasar, S.A., Electric Energy Conversion and Transmission, Macmillan Publishing Company, New York, 1985.

Padiyar K.R., Power System Dynamics : Stability and Control, BS Publishers, India.

Sauer, P.W., and Pai, M. A., Power System Dynamics and Stability, Prentice-Hall, Englewood Cliffs, New Jersey, 1998.

Stagg, G. W., and El-Abiad, A. H., Computer Methods in Power System Analysis, McGraw-Hill Book Company, New York, 1968.

Stevenson York,, W.D., and Grainger, J.J., Power System Analysis, McGraw-Hill Book Company, New York, 1994.

Stott, B., "Decoupled Newton Load Flow", IEEE Trans. Power Apparatus and Systems, PAS-91, pp. 1955-1959, October 1972.

Stott, B., and Alsac, O., "Fast Decoupled Load Flow", IEEE Trans. Power Apparatus and Systems, PAS-93, pp. 859-869, May-June 1974.

Sullivan R., Power System Planning, McGraw-Hill Book Company, New York, 1977.

Taylor, C.W., Power System Voltage Stability, McGraw-Hill Book Company, New York, 1993.

Tinney, W.F., "Compensation Methods for Network Solutions by Optimally. Ordered Triangular Factorization", IEEE Trans., PAS-91, pp. 123-127, January 1972.

Wallach, Y., Calculations and Programs for Power System Networks, Prentice-Hall, Englewood Cliffs, New Jersey, 1986.

Weedy, B. M., Electric Power Systems, Third Edition, John Wiley & Sons, Inc., New York, 1979.

Westinghouse Electric Corporation, Electric Transmission and Distribution Reference Book, East Pittsburgh, Pennsylvania, 1964.

Wood, A. J., and Wollenberg, B. F., Power Generation Operation and Control, John Wiley & Sons, Ind., New York, 1974.

Yu, Yao-nan, Electric Power Systems Dynamics, Academic Press, New York, 1983.

Index